Introduction to Organic Spectroscopy

Introduction to Organic Spectroscopy

JOSEPH B. LAMBERT
Northwestern University
HERBERT F. SHURVELL
Queen's University
DAVID A. LIGHTNER
University of Nevada Reno
R. GRAHAM COOKS
Purdue University

Macmillan Publishing Company
New York
Collier Macmillan Publishers
London

To

LAWRENCE VERBIT
1935–1983

Macmillan Publishing Company
866 Third Avenue, New York, New York 10022

Collier Macmillan Canada, Inc.

Library of Congress Cataloging-in-Publication Data

Introduction to organic spectroscopy.

Includes bibliographies and index.
1. Spectrum analysis. 2. Chemistry, Organic.
I. Lambert, Joseph B.
QD272.S6I54 1987 547.3′0858 86–31304
ISBN 0–02–367300–1

Printing: 1 2 3 4 5 6 7 8 Year: 6 7 8 9 0 1 2 3 4 5

ISBN 0-02-367300-1

Preface

Perhaps the most revolutionary event in organic chemistry during the past 30 years has been the development of spectroscopic methods for analyzing molecular structure. These methods were imported from physics and transformed by chemists into entirely new procedures for providing structural information. Prior to the advent of spectroscopic methods, structural determination was a complex and often impossible task based primarily on modifying molecules through reactions and comparing the products with known structures. In contrast, spectroscopic experiments for the most part are simple and even verge on the routine.

Structural analysis of an organic compound typically requires combining the information obtained from nuclear magnetic resonance (NMR), infrared, and ultraviolet–visible spectroscopy and mass spectrometry. In this textbook we treat each of these areas from the point of view of the beginner, repeating if necessary material that may have been covered in a course on elementary organic chemistry. The level is brought to the point at which, we hope, the student can apply the principles in general to structural problems. We have tried to present the material as it is actually used today. Thus NMR is treated as an integral mixture of proton and carbon-13 methods, and Fourier transform techniques are introduced at an early stage. Problems are given for each type of spectroscopy. Answers are provided for most problems, and others are worked in the text. In addition, a set of problems is given that combines the use of all four types of spectroscopy.

Spectroscopy is taught in many different formats. It is intended that the material in this text be covered in a course that is approximately one semester in length. For instructors who teach in the quarter system, who teach spectroscopy as an adjunct to another course, or who require less material for any other reason, the following sections may be omitted without loss of the integrity of the introductory material: NMR advanced topics (Chapter 5); Raman and special infrared techniques (Sections 6-6, 6-8, 6-9, 7-4, and 8-6c); chiroptical methods (Sections 9-5, 10-1b, 10-4, and 10-6d) and ORD–CD portions of other sections; some mass spectral theory and mixture analysis (Sections 11-5, 11-6, and 12-5).

The authors wish to thank the following individuals for assistance in preparing this manuscript: Elisabeth Belfer, Y. S. Byun, Sonya Marquie, Arlene Rothwell, Carol Slingo, and Irene Shurvell. We also acknowledge Paul Salverda, Gary Taniguchi, and Mike Wilson of Hewlett-Packard for critical reading of portions of the manuscript.

J. B. L.
H. F. S.
D. A. L.
R. G. C.

Contents

1 Introduction

1-1 Structural Determination

What is meant by the structure of an organic molecule? In different contexts, organic chemists tend to accept different levels of meaning. In its most fundamental sense, structure connotes the order in which atoms are bonded to one another. For example, molecules **1-1, 1-2**, and **1-3** all have molecular formula $C_8H_{14}O_2$, but

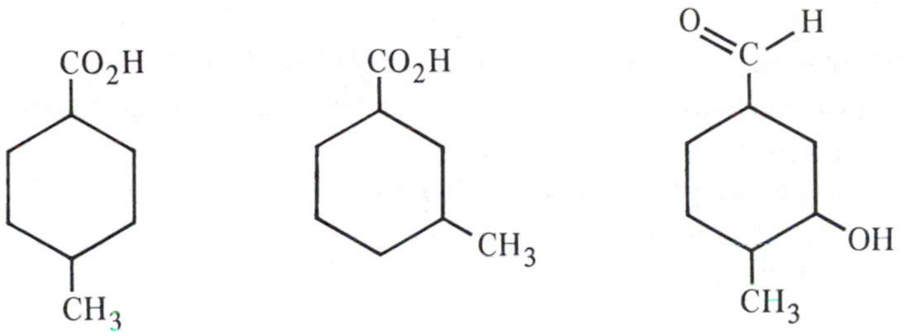

CO_2H CO_2H O=C–H

CH_3 CH_3 OH CH_3

1-1 **1-2** **1-3**

they clearly have different structures. Molecules **1-1** and **1-2** are *positional isomers* of each other, and molecule **1-3** is a *functional isomer* of both. The three molecules differ in their atomic connectivity and hence have different molecular structures. Because they have the same molecular formula but different connectivities, they are said to be *structural isomers*. Thus, in one sense, structural determination identifies atoms and functional groups and orders these components into a molecule.

The two substituents in molecule **1-1** can be located on the same side or on opposite sides of the six-membered ring, as in **1-4** and **1-5**. Because these two forms have the same constitution or connectivity but

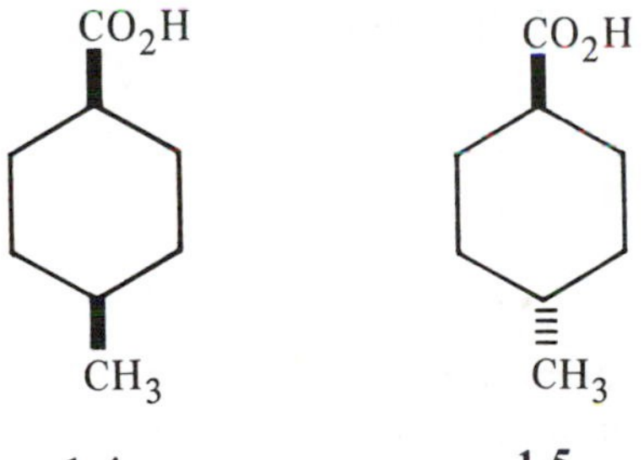

CO_2H CO_2H

CH_3 CH_3

1-4 1-5

differ in the arrangement of atoms in space, they are said to be *stereoisomers*. Specifically, **1-4** and **1-5** are *diastereoisomers* of the type termed geometrical isomers. *Enantiomers* are another important class of stereoisomers. A pair of enantiomers are nonsuperimposable mirror images of each other. The trans form of

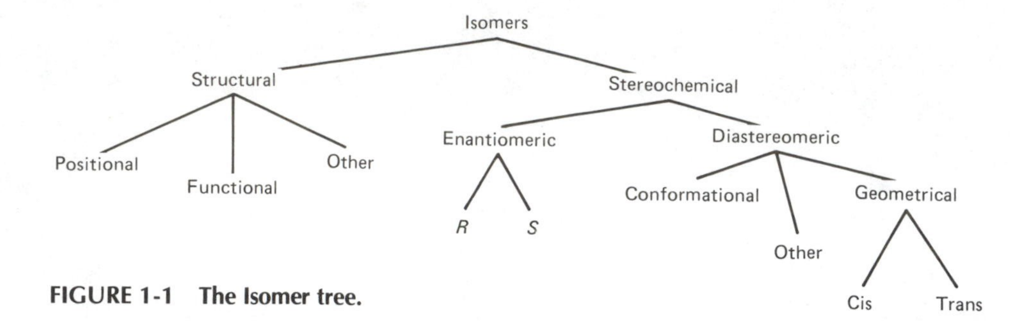

FIGURE 1-1 The Isomer tree.

molecule **1-2**, but not the cis form, exists in two enantiomeric forms. Thus in a second sense, structural determination includes specifying the static spatial arrangement of atoms within a molecule.

The six-membered ring of molecules **1-1**–**1-5** assumes the chair conformation. For molecule **1-4**, there are two forms, depending on whether the methyl group is equatorial (**1-6**) or axial (**1-7**). Because these two mole-

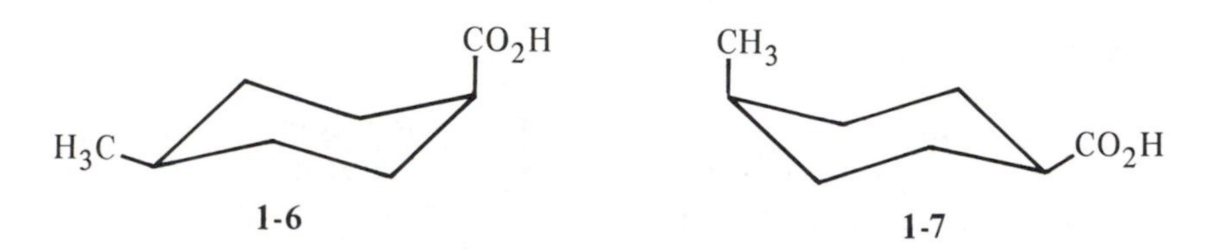

cules may be interconverted by a series of rotations about single bonds, they are said to be *conformational isomers*. In a third sense, structural determination involves identifying the rotational arrangements of atoms around single bonds.

In the most detailed sense, the structure of a molecule includes all its bond lengths, valence angles, and torsional angles, which in turn define connectivity, stereochemistry, and conformation. For chiral molecules, stereochemical structure includes absolute configuration (*R* or *S*). An incomplete tree of organic isomers is given in Figure 1-1.

The organic chemist seldom has recourse to a complete structural determination by X-ray crystallography, electron diffraction, or microwave spectroscopy. The appropriate equipment may be lacking, the molecules may be too complex or too impure for these methods, or the molecules may exist in a state of matter, such as the liquid, that is inappropriate for these methods (X-ray crystallography requires solids; the other two methods require gases). A more accessible and straightforward approach to structural determination involves acquisition of the four types of spectra described in this volume: nuclear magnetic resonance, vibrational, electronic, and mass. These spectra can be obtained on almost all types of organic materials, the equipment is widely available, and each experiment often takes only a few minutes. The spectroscopist must construct as much of the molecule as possible from the wealth of information provided by these spectroscopies, including the molecular formula, the types of functional groups, the number and types of hydrogens and carbons, and the absolute configuration of chiral molecules.

1-2

Molecular Weight and Formula

In many cases, the first order of business in structural determination is obtaining the molecular weight and formula. There are numerous experimental methods for molecular weight determination, including freezing or melting point lowering, boiling point elevation, and vapor pressure measurement. These methods do not require expensive equipment and are described in textbooks of experimental physical chemistry.

Molecular weight is obtained more reliably and easily, however, by mass spectrometry. High resolution mass spectrometry provides the exact molecular weight to six or seven significant digits. Because the atomic weights or carbon, hydrogen, oxygen, and nitrogen are nonintegral, the numbers after the decimal point are reliably characteristic of the identity of the atom. The exact molecular weight, then, is a nearly unique combination of these atomic weights. Reference to tables or computer listings of exact molecular weights gives the correspondence to the molecular formula.

More traditionally, chemists have used quantitative elemental analysis to derive the *empirical formula* (the lowest integer ratio of elements present), which is related to the *molecular formula* by an integral factor. Elemental analysis is a necessary step in a structure proof, and almost all journals require such analyses for new compounds. Division of an atomic percentage X by the atomic weight M provides the relative number of atomic equivalents ($E = X/M$) of that element in the compound. If all the elements are accounted for, a similar process of division provides the relative number of atomic equivalents for each element. These numbers E, E', E'' are converted to the simplest ratio by division by the smallest: E/E'', E'/E'', 1.0. These numbers are then multiplied by the lowest integer n that yields whole numbers for all numbers: nE/E'', nE'/E'', n. The procedure is illustrated as follows.

	Carbon	Hydrogen	Oxygen
Elemental analysis (X)	48.57	8.19	43.24
Relative atoms ($E=X/M$)	$\frac{48.57}{12.01} = 4.04$	$\frac{8.19}{1.008} = 8.13$	$\frac{43.23}{16.00} = 2.70$
Simplest ratio (E/E'')	$\frac{4.04}{2.70} = 1.50$	$\frac{8.13}{2.70} = 3.01$	$\frac{2.70}{2.70} = 1.00$
Integer ratio (nE/E'')	$2 \times 1.50 = 3$	$2 \times 3.01 \approx 6$	$2 \times 1.00 = 2$
Empirical formula	C_3	H_6	O_2

The result of such calculations is the empirical formula. The molecular formula may be a multiple of this formula, so that a definitive result is not obtained. The procedure is poor at distinguishing small differences in carbon or hydrogen, it is subject to experimental error in the percentages, and it is very sensitive to the purity of the sample. For these reasons, the spectroscopic method, using high resolution mass spectrometry, is preferred. Elemental analysis nonetheless is retained as an important measure of purity.

The molecular formula itself provides structural information beyond the simple facts of elemental identity. Introduction of a ring or a double bond into an alkane structure reduces the number of hydrogen atoms by two. Thus, the formula for the homologous series of straight chain alkanes is C_nH_{2n+2}, whereas that for alkenes and monocyclic alkanes is C_nH_{2n}. Detailed examination of the molecular formula can provide the number of sites of unsaturations. Each ring is one site, each double bond is one site, and each triple bond is two sites.

The *unsaturation number* (U) is given by eq. 1-1, in which C is the number of the tetravalent elements

$$U = C + 1 - \tfrac{1}{2}(X - N) \qquad (1\text{-}1)$$

(carbon, silicon, etc.), X is the number of monovalent elements (hydrogen and the halogens), and N is the number of trivalent elements (nitrogen, phosphorus, etc.). The formula is an elaboration on the formula for the alkane series, whose unsaturation number is zero. Divalent elements such as oxygen and sulfur extend chains but do not alter the number of unsaturations, so they do not appear in the formula. Trivalent elements add a hydrogen, and univalent elements subtract off a hydrogen, as indicated in the second term. The formula for molecules **1-1**–**1-3** is $C_8H_{14}O_2$, so their unsaturation number is $8 + 1 - \frac{1}{2}(14) = 2$, one each for the ring and the π bond of the carbonyl group.

How an unknown material was produced provides important input concerning molecular structure. The results of organic reactions often are predictable or at least can be circumscribed, so that the organic chemistry can provide considerable information about the structure before elemental and spectroscopic analysis. Hydride reduction of a molecule containing a cyclohexane ring ought to retain the ring. Grignard addition to an ester lacking nitrogen and halogen should not alter the nitrogen and halogen content. An anticipated elimination reaction should provide a molecule with a new double bond and without the leaving group. Although these deductions may seem self-evident, they usually provide the foundation on which spectroscopy builds the remainder of the molecule.

1-3

Contributions of Different Forms of Spectroscopy

Elucidation of the structure of an organic molecule draws on the results of several spectroscopic methods. In this volume we examine the four most widely used and most successful forms of molecular spectroscopy. Each provides its own special kind of data that apply to molecular structure. The following four sections briefly describe the methods that are the subjects of the four parts of this volume.

1-3a Nuclear Magnetic Resonance Spectroscopy

Nuclear magnetic resonance (NMR) spectroscopy provides information about the types and numbers of particular nuclei. Thus it can show that ethanol, CH_3CH_2OH, has two types of carbons in the ratio 1/1 and three types of hydrogens in the ratio 3/2/1. The experiment involves changing the spin states of magnetically active nuclei. The most commonly examined nuclei are hydrogen and carbon-13. Examination of the hydrogens and carbons in an organic molecule by NMR provides a characterization of each nucleus according to a physical parameter termed the chemical shift. Partial structures or even the complete carbon connectivity of organic molecules is derived from analysis of chemical shifts. The interaction between nuclei, as measured by a coupling constant, provides further information about subunits of a molecule. Because these coupling constants also depend on the distance between the coupled nuclei in the molecule, stereochemical information is obtainable. Furthermore, because chemical reactions move nuclei from one position to another within a molecule or even to a separate molecule, NMR is used to follow the course of many kinds of reactions.

The NMR experiment may be applied to molecules in any state of matter, but routine applications are usually carried out on liquid solutions. The NMR method may be applied to submilligram quantities, which need not be completely pure. Thus, NMR is the most general approach to organic structural elucidation, as described in Part I of this volume. Although not the specific subject of this volume, inorganic and bioorganic molecules also are perfectly amenable to study by NMR.

1-3b Vibrational Spectroscopy

Infrared (IR) and Raman spectra come from minute vibrations that occur naturally within molecules. The spectra provide a variety of information on molecular structure, symmetry, and many other problems of interest to the organic chemist. Infrared spectroscopy is more widely used than Raman because of the relative ease of sample preparation and operation of the instruments. Small grating infrared spectrometers are inexpensive, compact, and so simple to use that any undergraduate student can obtain a spectrum after just a few minutes of instruction. Raman spectrometers are more expensive and complicated. Nevertheless, Raman spectroscopy is an important technique complementary to infrared and is included in Part II of this book.

The infrared and Raman spectra of any compound are unique and can be used as fingerprints for identification purposes. Moreover, vibrations of certain functional groups always give rise to features in the infrared and Raman spectra within well-defined ranges, regardless of the molecule containing the functional group. Consequently, these methods are particularly useful in the identification of functionalities such as carbonyl or nitrile. Whereas NMR provides information about the carbon and hydrogen skeleton of the molecule, IR and Raman focus on important substructures.

The development of the laser brought about a revival in interest in Raman spectroscopy around 1970. More recently, developments in small computer technology have enabled the theoretical advantages of Fourier transform infrared spectroscopy to become a practical reality. Regardless of how the vibrational spectrum of a compound is obtained, however, there always remains the process of spectral interpretation. Part II of this volume is concerned with these matters.

1-3c Ultraviolet-Visible and Circular Dichroism Spectroscopy

Electronic absorption spectroscopy measures the energy and probability of promoting a molecule from its ground electronic state to an electronically excited state. The excitation involves moving an electron from an occupied molecular orbital to a higher, unoccupied orbital. Since an organic molecule typically has numerous occupied and unoccupied molecular orbitals, many different electronic excitations from the ground state are possible. The associated transition energies are found in the ultraviolet (UV) and visible (vis) regions of the electromagnetic spectrum. The electronic transitions of greatest importance in organic chemistry are the lowest energy excitations, that is, those involving promotion from the highest occupied bonding or nonbonding orbital to the lowest unoccupied molecular orbital. Information about these states may be obtained either directly by the UV–vis spectrum or by a difference procedure known as circular dichroism (CD) spectroscopy that is sensitive only to chiral molecules. Both UV–vis and CD spectroscopies, however, arise from the same photophysical process: promotion of a molecule from its ground electronic state to an excited electronic state. Therefore it is not surprising that these spectroscopic methods, along with a related technique known as optical rotatory dispersion (ORD), are employed to extract closely related structural information.

UV–vis and CD spectroscopies are used in a qualitative way to detect certain functional groups, called chromophores, from the position and intensity of an absorption band, as described in Part III of this volume. In addition, they can detect the interaction between neighboring chromophores, such as conjugated ketones, from spectral shifts, and they can follow chemical and photochemical reactions that lead to an alteration of the chromophore. Circular dichroism in particular can determine absolute configuration or conformation from chirality rules. These methods are also used for quantitative measurements, for example, to determine solute concentrations, to follow reaction kinetics, and to measure equilibrium constants. The sensitivities are typically quite high, with detectability up to 10^{-9} M.

1-3d Mass Spectrometry

Mass spectrometry (MS) examines ions in the gas phase that are produced by techniques such as collision of the molecule with electrons. MS provides molecular weights, both nominal (nearest integer) and exact (five or more significant figures). In addition to ions representative of the intact molecule, fragment ions often are generated, and these can be pieced together to formulate the molecular structure. The experiment involves generation of ions in the gas phase followed by measurement of their mass-to-charge ratios and relative abundances. These measurements are made by any one of a number of techniques based on properties of electric or magnetic fields. Isotopes, because they differ in mass, can be recognized by mass spectrometry, thus allowing quantitative and qualitative analysis of compounds enriched in stable isotopes.

Mass spectra can be taken on molecules in any state of matter, including fragile, thermally labile compounds in the solid state. Mixtures are commonly examined by mass spectrometry, especially in instruments that combine mass analysis with separations based on gas chromatography, liquid chromatography, or a second stage of mass spectrometry. The sensitivity of mass spectrometry is its strongest suit, quantities as small as femtomoles (10^{-12} mol) being detectable and work being done routinely at the submicrogram ($<10^{-6}$ g) level.

Chemical reactions invariably occur during the mass spectrometry experiment, and it is possible to follow their kinetics, to make thermochemical determinations, and to prepare new compounds in the mass spectrometer. Part IV of this volume describes the use of MS to obtain molecular weights, to analyze fragment ions, and to follow gas-phase reactions.

1-3e The Electromagnetic Spectrum

NMR, vibrational, and electronic spectroscopies involve absorption of electromagnetic energy corresponding to specific regions of the electromagnetic spectrum (Figure 1-2). Consequently, they are referred to

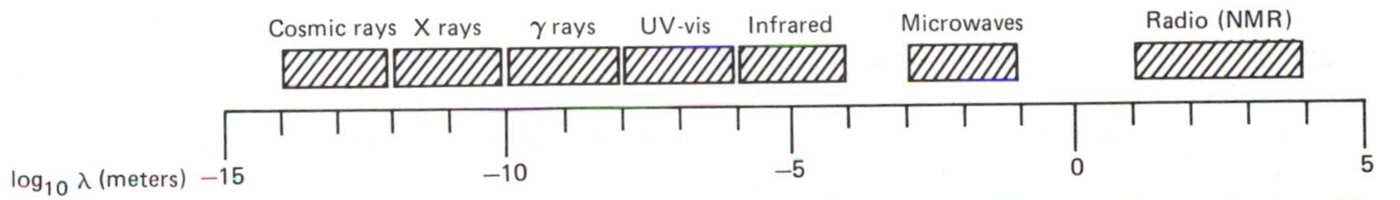

FIGURE 1-2 **Names given to various regions of the electromagnetic spectrum. The plot is linear in the logarithm of wavelength.**

as specific varieties of *absorption spectroscopy*. Mass spectrometry, however, involves measurement of mass numbers rather than energy absorption (hence it is a spectrometry rather than a spectroscopy) and does not appear in Figure 1-2.

Figure 1-2 is arranged with waves of longest wavelength on the right. Eq. 1-2 gives the relationship among frequency ν, wavelength λ, and the speed of light c (2.998×10^8 m s^{-1} in vacuum). Eq. 1-3 gives the

$$\nu = c/\lambda \tag{1-2}$$

$$\Delta E = h\nu \tag{1-3}$$

relationship between the frequency ν of the wave, Planck's constant h (6.624×10^{-34} J·s), and the amount of energy ΔE that is absorbed by a molecule. Thus energy is directly proportional to frequency, and both would be largest at the left of Figure 1-2. Each form of spectroscopy has its preferred unit for wavelength, frequency, and energy. The appropriate expressions are discussed in Parts I–IV.

PROBLEMS

1-1 Specify whether the following pairs of molecules are structural isomers or stereoisomers.

a. CH_3 CH_2

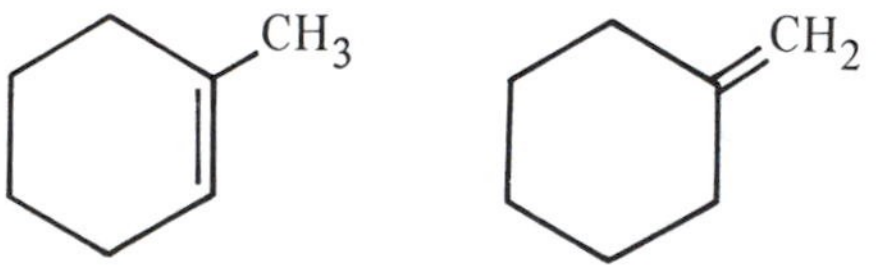

b. CH_3NH CH_3 H_2N CH_2CH_3

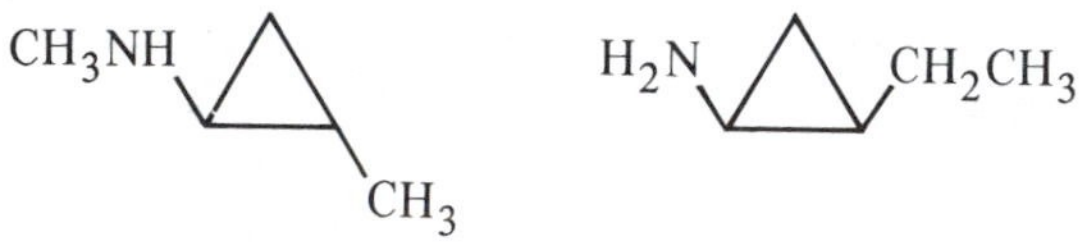

c. CH_3 CH_3 CH_3 CH_3

d. $C_6H_5CH_2OCCH_3$ (C=O) $C_6H_5OCCH_2CH_3$ (C=O)

e. H OH HO H

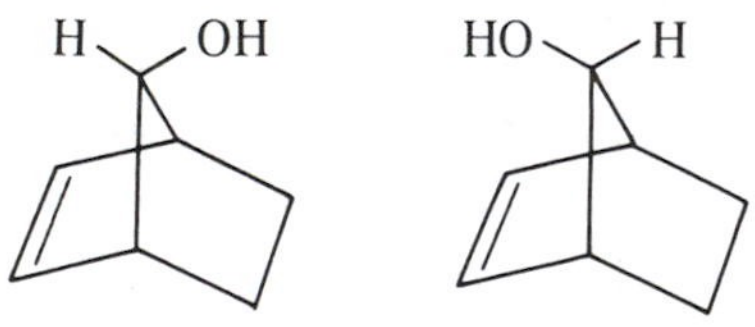

1-2 Carry out the following operations for each of the molecules in problem **1-1**.

a. Write down the molecular formula and calculate elemental percentages for C, H, N, and O (if present).

b. Determine the unsaturation number.

c. Specify the functional groups present.

1-3 For the molecules in problems **1-1a, 1-1d,** and **1-1e**, determine the number and relative proportions of different types of carbon atoms. Do the same with the hydrogens. Atoms are said to be different if they cannot be interconverted by a symmetry operation such as reflection through a mirror plane or rotation about an axis.

1-4 Determine the empirical formula of the molecules that gave the following elemental analyses.

a. C, 59.89; H, 8.12; O, 31.99.

b. C, 66.02; H, 10.34; N, 10.96; O, 12.68.

Part I

NUCLEAR MAGNETIC RESONANCE SPECTROSCOPY

2

Basic Concepts of Nuclear Magnetic Resonance

2-1

Magnetic Properties of Nuclei

The ideal technique for the determination of the structure of an organic molecule would be analogous to a microscope. Its lens would provide a view of the molecule with full identification of all atoms in their proper conformation. Nuclear magnetic resonance (NMR) probably comes closest today to being a lens on molecular structure. The experiment takes advantage of magnetic properties of certain nuclei, such as hydrogen-1 (the proton) or carbon-13. The NMR experiment is tuned to a particular nucleus and yields a portrait of all such nuclei found in the molecule under study. Unlike an imaginary molecular microscope, the NMR experiment does not connect all the atoms to give a full structure. The experimentalist must assemble the molecule in much the same way a puzzle is put together. NMR provides all the pieces, replete with molecular hooks, but we must figure out how they fit together.

The NMR experiment is made possible by the magnetic properties of various nuclei. In general, a moving charge creates a magnetic field and hence has a magnetic moment μ. Certain nuclei act as if they spin in a manner that is equivalent to circular motion (Diagram 2-1). Since they possess the charge of the constituent protons, these nuclei therefore have vectorial magnetic properties such as a magnetic moment.

Diagram 2-1

Charge moving in a circle

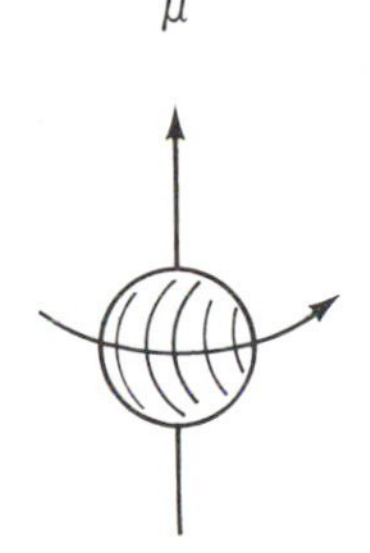

Spinning spherical nucleus

Nuclei that do not spin, including the common isotopes of carbon (^{12}C) and oxygen (^{16}O), are said to have a spin (represented by the letter I) of zero and are not of interest in the magnetic resonance experiment (Diagram 2-2). Of the spinning, magnetic nuclei, those with a spherical distribution of nuclear charge are assigned a spin of $\frac{1}{2}$, and those with a nonspherical, or quadrupolar, distribution are assigned a spin of 1 or more, in integral or half-integral units. Spin $\frac{1}{2}$ nuclei include ^{1}H, ^{13}C, ^{15}N, ^{19}F, ^{29}Si, and ^{31}P, whereas quadrupolar nuclei include ^{2}D, ^{7}Li, ^{11}B, ^{14}N, ^{17}O, and ^{23}Na. The precise value of the spin (I) is determined by the number of protons and neutrons in the nucleus.

Diagram 2-2

No spin

$I = 0$

Spinning sphere

$I = 1/2$

Spinning ellipsoid

$I = 1, 3/2, 2, \ldots$

The NMR experiment involves placing a small sample of a compound between the poles of a strong laboratory magnet, whose field is represented by B_0 in units of tesla (Diagram 2-3). In the absence of this applied field, all orientations of nuclei are equivalent (Diagram 2-4). The spin of magnetic nuclei, however, is quantized and can possess only certain well-defined values. In the B_0 field, a nucleus of spin $I = \frac{1}{2}$ can adopt only two orientations, said to be with the field and against the field. Because the z direction is defined as parallel to the B_0 field, these nuclear spin states are assigned quantum numbers of $I_z = +\frac{1}{2}$ for the favored arrangement parallel to the applied field and $-\frac{1}{2}$ for the less favored arrangement antiparallel to the field. Thus there are a few more $+\frac{1}{2}$ than $-\frac{1}{2}$ nuclei, according to Boltzmann's law (Diagram 2-4).

Diagram 2-3

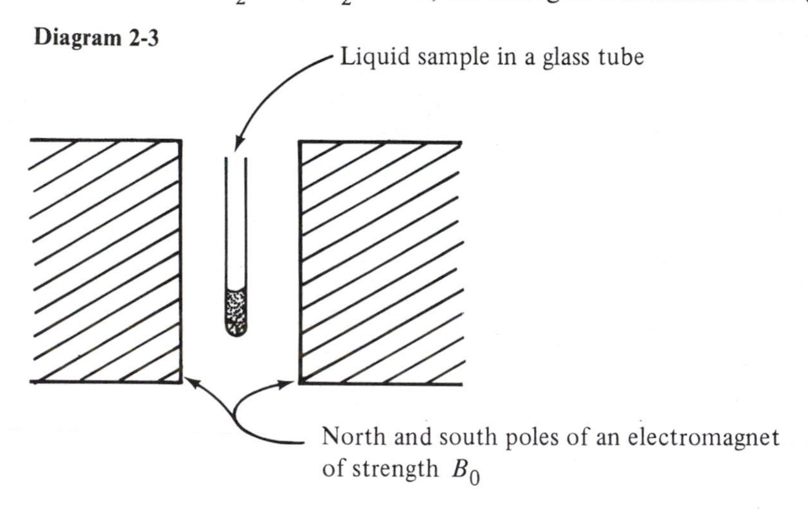

Diagram 2-4

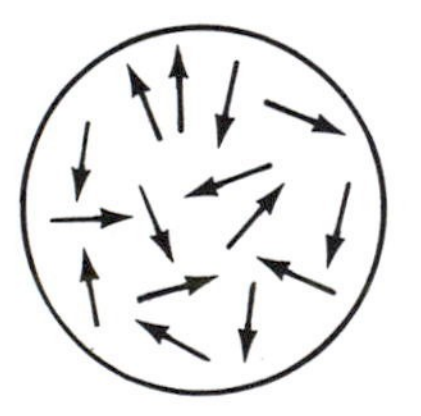

Nuclear magnets in the sample in the absence of B_0

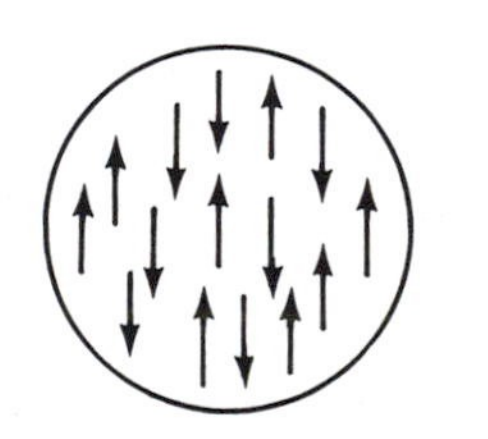

Nuclear magnets in the presence of B_0

For any given value of B_0, the two states of a spin $\frac{1}{2}$ nucleus have well-defined energies, say, E_2 for the less stable $I = -\frac{1}{2}$ state and E_1 for the more stable $I = +\frac{1}{2}$ state. The energy difference ΔE between the states is given by eq. 2-1, in which $\hbar$ is Planck's constant divided by 2π and γ is a collection of nuclear properties

$$\Delta E = E_2 - E_1 = \hbar \gamma B_0 \tag{2-1}$$

called the *gyromagnetic ratio*. This simple relationship is the fundamental equation of nuclear magnetic resonance. It shows that the energy separation between the states of a spin $\frac{1}{2}$ nucleus depends only on nuclear properties imbedded in the gyromagnetic ratio and on the B_0 field selected by the experimentalist. Since every magnetic nuclide, from hydrogen-1 to uranium-235, has a distinct gyromagnetic ratio, the NMR experiment may be fine-tuned to a specific nucleus. Furthermore, according to eq. 2-1, the energy difference between the

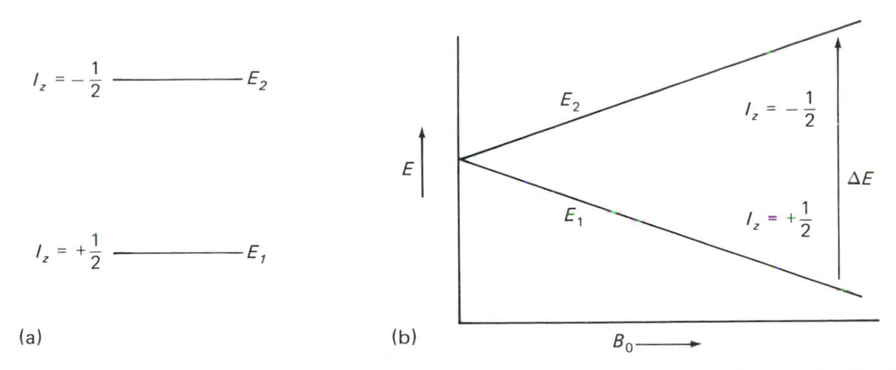

FIGURE 2-1 (a) Spin states of an $I = ½$ nucleus in a magnetic field of value B_0. (b) The dependence of ΔE on the value of B_0.

spin states is directly proportional to the B_0 field. The higher the field, the greater the energy separation of the spin states. These properties of nuclear magnets are illustrated in Figure 2-1.

The existence of these spin states may be detected by applying a second magnetic field, of value B_1, to the sample. The B_1 field is part of the electromagnetic spectrum and has a characteristic frequency ν (see page 6). This frequency may be adjusted until it achieves a value ν_0 that has the same energy ($h\nu_0$) as the energy difference between the spin states ($\hbar\gamma B_0$), eq. 2-1. Energy then can be absorbed by a $+\frac{1}{2}$ nucleus to convert it to a $-\frac{1}{2}$ nucleus. This absorption of energy is referred to as a *resonance* and may be detected electronically. Since $h\nu_0 = \hbar\gamma B_0$ at resonance, the resonance frequency ν_0 is $\gamma B_0/2\pi$ (Diagram 2-5).

Diagram 2-5

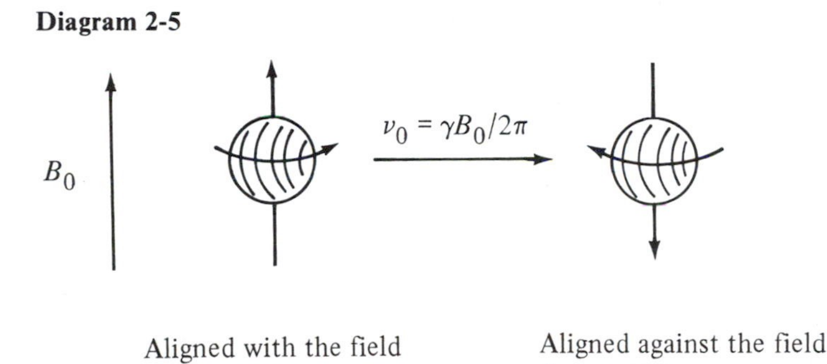

Values of B_0 range from 1.4 to 14 tesla. The B_1 fields have frequencies in the radiofrequency (rf) range, from 60 MHz when B_0 is 1.4 tesla to 600 MHz when B_0 is 14 tesla for a proton resonance (1 MHz is one million hertz or cycles per second). Other nuclei resonate at other distinct frequencies, for example, 25.2 MHz for ^{13}C and 10.1 MHz for ^{15}N at 2.35 tesla, for which ^{1}H resonates at 100 MHz.

These numbers mean that the energy states (E_1 and E_2) for protons are much farther apart than those for carbon-13 and nitrogen-15 at the same field. Absorption of energy is easier to observe when the states are farther apart, so the NMR experiment is particularly straightforward with protons. Diagram 2-6 illustrates the situation, but is not to scale. As ΔE increases, the relative energy of E_2 becomes higher, and fewer nuclear

Diagram 2-6

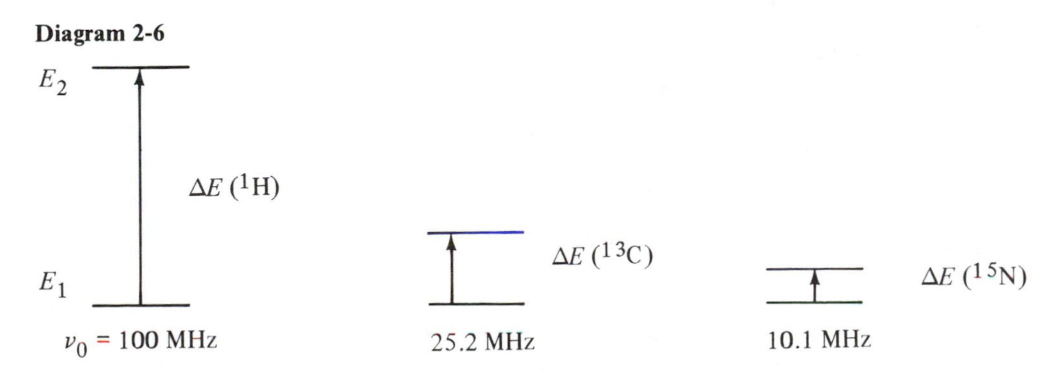

spins are present in E_2. More nuclei then are available in E_1 to absorb energy and undergo spin flips. The larger absorption means a more easily observed NMR signal.

The NMR experiment thus consists of immersing magnetic spins in a strong field B_0 to distinguish the nuclei according to their values of I_z, followed by application of a B_1 field at the frequency corresponding to the energy difference between spin states. This energy difference depends on the identity of the nucleus, according to its gyromagnetic ratio (eq. 2-1), and on the value of the B_0 field (Figure 2-1). How this subtle physics experiment can be exploited as a general tool to elucidate molecular structure is described in the next section.

2-2

The Chemical Shift

The above discussion seems to imply that all examples of a given nucleus such as the proton should have the same resonance frequency ν_0, determined only by γ and B_0. Examination of molecules with several types of protons, however, reveals that the exact resonance frequency depends on the nature of the electron cloud surrounding the nucleus. The actual field experienced by the nucleus, B_{local}, may be increased or decreased by the interaction between the electron cloud and the applied field B_0 (eq. 2-2). The quantity σ, called the

$$B_{local} = B_0(1 - \sigma) \tag{2-2}$$

shielding, is a measure of the ability of the electrons to alter B_0. Shielding varies from compound to compound and even within individual compounds. The resonance frequency of the nucleus therefore depends on its electronic environment, as well as on B_0 and γ (eq. 2-3). Nuclei are said to be

$$\nu_0 = \gamma B_0(1 - \sigma)/2\pi \tag{2-3}$$

shielded or screened by the surrounding electrons. Variation of the resonance frequency with electronic structure has been termed the *chemical shift*. The terms resonance frequency and chemical shift have in fact become synonymous in most contexts.

From eq. 2-3, it can be seen that electronic screening or shielding results in a lower resonance frequency ν_0 at constant B_0, since σ enters the equation with a negative sign. Conversely, as shielding decreases, the resonance frequency increases. For example, introduction of an electron-withdrawing group into a molecule reduces the electron cloud around a proton, so that there is less shielding and the resonance frequency increases. These factors are discussed in detail in Chapter 3, but they may be illustrated here by the resonance positions for the methyl protons in monosubstituted methanes.

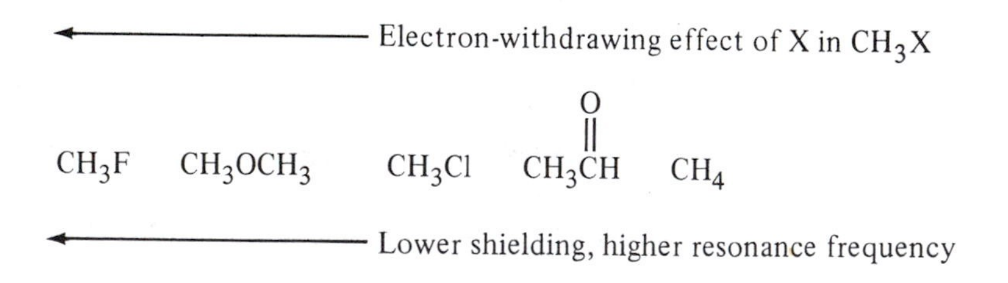

Figure 2-2 shows the NMR spectra of the protons (^1H) and the carbons (^{13}C at its 1.1% natural abundance) in methyl acetate ($CH_3CO_2CH_3$). Although 98.9% of naturally occurring carbon is nonmagnetic ^{12}C, the NMR experiment may be carried out on the small amount of ^{13}C, which is magnetic ($I = \frac{1}{2}$). The B_1 frequency in Figure 2-2 is varied so that each type of proton or carbon in turn comes into resonance and gives a spike-like signal. It can be seen, as expected, that there are two types of proton but three types of carbon. The methyl groups produce both proton and carbon resonances, but the carbonyl group of course has only a carbon resonance (Diagram 2-7).

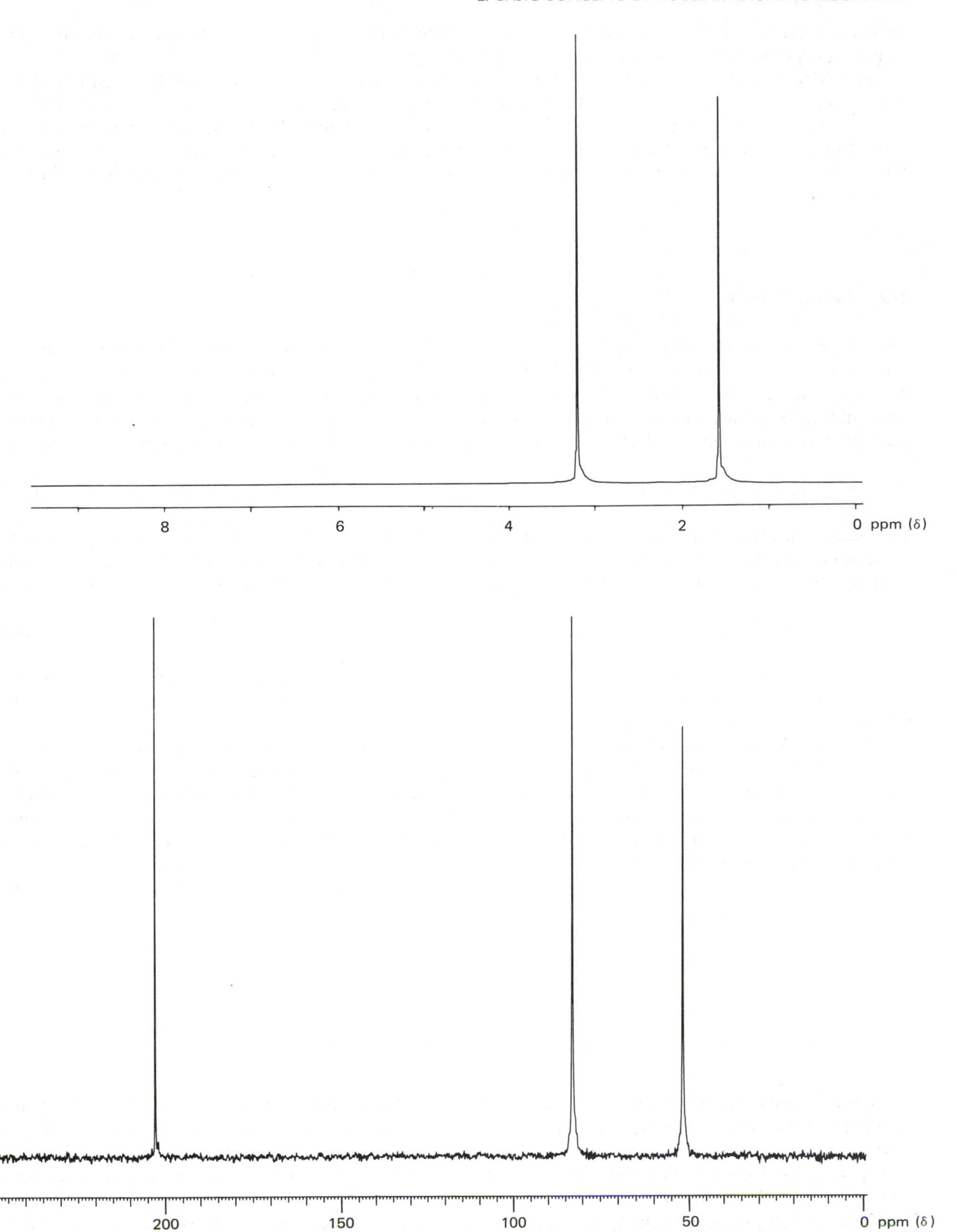

FIGURE 2-2 The 90 MHz proton spectrum (upper) and the 22.6 MHz carbon-13 spectrum (lower) of methyl acetate without solvent. The frequency units on the horizontal axis are defined in Section 2-4. Higher frequencies are always on the left.

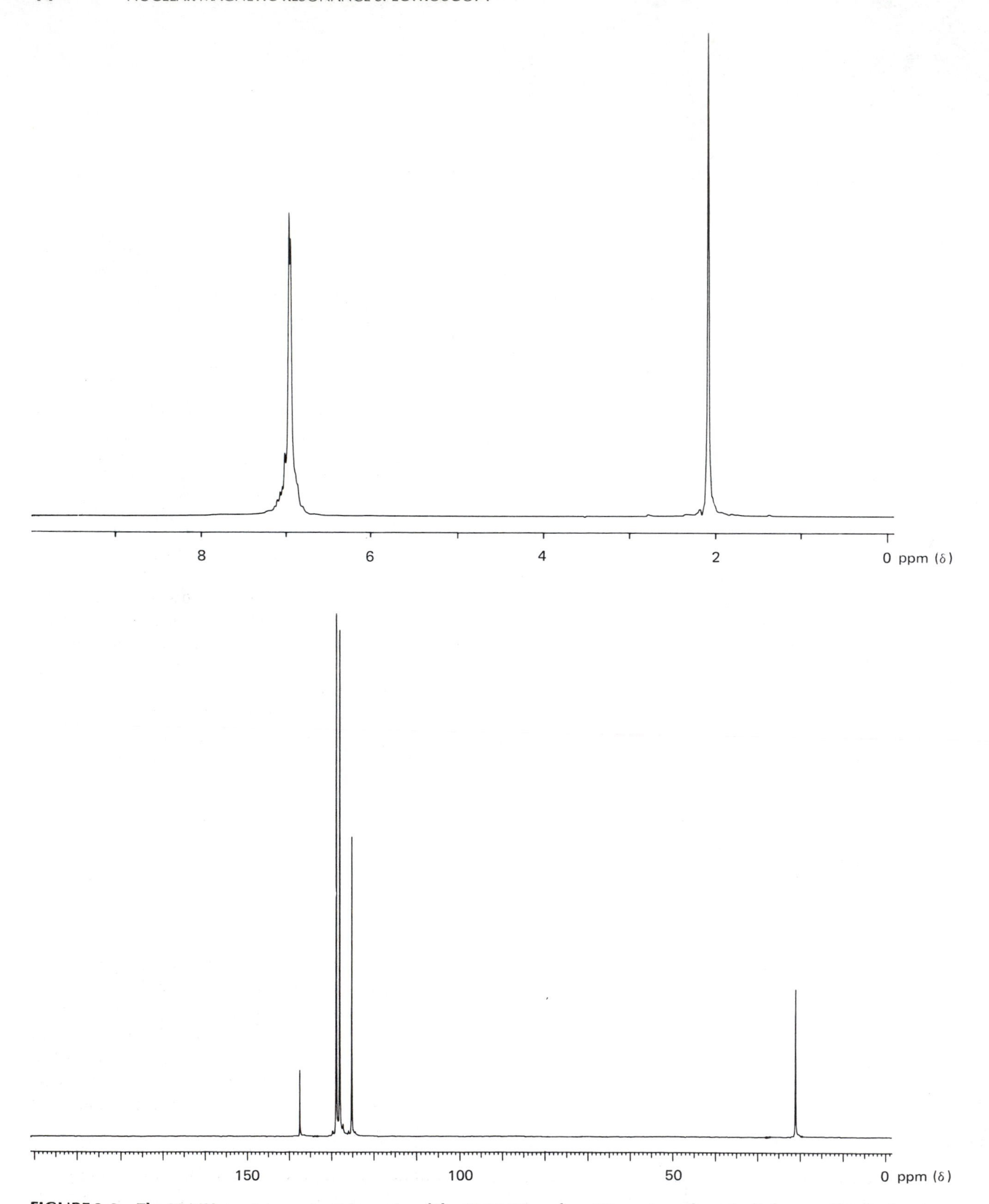

FIGURE 2-3 The 90 MHz proton spectrum (upper) and the 22.6 MHz carbon-13 spectrum (lower) of toluene without solvent.

Diagram 2-7

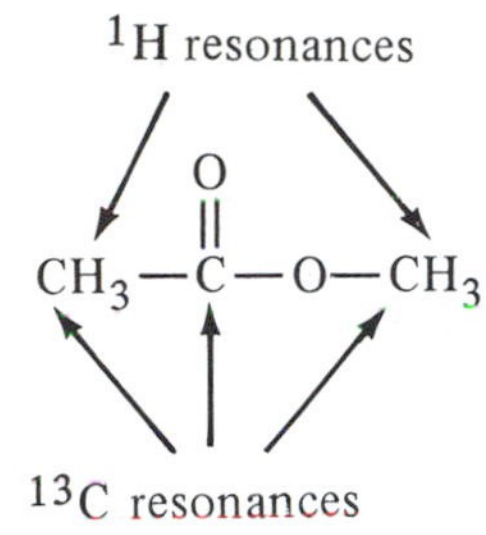

The proton resonances may be assigned by consideration of electronegativities of the groups attached to each methyl group. The ester oxygen is more electron withdrawing than the carbonyl group, so the $—OCH_3$ resonance is further to the left (higher frequency, less shielding) than the $CH_3(C{=}O)$ resonance. The molecule also would be expected to give two oxygen-17 resonances, for the carbonyl and ester oxygens. Again, the principal oxygen nuclide (^{16}O) is nonmagnetic, but the experiment may be carried out on the trace nuclide (^{17}O).

The success of the NMR experiment results from the narrow width of the resonances and the large range of the resonance frequencies. In frequency units (Hz or cps) each peak is typically less than 0.5 Hz wide and can be less than 0.1 Hz. The full frequency range for protons is more than 1000 Hz at 1.41 tesla, and most other nuclei, including ^{13}C, ^{15}N, ^{17}O, and ^{19}F, have ranges of many thousand hertz.

Not every chemically distinct proton has a different chemical shift. Inevitably there is some overlap, as in the aromatic protons in toluene (Figure 2-3). Although the methyl resonance (on the right) is well shifted from the aromatic resonance, the ortho, meta, and para protons do not have separately resolved resonances. The electronic environments of these protons are not sufficiently different to cause differential shielding. This problem occurs often in proton spectra. The carbon-13 spectrum of the same material, however, shows five resonances. In addition to the methyl resonance on the right, there are four aromatic peaks, for the ortho, meta, para, and ipso carbons. The ipso position of course carries no proton and would not have produced a resonance under any circumstance in the proton spectrum.

In this fashion we can see how NMR does resemble a molecular microscope. At one frequency we have a view of all the various protons, although sometimes they stand in each other's way and overlap. At other frequencies we can obtain a view of all the carbons, nitrogens, or oxygens. These resonances must be interpreted according to their frequency values, in order to try to assemble a molecular structure from the spectrum. This process is the subject of Chapter 3.

2-3

The Coupling Constant

The form of a resonance is altered when there are other nearby magnetic nuclei. In dichlorofluoromethane ($CHFCl_2$), for example, both the proton and the fluorine nucleus have spin of $\frac{1}{2}$. Both of the spin $\frac{1}{2}$ nuclei exist in two I_z spin states, $+\frac{1}{2}$ and $-\frac{1}{2}$. The proton then has one magnetic environment when ^{19}F is $+\frac{1}{2}$ and a slightly different one when ^{19}F is $-\frac{1}{2}$. As a result, the resonance of the proton contains two peaks (Diagram 2-8 and Figure 2-4). Similarly, the fluorine nucleus exists in two distinct magnetic environments because the proton has two spin states, so the ^{19}F spectrum also is a doublet (Figure 2-4). Quadrupolar nuclei such as chlorine often act as if they are nonmagnetic and may be ignored in this context. Thus the proton resonance of dichloromethane (CH_2Cl_2) is a singlet; the protons are not split by chlorine.

This influence of neighboring spins on the multiplicity of peaks is termed *spin–spin coupling*. The distance between the two peaks for the resonance of one nucleus split by another is a measure of how strongly the nuclear spins perturb each other and is given the symbol J, the *coupling constant*. In $CHFCl_2$ the coupling between H and F is strong and J is 50.7 Hz, a relatively large value. Hertz is actually a unit of frequency, but in this context it often is considered equivalent to a unit of energy (the measure of the spin–spin interaction).

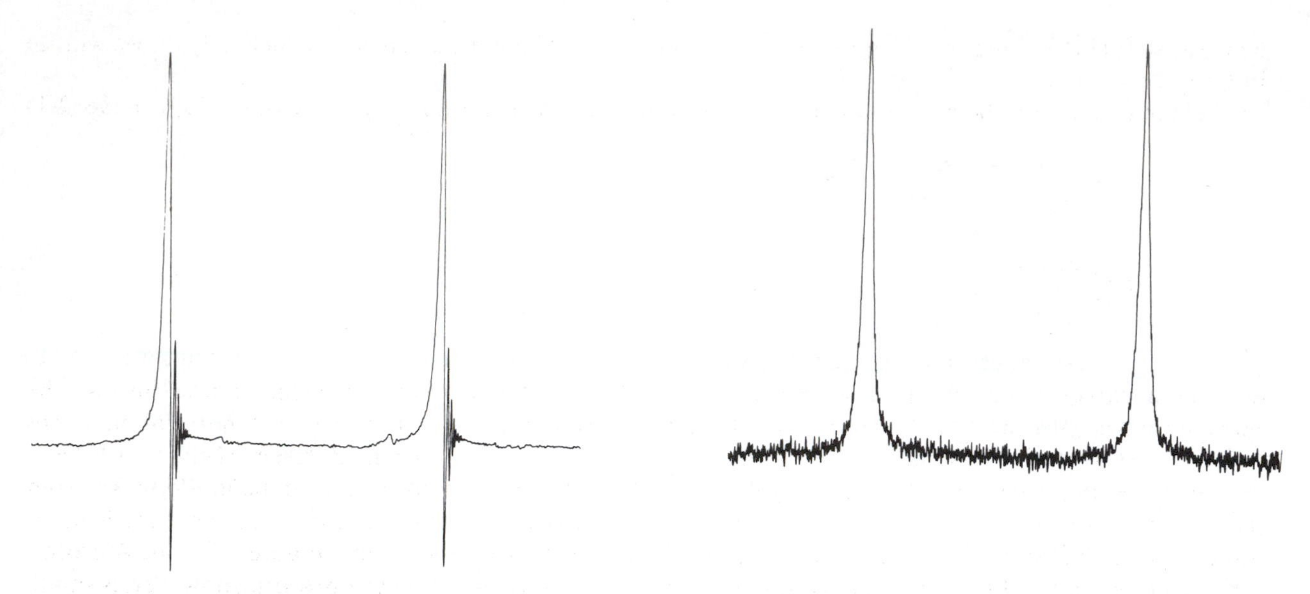

FIGURE 2-4 The 90 MHz proton spectrum (left) and the 84.6 MHz fluorine-19 spectrum (right) of $CHFCl_2$ without solvent.

Diagram 2-8

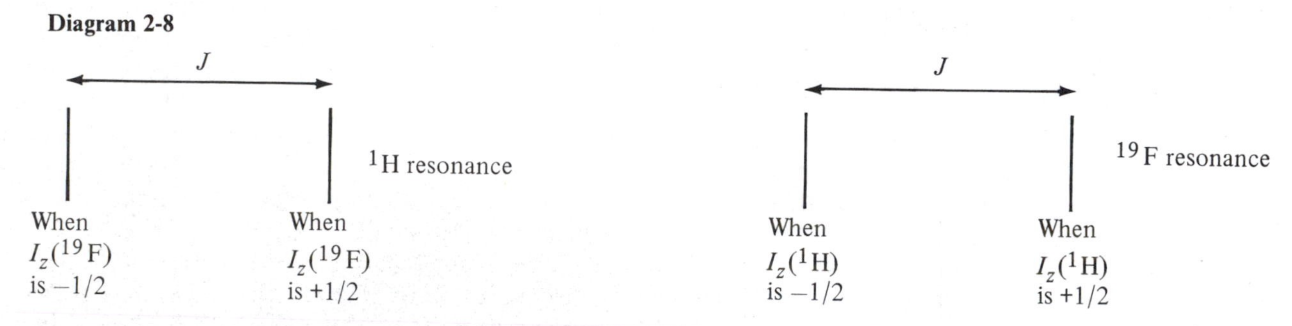

Frequency may be converted to energy simply through multiplication by Planck's constant, $\Delta E = h\nu$. When there are only two nuclei in a coupled system, the resulting spectrum is referred to as AX. Notice that the splitting in both the A and the X portions of the spectrum is the same (Figure 2-4) because J is a measure of the interaction between two nuclei and is the same when viewed from the perspective of either the proton or the fluorine.

How does one nucleus "know" what the spin states of neighboring nuclei are? Although there are several conceivable mechanisms, the most important is a rather indirect one. The neighboring nucleus very slightly polarizes the spins of its surrounding electrons, that is, it makes the electron spins line up slightly more in one direction with respect to the B_0 field. This spin polarization is passed through the electrons of other bonds and finally reaches the resonating nucleus (Diagram 2-9). Thus J normally is a through-bond interaction and

Diagram 2-9

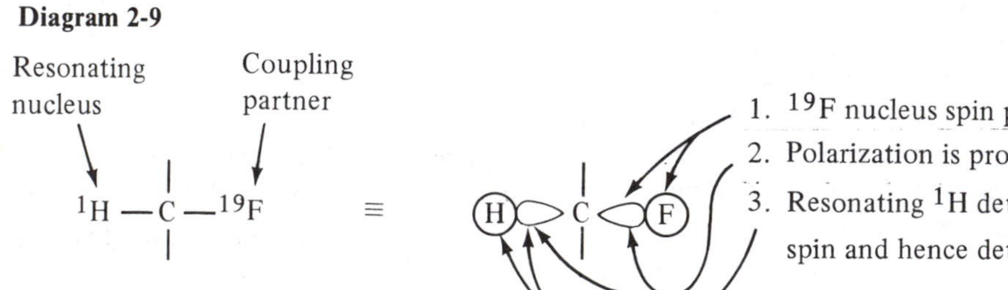

hence is useful in drawing conclusions about the arrangement of bonds, that is, stereochemistry, as we will see in Chapter 4.

There may of course be more than one nucleus in the vicinity of a resonating nucleus. Cyclopropene **(2-1)**

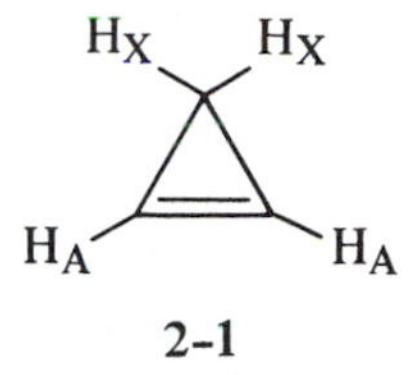

provides a classic example of an A_2X_2 spectrum. Each X proton is coupled to two identical A protons. The spins of the two A protons must be considered together. Both spins can be plus $(+\frac{1}{2}, +\frac{1}{2})$, both can be minus $(-\frac{1}{2}, -\frac{1}{2})$, or one can be plus and the other minus $(+\frac{1}{2}, -\frac{1}{2}$ or $-\frac{1}{2}, +\frac{1}{2})$. There is no difference between the (+−) and the (−+) arrangements. Thus an X proton has three different environments, depending on the spin states of the A protons: (++), (+−)/(−+), (−−). Therefore the X proton will have a resonance made up of three peaks in the ratio 1/2/1 (Diagram 2-10), as in the spectrum of cyclopropene (Figure 2-5). The A proton also is split into a 1/2/1 triplet for the same reason. The spin interaction for cyclopropene, however, is small, and J is only 1.8 Hz (the distance between any two adjacent peaks).

Diagram 2-10

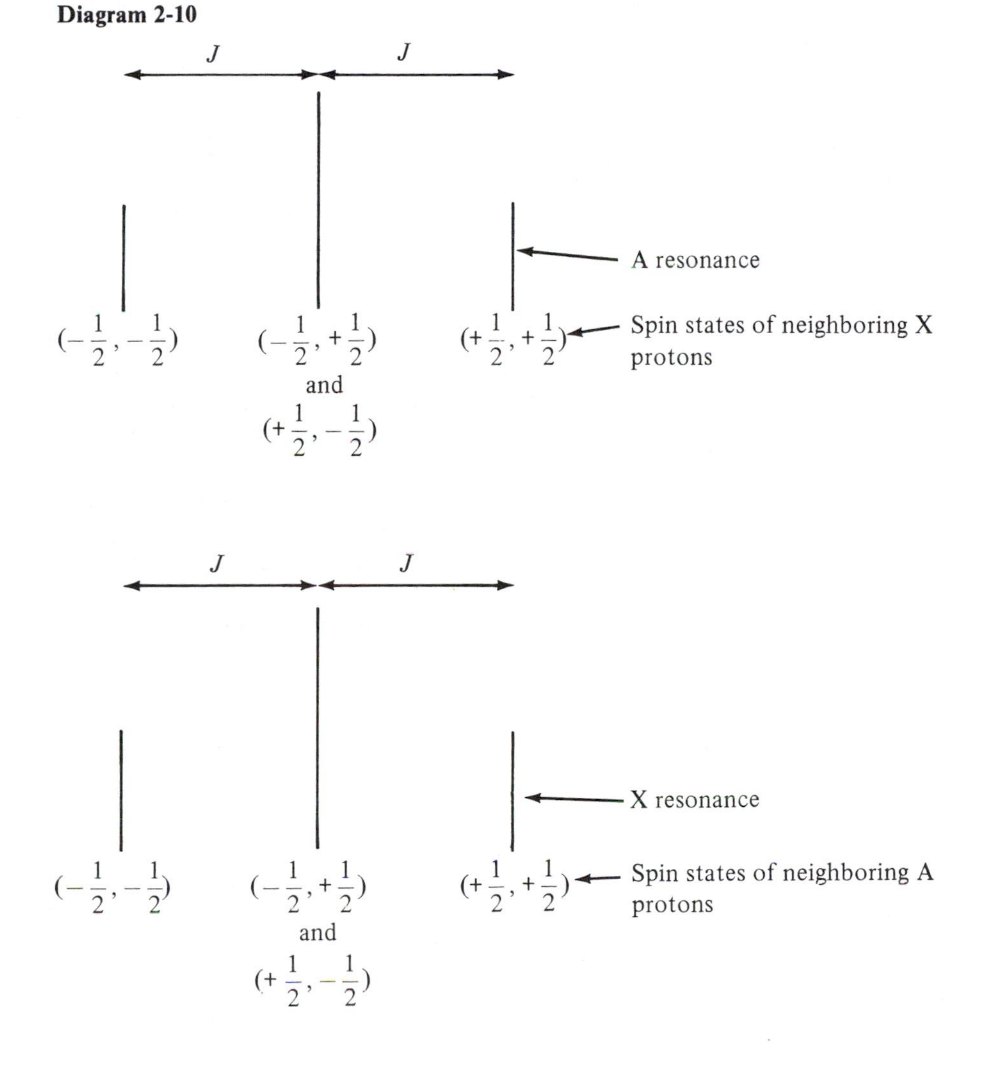

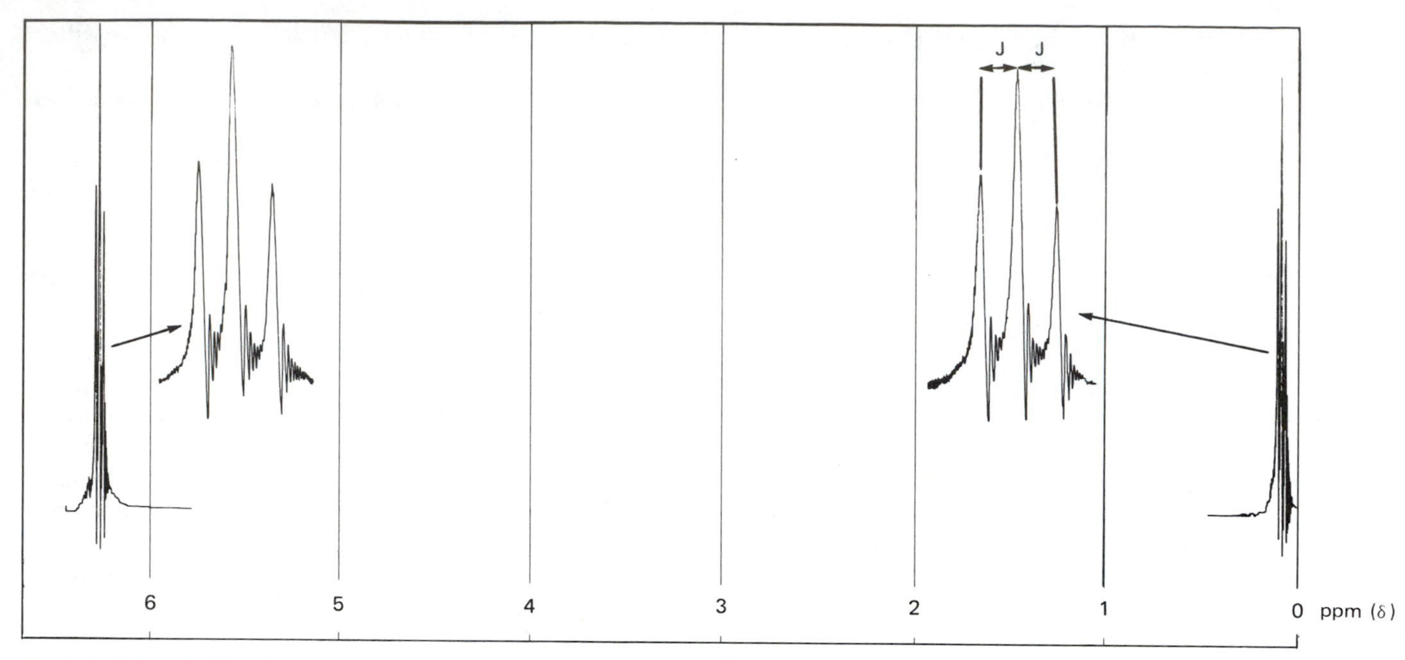

FIGURE 2-5 The 90 MHz proton spectrum of cyclopropene (neat) taken at −70°C. The spectrum is 600 Hz wide, and the insets have been expanded by a factor of 10. [Reprinted with permission from J. B. Lambert, A. P. Jovanovich, and W. L. Oliver, Jr., *J. Phys. Chem.*, **74**, 2221 (1970). Copyright 1970 American Chemical Society.]

As numbers of neighboring spins increase, so does the complexity of the spectrum. The ethyl groups in diethyl ether form an A_2X_3 spectrum (Figure 2-6). The methyl protons are split by the two methylene protons into a triplet, similar to the triplets of Figure 2-5. Because the methylene protons are split by three methyl protons, there are four peaks in the CH_2 resonance. This arrangement arises because the neighboring methyl protons can exist one way with all spins positive (+++), three ways with two spins positive (++−, +−+, −++), three ways with two spins negative (−−+, −+−, +−−), and one way with three spins negative

Diagram 2-11

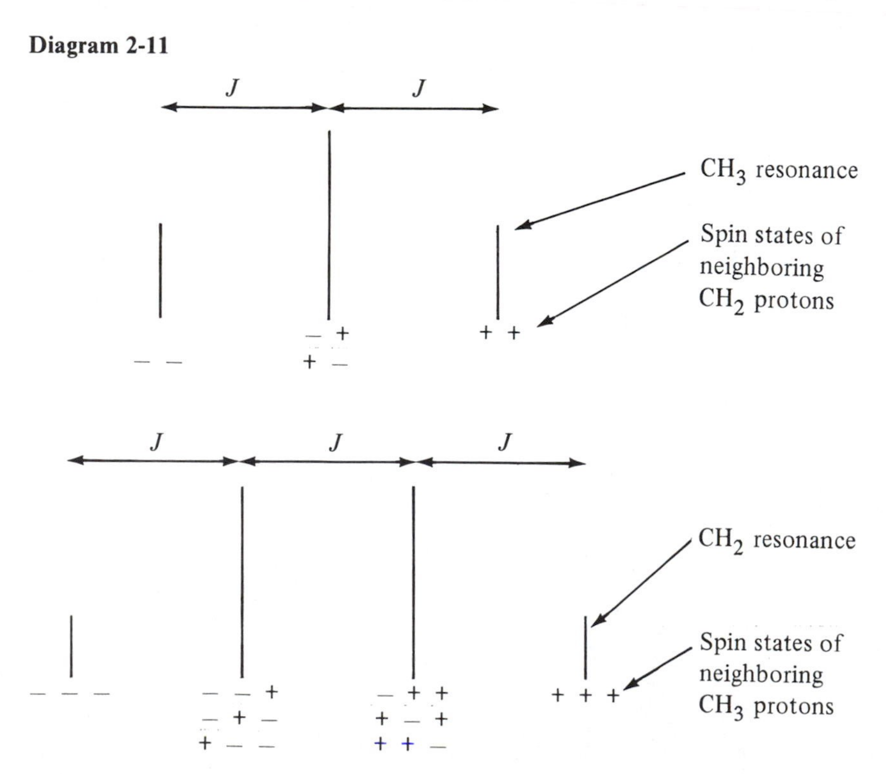

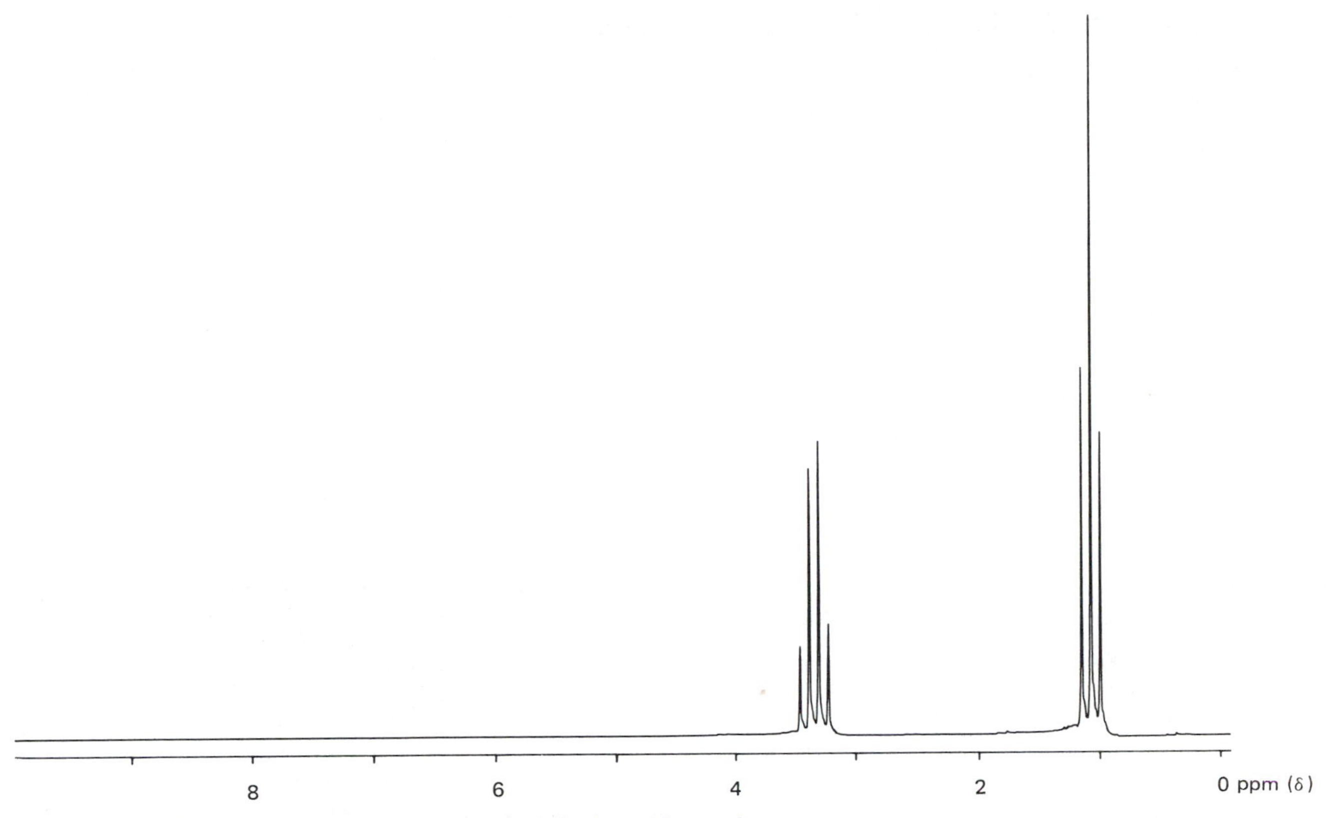

FIGURE 2-6 The 90 MHz proton spectrum of diethyl ether without solvent.

(———).The result is a 1/3/3/1 quartet (Diagram 2-11). The triplet–quartet pattern seen in Figure 2-6 is a reliable diagnostic for an ethyl group.

The splitting patterns of large spin systems may be deduced in a similar fashion, but they are less often encountered. If a nucleus is coupled to n nuclei, there are always $n + 1$ peaks, unless so-called second-order effects are present (Chapter 4). The intensity ratios correspond to the coefficients in the binomial expansion and may be obtained from Pascal's triangle, since we are dealing with statistical arrangements of independent events (the I_z identities of the neighboring spins). Pascal's triangle (Figure 2-7) is developed by summing two adjacent integers and placing the result one row lower and between the two integers. Zeros are imagined outside the triangle in order to obtain the figures along the edges. The first row (1) gives the resonance multiplicity when there is no neighboring spin, the second row (1/1) when there is one neighboring spin, and so on. We have already seen that two neighboring spins give a 1/2/1 triplet and three neighboring spins give a 1/3/3/1 quartet. Four neighboring spins are present for the CH proton in the arrangement —CH_2—CHX—CH_2— (X is nonmagnetic), and the CH resonance would be a 1/4/6/4/1 quintet. The CH resonance from an

1
1 1
1 2 1
1 3 3 1
1 4 6 4 1
1 5 10 10 5 1
1 6 15 20 15 6 1
1 7 21 35 35 21 7 1
1 8 28 56 70 56 28 8 1
1 9 36 84 126 126 84 36 9 1
1 10 45 120 210 252 210 120 45 10 1

FIGURE 2-7 Pascal's triangle.

TABLE 2-1 Common First-Order Spin–Spin Splitting Patterns

Spin System	Molecular Substructure	A Multiplicity	X Multiplicity
AX	$—CH^A—CH^X—$	doublet (1/1)	doublet (1/1)
AX_2	$—CH^A—CH_2^X—$	triplet (1/2/1)	doublet (1/1)
AX_3	$—CH^A—CH_3^X$	quartet (1/3/3/1)	doublet (1/1)
AX_4	$—CH_2^X—CH^A—CH_2^X—$	quintet (1/4/6/4/1)	doublet (1/1)
AX_6	$CH_3^X—CH^A—CH_3^X$	septet (1/6/15/20/15/6/1)	doublet (1/1)
A_2X_2	$—CH_2^A—CH_2^X—$	triplet (1/2/1)	triplet (1/2/1)
A_2X_3	$—CH_2^A—CH_3^X$	quartet (1/3/3/1)	triplet (1/2/1)
A_2X_4	$—CH_2^X—CH_2^A—CH_2^X—$	quintet (1/4/6/4/1)	triplet (1/2/1)

isopropyl group, $—CH(CH_3)_2$, would be a 1/6/15/20/15/6/1 septet. Several possible spin systems are described in Table 2-1.

Appreciable couplings are primarily observed between neighboring (*vicinal*) protons, over three bonds (H—C—C—H) (Table 2-2). Coupling over four or more bonds is usually too small to observe. For protons to couple over two bonds, they must be attached to the same carbon (H—C—H) and are said to be *geminal* to each other. Geminal protons often are related by a molecular symmetry element and hence are equivalent. As a first approximation, equivalent protons never split each other. For this reason, unsplit singlets are observed for numerous simple molecules, such as benzene, acetone, chloromethane, ethene, and ethane. Exceptions to these general rules are discussed in Chapter 4.

Because 99% of carbon is the nonmagnetic ^{12}C, for all practical purposes protons are not split by carbon in proton spectra. The 1% of protons that are attached to and coupled to ^{13}C give resonances that are obscured by the resonances of all the protons attached to ^{12}C. In carbon spectra, however, the 1% of ^{13}C is examined directly, since ^{12}C gives no spectrum. When carbons are attached to protons, the ^{13}C resonances are split in the usual way. The most important splittings are those of the ^{13}C by protons that are directly attached. Thus a methyl carbon (CH_3) is split into a quartet, a methylene carbon (CH_2) into a triplet, and a methinyl carbon (CH) into a doublet; a quaternary carbon is not split. Figure 2-8 (upper) shows the ^{13}C spectrum of 3-hydroxybutyric acid, $CH_3CH(OH)CH_2CO_2H$, which contains a carbon resonance with each type of multiplicity. From right to left may be seen a quartet ($^{13}CH_3$), a triplet ($^{13}CH_2$), a doublet (^{13}CH), and a singlet ($^{13}CO_2H$).

Why do the carbon spectra in Figures 2-2 and 2-3 not show these splittings? There is an instrumental technique, termed *decoupling*, whereby spin–spin splittings may be removed. A third magnetic field, B_2, is applied at the resonance frequency of the neighboring nucleus. This field shuttles the spins back and forth between the various spin states, for example, $+\frac{1}{2} \rightleftharpoons -\frac{1}{2}$. If this process is very rapid, the resonating nucleus no longer detects individual spin states of the neighboring nucleus, but only the average, as if there were no coupling. Under such instrumental conditions, the resonating nucleus is said to be decoupled from the

TABLE 2-2 Common Types of Coupling Between Protons

Coupled Protons	Number of Bonds	Size of *J*	Example
H—C—C—C—H	4	small	$CH_3OC(=O)H$
H—C—C—H	3	normally observable	$ClCH_2CH_3$
H—C—H	2	not normally observable between protons related by symmetry	CH_2Cl_2

neighboring nucleus. For example, in Figure 2-2 (lower), all proton frequencies were subjected to the B_2 irradiation while B_1 was used to detect carbon resonances. As a result, both carbon resonances came out as singlets, decoupled from the protons. Figure 2-3 (lower) and Figure 2-8 (lower) also show proton-decoupled ^{13}C spectra. This instrumental procedure is quite useful because it results in one singlet for each type of

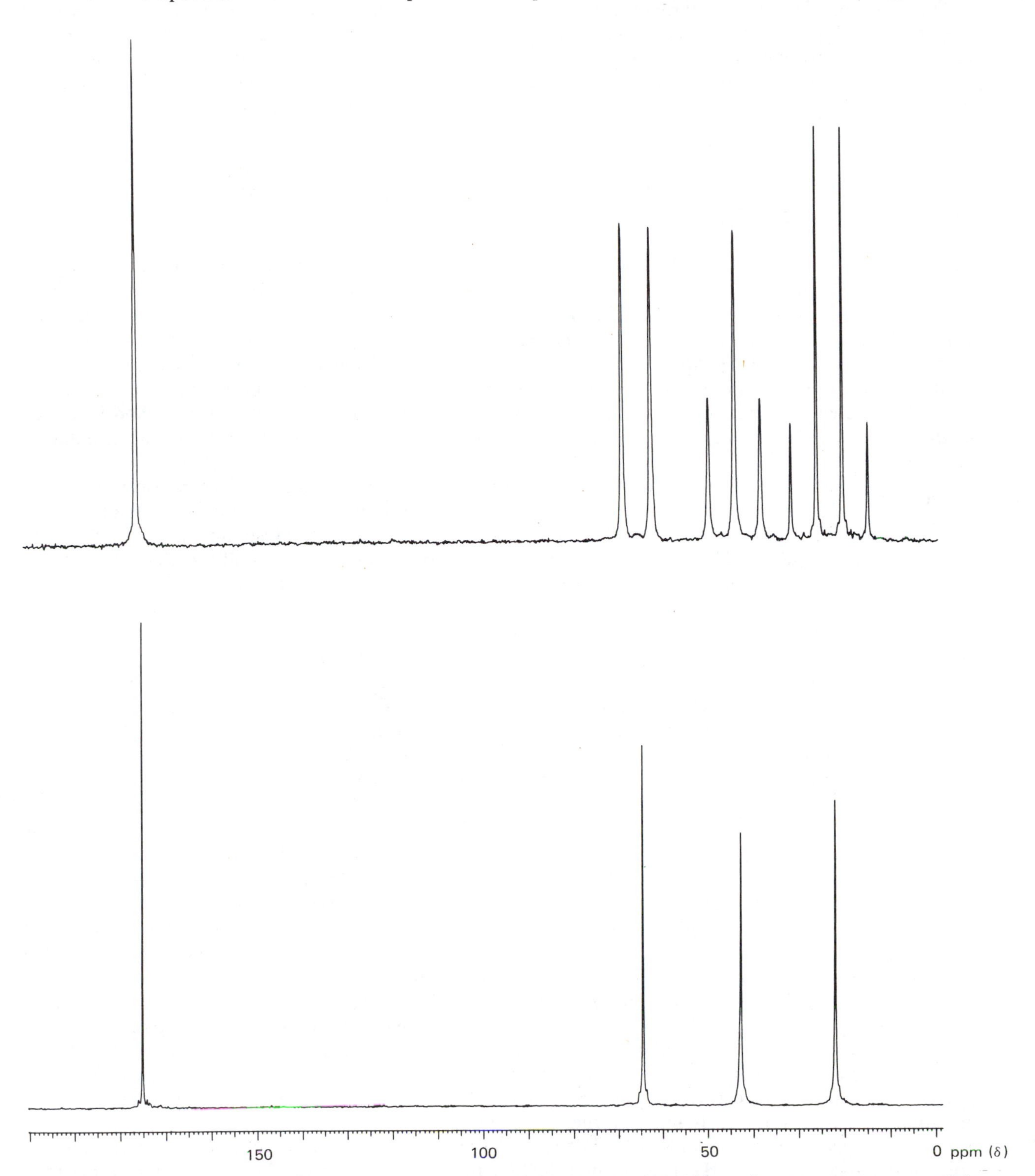

FIGURE 2-8 Upper: The 22.6 MHz carbon-13 spectrum of 3-hydroxybutyric acid without solvent. Lower: Spectrum of the same compound with decoupling at all proton frequencies.

carbon, a particularly simple map of the carbon skeleton of a molecule. The decoupling of protons from protons is discussed in Chapter 5. A fully coupled ^{13}C spectrum, as in the upper part of Figure 2-8, also is useful because it reveals from the splitting pattern whether a given carbon is methyl, methylene, methine, or quaternary. Thus a typical ^{13}C experiment often involves recording both coupled and decoupled spectra.

The spin–spin splitting pattern of a nucleus provides information about the number and type of neighboring magnetic nuclei. In a structural problem, the proton–proton splittings help us to assemble pieces of an unknown molecule. It is important to appreciate that magnetically equivalent nuclei do not split each other. The methyl proton resonance in an ethyl group is a triplet because of coupling to the neighboring methylene protons (not a quartet because the methyl group contains three protons).

2-4

Continuous Wave Spectra

Production of proton spectra such as those seen in Figures 2-2 and 2-3 (upper) requires a means to vary the frequency of the B_1 field so that all nuclei in turn may come into resonance. Several methods have been developed to carry out this operation. The most straightforward would be simply to hold B_0 constant and vary B_1. The B_0 field must be constant; otherwise, the resonance frequency for a given nucleus would be changing continuously. In the spectral representations of Figure 2-2 and in all others in Parts I and V of this book, frequency increases from right to left, as can be seen in the units underneath the spectral display. Increased frequency implies decreased shielding, by eq. 2-3, when B_0 is constant. This method is termed *continuous wave* and *frequency sweep*, since the frequency is swept continuously from one extreme to another.

Most older machines used a variation of this technique in which the B_1 frequency was held constant and the B_0 field was swept. Imagine in this approach that by proper variation of B_0 every nucleus actually has the same resonance frequency, the ν_0 of B_1. The field is altered so that all proton resonances, for example, are brought to this frequency. The method appears convoluted, but it was instrumentally very easy to alter the magnetic field with a sawtooth sweep. The procedure is termed *continuous wave* and *field sweep* and is still used on some current 60 and 90 MHz proton instruments.

When the frequency is held constant, the field increases from left to right in the spectral display. Consider the case of CH_3Cl compared with CH_4, with reference to eq. 2-3 [$\nu_0 = \gamma B_0(1 - \sigma)/2\pi$]. The chlorine atom removes electron density from the protons and deshields them (σ is smaller). To keep ν_0 the same for both CH_3Cl and CH_4, the B_0 field must be lowered in the case of CH_3Cl. At the lower B_0 field the less shielded CH_3Cl protons can resonate at the same frequency as the CH_4 protons. In terms of eq. 2-3, to keep ν_0 constant B_0 must be lowered when shielding (σ) is smaller, that is, a decrease B_0 offsets an increase in $(1 - \sigma)$. In a spectral display the right side corresponds to high shielding, low frequency, and high field.

Visually there is no way to distinguish frequency-sweep and field-sweep spectra. Because field-sweep instruments dominated the market early in the history of NMR, certain terminology has endured that has little relevance in many current spectral displays. For example, because the field increases from left to right, the left end of the spectrum is termed low field (or downfield) and the right end high field (or upfield). Thus in the spectrum of toluene (Figure 2-2, upper), the aromatic proton resonance is said to be at low field and the methyl proton resonance at high field, even if the spectrum was obtained by sweeping the frequency at constant field. It would be more accurate to say that the aromatic peak is at high frequency and the methyl peak at low frequency, but such terminology has not come into use.

One system of units for the NMR spectrum might be absolute frequency. At a B_0 of 1.41 tesla, protons resonate at a ν_0 of about 60 MHz. A scale involving numbers like 60.007238 MHz is cumbersome, so that a simpler scale for chemical shifts was developed during the field-sweep era. A reference material was chosen and assigned a position of 0.0. Changes in chemical shifts from this position are measured in parts per million (ppm) of the magnetic field. Thus in the proton spectra of Figures 2-2, 2-3, 2-5, and 2-6 the units are ppm. The ppm unit is used for both field- and frequency-sweep experiments.

Since protons resonate at 60.0 MHz at 1.41 tesla, 1 ppm of the frequency is 60 Hz in energy (frequency) units at this field. At 300 MHz, 1 ppm equals 300 Hz for the ^1H, and so on. The use of ppm rather than absolute field (tesla) or absolute frequency (Hz) means that the unit of chemical shift is independent of the choice of B_0. Two resonances that are 1 ppm apart at 1.41 tesla (60 MHz) are still 1 ppm apart at 2.36 tesla (100 MHz). The unit of ppm is also used for ^{13}C spectra (Figures 2-2, 2-3, and 2-8), but in general the scale is much larger.

Both the ^{1}H and the ^{13}C resonances of tetramethylsilane (Me_4Si or TMS) occur at a very high field (low frequency), almost at the extreme of observed resonances, because silicon is electron donating and increases shielding. In addition, TMS is volatile, soluble in most organics, and almost inert chemically and gives a strong signal. For these reasons, TMS has been selected to serve as the reference for ^{1}H and ^{13}C resonances. The distance δ in parts per million (ppm) of a resonance from that of TMS is the standard expression for the chemical shift. The distance measured in Hz may be converted to ppm by eq. 2-4. By convention, δ is positive

$$\delta\ (\text{ppm}) = \frac{\text{distance from TMS in Hz}}{\text{value of } B_1 \text{ in MHz}} \tag{2-4}$$

in the downfield (up-frequency) direction, from right to left in the spectra found in this text. Thus a resonance that is 84 Hz downfield of TMS at 60 MHz has a δ of 1.4 (84/60). Resonances at higher field than TMS have negative δ. Since TMS is insoluble in water, other internal standards are used for aqueous solutions. The sodium salts of 3-(trimethylsilyl)-1-propanesulfonic acid [$(CH_3)_3Si(CH_2)_3SO_3Na$] or of 3-(trimethylsilyl)-propionic acid [$(CH_3)_3SiCH_2CH_2CO_2Na$] are water soluble. The respective trimethylsilyl groups in these compounds resonate near TMS and can be used as an aqueous internal standard with δ values of 0.0. The methylene resonances are small and may be ignored. Diagram 2-12 shows a typical scale for a ^{1}H spectrum.

Diagram 2-12

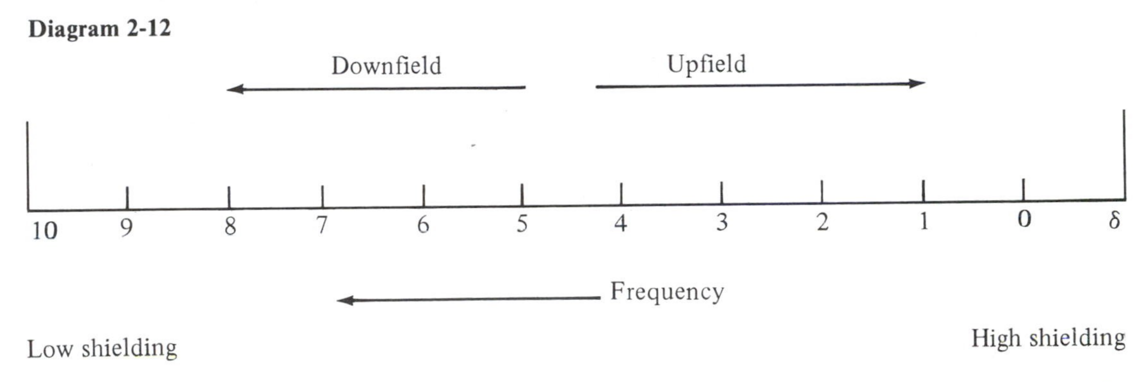

The coupling constant J is the measure of an interaction between two nuclei and is therefore independent of B_0. A J of 6.0 Hz at 60 MHz (0.1 ppm) is still 6.0 Hz at 300 MHz (0.02 ppm). Hence J is always expressed in frequency or energy units (Hz). If it is not known whether a particular separation is a chemical shift or a coupling constant, the problem is readily solved by taking the spectrum at two fields. A chemical shift separation in Hz changes with field, but a coupling constant remains the same.

2-5 Quantitation

Absorption of energy when nuclear spins flip over from $+\frac{1}{2}$ to $-\frac{1}{2}$ is directly proportional to the number of spins present. Thus a methyl group will have three times the proton signal of a single proton. This difference in intensity can be measured through electronic or mechanical integration and can be exploited to help work out the structure of unknown molecules. Figure 2-9 shows what an electronic integration looks like, for the spectrum of ethyl crotonate ($CH_3CH{=}CHCO_2C_2H_5$). The continuous line above the peaks is displaced vertically whenever there is a signal. A pencil and a ruler can provide intensity information from the displacement, showing here, for example, that the doublet at δ 5.8, the methylene quartet at δ 4.2, and the methyl triplet at δ 1.3 are in the ratio 1/2/3. Such conclusions about intensities are arrived at by the following procedure. The vertical displacement for each resonance is measured with a ruler. The resonance from a single proton (CH) is assigned an integral of unity and its vertical displacement is divided into that of all other resonances to give the number of protons for the remaining resonances. The procedure is illustrated in Diagram 2-13 for three resonances from Figure 2-9. If a resonance of unit intensity is not known, often a methyl resonance may be identified and assigned an integral of 3.0. For many instruments today the integral is provided by a digital readout rather than by the continuous line seen in Figure 2-9.

Diagram 2-13

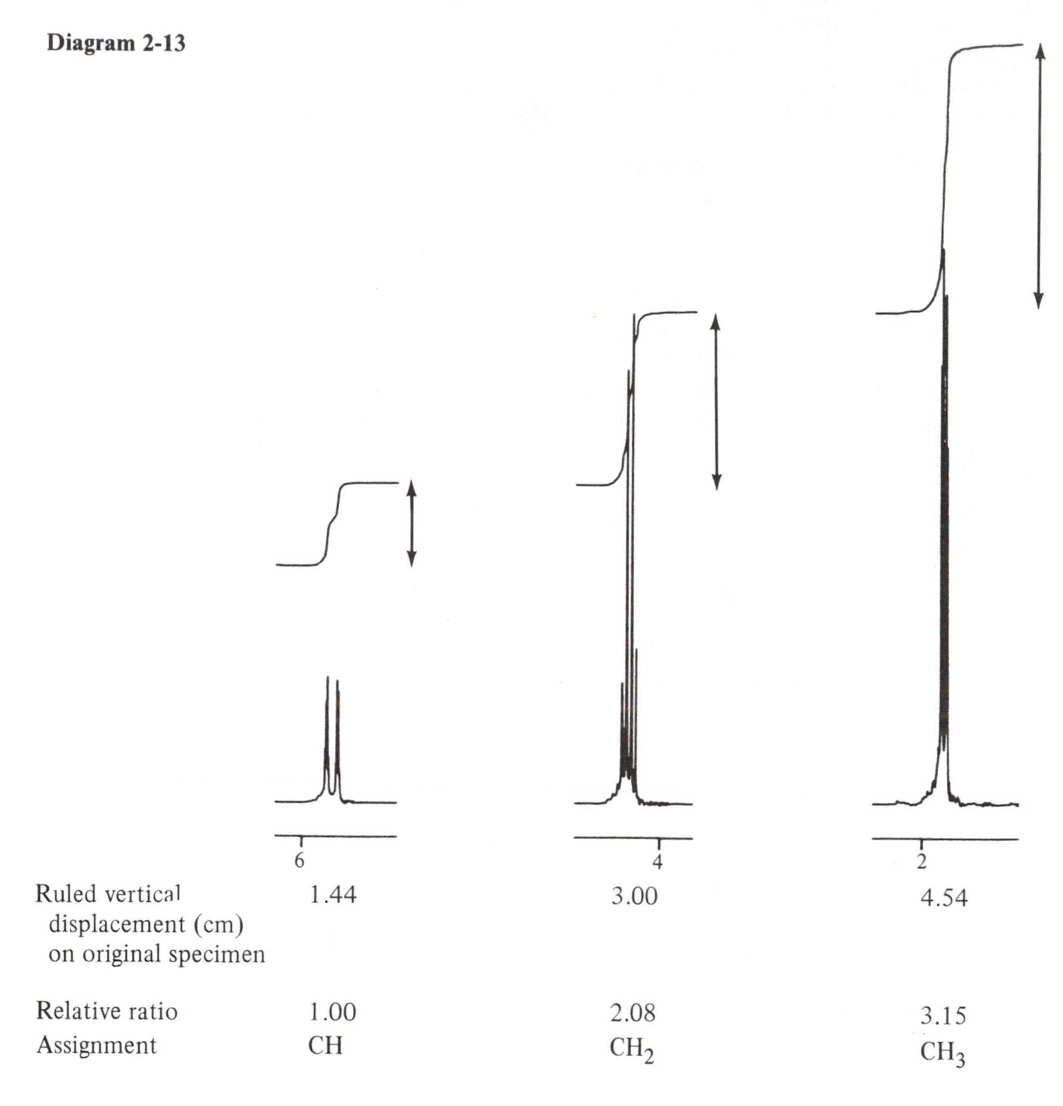

Routine spectral examination includes recording both the resonances and the integral, so that number as well as type of proton or carbon may be examined. Each of the peaks in the spectrum of ethyl crotonate (Figure 2-9) may be assigned by examination of the integral and the splitting pattern. The triplet at highest field (δ 1.3) has an integral of three and must come from a methyl group. Its coupling corresponds to that in the quartet in the middle of the spectrum at δ 4.2, whose integral is two. This latter multiplet must come from a methylene group. The mutually coupled methyl triplet and methylene quartet form the resonances for the ethyl group attached to oxygen in ethyl crotonate. The methylene resonance is further downfield than the methyl resonance because CH_2 is closer than CH_3 to the electron-withdrawing oxygen.

The remaining resonances in the spectrum come from protons that are coupled to more than one type of proton and deserve closer examination. The lowest field resonance (δ 7.0) has an intensity of unity and comes from one of the two alkene (CH=) protons. This resonance is split into a doublet by the other alkene proton, and then each member of the doublet is further split into a quartet by coupling to the methyl group on carbon. Stick diagrams are useful in analyzing complex multiplets, as in Diagram 2-14. The resonance at δ

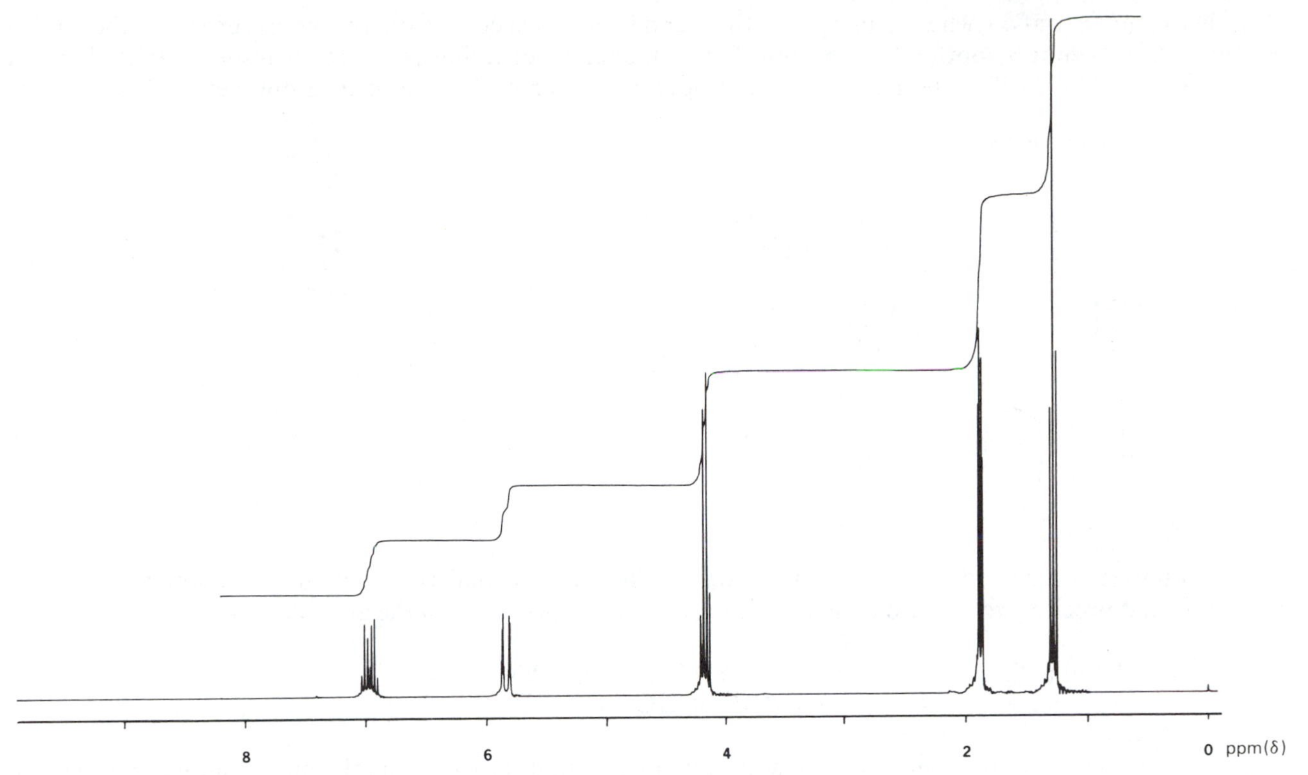

FIGURE 2-9 The 270 MHz proton spectrum of ethyl crotonate without solvent. The continuous line above the resonance is the integral.

Diagram 2-14

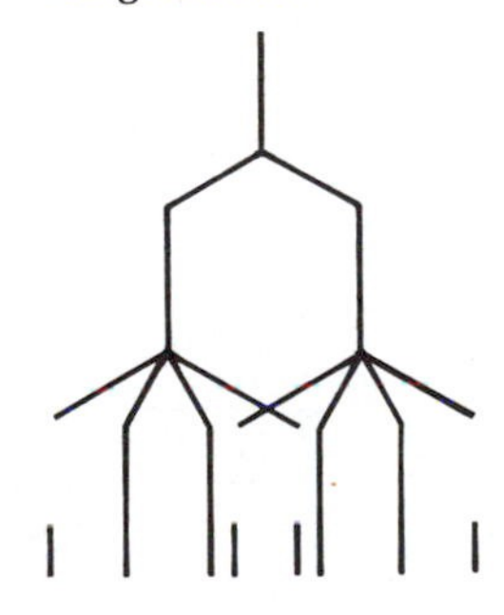

5.8 of unit integral is quite similar. It is split into a doublet by the other alkene proton. Although each member of the doublet is still split by the methyl group into a quartet, the splitting is so small that the resonance appears essentially as a doublet (Diagram 2-15). The significance of these differences in the magnitude of couplings is discussed in Chapter 4.

Diagram 2-15

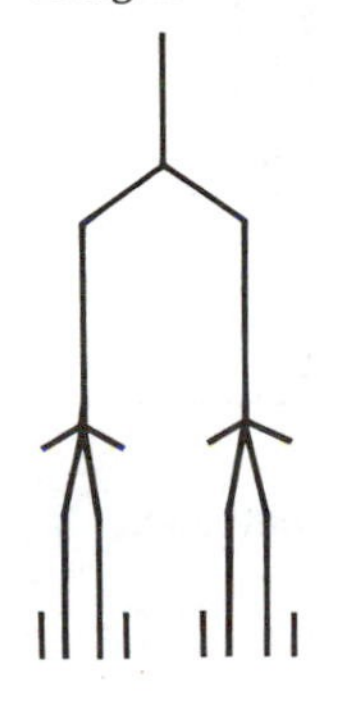

The resonance at δ 1.9 has an integral of three and hence must come from the methyl group on the double bond. As it is split by both alkene protons but with unequal couplings, four peaks result. A stick diagram (Diagram 2-16) clarifies the analysis. This grouping of four peaks is termed a doublet of doublets; the

Diagram 2-16

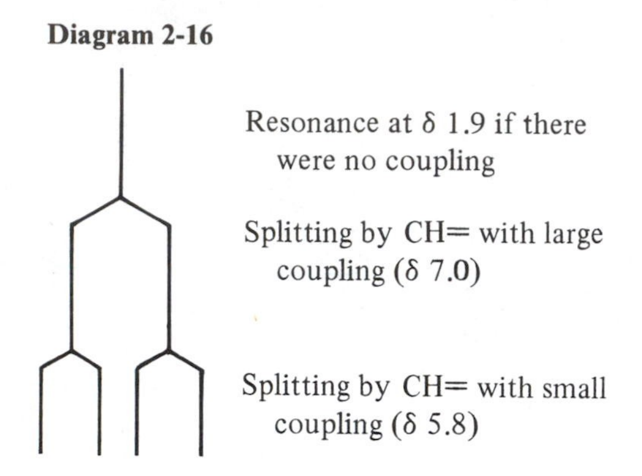

term quartet is reserved for a 1/3/3/1 multiplet. The two unequal couplings in the resonance at δ 1.9 correspond precisely to the two quartet splittings found, respectively, in the alkenic resonances at low field.

Final assignments: δ 1.9 (5.8, 7.0) 4.2 1.3
$CH_3CH{=}CHCO_2CH_2CH_3$

The exact values of the chemical shifts and coupling constants give further information about the structure and will be considered in Chapters 3 and 4. In the first approach to the integral of an unknown compound, it is necessary to select one resonance and assign a certain number of protons to it. The integral gives only relative intensities. Once a resonance is assigned an absolute intensity, all the other intensities follow.

Mixtures as well as pure compounds may be analyzed by NMR spectroscopy. In this case, the integral is particularly useful in figuring out how much of each component is present. Figure 2-10 is the ^1H spectrum of a mixture of bromobenzene (C_6H_5Br), dichloromethane (CH_2Cl_2), and iodoethane (ethyl iodide, CH_3CH_2I).

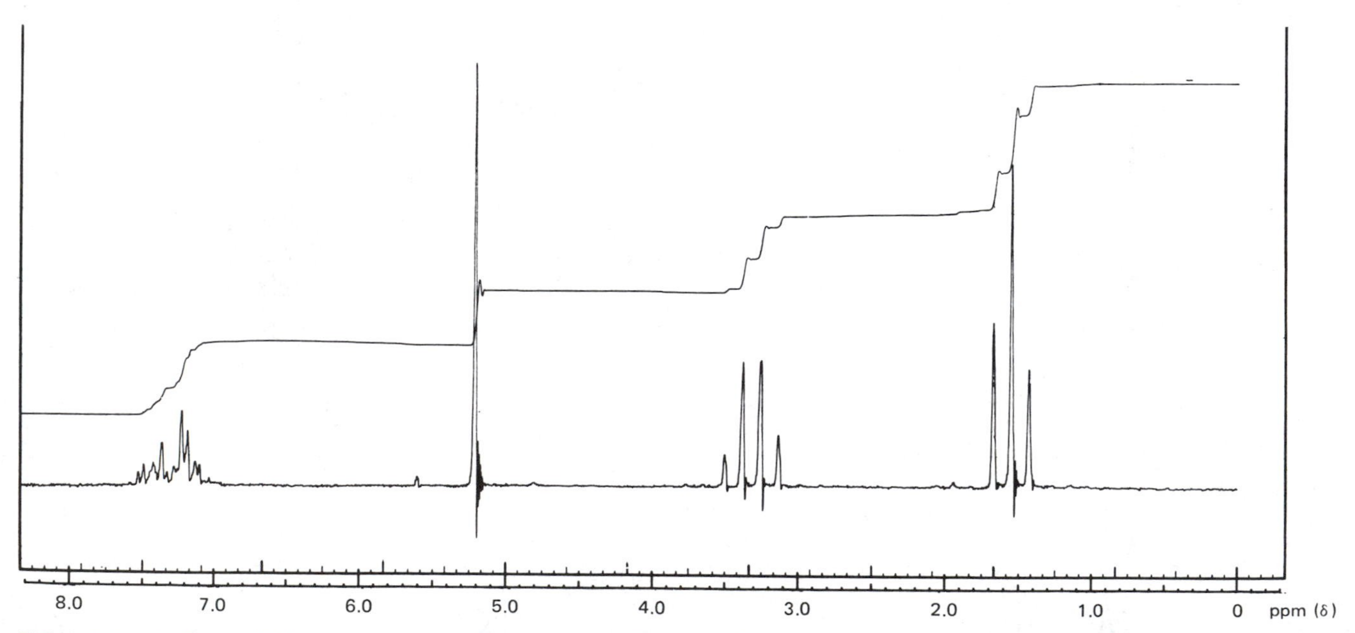

FIGURE 2-10 The 60 MHz spectrum of a mixture of bromobenzene, dichloromethane, and iodoethane.

The peaks may be assigned by consideration of their splitting patterns and by comparison with the spectra of the pure compounds. The multiplet at lowest field is from C_6H_5Br, the singlet is from CH_2Cl_2, and the quartet and triplet are from CH_3CH_2I. The complexity of the aromatic resonance arises from the slight nonequivalence of the individual protons, in contrast to the overlap seen in the spectrum of toluene (Figure 2-3). From the integrals, after allowance for the number of protons causing a given resonance, the $C_6H_5Br/CH_2Cl_2/CH_3CH_2I$ ratio may be calculated to be 18/34/48. The error on the integration permits an accuracy of no better than about ±2–3%.

2-6

Experimental Methods

Although there is a wide variety of NMR instruments available, certain components are common to all.

1. The laboratory magnet to supply the B_0 field.
2. The radiofrequency unit to transmit the B_1 field and to receive the NMR signal.
3. A probe for positioning the sample in the magnet.
4. Hardware for stabilizing the B_0 field and optimizing the signal.

2-6a The Magnet

Early NMR magnets were generally of the electromagnet type. Although a generation of chemists used them, they had low sensitivity and poor stability. Permanent magnets cost less and are simpler to maintain but still have low sensitivity. Iron core magnets of these types are available today on routine instruments operating at 1.4–2.3 tesla and are used primarily for proton work at 60–100 MHz. Iron core magnets resemble a thick horseshoe magnet whose ends (poles) have been bent toward each other. The sample is placed between the pole faces. Most research-grade instruments use a superconducting magnet and operate at 3.5–14 tesla (150–600 MHz for protons). These magnets provide much higher sensitivity and stability. Possibly most important, the very high fields produced by superconducting magnets result in better spectral dispersion, so that there is less overlap of resonances. The superconducting magnet resembles a solid cylinder with a hole drilled out of the center all the way along the axis. The sample is placed in the hole, with the axes of the sample and cylinder coincident.

2-6b The Radiofrequency Unit

Usually separate from the magnet is a console that contains, among other components, the radiofrequency (rf) unit. In a continuous wave (CW) experiment, the B_1 field is transmitted by an oscillator or frequency synthesizer in the rf unit and is relatively weak (10^{-4} gauss). When the Fourier transform (FT) method is used (Section 2-11), the B_1 field is supplied by a powerful pulse generator and is quite strong (10–400 gauss), in order to cover all frequencies of the resonating nucleus. In multinuclear spectrometers, the B_1 field must be tunable to frequencies of all desired nuclei. The rf unit also includes a receiver that detects and amplifies the NMR signal. In addition, the console contains various means to display the signal, including possibly an oscilloscope for immediate observation, an XY or strip-chart recorder for permanent record, or a computer for signal storage, further manipulation, and later examination.

2-6c The Probe and the Sample

The sample is placed in the most homogeneous region of the magnetic field by means of an adjustable probe. The probe contains a holder for the sample, mechanical means for adjusting its position in the field, electronic leads for supplying the B_1 and B_2 (decoupling) fields and for receiving the signal, and devices for improving magnetic homogeneity. The B_1 field is usually provided by a wire that coils around the sample tube (Figure 2-11).

Liquid samples are contained in a cylindrical tube, usually 5 mm in outer diameter for protons. A volume of 300–400 μl is required to fill the tube to about 35 mm, thereby exceeding the height of the rf coil. End effects

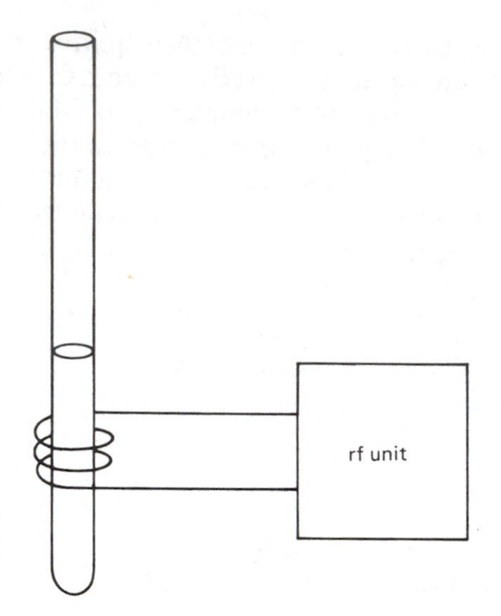

FIGURE 2-11 The NMR sample is contained in a cylindrical tube and placed in the transmitter coil that provides B_1 within the probe.

may become important if the cylinder of sample is too short. If sample is in short supply, a 2 mm microtube can be used that requires a much smaller volume. If there is plenty of sample and if sensitivity is the main problem, tubes of 25 mm or even larger diameter are available for specially constructed widebore magnets. Because of the lower sensitivity of the ^{13}C experiment, samples for carbon analysis are usually placed in 8 or 10 mm tubes. Sensitivity goes up as the square of the tube diameter, since the receiver coils contain a larger volume of sample.

A liquid sample may be examined without solvent. Normally, however, the sample is dissolved in a suitable solvent, which is chosen not only for its solubility characteristics but also for its magnetic properties. In particular, it must not have any resonances in the regions of interest. For protons, CCl_4, $CDCl_3$, D_2O, acetone-d_6, and methanol-d_4 are traditional solvents. If the spectrum is to be taken above or below room temperature, the solvent also must be chosen so that it does not boil or freeze during the experiment.

2-6d Optimizing the Signal

The entire range of proton frequencies is contained in about 15 parts per million (ppm) of the B_0 field at constant frequency. Consequently, strenuous efforts must be made to permit differentiation of resonances that have small separations. Corrections may be made for small gradients in B_0 by the use of *shim coils.* For example, the field along the z direction (from one pole to the other in an electromagnet or along the axis of the tube in a superconducting magnet) might be slightly higher at one point than at another. Such a gradient may be compensated for by applying a small current through a shim coil built into the probe. The shim coil is adjusted empirically or through computer control until an observed resonance has a minimum linewidth or maximum height. Similar coils correct inhomogeneities in the x and y directions and in higher order gradients (x^2, xy, and so on).

The sample also is spun along the axis of the cylinder at a rate of 20–50 Hz in order to improve homogeneity within the tube. Spinning improves resolution because a molecule at a particular location in the tube experiences a field that is averaged over a circular path. In an electromagnet, the axis of the cylinder is the y direction, in a superconducting magnet the z direction. Spinning does not average gradients along the axis of the cylinder, so shimming is required primarily for y gradients in an electromagnet or z gradients in a superconducting magnet.

Electromagnets in particular are subject to field drift, even over the time period of a NMR experiment. Drift can be avoided by electronically locking the field to the resonance of a substance contained in the sample. Because deuterated solvents are used quite commonly, an *internal lock* normally is at the deuterium frequency and uses deuterons of the solvent, such as $CDCl_3$ or CD_3COCD_3. The deuterium lock is convenient for both ^1H and ^{13}C spectra. Superconducting magnets are so stable that this field-frequency lock is not always necessary. In some instruments, the field is locked to a sample contained in a separate tube located permanently elsewhere in the probe. This *external lock* is usually found only in spectrometers that are designed for a highly specific use, such as taking only ^1H spectra.

Sensitivity is a problem for ^{1}H spectra of poorly available or poorly soluble materials, for most ^{13}C spectra, and for all ^{15}N spectra at natural abundance. Computer techniques have been developed for routine improvement of sensitivity through multiple scanning or signal averaging. In this procedure, the entire NMR spectrum is digitized and stored in the memory of the spectrometer's computer. The spectrum then is recorded a second time and stored in the same locations. Any signal present will be reinforced, but noise will tend to cancel out. If n such measurements are carried out and stored, the theory of random processes tells us that the signal amplitude is proportional to n but the noise is proportional to $\sqrt{n}$. The signal to noise ratio (S/N) therefore increases by $n/\sqrt{n}$, or $\sqrt{n}$. Thus 100 sweeps added together will theoretically enhance the S/N by a factor of about 10 (the actual result is somewhat smaller). Figure 2-12 shows progressive averaging for the methylene quartet of ethylbenzene. Multiple scanning is necessary for most nuclei other than ^{1}H, ^{19}F, and

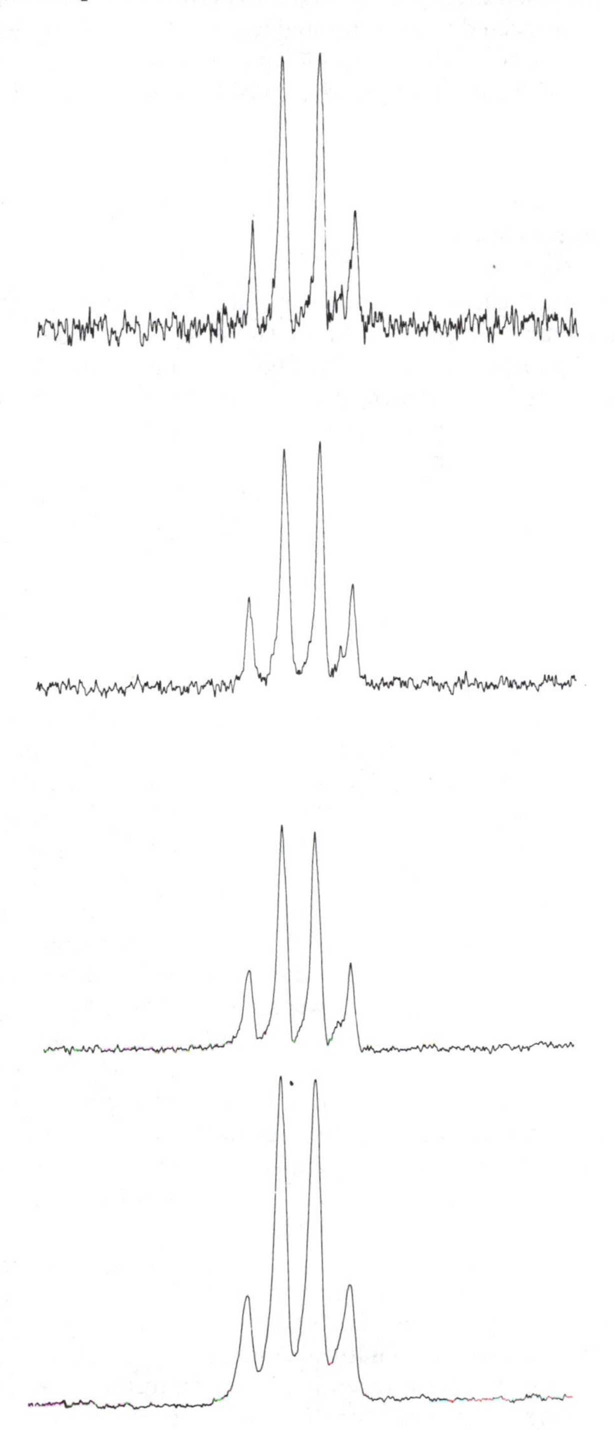

FIGURE 2-12 The methylene quartet of ethylbenzene (1% solution in CCl_4) with accumulations (from the top) of 1, 4, 16, and 128 scans.

^{31}P. In particular, ^{13}C and ^{15}N spectra routinely require signal averaging. With Fourier transform methods (Section 2-11), several hundred scans can be obtained in a few minutes. All ^{13}C spectra illustrated in this chapter used the multiple scan method.

2-6e Accessories

Variable temperature operation makes possible the measurement of reaction kinetics (Section 2-9). In addition, poorly soluble materials may be examined at high temperatures and unstable compounds at low temperatures. Thus many spectrometers have probes that are designed to control temperature.

Multiple resonance techniques such as spin decoupling can provide a host of useful spectral results. Probes must be constructed with additional leads to bring the B_2 field to the sample. It sometimes is useful to apply more than one rf pulse to a sample, in which case equipment must be available to generate pulse sequences. Some uses of these techniques are considered in Chapter 5.

2-7

Common Nuclides

What nuclei are useful in chemical problems? The answer depends on one's area of specialty. Certainly for the organic chemist, the most common elements are carbon, hydrogen, oxygen, and nitrogen. The biochemist would add phosphorus to the list. The inorganic chemist potentially uses the entire periodic table, so for one person iron might be important; for another, mercury or boron. When can the chemist hope to be able to use NMR to study these various elements? Several different properties determine how accessible a given nucleus is to the NMR experiment.

Spin. The spin of a nucleus is determined by how the protons and neutrons couple. As mentioned earlier, many common nuclides (^{12}C, ^{16}O) have zero spin and hence are not available to the NMR experiment. These are the nuclides with both an even mass number (protons plus neutrons) and an even atomic or charge number (protons). In a way, it is fortunate that the common carbon and oxygen nuclides are magnetically invisible. Otherwise, proton spectra would be made exceedingly complex because of additional spin–spin splittings. By and large, spin $\frac{1}{2}$ nuclei exhibit more favorable NMR properties than quadrupolar nuclei. Nuclei with odd mass numbers have half-integral spins ($\frac{1}{2}$, $\frac{3}{2}$, etc.), whereas those with even mass and odd charge have integral spins (1, 2, etc.). We will concentrate on spin $\frac{1}{2}$ nuclei in this text.

Natural abundance. Nature provides us with nuclides in varying amounts. Whereas ^{19}F and ^{31}P are 100% abundant and ^{1}H nearly so, ^{13}C is present only to the extent of 1.1% and the useful nitrogen (^{15}N) and oxygen (^{17}O) nuclides occur to the extent of much less than 1%. Clearly, the NMR experiment is easier for nuclides with higher natural abundance. Because so little ^{13}C is present, there is a very small probability of finding two ^{13}C atoms in the same molecule ($1.1\% \times 1.1\%$, or 0.00012). Thus J coupling is not easily observed between two ^{13}C nuclei in ^{13}C spectra, in considerable contrast to ^{1}H spectra.

Sensitivity. Nuclides have different natural sensitivities to the NMR experiment, as determined by the distance ΔE between the $+\frac{1}{2}$ (E_1) and $-\frac{1}{2}$ (E_2) spin states (Diagram 2-6). The larger the energy difference, the more nuclei are present in the lower state and hence are available to absorb energy. The proton is one of the most sensitive nuclides, whereas ^{13}C and ^{15}N unfortunately are rather weak. Natural sensitivity should be distinguished from natural abundance. Tritium (^{3}H) is very useful to the biochemist as a label and has high NMR sensitivity, but occurs in zero natural abundance. The biochemist, however, can use naturally occurring deuterium instead, although its inherent sensitivity is very low. Similarly, many nuclei that might be of interest to the inorganic chemist, such as iron and potassium, have poor sensitivities. Others, such as cobalt, have quite good sensitivity. Thus it is important to know something about the natural sensitivity of a nucleus before designing a NMR experiment.

Receptivity. The accessibility of a given nucleus is determined largely by the two factors natural abundance and natural sensitivity. Combining these two factors as their product therefore gives a more complete idea of how amenable the nucleus is to the NMR experiment. Because people are quite familiar with the ^{13}C experiment, the product of natural abundance and natural sensitivity for one nucleus is compared to that for ^{13}C by division. The term receptivity is given to the ratio (abundance times sensitivity for a given nuclide, divided by the same product for ^{13}C), so the receptivity of ^{13}C is 1.00. For example, the ^{15}N nucleus is about 50

TABLE 2-3 NMR Properties of Common Nuclei

Nuclide	Spin	Natural Abundance (%)	Natural Sensitivity (for equal numbers of nuclei) (vs. ^1H)	Receptivity (vs. ^{13}C)	NMR Frequency (2.35 tesla)	Reference Substance
Proton	$\frac{1}{2}$	99.985	1.00	5680	100.00	$(CH_3)_4Si$
Deuterium	1	0.015	0.00965	0.0082	15.35	$(CD_3)_4Si$
Boron-11	$\frac{3}{2}$	80.42	0.0165	754	32.07	$Et_2O{\cdot}BF_3$
Carbon-13	$\frac{1}{2}$	1.108	0.0159	1.00	25.15	$(CH_3)_4Si$
Nitrogen-14	1	99.63	0.00101	5.69	7.23	$NH_3(l)$
Nitrogen-15	$\frac{1}{2}$	0.37	0.00104	0.0219	10.14	$NH_3(l)$
Oxygen-17	$\frac{5}{2}$	0.037	0.0291	0.0611	13.56	H_2O
Fluorine-19	$\frac{1}{2}$	100	0.833	4730	94.09	CCl_3F
Silicon-29	$\frac{1}{2}$	4.70	0.00784	2.09	19.87	$(CH_3)_4Si$
Phosphorus-31	$\frac{1}{2}$	100	0.0663	377	40.48	85% H_3PO_4
Sulfur-33	$\frac{3}{2}$	0.76	0.00226	0.0973	7.68	CS_2
Chlorine-35	$\frac{3}{2}$	75.53	0.0047	20.2	9.80	NaCl (aq)
Cobalt-59	$\frac{7}{2}$	100	0.277	1570	23.73	$K_3Co(CN)_6$
Platinum-195	$\frac{1}{2}$	33.8	0.00994	19.1	21.46	Na_2PtCl_6
Mercury-199	$\frac{1}{2}$	16.84	0.00567	5.42	17.91	$(CH_3)_2Hg$

times less receptive than ^{13}C, when both abundance and sensitivity are taken into account, so its receptivity is about 0.02.

Table 2-3 lists the spins, natural abundances, natural sensitivities, receptivities, and resonance frequencies at 2.35 tesla for some common nuclei. More complete tables may be found in the references listed at the end of the chapter. For each nuclide a compound has been selected whose resonance is taken to be $\delta = 0.00$. Some common reference substances are given in the last column. In this text we will continue to emphasize ^1H and ^{13}C spectra, but from time to time we will mention other nuclei contained in this table.

2-8

Relaxation

Just how many more $+\frac{1}{2}$ nuclei are there than $-\frac{1}{2}$ nuclei? The Boltzmann law defines the ratio of populations for two species that are in equilibrium. The Boltzmann expression for the ratio of spin populations is given by eq. 2-5, in which n_+ is the population of $+\frac{1}{2}$ nuclei, n_- is the population for $-\frac{1}{2}$ nuclei, ΔE is

$$\frac{n_+}{n_-} = \exp(\Delta E/kT) \tag{2-5}$$

the energy difference between the two spin states, k is the Boltzmann constant, and T is the temperature. From eqs. 2-1 through 2-3, we know the value of ΔE in terms of nuclear magnetic properties, so the ratio takes the form of eq. 2-6. If appropriate values are inserted, we learn that there are only about six more $+\frac{1}{2}$

$$\frac{n_+}{n_-} = \exp[\gamma \hbar B_0 (1 - \sigma)/kT] \tag{2-6}$$

than $-\frac{1}{2}$ nuclei per million nuclei for protons at 60 MHz (B_0 of 1.41 tesla). The excess increases at higher field, since ΔE is larger.

When the B_1 field is applied at the resonance frequency, nuclear spin flips occur in both directions. The $+\frac{1}{2}$ spins absorb energy and become $-\frac{1}{2}$; the $-\frac{1}{2}$ spins emit energy and become $+\frac{1}{2}$. Because there are more $+\frac{1}{2}$ than $-\frac{1}{2}$ nuclei, the net effect is absorption. After a period of time, however, the excess of $+\frac{1}{2}$ nuclei becomes exhausted and the rates of absorption and emission are equal. Under these conditions, the sample is said to be *saturated*, and no further resonance can be observed. Nonetheless, nuclear magnetic resonances tend to persist, so that there must be natural mechanisms whereby nuclear spins may be returned to the equilibrium condition with an excess of $+\frac{1}{2}$ nuclei. Such a process is called *relaxation*.

NMR relaxation derives from the existence of local oscillating fields in the sample that correspond to the resonance frequency. These fields do not come directly from laboratory operations involving B_0 or B_1. The primary source of oscillating fields is nuclear magnets in motion. As a molecule tumbles in solution in the B_0 field, each nuclear magnet is in motion and sets up a magnetic field in opposition to B_0 by Lenz's law. If the motion is at just the right frequency, it corresponds in energy to the excess energy of the $-\frac{1}{2}$ spin state. Therefore, excess spin energy can be transferred to the motional energy of the neighboring nuclei. Such a process is called *spin–lattice relaxation*, since the term lattice is used to describe the general environment of the nuclear spin. Thus after a nucleus is excited from $+\frac{1}{2}$ to $-\frac{1}{2}$ by the NMR experiment, relaxation occurs ($-\frac{1}{2} \rightarrow +\frac{1}{2}$), and spin energy is passed to the lattice with a minuscule rise in temperature. The nucleus then becomes available for another spin flip to continue the resonance experiment.

For effective relaxation, there must be magnetic nuclei close to the resonating nucleus. For ^{13}C, the attached protons provide spin–lattice relaxation. A carbon attached to four other carbons relaxes very slowly because almost 99% of the attached carbons are the nonmagnetic ^{12}C. The motion of ^{12}C in solution provides no relaxation. Protons must reach out to nearest neighbor protons for their relaxation. Detailed analysis of relaxation can give information about distance between atoms and about motion of atoms in solution. For the present, we need view relaxation only as the means to keep the NMR experiment going.

Relaxation processes also are responsible for generating the initial excess of $+\frac{1}{2}$ nuclei when the sample is first placed in the probe. Outside the laboratory field, all spin states are equally populated. When the sample then is immersed in the B_0 field, spin–lattice relaxation drives the sample to the equilibrium condition with more $+\frac{1}{2}$ than $-\frac{1}{2}$ nuclei.

If there are no nearby spins, the time for relaxation can be quite long, up to 100 seconds or more for ^{13}C. Consequently, it is possible to saturate the signal with attendant loss of signal. Also, when a very high level of power is used for the B_1 field, even when there are nearby nuclei, the sample may become saturated. Relaxation simply cannot keep up with the effects of a strong B_1 field. In practice, the value of B_1 is chosen carefully. Too low a value gives a weak signal; too high a value causes saturation.

2-9 Dynamic Effects

According to the principles outlined in the previous sections, the proton spectrum of methanol should contain a doublet of integral 3 for the CH_3 coupled to OH and a quartet of integral 1 for the OH coupled to CH_3. Under conditions of high purity and low temperatures, such a spectrum is observed (Figure 2-13, lower). The presence of a small amount of acidic or basic impurity, however, can catalyze the intermolecular exchange of the hydroxyl proton. When this proton becomes detached from the molecule by any mechanism, information about its spin states is no longer available to the rest of the molecule. If the rate of exchange is very fast, no coupling is observed between the exchanging hydroxyl proton and the other nuclei in the molecule. Thus at high temperatures (Figure 2-13, upper), the proton spectrum of methanol contains only two singlets. If the exchange rate is slowed by lowering the temperature or by decreasing the amount of acidic or basic catalyst, the coupling constant continues to be washed out unless the exchange rate reaches a critical value at which the proton resides sufficiently long on oxygen to permit the methyl group to detect the unaveraged spin states. For coupling to be observed, the rate of exchange in reciprocal seconds (s^{-1}) must be slower approximately than the magnitude of the coupling in Hz (the unit Hz is the same as s^{-1}). Thus a proton could exchange a few times per second and still maintain coupling. As can be seen from Figure 2-13, the transition from slow to fast exchange can be accomplished for methanol over an 80°C temperature range.

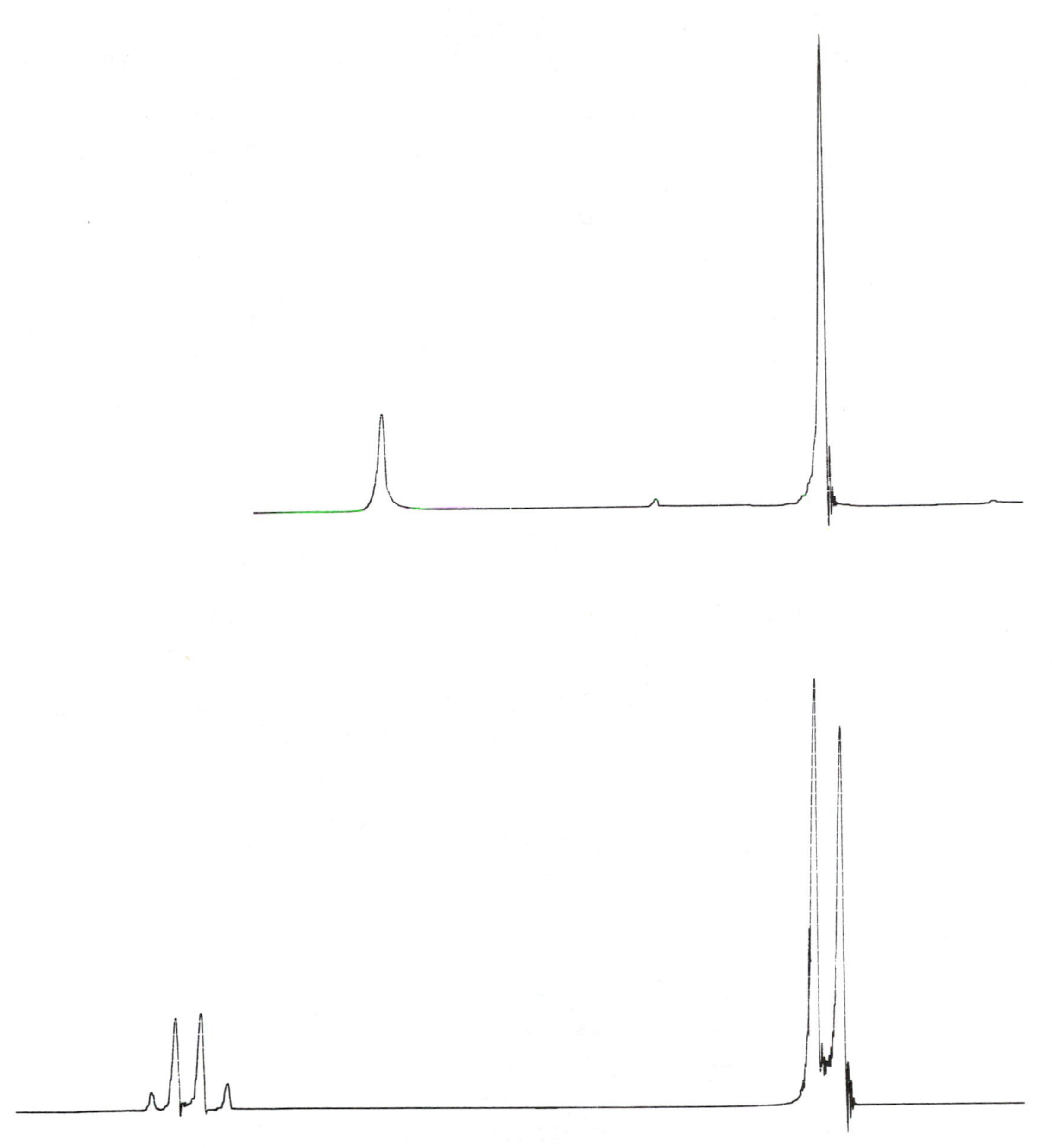

FIGURE 2-13 The 60 MHz proton spectrum of CH_3OH at +50°C (upper) and at −30°C (lower).

Under most spectral conditions, there are minor amounts of acid or base impurities, so hydroxyl protons do not usually exhibit couplings to other nuclei. The integral is still unity for the OH group because the amount of catalyst is very small. Similar exchange phenomena can wash out couplings to amino hydrogens, so NH_2 resonances usually are singlets.

Related to the averaging of coupling constants is the averaging of chemical shifts. A mixture of acetic acid and benzoic acid contains only one resonance for the CO_2H protons from both molecules. In this case the carboxyl protons of the two compounds interchange so rapidly that the spectrum exhibits only the average of the two. Moreover, if water is the solvent, exchangeable protons such as OH do not give separate resonances. For example, the proton spectrum of acetic acid in water contains two, not three, peaks. The water and carboxyl protons appear as a single, weighted-average resonance. Theoretically, if the rate of exchange between OH and solvent could be slowed sufficiently, separate resonances could be observed.

Intramolecular (unimolecular) reactions also can influence the appearance of the NMR spectrum. The molecule cyclohexane, for example, contains distinct axial and equatorial protons, yet the spectrum at room

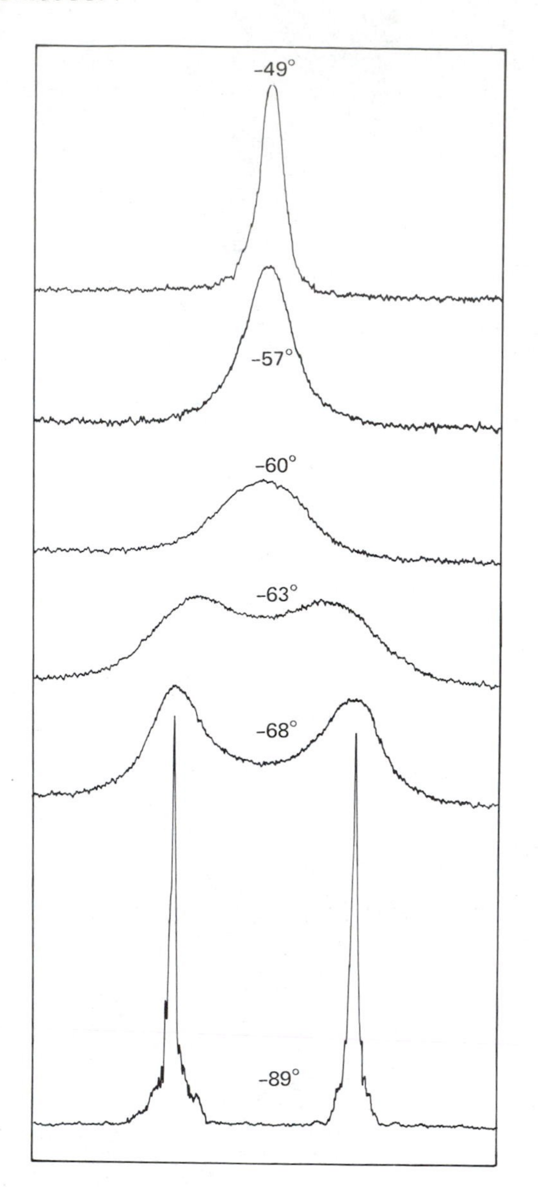

FIGURE 2-14 The 60 MHz proton spectrum of cyclohexane-d_{11} as a function of temperature. [Reproduced with permission from F. A. Bovey, F. P. Hood III, E. W. Anderson, and R. L. Kornegay, *J. Chem. Phys.*, **41**, 2042 (1964).]

temperature exhibits only one sharp singlet. The process of ring reversal intramolecularly interconverts axial and equatorial positions. When the rate of this process is greater (in s^{-1}) than the chemical shift difference between axial and equatorial protons (in Hz), the NMR experiment cannot distinguish that there are chemically distinct protons and only one peak is observed. At lower temperatures, however, the process of ring reversal is slowed considerably. By −100°C, the NMR experiment can distinguish the two proton types, so two resonances are observed. The transition from fast to slow exchange is shown in Figure 2-14 for cyclohexane in which all protons but one have been replaced by deuterium in order to simplify the spectrum (eq. 2-7). Couplings to deuterium may be removed by double irradiation at the deuterium frequency.

(2-7)

When part of a molecule is altered reversibly during the NMR experiment, either by an acid-catalyzed intermolecular exchange or by a unimolecular reorganization, the spectrum can be sensitive to the rate of this process. Hence one speaks of the effects of kinetic or dynamic processes. NMR provides one of the few methods for examining the rates of reactions when the system is at equilibrium. Normally, kinetics are followed by the transformation of one substance into another. In NMR spectroscopy, kinetic effects are observed by the averaging of chemical shifts or coupling constants. A wide variety of rate processes has been studied by these techniques. Some of them are considered further in Chapter 5.

2-10

Spectra of Solids

All of the examples and spectra illustrated thus far have been for liquid samples. Certainly it would be useful to be able to take NMR spectra of solids, so why have we avoided discussion of solid samples? Until recently, it was not possible to obtain high resolution spectra of solids. Proton or carbon spectra of solid samples were broad and unresolved. Only modest amounts of information could be obtained from such spectra. There were two primary reasons for the poor resolution of solid samples. We have already spoken of the indirect spin–spin interaction between nuclei that occurs through bonds and gives rise to the J coupling. Nuclear magnets also can couple directly through the interaction of their nuclear dipoles. This *dipole–dipole coupling* occurs through space and gives rise to a D coupling that is much larger than the J coupling. In solution, dipoles are continuously reorienting through molecular tumbling. Just as two bar magnets have no net interaction when averaged over all mutual geometries, two nuclear magnets have no net dipolar interaction because of the randomizing process of tumbling. The indirect J coupling is not averaged to zero because tumbling cannot destroy a through-bond mechanism. In the solid phase, nuclear dipoles are held rigidly in position. The dominant interaction between nuclei therefore is the D coupling, which is on the order of several hundred to a few thousand hertz. Such splittings tend to overpower J couplings and even most chemical shifts, so that all one observes is a broad, almost featureless band.

Just as the J coupling may be removed by decoupling, the D coupling may be eliminated by use of a strong B_2 field. Power levels for D decoupling must be higher than those for J decoupling, since D is two to three orders of magnitude larger than J. Such B_2 fields are now available and can be adapted to reduce the linewidth of the spectra of solids.

Unfortunately, there is a second factor that also contributes to line broadening for solids, *anisotropy of the chemical shielding*. In solution, the observed chemical shift is the average of the shielding around a nucleus for all orientations in space, as the result of tumbling. In the solid, shielding is not always the same for a given type of nucleus. Consider the carbonyl carbon in acetone. When the B_0 field is parallel to the C=O bond, the ^{13}C nucleus experiences one shielding. When the B_0 field is perpendicular to the C=O bond, the nucleus experiences quite a different shielding (Figure 2-15). Recall that shielding arises because electrons in motion set up fields opposite to the direction of the applied field. The ability of electrons to circulate varies according to their orientation toward the field. Double irradiation cannot average the anisotropy of chemical shielding to zero, since the effect is entirely geometrical. The problem was solved by spinning the sample at very high

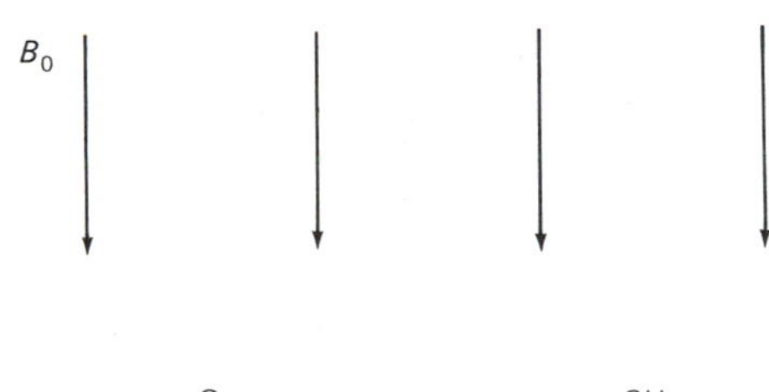

FIGURE 2-15 The shielding of the carbonyl carbon of acetone varies with the orientation of the molecule in the B_0 field.

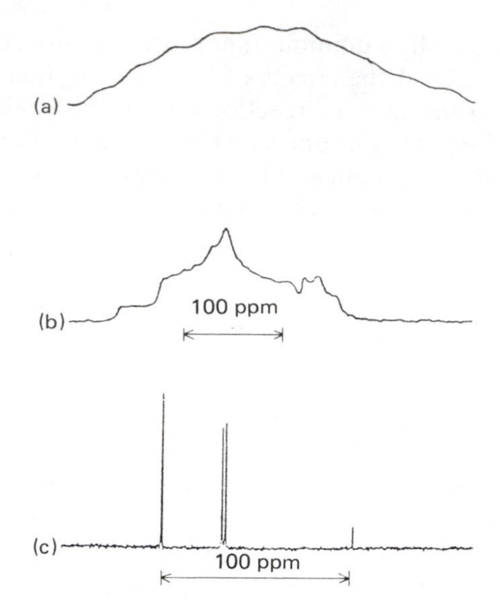

FIGURE 2-16 The ^{13}C spectrum of polycrystalline β-quinolmethanol clathrate (a) without decoupling, (b) with decoupling, and (c) with decoupling and magic angle spinning. [Reproduced with permission from T. Terao, *JEOL News,* **19,** 12 (1983).]

speeds. Shielding anisotropies are a few hundred to a couple of thousand hertz, so spinning must exceed this range in order to average all orientations. Typical values are 2–4 kHz. Spinning must be carried out at a special angle with respect to the applied field for maximum results (the best angle between B_0 and the axis of spinning is 54°44′). The technique therefore has been given the name *magic angle spinning*.

The combination of strong irradiation to eliminate dipolar couplings and magic angle spinning to eliminate chemical shielding anisotropy results in spectra of solids that are almost as high resolution as those of liquids. Figure 2-16 shows the ^{13}C spectrum of polycrystalline β-quinolmethanol clathrate. In (a) the broad, featureless spectrum is typical of solids. Strong double irradiation in (b) eliminates dipolar couplings and brings out some features. In (c) magic angle spinning as well as decoupling gives a high resolution spectrum.

For spectra in the solid, relaxation times are extremely long because motion of nuclei that is necessary for spin–lattice relaxation is absent. Particularly for ^{13}C, spectra could take exceedingly long amounts of time to record because nuclei must be allowed to relax for several minutes between pulses. One way around this problem takes advantage of the high natural sensitivity of the ^{1}H experiment compared with that of ^{13}C (Table 2-3). The same double irradiation process that eliminates J and D couplings also can be used to transfer some of the proton's favorable amplitude to the carbon atoms. Such a process is called *cross-polarization* and is standard for almost all solid spectra of ^{13}C.

The higher resolution and sensitivity of the magic angle spinning and cross polarization experiment (MAS–CP) opened vast new areas to NMR. Inorganic and organic materials that do not dissolve can now be subjected to NMR analysis. Synthetic polymers and coal were two of the first materials to be examined. Biological and geological materials such as wood, humic acids, and biomembranes became general subjects for NMR study. Amber is one such geomaterial of organic source. It is polymeric, noncrystalline, and nearly insoluble in all solvents. Its structure is largely unknown. Figure 2-17 shows that high resolution spectra can be obtained from amber with MAS–CP and that numerous specific functional groups can be picked out through chemical shift analyses of the type described in the next chapter.

2-11

Fourier Transform Spectra

It takes about 5 minutes to record a spectrum such as that in Figure 2-2. Quite a bit of that time is wasted just sweeping through the noise in regions that have no peaks. About 1970, a new method was developed that avoids this instrumental inefficiency. Instead of continuously sweeping the field or the frequency, the new procedure uses a strong burst of radiofrequency energy that encompasses all possible resonance frequencies. This burst, or *pulse*, is centered at a frequency, called the carrier, that is usually set slightly apart from the band of resonance frequencies. The pulse lines up nuclear magnets in the usual fashion, and this

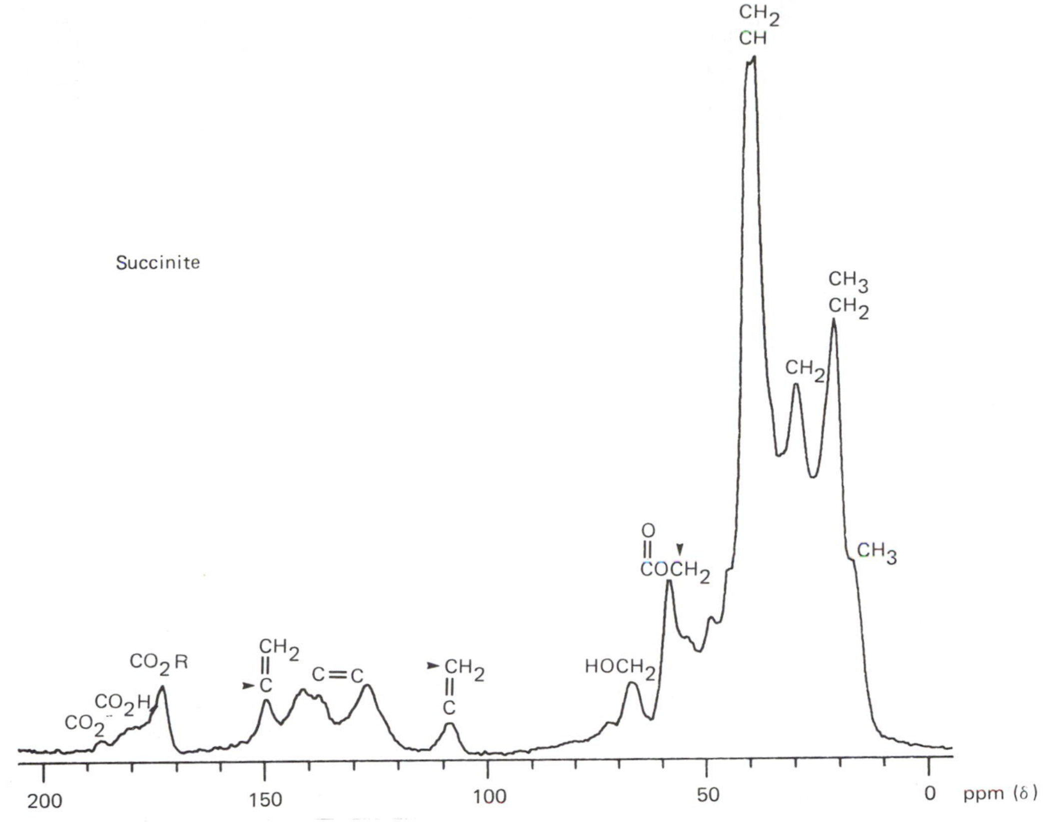

FIGURE 2-17 The ^{13}C spectrum of Baltic amber, with cross polarization and magic angle spinning. [Reproduced with permission from the J. B. Lambert and J. S. Frye, *Science*, **217**, 55 (1982). Copyright 1982 by the American Association for the Advancement of Science.]

magnetization, as lining up of magnets is called, decays with time, after the pulse is shut off. Since the carrier and resonance frequencies are not quite the same, the plot of magnetization as a function of time is an interference pattern, as illustrated in Figure 2-18a for acetone. This plot is called the *free induction decay* (FID) of magnetization and actually contains the NMR spectrum, but along a time rather than a frequency coordinate. The difference between each maximum is $1/(\nu_0 - \nu_c)$, in which ν_0 is the acetone resonance frequency and ν_c is the carrier frequency of the strong pulse. The signal dies off in an exponential fashion. The rate of decay is related to the linewidth. Thus the FID contains all frequency and linewidth information present in a usual NMR spectrum. The *time domain* representation of Figure 2-18 may be converted to the *frequency domain* used in all earlier spectral displays by the mathematical operation known as *Fourier transformation.*

If there is more than one line in the frequency domain spectrum, the free induction decay in the time domain is more complex. For two peaks, as in the ^1H spectrum of methyl acetate, the interference pattern, that is, the FID, contains two frequencies (Figure 2-18b) and after Fourier transformation (Figure 2-2) generates two peaks. The FID of a multipeak spectrum is very complex (see Figure 2-18c for the FID of the ^{13}C spectrum of 3-hydroxybutyric acid), but transforms smoothly to the traditional form (Figure 2-8). Chemists rarely examine the time domain FID but go right through Fourier transformation to get the more familiar frequency domain spectrum. The process of Fourier transformation takes only a few seconds on modern instrumentation.

The Fourier transform (FT) method seems much more complex than the continuous wave (CW) procedure, so why do chemists prefer it? The principal reason is sensitivity. Whereas the CW method wastes time sweeping through regions with no peaks, the FT method obtains signals only from points in the spectrum that have resonances that can interfere with the carrier frequency. Thus the spectrum is obtained much more quickly and with higher sensitivity. Especially for very small samples (a chemist may have less than a milligram of a precious material whose structure is unknown), it is very important to have the highest

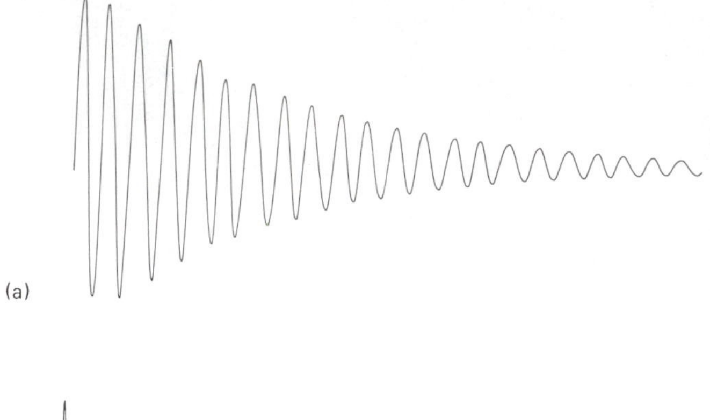

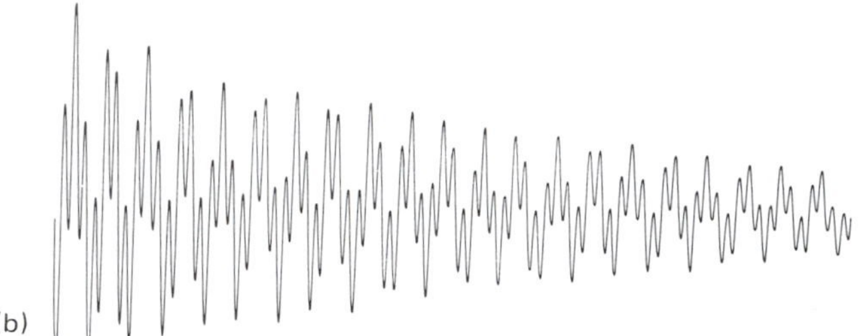

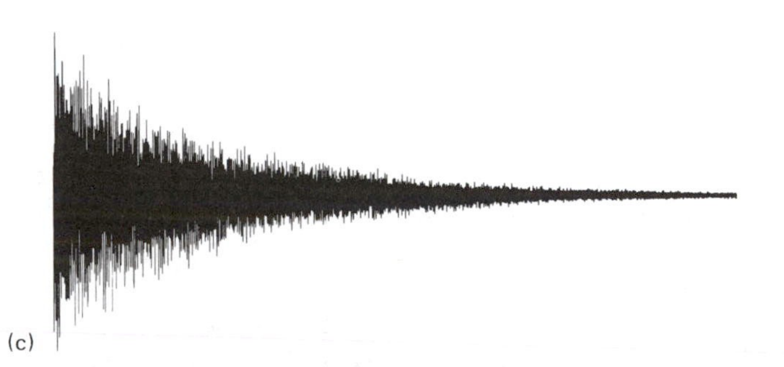

FIGURE 2-18 The free induction decay for the ^{1}H spectrum of (a) acetone and (b) methyl acetate. (c) The free induction decay for the ^{13}C spectrum of 3-hydroxybutyric acid. All samples were without solvent.

sensitivity possible. For cases in which there is plenty of material, say 50 mg or more, the CW spectrum is quite acceptable for protons. For poorly available or poorly soluble samples, however, the FT procedure becomes a necessity. Observation of ^{13}C at its natural abundance (1.1%) almost always requires the FT procedure. A second advantage of the FT method is that the rapid acquisition of data enables signal enhancement through repeat scanning to be managed very easily. Figure 2-17, for example, was the result of averaging 55,000 scans.

Once the Fourier transformation is carried out, the resulting frequency domain spectrum is very similar to an analogous CW spectrum of the same material. In fact, within this chapter, the FT method was used to record all ^{13}C spectra, as well as the 270 MHz ^{1}H spectrum of ethyl crotonate in Figure 2-9. As a rule, FT is used for all nuclei other than ^{1}H and for ^{1}H at frequencies above 100 MHz.

PROBLEMS

2-1 What is the multiplicity for the resonance of the underlined protons? Justify your answer in **a** by listing the spin states for the neighboring nuclei.

a. $ClCH_2C\underline{H}ClCH_2Cl$ **b.** $BrC\underline{H}(CH_3)_2$ **c.** cyclopropane **d.** H O Cl H $\underline{H}$

2-2 Predict the first-order multiplicities for the proton resonance and the carbon resonance in the absence of decoupling for each of the following compounds. For the carbon spectra, give only the multiplicities caused by coupling to attached protons. For the proton spectra, give only the mutiplicities caused by coupling to vicinal protons (HCCH).

a. $CH_3CH_2CH_2OC(=O)CH_3$ **b.** C_6H_5–CH_2CH_2Br **c.** (six-membered ring containing O)

d. $N(CH_2CH_3)_3$ **e.** $(CH_3)(H)C{=}C(H)(CO_2CH_3)$

2-3 (i) From the elemental formula, calculate the unsaturation number. (ii) Then calculate the relative sizes of the integrals for each group of protons in the following 90 MHz spectra. To convert to absolute numbers, it is necessary to select one group to be of known integral size, for example, a methoxy group (3) or a methine proton (1). (iii) Assign a structure to each compound. The first problem is worked as an illustration.

EXAMPLE

$C_7H_{16}O_3$

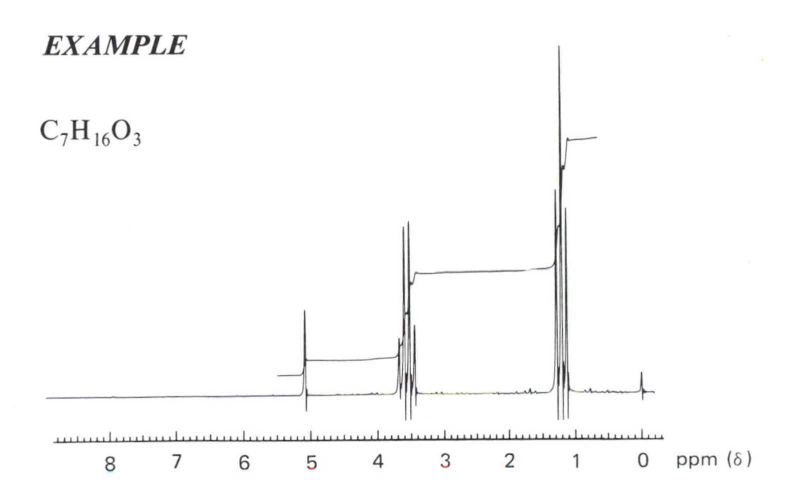

Answer: (i) $U = 8 - \frac{1}{2}(16) = 0$

(ii) Integrals

Resonance	Height (cm)	Relative Area (integers)
δ 5.1	0.20	1.00 (1)
δ 3.6	1.18	5.90 (6)
δ 1.2	1.76	8.80 (9)

(iii) *Interpretation*: The resonance at δ 0.0 is the reference TMS. The unknown has no rings or double bonds. The triplet and quartet must come from an ethyl group. The ratio of areas is $1.76/1.18 = 1.49$ or 3/2, in agreement with this assignment. Since the respective areas of the CH_2 and CH_3 resonances are 6 and 9 times the size of the resonance at δ 5.1, there must be three equivalent ethyl groups, $(CH_2CH_3)_3$. Since there are three oxygen atoms, the most likely substructure is three ethoxy groups, $(OCH_2CH_3)_3$. All that remains from the elemental formula to attach is one proton and one carbon. All three ethoxy groups and the last proton can be attached to the one carbon, so the structure is $HC(OCH_2CH_3)_3$.

a. $C_5H_8O_2$

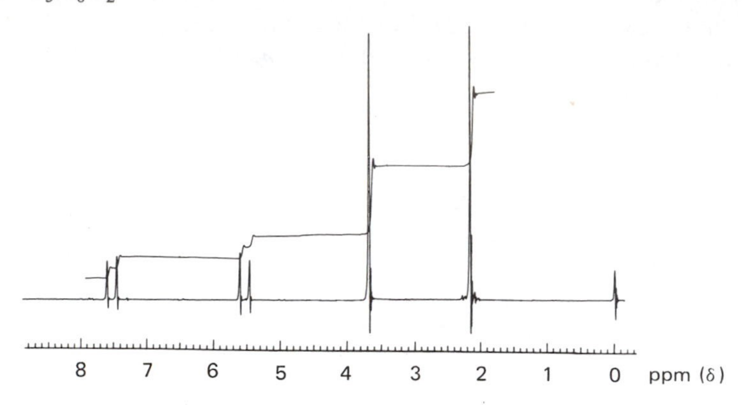

b. $C_9H_{11}O_2N$ [*Hint*: The lowest field resonances (δ 6.5–8.0) come from a para-disubstituted phenyl ring.]

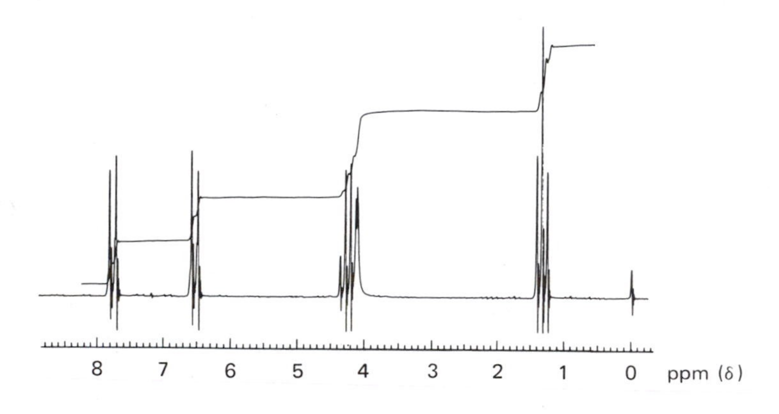

c. C_5H_9ON [*Hint*: The resonance at δ 3.5 comes from the partial overlap of a triplet and a quartet.]

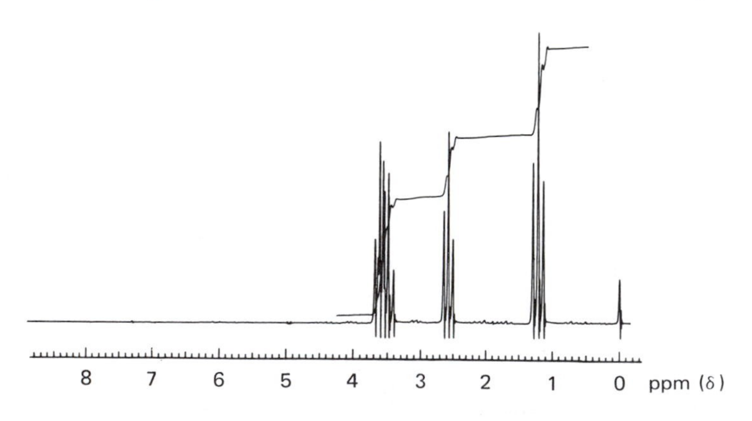

2-4 For the example in problem **2-3**, calculate the distance of each resonance from TMS in Hz.

REFERENCES/BIBLIOGRAPHY

2-1 J. A. Pople, W. G. Schneider, and H. J. Bernstein, *High-resolution Nuclear Magnetic Resonance*, McGraw-Hill, New York, 1959.
2-2 J. D. Roberts, *Nuclear Magnetic Resonance*, McGraw-Hill, New York, 1959.
2-3 J. W. Emsley, J. Feeney, and L. H. Sutcliffe, *High Resolution Nuclear Magnetic Resonance Spectroscopy*, Pergamon Press, Oxford, 1965.
2-4 F. A. Bovey, *Nuclear Magnetic Resonance Spectroscopy*, Academic Press, New York, 1969.
2-5 L. M. Jackson and S. Sternhell, *Application of Nuclear Magnetic Resonance Spectroscopy in Organic Chemistry*, 2nd ed., Pergamon Press, Oxford, 1969.
2-6 W. W. Paudler, *Nuclear Magnetic Resonance*, 2nd ed., Allyn and Bacon, Boston, 1974.
2-7 N. F. Chamberlain, *The Practice of NMR Spectroscopy*, Plenum Press, New York, 1974.
2-8 T. L. James, *Nuclear Magnetic Resonance in Biochemistry*, Academic Press, New York, 1975.
2-9 J. B. Lambert, H. F. Shurvell, L. Verbit, R. G. Cooks, and G. H. Stout, *Organic Structural Analysis*, Macmillan, New York, 1976.
2-10 D. Shaw, *Fourier Transform N.M.R. Spectroscopy*, 2nd ed., Elsevier, Amsterdam, 1984.
2-11 E. D. Becker, *High Resolution NMR*, 2nd ed., Academic Press, New York, 1980.
2-12 H. Günther, *NMR Spectroscopy*, Wiley, Chichester, 1980.
2-13 R. K. Harris, *Nuclear Magnetic Resonance Spectroscopy*, Pitman, London, 1983.
2-14 Quantitative analysis: J. N. Shoolery, *Progr. NMR Spectrosc.,* **11**, 79 (1977); D. E. Leyden and R. H. Cox, *Analytical Applications of NMR*, Wiley, New York, 1977.
2-15 Fourier transform NMR: T. G. Farrar and E. D. Becker, *Pulse and Fourier Transform NMR*, Academic Press, New York, 1971; A. G. Redfield and R. K. Gupta, *Advan. Magn. Reson.*, **5**, 82 (1971); D. A. Netzel, *Appl. Spectrosc.*, **26**, 430 (1972); D. G. Gillies and D. Shaw, *Ann. Rep. NMR Spectrosc.*, **5A**, 560 (1972); E. Fukushima and S. B. W. Roeder, *Experimental Pulse NMR: A Nuts and Bolts Approach*, Addison-Wesley, Reading, MA, 1981; R. Benn and H. Günther, *Angew. Chem., Int. Ed. Engl.,* **22**, 350 (1983); *Nuclear Magnetic Resonance*, A Specialist Periodical Report, The Chemical Society, London.
2-16 Solid state magic angle spinning: E. R. Andrews, *Progr. NMR Spectrosc.,* **8**, 1 (1971); M. Mehring, *NMR Basic Princ. Progr.,* **11**, 1 (1976); C. A. Fyfe, *Solid State NMR for Chemists*, C. F. C. Press, Guelph, Ontario, 1983.

3

The Chemical Shift

3-1

Factors That Influence Proton Shifts

We have seen that spin $\frac{1}{2}$ nuclei are aligned with or against the applied magnetic field during the NMR experiment. The field, however, is not successful in bringing about perfect alignments, so nuclei actually precess with or against the direction of the magnetic field (Figure 3-1). That is, the direction of the magnetic vector circles around the B_0 direction. Precession occurs at an angular frequency ω_0 in radians per second that is related to the resonance frequency by the standard expression $\omega_0 = 2\pi\nu_0$. The process is sometimes referred to as Larmor precession and ω_0 as the Larmor frequency.

Electrons surrounding an isolated atom possess charge and hence also precess in the B_0 field. By Lenz's law, the magnetic field that arises from electron precession opposes the direction of the B_0 field. This opposing field, experienced at the nucleus, is the cause of the shielding phenomenon and the chemical shift. Any material such as the electron cloud that gives rise to a magnetic field that opposes B_0 is said to be *diamagnetic*, so this contribution to the shielding is termed diamagnetic shielding or σ^d. For the effect to be maximal, precession must be unhindered, as when the electron cloud is spherical. This condition is found in free atoms and some molecules with tetrahedral symmetry, such as the carbon in CH_4 and the nitrogen in $^+NH_4$. Because hydrogen atoms have only s electrons, which are reasonably spherical in distribution, proton chemical shifts are dominated by diamagnetic shielding. To compensate for the effect of diamagnetic shielding, the B_0 field must be raised to bring the nucleus into resonance, resulting in an upfield shift at constant frequency, to the right on normal spectral displays. At

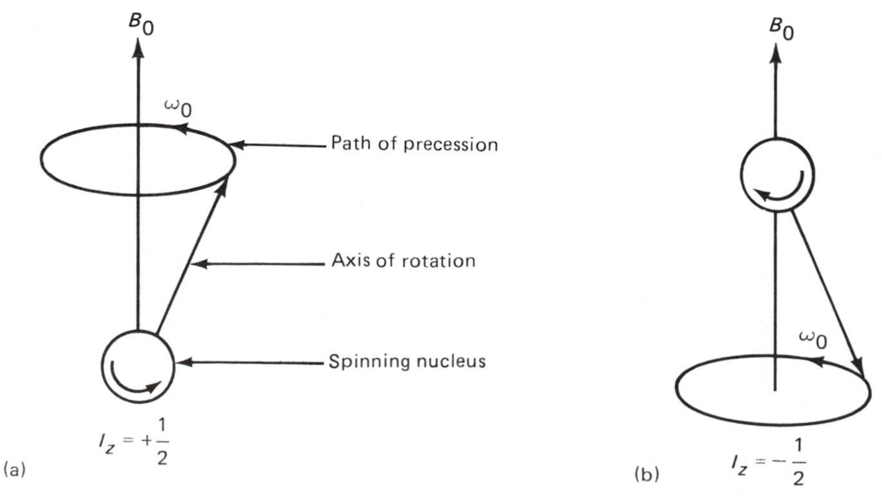

FIGURE 3-1 **Precession of spin ½ nuclei with (a) and against (b) the applied magnetic field B_0.**

constant B_0 field, the resonance frequency must be reduced to obtain the resonance condition because the shielding electrons reduce the effective field at the nucleus (eq. 2-2).

Diamagnetic shielding depends strongly on the electron density at the nucleus. An electronegative substituent within a molecule, that is, one that pulls electron density away from a proton, reduces σ^d for that proton and shifts its resonance position to lower field (higher frequency). Thus the physical organic concepts of electron withdrawal and electron donation are important in understanding proton chemical shifts. For example, when a proton in methane is replaced by a chlorine atom, the resonance position moves from δ 0.2 to 2.7 for chloromethane (CH_3Cl), a shift of 2.5 ppm. Progressive replacement of additional hydrogens by chlorine results in further downfield shifts, to δ 5.3 for CH_2Cl_2 and δ 7.3 for $CHCl_3$. Other electron-withdrawing substituents such as nitro, cyano, carbonyl, fluorine, alkoxyl, hydroxyl, or amino result in similar downfield shifts that are proportional to the group or atomic electronegativities. Conversely, electropositive groups such as organosilicon or organomagnesium enhance the diamagnetic shielding effect and result in an upfield shift, as exemplified by the high-field resonance position of TMS. We will see in Section 3-3 how specific shifts are related to functionality.

Electron density is influenced by resonance as well as by inductive effects, as seen in alkenes and aromatic compounds. A methoxy group can donate electrons by resonance, so the positions of the CH_2 protons in the vinyl ether **3-1** or of the para protons of anisole **3-2** are found at a higher field, respectively, than those in

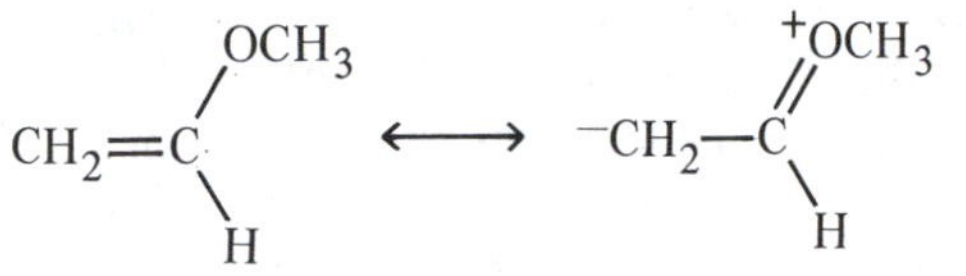

$CH_2{=}C(OCH_3)H \longleftrightarrow {}^-CH_2{-}C(={}^+OCH_3)H$

3-1

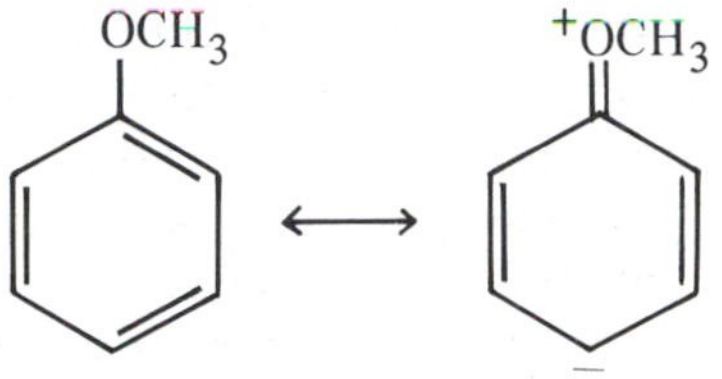

3-2

ethene or benzene. By the same token, the positions of the CH_2 protons in methyl vinyl ketone **3-3** or of the para protons in nitrobenzene **3-4** are shifted to lower fields, respectively, than in ethene or benzene, because of electron withdrawal by resonance.

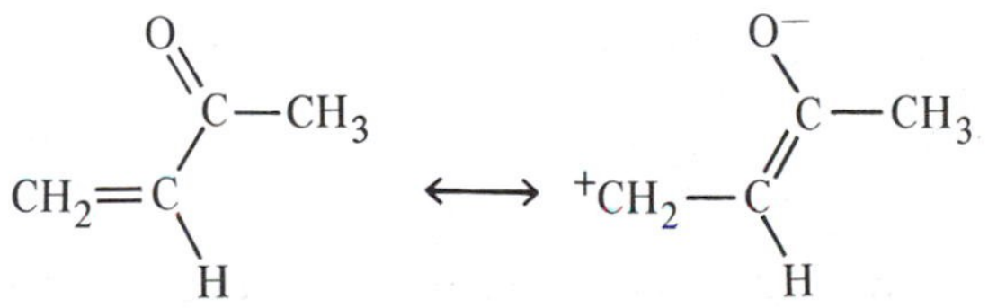

$CH_2{=}C(H){-}C({=}O){-}CH_3 \longleftrightarrow {}^+CH_2{-}C(H){=}C(O^-){-}CH_3$

3-3

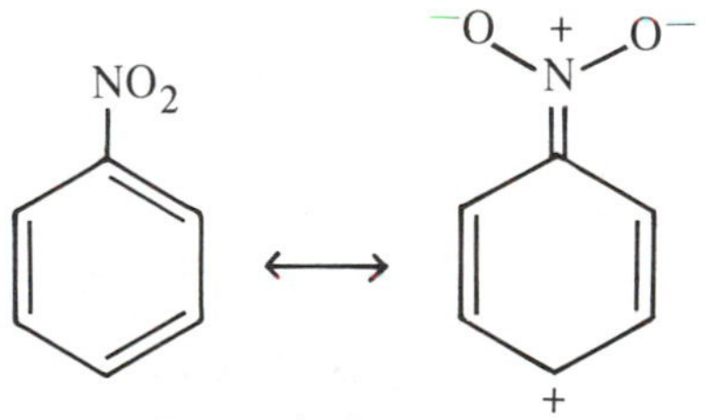

3-4

Hybridization also influences the electron density around protons. As the s character increases from 25% (sp^3) to 33% (sp^2) to 50% (sp), bonding electrons move closer to carbon, thereby deshielding the protons. For this reason, alkene protons resonate about 4 ppm lower field than do alkane protons.

The effects discussed so far have to do with changes in induction, resonance, or hybridization at the atom to which the resonating proton is attached. They arise from perturbation of the electron density at the proton because of electron donation or withdrawal. A second major factor in determining proton shifts comes from magnetic properties of substituents. As we shall see, a spherical or isotropic substituent has no perturbing effect, but an ellipsoidal or cylindrical substituent can have a major effect on chemical shifts even in the absence of changes in electron density.

To see why an isotropic substituent has no additional effect, consider a proton attached to a spherical distribution of electrons, for example, a methyl substituent. Electrons in the substituent precess in the applied field and give rise to an induced field that opposes B_0. The induced field may be represented by lines of force. If the bond from the resonating proton to the spherical substituent is parallel to the direction of B_0, as in **3-5**, the lines of force from the induced field oppose the applied field at the proton. At constant frequency, the B_0 field must then be raised, so that the proton resonates at higher field. If the bond from the proton to the group is perpendicular to B_0, as in **3-6**, the proton is in that portion of the induced field that reinforces B_0. In **3-6**, the induced and applied lines of force at the atom are parallel, so the proton is less shielded and resonates at lower field. Because the group is isotropic, the two arrangements (**3-5**, **3-6**) are equally probable and precisely

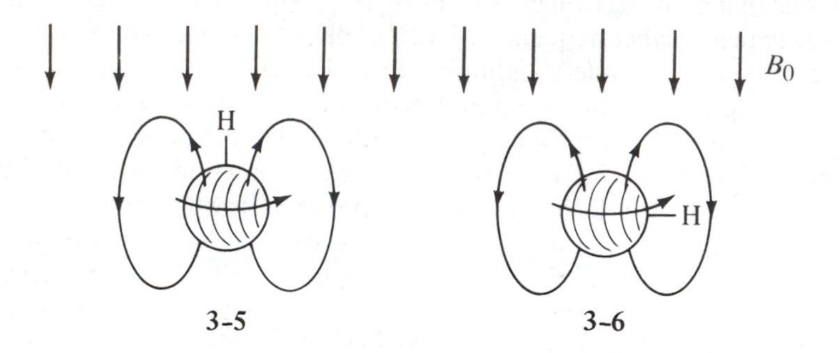

cancel. Any other orientation can be considered to be a weighted average of these two. Averaging over all orientations, since the molecule is tumbling freely, results in no net shielding or deshielding of the proton by the spherical substituent, aside from the effects of electron donation or withdrawal described above.

Electron clouds with an ellipsoidal shape may be classified either as oblate (dish-like) or prolate (rod-like) and are said to be anisotropic (different in different directions). The oblate ellipsoid is a good model for aromatic rings and the prolate ellipsoid for single or triple bonds. For a proton situated at the edge of an oblate ellipsoid such as a benzene ring, there again are two extreme arrangements. When the flat portion of the ring is perpendicular to the applied field (**3-7**), a proton at the edge resonates at lower field because it is in

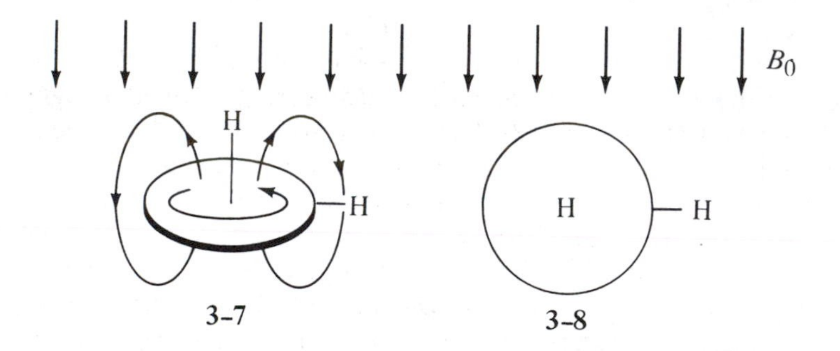

the reinforced region of the field. A proton situated over the ellipsoid resonates at higher field because it is in the opposed region of the field. For a benzene ring, oriented as in **3-7**, the induced field is particularly large because the aromatic electrons circulate easily. If the plane of the ring is rotated so that it is parallel to the field (**3-8**), there is little or no induced circulation of current. Current does not easily flow from one benzene face to the other. Thus arrangement **3-7** has a greater ability to induce a current, or a greater *diamagnetic susceptibility*, since current flow can take place easily in each benzene face. A group that has different currents induced by B_0 from different orientations in the field is said to have *diamagnetic anisotropy*. For the oblate ellipsoid, a proton at the edge of the ring is deshielded and one at the center is shielded. Any other orientation may be expressed as the weighted average of these two extremes. Thus the protons on benzene resonate at a particularly low field because of the diamagnetic anisotropy of the ring. The effect is not averaged to zero by tumbling because one configuration does not completely cancel the other.

Numerous interesting aromatic molecules have been prepared that explore the effects of diamagnetic anisotropy. Although the protons on benzene are shifted downfield, the indicated protons on [2,2]metacyclophane (**3-9**) are shifted upfield. These protons are constrained to positions over the neighboring benzene ring and hence are shielded. The aromatic 10 π electron system of methano[10]annulene (**3-10**) also can support a ring current. The methylene protons are located over the π cloud and hence are shielded to a position (δ -0.5) even higher field than TMS. The effects of shielding are often used as criteria for aromaticity in cyclic π systems.

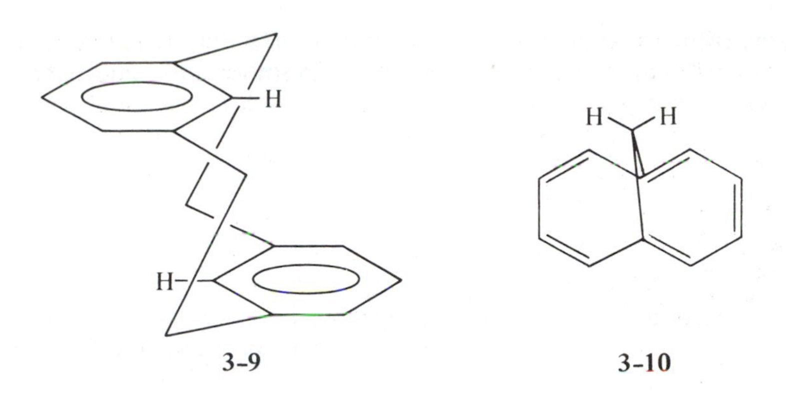

The two arrangements of a prolate ellipsoid (**3-11** and **3-12**) may be considered in a similar fashion. In this

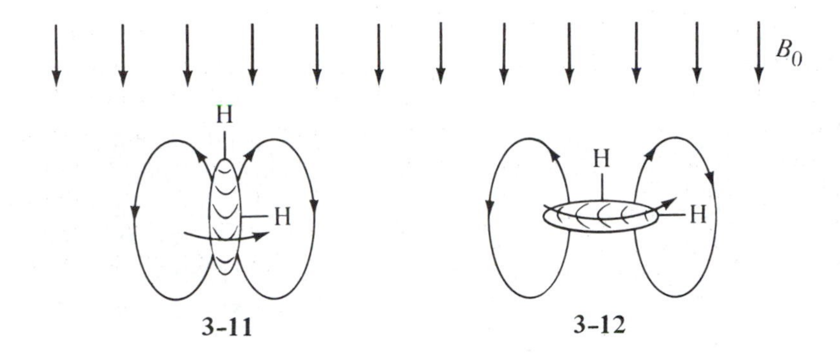

case, it is not clear which arrangement has a stronger induced current. If the diamagnetic susceptibility is greater when the long axis is parallel to the field (**3-11**), then a proton at the end is shielded and one at the side is deshielded. If the susceptibility is greater when the long axis is perpendicular to the field (**3-12**), the reverse is true. The prolate ellipsoid is a good model for a chemical bond, and, as we shall see, either case (**3-11** or **3-12**) can dominate, depending on the nature of the bond.

The π electrons of acetylene (ethyne) are cylindrically arranged about the carbon–carbon triple bond and are particularly susceptible to circulation when in arrangement **3-11**. Imagine electrons circulating freely around the cylinder. The acetylenic proton is attached to the end of this array of electrons and hence is shielded. For this reason, the acetylene resonance (δ 2.7) falls between those of ethane (δ 0.8) and ethene (δ 5.3). Consideration of hybridization alone would have predicted that acetylene have the lowest field resonance. The model of configuration **3-11** also indicates that a proton situated at the side of the triple bond would be deshielded. A peri relationship between a triple bond and a proton on naphthalene would exemplify this situation. Other triple bonds, such as in the nitrile group, have properties like those of acetylene. Since the triple bond can either shield or deshield a proton, at some point in space the effect must be null. This point occurs along a cone that is 55° 44′ up from the axis of the bond (Figure 3-2). A similar null surface must exist for benzene rings.

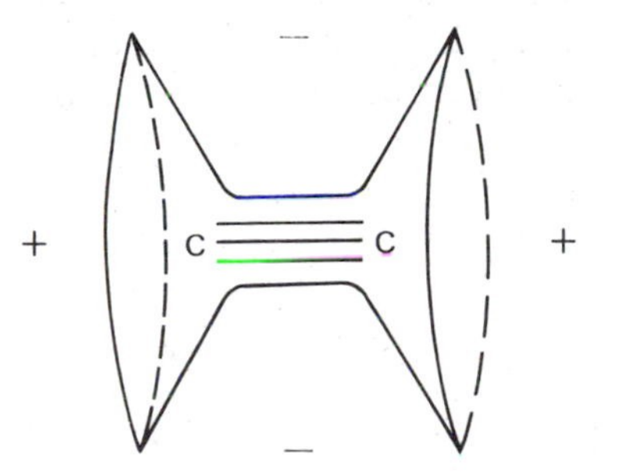

FIGURE 3-2 The diamagnetic anisotropy of the triple bond, showing the shielded (+) and deshielded (–) regions. For the single bond, the same model is used with the signs reversed.

Circulation of charge within a single bond is less effective than that in a triple bond and occurs more strongly in arrangement **3-12**. Thus a proton along the side of a single bond is more shielded than one along

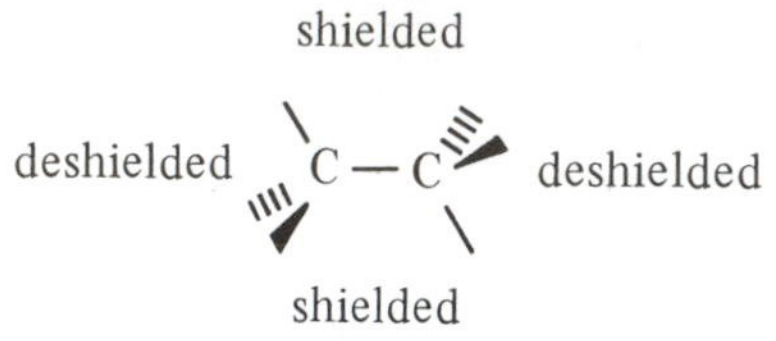

shielded

deshielded C—C deshielded

shielded

the end. The axial and equatorial protons in cyclohexane, when ring reversal is frozen out, exemplify these arrangements (**3-13**). The two illustrated protons are equivalently positioned with respect to the 1,2 and 6,1

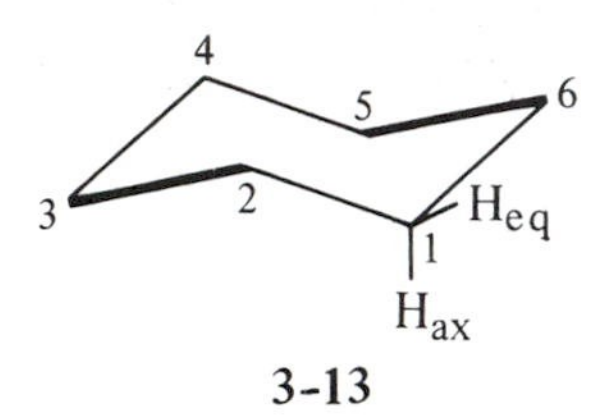

3-13

bonds, which thus produce no differential effect. The axial proton is in the shielding region of the 2,3 and 5,6 bonds (darkened), whereas the equatorial proton is in their deshielding region. Therefore H_{ax} resonates at a higher field than H_{eq} by about 0.5 ppm. The downfield effect for methine compared with methylene protons and for methylene compared with methyl protons, i.e., $\mathbf{CH_3X}$ to $\mathbf{CH_3CH_2X}$ to $\mathbf{(CH_3)_2CHX}$, or in general for any series methyl to methylene to methine for a common X group, has been attributed to the anisotropy of the additional C—C bonds. The cyclopropane ring also has unusual shielding properties that may be explained

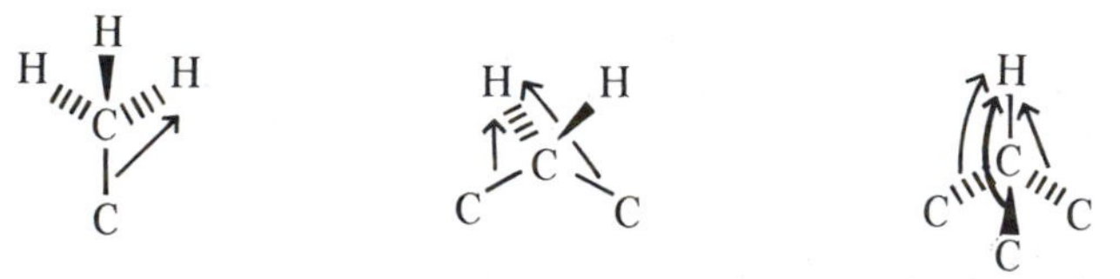

methyl deshielded by 1 bond

methylene deshielded by 2 bonds

methine deshielded by 3 bonds

in terms of C—C bond anisotropy. The methylene group in unsubstituted cyclopropane is situated directly opposite a C—C bond (**3-14**) and is therefore shielded with respect to an ordinary methylene group (**3-15**).

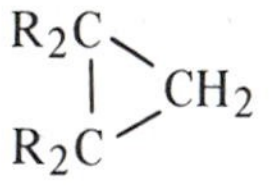

3-14

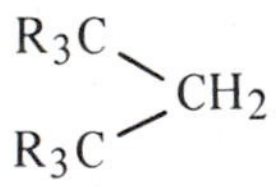

3-15

The anisotropy of double bonds is more difficult to assess, because they have three nonequivalent axes. In general, protons situated over double bonds are shielded. Thus the **CHOH** proton is over the double bond in **3-16** but not in **3-17**, so the resonance in **3-16** is at higher field. An aldehydic proton is in the severely de-

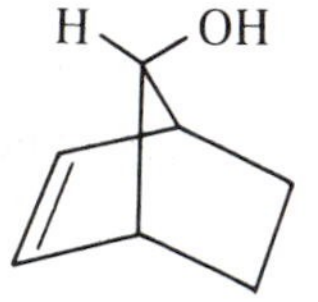

3-16

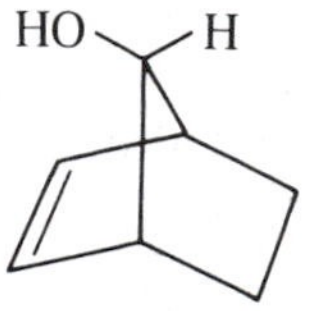

3-17

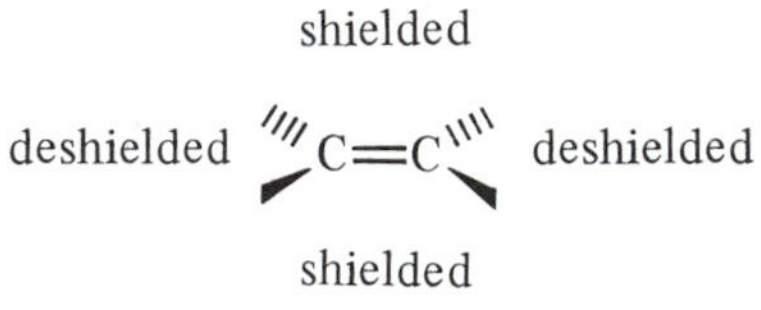

shielded

deshielded C=C deshielded

shielded

shielded region at the end of a carbon–oxygen double bond, so it resonates at a very low field (δ 9.7).

shielded

deshielded R, H C=O

shielded

A proton involved in a hydrogen bond experiences a strong downfield shift. Thus the hydroxyl proton in free ethanol resonates at δ 0.7, but in neat ethanol, with more hydrogen bonding, at δ 5.3. One explanation considers that the hydrogen bonding lone pair is a prolate ellipsoid of the type **3-12**. The proton is located at the end of this ellipsoid and hence is deshielded. In general, carboxylic acid protons (CO_2H) resonate at extremely low field because of hydrogen bonding within dimers.

O deshielded

Most proton chemical shifts may be explained by the two general effects described above. (1) Electron withdrawal or donation by induction, resonance, or hybridization alters the electron density at the resonating proton. Higher electron density results in an upfield shift, lower density in a downfield shift. (2) Diamagnetic anisotropy influences the positions of protons in the vicinity of nonspherical substituents and is largely responsible for the proton shifts of aromatics, acetylenes, aldehydes, cyclopropanes, hydrogen-bonded species, some double-bonded compounds, and cyclohexanes, and for branching effects in alkanes. Both these factors are considered more fully in Section 3-3.

3-2

Factors That Influence Shifts of Nuclei Other Than the Proton

Diamagnetic shielding (σ^d) is maximal when the electron cloud is spherically symmetrical. Electronic precession in the magnetic field, however, is impeded when the cloud is nonspherical. Because protons possess only s electrons, which have spherical symmetry, the diamagnetic term is dominant. When molecules have unsymmetrical electronic distributions, a second contribution to the shielding becomes important. This contribution is called the *paramagnetic term* (σ^p) and is in the opposite direction to σ^d. Electronic asymmetries arise because of the presence of excited states, which are described in more detail in Part III. Promotion of an electron from a ground state molecular orbital, such as those containing σ electrons or a lone pair, to an excited molecular orbital, as offered perhaps by antibonding σ or π orbitals, produces electronic currents that reinforce the applied magnetic field and hence are called paramagnetic rather than diamagnetic. For nuclei other than the proton, such as ^{13}C, ^{15}N, and ^{17}O, the presence of p electrons provides accessible excited states and hence large paramagnetic shifts. In this context, the term paramagnetic should not be confused with its usage to describe molecules with unpaired electrons.

The paramagnetic term can be quite large. Whereas for protons σ^d covers a range of only a few parts per million, σ^p can extend over a range of many hundreds or even thousands of ppm for other nuclei. Because σ^p dominates the shielding of nuclei other than the proton, we can expect their ranges of chemical shifts to be extremely large.

In the analysis of chemical shifts dominated by σ^p, it is important to think in terms that are somewhat different from those used for protons. Various contributions to σ^p are collected in eq. 3-1. This expression

$$\sigma^p \propto \frac{1}{\Delta E}\left\langle r^{-3}\right\rangle Q_{ij} \qquad (3\text{-}1)$$

includes three factors: the average excitation energy ΔE, a radial term $\langle r^{-3}\rangle$, and a molecular orbital description of p bonding Q_{ij}. We will describe the meaning of each of these factors in terms of molecular structure.

Since paramagnetic shielding depends on the availability of excited electronic states, the lower the energy of the excited state, the better able it is to provide shielding. The quantity ΔE in eq. 3-1 is the difference between ground and excited states. It is called the *average excitation energy* because it represents an average over all available excited states. As seen from eq. 3-1, σ^p is inversely proportional to ΔE, that is, the larger the energy separation, the smaller the paramagnetic term. Saturated molecules, such as alkanes, typically have no low-lying excited states, so that alkane carbon resonances are found at very high field (small σ^p). Remember that paramagnetic shielding causes downfield shifts, whereas diamagnetic shielding causes upfield shifts. Similarly, the nitrogens in aliphatic amines and the oxygens in aliphatic ethers have no low-lying excited states, so that their ^{15}N and ^{17}O resonances also are found at high field.

At the opposite extreme are carbonyl carbons, C=O, which have a low-lying excited state associated with circulation of charge from the oxygen lone pair to the antibonding π orbital, hence called an $n \rightarrow \pi^*$ transition. Typical ^{13}C resonances of carbonyl groups are found about 200 ppm downfield of TMS. The ^{15}N resonance of the nitrogen analogue of carbonyl, the nitroso group N=O, is found 900 ppm downfield of the ammonia resonance. Excited states that involve only π electrons, such as promotion of electrons from bonding to antibonding π orbitals (a $\pi \rightarrow \pi^*$ transition) are too localized and do not give rise to significant paramagnetic currents. Thus the chemical shifts of alkene carbons, for example, do not have an important contribution from the ΔE term. A complete understanding of paramagnetic shifts thus requires a good appreciation of the molecular orbital structure of the molecule. We will try to indicate general trends without invoking complex theory.

The second term in eq. 3-1 represents the average of the inverse cube of the distance r between the nucleus and the bonding electrons (2p electrons are the most important). The average is necessary because electronic location cannot be specified exactly. Because σ^p is inversely proportional to the cube of the distance, the further the electrons, the less important the paramagnetic shielding. Electrons are closer to the nucleus as one proceeds from the left to the right in a given row in the periodic table. Thus oxygen, for example, has closer 2p electrons than carbon. This same phenomenon is responsible for the higher electronegativity of oxygen. As a result, ^{17}O shifts are about three times as large as ^{13}C shifts because the radial term in eq. 3-1 is about three times as large for oxygen. Thus a plot of the ^{17}O shifts of aliphatic ethers vs. the ^{13}C shifts of analogous alkanes is linear with a slope of about three. The linearity shows that oxygen and carbon shifts are sensitive to the same structural factors, and the slope indicates that oxygen is more sensitive to these factors because its 2p electrons are closer.

The radial term in eq. 3-1 gives rise to an electronegativity dependence that is very reminiscent of diamagnetic shielding. When electrons are donated to a carbon atom, the electrons around the carbon atom must move further away to reduce electrostatic repulsion, thereby increasing r and decreasing σ^p. The result is an upfield shift. Similarly, electron withdrawal permits electrons to move nearer to the nucleus, increasing σ^p and causing a downfield effect. Thus placing a series of electron-withdrawing atoms on carbon results in progressively lower field shifts, as in the series CH_3Cl (25 ppm below TMS), CH_2Cl_2 (δ 54), $CHCl_3$ (δ 78), and CCl_4 (δ 97). The situation is qualitatively similar to that for protons, but the numbers are much larger because the shift arises from σ^p rather than from σ^d. We can conclude that substituent effects generally follow the electronegativity of groups attached to carbon.

The third factor in eq. 3-1 is related to charge densities and bond orders but can be considered to be a measure of multiple bonding. The greater the multiple bonding, the larger the contribution to σ^p, and the lower field the chemical shift. This term is not entirely independent of the radial term, since multiple bonding and electron distance are related. Nonetheless, the Q term provides an empirical rationalization for the series ethane (δ 6), ethene (δ 123), and the central sp-hybridized carbon of allene (δ 214). Arenes are similar to alkenes (benzene, δ 129). The molecular orbital interpretation for alkynes is more complex, so that their chemical shifts are at an intermediate position (acetylene is at δ 70), still, however, determined by the Q term.

Interpretation of the chemical shifts of most elements is accomplished by analysis of the three factors in eq. 3-1: accessibility of excited states, distance of the electron cloud, and multiple bonding. For carbon, we have seen that the shifts of alkanes, alkenes, arenes, alkynes, and carbonyl groups and the substitution effects of electron-donating or electron-withdrawing groups may be interpreted in this fashion. There are exceptions, the most prominent being that associated with heavy atoms. The series CH_3Br (δ 10), CH_2Br_2 (δ 22), $CHBr_3$ (δ 12), and CBr_4 (δ −29) defies any explanation in terms of electronegativity, as was just used successfully for the same chlorine series. The iodine series is monotonic in an upfield shift with increased number of iodine atoms, in complete opposition to the expectations of electronegativity. This so-called *heavy atom effect* has been attributed to *spin–orbit coupling*. The nuclear spins are shielded (upfield shift) by currents associated with electrons in orbits that are unsymmetrical and in motion. These anomalous upfield effects can be expected for any nucleus, other than proton, with heavy atom substituents.

3-3

Chemical Shift and Structure (Proton)

Assignment of structure from NMR spectra requires knowledge of the relationship between chemical shift and functional group. In many cases, both carbon and proton spectra must be recorded and analyzed in order to assign organic structures. In this section we discuss qualitative functional group analysis for protons, and in the next section there is a similar analysis for carbons. In Section 3-5, we examine a few quantitative relationships. Figure 3-3 illustrates the resonance ranges for common proton functionalities.

Saturated Alkanes. Cyclopropane has the highest field position (δ 0.22) of any simple hydrocarbon, because of the anisotropy of the carbon–carbon bonds. Unsubstituted methane has essentially the same chemical shift (δ 0.23). Progressive addition of carbon–carbon bonds results in a downfield shift of about 0.3–0.5 ppm per bond, as in the series ethane (CH_3CH_3, δ 0.86), propane ($CH_3\mathbf{CH_2}CH_3$, δ 1.33), and isobutane ($(CH_3)_3\mathbf{CH}$, δ 1.56). This effect is very general and has been attributed to the anisotropy of the added bonds. Cyclic structures other than cyclopropane have resonance positions similar to those of open chain systems. Thus cyclohexane resonates at δ 1.42. Spin–spin splitting between adjacent, nonequivalent methylene groups brings about broad, overlapping resonances. Molecules such as steroids that have numerous hydrocarbon segments that are structurally similar have an envelope of resonances in the δ 0.8–2.0 region that is punctuated only by a few sharp methyl peaks that emerge from the background. For complex molecules of this type, the decoupled ^{13}C spectrum, which lacks spin–spin splitting, can be more useful than the ^1H spectrum in structure elucidation.

Functionalized, Saturated Alkanes. The presence of a functional group alters the resonance position according to the electronegativity of the group and its diamagnetic anisotropy. Ethane (δ 0.86) is a useful point of reference for methyl groups. Replacement of one methyl group in ethane with hydroxyl yields the

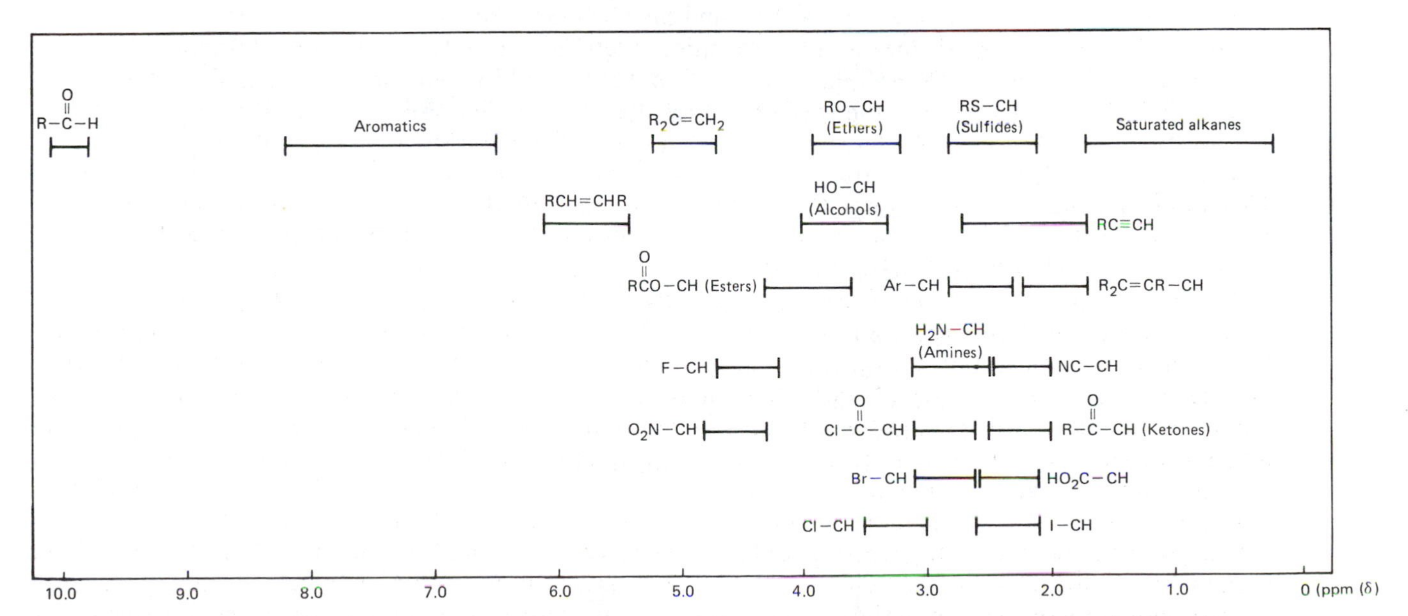

FIGURE 3-3 Proton chemical shift ranges for common structural units. The symbol CH represents methyl, methylene, or methine; R represents a saturated alkyl group. The indicated ranges are for common examples; the actual range for a given functional group can be larger.

molecule methanol, whose resonance position is δ 3.38. The electron-withdrawing effect of the oxygen atom is the primary cause of this large downfield shift. Just as in unfunctionalized alkanes, methylene groups ($CH_3\mathbf{CH_2}OH$, δ 3.56) and methine groups ($(CH_3)_2\mathbf{CH}OH$, δ 3.85) are found at lower field. We will illustrate most other functional groups with only the methyl resonance position. It should be borne in mind that methylene and methine positions will be, respectively, about 0.4 and 0.7 ppm further downfield. Furthermore, there is considerable variation from one example to another, depending on the remainder of the structure, so that resonances for a given structural type can range over 1 ppm. Ether resonances are similar to those for alcohols ($\mathbf{CH_3}OCH_3$, δ 3.24), but ester alkoxy groups usually resonate further downfield ($\mathbf{CH_3}O(CO)CH_3$, δ 3.67).

Because nitrogen is not so electron withdrawing as oxygen, amines resonate somewhat further upfield than ethers, δ 2.42 for methylamine ($\mathbf{CH_3}NH_2$ in aqueous solution). For the same reason, sulfides are at even higher field, δ 2.12 for dimethyl sulfide ($\mathbf{CH_3}SCH_3$). Halogen resonances move downfield according to the electronegativity of the atom, δ 2.15 for CH_3I, 2.69 for CH_3Br, 3.06 for CH_3Cl, and 4.27 for CH_3F. In every case, undoubtedly there are contributions from the anisotropy of the C—X bonds, but this factor is hard to assess and is somewhat diminished by free rotation in open chain systems. Multiple substitution by halogen can result in large downfield shifts, as seen in the resonance position of the common NMR solvent chloroform ($CHCl_3$, δ 7.27). Pseudohalogen substituents also cause downfield shifts, as for cyano in acetonitrile (CH_3CN, δ 2.00) and nitro in nitromethane (CH_3NO_2, δ 4.33). Electron-donating substituents like silicon in TMS cause upfield shifts.

Methyl groups on double bonds are usually found in the region δ 1.7–2.5, as for isobutylene ($(\mathbf{CH_3})_2C{=}CH_2$, δ 1.70) and toluene ($C_6H_5\mathbf{CH_3}$, δ 2.31). Methyls on carbonyl groups also are found in this region. The exact position varies with the type of carbonyl. The ketone acetone ($CH_3(CO)CH_3$) is found at δ 2.07, acetic acid ($\mathbf{CH_3}CO_2H$) at δ 2.10, acetaldehyde ($\mathbf{CH_3}CHO$) at δ 2.20, and acetyl chloride (CH_3COCl) at δ 2.67.

Protons on Atoms Other Than Carbon. Hydroxyl and amino protons are strongly influenced by the acidity and hydrogen-bonding properties of the solution. At infinite dilution in CCl_4, the OH resonance of methanol might be found at about δ 0.5. Under normal conditions, higher degrees of hydrogen bonding result in resonances in the δ 2–4 range. Phenolic hydroxyl groups are found at lower field, δ 4–8. Because of the variable location of these resonances, one must be extremely wary of spectral assignments for hydroxyl-containing molecules. A convenient experimental procedure to identify hydroxyl resonances in organic solvents such as CCl_4 is to add a couple of drops of D_2O to the NMR tube. Shaking the tube briefly will result in exchange of the hydroxyl protons with deuterium, which is in excess. The aqueous layer separates out, usually to the top in halogenated solvents, and is located above the receiver coil (see Figure 2-11). Consequently, hydroxyl resonances disappear and may be identified by their absence.

Amino resonances similarly are variable in location. Methylamine ($CH_3\mathbf{NH_2}$) in aqueous solution is found at about δ 2.0, but the range for aliphatic amines is δ 0.5–3.0 and for aromatic amines (anilines) δ 3–5.

Alkynes. The anisotropy of the triple bond results in a relatively high-field position for protons on sp-hybridized carbons. For acetylene itself, the location is δ 2.88, and the range is about δ 1.82–2.93.

Alkenes. The high electronegativity of sp^2 carbon results in a low-field position for protons on alkene carbons. The range, however, is large (δ 4.5–7.0), as the exact resonance position depends on the nature of the substituents on the double bond. The value for ethene is δ 5.28. 1,1-Disubstituted alkenes, including exo-methylene groups on rings (=CH_2), resonate at somewhat higher field, δ 4.73 for isobutylene ($(CH_3)_2C{=}\mathbf{CH_2}$). The CH_2 part in a vinyl group, $—CH{=}\mathbf{CH_2}$, also is usually found at higher field than δ 5. 1,2-Disubstituted alkenes, as found, for example, in endocyclic ring double bonds (—CH=CH—), and trisubstituted double bonds generally are found at lower field than δ 5, as in *trans*-2-butene ($CH_3\mathbf{CH}{=}CHCH_3$, δ 5.46). Angle strain on the double bond moves the resonance position to lower field, as in norbornene (**3-18**, δ 5.94). Conjugation usually moves the resonance to lower field, as in 1,3-cyclohexadiene (**3-19**, δ 5.78). The double bonds in 1,3-cyclopentadiene (**3-20**) are both strained and conjugated, so the

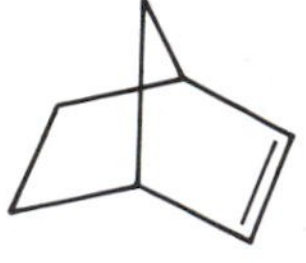

3-18 **3-19** **3-20**

position is at even lower field, δ 6.42. The phenyl ring in styrene ($C_6H_5CH{=}CH_2$) is electron withdrawing by induction and resonance, so the positions of all the alkenic protons are moved downfield, to δ 6.66 for CH and to δ 5.15–5.63 for CH_2.

Carbonyl groups are strongly electron withdrawing by induction and resonance, so double bonds conjugated with esters, for example, have very low field resonances, δ 6.83 for *trans*-$CH_3CH_2O_2CCH{=}CHCO_2CH_2CH_3$. The series **3-21**–**3-23** shows the important influence of resonance effects on alkene chemical

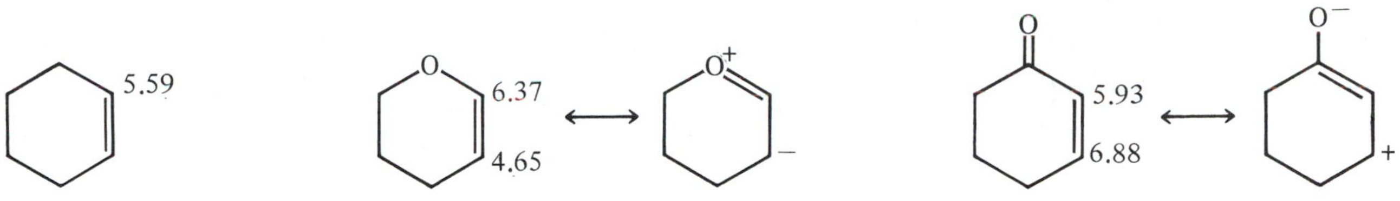

5.59

3-21

6.37 4.65

3-22

5.93 6.88

3-23

shifts. Whereas the alkenic protons of cyclohexene (**3-21**) resonate at a normal δ 5.59, the electron-donating oxygen atom in **3-22** moves the α proton downfield and the β proton upfield. In contrast, the electron-withdrawing carbonyl group in **3-23** moves the β proton downfield; in this case the inductive effect of the carbonyl group is strong enough to cause a small downfield shift for the α proton. The term *vinylic* should not be used generically for protons on a double bond. Strictly speaking, a vinyl group is $—CH{=}CH_2$, and *alkenic* should be the general term.

Aromatic Molecules. Diamagnetic anisotropy of the benzene ring augments the already deshielding influence of the sp^2 carbon atoms to yield a very low-field position for benzene, δ 7.27. Substituent effects are similar to those found in alkenes. In toluene ($C_6H_5CH_3$), the effect of the methyl group is small, and all five aromatic protons resonate at about δ 7.10. This narrow range is typical for saturated hydrocarbon substituents on benzene (see Figure 2-3 for the ^1H spectrum of toluene). Conjugating substituents, on the other hand, result in a large spread in aromatic resonances and considerable multiplicity because of spin–spin splitting (Figure 3-4). The electron-withdrawing nitro group (**3-24**), for example, pushes all resonances downfield, but the largest effects are seen at the ortho and para positions because of conjugation. By contrast,

NO_2 8.22 7.48 7.61

3-24

OCH_3 7.69 7.24 7.63

3-25

the methoxy group in anisole (**3-25**) donates electrons by conjugation, so the ortho and para positions are higher field than benzene. Similar sorts of effects are seen in heterocycles. The nitrogen atom in pyridine (**3-26**) is strongly electron withdrawing, so the α position is much further downfield than is the β position.

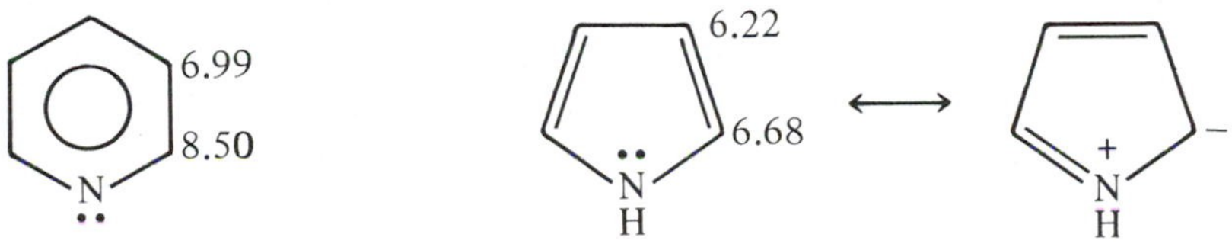

6.99 8.50 N

3-26

6.22 6.68 N H

3-27

In pyrrole (**3-27**), electron donation from the lone pair on nitrogen, necessary for aromaticity, partially offsets the inductive effect of the nitrogen, so the resonance positions of the α and β protons are closer together.

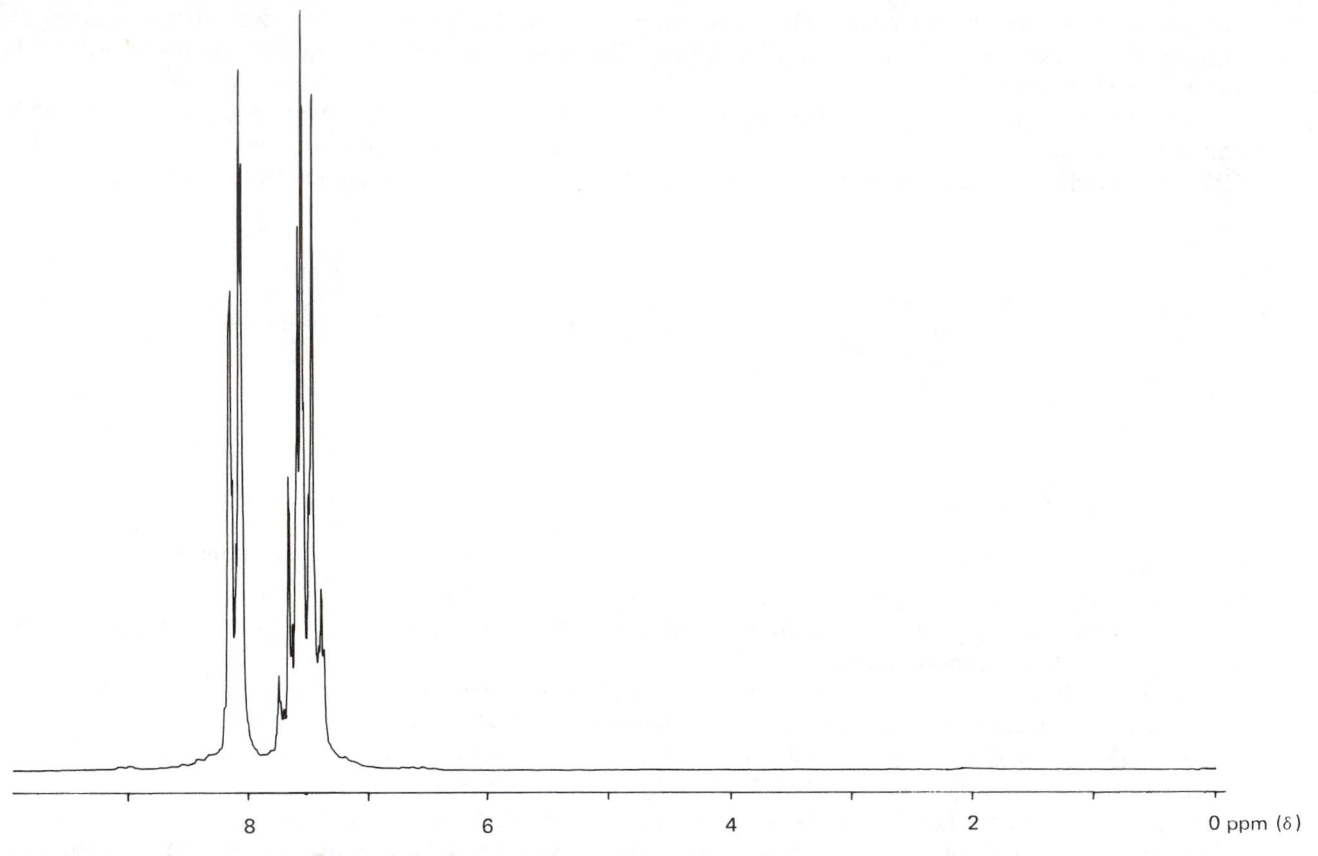

FIGURE 3-4 The proton spectrum of nitrobenzene at 90 MHz without solvent.

Aldehydes. The aldehydic proton resides in the deshielding region of the anisotropic carbonyl group, so the resonance position is extremely far downfield, δ 9.80 for acetaldehyde (CH_3**CHO**). The aldehyde range is relatively small, generally very close to δ 10 ± 0.3.

Carboxylic Acids. Most carboxylic acids exist as hydrogen-bonded dimers or oligomers, even in dilute solution. Because of the anisotropy of the lone pair on oxygen, the acid protons resonate at the lowest field general range for protons, δ 10–13 (for acetic acid, CH_3CO_2H, the position is δ 11.37). Other highly hydrogen-bonded protons also may be found in this range, such as the enol proton of acetylacetone.

3-4

Chemical Shift and Structure (Carbon)

Figure 3-5 illustrates general ranges for ^{13}C chemical shifts. As with protons, we will examine these shifts according to functional group.

Saturated Alkanes. The absence of low-lying excited states and of π bonding minimizes the paramagnetic term and places alkane chemical shifts at very high field. Methane itself resonates at δ −2.1, exceeded only by the resonance position of cyclopropane, δ −2.6. The ethane (CH_3), propane (CH_2), isobutane (CH) series follows a steady downfield trend, δ 5.9, 16.1, 25.2, similar to the methyl, methylene, methine series in proton shifts. Each additional methyl group adds about 9 ppm to the chemical shift of methane. Because the methyl groups are attached directly to the resonating carbon, this phenomenon has been termed the *α effect* and is not restricted to methyl (**3-28**, in which the resonating atom is shown). Any group X that replaces hydrogen on a carbon atom causes a relatively reproducible shift. Quantitative aspects of the α effect are considered in Section 3-5.

Replacement of hydrogen at a β position (**3-29**) also has a reproducible effect. Thus the resonance position

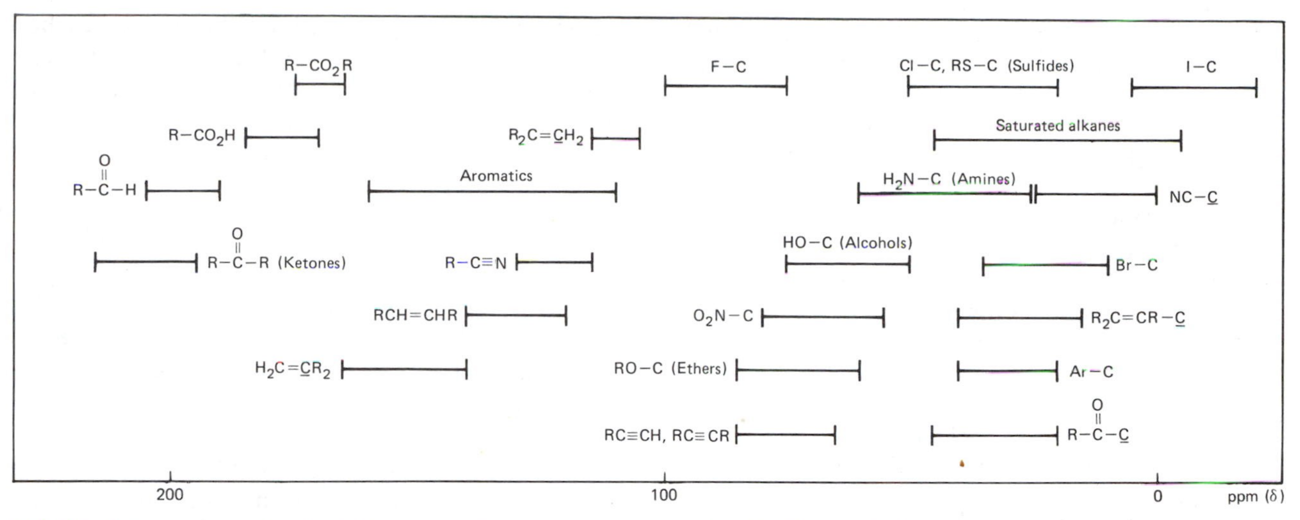

FIGURE 3-5 Carbon chemical shift ranges for common structural units. The symbol C represents methyl, methylene, methine, or quaternary carbon; R represents a saturated alkyl group. The indicated ranges are for common examples; the actual range for a given functional group can be larger.

α effect

$$-\overset{|}{\underset{|}{C}}-H \longrightarrow -\overset{|}{\underset{|}{C}}-X$$

3-28

β effect

$$-\overset{|}{\underset{|}{C}}-Y-H \longrightarrow -\overset{|}{\underset{|}{C}}-Y-X$$

3-29

of cyclohexane is at δ 27.8, or about 11 ppm downfield from the position for propane (16.1). A given cyclohexane methylene carbon has two α methylene groups, like propane, but also has two β methylene groups. The 11 ppm downfield shift is the result of the *β effect* of these two additional carbons. There are also *γ effects* (**3-30**), which vary with stereochemical relationships and are discussed in greater detail in Section 3-5. Thus much of ^{13}C chemical shift analysis is a process of counting substituents.

γ effect

$$-\overset{|}{\underset{|}{C}}-Z-Y-H \longrightarrow -\overset{|}{\underset{|}{C}}-Z-Y-X$$

3-30

Because of the large α and β effects, the alkane chemical shift range is rather large. Methyl resonances are found in the δ 5–15 range, depending on the number of β substituents; methylene resonances are in the δ 15–30 range; and methine resonances are in the δ 25–45 range.

Functionalized, Saturated Alkanes. Replacement of hydrogen on carbon with a heteroatom or with an unsaturated group usually results in downfield α effects because of changes in the radial term. Strong electron-withdrawing groups have large α effects. In the halogen series CH_3X, the methyl chemical shifts are δ 75.4 for F, 25.1 for Cl, 10.2 for Br, and −20.5 for I. Multiple substitution results in larger effects, δ 77.7 for $CHCl_3$. Recall that the α effect of heavy atoms such as iodine and bromine is influenced by a spin–orbit mechanism and hence does not follow a simple order of electronegativity. Values given here are for methyl resonances. The range for a given halogen substituent will extend from this methyl value to some 20 ppm downfield, depending on the summation of the β and γ effects of the various carbon substituents. Multiple functionalization could extend the range even further.

For oxygen substituents, methanol resonates at δ 49.2, and the range for hydroxy-substituted carbons

($—CH_2OH$) is δ 49–70. Dimethyl ether resonates at δ 59.5, and the range for alkoxy-substituted carbons ($—CH_2OR$) is δ 59–75. The ether range is translated a few parts per million further downfield because each ether must have one additional β effect with respect to the analogous alcohol. The lower electronegativity of nitrogen moves the amine range somewhat upfield. Methylamine in aqueous solution resonates at δ 28.3, with the range for amines extending some 30 ppm downfield. The amine range is larger than the alcohol range because nitrogen can carry three substituents, with the possibility of more β and γ effects. Dimethyl sulfide resonates at δ 19.5, acetonitrile at δ 0.3, and nitromethane at δ 57.3. The respective ranges for thioalkoxy, cyano, and nitro substitution would extend some 25 ppm downfield from these values. The high-field position for cyano substitution is unusual.

A methyl group is not strongly affected by substitution on a double bond. Thus the position for the methyls of *trans*-2-butene is δ 17.3, and that for the methyl of toluene is δ 21.3. The range of carbons on double bonds is about δ 15–40. Methyls on carbonyl groups are at slightly lower field, δ 30.2 for acetone, with a range of about δ 21–45.

Alkynes. An alkyne carbon that carries a hydrogen substituent (≡CH) generally resonates in the narrow range δ 67–70. An alkyne carbon that carries a carbon substituent (≡CR) resonates at lower field (δ 74–85 range) because of α and β effects from the R substituent. The somewhat similar nitriles resonate in the δ 117–130 range (acetonitrile is δ 117.2). The $n \rightarrow \pi^*$ transition pushes the range downfield.

Alkenes and Arenes. Because of the lower importance of diamagnetic anisotropy in carbon chemical shifts, alkene and arene ranges overlap in the δ 100–170 region. In the highest field portion of this range are carbons that bear no substituents ($=CH_2$). In isobutylene, the position is δ 107.7, and the range is about δ 104–115. Carbons that have one substituent (=CHR), like those in *trans*-2-butene (δ 123.3), resonate in the δ 120–140 range. Finally, disubstituted carbons (=CRR′), like that in isobutylene (δ 146.4), resonate at lowest field, δ 140–165. Polar substituents on double bonds, particularly those in conjugation with the bond, can alter the resonance position appreciably. α,β-Unsaturated ketones, such as **3-31** and **3-32**, have high-field α reso-

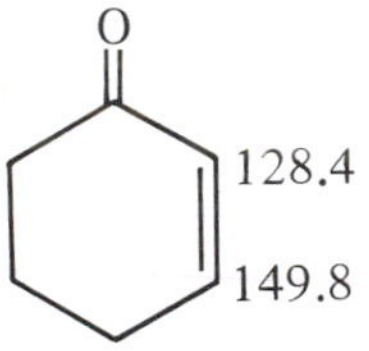

O 128.4 149.8
3-31

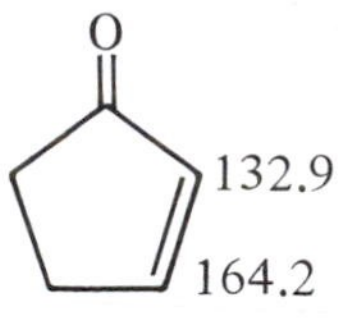

O 132.9 164.2
3-32

nances and low-field β resonances, for example. Further substitution on **3-31** or **3-32** could result in still lower field resonances.

Benzene resonates at δ 128.7, not far from the position for *trans*-2-butene, in contrast to the situation in proton NMR. Alkyl substitution, as in toluene (**3-33**), has its major effect on the ipso carbon. Because the

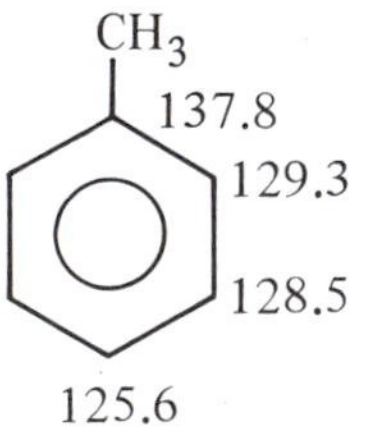

CH_3 137.8 129.3 128.5 125.6
3-33

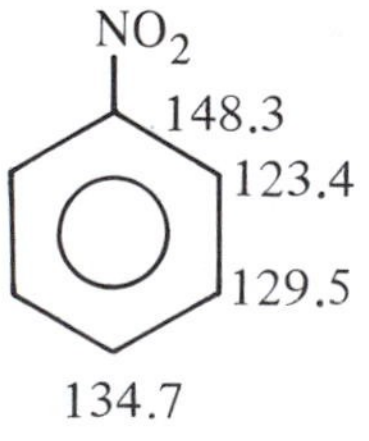

NO_2 148.3 123.4 129.5 134.7
3-34

ipso carbon has no attached proton, its relaxation time is much longer than those of the other carbons and its intensity is usually lower. Conjugating substituents like nitro (**3-34**) have stronger perturbations on the aromatic resonance positions, as the result of a combination of traditional α, β, and γ effects and changes in electron density through resonance. A similar interplay of effects is seen in the resonance positions of pyridine (**3-35**) and pyrrole (**3-36**).

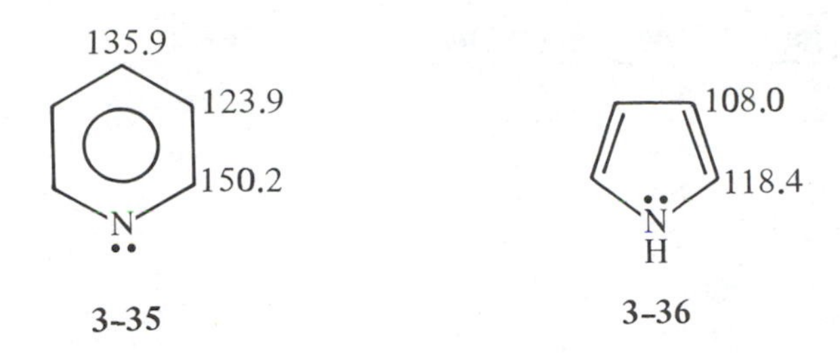

Carbonyl Groups. This functional group is entirely missing in proton NMR, so carbon NMR provides a new approach to the analysis of carbonyl compounds. The entire chemical shift range, about δ 160–220, is well removed from that of almost all other functional groups. Like aromatic ipso carbons and nitriles, carbonyl carbons other than aldehydes carry no attached protons and hence relax slowly and tend to have weak resonances.

Aldehydes resonate toward the middle of the carbonyl range, about δ 190–205, with acetaldehyde (CH_3CHO) at δ 199.6. Unsaturated aldehydes, in which the double bond is in conjugation with the carbonyl group, are shifted upfield. Ketones are at the high-field portion of the carbonyl range, δ 205.1 for acetone. Their overall range is about δ 195–215, again with unsaturated ketones at the high-field end.

Carboxyl functionalities are at the low-field portion of the carbonyl range, and there is a good separation for carboxylate (CO_2^-), carboxyl (CO_2H), and ester (CO_2R) resonances. The series sodium acetate, acetic acid, methyl acetate, for example, has respective resonances at δ 181.5, 177.3, and 170.7. The range for esters is about δ 165–175 and that for acids is δ 170–185. Acid chlorides are at somewhat higher field, δ 168.7 for acetyl chloride ($CH_3(CO)Cl$), and anhydrides are at lower field, δ 197.7 for acetic anhydride ($CH_3(CO)O(CO)CH_3$). Amides and lactones overlap the acid–ester range, δ 172.7 for acetamide ($CH_3(CO)NH_2$) and δ 176.5 for the six-membered lactone. These functional groups have ranges of about 25 ppm centered about these values, with specific positions depending on the number of β and γ substitutions and on the presence of conjugating substituents.

3-5

Empirical Correlations

Quantitative correlations between structure and chemical shift have been developed for common but relatively simple structural units. The earliest, called Shoolery's rule, provides the chemical shift of protons in a Y—CH_2—X group from substituent parameters Δ_i added to the chemical shift of methane (eq. 3-2, Table

$$\delta = 0.23 + \Delta_Y + \Delta_X \tag{3-2}$$

3-1). The calculation is reasonably successful for CH_2XY, but less so for CHXYZ. For example, the calculated shift for CH_2Cl_2 is $0.23 + (2 \times 2.53) = 5.29$ (obs. δ 5.30); for $CHCl_3$, $0.23 + (3 \times 2.53) = 7.82$ (obs. 7.27). Shoolery's rule is useful as an indication of resonance location for simple methylene groups.

TABLE 3-1 Substituent Parameters for Shoolery's Rule (R = H or Alkyl)

Substituent	Δ_i	Substituent	Δ_i
CH_3	0.68	C_6H_5	1.83
$CR{=}CR_2$	1.32	I	2.19
$C{\equiv}CR$	1.44	Br	2.33
CO—R	1.50	OR	2.36
NR_2	1.57	Cl	2.53
CN	1.59	O(CO)R	3.01
SR	1.64	NO_2	3.36

TABLE 3-2 Substituent Parameters for the Tobey–Simon Rule

Substituent	Z_{gem}	Z_{cis}	Z_{trans}
H	0.0	0.0	0.0
Alkyl	0.44	−0.26	−0.29
CH_2O, CH_2I	0.67	−0.02	−0.07
CH_2S	0.53	−0.15	−0.15
CH_2Cl, CH_2Br	0.72	0.12	0.07
CH_2N	0.66	−0.05	−0.23
C=C	0.50	0.35	0.10
C≡N	0.23	0.78	0.58
C=C (isolated)	0.98	−0.04	−0.21
C=C (conjugated)	1.26	0.08	−0.01
C=O (isolated)	1.10	1.13	0.81
C=O (conjugated)	1.06	1.01	0.95
CO_2H (isolated)	1.00	1.35	0.74
CO_2R (isolated)	0.84	1.15	0.56
CHO	1.03	0.97	1.21
OR (R aliphatic)	1.18	−1.06	−1.28
OCOR	2.09	−0.40	−0.67
Aromatic	1.35	0.37	−0.10
Cl	1.00	0.19	0.03
Br	1.04	0.40	0.55
NR_2 (R aliphatic)	0.69	−1.19	−1.31
SR	1.00	−0.24	−0.04

Tobey and Pascual, Meier, and Simon developed a similar relationship (eq. 3-3) for the chemical shift of

$$\delta = 5.28 + Z_{gem} + Z_{cis} + Z_{trans} \qquad (3\text{-}3)$$

a proton on a double bond. Substituent constants Z_i (Table 3-2) for all groups on the double bond are added to the chemical shift of ethene. For example, by this means the stereochemistry for the crotonaldehyde ($CH_3{-}CH_a{=}CH_b{-}CHO$) with resonances at δ 6.87 and 6.03 may be assigned. The calculated shifts for the cis stereochemistry are δ 6.93 (H_a, 5.28 + 0.44 + 1.21) and 6.02 (H_b, 5.28 + 1.03 − 0.29), and for the trans stereochemistry δ 6.69 (H_a, 5.28 + 0.44 + 0.97) and 6.05 (H_b, 5.28 + 1.03 − 0.26). The cis stereochemistry is supported, and the lower field resonance is unambiguously assigned to H_a. The empirical parameters Z_i in the Tobey–Simon rule incorporate the effects of resonance and induction mentioned in Section 3-3, but steric effects can cause deviations that weaken the assignment.

Aromatic proton resonances may be treated in a similar way, provided no two substituents are ortho to each other (steric effects again). The shift of a particular proton is obtained (eq. 3-4) by adding substituent

$$\delta = 7.27 + \Sigma S_i \qquad (3\text{-}4)$$

parameters to the shift of benzene (Table 3-3, gathered from many sources by Jackman and Sternhell). For example, *p*-chlorobenzaldehyde (**3-37**) gives an aromatic AB quartet (with some further complications) with resonances centered at δ 7.75 and 7.50. The calculated position for H_a is δ 7.79 (7.27 + 0.58 − 0.06) and for H_b is δ 7.50 (7.27 + 0.02 + 0.21). The observed resonances for *p*-methoxybenzoic acid (**3-38**) are at δ 8.08 and 6.98, and the calculated positions are δ 7.98 (H_a) and 6.98 (H_b). Spectral assignments therefore can be made with confidence. Conversely, eq. 3-4 may be used for specifying the substituent pattern by calculating the shifts for all possibilities and comparing them with the observed values.

TABLE 3-3 Substituent Parameters for Aromatic Proton Shifts

Substituent	S_{ortho}	S_{meta}	S_{para}
CH_3	−0.17	−0.09	−0.18
CH_2CH_3	−0.15	−0.06	−0.18
NO_2	0.95	0.17	0.33
Cl	0.02	−0.06	−0.04
Br	0.22	−0.13	−0.03
I	0.40	−0.26	−0.03
CHO	0.58	0.21	0.27
OH	−0.50	−0.14	−0.4
NH_2	−0.75	−0.24	−0.63
CN	0.27	0.11	0.3
CO_2H	0.8	0.14	0.2
CO_2CH_3	0.74	0.07	0.20
$COCH_3$	0.64	0.09	0.3
OCH_3	−0.43	−0.09	−0.37
$OCOCH_3$	−0.21	−0.02	—
$N(CH_3)_2$	−0.60	−0.10	−0.62
SCH_3	−0.03	0.0	—

CHO, H_a, H_a, H_b, H_b, Cl

3-37

CO_2H, H_a, H_a, H_b, H_b, OCH_3

3-38

Frequently in spectral analysis, the investigator is faced with differentiating a pair of closely related structures. There is a strong temptation to try to obtain a definitive answer on the basis of small differences in proton chemical shifts. Such simplistic approaches should be resisted. Too many factors influence proton shifts, and without the use of numerous model compounds incorrect conclusions are often possible from chemical shift data alone.

Carbon-13 chemical shifts lend themselves conveniently to empirical relationships because shifts may be measured easily and tend to have well-defined substituent effects. For saturated, acyclic hydrocarbons, eq. 3-5 provides an empirical measure of chemical shifts, developed by Grant. To the chemical shift of

$$\delta = -2.5 + \Sigma A_i n_i \qquad (3\text{-}5)$$

methane (δ −2.5) is added a substituent parameter A_i for each carbon atom in the molecule, up to a distance of five bonds from the resonating carbon. There is a different substituent parameter for carbons that are α (9.1), β (9.4), γ (−2.5), δ (0.3), or ϵ (0.1) to the resonating carbon. If there is more than one α carbon, the substituent parameter is multiplied by the appropriate number n_i, and a similar factor is applied for multiple substitution at all β, γ, and δ positions. Thus the methylene resonance position for propane is calculated to be $-2.5 + (9.1 \times 2) = 15.7$ (obs. δ 16.1).

$$\underset{\alpha}{\overset{(9.1)}{CH_3}}-CH_2-\underset{\alpha}{\overset{(9.1)}{CH_3}}$$

TABLE 3-4 Substituent Parameters for Methyl Substitution on Cyclohexane

Stereochemistry	α	β	γ	δ
Equatorial	5.6	8.9	0.0	−0.3
Axial	1.1	5.2	−5.4	−0.1

Corrections must be applied if there is branching, as eq. 3-5 applies only to straight chains. The resonance position of a methyl group is corrected for the presence of an adjacent tertiary (CH) carbon by addition of the factor −1.1 and for an adjacent quaternary carbon by −3.4. Methylene carbons have corrections of −2.5 and −7.2, respectively, for adjacent tertiary and quaternary carbons. Methine carbons have corrections of −3.7, −9.5, and −1.5 for adjacent secondary, tertiary, or quaternary carbons. Finally, quaternary carbons have corrections of −1.5 and −8.4 for adjacent primary and secondary centers (the numbers are not known accurately for tertiary and quaternary carbons adjacent to a resonating quaternary carbon). For example, the methyl group in isobutane ($(CH_3)_3CH$) is adjacent to a tertiary center, so the calculated shift is $-2.5 + 9.1 + (2 \times 9.4) - 1.1 = 24.3$ (obs. δ 24.3).

(9.4)
β
(−1.1) CH_3 (9.4)
Me/3° | β
CH_3—CH—CH_3
α
(9.1)

It is interesting that the γ effect of a carbon substituent is negative (−2.5). Because there is dihedral ambiguity in the γ effect, this value can vary according to stereochemistry and −2.5 actually is a weighted average for a mixture of open-chain gauche and anti conformations. The α and β effects are determined by fixed geometries and have no stereochemical component. Hydrocarbons with unusually large contributions from gauche stereochemistries may give poor results from eq. 3-5.

The fixed stereochemistry represented by cyclohexane requires an entirely new set of empirical parameters that depend on the axial or equatorial nature of the substituent, as well as on the distance from the resonating carbon. Table 3-4 lists the parameters for the ring carbons, which are added to the value for cyclohexane (δ 27.7). Corrections again are needed for branching. For two α methyls that are geminal, the correction is −3.4; for two β methyls that are geminal, it is −1.2. Thus the calculated resonance position for C2 of 1,1,3-trimethylcyclohexane is $27.7 + 5.2 + 2\,(8.9) - 1.2 = 49.5$ (obs. 49.9). A pair of vicinal, diequatorial

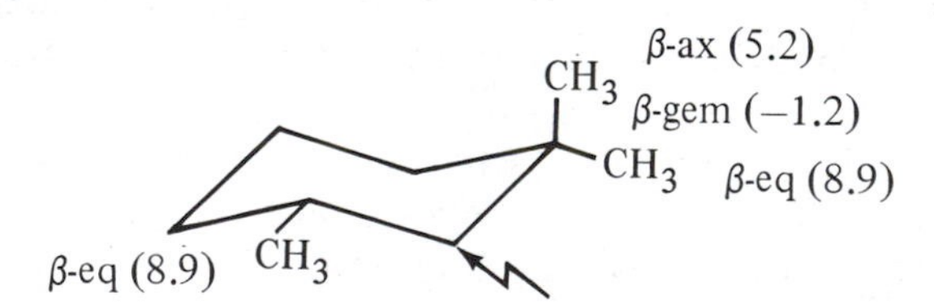

methyls (α, β) require a correction of −2.3, and vicinal, axial–equatorial methyls −3.1. The parameter for a γ-axial methyl is large and negative, reflecting the gauche stereochemistry between perturbing and resonating carbon. This large, negative shift has been termed the γ *gauche effect*.

Useful though these empirical relationships are, they refer only to saturated hydrocarbons. Any type of heteroatom or unsaturation would require another whole table of substituent parameters. Table 3-5 lists the α, β, and γ effects for a few functional groups. These numbers represent the effect on a resonating carbon of replacing a hydrogen atom at the respective position with a group X. With the exception of the heavy atom iodide, the α effects are determined largely by considerations of electronegativity. It is interesting that the β effects are all positive and of similar magnitude and that the γ effects are all negative and of similar magnitude. Although the details are not well understood, it is clear that simple polar considerations do not dominate the β and γ effects. The values in Table 3-5 offer guidelines for interpreting simple, functionalized alkanes. To utilize these values, add the substituent parameters in Table 3-5 to the values for the carbon in the

TABLE 3-5 Carbon Substituent Parameters for Functional Groups

X	α	β	γ
I	−6.0	11.3	−1.0
CN	4	8	−2
Br	20.0	10.6	−3.1
CO_2H	20.9	2.5	−2.2
NH_2	29.3	11.3	−4.6
Cl	31.2	10.5	−4.6
OH (primary)	48.3	10.2	−5.8
NO_2	61	3	−2

analogous hydrocarbon. Thus for the 1 carbon of 1,3-dichloropropane, start with δ 15.6, the value for the 1 carbon of propane: $15.6 + 31.2 - 4.6 = 42.2$ (obs. 42.2). Addition of a functional group to an already

(31.2) (−4.6)
α γ
$ClCH_2CH_2CH_2Cl$

functionalized molecule can be treated in a similar fashion. Thus the chemical shift of the 4 carbon of 4-hydroxy-2-butanone could be calculated from the chemical shift (δ 7.6) of the 4 carbon of butanone: $7.6 + 48.3 = 55.9$.

(48.3)
α
$HOCH_2CH_2C(=O)CH_3$

Aromatic systems have easily defined effects for substitution α (ipso), β (ortho), γ (meta), and δ (para) to a resonating carbon (Table 3-6, in which the values represent the effect of replacing H with X; collected by Stothers from several sources). In this fully conjugated system, normal α, β, and γ effects are altered by polar and resonance effects characteristic of aromatic systems. An empirical correlation for alkenic carbons, similar to the Tobey–Simon rule for protons, has been developed.

TABLE 3-6 Carbon Substituent Parameters for Aromatic Systems

X	ipso	ortho	meta	para
CH_3	8.9	0.7	−0.1	−2.9
CN	−19.0	1.4	−1.5	1.4
F	35.1	−14.1	1.6	−4.4
Cl	6.4	0.2	1.0	−2.0
Br	−5.4	3.3	2.2	−1.0
I	−32.0	10.2	2.9	1.0
NH_2	19.2	−12.4	1.3	−9.5
OH	26.9	−12.6	1.8	−7.9
OCH_3	30.2	−15.5	0.0	−8.9
CO_2CH_3	1.3	−0.5	−0.5	3.5
CHO	9.0	1.2	1.2	6.0
$CO{-}CH_3$	7.9	−0.3	−0.3	2.9
NO_2	19.6	−5.3	0.8	6.0

All of these correlations provide useful initial approaches to understanding proton and carbon chemical shifts. They are only as good as the data set used in their creation, and they all assume that the effects of multiple substitution are additive. Unconsidered or nonadditive factors, such as steric effects or conformational changes, can cause unexpected deviations between calculated and observed chemical shifts. Due judgment must be exercised in all uses of these empirical correlations.

3-6

Summary and Tables of Chemical Shifts

The complete NMR analysis of an unknown organic material should involve examination of both the ^{1}H and the ^{13}C spectra, as well as of other nuclei in special circumstances. Analysis of the resonances is then carried out with the benefit of knowledge of molecular formula or structural information based on synthetic precursors. The proton spectrum is divided into regions for saturated hydrocarbon structures (approximately δ 0–2), saturated but functionalized structures (δ 2–5, but even lower field if a proton is near several functional groups), the alkenic region (δ 4–7), and the aromatic region (δ 6–9). In addition, acetylenic, aldehydic, and carboxylic protons have well-characterized regions.

The carbon spectrum is divided into similar regions for saturated hydrocarbon structures (δ 0–40), saturated but functionalized structures (δ 30–90, higher field for cyano, bromo, and iodo functional groups, lower field for multiply functionalized carbons), unsaturated structures including both alkenes and arenes (δ 100–170), and a type not found in proton spectra, carbonyl structures (δ 160–220). Again, there are special regions for acetylenic and cyano carbons.

Representative chemical shifts for the major regions are given in Tables 3-7 to 3-11. Most of these values came from the compilations of Jackman and Sternhell and of Stothers and from the Aldrich Library of NMR Spectra. Analysis of each region, comparison with chemical shift tables, and application of molecular formula or knowledge of synthetic precursors then provide partial structures or sometimes the entire structure of the unknown.

The observed chemical shift of a particular nucleus depends not only on structure but also on medium. Choice of solvent can have a major effect on the chemical shift and should always be taken into account. Hydrocarbons such as cyclohexane, pentane, and hexane interact with solutes only through van der Waals interactions and have a minimal effect on the chemical shift. Polar molecules such as acetone, chloroform, diethyl ether, dimethyl sulfoxide, or acetonitrile, or even nonpolar molecules with polar substituents, such as carbon tetrachloride, can interact with a polar solute through electrostatic interactions. Because these same solvent molecules interact much less with the reference, TMS, there is a net effect on solute proton chemical shifts, sometimes several tenths of a part per million.

Solvent effects also can be exploited to separate overlapping resonances. The use of solvents with large diamagnetic anisotropy can be particularly effective in influencing proton chemical shifts. Flat molecules such as benzene, toluene, nitrobenzene, and nitroethane, because of their anisotropy, interact with many solutes differently from the way they do with TMS and hence give rise to significant solvent shifts. Rod-shaped molecules such as acetonitrile, carbon disulfide, or sulfur dioxide also are anisotropic, but their diamagnetic susceptibility has the opposite sense from that of the flat molecules. Consequently, the solvent shifts of rod-like solvents are in the opposite direction to those of flat solvents. Thus the position of the CH_2 protons in propargyl chloride ($HC{\equiv}CCH_2Cl$) moves from δ 3.87 in noninteracting cyclohexane to 4.17 in the rod-like acetonitrile and to 3.42 in the disc-like benzene.

Solvent effects do not shift all protons or carbons in a molecule to the same extent. The influence of solvent should be kept in mind in any analysis of chemical shifts and can be exploited to remove spectral overlap, particularly when high-field instrumentation is not available. Thus the α and β protons of pyrrole overlap completely in benzene but give well-separated resonances in hexane or acetone.

TABLE 3-7 Methylene Groups (CH_3CH_2X)

X	$^{1}H(CH_2)$	$^{13}C(CH_2)$
CH_3	1.17	16.1
$CH{=}CH_2$	2.00	
SCH_3	2.09	
$C{\equiv}CH$	2.14	
$(CO)N(CH_3)_2$	2.23	
CO_2CH_3	2.28	27.2
CN	2.35	10.4
$(CO)CH_3$	2.47	36.4
C_6H_5	2.63	29.3
$N(CH_3)_2$	2.63	
I	3.16	0.2
Br	3.37	28.3
OCH_3	3.37	
Cl	3.47	39.9
OH	3.59	57.3
$O(CO)CH_3$	4.05	60.1
NO_2	4.29	70.4

TABLE 3-8 Ring Systems

		^{1}H	^{13}C
Cyclopropane		0.22	−2.6
Cyclobutane		1.98	23.3
Cyclopentane		1.51	26.5
Cyclohexane		1.43	27.8
Cycloheptane		1.53	29.4
Cyclopentanone	(α)	2.06	37.0
	(β)	2.02	22.3
Cyclohexanone	(α)	2.22	40.7
	(β)	1.8	26.8
	(γ)	1.8	24.1
Oxirane		2.54	40.5
Tetrahydrofuran	(α)	3.75	69.1
	(β)	1.85	26.2
Oxane (tetrahydropyran)	(α)	3.52	68.0
	(β)	1.51	26.6
	(γ)	—	23.6
Pyrrolidine	(α)	2.75	47.4
	(β)	1.59	25.8
Piperidine	(α)	2.74	47.5
	(β)	1.50	27.2
	(γ)	1.50	25.5
Thiirane		2.27	18.9
Tetrahydrothiophene	(α)	2.82	31.7
	(β)	1.93	31.2
Sulfolane	(α)	3.00	51.1
	(β)	2.23	22.7

TABLE 3-9 Alkenes

		^{1}H	^{13}C
$CH_2{=}CHCN$	(α)	5.5–6.4	107.7
	(β)		137.8
$CH_2{=}CHC_6H_5$	(α)	6.66	112.3
	(β)	5.15, 5.63	135.8
$CH_2{=}CHBr$	(α)	6.4	115.6
	(β)	5.7–6.1	122.1
$CH_2{=}CHCO_2H$	(α)	6.5	128.0
	(β)	5.9–6.5	131.9
$CH_2{=}CH(CO)CH_3$	(α)	5.8–6.4	138.5
	(β)		129.3
$CH_2{=}CHO(CO)CH_3$	(α)	7.28	141.7
	(β)	4.56, 4.88	96.4
$CH_2{=}CHOCH_2CH_3$	(α)	6.45	152.9
	(β)	3.6–4.3	84.6
$\overset{4}{C}H_3\overset{3}{C}H{=}CCH_3{-}\overset{2}{C}H{=}\overset{1}{C}H_2$	(1)	5.02	
	(2)	6.40	
	(4)	5.70	
$(E){-}CH_3CH{=}C(CH_3)CO_2CH_3$	(α)	—	128.6
	(β)	6.73	136.4
$(CH_3)_2C{=}CHCO_2CH_3$	(α)	—	114.8
	(β)	5.62	155.9
Cyclopentene		5.60	130.6
Cyclohexene		5.59	127.2
1,3-Cyclopentadiene		6.42	132.2, 132.8
1,3-Cyclohexadiene		5.78	126.3
2-Cyclopentenone	(α)	6.10	132.9
	(β)	7.71	164.2
2-Cyclohexenone	(α)	5.93	128.4
	(β)	6.88	149.8
exo-Methylenecyclohexane	($=CH_2$)	4.55	106.5
	($C=$)	—	149.7
Allene	($=CH_2$)	4.67	74.0
	($=C=$)	—	213.0

TABLE 3-10 Aromatics

	^1H			^{13}C			
	o	*m*	*p*	*i*	*o*	*m*	*p*
$C_6H_5CH_3$	7.16	7.16	7.16	137.8	129.3	128.5	125.6
$C_6H_5CH{=}CH_2$	7.24	7.24	7.24	138.2	126.7	128.9	128.2
$C_6H_5SCH_3$	7.23	7.23	7.23	138.7	126.7	128.9	124.9
C_6H_5F	6.97	7.25	7.05	163.8	114.6	130.3	124.3
C_6H_5Cl	7.29	7.21	7.23	135.1	128.9	129.7	126.7
C_6H_5Br	7.49	7.14	7.24	123.3	132.0	130.9	127.7
C_6H_5OH	6.77	7.13	6.87	155.6	116.1	130.5	120.8
$C_6H_5OCH_3$	6.84	7.18	6.90	158.9	113.2	128.7	119.8
$C_6H_5O(CO)CH_3$	7.06	7.25	7.25	151.7	122.3	130.0	126.4
$C_6H_5(CO)CH_3$	7.91	7.45	7.45	136.6	128.4	128.4	131.6
$C_6H_5CO_2H$	8.07	7.41	7.47	130.6	130.0	128.5	133.6
$C_6H_5(CO)Cl$	8.10	7.43	7.57	134.5	131.3	129.9	136.1
C_6H_5CN	7.54	7.38	7.57	109.7	130.1	127.2	130.1
$C_6H_5NH_2$	6.52	7.03	6.63	147.9	116.3	130.0	119.2
$C_6H_5NO_2$	8.22	7.48	7.61	148.3	123.4	129.5	134.7

	^1H			^{13}C		
	α	β	**9,10**	α	β	**9,10**
Naphthalene	7.81	7.46	—	128.3	126.1	—
Anthracene	7.91	7.39	8.31	130.3	125.7	132.8
Furan	7.40	6.30	—	142.8	109.8	—
Thiophene	7.19	7.04	—	125.6	127.4	—
Pyrrole	6.68	6.22	—	118.4	108.0	—
Pyridine	6.99	8.50	—	150.2	123.9	—

TABLE 3-11 Carbonyl Compounds

	$^1H(CH_3)$	^1H(other)	$^{13}C(C{=}O)$
$H(CO)CH_3$		9.80 (HCO)	199.6
$H(CO)OCH_3$	3.79	8.05 (HCO)	
$CH_3(CO)Cl$	2.67		168.6
$CH_3(CO)OCH_2CH_3$	2.02(CH_3CO)	4.11(CH_2),1.24(CH_3C)	169.5
$CH_3(CO)N(CH_3)_2$	2.10(CH_3CO)	6.94, 7.04 (CH_3N)	169.6
CH_3CO_2H	2.12	1.37 (HO)	177.3
$CH_3CO_2^-\ Na^+$			181.5
$CH_3(CO)C_6H_5$	2.62		196.0
$CH_3(CO)CH{=}CH_2$	2.32	5.8–6.4($CH{=}CH_2$)	197.2
$CH_3(CO)CH_3$	2.19		205.1
Cyclopentanone		1.9–2.3	213.9
Cyclohexanone		1.7–2.5	208.8
2-Cyclopentenone		6.10,7.71($CH{=}CH_2$)	208.1
2-Cyclohexenone		5.93,6.88($CH{=}CH_2$)	197.1

PROBLEMS

3-1 Should the α or the β proton of naphthalene resonate at lower field? Why? Compare both resonance positions with that of benzene.

H_α H_β

3-2 The —OH proton resonance is found at δ 5.80 for phenol in dilute $CDCl_3$ and at δ 10.67 for *o*-nitrophenol in dilute $CDCl_3$. Explain.

3-3 The α carbons of pyridine resonate at δ 150.2, those of pyrrole at δ 118.4. Explain.

3-4 From Shoolery's rule, calculate the expected resonance position for the CH_2 resonance in **(a)** CH_3CH_2I, **(b)** $NC—CH_2CH{=}CH_2$, **(c)** $CH_3OCH_2C_6H_5$, **(d)** $CH_3C{\equiv}CCH_2Br$.

3-5 The Aldrich Library of NMR Spectra (Vol. I, p. 71) contains the proton spectrum of a mixture of the *cis*- and *trans*-1,2-dibromoethenes. The resonance positions are δ 6.65 and 7.03. Which comes from the cis isomer and which from the trans isomer?

3-6 A trisubstituted benzene possessing one bromine and two methoxy substituents exhibits three aromatic resonances, at δ 6.40, 6.46, and 7.41. What is the substitution pattern?

3-7 Calculate the expected ^{13}C resonance positions for all the carbon atoms in the following molecules.

a. $CH_3CH_2CH(CH_3)CH(CH_3)_2$ **b.** $CH_3CH_2CHCH_3$ (with NO_2 on the CH) **c.** $ICH_2CH_2CH_2Br$

d. $(CH_3)_3CCN$

3-8 Derive the structure of the compounds that have the following proton and carbon-13 spectra. The 1/1/1 triplet at δ 78 in the ^{13}C spectra is from the solvent $CDCl_3$.

a. $C_4H_6O_2$ (A ^1H resonance of unit integral at δ 12 is not shown.)

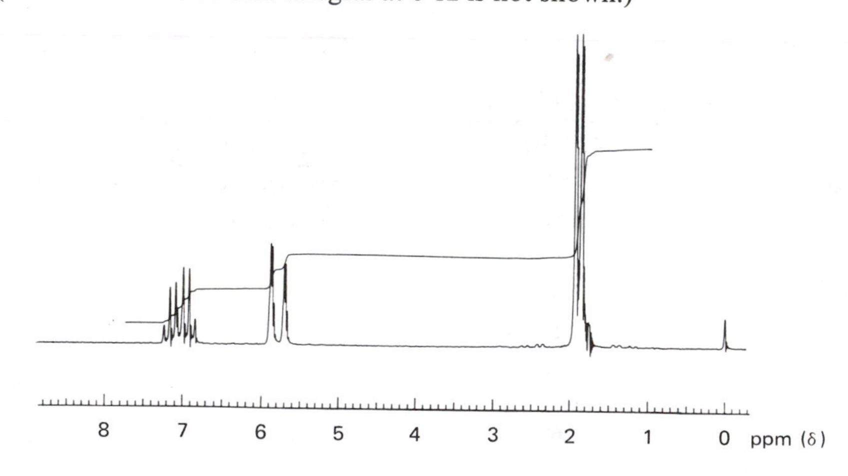

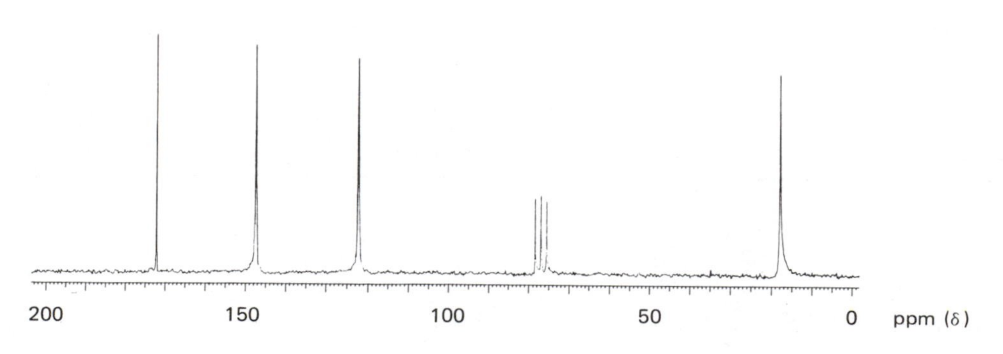

b. $C_4H_8Cl_2$

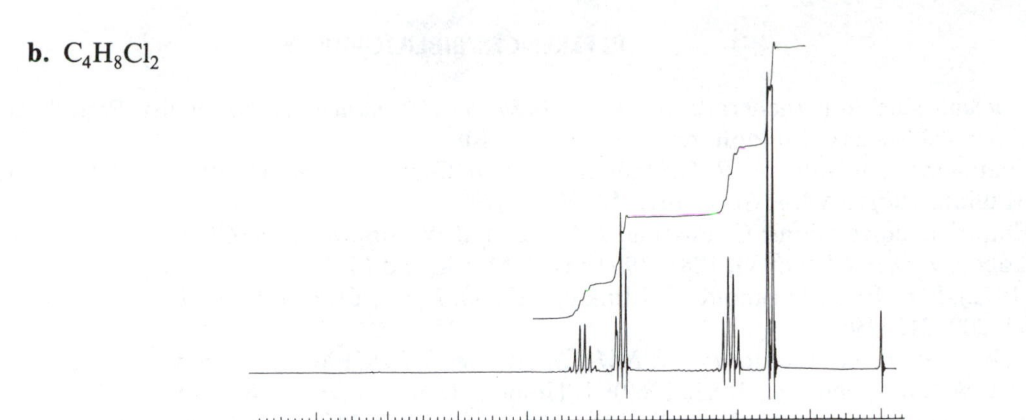

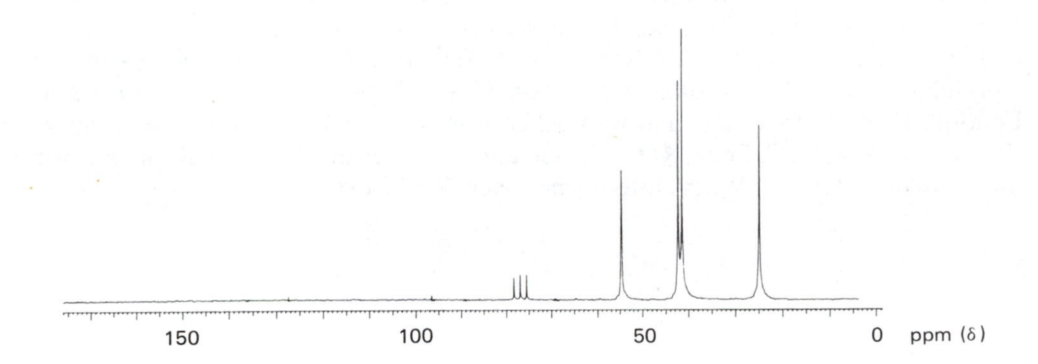

c. C_5H_9OCl

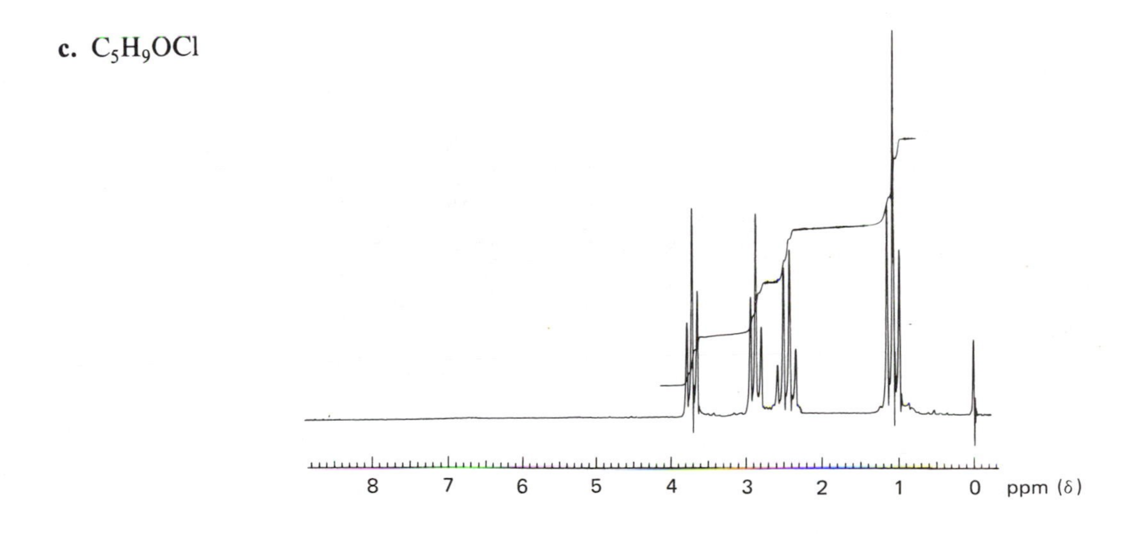

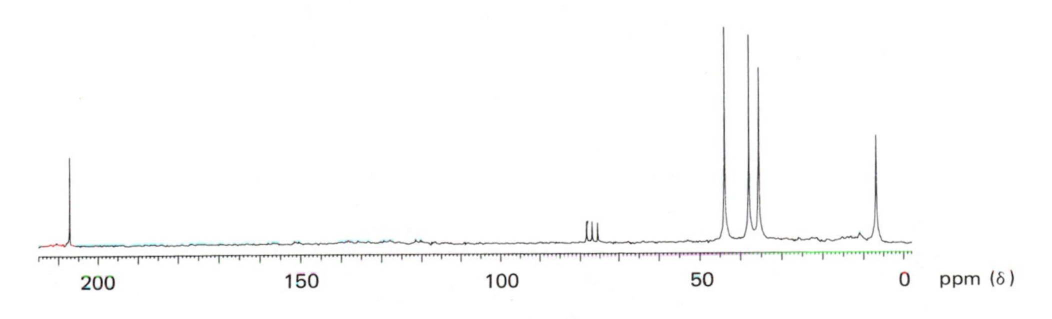

REFERENCES/BIBLIOGRAPHY

3-1 Nuclear shielding (general): *Nuclear Magnetic Resonance*, A Specialist Periodical Report, The Chemical Society, London, reviewed in each issue.

3-2 Diamagnetic anisotropy: R. C. Haddon, *Fortsch. Chem. Forsch.*, **16**, 105 (1971); C. W. Haigh and R. B. Mallion, *Progr. NMR Spectrosc.*, **13**, 303 (1979).

3-3 Empirical correlations: C. Pascual, J. Meier, and W. Simon, *Helv. Chim. Acta*, **49**, 164 (1966); S. W. Tobey, *J. Org. Chem.*, **34**, 1281 (1969); G. J. Martin and M. L. Martin, *Progr. NMR Spectrosc.*, **8**, 163 (1972); E. C. Friedrich and K. G. Runkle, *J. Chem. Educ.*, **61**, 830 (1984); D. W. Brown, *J. Chem. Educ.*, **62**, 209–212 (1985).

3-4 Solvent effects: P. Laszlo, *Progr. NMR Spectrosc.*, **3**, 231 (1967); J. Ronayne and D. H. Williams, *Ann. Rev. NMR Spectrosc.*, **2**, 83 (1969); J. Homer, *Appl. Spectrosc. Rev.*, **9**, 1 (1975); *Nuclear Magnetic Resonance*, A Specialist Periodical Report, The Chemical Society, London, reviewed in each issue.

3-5 Proton NMR: references 2-3, 2-4, 2-5, 2-9, 2-11, and 2-12.

3-6 Carbon-13 NMR: J. B. Stothers, *Carbon-13 NMR Spectroscopy*, Academic Press, New York, 1973; N. K. Wilson and J. B. Stothers, *Top. Stereochem.*, **8**, 1 (1974); G. E. Maciel, *Top. Carbon-13 NMR Spectrosc.*, **1**, 53 (1974); G. L. Nelson and E. A. Williams, *Progr. Phys. Org. Chem.*, **12**, 220 (1976); R. J. Abraham and P. Loftus, *Proton and Carbon-13 NMR Spectroscopy: an Integrated Approach*, Heyden, London, 1978; E. Breitmaier and W. Voelter, *Carbon-13 NMR Spectroscopy*, 2nd ed., Verlag Chemie, Weinheim, 1978; G. C. Levy, R. L. Lichter, and G. L. Nelson, *Carbon-13 Nuclear Magnetic Resonance Spectroscopy*, 2nd ed., Wiley–Interscience, New York, 1980.

4

The Coupling Constant

4-1

Equivalences and Second-Order Spectra

It would be an easy matter to obtain coupling constants if all spectra were of the first-order type described in Chapter 2. For a spectrum to be first order, the chemical shift difference ($\Delta\nu$) between two coupled nuclei must be larger than their coupling constant by about a factor of 10 ($\Delta\nu / J > 10$). Coupling constants may then be read directly from peak spacings. Unfortunately, chemical shift differences often are small, particularly for protons and at low field strengths, so that second-order effects are observed. These include deviation of intensities from the binomial pattern, observation of spacings that do not correspond to coupling constants, and appearance of additional peaks. For example, as the AX (first-order) pattern goes to AB (second-order), the inner peaks in each doublet grow larger at the expense of the outer peaks (Figure 4-1). As in this example, intensity distortions tend toward maximum peak heights at the center of the spectrum. Thus a second-order multiplet appears to lean toward the resonances of its coupling partner.

For spectra to be first order, it is also necessary that all chemically equivalent nuclei be magnetically equivalent. In order to define these terms, we first must examine in detail the influence of molecular symmetry on spectral characteristics. Then we will be in a position to complete the discussion of first- and second-order spectra.

Nuclei are *chemically equivalent* if they can be interchanged by a symmetry operation of the molecule. For example, the two protons in 1,1-difluoroethene (**4-1**) or in difluoromethane (**4-2**) may be interchanged by a

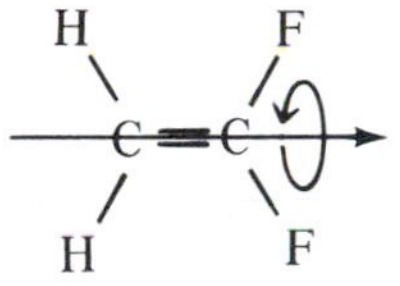

H F H F

4-1

H F H F

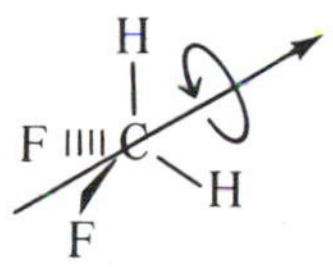

4-2

180° rotation. Nuclei that are interchangeable by rotational symmetry are said to be *homotopic*.

Nuclei that are related by other symmetry elements such as a plane are called *enantiotopic*, if there is no rotational axis of symmetry. For example, the protons in bromochloromethane (**4-3**) are chemically

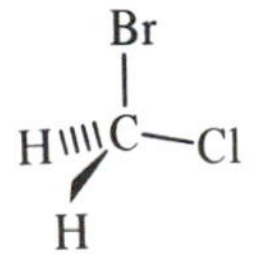

Br H C Cl H

4-3

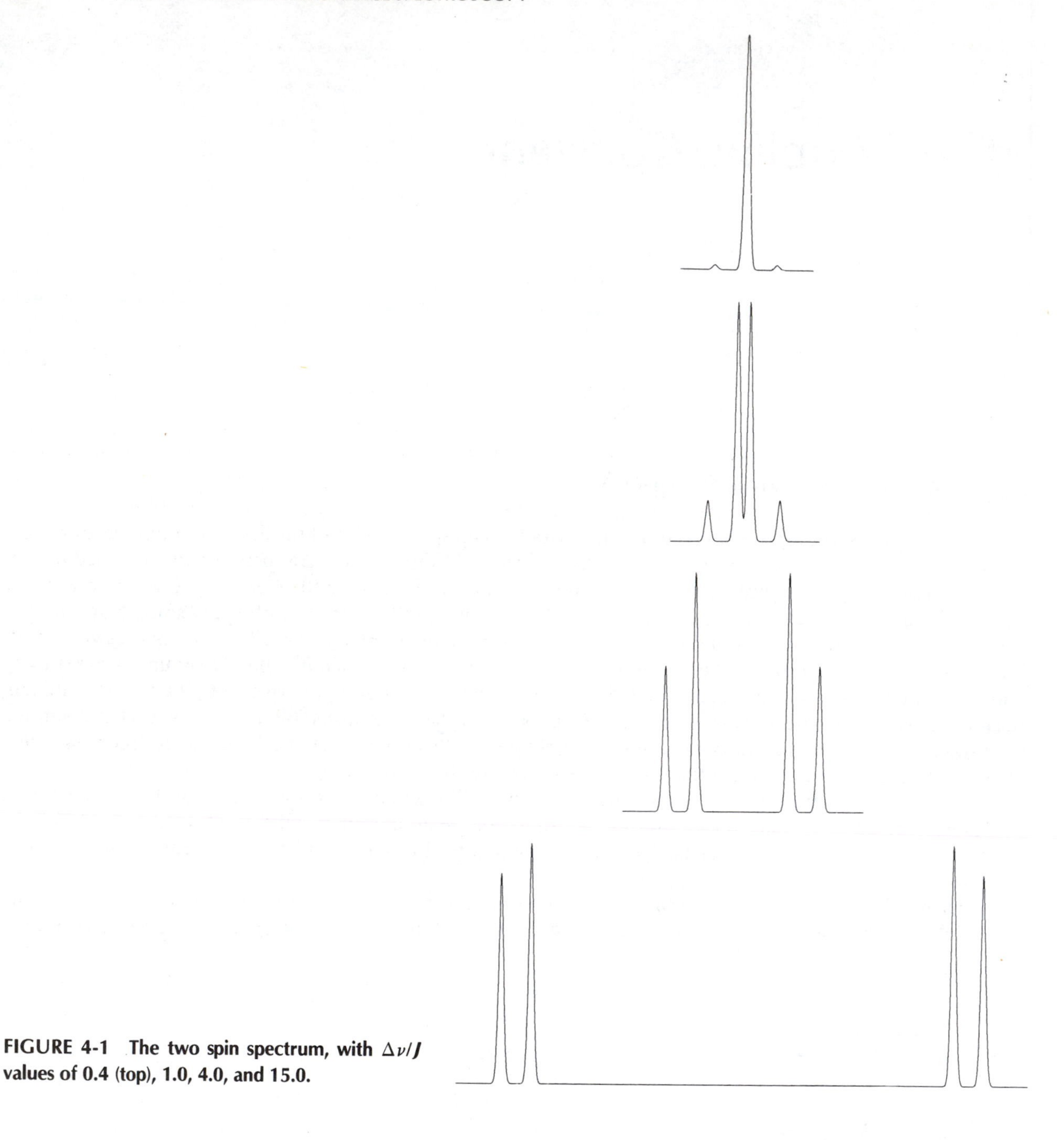

FIGURE 4-1 The two spin spectrum, with $\Delta\nu/J$ values of 0.4 (top), 1.0, 4.0, and 15.0.

equivalent and enantiotopic because they are related by the plane of symmetry containing C, Br, and Cl. The term enantiotopic was invented because replacement of one proton by another atom or group such as deuterium produces the enantiomer (**4-3a**) of the molecule that results when the other proton is replaced by the same group (**4-3b**). Homotopic nuclei handled in this fashion produce identical molecules (superimposable mirror images). Enantiotopic or homotopic protons need not be on the same carbon atom. Thus

Br, H, C, Cl, D

4-3a

Br, D, C, Cl, H

4-3b

the alkenic protons in cyclopropene (**4-4**) are homotopic, but those in 3-methylcyclopropene (**4-5**) are enantiotopic.

H H H H

4-4

H CH3 H H

4-5

Two nuclei that are enantiotopic have identical chemical properties, except in chiral environments that can be created by optically active solvents or by enzymes. Under these conditions of altered external symmetry, enantiotopic nuclei may become chemically nonequivalent and exhibit coupling in the NMR spectrum or differences in chemical properties such as acidity. Chemically equivalent nuclei are represented by the same letter in the spectral shorthand (A_2X_2 and AX_2 for the ring protons of **4-4** and **4-5**, respectively).

A set of chemically equivalent nuclei are said to be *magnetically equivalent* only if they are equally coupled to any other given nucleus in the molecule outside the set. Thus the two protons in difluoromethane (**4-2**) are magnetically equivalent because they have the same coupling to each fluorine. Similarly, the two fluorine atoms in **4-2** are magnetically equivalent, as are the two alkenic protons in cyclopropene (**4-4**).

Nuclei that are not equally coupled to another atom are said to be magnetically nonequivalent, even if they are chemically identical. 1,1-Difluoroethene (**4-1**) provides a classic example of magnetic nonequivalence. Although the two protons are chemically equivalent, they have different coupling constants to a given fluorine. The upper proton is coupled to the upper fluorine by a J_{cis}, whereas the lower proton is coupled to the same fluorine by a J_{trans}. The notation for such a set of nuclei is AA′XX′, indicating that an A nucleus has two different couplings with the X nuclei, J_{AX} and $J_{AX'}$. Any spin system that contains magnetically nonequivalent nuclei is second order by definition. Furthermore, it remains second order at the highest accessible fields, since the property results from symmetry considerations alone. The AA′ portion of an AA′XX′ spectrum contains ten peaks (Figure 4-2), compared with three in the A_2 portion of the first-order A_2X_2 spectrum. Second-order spectra are discussed further in Sections 4-7 and 4-8.

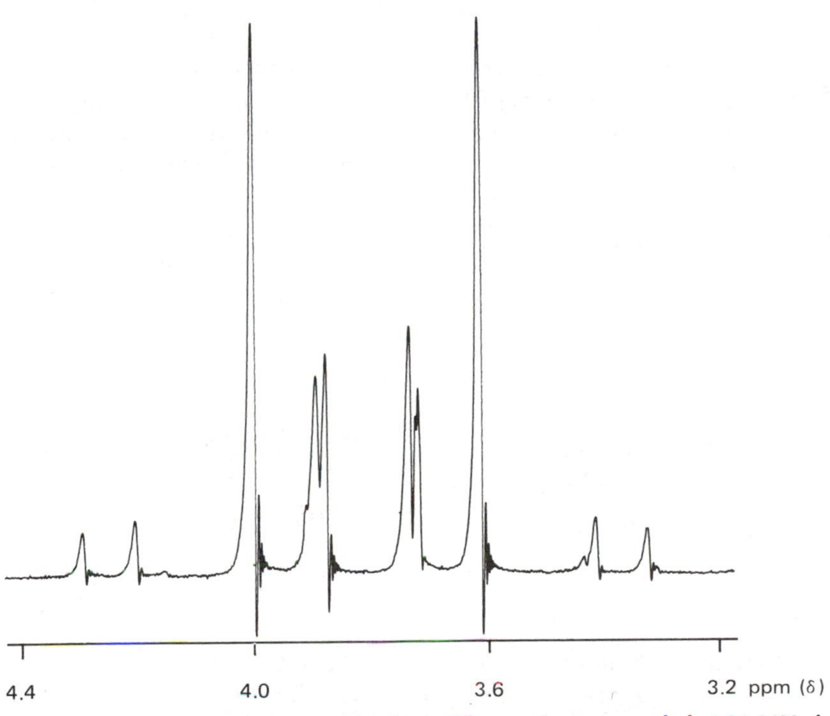

FIGURE 4-2 The proton spectrum of 1,1-difluoroethene, recorded at 90 MHz in $CDCl_3$.

Thus for a spectrum to be first order, it is necessary that $\Delta\nu/J$ be greater than about 10 and that all members of a chemically equivalent set of nuclei be magnetically equivalent. The ring protons of *p*-nitrotoluene (**4-6**) fail both tests. The two protons ortho to methyl have different couplings (J_{ortho}, J_{para}) to

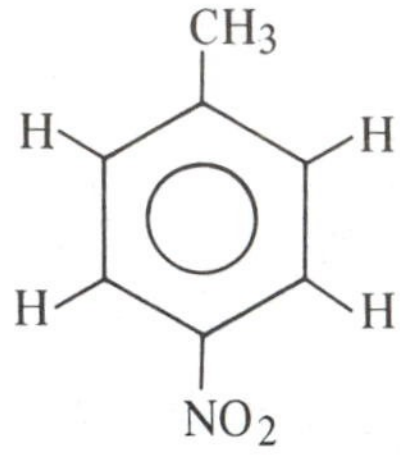

4-6 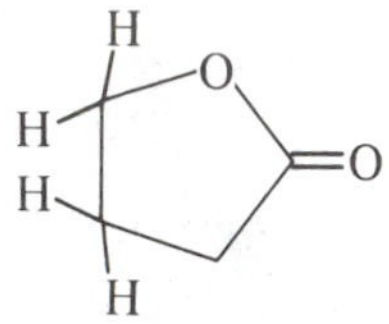 **4-7**

the lower left proton that is ortho to nitro. Since the chemical shifts are close together, the spin system is called AA′BB′ rather than AA′XX′. The protons next to oxygen in butyrolactone (**4-7**) are magnetically nonequivalent, since they are unequally coupled to the adjacent protons (J_{cis}, J_{trans}).

Magnetically equivalent nuclei of course must be chemically equivalent and hence have the same chemical shift, that is, they must be *isochronous*. Although most cases of chemical nonequivalence are self-evident, there is one class that is relatively subtle. Consider the methylene protons in ethylbenzene (**4-8**) and in its β-bromo-β-chloro derivative (**4-9**). The Newman projection **4-8a** shows that H_A and H_B are chemically

$C_6H_5—CH_2—CH_3$ **4-8** $C_6H_5—CH_2—CHBrCl$ **4-9**

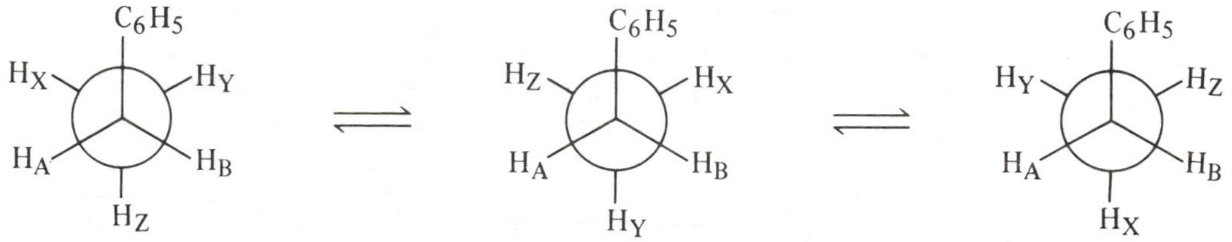

4-8a ⇌ **4-8b** ⇌ **4-8c**

equivalent and enantiotopic by reason of a plane of symmetry. Although H_A and H_B have different couplings to H_X in **4-8a**, they are magnetically equivalent on the average. Rapid C—C rotation interchanges the positions of H_A and H_B and results in only one average coupling constant. If the chemical shift between the methylene and the methyl protons is sufficiently large, the aliphatic portion of the spectrum would be entirely first order, A_2X_3.

The Newman projections for **4-9** give a contrasting situation. Each rotamer, **4-9a**, **4-9b**, and **4-9c**, is distinct, whereas the rotamers for **4-8** are identical. In none of the three are H_A and H_B related by a symmetry operation. Even with rapid C—C rotation, H_A and H_B have different chemical shifts and exhibit a coupling constant. The spin system is ABX.

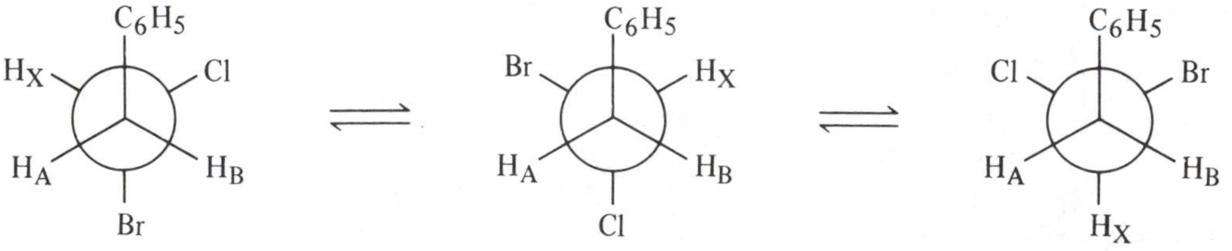

4-9a ⇌ **4-9b** ⇌ **4-9c**

The AB protons in **4-9** exemplify a particular type of nonequivalent nuclei that are termed *diastereotopic*. Replacement of H_A by deuterium in **4-9** gives **4-9d**, a diastereomer, not the enantiomer, of **4-9e**,

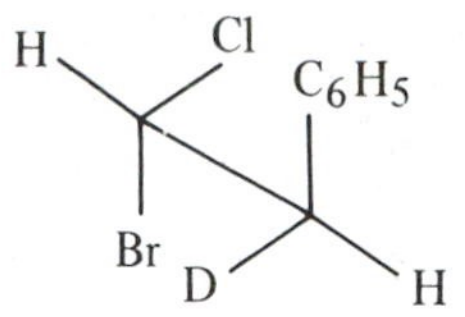

4-9d 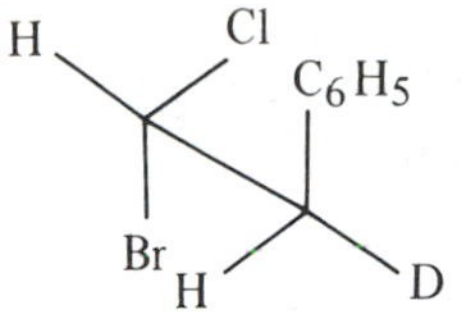 **4-9e**

which is formed on replacement of H_B by deuterium. Diastereoisomerism results because the addition of deuterium adds a second chiral center to the already existing center (—CHClBr). In general, the protons of a methylene group are diastereotopic when there is a chiral center elsewhere in the molecule. Accidental degeneracy, however, is frequently the case ($\Delta\nu = 0$), so that diastereotopic protons can appear to be equivalent in the spectrum. Diastereotopic groups can be interconverted by intramolecular processes in certain cases and hence can become enantiotopic on the average. Methyl groups in an isopropyl group can be diastereotopic when a chiral center is present, as in α-thujene (**4-10**), so that the methyl protons and the

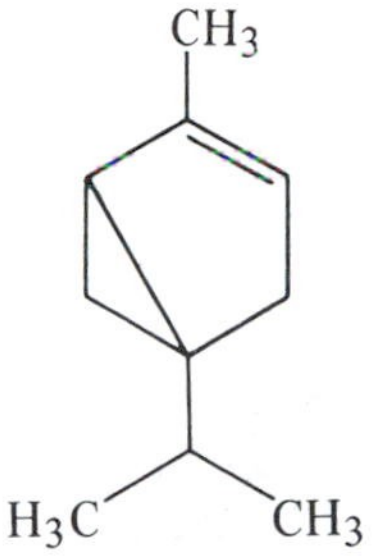

4-10 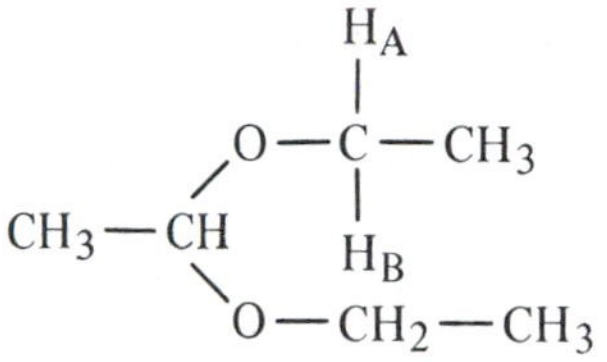 **4-11**

methyl carbons can give doubled resonances. It is not necessary that a molecule contain a chiral center for methylene protons to be diastereotopic. The diethyl acetal of acetaldehyde (**4-11**) contains diastereotopic protons (H_A and H_B), even though there is no chiral center. There is no symmetry operation that interconverts the two protons. Replacement of H_A by deuterium creates two chiral centers, and the resulting molecule is a diastereomer of the molecule in which H_B is replaced by deuterium.

The term diastereotopic may be applied to a broad range of cases, since diastereomers are defined as stereoisomers, including geometric, that are not enantiomers. Thus the methylene protons in *cis*-1,2-dichlorocyclopropane (12) are diastereotopic because **4-12a** and **4-12b** are diastereomers. By the same token, the axial and equatorial protons on a single carbon atom in ring-frozen cyclohexane are diastereotopic.

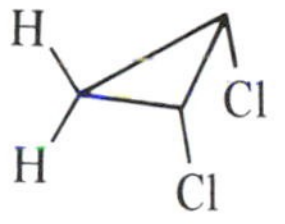

4-12 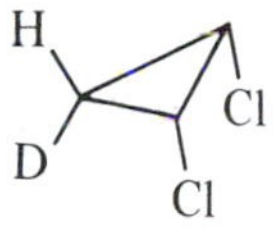 **4-12a** 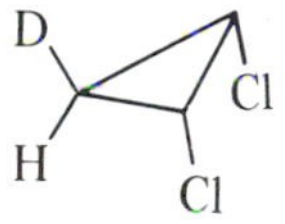 **4-12b**

4-2 Signs and Mechanisms: One Bond Couplings

As described in Chapter 2, spin–spin coupling arises because information about nuclear spin is transferred from nucleus to nucleus by the electron cloud. Exactly how does this process take place? Several mechanisms have been considered, but the most widely accepted is the *Fermi contact mechanism*. According to this model, an electron in a bond X—Y (both X and Y magnetic) spends a finite amount of time at the same point in space as, say, nucleus X. The electron and the nucleus are then said to be in contact. If nucleus X has a spin of $I_z = +\frac{1}{2}$, then the spin of the electron must be opposite ($-\frac{1}{2}$), so that the two spins can occupy the same space at the same time. In this way the nuclear spin polarizes the electron spin. This electron in turn shares an orbital in the X—Y bond with a second electron, which must have a spin of $+\frac{1}{2}$ when the spin of the first electron is $-\frac{1}{2}$. This second electron ($+\frac{1}{2}$) occupies the same point in space as nucleus Y only when Y has a spin of $-\frac{1}{2}$. Thus whenever X has a spin of $+\frac{1}{2}$, a spin of $-\frac{1}{2}$ will be slightly favored for Y, as shown in Diagram 4-1 for a $^{13}C-^{1}H$ coupling. Since the bonding electrons are used to pass the spin information, the contact term is not averaged to zero by molecular tumbling.

Diagram 4-1

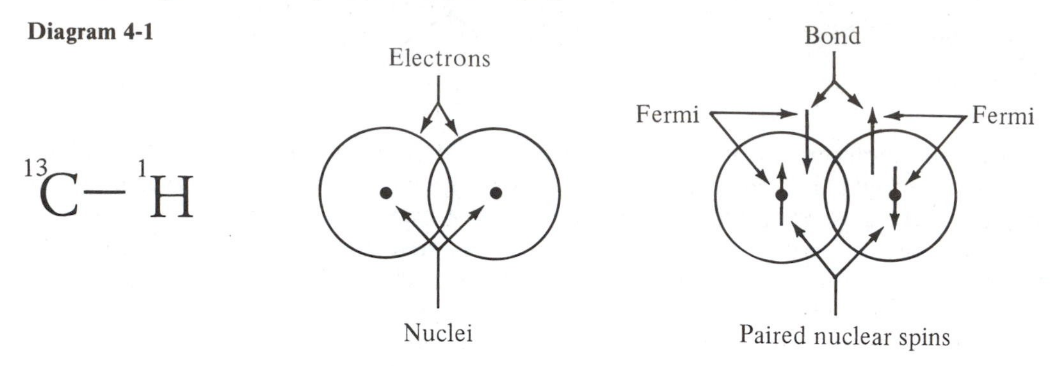

When the preferred arrangement of two coupled spins is opposite, as in the above model for coupling across X—Y, the coupling constant J by convention has a positive sign. A negative coupling occurs when the preferred arrangement of the coupled spins is parallel. An example may be found in the coupling between X and Z in an X—Y—Z fragment, a two bond coupling such as in H—C—H. When nucleus X is $+\frac{1}{2}$, the preferred spin for an electron in the X—Y bond that is at the same point in space as the nucleus must be $-\frac{1}{2}$. This electron in turn is spin-paired with the second electron ($+\frac{1}{2}$) in the bond. This electron resides in an orbital about nucleus Y. By Hund's rule, an electron in another, similar orbital about Y preferentially has the same spin ($+\frac{1}{2}$). Therefore the electron in the X—Y bond polarizes the spin of an electron in the Y—Z bond. The other electron in the Y—Z bond necessarily has the opposite spin ($-\frac{1}{2}$). This electron in turn spends time at the Z nucleus, which consequently prefers the opposite spin ($+\frac{1}{2}$). A $+\frac{1}{2}$ X nucleus thus favors a $+\frac{1}{2}$ Z nucleus. These spin polarizations may be summarized in Diagram 4-2 for the H—C—H group. Thus X and Z are stabler when their spins are parallel, and the coupling constant by convention has a negative sign. Note that the spin of the middle nucleus Y (carbon in H—C—H) does not enter into this process.

Diagram 4-2

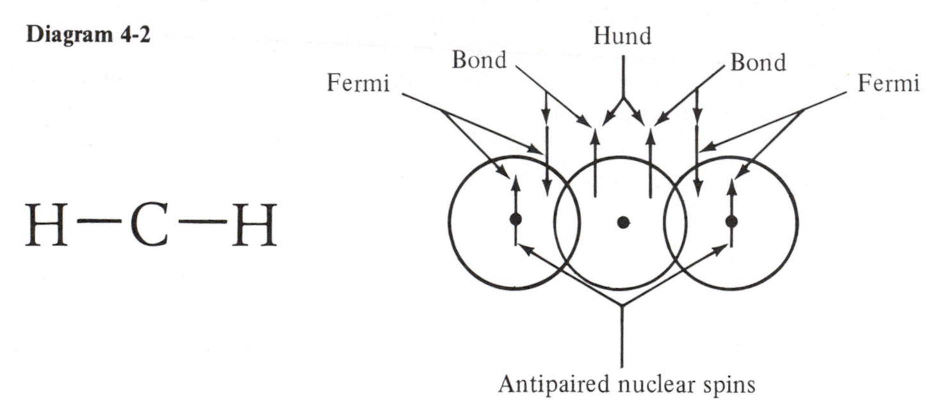

Although this qualitative argument, called the Dirac vector model, has little theoretical basis and many exceptions, it is useful primarily as a mnemonic device. Many two bond couplings do, in fact, have a negative sign. If the argument is pursued to the next stage, as in the three bond coupling over the pathway W—X—Y—Z, a positive sign is predicted by using Hund's rule twice. Couplings over an odd number of bonds are expected in general to be positive and couplings over an even number of bonds to be negative.

High resolution NMR spectra normally are not dependent on the absolute signs of the coupling constants. Reversal of the sign of every coupling constant in a spin system would result in an identical spectrum. Many spectra, however, are dependent on the relative signs of the component couplings. For example, the ABX spectrum is determined in part by three coupling constants, J_{AB}, J_{AX}, and J_{BX}. Different spectra can be obtained when J_{AX} and J_{BX} have the same sign (both positive or both negative) from when they have opposite signs, even if the magnitudes remain the same.

The usual convention for referring to a coupling constant is to denote the number of bonds between the coupling nuclei by a superscript on the left and any other descriptive material by a subscript on the right or parenthetically. A two bond (*geminal*) coupling between protons then is ${}^2J_{H-C-H}$ or ${}^2J(H-C-H)$, and a three bond (*vicinal*) coupling between a proton and a carbon is ${}^3J_{H-C-C-C}$. Beyond three bonds couplings between protons are said to be *long range*.

Carbon-13 spectra are dominated by the one bond ${}^{13}C-{}^1H$ coupling when the decoupler is not used. This coupling in fact can provide useful information and illustrates several important principles. Because a p orbital has a node at the nucleus, only electrons in s orbitals can contribute to the Fermi coupling mechanism. For protons, all electrons reside in s orbitals, but for other nuclei only that proportion of an orbital that has s character contributes to the coupling. When a proton is attached to an sp^3 carbon atom (25% s character), ${}^1J({}^{13}C-{}^1H)$ should be half as large as that for a proton attached to an sp hybridized carbon (50%). The alkenic coupling should be intermediate. The values for ${}^1J({}^{13}C-{}^1H)$ for methane (sp^3), ethene (sp^2), benzene (sp^2), and ethyne (sp) are 125, 157, 159, and 249 Hz, respectively. These points define a linear relationship between the percentage of s character of the carbon orbital and the one bond coupling (eq. 4-1). The

$$\%\,s(C-H) = 0.2J({}^{13}C-{}^1H) \tag{4-1}$$

zero intercept of this equation indicates that there is no coupling when there is no s character, in agreement with the Fermi contact model.

The *J*-s relationship of eq. 4-1 can be used as an empirical source of hybridization information. The coupling constant in cyclopropane (160 Hz) demonstrates that the carbon orbital to hydrogen is approximately sp^2 hybridized. The existence of ${}^{13}C-{}^1H$ coupling constants for hydrocarbons over the entire range from 125 to 260 Hz clearly shows that fractional hybridization of any amount is permitted. The coupling constants and s characters for the C—H bonds indicated below are 144 Hz (29%, $sp^{2.4}$) in **4-13**, 160 Hz (32%, sp^2) in cubane **4-14**, and 179 (36%, $sp^{1.8}$) in norbornane **4-15**. Although the *J*-s relationship has

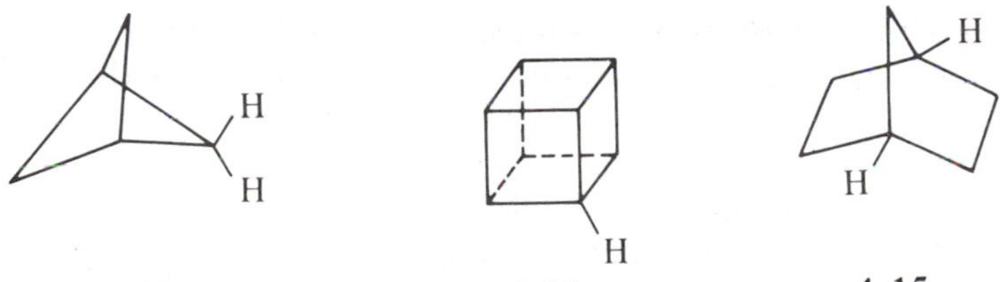

4-13 **4-14** **4-15**

worked well for hydrocarbons, there is some question as to its applicability to polar molecules. Variations in the effective nuclear charge, in addition to hybridization effects, may alter the coupling constant.

Just as the resonance frequency of a nucleus is proportional to its gyromagnetic ratio, the coupling constant between two nuclei is proportional to the product of their gyromagnetic ratios, $J(X-Y) \propto \gamma_X\gamma_Y$. The gyromagnetic ratio γ was defined in eq. 2-3. Thus because nitrogen-15 has a very small gyromagnetic ratio, its couplings are smaller than analogous ones to carbon-13. Furthermore, $\gamma({}^{15}N)$ is negative, whereas $\gamma({}^{13}C)$ is positive, so one bond couplings between hydrogen and nitrogen-15 should have a negative sign. This sign does not indicate an exception to the Dirac vector model but only reflects the negative sign of the gyromagnetic ratio.

A *J*-s relationship has been developed between the ${}^{15}N-{}^1H$ coupling constant and the hybridization of the nitrogen orbital to hydrogen, based on the couplings in ${}^+NH_4$ (sp^3, −73 Hz), $Ph_2C{=}NH_2{}^+$ (sp^2, −93 Hz), and $CH_3-C{\equiv}NH^+$ (sp, −130 Hz), but applications are limited.

Whenever the nitrogen atom possesses a lone pair, an added contribution to the contact term from this orbital has the opposite sign from the contributions of the other σ orbitals and renders any considerations of s character meaningless. Thus $J({}^{15}N-{}^1H)$ for the neutral imine $Ph_2C{=}NH$ is −51 Hz, yet the hybridization should still be sp^2. Similar effects are present in phosphorus couplings. For Me_2P-PMe_2, ${}^1J({}^{31}P-{}^{31}P)$ is −180 Hz (the negative contribution comes from the phosphorus lone pairs, since $\gamma({}^{31}P)$ is positive), whereas for $Me_2P({=}S)-P({=}S)Me_2$ without lone pairs the one bond coupling is +19 Hz.

When neither nucleus is hydrogen, the coupling constant is determined by the product of the s characters of the orbitals from both nuclei that form the bond. Early studies on ${}^1J({}^{13}C-{}^{19}F)$ and ${}^1J({}^{13}C-{}^{15}N)$ found no simple relationship with hybridization, presumably because of effects of the fluorine and nitrogen lone pairs. The one bond coupling between two carbons, however, has limited applications to considerations of hybridization and is extremely useful in mapping carbon connectivities in complex molecules. This latter use is considered in Section 5-6d.

4-3

Geminal Couplings

The two bond (geminal) H—C—H coupling often may be measured directly from the AB portion of ABX spectra. For many important compounds, however, couplings must be measured between chemically and magnetically equivalent nuclei. Thus the protons of dichloromethane are coupled, but the spectrum is a singlet because both protons have the same chemical shift. Such nuclei give no spectral splittings, so recourse must be made to alternative methods to measure J. For example, one of the protons in the methylene group can be replaced by deuterium ($CHDCl_2$). The H—D coupling then is readily measured from the proton spectrum. Since a coupling constant is proportional to the product of the gyromagnetic ratios of the coupling nuclei, J_{H-H} in general may be derived from J_{H-D} (eq. 4-2).

$$J_{H-H} = \frac{\gamma_H}{\gamma_D} J_{H-D} = 6.51 J_{H-D} \tag{4-2}$$

The geminal H—C—H coupling was generally enigmatic until 1965, when Pople and Bothner-By published a molecular orbital approach to the problem. By concentrating on substituent effects, they were able to make several important generalizations. The $^2J_{H-C-H}$ in methane is -12.4 Hz, in ethene $+2.3$ Hz. Both these couplings had been measured from deuterated samples. The 15 Hz difference between these two quantities is due to significant differences in structure and hybridization. Rather than attempt to explain these differences, Pople and Bothner-By used these molecules as points of comparison for related molecules. A substituent that withdraws electrons from the CH_2 group by induction (σ withdrawal) was found to make ^2J more positive. Similarly, a σ-donating substituent makes ^2J less positive. In contrast, a substituent that withdraws electrons hyperconjugatively (π withdrawal) makes ^2J less positive, but a π donor makes it more positive. These effects are best understood by examples.

The coupling constant in an H—C—H fragment with an sp^2 carbon is compared with the coupling in ethene (+2.3 Hz). For most hydrocarbon alkenes, ^2J is close to this value or even smaller. As a result, geminal couplings are difficult to observe in exomethylene ($C{=}CH_2$) or terminal vinyl ($CH{=}CH_2$) groups. In formaldehyde, the substituent on the methylene group has been changed from $=CH_2$ to $=O$. The oxygen atom withdraws electrons by the σ mechanism and makes the coupling constant more positive. Moreover, a nonbonding pair of electrons on oxygen has the correct symmetry for hyperconjugative donation (**4-16**),

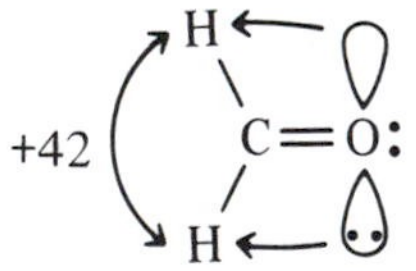

4-16 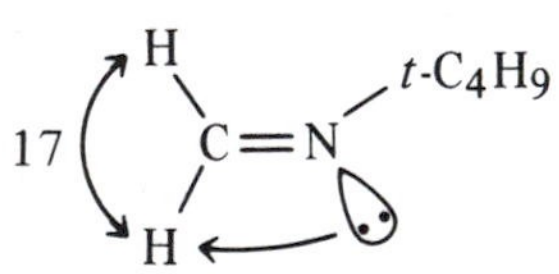 **4-17**

which also makes ^2J more positive. Reinforcement of the α and π effects gives rise to the very large and positive coupling constant observed in formaldehyde ($^2J = +42$ Hz). Similar considerations apply to imines such as **4-17**, but the lower electronegativity of nitrogen and less effective π overlap decrease the effects ($^2J =$ 17 Hz for **4-17**). If known, signs of J are given. The absence of a sign in the figures implies that it is unknown, rather than positive.

In an allene, the $=CH_2$ group of ethene has been replaced by $=C{=}CR_2$. There is only a small inductive change. The more distant double bond can accept electrons from orbitals of π symmetry in the methylene group (**4-18**). The result of this hyperconjugative electron withdrawal is a more negative coupling constant, -9 Hz in 1,1-dimethylallene.

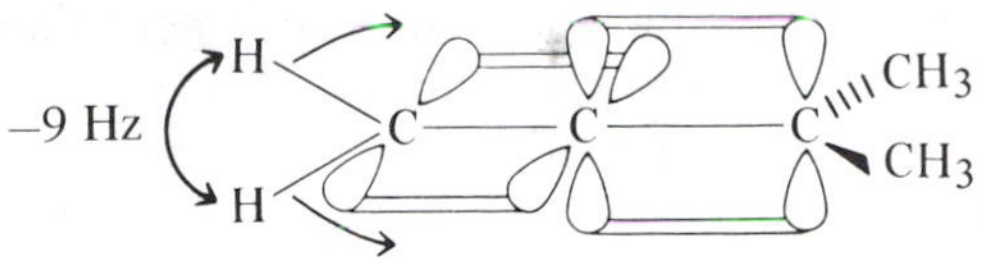

4-18

Geminal couplings through an sp^3 carbon are compared with the −12.4 Hz value in methane. Simple σ withdrawal of electrons makes ^2J more positive (less negative), as in CH_3OH (−10.8 Hz), CH_3I (−9.2 Hz), and CH_2Br_2 (−5.5 Hz); σ donation makes ^2J less positive, as in CH_3SiMe_3 (TMS, −14.1 Hz). A lone pair of electrons on oxygen can serve as a π donor to the CH_2 group (**4-19**) and thereby make ^2J more positive. In

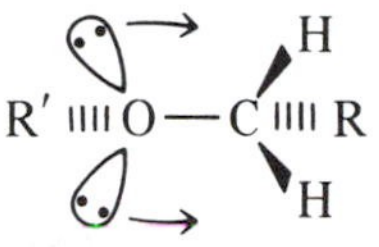

4-19

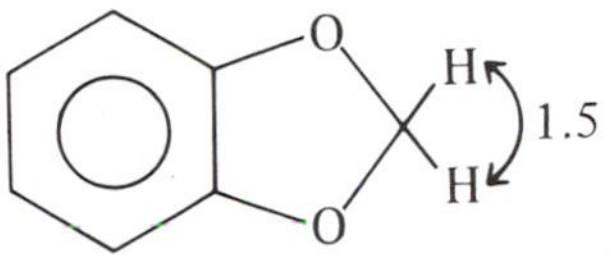

4-20

open chain systems such as **4-19**, there is free rotation about the O—C bond, so strong π overlap is inhibited. In a planar ring system such as **4-20**, the filled lone pair orbitals are frozen into the optimal orientation for overlap, and in this case there are two contributing oxygen atoms. As a result, ^2J is 1.5 Hz in **4-20**, an increase of more than 10 Hz from the value in methane.

An adjacent double bond can withdraw π electrons from a CH_2 group and thereby make ^2J less positive. In open chain systems, free rotation dilutes overlap, so the effect is minimal, $^2J = -14.9$ Hz in acetone. Overlap again is optimal in planar systems. The negative contributions of two carbonyl groups produce a coupling constant of −21.5 Hz in **4-21**, and of two benzene rings a coupling of −22.3 Hz in **4-22**. Overlap with triple bonds has no such angular dependence, so an alkynic or nitrilic substituent is a very effective π withdrawer, as in $CH_3C{\equiv}N$ ($^2J = -16.9$ Hz) and $CH_2(CN)_2$ ($^2J = -20.4$ Hz).

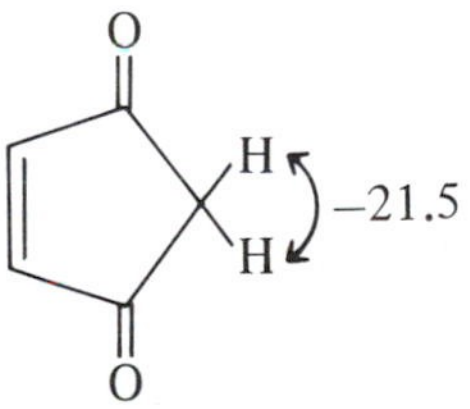

4-21

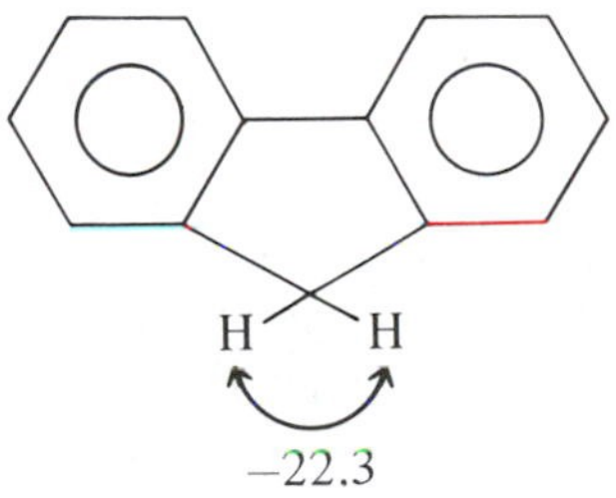

4-22

Geminal coupling constants between protons and other nuclei have also been studied. The H—C—^{13}C coupling responds to substituents in much the same manner as does the H—C—H coupling. Most such couplings are relatively small: **H—CH_2—CH_3** (−4.8 Hz), **H—CCl_2—$CHCl_2$** (+1.2 Hz). Couplings from hydrogen to sp^2 carbon (**H—CH_2—C=**) are typically −4 to −7 Hz. When the intervening carbon is sp^2 (**H—(C=)—C**) the coupling becomes larger and positive (+20–50 Hz). Unlike the H—C—H coupling, the coupling pathway can include a double bond (**H—C=C**), and factors not previously discussed can become important. Thus $^2J_{H-C-C}$ is 16.0 Hz in *cis*-dichloroethene (**4-23**) but 0.8 Hz in the trans isomer (**4-24**). The

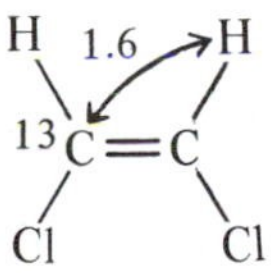

4-23

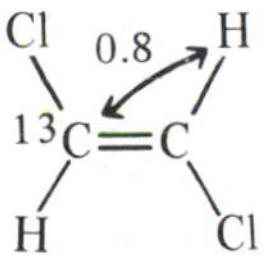

4-24

value in ethene itself is −2.4 Hz. This large difference can be exploited to document alkenic stereochemistries. Further examples may be found in Table 4-4 at the end of this chapter.

The two bond coupling between hydrogen and nitrogen-15 is critically dependent on the presence and orientation of the nitrogen lone pair. With reliable reproducibility, the H—C—^{15}N coupling in an imine is large and negative between nitrogen and a cis aldehydic proton, but small and positive for a trans aldehydic proton, as in **4-25**. Thus $^2J_{H-C-N}$ is a useful structural diagnostic for syn–anti isomerism in imines, oximes,

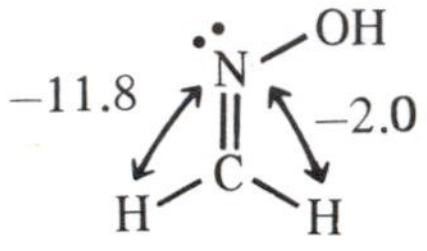

−11.8 −2.0 −10.8 −3.0

4-25 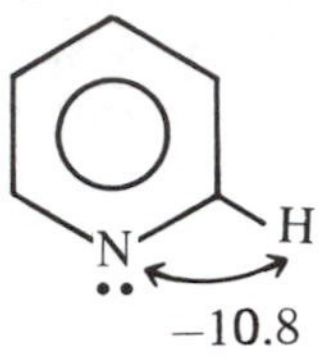 **4-26** 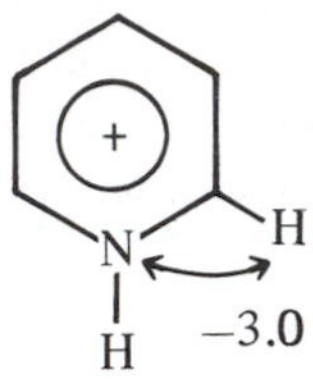 **4-27**

and related compounds. Values are typically small and negative in saturated amines (−1.0 Hz in CH_3NH_2). The cis relationship between the nitrogen lone pair and a coupling hydrogen is found in other structural situations, such as pyridine (**4-26**), and here too the coupling is large and negative (−10.8 Hz in **4-26**). Removal of the lone pair by protonation reduces the coupling substantially (−3.0 Hz in protonated pyridine, **4-27**).

Two bond couplings between ^{15}N and ^{13}C follow a similar pattern and can also be used for structural and stereochemical assignments. The carbon on the same side as the lone pair in imines again has a large and negative coupling (−11.6 Hz in **4.28**). The isomer in which the methyl is syn to hydroxyl (anti to the lone pair)

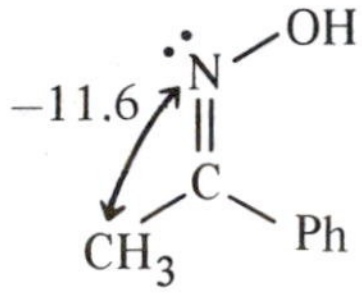

−11.6 +2.7 −9.3 +2.5 −6

4-28 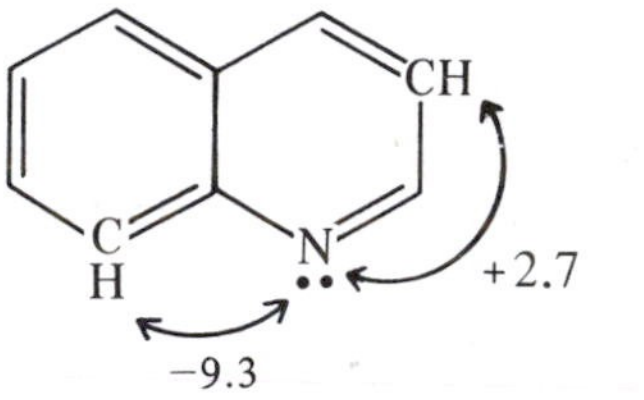 **4-29** 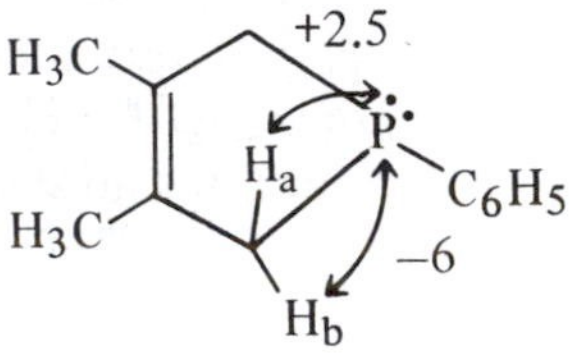 **4-30**

has a coupling of only 1.0 Hz. The two indicated carbons in quinoline (**4-29**) have couplings, respectively, of −9.3 and +2.7 Hz, as one is syn and the other anti to the nitrogen lone pair.

Couplings between ^{31}P and hydrogen also have been exploited stereochemically. The maximum positive value of $^2J_{H-C-P}$ is observed when the H—C bond and the phosphorus lone pair are eclipsed (syn), and the maximum negative value when they are orthogonal or anti. The situation is similar to that for couplings between protons and ^{15}N, but signs are reversed as a result of the opposite signs of the gyromagnetic ratios of ^{15}N and ^{31}P. The heterocycle **4-30** exhibits a coupling of +25 Hz between ^{31}P and H_a (syn) and of −6 Hz between ^{31}P and H_b (anti).

Geminal H—C—F couplings are usually close to +50 Hz for an sp^3 carbon and +80 Hz for an sp^2 carbon: CH_3CH_2F (47.5 Hz) and $CH_2{=}CHF$ (84.7 Hz). Geminal F—C—F couplings are quite large (+150–250 Hz) for saturated carbon (240 Hz for 1,1-difluorocyclohexane), but less than 100 Hz for unsaturated carbon (35.6 Hz for $CH_2{=}CF_2$).

4-4 Vicinal Couplings

The vicinal proton–proton coupling constant has provided probably the most important stereochemical application of NMR spectroscopy. In 1959, Karplus utilized valence bond calculations to define the mathematical relationship between $^3J_{H-C-C-H}$ and the H—C—C—H dihedral angle ϕ. The simple form of

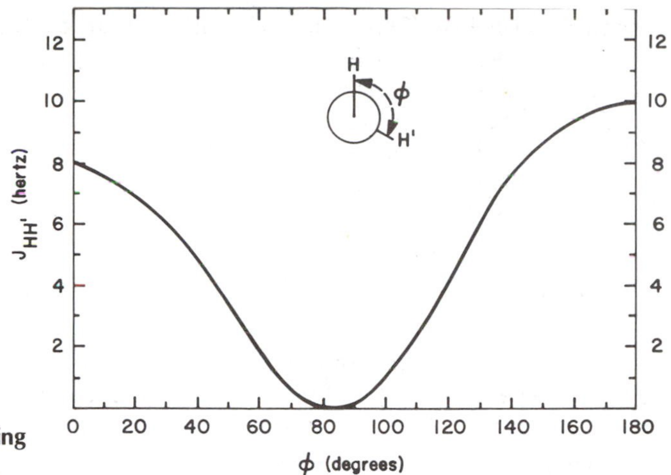

FIGURE 4-3 The vicinal H—C—C—H coupling constant as a function of the dihydral angle (ϕ).

the relationship (eq. 4-3, Figure 4-3) has provided stereochemists with a general and easily applied tool. The

$$^3J = \begin{cases} A\cos^2\phi + C & (\phi = 0\text{–}90^\circ) \\ A'\cos^2\phi + C' & (\phi = 90\text{–}180^\circ) \end{cases} \tag{4-3}$$

cosine-squared form of the dependence indicates that ^3J is largest when the protons are antiperiplanar ($\phi =$ 150–180°), also large when eclipsed (0–30°), and small when staggered (60–120°).

The shape of the Karplus curve derives from the ability of spin information to be passed through bonds. When orbitals are parallel, as at $\phi = 0°$, good overlap permits a strong interaction between nuclei and hence a large J (Diagram 4-3). Overlap is also very effective at $\phi = 180°$. The backside of the sp^3 or sp^2 orbital, rarely drawn, in one carbon is parallel to the large portion on the other carbon and provides an excellent coupling pathway. At $\phi = 90°$ the orbitals are orthogonal, as overlap is at a minimum, and J is small.

Diagram 4-3

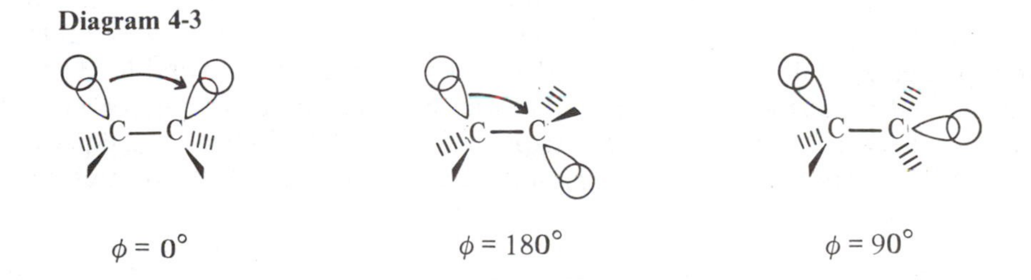

If the constants A and C can be evaluated, quantitative work is possible. The inequality of A and A' means that J is different at the two maxima of $\phi = 0$ and 180°. The additive constants C and C' are usually neglected, since they are thought to be less than 0.3 Hz. The multiplicative constants A and A' vary from system to system in the range 8–14 Hz. Because of this variation, quantitative applications cannot be transferred from one system to another, unless there is good reason to expect that A is unvaried.

Before giving specific examples, let us examine the general areas of applicability of the Karplus equation. In chair cyclohexanes, J_{aa} is large (8–12 Hz) because ϕ_{aa} is close to 180° (Figure 4-3), whereas J_{ee} (0–4 Hz) and J_{ae} (1–5 Hz) are small because ϕ_{ee} and ϕ_{ae} are close to 60°. An axial proton that has an axial proton neighbor can easily be identified by its large J_{aa}.

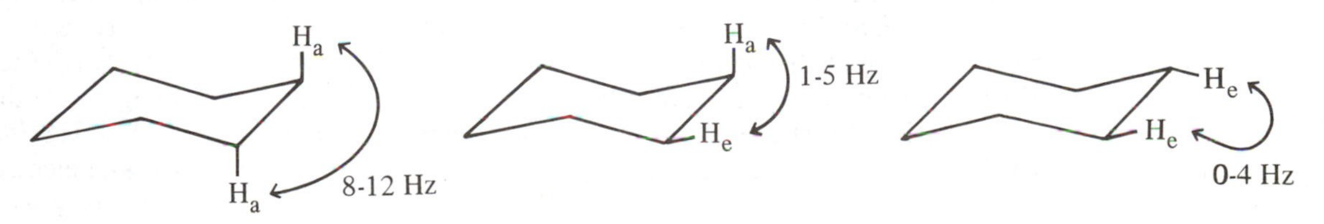

In three-membered rings (**4-31**), J_{cis} ($\phi = 0°$) is larger than J_{trans} ($\phi = 120°$). For the parent cyclopropane,

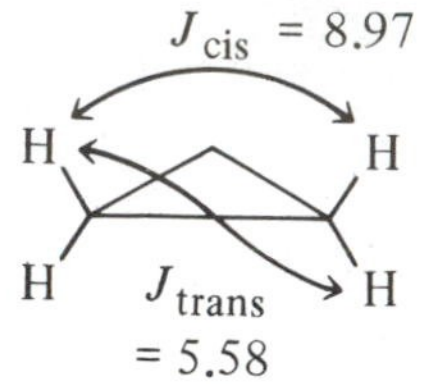

J_{cis} = 8.97

J_{trans} = 5.58

4-31

$J_{cis} = 8.97$ Hz and $J_{trans} = 5.58$ Hz. In five-membered rings, either J_{trans} or J_{cis} can be the larger, since the dihedral angles are toward the center of the Karplus curve. In alkenes, J_{trans} ($\phi = 180°$) is always larger than J_{cis} ($\phi = 0°$), for example, 11.3 and 18.2 Hz in acrylonitrile. Because ortho protons in aromatic rings have a dihedral angle of 0°, $^3J_{ortho}$ is generally quite large (6–10 Hz in benzenes) and can be distinguished from J_{meta} (1–4 Hz) and J_{para} (0–1.5 Hz).

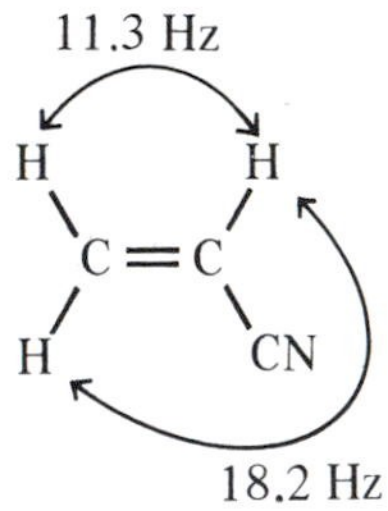

11.3 Hz

18.2 Hz

Qualitative Karplus considerations lend themselves readily to stereochemical assignments in six-membered rings. For example, the proton spectrum of a 1,4-dinitro-2,3,5,6-tetraacetoxycyclohexane is δ 2.02 (s, 6H), 2.12 (s, 6H), 5.03 (d of d ($J = 3.5$, 11 Hz), 2H), 5.89 (d of d ($J = 3.5$, 11 Hz), 2H), 6.22 (t ($J = 3.5$ Hz), 2H). [This shorthand notation is commonly used to report NMR spectra in experimental sections; the δ value for each set of protons is followed parenthetically by the multiplicity (s for singlet, d for doublet, t for triplet, m for multiplet) and by the integrated number of protons. The phrase d of d indicates a doublet of doublets, or four peaks.] The two acetoxy peaks suggest but do not prove that there are only two types of acetoxy groups. A doublet of doublets with coupling constants of 3.5 and 11 Hz can arise only from an axial proton flanked by one axial and one equatorial proton. There must be two distinct protons of this type adjacent to each other (δ 5.03, 5.89). The triplet with $J = 3.5$ Hz indicates an equatorial proton. Each of these types of protons must occur twice in the molecule. Of the various possibilities, only **4-32** is consistent with these observations.

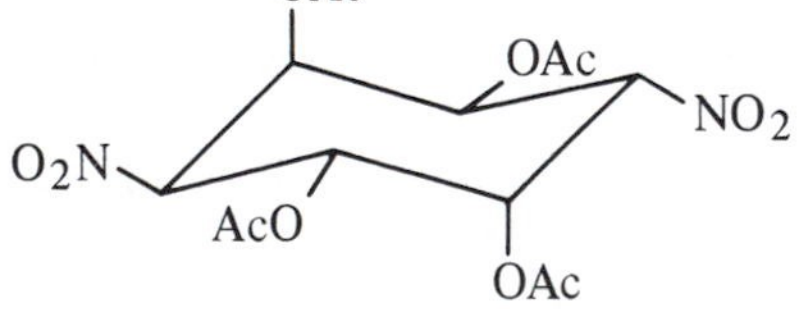

4-32

Despite the potentially general application of the Karplus equation to dihedral angle problems, there are limitations that must be considered. The vicinal H—C—C—H coupling constant is dependent on the C—C bond length, the H—C—C valence angle, the electronegativity of substituents, and the orientation of substituents, in addition to the H—C—C—H dihedral angle. In a properly controlled series of molecules, however, bond length and valence angle factors should remain constant. By far the most difficult additional factor to deal with is the effect of substituent electronegativity. If a hydrogen on ethane is replaced by a more electronegative substituent, the coupling constant decreases, from 8.0 Hz in CH_3CH_3 to 7.2 Hz in CH_3CH_2Cl, for example. Several approaches have been developed to take electronegativity into account.

These include derivation of the mathematical dependence of J on electronegativity, empirical allowance by the use of chemical shifts that depend in a similar fashion on electronegativity, and elimination of the problem through the use of the ratio (the R value) of two coupling constants that respond to the same dihedral angle and have the same multiplicative dependence on substituent electronegativity. These more sophisticated versions of the Karplus method have been quite successful in obtaining quantitative results.

Considerable data now confirm that vicinal couplings over any H—C—X—H pathway, in which X is an atom other than carbon, follow a Karplus-like relationship in their dependence on dihedral angle (X = O, N, S, Se, Te, Si). The H—C—C—C, H—C—C—F, and C—C—C—C couplings also follow Karplus relationships, although specific A values must be derived for each case. The F—C—C—F and H—C—C—P couplings appear not to follow the Karplus pattern. The presence of lone pairs on the coupling nuclei may introduce another contribution to the contact term that alters the stereochemical dependence.

4-5

Long Range Couplings

Ordinarily, spin–spin coupling takes place through the σ framework of a molecule and falls off very rapidly with the number of intervening bonds. Except in certain special cases, σ couplings greater than 1 Hz are rarely observed over more than three bonds. Contributions from π electrons and from specific σ geometries, however, can become important over these longer pathways. One such case is the allylic coupling **HC—C=CH**, with a range of about +1 to −3 Hz and typical values close to −1 Hz. The larger values are observed when the saturated C—H_a bond (**4-33**) is parallel to the π orbitals. This σ–π overlap enables the

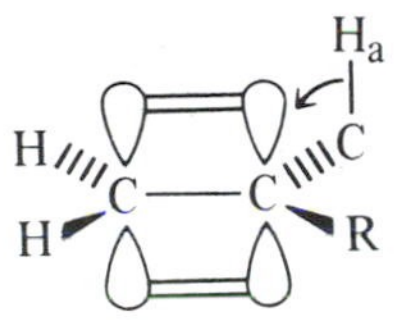

4-33 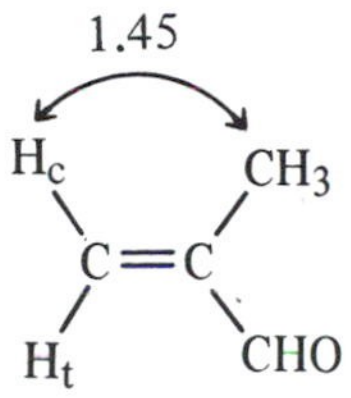 4-34 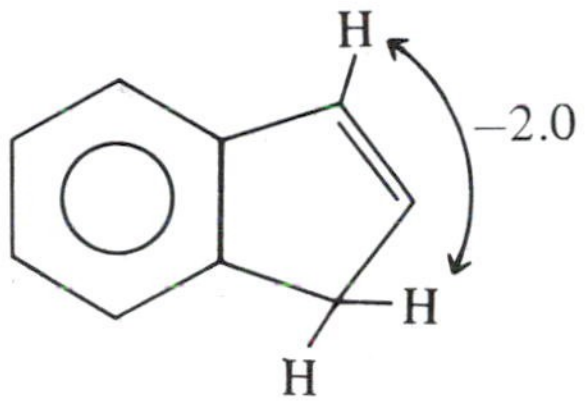 4-35

coupling to be transmitted more effectively. When the C—H_a bond is orthogonal to the π orbitals, there is no σ–π contribution and couplings are small (<1 Hz). In acyclic systems, the dihedral angle is averaged over both favorable and unfavorable arrangements, so an average $^4J_{\text{allylic}}$ is found, as in 2-methylacrolein (**4-34**, ^4J = 1.45 Hz). Ring constraints can freeze bonds into the favorable arrangement, as in indene (**4-35**, ^4J = −2.0 Hz).

The homoallylic coupling, **HC—C=C—CH**, depends on the orientation of two C—H bonds with respect to the π framework. For acyclic systems such as *cis*- or *trans*-2-butene, ^5J typically is +2 Hz. When both protons are properly aligned, as in the nearly planar 1,4-cyclohexadiene (**4-36**), ^5J can be quite large. In **4-36**, the coupling is 9.63 Hz between the cis homoallylic protons and 8.04 Hz between the trans protons. It is

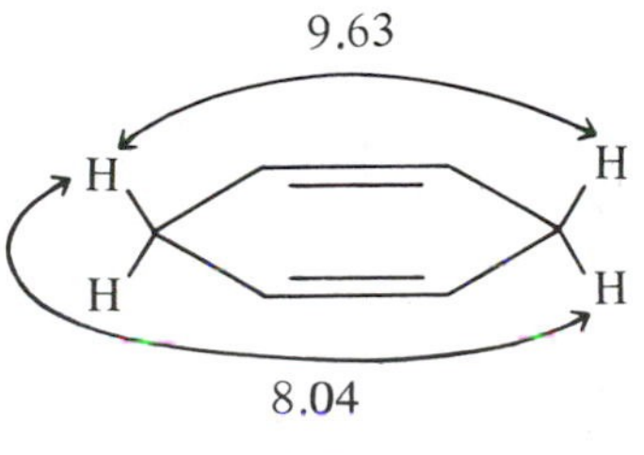

4-36 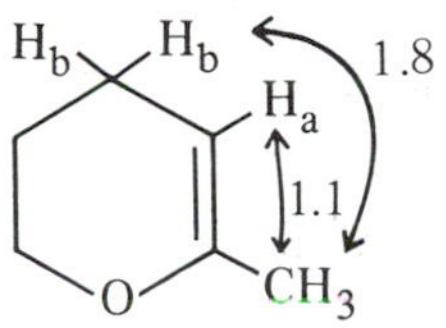 4-37

not unusual for the homoallylic coupling to be larger than the allylic coupling. Thus $^4J(CH_3—H_a)$ is 1.1 Hz and $^5J(CH_3—H_b)$ is 1.8 Hz in **4-37**.

Long range couplings can be large in alkynic and allenic systems, although such molecules are not frequently encountered in practice. In allene itself, ^4J is -7 Hz, since the rigid geometry is optimal for σ–π overlap. In 1,1-dimethylallene, ^5J decreases to (+)3 Hz, since σ–π overlap is averaged by methyl rotation. In both methylacetylene (propyne, ^4J) and dimethylacetylene (2-butyne, ^5J), the long range coupling is

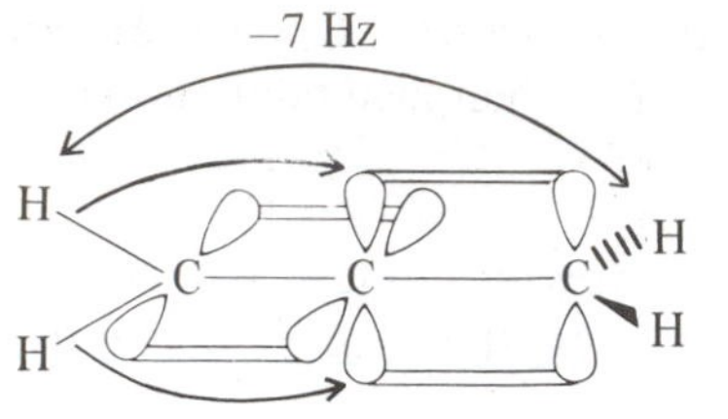

−7 Hz

allene

3 Hz

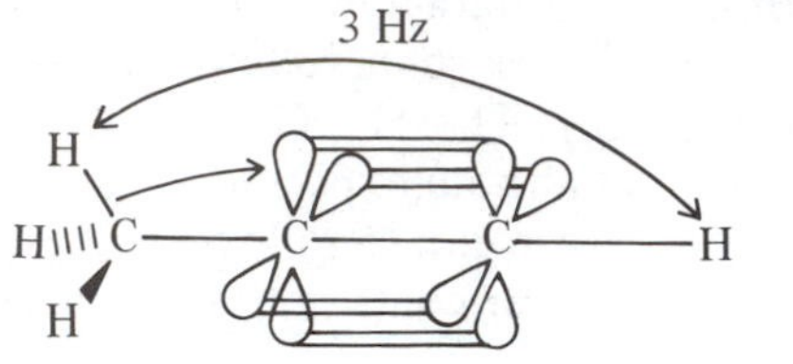

propyne

about 3 Hz. The triple bond imposes no steric limitations on σ–π overlap or on conjugation with other triple bonds, so appreciable long range couplings (>1 Hz) have been observed even over five to seven bonds in polyalkynes.

Conjugation of double bonds is a stereochemically more complicated situation. In butadiene, there are two four-bond couplings (−0.86, −0.83 Hz) and three five-bond couplings (+0.60, +1.30, +0.69 Hz). In aromatic rings, the meta coupling is a ^4J and the para coupling is a ^5J. Both may be distinguished from $^3J_{\text{ortho}}$ by their smaller magnitudes. In benzene itself, $^3J_{\text{ortho}}$ is 7.54 Hz, $^4J_{\text{meta}}$ is 1.37 Hz, and $^5J_{\text{para}}$ is 0.69 Hz. Protons on saturated carbon atoms attached to an aromatic ring couple with all three types of protons. These couplings, frequently called benzylic, depend on a σ–π interaction between the substituent C—H bonds and the π cloud, much like the allylic coupling.

The planar-zigzag or W pathway between coupling protons often results in large long range couplings. This geometry is seen, for example, in the 1,3 diequatorial arrangement between protons in **4-38** (^4J = 1.7 Hz)

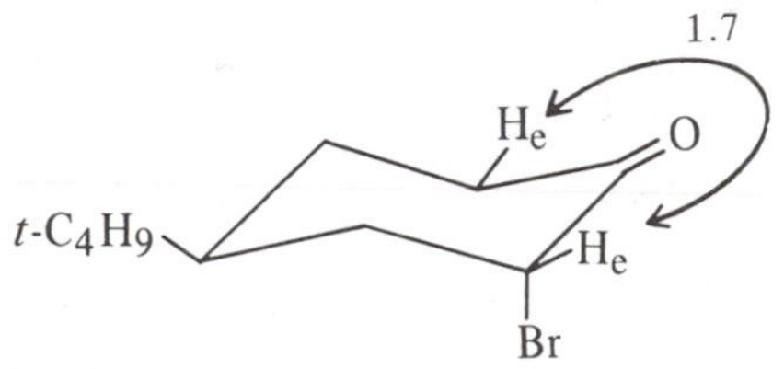

4-38

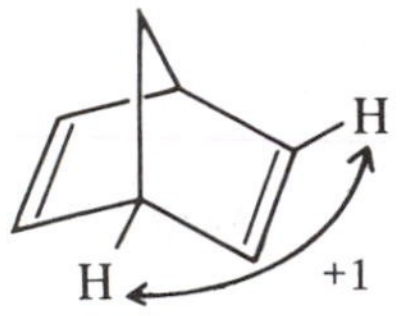

4-39

or in the allylic arrangement in norbornadiene (**4-39**, ^4J = +1 Hz). In the planar-zigzag arrangement, there is favorable overlap between the antiperiplanar H—C and C—C bonds (**4-40**) (analogous to the optimal vicinal

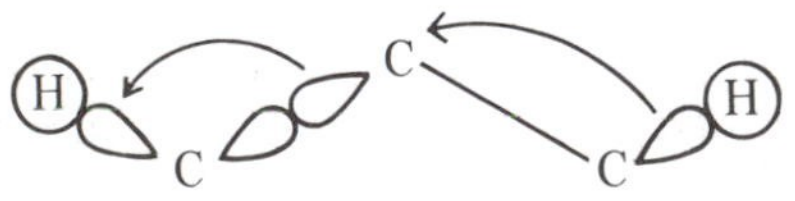

4-40

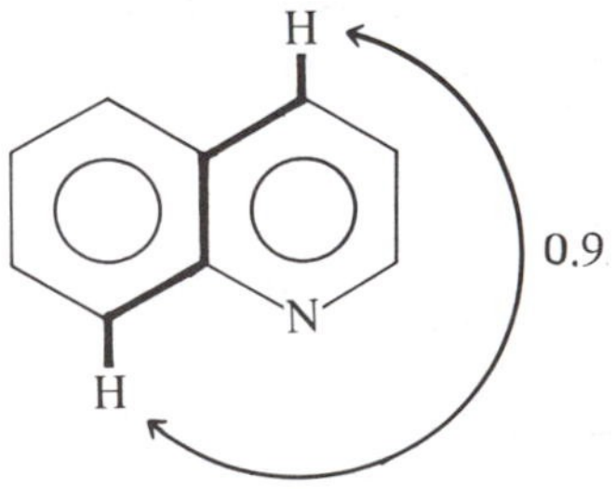

4-41

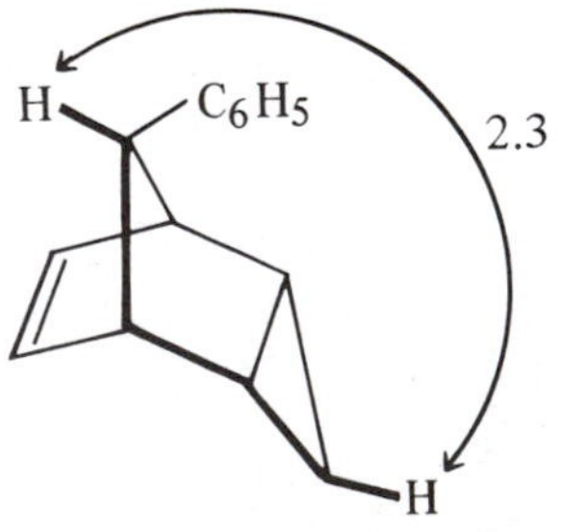

4-42

coupling at $\phi = 180°$), and this interaction is coupled to a normal geminal H—C—C arrangement. The zigzag pathway is entirely within the σ framework but is important for many π systems, including the aromatic meta coupling. Five-bond zigzag pathways similarly can give rise to unusual long range couplings. The +1.30 Hz coupling in 1,3-butadiene is one such example. Others include quinoline (**4-41**, $^5J =$ 0.9 Hz between the indicated protons) and the norbornyl system **4-42** ($^5J = 2.3$ Hz).

Although coupling information is always passed by electron-mediated pathways, in some cases part of the through bond pathway may be skipped, as in σ–π allylic or benzylic overlap. Two nuclei that are within van der Waals contact in space over any number of bonds can interchange spin information, if at least one nucleus possesses lone pair electrons. These so-called through space couplings are found most commonly in H—F and F—F pairs. The six-bond CH_3—F coupling in the structurally similar **4-43** and **4-44** is <0.5 Hz and

CH_3 F CH_3 F

<0.5 8.3

4-43 **4-44**

8.3 Hz, respectively. In **4-43**, the H—F distance is 2.84 Å, whereas in **4-44** it is 1.44 Å. In the latter case the nuclei are well within the sum of the van der Waals radii (about 2.55 Å), and coupling information is probably passed from the proton through the lone pair electrons to the fluorine nucleus. Such a mechanism very likely is quite important in the geminal F—C—F coupling also.

4-6

Measurement of the Coupling Constant

Spectral analysis is the derivation of the chemical shift ν_i of each nucleus and the coupling constant J_{ij} between each pair of nuclei from an observed spectrum. If all the ν_i and J_{ij} are known, the spectrum may be precisely reproduced by computer calculation. Analysis of first-order spectra is a straightforward process of measuring the distance between certain pairs of peaks. Thus an AX spectrum consists of two doublets. The doublet spacing is J_{AX} and the midpoints of the doublets are ν_A and ν_X. Because most nuclei other than the proton have enormous ranges of chemical shifts and because these nuclei often are in low natural abundance and hence do not show coupling to each other, second-order analysis is primarily an activity for proton spectroscopists. Occasionally, boron spectra and those of a few other nuclei exhibit second-order behavior. Even for protons, the exercise is becoming less important, since many spectra are now being recorded at 400 MHz or higher. Magnetic nonequivalence, however, is independent of field and can produce second-order spectra with the most expensive superconducting magnets. Even with the best equipment, spectral overlap is still common in proton spectra, so that the process of spectral analysis must be learned in order to obtain reliable spectral parameters.

The second-order, two spin system (AB) contains four lines, the inner peaks of which are more intense (Figure 4-4). The coupling constant (J_{AB}) is obtained directly from the doublet spacings, but no specific peak

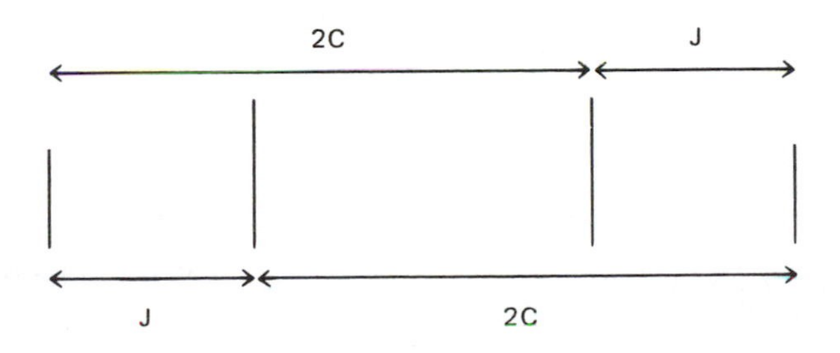

FIGURE 4-4 The AB spin system.

positions correspond to the chemical shifts. Because the A peaks have unequal intensities, ν_A is defined by the weighted average position. The difference $\Delta\nu_{AB}$ between the A and B chemical shifts may be obtained from eq. 4-4, in which $2C$ is the spacing between alternate peaks in Figure 4-4. The actual values of ν_A and ν_B

$$\Delta\nu_{AB} = (4C^2 - J^2)^{1/2} \tag{4-4}$$

are readily determined by adding and subtracting $\frac{1}{2}\Delta\nu_{AB}$ to the midpoint of the spectrum. The ratio of intensities of the inner to outer peaks is given by eq. 4-5.

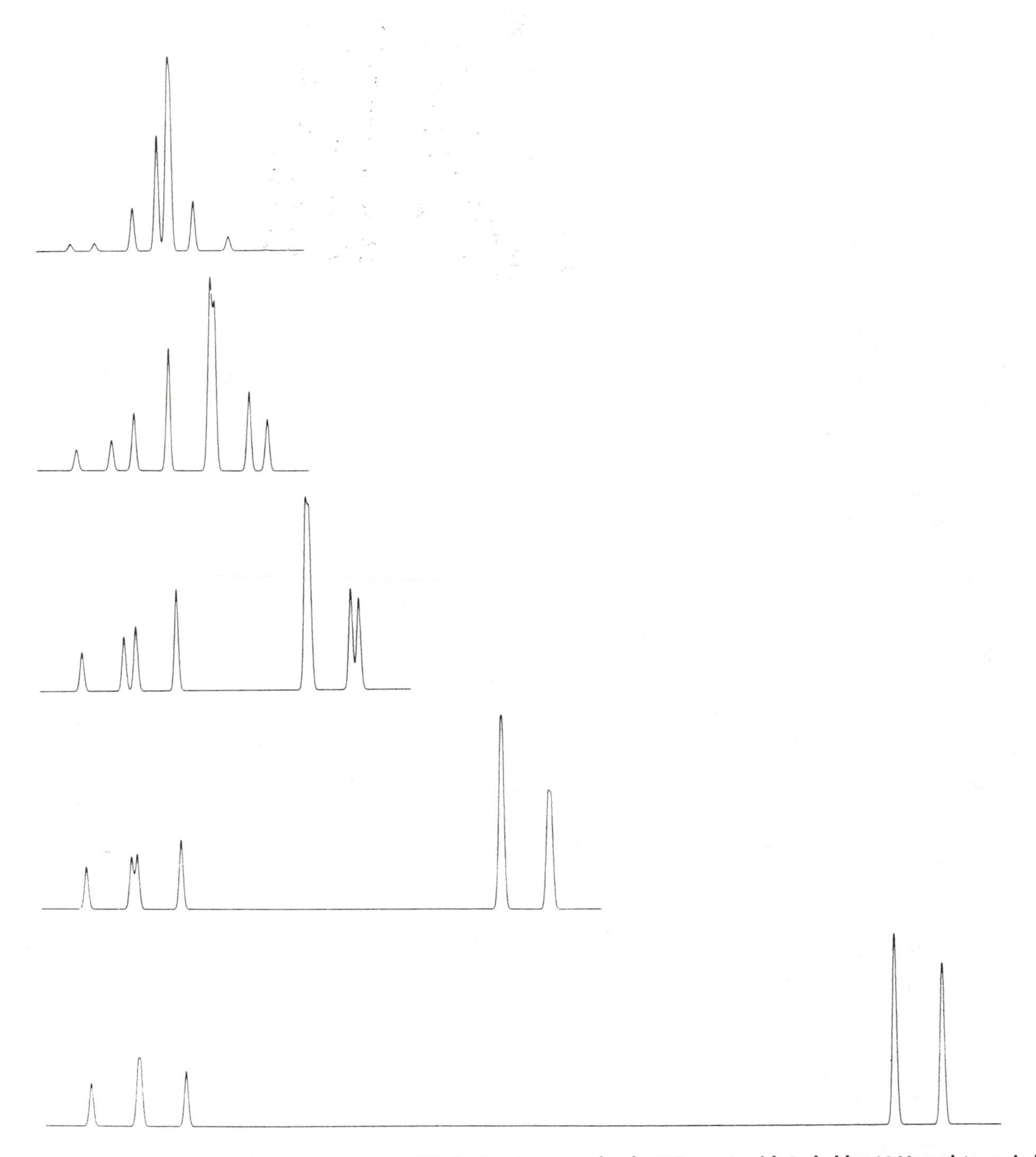

FIGURE 4-5 The transition from first order (AX_2, bottom) to second order (AB_2, top), with J_{AB} held at 10 Hz and $\Delta\nu$ varied: 10 (top), 20, 40, 80, and 160 Hz.

$$\frac{\text{inner}}{\text{outer}} = \frac{1 + J/2C}{1 - J/2C} \qquad (4\text{-}5)$$

Analysis of three spin systems can be carried out by inspection in only a few instances, such as the first-order AX_2 and AMX systems. The second-order AB_2 spectrum can contain up to nine peaks (Figure 4-5). Four of these peaks result from spin flips of the A proton, four from spin flips of the B protons, and one from simultaneous spin flips of both A and B protons. The ninth peak is called a *combination line* and is ordinarily forbidden.

When eight AB_2 peaks are observed (Figure 4-5, numbering from left to right), all the spectral parameters may be obtained from the various peak positions. The chemical shift of the A proton corresponds to ν_3, the position of the third peak. The average of ν_5 and ν_7 gives the chemical shift of the B proton (eq. 4-6). The

$$\nu_B = \tfrac{1}{2}(\nu_5 + \nu_7) \qquad (4\text{-}6)$$

coupling constant may be derived from a linear combination of four peak positions (eq. 4-7). Thus the entire

$$J_{AB} = \tfrac{1}{3}(\nu_1 - \nu_4 + \nu_6 - \nu_8) \qquad (4\text{-}7)$$

spectrum may be analyzed without reference to ν_9, the unobserved combination peak. If not all eight of the remaining peaks are observed (ν_5 and ν_6 often are degenerate), confusion as to numbering may arise, and the spectrum would probably best be analyzed by computer techniques.

The ABX, ABC, AA′XX′, and AA′BB′ systems are commonly encountered second-order spectra whose analysis has been treated in specialized texts found in the bibliography at the end of the chapter. In many such cases, hand analysis proves difficult or impossible, and recourse is made to computer techniques. Programs are available for many types of computers, including those that come with moderately priced spectrometers. The first step usually is a trial and error procedure of guessing the chemical shifts and coupling constants in order to match the observed spectrum through computer simulation. This method is relatively successful in the majority of cases with four or fewer spins. The chemical shifts are varied initially, until the widths of the observed and calculated spectra agree approximately. Then the coupling constants or their sums and differences are varied systematically one at a time until a reasonable match is obtained.

Refinements of hand or of trial and error analyses utilize iterative procedures. The program of Castellano and Bothner-By (LAOCN3) iterates on peak positions in order to adjust chemical shifts and coupling constants for best agreement between observed and calculated frequencies. The program of Swalen and Reilly (NMREN/NMRIT) calculates the energies of each spin state from observed peak positions and then iterates by adjusting the spectral parameters until the theoretical and observed energies of the spin states agree. Both iterative methods require considerable knowledge of the spin system before they are successful. This knowledge is generally gained by the trial and error procedure. Analysis is carried as far as possible by this means, and iteration then serves more as a refinement. Iteration, however, can succeed in cases in which trial and error fails. The more recent program of Stephenson and Binsch (DAVINS) dispenses with the trial and error stage and operates directly on unassigned peak positions.

4-7

Second-Order Effects

Most NMR analyses are carried out simply by inspection, without recourse to quantitative techniques. Unfortunately, such procedures often can lead to incorrect conclusions, because of erroneous interpretation of second-order phenomena. This section is concerned with qualitative aspects of second-order effects, since this approach is most useful for day-to-day spectral analyses. There are three major classes of second-order effects that can lead to misinterpretations: (1) spacings do not correspond to coupling constants, even when the spectrum contains an apparently first-order pattern; (2) certain unequal couplings appear to be equal, so that there are fewer lines than first-order considerations would predict (*deceptive simplicity*); and (3) certain

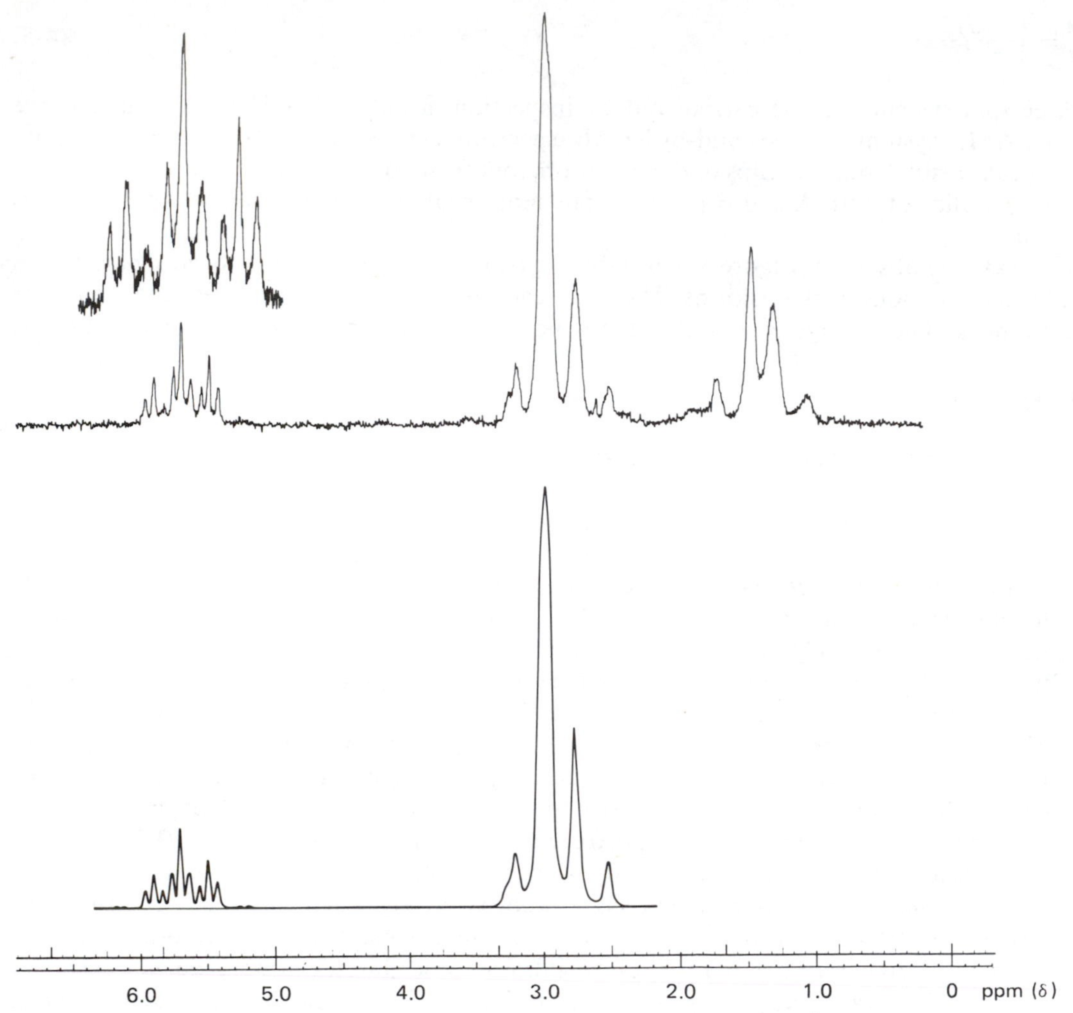

FIGURE 4-6 The 60 MHz proton spectrum of thiane-3,3,5,5-d_4 in FSO_3H at −30° C. [Reproduced with permission from J. B. Lambert, R. G. Keske, and D. K. Weary, *J. Am. Chem. Soc.*, **89**, 5922 (1967). Copyright 1967 American Chemical Society.]

zero or nearly zero couplings appear to give rise to significant splittings, so that there are more lines than first-order considerations would predict (*virtual coupling*).

When spectra are second order, peak spacings do not necessarily correspond to coupling constants, even when intuition seems to be clear. For example, the spectrum of protonated thiane-3,3,5,5-d_4 (**4-45**, Figure 4-6) contains a triplet of triplets for the proton on sulfur. The large splitting (12.4 Hz) corresponds to the

H_S S^+ D_2 D_2 H H H_{eq} H_{ax}

4-45

coupling between H_S and H_{ax}, the small splitting (4.0 Hz) to the coupling between H_S and H_{eq}. These spacings, however, are not true coupling constants. In order to reproduce a spectrum with these spacings by computer methods, coupling constants of 14.1 and 2.3 Hz are required. The spectrum gives only the sum of the coupling constants by inspection (16.4 Hz). This situation is common when a proton is coupled to two or more protons that in turn are closely coupled (small $\Delta\nu/J$). Under such conditions, the spin states of the closely coupled nuclei are mixed, so that the first nucleus cannot distinguish them properly.

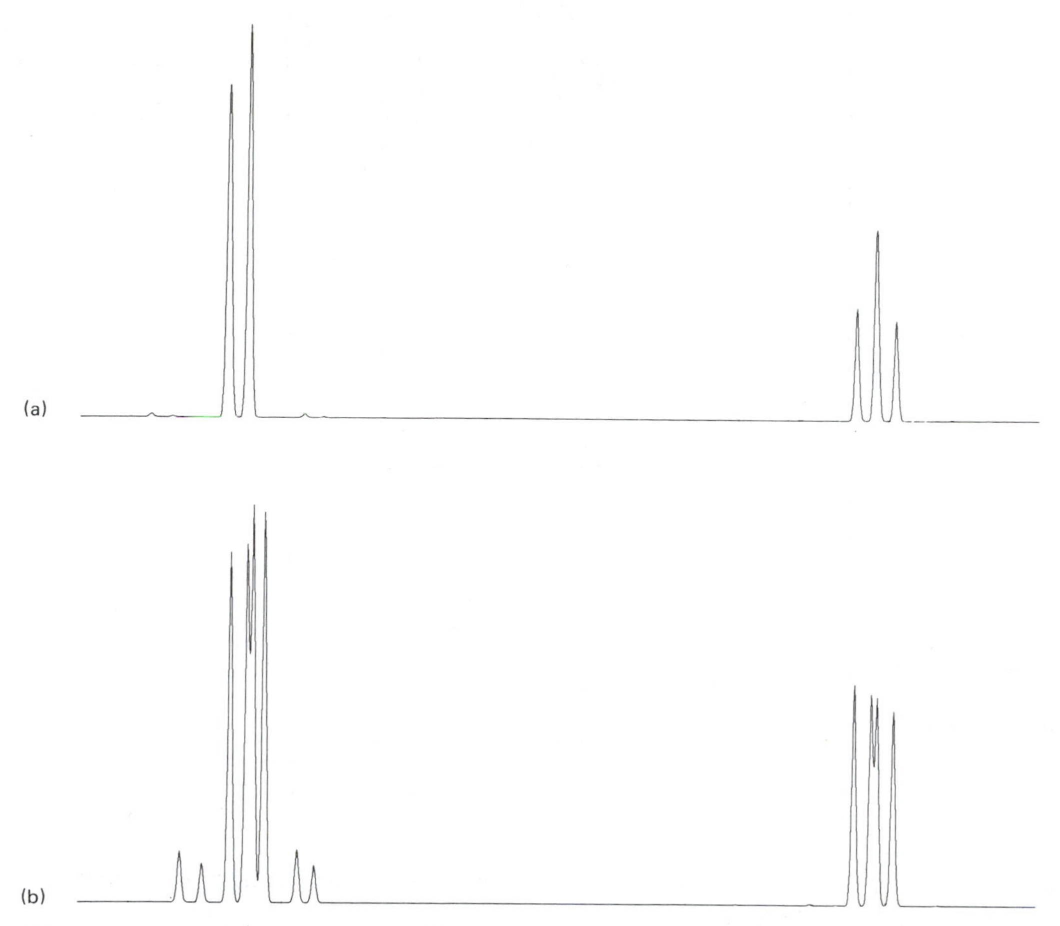

FIGURE 4-7 (a) A deceptively simple ABX spectrum: $\nu_A = 0.0$ Hz, $\nu_B = 3.0$ Hz, $\nu_X = 130.0$ Hz; $J_{AB} = 15.0$ Hz, $J_{AX} = 5.0$ Hz, $J_{BX} = 3.0$ Hz. (b) The same parameters, except $\nu_B = 8.0$ Hz. The larger value of $\Delta\nu_{AB}$ removes the deceptive simplicity and produces a typical ABX spectrum.

Lines often coincide in such a way that the spectrum assumes a simpler appearance than seems consistent with actual spectral parameters. The ABX spectrum offers many examples of such deceptive simplicity. If the A and B nuclei are closely coupled, the X nucleus receives mixed spin information with the result that the X resonance is an apparently first-order 1/2/1 triplet, as if $J_{AX} = J_{BX}$. In fact, the couplings can be quite unequal, but the spectrum is sensitive only to their sum. Analysis of such spectra often is impossible. Deceptive simplicity sometimes can be removed by operating at a higher or lower field (Figure 4-7).

Second-order effects also can cause spectra to be apparently too complex. Consider the ABX case for which A and B are closely coupled, J_{AX} is large, but J_{BX} is zero. The X spectrum, it would seem, should be a doublet from coupling only to A. Since A and B are closely coupled, however, the spin states of A and B are mixed, and X acts as if it were coupled to both nuclei. For example, in a slightly larger but equivalent spin system, the methine and methylene protons of **4-46** are closely coupled. Although the methyl group is

$HO_2C—CH_2—C(H)(CH_3)—CH_2—CO_2H$

4-46

$HO_2C—C(H)(CH_3)—CH_2—CO_2H$

4-47

coupled only to the methinyl proton, its resonance is much more complicated than a simple doublet. In **4-47**,

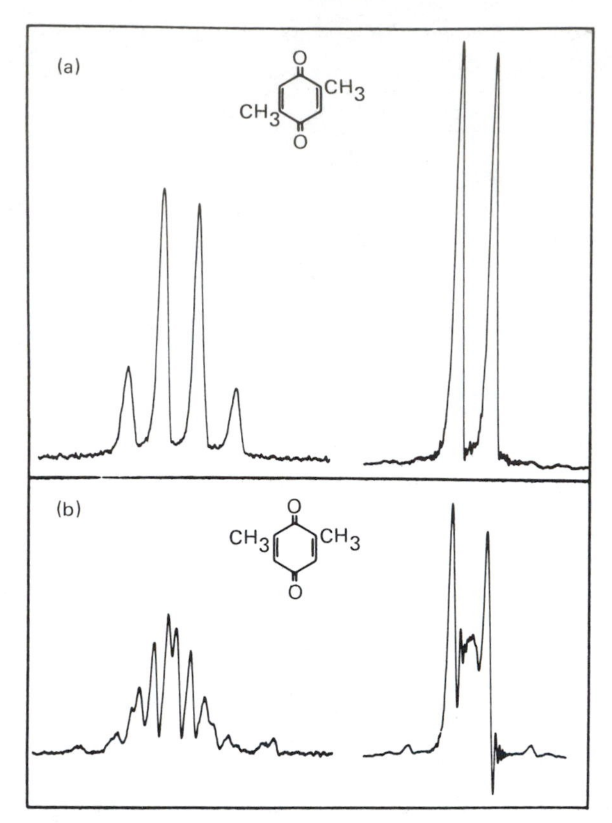

FIGURE 4-8 The 60 MHz proton spectra of the 2,5- and 2,6-dimethylquinones (4-48, 4-49). [Reproduced with permission from E. D. Becker, *High Resolution NMR*, 2nd ed., Academic Press, Orlando, FL, 1980, p. 166.]

the carboxyl group shifts the methine resonance downfield, so that methine and methylene protons no longer are closely coupled. Consequently, the methyl resonance is a clean, first-order doublet.

The dimethylquinones provide a further example. The spectrum of the 2,5-dimethyl compound (**4-48**, Figure 4-8) contains a first-order methyl doublet and an alkene quartet. The spectrum of the 2,6 compound

O, CH_3, CH_3, O — **4-48**

O, CH_3, CH_3, O — **4-49**

(**4-49**) is much more complicated. The alkenic protons in both molecules are equivalent (AA′). In **4-48** they are coupled only to the methyl protons, but in **4-49** they are closely coupled to each other because of the zigzag pathway. The multiplicity of the methyl resonance therefore is perturbed not only by the adjacent alkenic proton but also by the one on the opposite side of the ring. The zero AA′ coupling in **4-48** removes this second-order effect.

Although special names have been given to specific situations, the common source of all these phenomena is the second-order nature of the spectra. Errors in interpretation can be avoided if the frequent occurrence and subtle appearance of second-order effects in proton spectra are appreciated.

4-8

Aids in Spectral Analysis

Complete or even partial analysis of complex spins is time consuming and sometimes impractical or impossible. Consequently, numerous methods have been developed to make the task easier.

4-8a Higher Fields

Since the chemical shift is directly proportional to the field strength, the ratio $\Delta\nu/J$ in a second-order spectrum can be increased by raising the field. This advantage, together with increased sensitivity, has led to the development of magnets with very high fields. Routine spectra are still taken at 60–90 MHz on iron core magnets, but superconducting magnets operating at 100–300 MHz have superseded the lower field magnets in many laboratories. For research purposes, particularly with biomolecules that have numerous overlapping resonances, spectra at 400–600 MHz have become increasingly common. Figure 4-9 illustrates the advantages of increased field for chemical shift dispersion. Although higher fields serve to decrease second-order effects caused by small $\Delta\nu/J$ ratios, second-order properties caused by magnetic nonequivalence, as in the AA′XX′ spectrum, cannot be altered.

4-8b Lanthanide Shift Reagents

The unpaired spins of certain lanthanide reagents can give rise to unusual shielding effects in molecules with which they form acid–base complexes. The chemical shift enhancement is considered to derive from a *pseudocontact mechanism*, by which spin information is passed through space within the complex rather

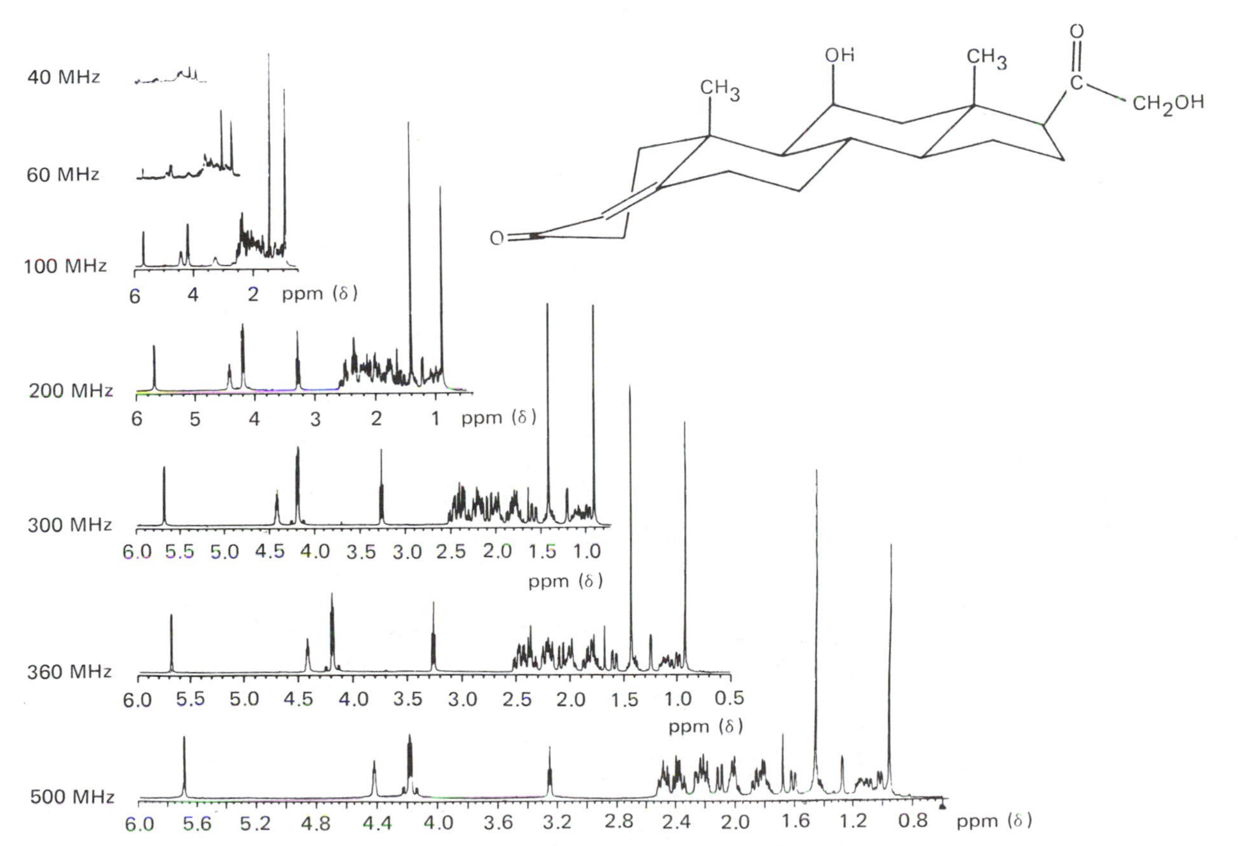

FIGURE 4-9 The proton spectrum of corticosterone from 40 to 500 MHz. [Reproduced with permission of Nicolet Instruments Ltd., Budbrooke Road, Warwick, England CV34 5XH.]

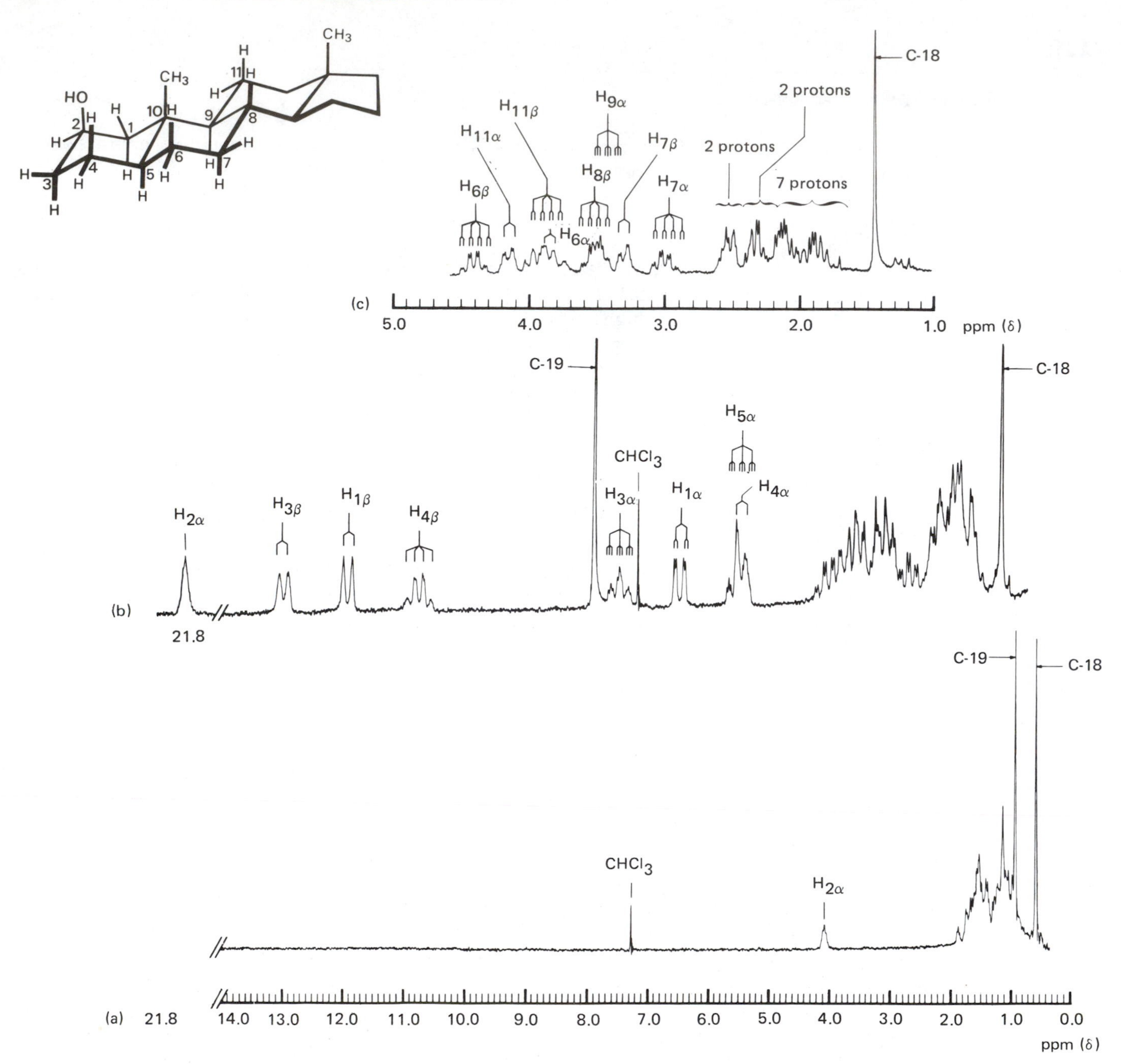

FIGURE 4-10 The 100 MHz proton spectrum of androstan-2β-o1 with (b) and without (a) added Eu(dpm)$_3$·2pyr (4-50). Spectrum c is a 220 MHz blowup of the region δ 1–5. [Reproduced with permission from P. V. Demarco, T. K. Elzey, and E. Wenkert, *J. Am. Chem. Soc.*, **92**, 5737 (1970). Copyright 1970 American Chemical Society.]

than through the bonding electrons. The presence of unpaired electron spins in the same region of space as the resonating protons causes very large shielding effects (hence the term pseudocontact). This mechanism has both a distance and an angular dependence. Because the pseudocontact mechanism drops off rapidly with distance, the shifts are largest close to the point of complexation. Europium shift reagents have come to be used primarily to minimize second-order spectral effects by increasing $\Delta\nu/J$. The spectrum of androstan-2β-ol is given in Figure 4-10 with and without tris(dipivalomethanato)europium(III)·2(pyridine) (**4-50**). This reagent, when used without the two molecules of pyridine, is referred to as Eu(dpm)$_3$. Another commonly used shift reagent is 1,1,1,2,2,3,3-heptafluoro-7,7-dimethyloctanedionatoeuropium(III), Eu(fod)$_3$ (**4-51**), which has better solubility properties than Eu(dpm)$_3$ and complexes with less basic sites.

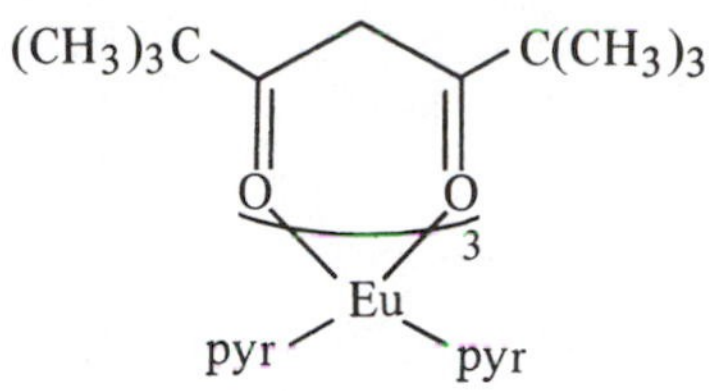

4-50

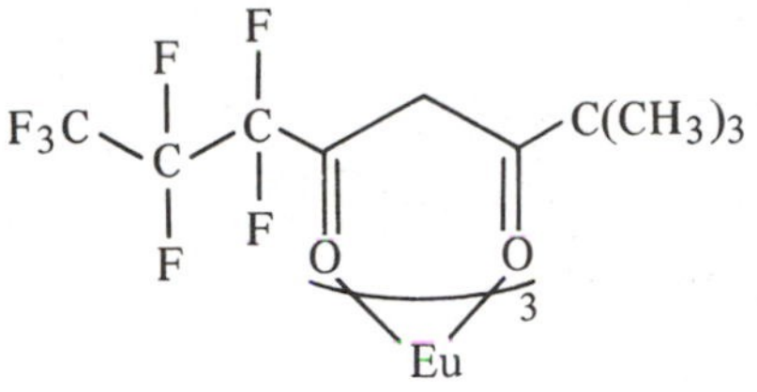

4-51

Shift reagents are available with a wide range of ligands and with most of the rare earths (praseodymium, holmium, and dysprosium, among others). Useful shift information has been obtained with almost all organic functional groups: alcohols, amines, ketones, esters, sulfoxides, ethers, thioethers, nitriles, oximes, lactones, aldehydes, and so on. Lanthanide shifts also have been observed for nuclei other than ^{1}H, including ^{19}F and ^{13}C.

4-8c Multiple Resonance

When only certain protons in a complicated spin system are of interest, the coupling properties of the unwanted nuclei often may be removed by multiple irradiation. For example, irradiation at the X frequency of an ABX system reduces the spectrum to an AB quartet. This technique is described more fully in Chapter 5.

4-8d Deuterium Substitution

Alternatively, the unwanted proton can be replaced synthetically by deuterium. In **4-52**, the vicinal H—C—C—F couplings were needed to obtain conformational information. When X was H, the resulting five spin

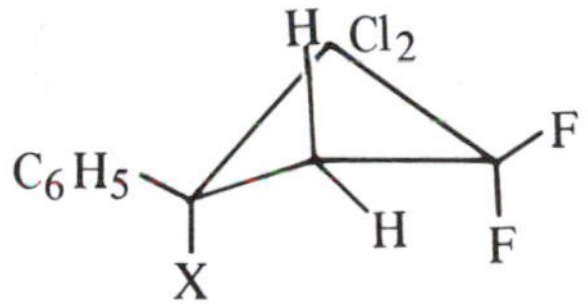

4-52

ABCXY spectrum was too complicated for analysis. Because the benzyl proton was too close in frequency to the methylene protons for double irradiation, the molecule was prepared with X = D. The smaller ABXY spin system was readily analyzed. Irradiation at the deuterium frequency was necessary to remove small couplings from the benzyl deuterium (Figure 4-11). Tetrahydrofuran has only two α,β couplings, which are buried in an $A_2A'_2B_2B'_2$ spin system. Synthesis of the 2,2,3,3-d_4 derivative (**4-53**) reduces the problem to the analysis of a four spin AA′BB′ system, from which the vicinal couplings are readily obtained.

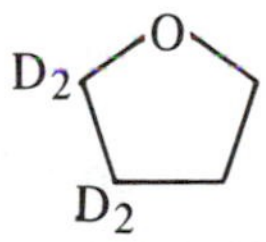

4-53

4-8e Satellite Spectra

Naturally abundant carbon-13 provides a convenient method to measure couplings between chemically equivalent protons on different carbon atoms. For 1.1% of the molecules (uncorrected for statistical factors), the spin system is H—^{12}C—^{13}C—H. The proton on ^{12}C resonates at almost the same position as the molecules with no ^{13}C. The large one bond ^{13}C—^{1}H coupling produces multiplets, called *satellites*, on either

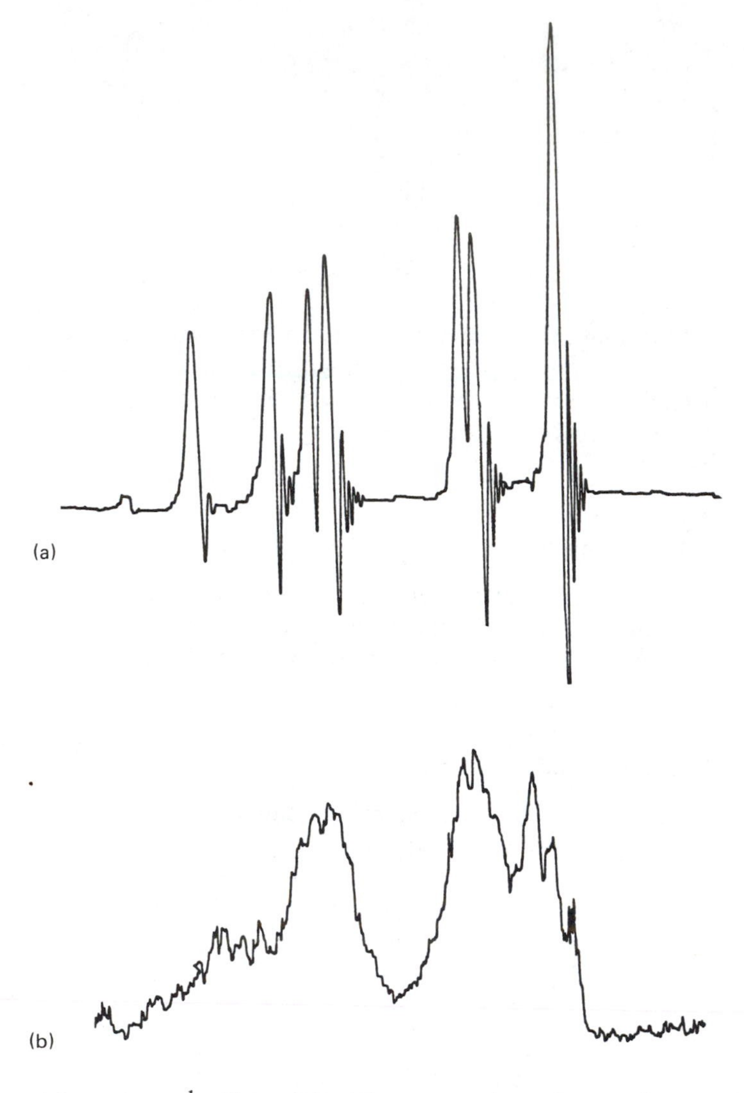

FIGURE 4-11 The 60 MHz proton spectrum of 1,1-difluoro-2,2-dichloro-3-phenylcyclobutane-3-*d* with (a) and without (b) deuterium irradiation.

side of the centerband and separated from it by $\frac{1}{2}J(^{13}C—^{1}H)$. The separation of each sideband from the centerband serves as an effective chemical shift difference, so that the H—H coupling between H—^{12}C and H—^{13}C is observed in the satellite. Figure 4-12 shows the satellite spectrum of the alkenic protons on cyclopropene. The satellite is a doublet of triplets, since the alkenic proton on ^{13}C is coupled to the other alkenic proton and to the two methylene protons.

Other satellite spectra can be used for analytical purposes. Nitrogen (^{15}N), silicon (^{29}Si), selenium (^{77}Se), cadmium (^{111}Cd, ^{113}Cd), tin (^{117}Sn, ^{119}Sn), tellurium (^{125}Te), platinum (^{195}Pt), and mercury (^{199}Hg) have spin $\frac{1}{2}$ isotopes that give useful satellite resonances in the proton spectrum.

4-8f Multiple Pulse Sequences

More advanced NMR techniques include the use of several pulses to achieve specific aims, such as better sensitivity, examination of multiple quantum relationships, and measurement of relaxation times. These methods are discussed in Chapter 5.

4-8g Two-Dimensional NMR

In very complex spectra, it is useful sometimes to examine two frequency domains simultaneously. These experiments result in two-dimensional plots, in which, for example, ^{1}H frequency is plotted versus ^{13}C frequency, or chemical shift is plotted against coupling constant. The methods are of particular advantage in the analysis of the spectra of extremely large molecules. A discussion is found in Chapter 5.

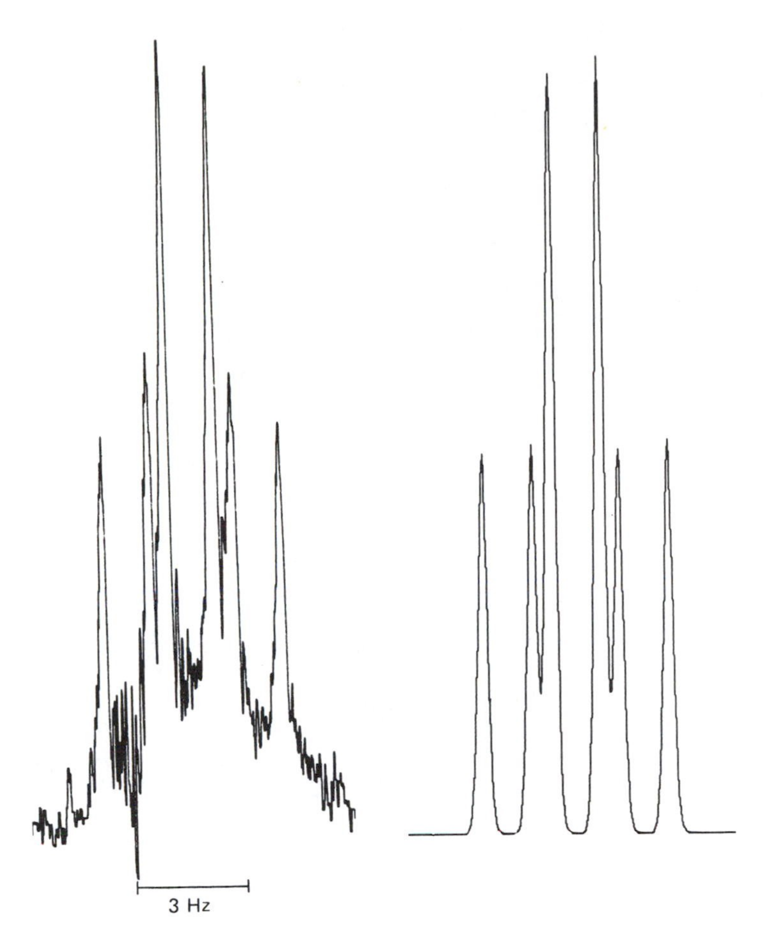

FIGURE 4-12 The 90 MHz spectrum of cyclopropene, showing the observed (left) and calculated (right) downfield ^{13}C satellite of the alkenic protons. [Reproduced with permission from J. B. Lambert, A. P. Jovanovich, and W. L. Oliver, Jr., *J. Phys. Chem.*, **74**, 2221 (1970). Copyright 1970 American Chemical Society.]

4-8h Liquid Crystal Solvents

Molecules such as *p,p'*-di-*n*-hexyloxyazoxybenzene (**4-54**) exist as nematic liquid crystals in certain temper-

$n\text{-}C_6H_{13}O\text{-}C_6H_4\text{-}\overset{+}{N}(O^-)\text{=}N\text{-}C_6H_4\text{-}O\text{-}n\text{-}C_6H_{13}$

4-54

ature ranges. The properties of this mesophase are intermediate between those of solids and liquids. Large numbers of these rod-shaped molecules align themselves along their long axis to form a *domain*. A domain can be oriented by a magnetic field so that the long molecular axis is parallel to B_0. Small organic molecules dissolved in a liquid crystal also will be aligned. The ordering is not so rigid as in the solid phase. For example, benzene as a solute is aligned with its plane parallel to the B_0 field, but it rotates freely around the sixfold axis perpendicular to this plane.

Because the solute molecules maintain some degrees of freedom, the NMR spectrum is not the broad, almost featureless mass often given by solids. The peaks are relatively sharp, but cover a much larger range of frequencies than do ordinary liquid phase (isotropic) spectra. Because molecules are no longer tumbling freely, the spectrum is determined by the direct dipole–dipole couplings (D) as well as by the indirect spin–spin couplings (J). Not only do liquid crystal spectra provide an approach for measuring D, but analysis of the various couplings gives the absolute signs of J.

In liquid crystal spectra, coupling constants are observed between chemically equivalent nuclei. Thus an isolated methyl group, as in acetonitrile, gives a triplet resonance, since a given proton is coupled to the other two. Because D has a known mathematical relationship with the internuclear distance and with the angle between the B_0 field and the internuclear vectors, liquid crystal spectra also may be analyzed to obtain hard structural data.

4-9

Tables of Coupling Constants

The following tables (Tables 4-1 to 4-5) summarize values of coupling constants by class of structure, extracted from the references found at the end of the chapter. Further examples may be obtained by examination of these references.

TABLE 4-1 One Bond Couplings (Hz)

$^{13}C—^{1}H$	CH_3CH_3	125	$^{13}C—^{19}F$	CF_3CF_3	259
	$(CH_3)_4Si$	118		CF_3I	345
	CH_3Li	98		C_6F_6	362
	$(CH_3)_3N$	132	$^{13}C—^{31}P$	CH_3PH_2	9.3
	CH_3CN	136		$(CH_3)_3P$	−13.6
	$(CH_3)_2S$	138		$(CH_3)_4P^+ I^-$	56
	CH_3OH	142	$^{13}C—^{15}N$	CH_3NH_2	−4.5
	CH_3F	144		$C_6H_5NH_2$	−11.43
	CH_3Cl	150		$CH_3(CO)NH_2$	−14.8
	CH_2Cl_2	177		$CH_3C{\equiv}N$	−17.5
	$CHCl_3$	208		Pyridine	+0.62
	Cyclohexane	125		Pyridinium	−11.85
	Cyclobutane	136		$CH_3HC{=}N{—}OH$ (E, Z)	−4.0,−2.3
	Cyclopropane	162	$^{15}N—^{1}H$	CH_3NH_2	−64.5
	Tetrahydrofuran (α, β)	145,133		$C_6H_5NH_2$	−78.5
	Norbornane (Cl)	142		CH_3CONH_2	−89
	Bicyclo[1.1.1]pentane (Cl)	164		Pyridinium	−90.5
	Cyclohexene (Cl)	157		$HC{\equiv}N^+H$	−134
	Benzene	159		$(C_6H_5)_2C{=}NH$	−51.2
	1,3-Cyclopentadiene (C2)	170	$^{15}N—^{15}N$	Azoxybenzene	12.5
	$CH_2{=}CHBr$ (gem)	197		Phenylhydrazine	6.7
	Acetaldehyde (CHO)	172	$^{15}N—^{31}P$	$C_6H_5NHP(CH_3)_2$	+53.0
	Pyridine (α,β,γ)	180,157,160		$C_6H_5NH(PO)(CH_3)_2$	−0.5
	Allene	158		$[(CH_3)_2N]_3P{=}O$	−26.9
	Propyne (≡CH)	248	$^{31}P—^{1}H$	$C_6H_5(C_6H_5CH_2)(PO)H$	474
	$(CH_3)_2(C^+H$ (^+CH)	164	$^{31}P—^{31}P$	$(CH_3)_2P{—}P(CH_3)_2$	−179.7
				$(CH_3)_2(PS)(PS)(CH_3)_2$	18.7

TABLE 4-2 Geminal Proton-Proton (H—C—H) Couplings (Hz)

CH_4	−12.4	$CH_2{=}CH_2$	+2.3
$(CH_3)_4Si$	−14.1	$CH_2{=}O$	+40.22
$C_6H_5CH_3$	−14.4	$CH_2{=}NOH$	9.95
$CH_3(CO)CH_3$	−14.9	$CH_2{=}CHF$	−3.2
CH_3CN	−16.9	$CH_2{=}CHNO_2$	−2.0
$CH_2(CN)_2$	−20.4	$CH_2{=}CHOCH_3$	−2.0
CH_3OH	−10.8	$CH_2{=}CHBr$	−1.8
CH_3Cl	−10.8	$CH_2{=}CHCl$	−1.4
CH_3Br	−10.2	$CH_2{=}CHCH_3$	2.08
CH_3F	−9.6	$CH_2{=}CHCO_2H$	1.7
CH_3I	−9.2	$CH_2{=}CHC_6H_5$	1.08
CH_2Cl_2	−7.5	$CH_2{=}CHCN$	0.91
Cyclohexane	−12.6	$CH_2{=}CHLi$	7.1
Cyclopropane	−4.3	$CH_2{=}C{=}C(CH_3)_2$	−9.0
Aziridine	+1.5		
Oxirane	+5.5		

TABLE 4-3 Vicinal Proton-Proton (H—C—C—H) Couplings (Hz)

Compound	J	Compound	J
CH_3CH_3	8.0	$CH_2{=}CH_2$ (cis,trans)	11.5,19.0
$CH_3CH_2C_6H_5$	7.62	$CH_2{=}CHLi$ (cis,trans)	19.3,23.9
CH_3CH_2CN	7.60	$CH_2{=}CHCN$ (cis,trans)	11.75,17.92
CH_3CH_2Cl	7.23	$CH_2{=}CHC_6H_5$ (cis,trans)	11.48,18.59
$(CH_3CH_2)_3N$	7.13	$CH_2{=}CHCO_2H$ (cis,trans)	10.2,17.2
CH_3CH_2OAc	7.12	$CH_2{=}CHCH_3$ (cis,trans)	10.02,16.81
$(CH_3CH_2)_2O$	6.97	$CH_2{=}CHCl$ (cis,trans)	7.4,14.8
CH_3CH_2Li	8.90	$CH_2{=}CHOCH_3$ (cis,trans)	7.0,14.1
$(CH_3)_2CHCl$	6.4	$ClHC{=}CHCl$ (cis,trans)	5.2,12.2
$ClCH_2CH_2Cl$ (neat)	5.9	Cyclopropene (1–2)	1.3
$Cl_2CHCHCl_2$ (neat)	3.06	Cyclobutene (1–2)	2.85
Cyclopropane (cis,trans)	8.97,5.58	Cyclopentene (1–2)	5.3
Oxirane (cis,trans)	4.45,3.10	Cyclohexene (1–2)	8.8
Aziridine (cis,trans)	6.0,3.1	Benzene	7.54
Cyclobutane (cis,trans)	10.4,4.9	C_6H_5Li (2–3)	6.73
Cyclopentane (cis,trans)	7.9,6.3	$C_6H_5CH_3$ (2–3)	7.64
Tetrahydrofuran (α–β: cis,trans)	7.94,6.14	$C_6H_5CO_2CH_3$ (2–3)	7.86
Cyclopentene (3–4: cis,trans)	9.36,5.72	C_6H_5Cl (2–3)	8.05
Cyclohexane (av.: cis,trans)	3.73,8.07	$C_6H_5OCH_3$ (2–3)	8.30
Cyclohexane (ax–ax)	12.5	$C_6H_5NO_2$ (2–3)	8.36
Cyclohexane (eq–eq and ax–eq)	3.7	$C_6H_5N(CH_3)_2$ (2–3)	8.40
Piperidine (av. α–β: cis,trans)	3.77,7.88	Naphthalene (1–2, 2–3)	8.28,6.85
Oxane (av. α–β: cis,trans)	3.87,7.41	Furan (2–3, 3–4)	1.75,3.3
Cyclohexanone (av. α–β: cis,trans)	5.01,8.61	Pyrrole (2–3, 3–4)	2.6,3.4
Cyclohexene (3–4: cis,trans)	2.95,8.94	Pyridine (2–3, 3–4)	4.88,7.67

TABLE 4-4 Carbon Couplings Other Than $^1J(^{13}C{-}^1H)$ (Hz)

Compound	J	Compound	J
$CH_3\mathbf{CH_3}$	−4.8	$\mathbf{CH_3CH_3}$	34.6
$\mathbf{CH_3}CH_2Cl$	2.6	$\mathbf{CH_3CH_2}OH$	37.7
$Cl_2CH{-}\mathbf{C}HCl_2$	+1.2	$\mathbf{CH_3CHO}$	39.4
Cyclopropane (^2J)	−2.6	$\mathbf{CH_3C}{\equiv}N$	56.5
$\mathbf{(CH_3)_2CH}CH_2CH(\mathbf{CH_3})_2$	5.	$\mathbf{CH_3CO_2}C_2H_5$	58.8
$\mathbf{(CH_3)_2C}{=}O$	5.9	$\mathbf{CH_2{=}CH_2}$	67.2
$CH_3(CO)\mathbf{H}$	26.7	$\mathbf{CH_2{=}CH}CN$	74.1
$\mathbf{CH_3}CH{=}\mathbf{C}(CH_3)_2$	4.8	C_6H_5CN (ipso)	80.3
$\mathbf{CH_2{=}CH_2}$	−2.4	$C_6H_5NO_2$ (1,2)	55.4
$\mathbf{CHCl{=}CHCl}$ (cis,trans)	16.0,0.8	$\mathbf{HC{\equiv}CH}$	170.6
$\mathbf{CH_2}{=}CHBr$ (cis,trans)	−8.5,+7.5	$(\mathbf{CH_3CH_2})_3\mathbf{P}$	+14.1
Benzene (^2J(CH),^3J(CH))	+1.0,+7.4	$(CH_3\mathbf{CH_2})_4\mathbf{P}^+$ Br^-	−4.3
$\mathbf{CH_3C}{\equiv}CH$ (CH_3, ≡CH)	−10.6,+50.8	$(CH_3O)_3\mathbf{P}$	+10.05
$\mathbf{CF_3CF_3}$	46.0	$(CH_3O)_3\mathbf{P}{=}O$	−5.8
$\mathbf{CH_3(CO)F}$	59.7		
$Cl_2\mathbf{C{=}CF_2}$	44.2		

TABLE 4-5 Nitrogen-15 Couplings beyond One Bond (Hz)

CH_3NH_2	−1.0	$CH_3CH_2CH_2NH_2$	1.2
Pyrrole (HNCH)	−4.52	CH_3CONH_2	9.5
Pyridine (NCH)	−10.76	$CH_3C{\equiv}N$	3.0
Pyridinium (HNCH)	−3.01	Pyridine (NCC)	+2.53
$(CH_3)_2NCHO$ (CH_3,CHO)	+1.1,−15.6	Pyridinium (HNCC)	+2.01
$H{-}C{\equiv}N$	8.7	Aniline (NCC)	−2.68
$H_2N(CO)CH_3$	1.3	Pyrrole (HNCC)	−3.92
Pyrrole (HNCCH)	−5.39	$CH_3CH_2CH_2NH_2$	1.4
Pyridine (NCCH)	−1.53	Pyridine (NCCC)	−3.85
Pyridinium (HNCCH)	−3.98	Pyridinium (HNCCC)	−5.30
$CH_3{-}C{\equiv}N$	−1.7	Aniline (NCCC)	−1.29

PROBLEMS

4-1 Characterize the indicated protons as being homotopic, enantiotopic, or diastereotopic; magnetically equivalent or nonequivalent.

a. 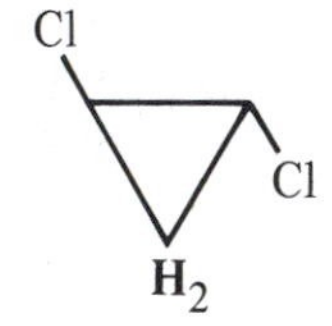 Cl, Cl, H_2

b. 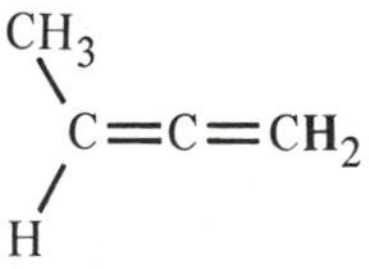 CH_3, C=C=CH_2, H

c. H, H, $CH(CH_3)(C_2H_5)$

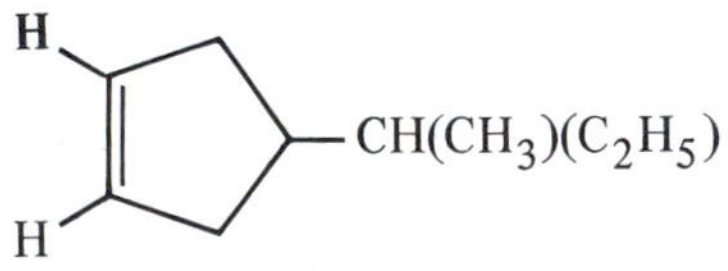

d. 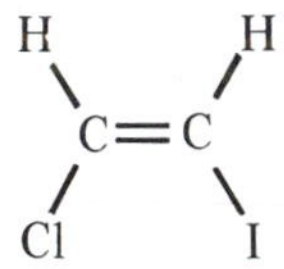 H, H, C=C, Cl, I

e. 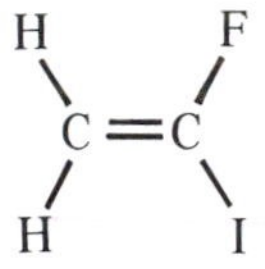 H, F, C=C, H, I

f. O, CH_3, H, CH_3, $CH(CH_3)_2$

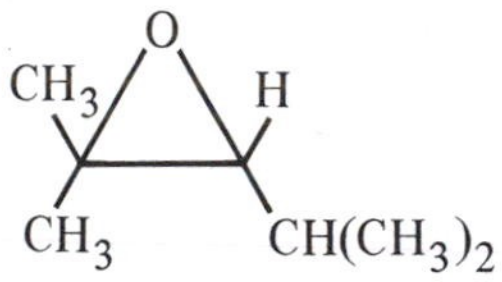

(two answers—one for each geminal pair of methyls)

g. 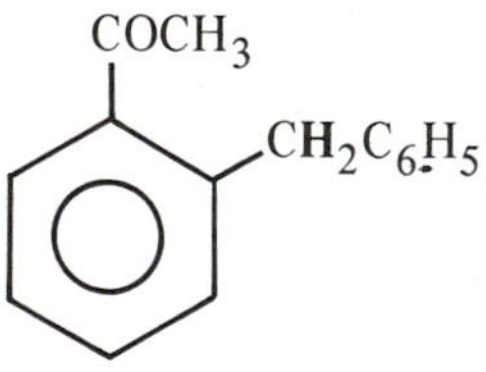 $COCH_3$, $CH_2C_6H_5$

h. 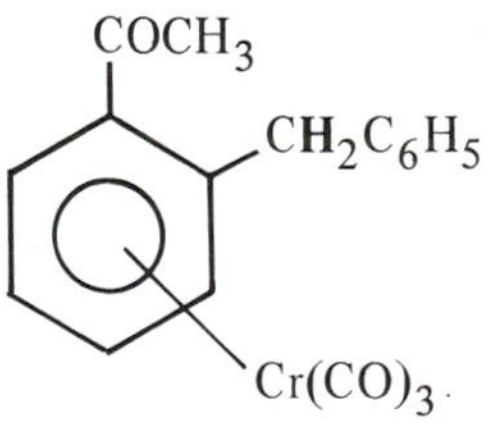 $COCH_3$, $CH_2C_6H_5$, $Cr(CO)_3$

i. 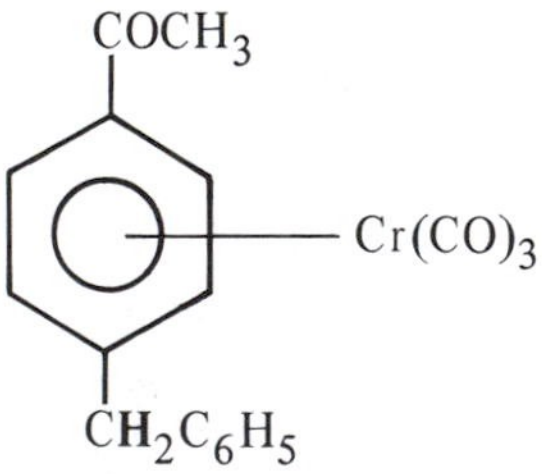 $COCH_3$, $Cr(CO)_3$, $CH_2C_6H_5$

4-2 What is the spin notation for each of the following molecules (AX, AMX, AA′XX′, etc.)? Consider only major isotopes.

a. 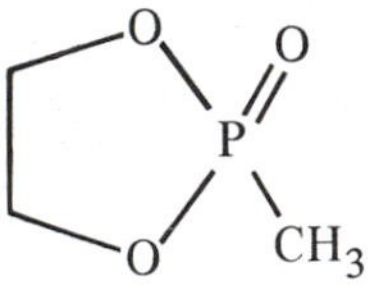 O, O, P, O, CH_3

b. $Ph(CH_3)P{-}P(CH_3)Ph$

(Ignore the aromatic protons.)

c. 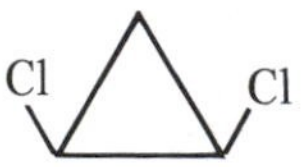 Cl, Cl

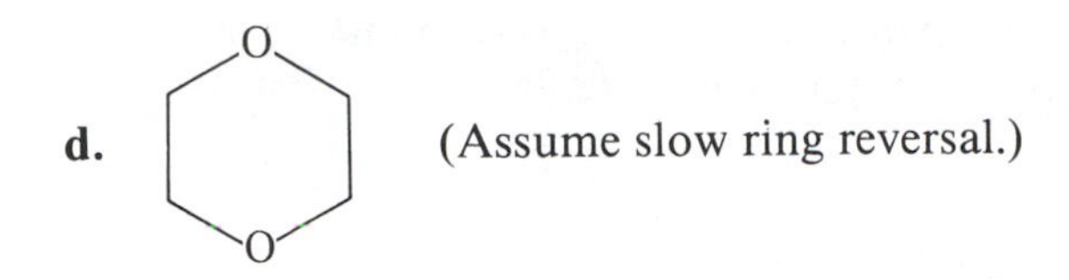

d. (Assume slow ring reversal.)

4-3 Using the Dirac vector model, rationalize why the three bond H_A—C—C—H_B coupling is usually positive.

4-4 Elimination of two moles of HBr from the molecule below should give the indicated cyclopropane. The $^1J(^{13}C—^1H)$ for the bridge CH_2 group in the isolated product was measured to be 142 Hz. Explain in terms of product structures.

Br Br Br Br — KOH / EtOH → H H

4-5 There are two isomers of thiane 1-oxide, A and B. The observed geminal coupling constant between the α protons is −13.7 Hz in one isomer and −11.7 Hz in the other. Which coupling belongs to which isomer and why?

A B

4-6 Explain why the linewidth of the angular methyl group is larger in *trans*-decalins than in *cis*-decalins.

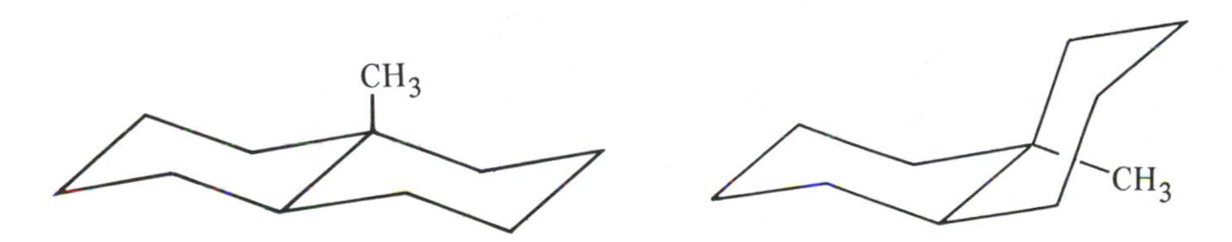

4-7 In cycloheptatriene (A), J_{23} is 5.3 Hz, whereas in its bistrifluoromethyl derivative (B), J_{23} is 6.9 Hz. Explain.

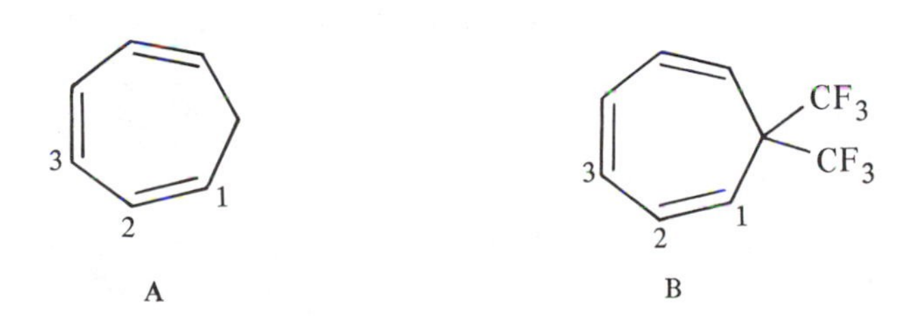

4-8 Analyze the following ^{1}H spectrum[1] of a thujic ester. The CH_3 resonances are not shown. Assign the resonances to specific protons and give very approximate coupling constants. Explain your chemical shift assignments.

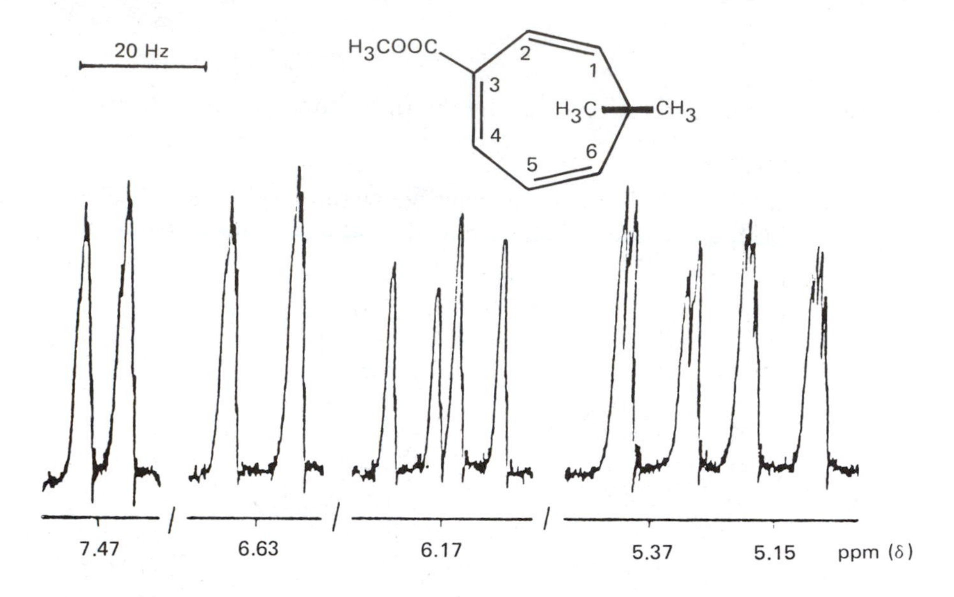

4-9 Bromination of 1-bromonaphthalene gives a dibromo isomer whose ^{1}H spectrum[2] is given below. What is the structure? Explain and give the spectral notation, AB, ABX, etc.

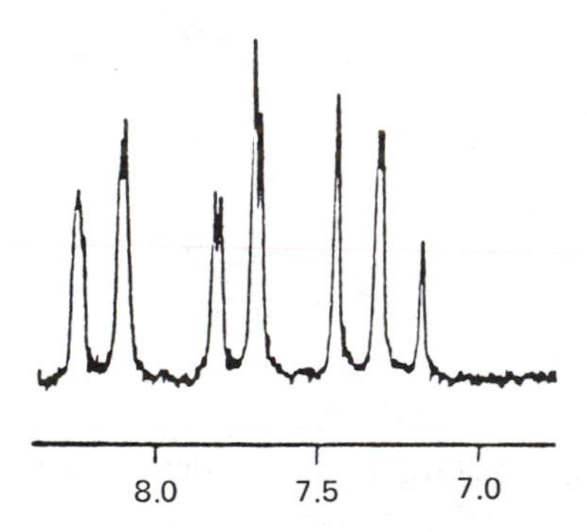

4-10 The following four spectra[3] are of lutidines (dimethylpyridines). From the chemical shifts and coupling patterns, deduce the placement of methyl groups on each molecule. Assume the spacings are first order.

[1]From H. Günther et al., *Org. Magn. Reson.*, **6**, 388 (1974). Copyright 1974 John Wiley & Sons Ltd. Reprinted by permission of John Wiley & Sons Ltd.

[2]From H. Günther, *NMR Spectroscopy*, Wiley, Chichester, 1980, p. 194. Copyright 1980 John Wiley & Sons Ltd. Reprinted by permission of John Wiley & Sons Ltd.

[3]Reproduced with permission of Aldrich Chemical Company, Inc., from *The Aldrich Library of NMR Spectra*, 2nd ed., vol. IX, pp. 8–9.

a.

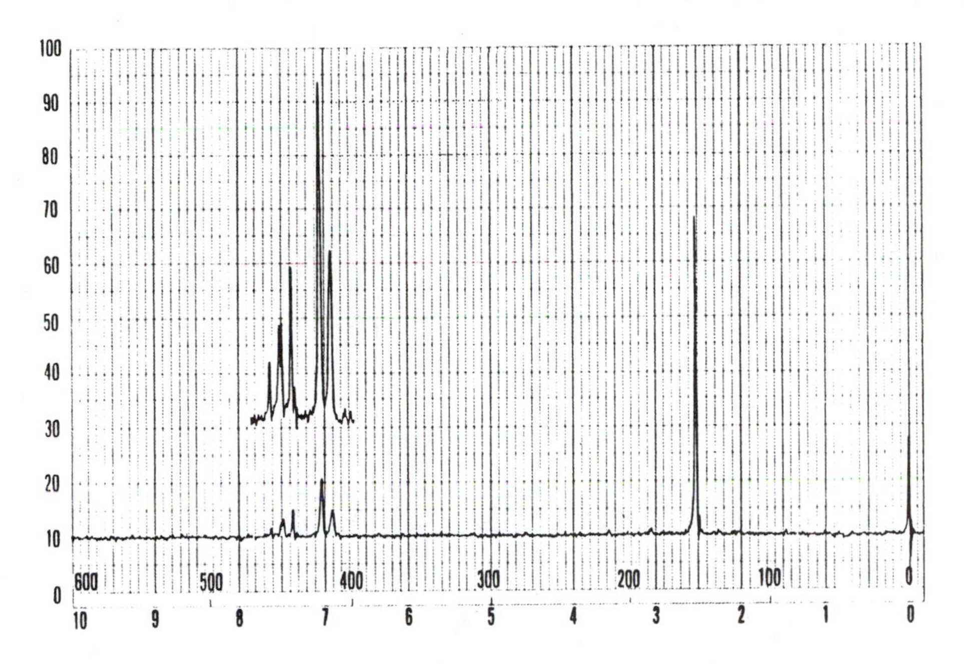
100
90
80
70
60
50
40
30
20
10
0
600
500
400
300
200
100
0
10
9
8
7
6
5
4
3
2
1
0

b.

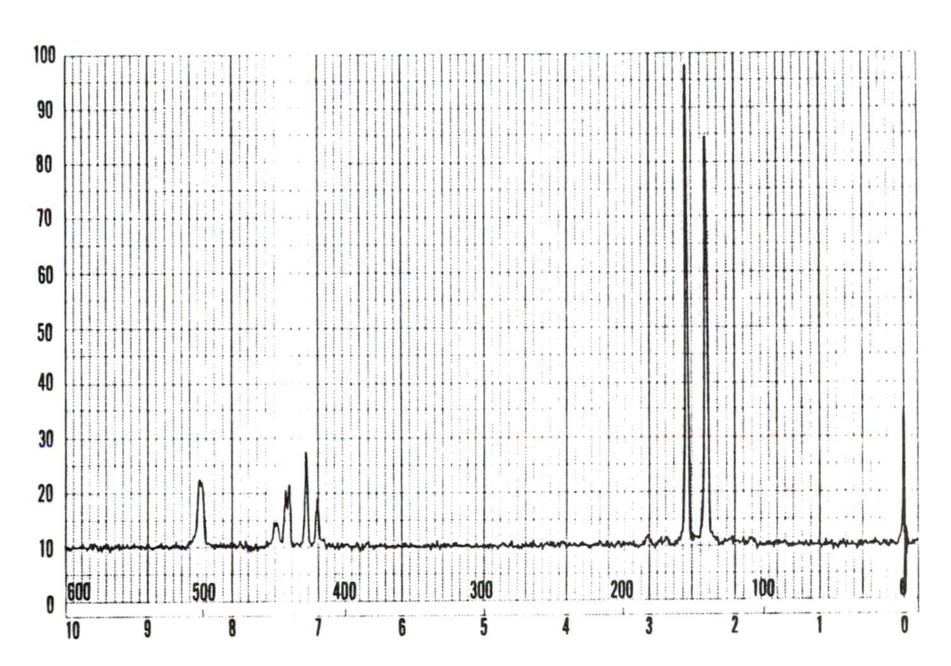
100
90
80
70
60
50
40
30
20
10
0
600
500
400
300
200
100
0
10
9
8
7
6
5
4
3
2
1
0

c.

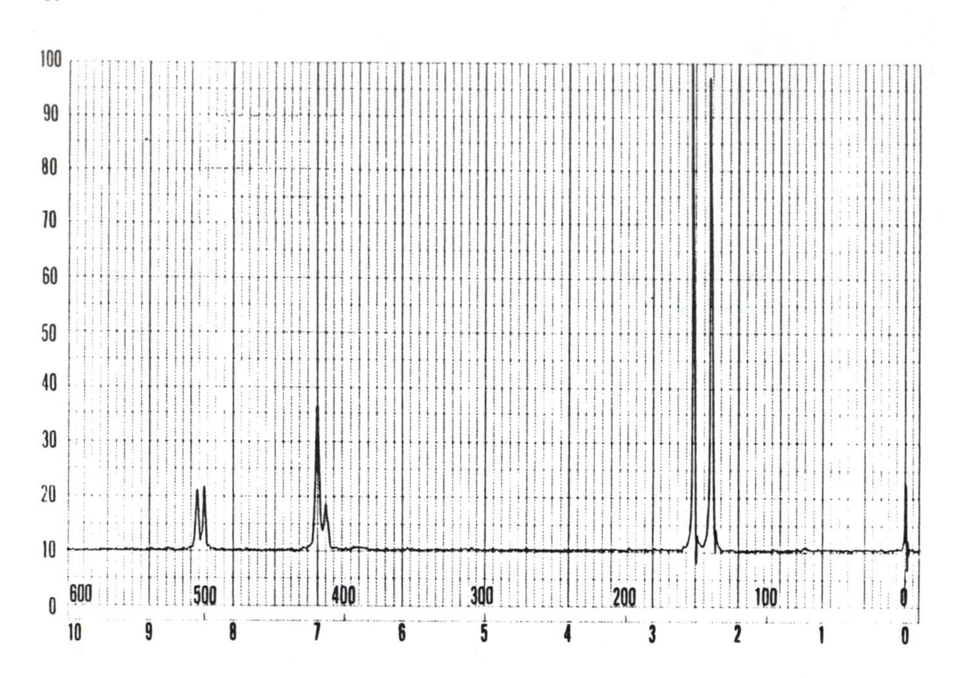
100
90
80
70
60
50
40
30
20
10
0
600
500
400
300
200
100
0
10
9
8
7
6
5
4
3
2
1
0

d.

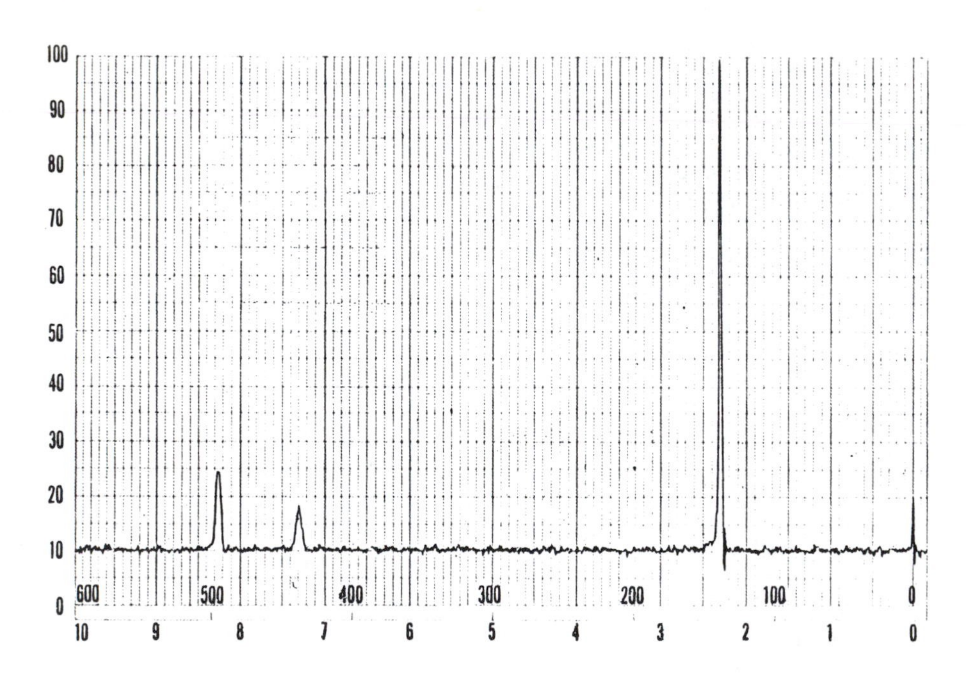
100
90
80
70
60
50
40
30
20
10
0
600
500
400
300
200
100
0
10
9
8
7
6
5
4
3
2
1
0

REFERENCES/BIBLIOGRAPHY

4-1 Coupling (general): *Nuclear Magnetic Resonance*, A Specialist Periodical Report, The Chemical Society, London, reviewed in each issue.

4-2 Magnetic equivalence: K. Mislow and M. Raban, *Top. Stereochem.*, **1**, 1 (1966); W. H. Pirkle and D. J. Hoover, *ibid.*, **13**, 263 (1982).

4-3 Directly bonded couplings: W. McFarlane, *Quart. Rev.*, **23**, 187 (1969); C. J. Jameson and H. S. Gutowsky, *J. Chem. Phys.*, **51**, 2790 (1969); J. H. Goldstein, V. S. Watts, and L. S. Rattet, *Progr. NMR Spectrosc.*, **8**, 103 (1971).

4-4 Geminal, vicinal, and long-range couplings (H—H): S. Sternhell, *Rev. Pure Appl. Chem.*, **14**, 15 (1964); A. A. Bothner-By, *Advan. Magn. Reson.*, **1**, 195 (1965); M. Barfield and B. Charkrabarti, *Chem. Rev.*, **69**, 757 (1969); S. Sternhell, *Quart. Rev.*, **23**, 236 (1969); V. F. Bystrov, *Russ. Chem. Rev.*, **41**, 281 (1972); J. Hilton and L. H. Sutcliffe, *Progr. NMR Spectrosc.*, **10**, 27 (1975); M. Barfield, R. J. Spear, and S. Sternhell, *Chem. Rev.*, **76**, 593 (1976).

4-5 Carbon-13 couplings: J. B. Stothers, *Carbon-13 NMR Spectroscopy*, Academic Press, New York, 1973; J. L. Marshall, D. E. Müller, S. A. Conn, R. Seiwell, and A. M. Ihrig, *Acc. Chem. Res.*, **7**, 333 (1974); D. F. Ewing, *Ann. Rep. NMR Spectrosc.*, **6A**, 389 (1975); R. E. Wasylishen, *Ann. Rep. NMR Spectrosc.*, **7**, 118 (1977); P. E. Hansen, *Org. Magn. Reson.*, **11**, 215 (1978); V. Wray, *Progr. NMR Spectrosc.*, **13**, 177 (1979); G. C. Levy, R. L. Lichter, and G. L. Nelson, *Carbon-13 Nuclear Magnetic Resonance Spectroscopy*, 2nd ed., Wiley-Interscience, New York, NY, 1980; V. Wray and P. E. Hansen, *Ann. Rep. NMR Spectrosc.*, **11A**, 99 (1981); P. E. Hansen, *ibid.*, **11A**, 65 (1981); P. E. Hansen, *Progr. NMR Spectrosc.*, **14**, 175 (1981); W. H. Pirkle and D. J. Hoover, *Top. Stereochem.*, **13**, 263 (1982); J. L. Marshall, *Carbon–Carbon and Carbon–Proton NMR Couplings*, Verlag Chemie, Deerfield Beach, FL, 1983.

4-6 Fluorine-19 couplings: J. M. Emsley, L. Phillips, and V. Wray, *Progr. NMR Spectrosc.*, **10**, 82 (1977).

4-7 Phosphorus-31 couplings: E. G. Finer and R. K. Harris, *Progr. NMR Spectrosc.*, **6**, 61 (1970).

4-8 Spectral analysis: J. D. Roberts, *An Introduction to the Analysis of Spin–Spin Splitting in High-Resolution Nuclear Magnetic Resonance Spectra*, W. A. Benjamin, New York, 1961; K. B. Wiberg and B. J. Nist, *The Interpretation of NMR Spectra*, W. A. Benjamin, New York, 1962; R. J. Abraham, *The Analysis of High Resolution NMR Spectra*, Elsevier, Amsterdam, 1971; R. A. Hoffman, S. Forsén, and B. Gestblom, *NMR Basic Princ. Progr.*, **5**, 1 (1971); C. W. Haigh, *Ann. Rep. NMR Spectrosc.*, **4**, 311 (1971); P. Diehl, H. Kellerhals, and E. Lustig, *NMR Basic Princ. Progr.*, **6**, 1 (1972); *Nuclear Magnetic Resonance*, A Specialist Periodical Report, The Chemical Society, London, reviewed in most issues.

4-9 Liquid crystals and oriented molecules: S. Meiboom and L. K. Snyder, *Science*, **162**, 1337 (1968); J. Bulthuis, *NMR in Liquid Crystalline Solvents*, Rodopi N. V., Amsterdam, 1974; J. W. Emsley and J. C. Lindon, *NMR Spectroscopy Using Liquid Crystal Solvents*, Pergamon, Oxford, 1975; L. Lunazi, *Determ. Org. Struc. Phys. Meth.*, **6**, 335 (1976).

5

Special Topics in Nuclear Magnetic Resonance

For the person who wants to use NMR for the structural analysis of simple compounds, the material in the first three chapters probably suffices as a general introduction. NMR, however, offers much more. In this chapter we will describe other areas of application and some of the advanced methods developed to solve structural problems.

5-1

Dynamic Processes

The use of spectral changes to obtain kinetic information (Section 2-9) has been referred to as the *dynamic nuclear magnetic resonance* (DNMR) method. In a typical DNMR experiment, the spectrum is recorded at a series of temperatures, although concentration sometimes is altered as well. The spectrum varies according to the value of the unimolecular rate constant and the distance between resonances, as in the case of exchange between axial and equatorial protons in cyclohexane (Figure 2-14). When exchange is very rapid, the spectrum gives only the average of the exchanging peaks. When exchange is slow, separate and distinct resonances are obtained. At intermediate temperatures, there is a well-defined gradation between these extremes. The point of maximum broadening is called the *coalescence temperature* (T_c). The rate constant at T_c for two unsplit peaks of equal intensity (as in Figure 2-14) is $\pi\Delta\nu/\sqrt{2}$ ($\Delta\nu$ is the distance between the two peaks at slow exchange). Thus NMR provides a very simple method for obtaining rates of unimolecular reactions. Computer programs are available for derivation of rates in more complex spin systems, including those with spin–spin coupling, unequal populations, and more than two exchange sites. These methods involve computer fitting of the spectra at several temperatures to produce a *complete lineshape analysis*. Rate constants at several temperatures then can be obtained.

Even when only one rate constant is obtained, as by the coalescence temperature method, the free energy of activation can be calculated by eq. 5-1, in which k_c is the rate constant at coalescence. Since the complete

$$\Delta G_c^{\ddagger} = 2.3RT_c[10.32 + \log(T_c/k_c)] \tag{5-1}$$

lineshape method produces rate constants as a function of temperature, Arrhenius plots may be constructed to give the activation energy (E_a) and pre-exponential factor (A) or the enthalpy and entropy of activation ($\Delta H^{\ddagger}$, $\Delta S^{\ddagger}$). As a check for systematic errors, it is a good idea to use both the coalescence temperature and complete lineshape methods. Barriers in the 4.5 to 27 kcal/mol range can be derived by measurements in the temperature range from −180 to +200°C.

Hindered rotation has been studied very thoroughly by DNMR methods. Bonds that are intermediate between single and double bonds often fall in the DNMR barrier range. These include amides, carbamates, thioamides, enamines, nitrosamines, alkyl nitrites, diazoketones, aminoboranes, and aromatic aldehydes. Steric congestion about single bonds can raise the barrier sufficiently to give DNMR effects, as in biphenyls, *tert*-butyl compounds, and 1-substituted trypticenes. Polyhalogenated alkanes such as 2,2,3,3-tetrachlorobutane also have barriers in the appropriate range. Molecules with adjacent atoms that both have lone pair electrons, such as hydrazines, disulfides, sulfenamides, and aminophosphines, also have measurable barriers.

Ring reversal is not limited to cyclohexane. Axial–equatorial interconversion is an example of hindered rotation and may be studied in a wide variety of systems, including heterocycles such as piperidine (**5-1**), unsaturated rings such as cyclohexene (**5-2**), fused rings such as *cis*-decalin (**5-3**), and rings of other than six members such as cycloheptatriene (**5-4**).

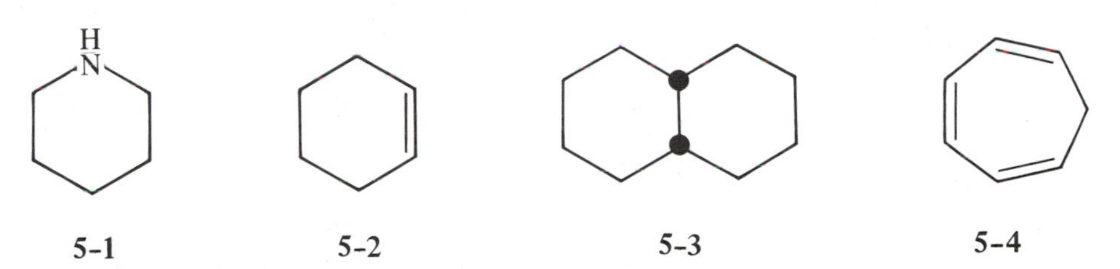

5-1 **5-2** **5-3** **5-4**

Trisubstituted atoms with a lone pair, such as amines, may undergo the process of atomic inversion, which often falls in the kinetic region able to be studied by DNMR methods. For example, the two methyl groups in the aziridine **5-5** become equivalent if inversion about nitrogen is rapid. This barrier is particularly high

5-5

(18–20 kcal/mol) because of angle strain in the three-membered ring. The effects of nitrogen inversion also have been observed in rings of other sizes and, at very low temperatures, in acyclic amines. When the nitrogen atom is substituted with an electron-withdrawing atom, as in oxazolidines and *N*-chloroamines, the barrier is raised substantially.

Inversion about other atoms also has been studied by NMR, including phosphorus, oxygen, and sulfur. Inversion about nitrogen in trigonal systems such as imines provides a method for studying syn–anti interconversion.

Numerous other types of unimolecular reactions may be studied by NMR, including trigonal bipyramidal reorganizations (axial–equatorial interchange in sulfur tetrafluoride, **5-6**), valence tautomerizations (Cope

5-6

rearrangement in 3,4-homotropilidene, **5-7**), and carbocation rearrangements (hydride shifts in the

5-7

norbornyl cation, **5-8**). Applications to inorganic and bioorganic chemistry are quite common now. A wide

5-8

range of exchanging nuclei also has been studied. Carbon-13 provides particularly simple spectra (a series of singlets) that are amenable to complete lineshape analysis. Other useful nuclei include ^{31}P, ^{2}H, ^{17}O, and metals such as ^{195}Pt and ^{199}Hg.

Chemically induced dynamic nuclear polarization (CIDNP) is a technique for the study of the effect of unpaired electron spins on nuclear resonances. It is especially useful for the study of reactions that produce free radicals. The subject is discussed in the references listed at the end of the chapter.

5-2

Relaxation Effects

The process of relaxation was discussed in Section 2-8. Because relaxation is closely related to the motion of molecules in solution, it can provide dynamic information that corresponds to a much faster time scale than the procedures described in the previous section. In this section we discuss the phenomenon of relaxation in greater detail and describe several applications to chemical problems.

5-2a Spin–Lattice Relaxation

Figure 5-1 illustrates the two states of a spin $\frac{1}{2}$ nucleus. Application of rf energy (B_1) brings about a spin flip. The NMR experiment is sustained because neighboring magnetic nuclei in motion (the *lattice*) can produce oscillating fields with the same frequency and phase as the excited nucleus. Energy may pass from the excited spins to the lattice, so that nuclei can return to the lower spin state and be available for another spin excitation by B_1. This process is called *spin–lattice relaxation.*

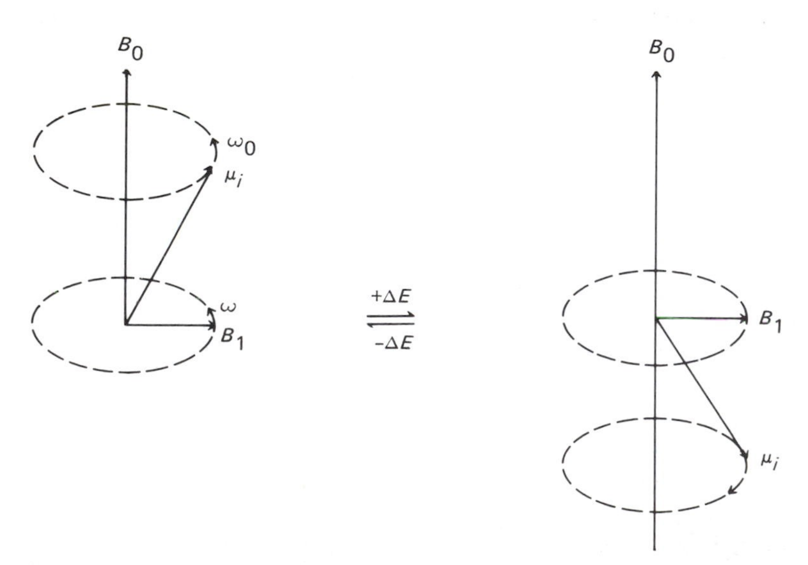

FIGURE 5-1 Left: Precession of a nucleus with magnetic moment μ_i about the direction of the B_0 field. Right: Absorption of energy results in precession against the field.

Magnetization (M) is the term given to the sum of the individual spins (for example, $M_z = \Sigma I_z$ is the magnetization along the direction of the external field). When the spins are at equilibrium in the magnetic field, M_z has a steady state value M_{z0}. The value of M_{z0} is not zero because there are more spins with $I_z = +\frac{1}{2}$ than with $I_z = -\frac{1}{2}$. When the B_1 field is applied, this equilibrium is disturbed by the process of spin flips shown in Figure 5-1, so that M_z no longer is at equilibrium. Spin–lattice relaxation is defined more rigorously as any process that returns M_z to its equilibrium value (M_{z0}). It is a first-order process with rate constant T_1^{-1}, and T_1 is called the *spin–lattice relaxation time.* The process also is termed *longitudinal relaxation* because changes in magnetization occur along the longitude (z axis) of the NMR experiment.

5-2b Measurement of T_1

The relaxation time T_1 has become the fourth important physical parameter to be obtained from the NMR experiment, after the chemical shift, the coupling constant, and the rate constant for dynamic processes. Its measurement, however, requires a specialized experiment known as *inversion recovery*.

At equilibrium, the net magnetization is along the z axis (Figure 5-2a). In the figure only the net magnetization M is shown (the vector sum of the nuclear magnets lined up with and against the field, with a slight excess lined up with the field). Application of a strong rf pulse (B_1) along the x direction applies a torque to the nuclear magnets. Because B_1 is perpendicular to the magnetization, the force it exerts moves the magnetization in a direction that is perpendicular to both original vectors (by a vector cross product right-hand rule). If the B_1 pulse is cut off after exactly the right time (on the order of microseconds), the net magnetization is left aligned along the y direction (Figure 5-2b). A pulse of this duration is called a 90° pulse because the magnetization is rotated 90°. In the normal NMR experiment, the detection coils are set up along the y direction, so that y signal may be detected. The normal Fourier transformation pulse experiment usually involves a 90° pulse followed by y detection.

If the B_1 pulse is left on twice as long as needed for a 90° pulse, the torquing effect of the B_1 field in the x direction continues to move the magnetization in a clockwise fashion, and the spins end up with their net orientation opposite to the B_0 field ($-z$ direction, Figure 5-2c). Because the spins have rotated through 180°, a pulse of this duration is said to be a 180° pulse. This arrangement is upside down with respect to the equilibrium situation, that is, more spins are aligned against than with the field. This unusual situation can be demonstrated by following the 180° pulse immediately with a 90° pulse (Figure 5-2d), the equivalent of a 270° pulse. The spins are rotated into the $-y$ direction, so the detector will pick up negative signals (upside down).

The nonequilibrium situation after the 180° pulse can be exploited to measure T_1. Figure 5-3 (bottom spectrum) shows the results of a 180° pulse followed almost immediately ($\tau = 1$ s) by a 90° pulse on the ^{13}C nuclei of 1-methylpiperidine. All the resonances are inverted. If the amount of time between the 180° and 90° pulses is quite long ($\tau = 100$ s), spin–lattice relaxation brings the system back to equilibrium, in which the net magnetization is again aligned along the B_0 field. The top spectrum shows the result of the pulse sequence

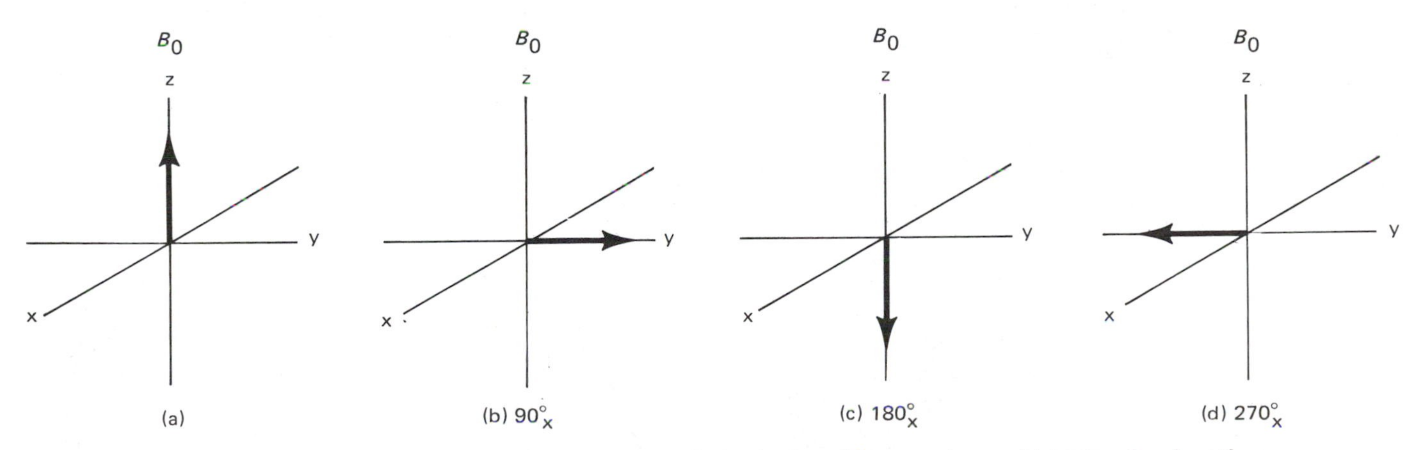

FIGURE 5-2 **(a) Magnetization is lined up along the z direction at the beginning of the experiment. (b) A 90° pulse along the x direction lines up the overall magnetization along the y direction. (c) A 180° pulse along the x direction lines up the overall magnetization along the $-z$ direction. (d) Overall magnetization after a 180° pulse followed by a 90° pulse.**

FIGURE 5-3 A stack of plots for the ^{13}C resonance of 1-methylpiperidine after the inversion recovery pulse sequence, 180°–τ–90°, with τ ranging from 1 s (bottom) to 100 s (top.) The peaks from left to right are from the α, CH_3, β, and γ carbons. [Reproduced with permission from J. B. Lambert and D. A. Netzel, *J. Am. Chem. Soc.*, **98**, 3784 (1976). Copyright 1976 American Chemical Society.]

180°–(100 s)–90°. All the resonances have normal intensity. If the amount of time (τ) between the pulses is intermediate, the result can be either negative peaks for short τ or positive peaks of reduced intensity for longer τ. The result of a series of experiments that varies τ from 1 to 100 s thus gives the stack of plots shown in Figure 5-3. The intensities for any given resonance provide a series of points for a first-order kinetic plot. The pulse interval τ is plotted vs. the logarithm of the intensities of each peak to give the plot, the slope of which is $(-1/T_1)$. The values of T_1 from Figure 5-3 are 17.0 (α CH_2), 17.8 (β CH_2), 15.3 (γ CH_2), and 14.0 (CH_3) s. Although other pulse sequences also can provide T_1, the inversion recovery sequence (180°–τ–90°) is the most common and probably the most accurate.

5-2c Mechanisms of Spin–Lattice Relaxation

The random oscillating fields that give rise to spin–lattice relaxation can come from a number of sources. As already mentioned, the major component of relaxation is caused by the motion of other magnetic nuclei or unpaired electrons, the *dipole–dipole relaxation* (T_1(DD)). Other sources of oscillating fields include interruption of the motion of rapidly rotating groups such as methyl by collisions (*spin rotation*), tumbling of molecules with anisotropic chemical shielding, averaging of coupling constants through chemical exchange or through interaction with a quadrupolar nucleus, and tumbling of quadrupolar nuclei (discussed separately in Section 5-5).

In practice, a carbon atom is almost always relaxed through dipolar interactions with its attached or nearest-neighbor protons. Eq. 5-2 provides a quantitative expression for dipolar relaxation of a carbon atom

$$\frac{1}{T_1(\mathrm{DD})} = n\gamma_{\mathrm{C}}^2\gamma_{\mathrm{H}}^2\hbar^2 r_{\mathrm{CH}}^{-6}\tau_c \qquad (5\text{-}2)$$

by hydrogen atoms. The symbol n is the number of attached or nearest neighbor hydrogens, $\hbar$ is Planck's constant divided by 2π, r is the distance from carbon to hydrogen, and τ_c is the effective correlation time for rotations (equivalent to the time for a molecule to rotate one radian). From this equation it can be seen that the relaxation time is shorter (that is, relaxation is faster) when there are more attached protons, when the internuclear distance is short, and when rotation in solution is slow. A quaternary carbon has a long relaxation time because it lacks an attached proton and r is large. A methyl relaxation time is $\frac{2}{3}$ as long as a methylene relaxation time, other things being equal, because of the difference in the number of attached protons. Large molecules tend to move more slowly in solution, so carbons in vitamin B_{12} relax more rapidly than those in cholesteryl chloride, and so on in the order naphthalene, toluene, methane.

The situation is made somewhat more complicated by the presence of nondipolar mechanisms of relaxation (T_1(other)), which add reciprocally according to eq. 5-3 (rate constants add directly and relaxation

$$\frac{1}{T_1} = \frac{1}{T_1(\text{DD})} + \frac{1}{T_1(\text{other})} \tag{5-3}$$

times add reciprocally). Spin rotation is important for very small molecules like carbon disulfide and for rapidly rotating groups like methyl within a molecule. Chemical shielding anisotropy is important for molecules with highly anisotropic groups such as triple bonds and becomes more important at high fields. Relaxation by averaging of coupling is rarely important except for nuclei attached to nuclei with large quadrupole moments, like the ipso carbon in bromobenzene. Quadrupolar nuclei themselves usually relax by the quadrupolar mechanism (Section 5-5), so that the spectra of ^{14}N and ^{17}O must be interpreted accordingly.

Eq. 5-2 is valid only in the so-called *extreme narrowing limit*, which fails for particularly large molecules. The optimal correlation times for dipolar relaxation lie in the range of about 10^{-7}–10^{-11} s. If motion is too rapid, τ_c is very short and T_1(DD) becomes long by eq. 5-2. If motion is too slow, the molecule is out of the extreme narrowing limit, and eq. 5-2 no longer is valid. Biopolymers often fall into this latter category.

5-2d Applications of Dipolar Relaxation Times

Different carbons within a molecule can move through solution at different rates, because of internal rotation or segmental motion. Knowledge of the dipolar relaxation time for each carbon, normalized for the number of attached carbons, thus gives a measure of the relative mobility of the carbons. Although the overall spin–lattice relaxation time can be used for this purpose, it is better to calculate T_1(DD) from T_1 by use of the nuclear Overhauser enhancement, as described in Section 5-4. The molecule decane (**5-9**) shows these effects. A longer value of nT_1 (in seconds) for a given carbon, by eq. 5-2, indicates a shorter effective

$CH_3CH_2CH_2CH_2CH_2CH_2CH_2CH_2CH_2CH_3$

nT_1 26.1 13.2 11.4 10.0 8.8

5-9

correlation time (other factors such as r are constant) and hence a more rapid motion. Thus the end of the decane chain moves more rapidly in solution, and motion is progressively hindered for carbons closer to the center of the molecule. In 1-bromodecane (**5-10**) the bromine atom serves as an anchor to one end of the

$CH_3CH_2CH_2CH_2CH_2CH_2CH_2CH_2CH_2CH_2Br$

nT_1 15.9 7.8 6.2 4.4 4.2 4.2 4.0 3.8 5.4 5.6

5-10

molecule, inhibiting motion. Thus the CH_3 end is more mobile than the $BrCH_2$ end. The values of nT_1 in general are shorter for 1-bromodecane, which is larger and moves more slowly in solution as a whole.

Molecules do not always rotate as perfect spheres in solution. Anisotropy of motion can be detected by examination of dipolar relaxation times of different atoms within the same molecule. In toluene (**5-11**), the

23 23 58 18 $-CH_3$ T_1 (s)

5-11

para carbon relaxes more quickly than the ortho and meta carbons. Preferred rotation about the ipso–para axis moves the ortho and meta carbons but not the para carbon. The faster motion of the ortho and meta carbons means a shorter effective correlation time and by eq. 5-2 a longer relaxation time. The ipso carbon relaxes very slowly because it has no attached protons.

The inversion recovery experiment can be exploited to examine certain components of a complex spectrum, provided that various carbon atoms have different relaxation times. Thus nuclei that relax more rapidly give resonances that pass through the null point and attain positive intensity at shorter τ in a stack plot such as Figure 5-3. The molecule prostacyclin (**5-12**) provides an instructive example (Figure 5-4) with

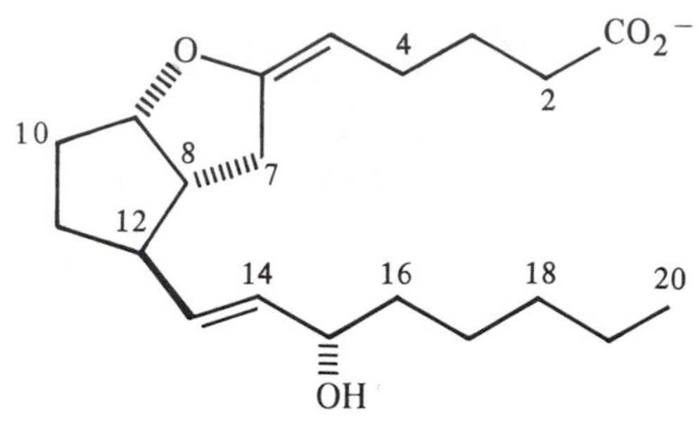

5-12

proton relaxation. The resonances of the allylic protons on C4 and C12 overlap. When $\tau = 0.6$ s, however, the H12 resonance is nulled and all the remaining peaks come from H4 and H4′. The protons on C4 relax each other more rapidly because of their geminal disposition, whereas the proton on C12 must be relaxed by more distant vicinal protons. When $\tau = 0.4$ s, the more slowly relaxing H8 resonances are nulled, but the overlapping doublet from the H7α peak remains (H8 is methinyl and H7α is part of a methylene group).

This technique gives *partially relaxed spectra* and may be used sometimes to simplify complex spectra. It also may be used to remove unwanted resonances, such as that from solvent. An inversion recovery experiment is carried out, and the τ value at which the solvent peak is nulled is chosen. The rest of the spectrum will have positive or negative intensities, depending on whether they relax more rapidly or more slowly than solvent. The experiment may be refined by applying the 180° pulse only at the resonance position of the solvent or of some other undesired nucleus. The τ value again is selected for nulling of this peak, so that the other peaks are not affected. The procedure is called *peak suppression.*

Relaxation times also may be used to obtain quantitative information about dynamic processes in solution. The mathematical analysis of methyl rotation has been carried out by Woessner, with the result that methyl rotation rates often can be measured from methyl carbon relaxation measurements. The method of *saturation transfer* is useful for measuring rates of slow dynamic processes and is discussed in Section 5-5f.

5-2e Spin-Spin Relaxation

A second type of relaxation is concerned with magnetization in the *xy* plane, perpendicular to the *z* axis. At equilibrium, chemically equivalent nuclei precess about the *z* direction with identical angular frequencies. Because there is a small excess of $+\frac{1}{2}$ nuclei, the net magnetization is directed along the $+z$ axis. At equilibrium, there is no magnetization in the *xy* plane ($M_x = M_y = 0$). In a Fourier transform experiment, a 90° pulse from the B_1 field tips the magnetization entirely into the *xy* plane (Figure 5-2b). In a CW experiment, the relatively weak field tips the net magnetization M slightly off the *z* axis, thereby developing a small component in the *xy* plane. In either case, *y* magnetization is induced and is detected by the receiver coil. When the B_1 field passes out of the region of resonance, the *x* and *y* components of magnetization decay back to their equilibrium values of zero, and the signal no longer is detectable. Because this phenomenon occurs in the plane perpendicular to the applied magnetic field, it is termed *transverse relaxation.* This relaxation is a first-order process with rate constant T_2^{-1}. Also called *spin–spin relaxation*, this phenomenon is defined as any process whereby the *xy* components of magnetization are brought back to their equilibrium values of zero.

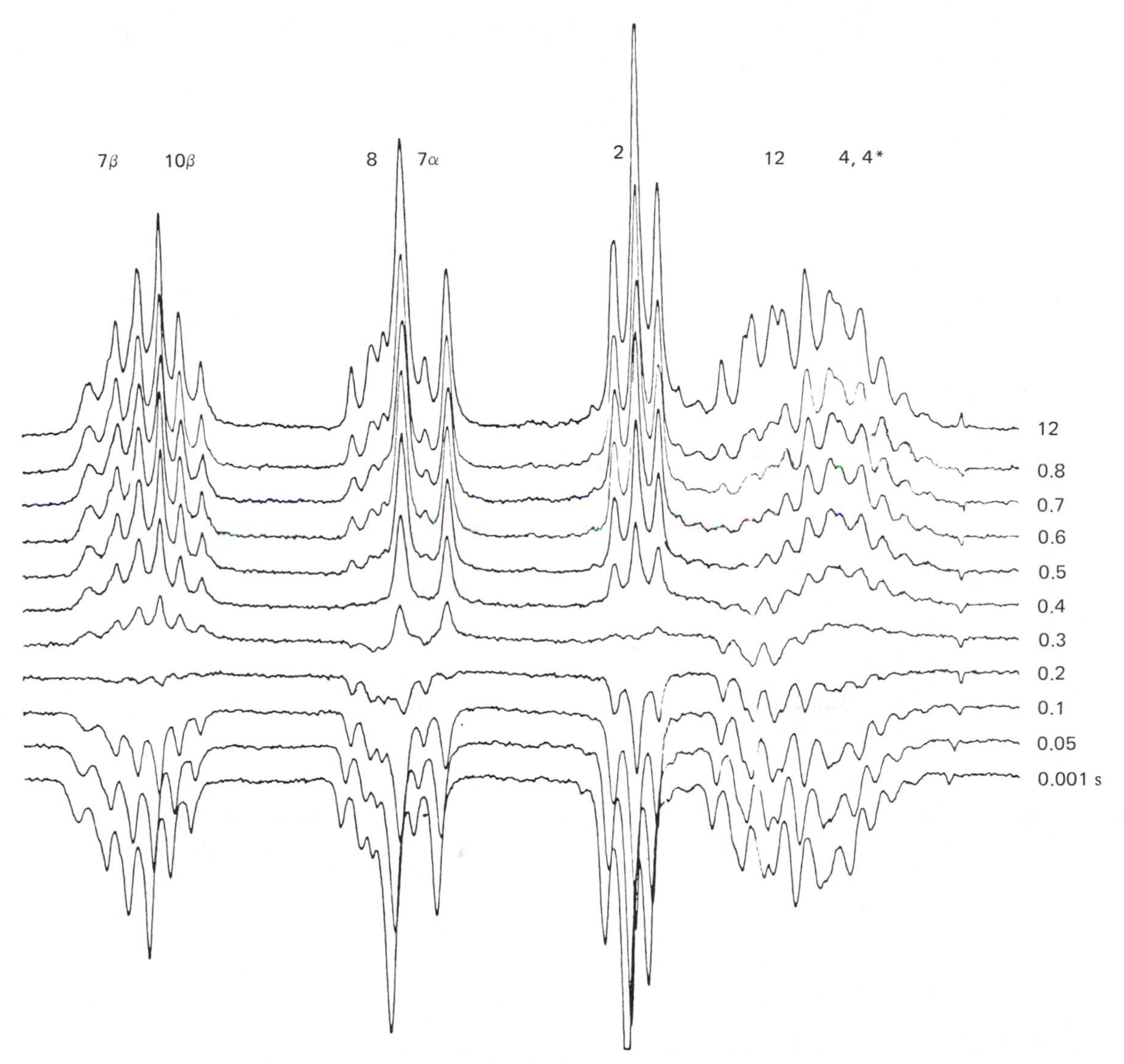

FIGURE 5-4 Stack plot of the proton spectrum of prostacyclin (PGI_2). [Reproduced with permission from G. Kotovych, G. H. M. Aarts, T. T. Nakashima, and G. Bigam, *Can. J. Chem.*, **58**, 978 (1980).]

5-3

Multiple Irradiation

The technique of multiple resonance utilizes additional sources of rf energy to bring about special effects in the NMR spectrum. Spectral simplification is one of the most useful such effects. The 90 MHz proton spectrum of methyldiphenylphosphine ($CH_3(C_6H_5)_2P$) contains a first-order doublet for the methyl group coupled to ^{31}P (Figure 5-5). If the ^{31}P resonance (32.1 MHz) is subjected to a second strong rf field ($\nu_2 = \gamma B_2$), the ^{31}P nuclei rapidly absorb and emit rf energy. The ^{31}P nuclei are shuttled back and forth between the $+\frac{1}{2}$ and $-\frac{1}{2}$ spin states so rapidly that the methyl protons no longer distinguish distinct energy levels. When *decoupled* from ^{31}P in this manner, the proton acts as if ^{31}P were a nonmagnetic nucleus (Figure 5-5).

When the observed nucleus (^1H in the above example) and the irradiated nucleus (^{31}P) are of different types, the experiment is termed *heteronuclear double resonance* and is represented by a notation such as $^1H\{^{31}P\}$, the nucleus in braces being irradiated. *Homonuclear double resonance* involves observation and irradiation of the same type of nucleus. Thus the quartet structure of the methylene resonance in

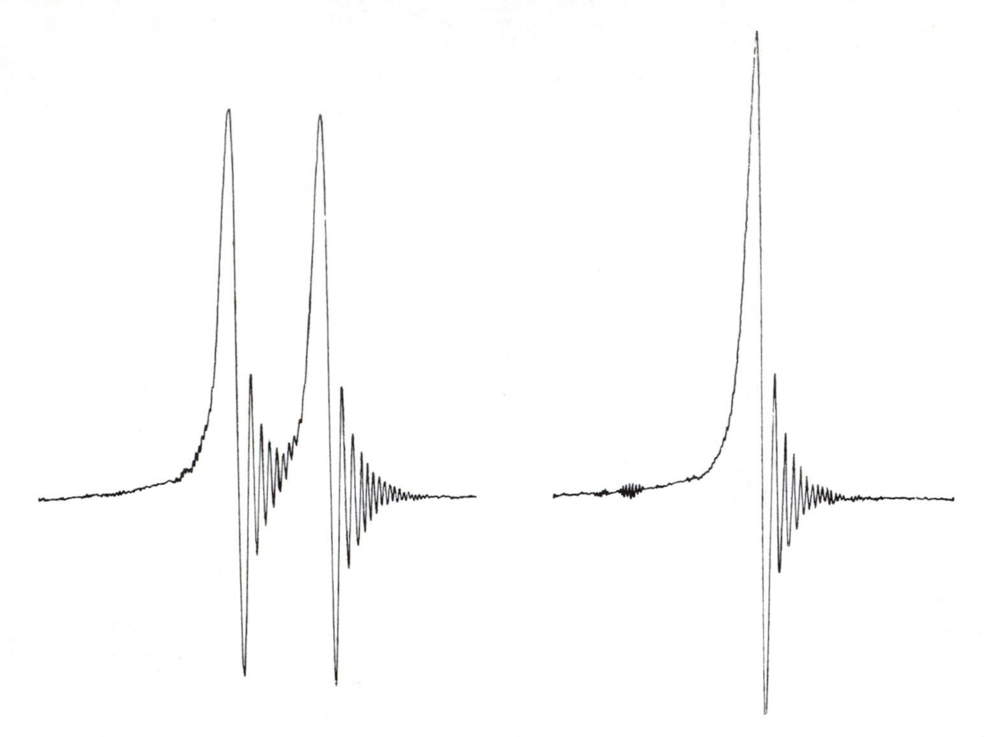

FIGURE 5-5 The proton spectrum (methyl resonance) of $CH_3(C_6H_5)_2P$ at 90 MHz, with (right) and without (left) simultaneous irradiation at the ^{31}P frequency (32.1 MHz).

ethylbenzene may be reduced to a singlet by irradiation at the frequency of the methyl protons. Figure 5-6 shows a typical spin decoupling experiment, in which ν_2 is set at δ4.62 (resonance frequency for H_5 of mannosan triacetate, **5-13**), and ν_1 is swept in the usual fashion. The two frequencies produce an interference

5-13

pattern at the point of coincidence. Spectral simplification is observed in both the H_4 and H_5 resonances. In this fashion, coupling between specific pairs of nuclei may be identified.

Double resonance experiments also may be classified according to the intensity or bandwidth of the irradiating field. A very wide range of frequencies may be irradiated by using either white noise or a broadband oscillator. This technique is commonly used to decouple ^{13}C from all protons (Figure 2-8). If the bandwidth of irradiation covers all the resonance lines of only a single nucleus, coupling is removed to this nucleus alone (Figure 5-6). If the irradiation bandwidth covers only part of the resonance, incomplete decoupling is observed elsewhere in the spectrum. Such techniques, known as *selective spin decoupling* or *spin tickling*, are useful in obtaining relative signs of coupling constants.

In a variation of the noise decoupling technique, the bandwidth of irradiation is moved away from the exact resonance frequencies. If *off-resonance decoupling* is applied to protons, observed ^{13}C resonances still have residual spin–spin coupling. Because multiplicities have not changed, off-resonance decoupling of protons is useful for determining, with minimal spectral overlap, whether carbons are methyl (quartet),

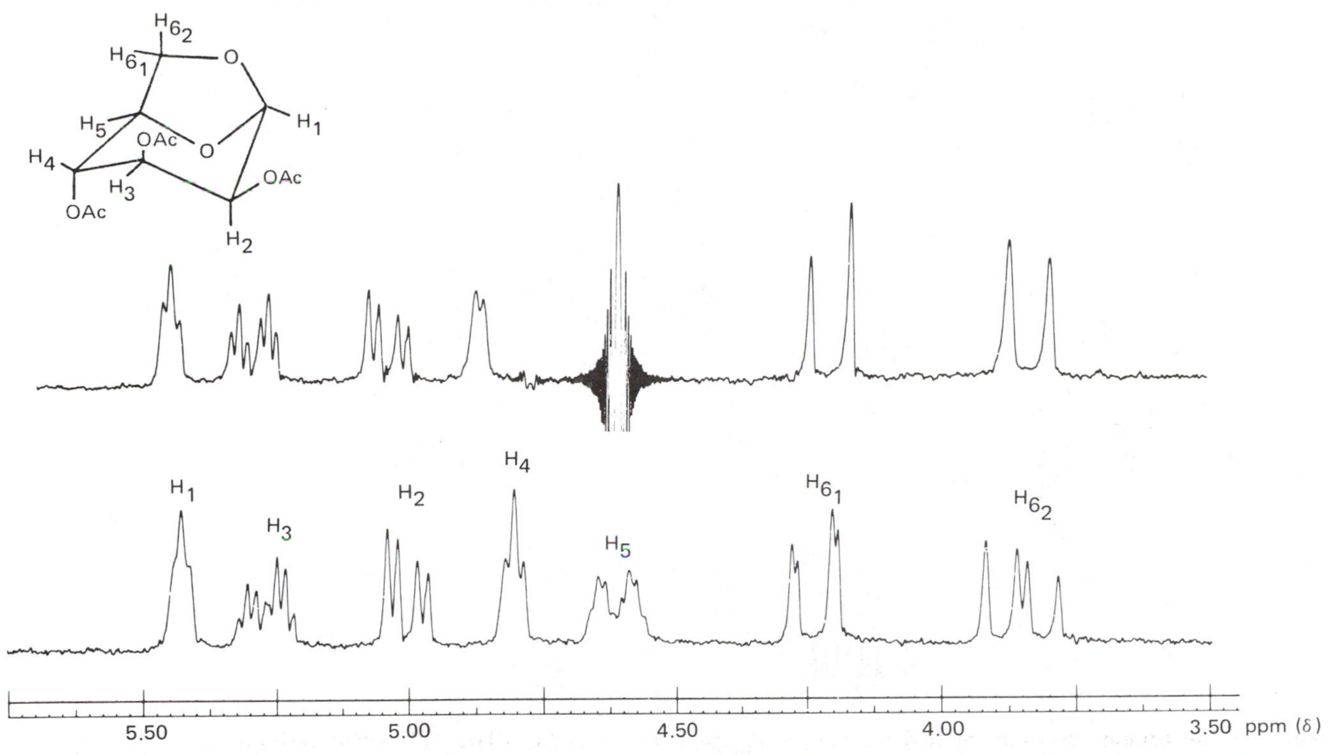

FIGURE 5-6 The 100 MHz proton spectrum of mannosan triacetate (5-13) in $CDCl_3$; without decoupling (lower) and with double irradiation at δ 4.62 (upper). [Reproduced with the permission of Varian Associates.]

methylene (triplet), methine (doublet), or quaternary (singlet). Figure 5-7 shows the spectrum for full decoupling and the residual multiplets for off-resonance decoupling of 3-methylpentane.

With complex molecules, it is useful to record the difference between coupled and decoupled spectra. Features that are not affected by decoupling are subtracted out and do not appear. Figure 5-8 shows the proton spectrum of 1-dehydrotestosterone. The complex region between δ 0.9 and δ 1.1 contains the resonances of four protons. One of these may be identified as 7α by irradiation of the vicinal 6α peak. Comparison of the coupled (a) and decoupled (b) spectra, however, shows little change as the result of double

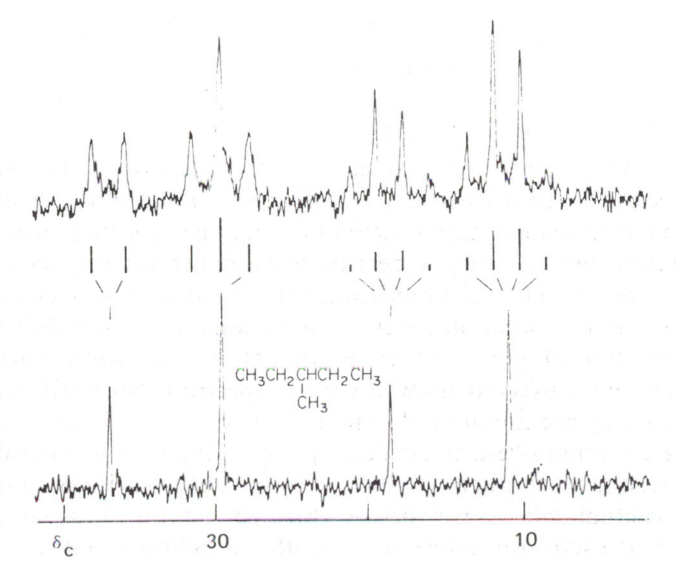

FIGURE 5-7 The fully decoupled (lower) and off-resonance decoupled (upper) ^{13}C spectra at 22.15 MHz of 3-methylpentane. The doublet, triplet, and quartet structures of the muliplets in the latter case readily identify the CH, CH_2, and CH_3 groups. [Courtesy of L. R. Johnson.]

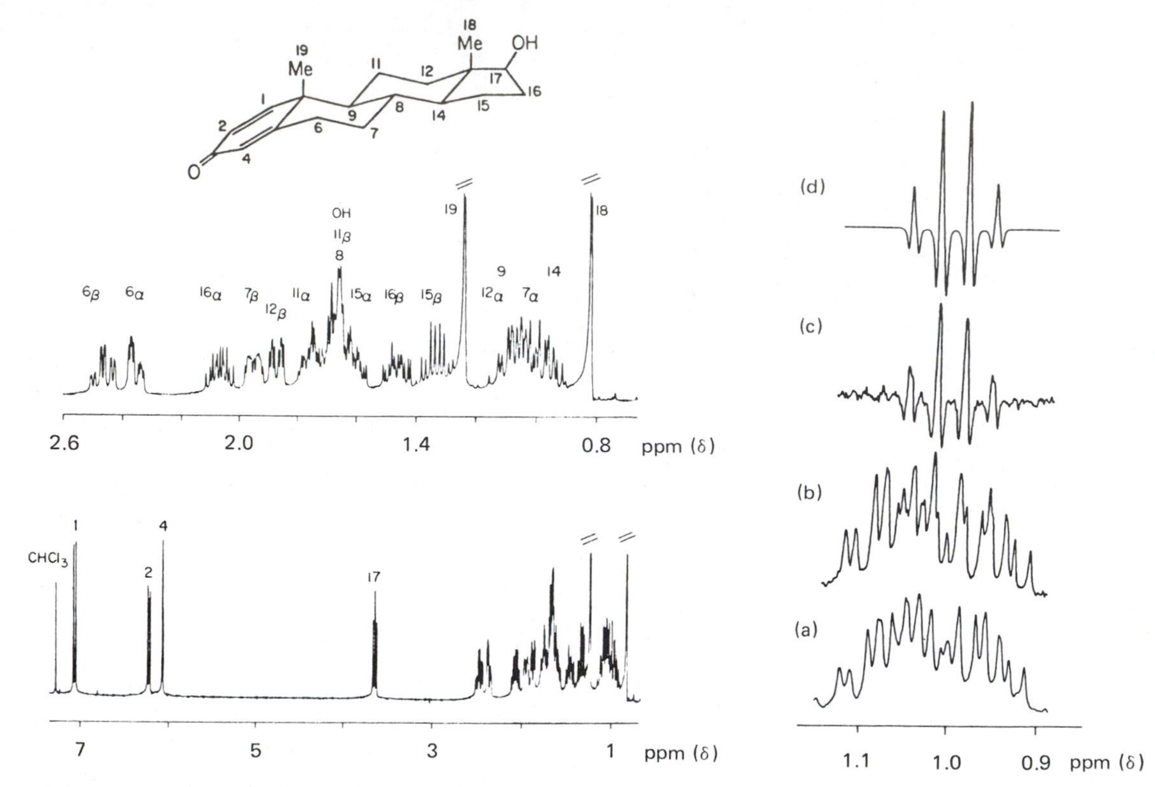

FIGURE 5-8 The 400 MHz spectrum of 1-dehydrotestosterone. The complete spectrum and an expansion of the high-field region are given on the left. On the right are given (a) the coupled spectrum for the δ 0.9–1.1 region, (b) the same region decoupled from the 6α proton, (c) the difference spectrum, and (d) the calculated difference spectrum. [Reproduced with permission from L. D. Hall and J. K. M. Saunders, *J. Am. Chem. Soc.*, **102**, 5703 (1980). Copyright 1980 American Chemical Society.]

irradiation, because of overlapping resonances from other protons. The *difference decoupling spectrum* subtracts (a) from (b) to give (c). The unaffected overlapping peaks are gone; the original resonances with coupling are present as negative peaks; the simpler decoupled resonances are present as positive peaks. The remaining peaks in this region may be identified by difference decoupling of the 7β, 12β, and 15β resonance positions. This procedure provides coupling relationships when spectral overlap is a serious problem.

5-4

Nuclear Overhauser Effect

Frequently two nuclei that are held close to each other by molecular constraints provide the dominant mechanism for mutual relaxation through dipole–dipole interactions. When one of these nuclei is doubly irradiated, the Boltzmann distribution of spins for the other nucleus is altered, and the intensity of its resonance is perturbed. The phenomenon was first observed by Overhauser between unpaired electrons and nuclei, but the *nuclear Overhauser effect*, when both spins are nuclei, is of more interest to the chemist. The phenomenon has great structural utility because the dipole–dipole mechanism for relaxation depends on the distance between the two spins. Qualitative and quantitative conclusions with regard to molecular structure or conformation are possible.

Figure 5-9 depicts the spin energy states for a pair of uncoupled nuclei of spin $\frac{1}{2}$. Nucleus A is irradiated, and nucleus B is observed. When the oscillating B_2 field is applied at the resonance frequency of A, states 2 and 4 are rapidly interconverted, as are 1 and 3. The process of cross relaxation (4 to 1 and 3 to 2) serves to return a B nucleus that is excited to state 4 from 3 back to state 1. Because 1 and 3 are in equilibrium by double

FIGURE 5-9 Spin states for a system of two uncoupled nuclei with double irradiation at the frequency of nucleus A and observation of B.

irradiation, the effective Boltzmann distribution for nucleus B can be determined by the energy difference between states 1 and 4, rather than 1 and 2 as is normally the case. The larger effective energy difference therefore conveys a larger population to the lower spin state. With a larger number of spins able to resonate, the B nucleus produces a signal with a larger intensity.

The maximum attainable increase in intensity depends on the ratio of the gyromagnetic ratios for the A and B nuclei (eq. 5-4), provided that all relaxation is by the dipole–dipole mechanism. Because the intensity is

$$\text{NOE(maximum)} = 1 + \frac{\gamma_A}{2\gamma_B} \tag{5-4}$$

increased by $\gamma_A/2\gamma_B$, this factor is referred to as η_{max}. The increase is less than maximum if other relaxation mechanisms are present or if the observed nucleus is relaxed by nuclei other than the one that is irradiated.

If both nuclei are protons, the intensities can increase by a maximum factor of only 1.5 ($\eta_{max} = 0.5$ or 50%). For the common case in which protons are irradiated and ^{13}C is observed, the NOE can be up to 2.988 ($\eta_{max} =$ 1.988 or about 200%). For irradiation of ^{1}H and observation of ^{31}P, the maximum increase is a factor of 2.24 ($\eta_{max} = 1.24$ or 124%). Certain nuclei have negative gyromagnetic ratios, so that by eq. 5-4 the nuclear Overhauser effect is negative and an inverted peak can result. For irradiation of ^{1}H and observation of ^{15}N, for example, $\eta_{max} = -4.92$ (the peak would be inverted with a maximum intensity 3.92 times that of the original peak, or 292%). If dipolar relaxation is only partial, the NOE can result in a completely nulled resonance when one nucleus has a negative gyromagnetic ratio. It should be emphasized that the Overhauser enhancement is completely independent of spectral changes that arise from the collapse of spin multiplets due to spin decoupling. The observation of the NOE does not require that nuclei A and B be spin-coupled, but only that A relax B through a dipolar mechanism.

The NOE is used for three distinct purposes. First, the increase in sensitivity improves the signal-to-noise ratio. Double irradiation of protons thus assists the ^{13}C spectrum both by collapsing multiplets to more intense singlets but also by enhancing the absolute intensity of the resulting singlets through the NOE. Because most carbons are relaxed almost entirely by attached protons, the Overhauser enhancement commonly attains the maximum value (η_{max}).

The second usage is to distinguish relaxation mechanisms. Because the NOE is determined by the proportion of dipolar relaxation, the size of the NOE can be used to determine how much of the spin–lattice relaxation comes from the dipole–dipole mechanism (eq. 5-5, in which η is the observed NOE minus one).

$$T_1(\text{DD}) = T_1(\text{obs}) \frac{\eta}{\eta_{max}} \tag{5-5}$$

Another way to look at it is that the percent of dipolar relaxation is 100 times η/η_{max}. By eq. 5-3, T_1(other) may be calculated once T_1(DD) is known.

In order to calculate T_1(DD) from eq. 5-5, it is necessary to measure the ^{13}C spectrum both with and without double irradiation, the difference in spectral intensities giving η. The experiment is normally carried out by a *gated decoupling* procedure (Figure 5-10). In the normal experiment (a), double irradiation is carried out continuously, so that both decoupling and NOE are observed. If double irradiation is gated off soon after the observed pulse is completed, as in (b), the NOE does not have sufficient time to develop, since it requires times on the order of the relaxation times. Thus η_{max} may be measured most readily by comparing the intensities of experiments (a) and (b). If there is no double irradiation during the observed pulse, there is no decoupling. By utilizing double irradiation during the long wait period between pulses, the NOE has time

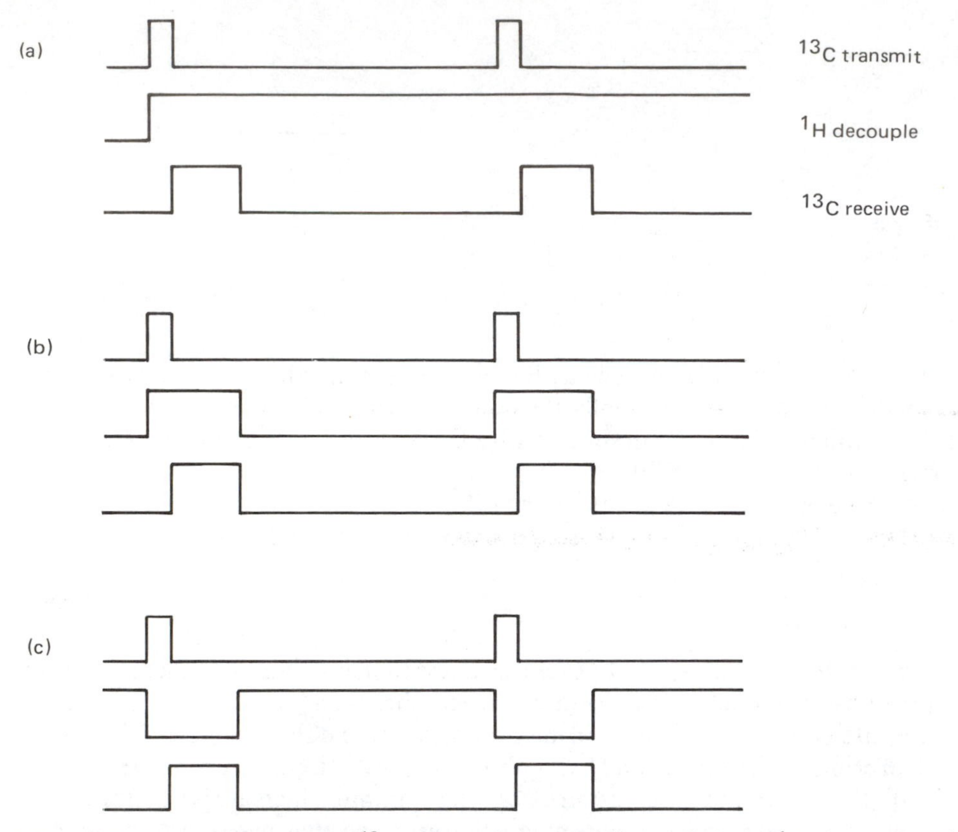

FIGURE 5-10 (a) Observation of ^{13}C with continuous double irradiation of ^{1}H (decoupling plus NOE). (b) Double irradiation applied during observation but gated off during wait period (decoupling, no NOE). (c) Double irradiation applied only during wait period (NOE, no decoupling). The pulse widths are not to scale. The scheme is shown for two cycles.

to develop, so that procedure (c) in Figure 5-10 yields a spectrum with NOE but without decoupling. Procedure (c) might be used to improve intensity while trying to analyze splitting patterns.

Finally, in the third application, the dependence of the NOE on geometry lends itself to analysis of structure and conformation. Two nuclei that are close together in space should exhibit an Overhauser effect. The adenosine derivative **5-14** (2′,3′-isopropylideneadenosine) can exist in a conformation (shown) with the purine ring lying over the sugar ring (syn) or in an extended form with the protons on C8 lying over the sugar

5-14

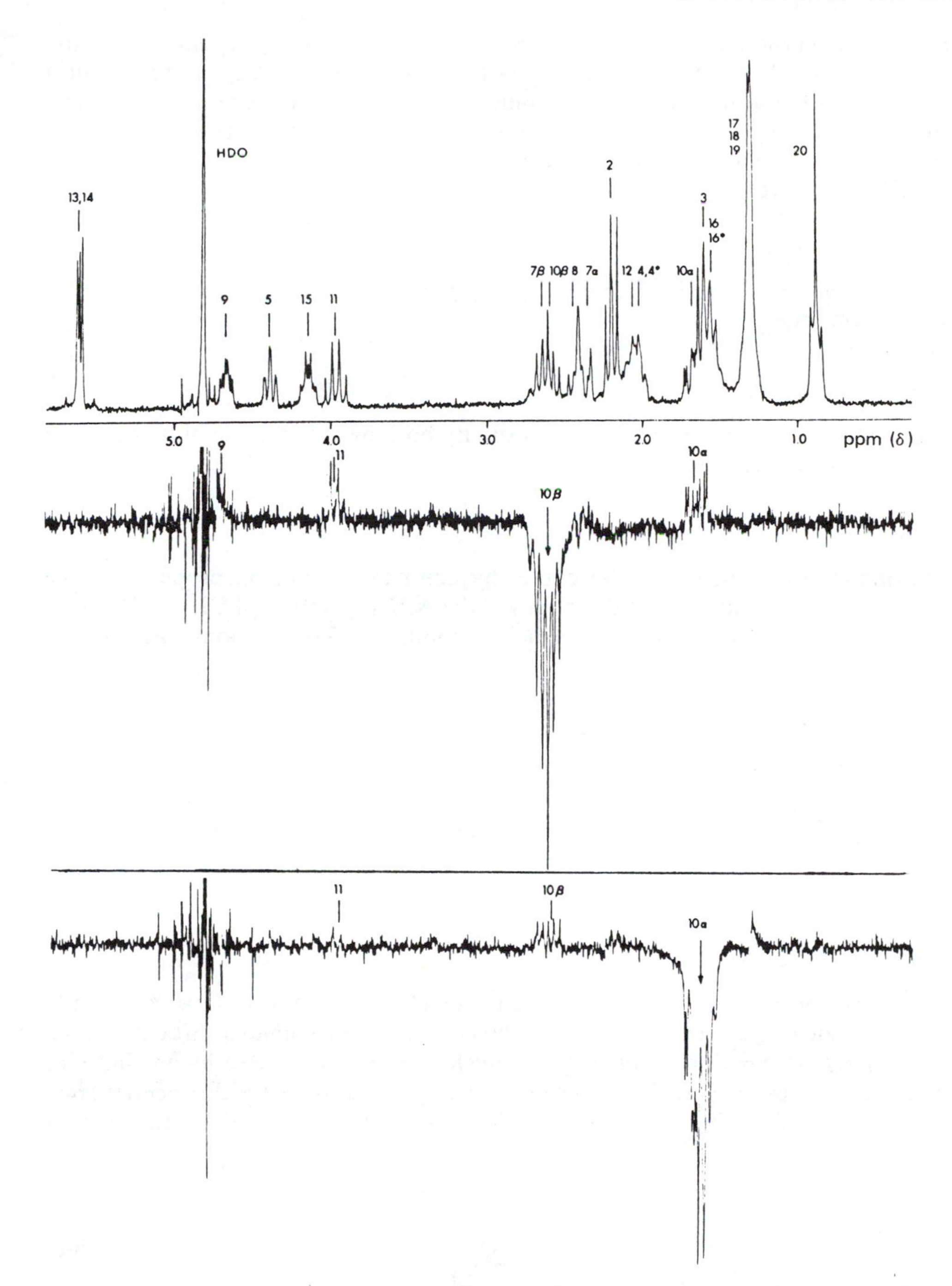

FIGURE 5-11 The proton spectrum of prostacyclin (PGI_2). In the center is the difference NOE spectrum for irradiation of the 10β proton, and at the bottom for irradiation of the 10α proton. [Reproduced with permission from G. Kotovych, G. H. M. Aarts, T. T. Nakashima, and G. Bigam, *Can. J. Chem.*, **58**, 978 (1980).]

(anti). Saturation of the H1′ resonance brings about a 23% enhancement of the H8 resonance. Similarly, saturation of H2′ produces less than a 5% enhancement. The proton at the 8 position, therefore, must be positioned most closely to H1′ but not far from H2′, as in the syn form.

When proton–proton NOEs are measured directly, enhancements of at least 10% are generally required before a positive assignment can be accepted. Smaller enhancements may be observed reliably if the normal and enhanced spectra are subtracted. The *difference NOE spectrum* for prostacyclin (**5-12**) is shown in Figure 5-11, with irradiation of one of the 10 protons. The small enhancements for H9, H11, and the other H10 resonances show that the irradiated proton must be in the 10β position by the following argument. Because the oxygens at 9 and 11 are down (alpha), their geminal protons must be up (beta). These protons in turn would be closest to the irradiated 10 proton, which then also must be beta. The lower experiment confirms the assignment, as irradiation of H10α gives rise to an enhancement only at 10β. The large negative peaks result from the process of spectral subtraction. By the difference procedure, enhancements of only 0.5%, as opposed to 10%, can be reliably observed.

5-5

Applications of Multiple Resonance

5-5a Signal Enhancement

We have already seen that double irradiation can increase sensitivity both by the collapse of mutiplets to fewer peaks and by the nuclear Overhauser effect.

5-5b Spectral Simplification

The complexities of a large spin system can sometimes be reduced by removing the coupling properties of one or more member nuclei. For example, irradiation of the X part of the ABX_3 spectrum of $CH{=}CHCH_3$ in **5-15** reduces the pattern to an AB quartet from which the vicinal AB coupling (15 Hz) demonstrates that the double bond is trans.

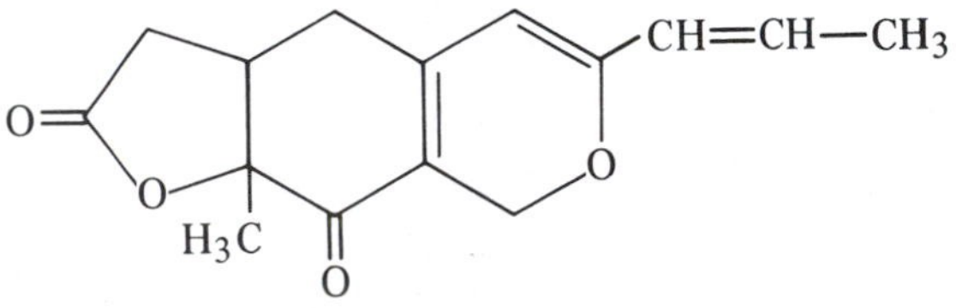

CH=CH—CH_3

O

O

O

H_3C

O

5-15

5-5c Elimination of Quadrupolar Effects

Quadrupolar nuclei that are surrounded by an unsymmetrical electron cloud have a special mechanism for spin–lattice relaxation. Line broadening associated with this relaxation can be eliminated through double resonance procedures. The charge distribution of quadrupolar nuclei can be represented by an ellipsoid, whereas that for spin $\frac{1}{2}$ nuclei resembles a sphere. For a nucleus with a spin of 1, there are three permissible orientations: with, normal to, or against the applied B_0 field (eq. 5-6). An unsymmetrical surrounding

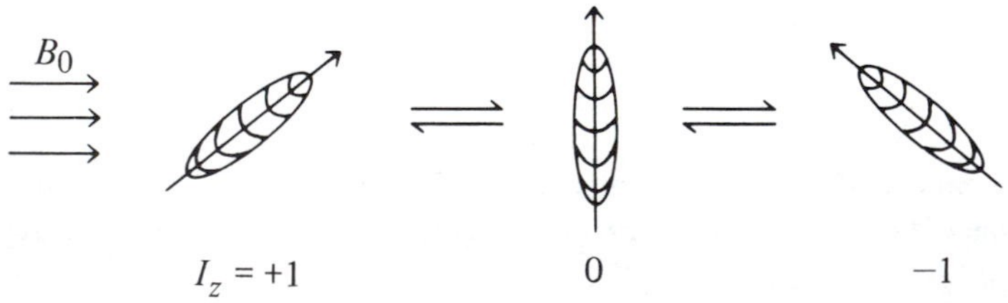

B_0 $\rightleftharpoons$ $\rightleftharpoons$ (5-6)

$I_z = +1$ 0 -1

electron cloud, tumbling in solution, gives rise to a fluctuating magnetic field, which exerts a torque on the quadrupolar nuclear charge. This electrostatic interaction alters the orientation of the nucleus in the magnetic field and thus causes a change in the I_z spin quantum number. Thus the electric field about a tumbling ^{14}N nucleus can bring about transitions among the three spin states ($+1 \rightleftharpoons 0 \rightleftharpoons -1$). If a nucleus has been excited to the -1 state by application of the B_1 field, the interaction between the quadrupolar nucleus and the electron cloud can induce relaxation. Since only the z component of magnetization is affected, the mechanism is a spin–lattice relaxation.

For quadrupolar relaxation to take place, both the nucleus and the electron cloud must have unsymmetrical charge distribution. Thus nuclei in highly symmetrical environments (^{14}N in $^{+}NH_4$, ^{10}B in $^{-}BH_4$, or ^{35}Cl in Cl^-) have no quadrupolar relaxation. Because ^{2}H has only s electrons, the electric field is relatively symmetrical, and the quadrupolar effect is small. For nuclei with a large quadrupole moment in highly unsymmetrical environments, T_1 can become extremely short with the result that linewidths are very broad. This problem is particularly important in ^{17}O and ^{14}N spectroscopy.

More important to the organic chemist often is the effect of quadrupolar nuclei on the resonances of other nuclei in the molecule. If relaxation is extremely rapid, a neighboring nucleus experiences only the average spin environment of the quadrupolar nucleus, so that no spin coupling is observed. Thus protons in chloromethane produce a sharp singlet even though ^{35}Cl and ^{37}Cl nuclei have spins of $\frac{3}{2}$. At the other extreme, deuterium has a weak quadrupole moment and relatively symmetrical electric fields, so that scalar (J) coupling of protons to neighboring deuterium atoms (spin of 1) can be observed.

The ^{14}N nucleus falls between these two extremes. In some cases, such as the interior nitrogen in biuret, $NH_2(CO)NH(CO)NH_2$, quadrupolar relaxation is rapid enough to produce a sharp singlet for the attached proton. The protons of the ammonium ion on the other hand give a sharp triplet with full coupling between ^{1}H and ^{14}N. When relaxation is at an intermediate rate, the three peaks can broaden out to the point of being unobservable. Irradiation at the ^{14}N frequency interconverts the three spin states rapidly, so that the spectrum appears as if ^{14}N were nonmagnetic. The ^{1}H{^{14}N} spectrum of pyrrole (Figure 5-12) shows the sharp resonances of the proton on nitrogen.

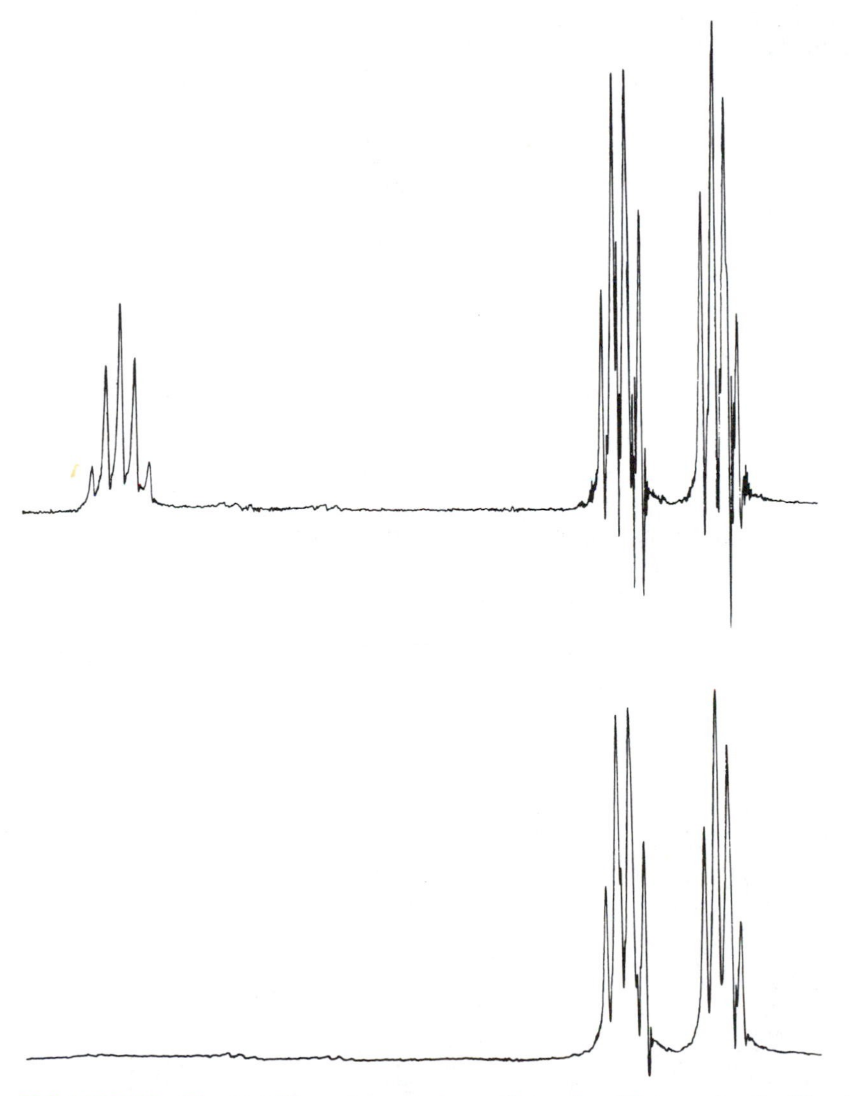

FIGURE 5-12 The 90 MHz proton spectrum of pyrrole with (upper) and without (lower) ^{14}N decoupling.

5-5d Connectivity

The collapse of vicinal couplings through double irradiation signifies that protons are attached to adjacent carbon atoms. Substructures of molecules can be identified by the appropriate double irradiation experiments, and sometimes a series of such experiments can map a significant portion of a molecule. These types of experiments are done more comprehensively by two-dimensional methods (Section 5-7).

5-5e Relative Signs of Coupling Constants

The relative signs of couplings, for example, in an AMX system, can be determined by selective spin decoupling. These experiments, although necessary in the historical development of NMR, are rarely carried out today. Descriptions may be found in the references at the end of the chapter.

5-5f Chemical Exchange

Double irradiation of a nucleus that can undergo chemical exchange between two sites provides useful dynamic information. When site A is doubly irradiated, the upper and lower spin states tend toward equal population (saturation). If a nucleus with identity A undergoes chemical exchange during double irradiation, and becomes a B form, the B nuclei will to some extent be saturated, provided that spin–lattice relaxation is not exceedingly fast. Thus irradiation at A will cause loss of intensity at B. By this means, partners in chemical exchange may be identified, mechanisms of exchange may be tested, and even the kinetics of exchange may be drived from peak intensity changes. This experiment is variously called *saturation transfer*, magnetization transfer, or the Forsén–Hoffman procedure.

5-6 Multiple Pulse Sequences

Many of the recent advances in NMR spectroscopy have resulted from manipulation of spin information through the use of several pulses. We have already seen (Section 5-2) that the use of a 180° pulse followed by a 90° pulse can be used to measure T_1. In this section we will describe several other uses of multiple pulse sequences.

5-6a Measurement of T_2

Many of the modern pulse sequences are patterned after experiments of Hahn, Carr, Purcell, Meiboom, and Gil in the 1950s. Figure 5-13 shows a pulse sequence that can refocus the effects of field inhomogeneity on

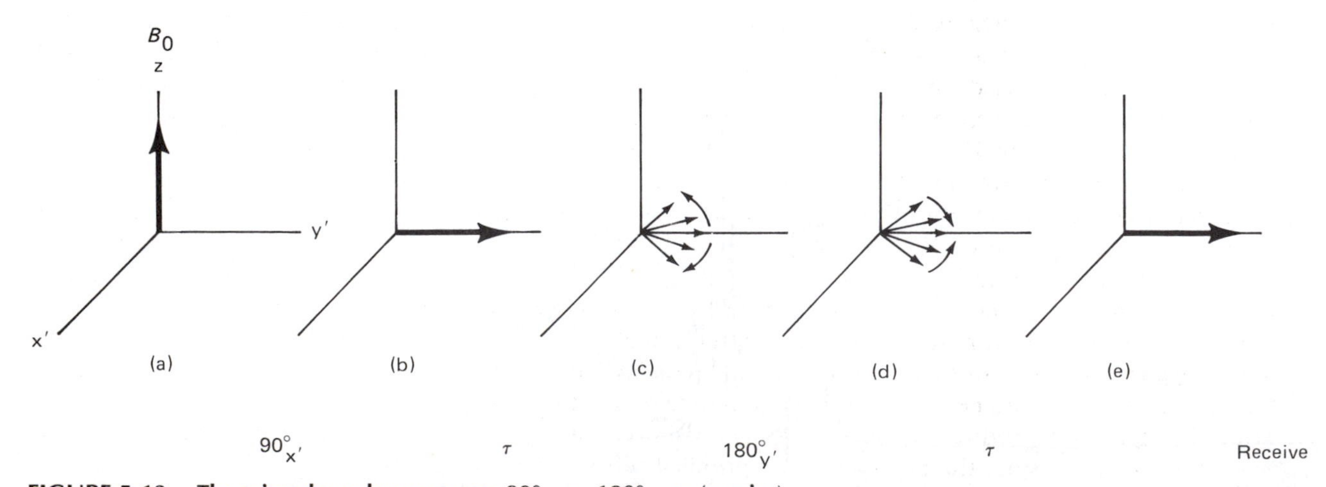

FIGURE 5-13. The spin echo pulse sequence, $90^\circ_{x'}-\tau-180^\circ_{y'}-\tau-$(receive).

magnetization in the xy plane. In diagram (a), the thick arrow shows that at the beginning there is an excess of z magnetization. A strong pulse along the x axis rotates the magnetization vector onto the y axis. Since nuclear magnets precess at the Larmor frequency ω_0, this vector moves in a circular fashion in the xy plane at this frequency, until it spirals back to the original state (a) through T_1 and T_2 relaxation. It is pictorially simpler to allow the xy axes to rotate about the z axis at ω_0, so that the magnetization vector remains stationary after the 90° pulse, as in (b). In this *rotating coordinate system*, the vector diminishes and moves back to the z axis by T_1 and T_2 processes. The rotating coordinates are labeled with primes to signify use of the alternative system.

If τ is short, the magnetization vector diminishes only slightly. Moreover, as in Figure 5-13c, if homogeneity is not perfect, nuclei with slightly different Larmor frequencies start to fan out in the xy plane, some faster, others more slowly than the frequency of the rotating coordinates. If a 180° pulse is applied along the y' axis, all vectors have the signs of their x' coordinate changed. Those still on the y' axis are not affected by the collinear pulse. Those with a $+x'$ component, caused by inhomogeneities, are flipped to the analogous $-x'$ position. Because they continue rotating in the same direction, they now are moving toward, rather than away from the y' axis, as in (d). Similarly, those with a $-x'$ component are flipped to $+x'$ and also move back to the y' axis. After an identical time τ, all the vectors come together (refocus) along the y' axis, as in (e). Sampling of magnetization at this time gives a value that is independent of magnet inhomogeneity. This pulse sequence, $90^\circ_{x'}-\tau-180^\circ_{y'}-\tau$, can be repeated to give a spin echo every 2τ period. The intensity of the echo dies off with a first-order rate constant $1/T_2$, in which T_2 is the true spin–spin relaxation time and does not depend on magnet inhomogeneity.

5-6b Spectral Editing

The model of Figure 5-13 assumes that the spin system contains no coupling. The A nucleus of an AX system gives two peaks at $\nu_0 \pm \frac{1}{2}J$. The two magnetization vectors corresponding to these two A peaks rotate, respectively, faster and more slowly than the Larmor frequency (ν_0) by $\frac{1}{2}J$. Spin echo experiments can be constructed that take advantage of the phase relationships between the different vectors of CH, CH_2, and CH_3 groups in the ^{13}C spectrum. Details of the pulse sequences are given in the references on spectral editing at the end of the chapter. One such sequence, known as DEPT (distortionless enhancement by polarization transfer), can provide separate spectra of the CH, CH_2, and CH_3 groups, as shown in Figure 5-14 for the trisaccharide gentamycin. It can be seen that there is only one type of methyl, at least four methylenes, and a large number of methines. Spectral editing of this type serves the same purpose as off-resonance decoupling but avoids problems of spectral overlap. Thus its utility increases as the complexity of the molecule increases.

5-6c Signal and Resolution Enhancement

A number of important nuclei, including ^{13}C and ^{15}N, have very low natural abundance and natural sensitivity. Pulse sequences have been devised to improve the sensitivity of these nuclei when they are coupled to another nucleus of high receptivity, usually the proton. Sequences of pulses are applied to both the insensitive, observed nucleus and to the sensitive, irradiated nucleus. The spin vectors are manipulated in such a way that the favorable polarization of the sensitive nucleus is transferred to the insensitive nucleus. A common sequence devised for this purpose is called INEPT, for insensitive nuclei enhanced by polarization transfer, and is most effective when the two nuclei are directly bonded. Figure 5-15 shows the enhancement in sensitivity achieved by INEPT for pyridine. The pulse sequence may be adjusted so that all the carbons are fully decoupled from protons. In the spectral display half the peaks are inverted as a result of the pulse sequence, but the decoupling procedure provides normally oriented singlets. INEPT has been used very successfully for observation of ^{15}N and ^{29}Si, as well as ^{13}C. The sequence DEPT also gives enhancement through polarization transfer, and some workers prefer DEPT because the pulse parameters are less sensitive to exact knowledge of the value of the coupling constant between the observed and irradiated nuclei.

Resolution may be improved through manipulation of pulse parameters. The process may be as simple as applying an exponential or linear decay function to the free induction decay, in order to alter the effective T_2. Less peak distortion can be achieved through more complex pulse sequences. The procedures can decrease linewidths and even produce peaks where only inflection points were present. Of course they cannot invent new information, so that the procedures are primarily cosmetic.

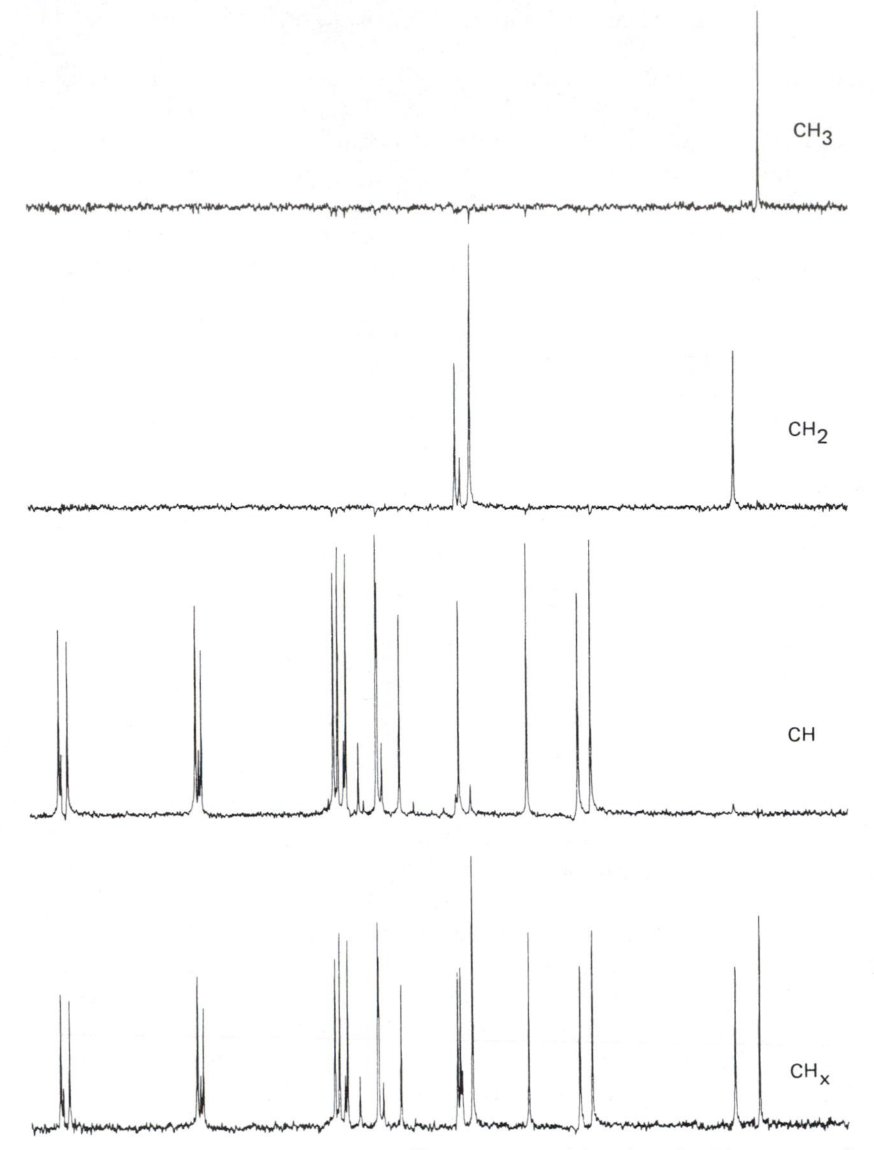

FIGURE 5-14 Spectral editing of the ^{13}C spectrum of the trisaccharide gentamycin, by the DEPT sequence, in $CHCl_3$ at 75.6 MHz. The bottom spectrum shows all carbons with attached protons. [Courtesy of Bruker Instruments, Inc.]

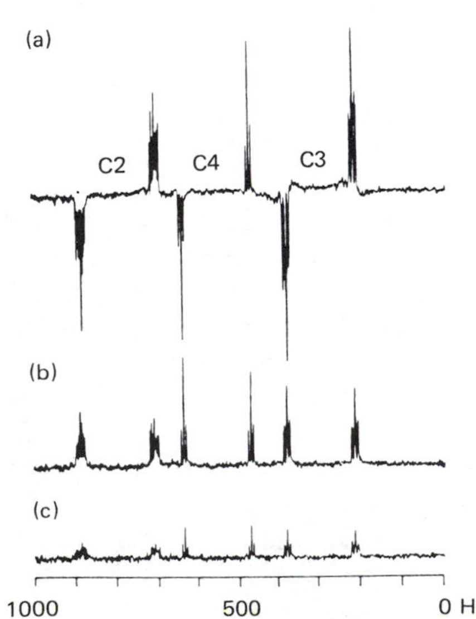

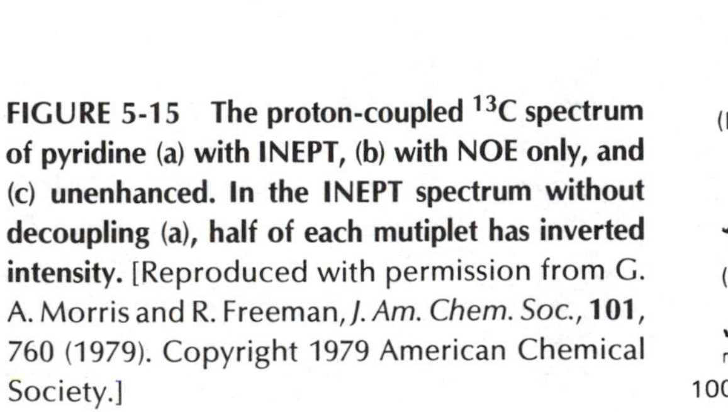

FIGURE 5-15 The proton-coupled ^{13}C spectrum of pyridine (a) with INEPT, (b) with NOE only, and (c) unenhanced. In the INEPT spectrum without decoupling (a), half of each mutiplet has inverted intensity. [Reproduced with permission from G. A. Morris and R. Freeman, *J. Am. Chem. Soc.*, **101**, 760 (1979). Copyright 1979 American Chemical Society.]

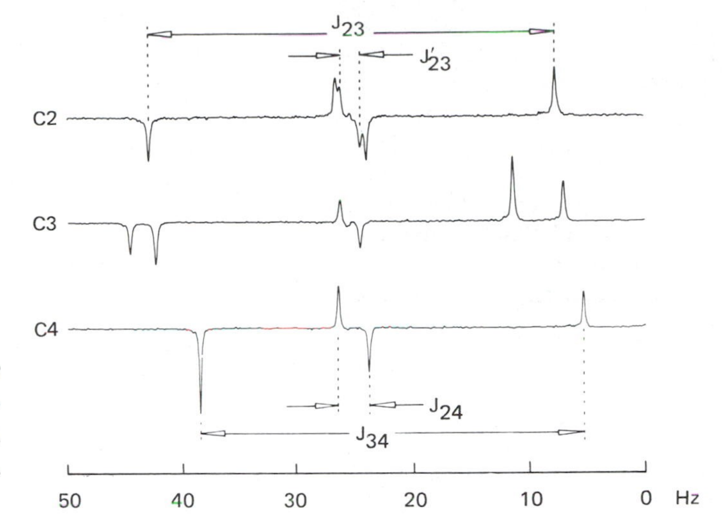

FIGURE 5-16 The INADEQUATE spectrum for the carbons of piperidine. [Reproduced with permission from A. Bax, R. Freeman, and S. P. Kempsell, *J. Am. Chem. Soc.*, **102**, 4849 (1980). Copyright 1980 American Chemical Society.]

5-6d Connectivity

There is a coupling constant between any two directly bonded carbon-13 nuclei that can be detected in the ^{13}C satellite of the ^{13}C spectrum. The satellites are on the order of 1% of the natural abundance spectrum, so for the most part they are obscured by the regular resonances. The pulse sequence INADEQUATE (incredible natural abundance double quantum transfer experiment) was developed to tune out the regular (single quantum) resonances and exhibit only the satellites (double quantum peaks). Figure 5-16 shows the double quantum spectra for the three carbons of piperidine. Each ^{13}C—^{13}C coupling constant is represented by a pair of peaks, one up, one down. Thus the INADEQUATE spectrum for C4 of piperidine has two such doublets, a large one for $^1J_{34}$ and a small one for $^2J_{24}$. For C3, there are two large doublets because the directly bonded couplings ($^1J_{23}$ and $^1J_{34}$) to the adjacent carbons are slightly different. There also is a small long range coupling $^3J_{23'}$ between C2 and the nonadjacent C3. The spectrum for C2 shows $^1J_{23}$, $^2J_{24}$, and $^3J_{23'}$.

Although longer range couplings are observable, the major parameters of interest are the directly bonded ^{13}C—^{13}C couplings. The single quantum center bands are entirely absent. The one bond couplings vary slightly for essentially every C—C bond. Thus a match of a $^1J(^{13}C—^{13}C)$ for two carbons suggests with high probability that the two carbons are connected. Even in complex molecules, there is sufficient variability of couplings that INADEQUATE can be used to map the complete connectivity of all carbons.

The major drawback to the INADEQUATE experiment is its low sensitivity. The experiment is done on 0.01% of the carbons. An alternative is provided by the RELAY experiment. This pulse sequence explores connectivity of carbons through vicinal proton–proton couplings. Both protons and carbons are pulsed, with the result that vicinally coupled protons transfer magnetization to carbon. In order to display the connectivity of two carbons bearing vicinally coupled protons, a two-dimensional display is required, as described in the next section.

5-7

Two-Dimensional NMR

In the one-dimensional FT experiment described up to this point, a 90° pulse is followed by a time-dependent period during which the free induction decay develops. Fourier transformation of the time variable into a frequency variable provides the familiar spectrum of δ values. If this 90° pulse is preceded by another 90° pulse (Figure 5-17), certain relationships between spins can evolve prior to the observation pulse. Thus during the constant period t_1, nuclei may undergo chemical exchange or mutual dipolar relaxation, or they may simply differentiate according to Larmor frequencies. When the Fourier transformation is carried out following the variable time period t_2, the spectrum reflects events that took place during the evolution period.

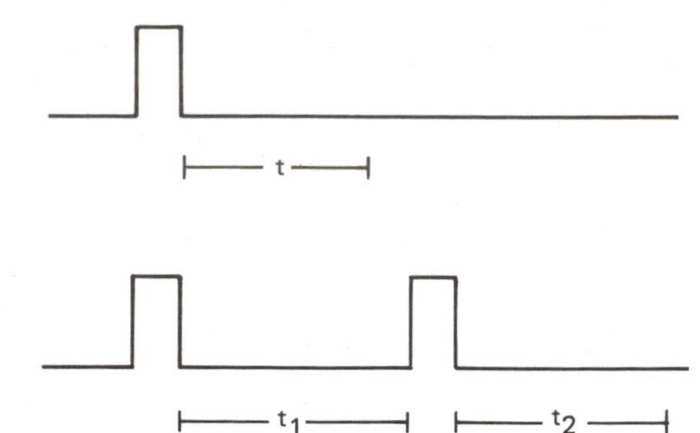

FIGURE 5-17 The pulse arrangement for a single cycle of one-dimensional (upper) and two-dimensional (lower) NMR. In this diagram each pulse is 90°. There are many other, more complex two-dimensional pulse sequences.

This experiment may be repeated n times, each with a different value of t_1. Each of these experiments would reflect the events during the evolution period in a slightly different fashion. Since t_1 now is a time variable, the total set of data may be transformed with respect to t_1, as well as t_2, with the result that two frequency domains, f_1 and f_2, are obtained. The exact nature of the frequency domains depends on the actual choice of pulses and on the range of frequencies. In this section we describe the experimental results of several two-dimensional NMR experiments. Such approaches have become very common in the analysis of complex organic and bioorganic molecules. Because of the wide variety of 2D experiments now available, it is not unusual to apply several of them to a single problem.

If one time domain is selected to correspond to resonance frequencies of one nucleus, and the other time domain to resonance frequencies of another nucleus, the result is a *chemical shift correlated 2D spectrum.* Figure 5-18 shows the ^1H–^{13}C 2D spectrum for the trisaccharide raffinose. Ordinarily, for steriods and many sugars, the saturated region is an envelope of undifferentiated resonances, with an occasional methyl spike. In the 2D spectrum, the proton peaks have been pulled into the second dimension provided by the ^{13}C shifts. The 2D experiment is equivalent to an entire set of heteronuclear decoupling experiments. The relationship between every coupled pair of protons and carbons is provided. Similar chemical shift correlated 2D spectra have been examined for ^{15}N, ^{31}P, and ^{183}W. Although it is not apparent from this discussion, correlation between chemical shifts requires the presence of a scalar (J) coupling between the two nuclei. The actual set of pulses is complex and involves pulsing both the ^1H and ^{13}C frequencies. Since quaternary carbons are not coupled to any protons, they do not appear in the 2D spectrum. Any peak in the stack plot of Figure 5-18 gives the ^{13}C shift to the right and the ^1H shift below. Thus each peak represents a coupled pair of nuclei. The one-dimensional spectra can be obtained by projection along an axis.

The set of t_1 values also may be chosen to correspond to coupling constants, although again a somewhat more complex set of pulses is required than that depicted in Figure 5-17. The double Fourier transformation results in a 2D array like that given in Figure 5-19 for the indicated glucose derivative. Such an experiment gives a *2D J-resolved spectrum*. The normal ^1H frequencies are found along the x (f_2) axis. The y (f_1) axis shows only proton–proton coupled multiplets. Thus the furthest downfield multiplet, from H3, is a quartet, with further splitting. All the f_1 multiplets are displayed symmetrically about a zero point. The projection (a) at the top of the spectrum provides the normal one-dimensional proton spectrum. By taking a projection at an angle (45°) that causes each of the members of the individual multiplets to overlap, as at the bottom (b), a display is obtained that in essence is a proton–proton decoupled proton spectrum. Resonances are present at each ^1H frequency, but are devoid of any couplings.

There are a number of advantages to 2D J-resolved spectra. The projection at the bottom is clearly an entirely new way to examine proton spectra. In addition, the pulse sequence, like the spin echo sequence, results in refocusing of spins whose frequencies spread out as the result of field inhomogeneities. The inset at the lower right of Figure 5-19 shows the one-dimensional spectra of the H4 and H5 resonances at the top (c and e) and the projection from the 2D J-resolved spectra (d and f, extracted from a). The much higher resolution of the 2D spectra is clearly evident. One drawback to this approach is that poor results are obtained when two protons are closely coupled (second order). Thus best results require use of the highest available field.

The simple 2D pulse sequence shown in Figure 5-17 is used to detect scalar (J) interactions between the entire set of spins of a single nuclide. Because a given spin will have the same resonance frequency ($f_1 = f_2$) in

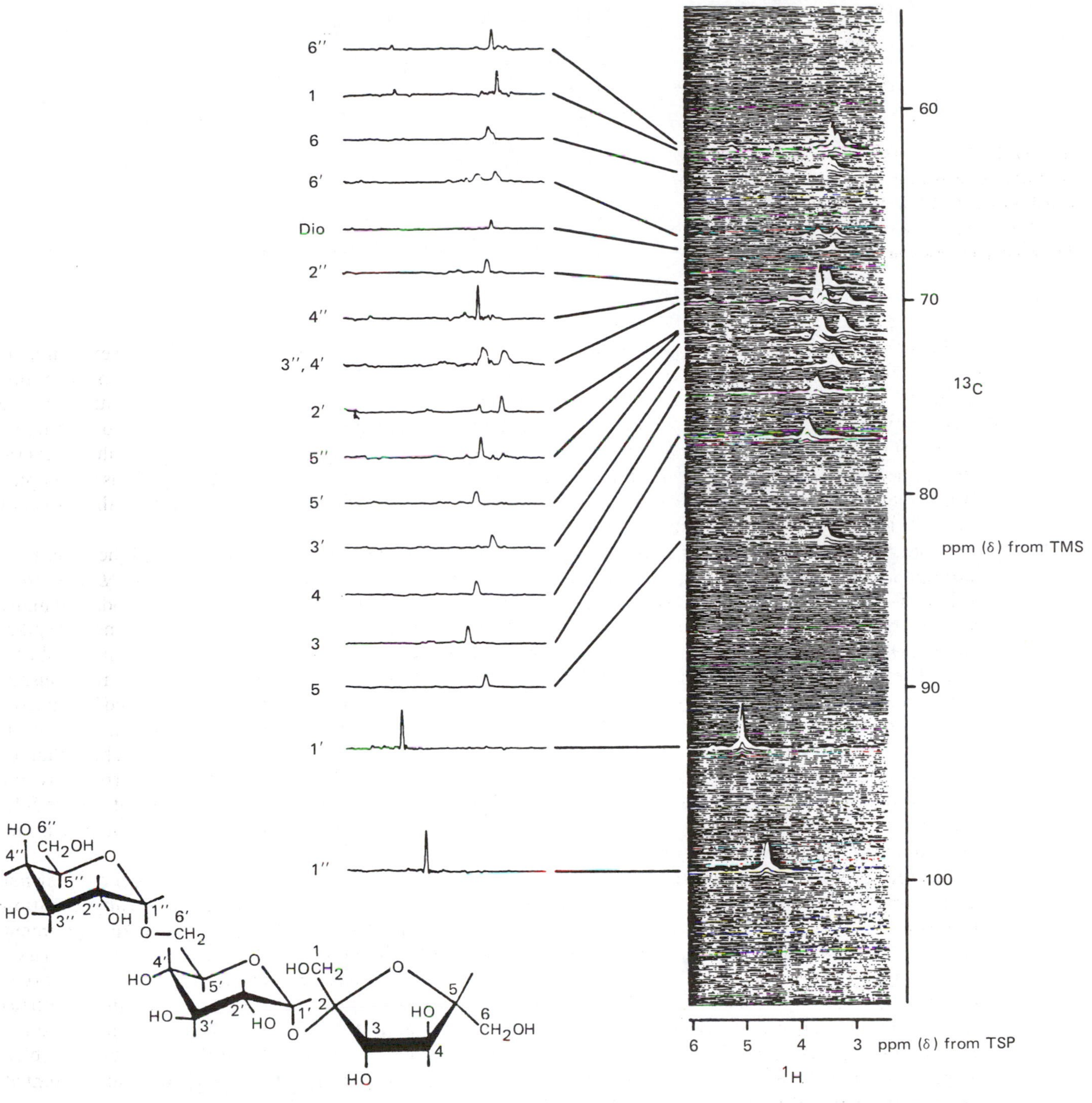

FIGURE 5-18 The $^1H/^{13}C$ two-dimensional spectrum of raffinose. In the 2D spectrum on the right, the *x* axis is the ^1H frequency and the *y* axis is the ^{13}C frequency. On the left, individual lines corresponding to a single ^{13}C frequency have been pulled out of the 2D display. [Reproduced with permission from G. A. Morris and L. D. Hall, *J. Am. Chem. Soc.*, **103**, 4703 (1981). Copyright 1981 American Chemical Society.]

both spectral domains, the normal (one-dimensional) spectrum is found on the diagonal of the 2D plot, as in Figure 5-20 for an annulene derivative. This particular mode of plotting is called a projection, rather than a stack, since only the outline of the base of the peaks is shown. Because the magnetization of a given nucleus Y (coupled to X) in domain f_1 is influenced by the presence of X, an off-diagonal peak is found at the point

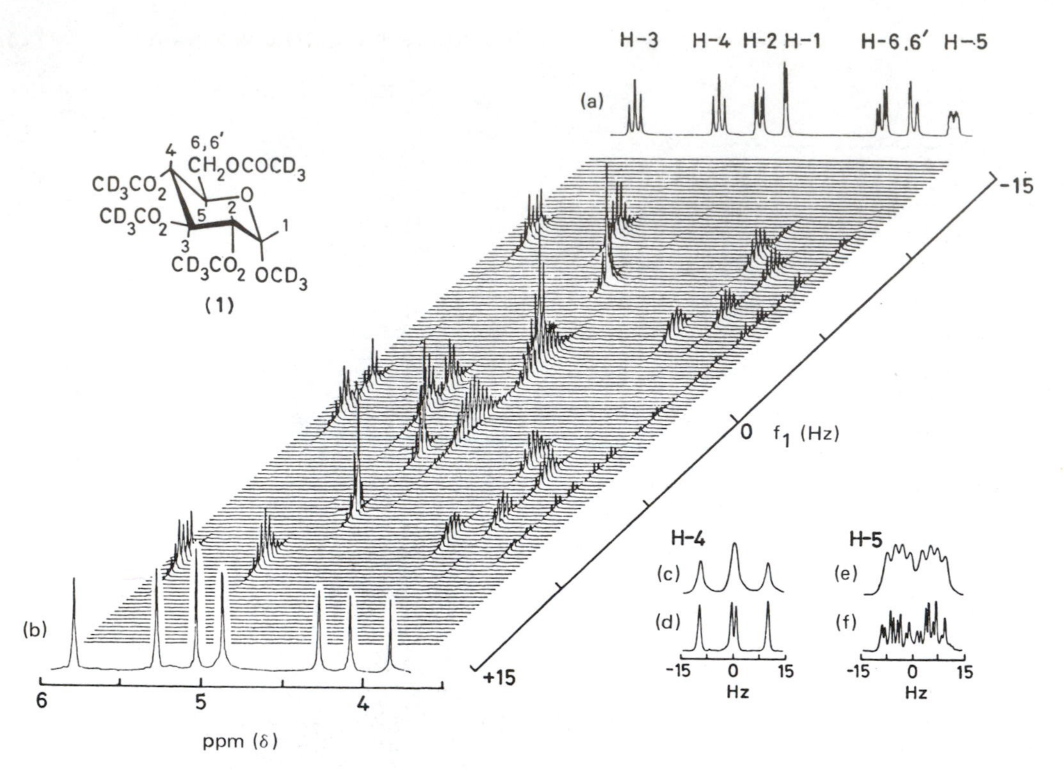

FIGURE 5-19 The 270 MHz two-dimensional *J*-resolved proton spectrum of 2,3,4,6-tetrakis-*O*-trideuterioacetyl-α-D-glucopyranoside. [Reproduced from L. D. Hall, S. Sukumar, and G. R. Sullivan, *J. Chem. Soc. Chem. Commun.*, 292 (1979).]

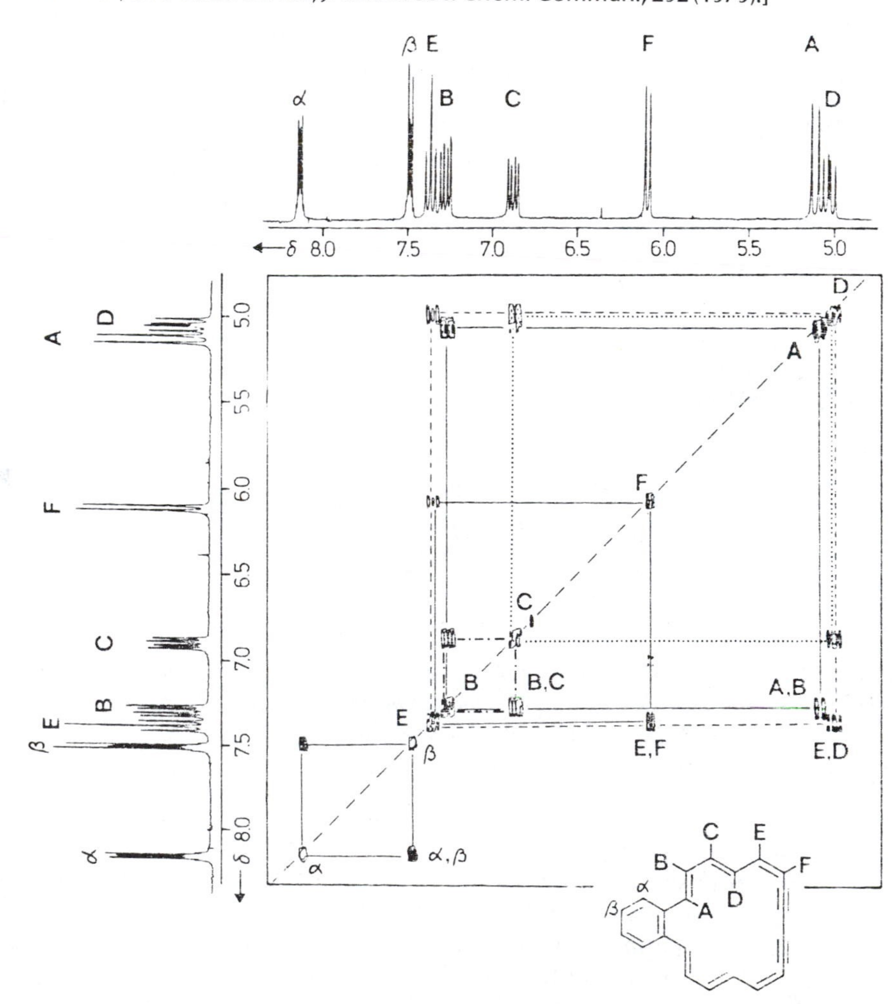

FIGURE 5-20 The COSY ¹H plot for 9,11-bisdehydrobenzo[18]annulene. [Reproduced with permission from R. Benn and H. Günther, *Angew. Chem., Intern. Ed. Engl.*, **22,** 350 (1983).]

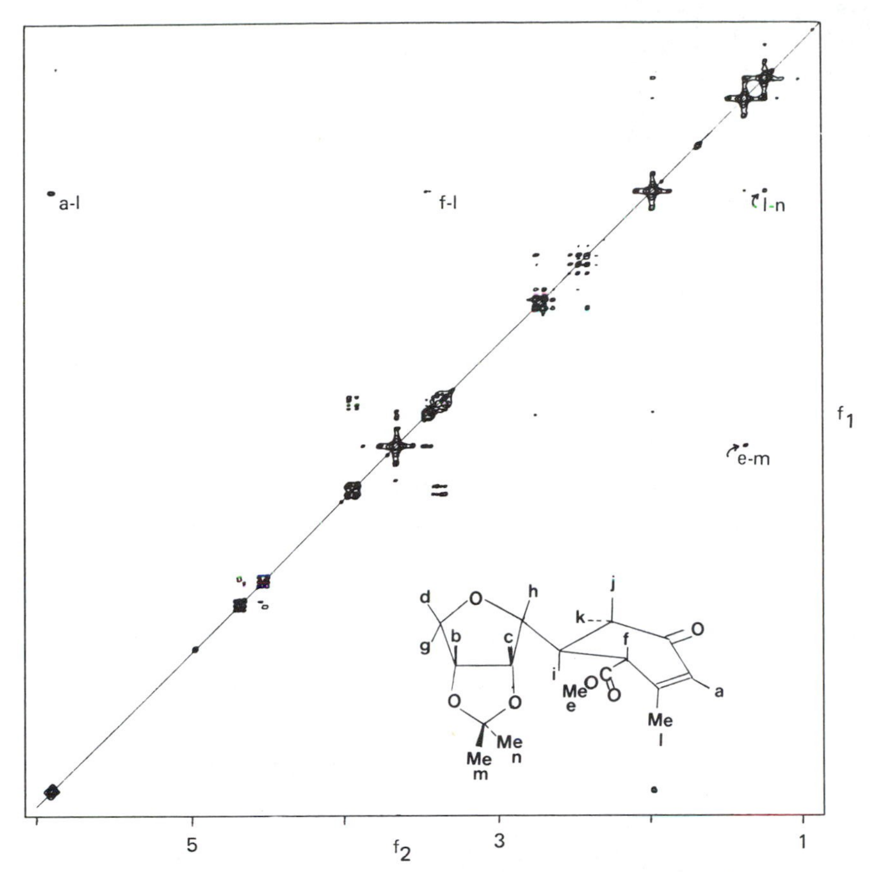

FIGURE 5-21 The ^1H NOESY spectrum for the indicated compound. [Courtesy of Bruker Instruments, Inc.]

with frequency X in one domain and frequency Y in the other domain. For example, in Figure 5-20, off-diagonal peaks are found at the locations marked (E, F), and so on, indicating coupling between two nuclei. Note that the projection along either frequency axis produces the same one-dimensional spectrum. This procedure is called COSY, for correlated spectroscopy, since nuclei with correlated scalar interactions give off-diagonal peaks. COSY in essence is a single experiment that locates all first-order couplings, just as a whole series of homonuclear decoupling experiments would. The COSY experiment currently is an essential aspect of the ^1H NMR analysis of a complex molecule. The SECSY (spin echo spectroscopy) variant provides the same information, but the one-dimensional spectrum is on a horizontal axis and the COSY off-diagonal resonances are located above and below.

In the structural analysis of complex molecules, it is useful to know not only which atoms are coupled to which but also which atoms are close together in space. The nuclear Overhauser experiment provides this information, and the NOESY (nuclear Overhauser effect spectroscopy) pulse sequence gives a complete 2D display of the Overhauser interactions. Again, the 1D spectrum is given on the diagonal, and the presence of Overhauser enhancements between pairs of protons is indicated off the diagonal, as in Figure 5-21. The methyl group l, for example, is demonstrated to be close in space to atoms a, f, and n.

Exchange between two sites also can be represented in two dimensions by the same pulse sequence as for NOESY but now referred to as EXTASY (exchange interaction spectroscopy). The two procedures are sensitive to different time ranges. Off-diagonal peaks are found here only for those sites that undergo chemical exchange, as in Figure 5-22 for the 1,2 methyl shift in the heptamethylbenzenium ion. Thus exchange is seen of site 3 with both 2 and 4, but site 1 exchanges only with site 2. The EXTASY sequence provides the same information as saturation transfer, but does so for all spins in one experiment.

The INADEQUATE and RELAY experiments are best carried out with 2D displays. Thus in the complete analysis of a complex molecule, several 2D experiments are needed, probably including ^1H–^{13}C chemical shift correlation, *J* correlation, COSY, NOESY, and one of the connectivity sequences such as

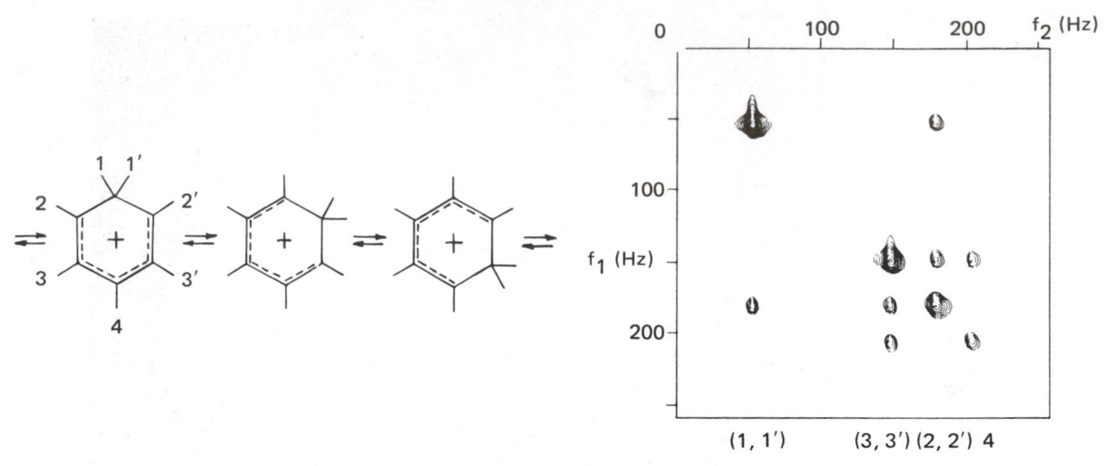

FIGURE 5-22 The ^{1}H EXTASY spectrum for the heptamethylbenzenium ion, given as a projection plot. [Reproduced with permission from R. H. Meier and R. R. Ernst, *J. Am. Chem. Soc.*, **101**, 6441 (1979). Copyright 1979 American Chemical Society.]

INADEQUATE. Spectral assignments and structural analysis are now possible for molecules with molecular weights in the tens of thousands.

5-8

NMR Imaging and Topical NMR

For a generation of chemists, the major purpose of NMR has been to solve structural and kinetic problems through the production of high resolution spectra. In 1973, Lauterbur described a method for using NMR to obtain images of objects. This area has grown in importance rapidly, and today it threatens to compete with X-ray methods for medical imaging. The major difference beween imaging and high resolution NMR spectroscopy is the former's use of field gradients. Figure 5-23 illustrates the principle. If a gradient on the

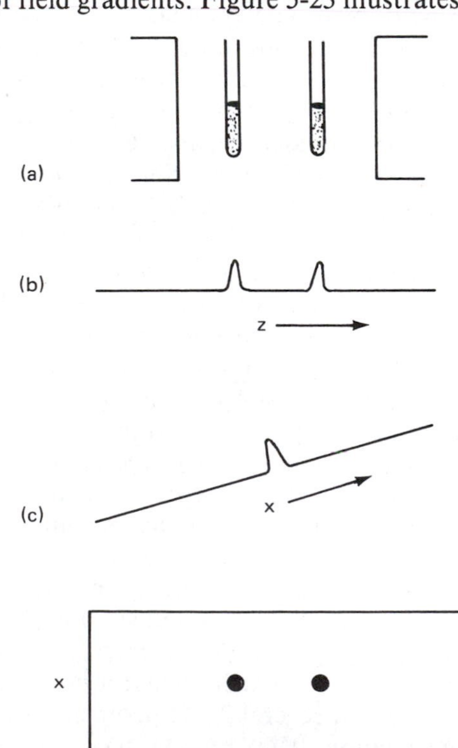

FIGURE 5-23 (a) Two NMR sample tubes immersed in a magnetic field. (b) The image from a gradient along the *z* direction (the axis from one pole face to the other, from left to right). (c) The image from a gradient along the *x* direction (the axis from front to back). (d) The image in the *xz* plane.

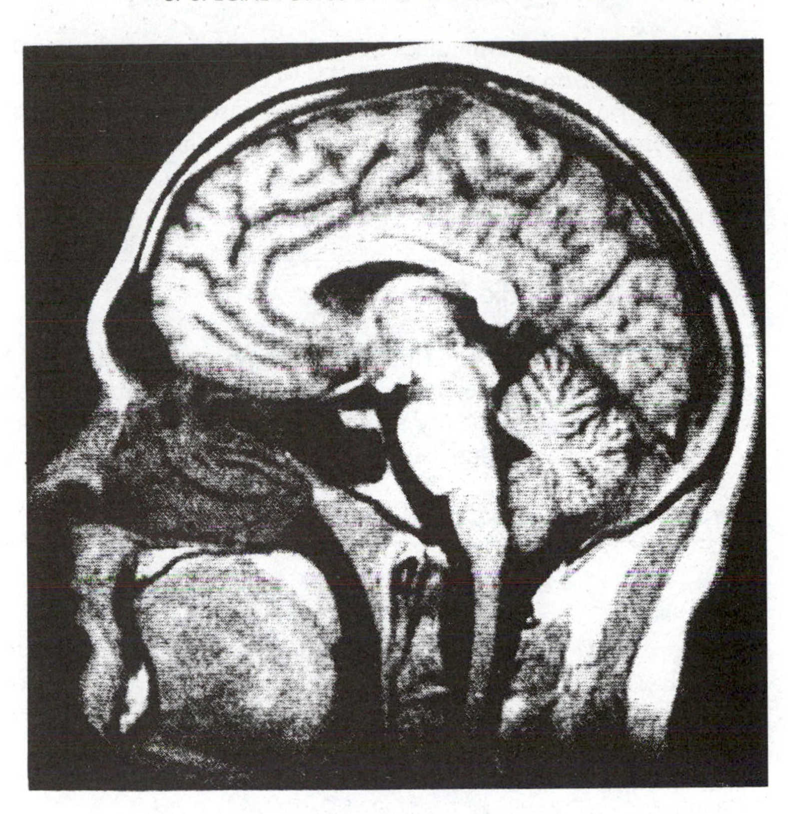

FIGURE 5-24 The proton NMR image of the human head in cross section (center-line sagittal section). [Reproduced with the permission of General Electric Medical Systems Group from R. E. Kinsinger, *Res. Develop.* (June 1984), p. 234.]

order of 10^{-2} T/m is applied to the sample along the z axis, in addition to the static B_0 field, the contents of the two NMR tubes experience slightly different field strengths. Consequently, protons in the tubes have different resonance frequencies. The different absorptions may be displayed spatially, so that a plot of magnetization as a function of the z coordinate gives the result in Figure 5-23b. If a gradient is applied along the x axis, the plot of absorption vs. magnetization reflects the different arrangement (Figure 5-23c). If the information is combined, a two-dimensional picture of the material in the probe is obtained (Figure 5-23d).

In the example of Figure 5-23, the tubes are assumed to be long compared with the static field, so that the experiment is independent of the y coordinate. If the object has structure along the y axis, an image can be obtained of a thin slice by alternating the z gradient in the plus and minus directions and allowing all points to cancel except the point at which the gradients coincide. To obtain two-dimensional information, the gradient is applied in the z direction of the plane defined by the alternating gradients for a time t_1 and then along the x direction for a time t_2. A double Fourier transformation, much like in normal 2D NMR, is then carried out on n values of t_1 and n values of t_2, to give the image of a plane made of $n \times n$ elements or pixels. Figure 5-24 shows the result for a cross section of the human head. Resolution is quite high.

NMR has several advantages over X-ray methods in the field of medical imaging. Foremost is the safety of radiofrequency energy compared with the considerable hazards of X-rays. For this reason, NMR is said to provide *noninvasive images*. In addition, the type of information is different. Proton NMR images are particularly revealing for soft tissues, which are less succesfully imaged with X-rays. Furthermore, NMR images can be obtained in several different fashions. Figure 5-24 shows the simple image of proton chemical shifts. Proton relaxation times, including both T_1 and T_2, also can be used to provide a different type of image. Other nuclei, such as ^{31}P, can be used as the sensor.

NMR imaging, also called NMR *tomography* or *zeugmatography*, has applications in numerous fields other than medicine. These include materials analysis and agriculture, since NMR can provide information about the internal structure of a cement block or of a seed without altering the object. In all these experiments, the object is placed in a field gradient. Because some such objects are extremely large, entirely new magnet systems have been constructed for these purposes, with pole gaps of sufficient size to admit the entire human body.

Topical NMR provides a different approach to the in vivo examination of organisms. Instead of bringing the organism to the NMR magnet, the magnet is brought to the organism. There are numerous variants of topical NMR. A magnetic probe of some sort, possibly as simple as a wire coil, is placed on the surface of an object to induce resonances in molecules close to the surface. These procedures have achieved reasonably high resolution examination of ^{1}H, ^{13}C, and ^{31}P resonances within living bodies. In this fashion, the in vivo concentration of ATP may be monitored following the administration of a drug, or the relative concentrations of different types of sugar phosphates may be monitored in a newborn human to detect abnormal conditions noninvasively.

5-9

The NMR Periodic Table

Although the discussion thus far has emphasized ^{1}H and ^{13}C NMR experiments, the impression should not be left that other nuclei do not give important information. Essentially every element in the periodic table has one or more magnetically active nuclides. Table 2-2 lists some of the most frequently encountered nuclei. Table 5-1 expands this list to include a number of other nuclei that are important in specific contexts.

TABLE 5-1 NMR Properties of Various Nuclei

Nuclide	Spin	Natural Abundance (%)	Natural Sensitivity (for equal numbers of nuclei) (vs. ^{1}H)	Receptivity (vs. ^{13}C)	NMR Frequency (2.35 tesla)
Lithium-6	1	7.42	0.00850	3.58	14.716
Lithium-7	$\frac{3}{2}$	92.58	0.293	1.54×10^{3}	38.864
Sodium-23	$\frac{3}{2}$	100.	0.0925	5.25×10^{2}	26.452
Potassium-39	$\frac{3}{2}$	93.1	0.000509	2.69	4.666
Magnesium-25	$\frac{5}{2}$	10.13	0.00268	1.54	6.122
Calcium-43	$\frac{7}{2}$	0.145	0.00640	5.27×10^{-2}	6.729
Aluminum-27	$\frac{5}{2}$	100.	0.206	1.17×10^{3}	26.057
Thallium-205	$\frac{1}{2}$	70.50	0.192	7.69×10^{2}	57.787
Tin-119	$\frac{1}{2}$	8.58	0.0518	25.2	37.291
Lead-207	$\frac{1}{2}$	22.6	0.00920	11.8	20.858
Selenium-77	$\frac{1}{2}$	7.58	0.00693	2.98	19.072
Tellurium-125	$\frac{1}{2}$	0.89	0.0176	0.89	26.170
Bromine-79	$\frac{3}{2}$	50.54	0.0788	2.26×10^{2}	25.054
Iodine-127	$\frac{5}{2}$	100.	0.0934	5.3×10^{2}	20.009
Xenon-129	$\frac{1}{2}$	26.44	0.0212	31.8	27.658
Vanadium-51	$\frac{7}{2}$	99.76	0.380	2.15×10^{3}	26.303
Chromium-53	$\frac{3}{2}$	9.55	0.000904	0.49	5.652
Tungsten-183	$\frac{1}{2}$	14.28	0.0000727	5.89×10^{-2}	4.166
Manganese-55	$\frac{5}{2}$	100.	0.175	9.94×10^{2}	24.745
Rhodium-103	$\frac{1}{2}$	100.	0.0000312	0.177	3.185
Nickel-61	$\frac{3}{2}$	1.19	0.00355	0.24	8.936
Palladium-105	$\frac{5}{2}$	22.23	0.00112	1.41	4.576
Copper-63	$\frac{3}{2}$	69.09	0.0931	3.65×10^{2}	26.528
Silver-107	$\frac{1}{2}$	51.82	0.0000663	0.195	4.048
Zinc-67	$\frac{5}{2}$	4.11	0.00285	0.665	6.257
Cadmium-111	$\frac{1}{2}$	12.75	0.00958	6.93	21.201

All of the alkali metals have magnetically active nuclides and may be used for example to study the properties of carbanions, particularly with regard to ion pairing and aggregation. The nuclei ^{6}Li and ^{7}Li have been used extensively in this context. Coupling between ^{13}C and ^{6}Li demonstrates covalency of the C—Li bond and provides the extent of aggregation, n in $(C{-}Li)_n$. The alkaline earth metals have been used not only for similar studies of organometallic compounds, but also extensively in biochemistry. Because both magnesium and calcium are present in certain large biomolecules, ^{25}Mg and ^{43}Ca can give important structural details when ^{1}H and ^{13}C offer little useful information.

The nuclides in Groups III–V, such as ^{27}Al, ^{119}Sn, and ^{77}Se, are invaluable in the structural analysis of molecules containing these elements. As can be seen in Table 5-1, several of these nuclei have a spin of $\frac{1}{2}$ and hence can give spectra with sharp peaks. The nuclides ^{29}Si and ^{119}Sn have come to provide key information in organosilicon and organotin chemistry. The halogen nuclides, other than ^{19}F, are not often used for structural analysis of haloorganic compounds, since ^{1}H and ^{13}C NMR spectra generally fulfill this role adequately. Chlorine and bromine nuclides in particular, however, have been used as probes of the structure of complex biomolecules. Although rare gas elements have not been studied extensively, ^{129}Xe has provided a useful standard for comparison of solvent effects, because the element is so noninteractive.

Metal nuclides are commonly used for structural elucidation of metal structures. Actual utility is very uneven. The best iron nuclide is still of such low receptivity that it is rarely used, but other elements such as platinum, cobalt, tungsten, and silver have highly receptive and useful nuclides. Cadmium-111, with a spin of $\frac{1}{2}$ and good receptivity, has found interesting applications as a substitute for the poor calcium and magnesium nucleides in studies of metal complexes of large biomolecules.

PROBLEMS

5-1 Give the spectral notation (AB, ABX, etc.) for each of the following substituted ethanes, first at fast C—C rotation, then at slow rotation: **(a)** CH_3CCl_3; **(b)** CH_3CHCl_2; **(c)** CH_3CH_2Cl; **(d)** $CHCl_2CH_2Cl$. At slow rotation there may be more than one rotamer.

5-2 Ring reversal in 7-methoxy-7,12-dihydropleiadene can be frozen out at −20° C. Two conformations are observed in the ratio 2/1. When the low field part of the 12-CH_2 AB quartet in the minor isomer is doubly irradiated, the intensity of the 7-methine proton is enhanced by 27%. Double irradiation of the same proton in the major isomer has no effect on the spectrum. What are the two conformational isomers and which is more abundant?

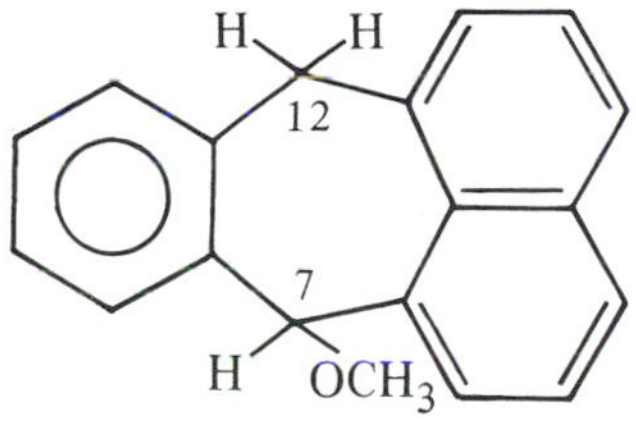

5-3 Permethyltitanocene reacts with an excess of nitrogen below −10° C to form a 1/1 complex. The methyl resonance of the complex above −50° C is a sharp singlet. Below −72° C the resonance splits

$$[C_5(CH_3)_5]_2Ti + N_2 \rightleftharpoons [C_5(CH_3)_5]_2TiN_2$$

reversibly into two peaks of not quite the same intensity. If the nitrogen is doubly labeled with ^{15}N, the ^{1}H-decoupled ^{15}N spectrum contains a singlet and an AX quartet ($J(^{15}N{-}^{15}N) = 7$ Hz), of not quite the same overall intensity at low temperatures. Explain these observations in terms of structures.

5-4 **a.** The resonance of the methylene protons of $C_6H_5CH_2SCHClC_6H_5$ in $CDCl_3$ is an AB quartet at room temperature. Why?

b. The AB spectrum coalesces at high temperatures to an A_2 singlet with a $\Delta G^{\ddagger}$ of 15.5 kcal/mol. The rate is independent of concentration in the range 0.0190–0.267 M. Explain in terms of a mechanism.

5-5 The off-resonance (upper) and fully (lower) decoupled $^{13}C\{^1H\}$ spectra of testosterone are given below.[1] Assign as many resonances as you can. Note ambiguities when appropriate.

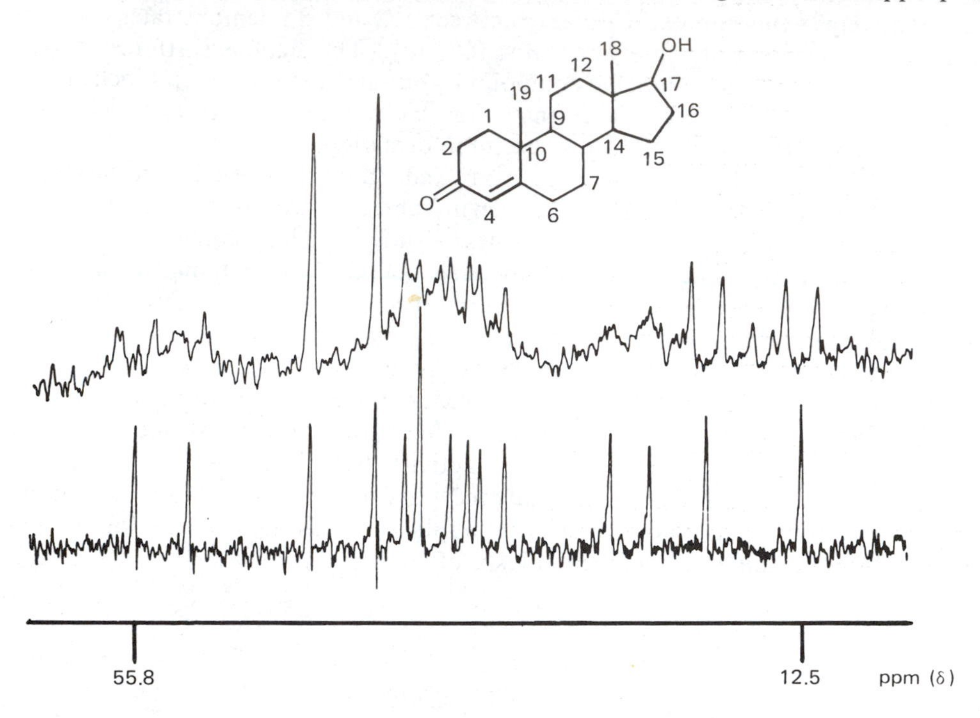

5-6 No coupling is observed between CH_3 and ^{14}N in acetonitrile ($CH_3—C{\equiv}N$), but there is a coupling in the corresponding isonitrile. Explain. This is not a distance effect. The phenomenon is general for nitriles and isonitriles.

5-7 Comment on the following ^{14}N linewidths.

$^+NMe_4$	<0.5 Hz
Me_3N	77
N H	172
$MeNO_2$	14
aniline	1300

5-8 The inversion–recovery (180°–τ–90°) spectral stack[2] for the aromatic carbons of *m*-xylene is given on the next page. Assign the resonances and explain the order of T_1 (look at the nulls).

5-9 1-Decanol has the following carbon T_1 values (s).

CH_3—	CH_2—	CH_2—	CH_2—	$(CH_2)_5$—	CH_2—OH
3.1	2.2	1.6	1.1	0.8–0.83	0.65

Explain the order and contrast the result with 1-bromodecane.

[1] Reprinted with permission from *J. Am. Chem. Soc.*, **91**, 7447 (1969). Copyright 1969 American Chemical Society.
[2] Reprinted with permission of R. Freeman from W. Bremser, H. P. W. Hill, and R. Freeman, *Messtechnik*, **79**, 14 (1971).

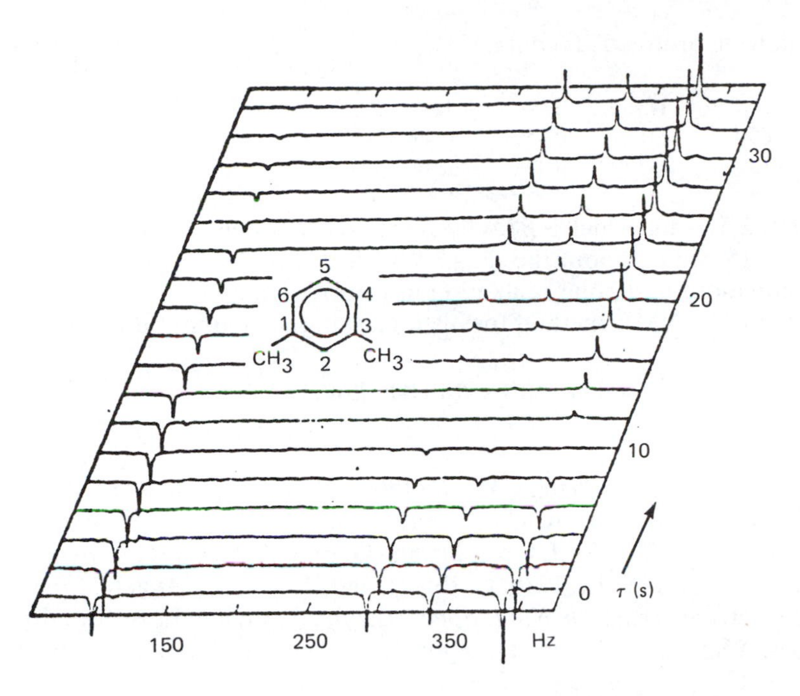

5-10 Would you expect axial or equatorial protons normally to have the shorter relaxation time? Why? Assign conformation and anomeric (C1) configuration to the following sugars, on the basis of the indicated relaxation times (T_1, s). Comment on (1) the relative size of T_1 for H1 in A and B (the basis for your assignment of anomeric configuration); (2) the size of T_1 for H4 in both anomers; (3) the relative size of T_1 for H3 in A and B; and (4) the relative size of T_1 for H2 in A and B.

A: CH_2OAc 1.3; H_4 3.1; H_5; OAc; AcO; H_2; OBz; H_1 2.0; H_3 2.7; Br; 3.6

B: CH_2OAc 1.3; H_4 2.7; H_5; OAc; AcO; H_2; OBz; H_1 4.1; H_3 3.4; Br; 2.1

Bz: PhC(=O)

5-11 In ribo-C-nucleosides the base is attached to C1′ by carbon. The α and β forms (C1 epimers) may be distinguished by T_1 studies.

β: NH_2; N; H; N; N; B; HO; O; 5′; 4′; H; H; H; 1′; H; 3′; 2′; HO; OH

α: HO; O; H; H; H; H; B; HO; OH

a. Consider the following proton T_1 (s) data.

	H1′	H3′	H5′	H5″
Isomer 1	1.60	1.31	0.45	0.45
Isomer 2	3.33	1.37	0.40	0.40

Which isomer (1 or 2) is α and which is β? Why are the H1′ T_1 values different for isomers 1 and 2 but the H3′, H5′, and H5″ values about the same? Why is T_1 for H5′ and H5″ smaller than the other values? Use the equation for dipolar relaxation in your reasoning.

b. Suggest another (not T_1) NMR method for distinguishing these α and β forms.

REFERENCES/BIBLIOGRAPHY

Dynamic Processes

5-1 General: G. Binsch, *Top. Stereochem.*, **3**, 97 (1968); *Dynamic Nuclear Magnetic Resonance Spectroscopy*, L. M. Jackman and F. A. Cotton, eds., Academic Press, New York, 1975; A. Steigel, *NMR Basic Princ. Progr.*, **15**, 1 (1978); J. I. Kaplan and G. Fraenkel, *NMR of Chemically Exchanging Systems*, Academic Press, New York, 1980; G. Binsch and H. Kessler, *Angew. Chem., Int. Ed. Engl*, **19**, 411 (1980); J. Sändstrom, *Dynamic NMR Spectroscopy*, Academic Press, London, 1982; *Nuclear Magnetic Resonance*, A Specialist Periodical Report, The Chemical Society, London, reviewed in each issue.

5-2 Carbon-13 applications: *Progr. NMR Spectrosc.*, **11**, 95 (1977).

5-3 Hindered rotation: H. Kessler, *Angew. Chem., Int. Ed. Engl.*, **9**, 219 (1970); W. E. Stewart and T. H. Siddall, *Chem. Rev.*, **70**, 517 (1970); M. Ōki, *Top. Stereochem.*, **14**, 1 (1983).

5-4 Ring reversal and cyclic systems: J. E. Anderson, *Quart. Rev.*, **19**, 426 (1965); H. Booth, *Progr. NMR Spectrosc.*, **5**, 149 (1969); J. B. Lambert and S. I. Featherman, *Chem Rev.*, **75**, 611 (1975); H. Günther and G. Jikeli, *Angew. Chem., Int. Ed. Engl.*, **16**, 599 (1977); E. L. Eliel and K. M. Pietrusiewicz, *Top. Carbon-13 Spectrosc.*, **3**, 171 (1979); F. G. Riddell, *The Conformational Analysis of Heterocyclic Compounds*, Academic Press, London, 1980; A. P. Marchand, *Stereochemical Applications of NMR Studies in Rigid Bicyclic Systems*, Verlag Chemie, Deerfield Beach, FL, 1982.

5-5 Atomic inversion: J. B. Lambert, *Top. Stereochem.*, **6**, 19 (1971); A. Rauk, L. C. Allen, and K. Mislow, *Angew. Chem., Int. Ed. Engl.*, **9**, 400 (1970).

5-6 Biomolecules: G. Govil and R. V. Hosur, *NMR Basic Princ. Progr.*, **20**, 1 (1982).

5-7 CIDNP: H. R. Ward, *Acc. Chem. Res.*, **5**, 24 (1972); R. G. Lawler, *ibid.*, **5**, 32 (1972); R. G. Lawler, *Progr. NMR Spectrosc.*, **9**, 145 (1973); *Chemically Induced Magnetic Polarization*, A. R. Lepley and G. L. Closs, eds., Wiley, New York, 1973; G. L. Closs, *Advan. Magn. Reson.*, **7**, 157 (1974); C. Richard and P. Granger, *NMR Basic Princ. Progr.*, **8**, 1 (1974); W. B. Moniz, C. F. Poranski, Jr., and S. A. Sojka, *Top. Carbon-13 Spectrosc.*, **3**, 361 (1979).

Relaxation Effects

5-8 General: *Nuclear Magnetic Resonance*, A Specialist Periodical Report, The Chemical Society, London, reviewed in each issue; D. A. Wright, D. E. Axelson, and G. C. Levy, *Magn. Reson. Rev.*, **3**, 103 (1979).

5-9 Carbon-13 relaxation: J. R. Lyerla, Jr., and G. C. Levy, *Top. Carbon-13 Spectrosc.*, **1**, 79 (1974); F. W. Wehrli, *ibid.*, **2**, 343 (1976).

5-10 Lanthanide shift reagents: A. F. Cockerill, G. L. O. Davies, R. C. Harden, and D. M. Rackham, *Chem. Rev.*, **73**, 553 (1973); B. C. Mayo, *Chem. Soc. Rev.*, **2**, 49 (1973); J. Reuben, *Progr. NMR Spectrosc.*, **9**, 1 (1973); *NMR Shift Reagents*, R. E. Sievers, ed., Academic Press, New York, 1973; O. Hofer, *Top. Stereochem.*, **9**, 111 (1976); B. D. Lockhart, *CRC Crit. Rev. Anal. Chem.*, **6**, 69 (1976); F. Inagaki and T. Miyazawa, *Progr. NMR Spectrosc.*, **14**, 67 (1980).

5-11 Nuclear Overhauser effect: J. H. Noggle and R. E. Schirmer, *The Nuclear Overhauser Effect*, Academic Press, New York, 1971; R. A. Bell and J. K. Saunders, *Top. Stereochem.*, **7**, 1 (1973); J. K. Saunders and J. W. Easton, *Determ. Org. Struct. Phys. Meth.*, **6**, 271 (1976).

5-12 Rates from relaxation times: J. B. Lambert, R. J. Nienhuis, and J. W. Keepers, *Angew. Chem., Int. Ed. Engl.*, **20**, 487 (1981).

Multiple Irradiation and Multipulse Techniques

5-13 General multiple resonance: R. A. Hoffman and S. Forsén, *Progr. NMR Spectrosc.*, **1**, 15 (1966); W. McFarlane, *Ann. Rev. NMR Spectrosc.*, **1**, 135 (1968); V. J. Kowalewski, *Progr. NMR Spectrosc.*, **5**, 1 (1969); W. McFarlane, *Determ. Org. Struct. Phys. Meth.*, **4**, 150 (1971); W. von Philipsborn, *Angew. Chem., Int. Ed. Engl.*, **10**, 472 (1971); W. McFarlane, *Ann. Rep. NMR Spectrosc.*, **5A**, 353 (1972); R. L. Micher, *Magn. Reson. Rev.*, **1**, 225 (1972); L. R. Dalton, *ibid.*, **1**, 301 (1972); W. McFarlane and D. S. Rycroft, *Ann. Rep. NMR Spectrosc.*, **9**, 320 (1979); *Nuclear Magnetic Resonance,* A Specialist Periodical Report, The Chemical Society, London, reviewed in most issues.

5-14 Difference spectroscopy: J. K. M. Sanders and J. D. Merck, *Progr. NMR Spectrosc.*, **15**, 353 (1982).

5-15 INEPT, INADEQUATE, and spectral editing: K. Hikichi and M. Ohuchi, *JEOL News*, **18A**, No. 1, pp. 2f, 1982; R. Benn and H. Günther, *Angew. Chem., Int. Ed. Engl.*, **22**, 350 (1983); C. J. Turner, *Progr. NMR Spectrosc.*, **16**, 311 (1984); G. A. Morris, *Magn. Reson. Chem.*, **24**, 371 (1986).

5-16 Two-dimensional spectra: R. Freeman and G. A. Morris, *Bull. Magn. Reson.*, **1**, 5 (1979); A. Bax, *Two Dimensional N.M.R. in Liquids*, D. Reidel, Hineman, MA, 1982; K. Hikichi, *JEOL News,* **19A**, No. 2, pp. 2f, 1983; R. Benn and H. Günther, *Angew. Chem., Int. Ed. Engl.*, **22**, 350 (1983); G. A. Morris, *Magn. Reson. Chem.*, **24**, 371 (1986).

NMR Imaging and Topical NMR

5-17 NMR imaging: *Nuclear Magnetic Imaging*, C. L. Partain et al., Saunders, Philadelphia, 1983; E. R. Andrew, *Acc. Chem. Res.*, **16**, 114 (1983).

5-18 Topical NMR: R. E. Gordon, P. E. Hanley, and D. Shaw, *Progr. NMR Spectrosc.*, **15**, 1 (1982).

The NMR Periodic Table

5-19 Fluorine-19 NMR: E. F. Mooney, *An Introduction to ^{19}F NMR Spectroscopy*, Heyden and Son, London, 1970; C. H. Dungan and J. R. Van Wazer, *Compilation of Reported Fluorine-19 NMR Chemical Shifts*, Wiley–Interscience, New York, 1970; J. W. Emsley and L. Phillips, *Progr. NMR Spectrosc.*, **7**, 1 (1971); R. Fields, *Ann. Rep. NMR Spectrosc.*, **5A**, 99 (1972); V. Wray, *ibid.*, **10B**, 1 (1980), **14**, 1 (1983).

5-20 Nitrogen-15 and nitrogen-14 NMR: *Nitrogen NMR*, M. Witanowski and G. A. Webb, eds., Plenum Press, London, 1973; G. C. Levy and R. L. Lichter, *Nitrogen-15 Nuclear Magnetic Resonance Spectroscopy*, Wiley, New York, 1979; M. Witanowski, L. Stefaniak, and G. A. Webb, *Ann. Rep. NMR Spectrosc.*, **7**, 117 (1977), **11a**, 1 (1981); G. J. Martin, M. L. Martin, and J. P. Gouesnard, *NMR Basic Princ. Progr.*, **18**, 1 (1981).

5-21 Boron-11 NMR: W. G. Henderson and E. F. Mooney, *Ann. Rev. NMR Spectrosc.*, **2**, 219 (1969); W. L. Smith, *J. Educ. Chem.*, **54**, 469 (1977); H. Nöth and B. Wrackmeyer, *NMR Basic Princ. Progr.*, **14**, 1 (1978); A. R. Siedle, *Ann. Rep. NMR Spectrosc.*, **12**, 177 (1982).

5-22 Phosphorus-31 NMR: *Top. Phosphorus Chem.*, **5** (1967); J. R. Van Wazer, *Determ. Org. Struct. Phys. Meth.*, **4**, 323 (1971); G. Mavel, *Ann. Rep. NMR Spectrosc.*, **5B**, 1 (1973); D. G. Gorenstein, *Progr. NMR Spectrosc.*, **16**, 1 (1983); D. G. Gorenstein, *Phosphorus-31 NMR*, Academic Press, New York, 1984.

5-23 Oxygen-17 NMR: W. G. Klemperer, *Angew. Chem., Int. Ed. Engl.*, **17**, 246 (1978); J.-P. Kintzinger, *NMR Basic Princ. Progr.*, **14**, 1 (1978); T. St. Amour and D. Fiat, *Bull. Magn. Reson.*, **1**, 118 (1980).

5-24 Deuterium NMR: H. H. Mantsch, H. Saitô, and I. C. P. Smith, *Progr. NMR Spectrosc.*, **11**, 211 (1977).

5-25 Tritium NMR: J. A. Elvidge, in *Isotopes: Essential Chemistry and Application*, J. A. Elvidge and J. R. Jones, eds., The Chemical Society, London, pp. 152f, 1980.

5-26 Sodium-23 NMR: P. Laszlo, *Angew. Chem., Int. Ed. Engl*, **17**, 254 (1978).

5-27 Silicon-29 NMR: J. Schraml and J. M. Bellama, *Determ. Org. Struc. Phys. Meth.*, **6**, 203 (1976); E. A. Williams and J. D. Cargioli, *Ann. Rep. NMR Spectrosc.*, **9**, 221 (1979); J. P. Kintzinger, *NMR Basic Princ. Progr.*, **17**, 1 (1981); H. C. Marsmann, *ibid.*, **17**, 65 (1981).

5-28 Halogen NMR: B. Lindman and S. Forsén, *NMR Basic Princ. Progr.*, **12**, 1 (1976).

5-29 Calcium and magnesium NMR: S. Forsén and B. Lindman, *Ann. Rep. NMR Spectrosc.*, **11A**, 183 (1981).
5-30 Cadmium-113 NMR: P. D. Ellis, *Science*, **221**, 1141 (1983).
5-31 Tin-119 NMR: P. J. Smith and A. P. Tupciauskas, *Ann. Rep. NMR Spectrosc.*, **8**, 292 (1978).
5-32 Thallium NMR: J. F. Hinton, K. R. Metz, and R. W. Briggs, *Ann. Rep. NMR Spectrosc.*, **13**, 211 (1982).
5-33 Transition metal NMR: R. G. Kidd, *Ann. Rep. NMR Spectrosc.*, **10A**, 1 (1980).
5-34 Nuclei other than ^{1}H (general): *Nuclear Magnetic Resonance Spectroscopy of Nuclei Other than Protons*, T. Axenrod and G. A. Webb, eds., Wiley–Interscience, New York, 1974; R. K. Harris and B. E. Mann, *NMR and the Periodic Table*, Academic Press, London, 1978; F. W. Wehrli, *Ann. Rep. NMR Spectrosc.*, **9**, 126 (1979); C. Brevard and P. Granger, *Handbook of High Resolution Multinuclear NMR*, Wiley–Interscience, New York, 1981; J. B. Lambert and F. G. Riddell, *The Multinuclear Approach to NMR Spectroscopy*, D. Reidel, Dordrecht, Holland, 1983; P. Laszlo, *NMR of Newly Accessible Nuclei*, Academic Press, London, 1983 (Vol. 1), 1984 (Vol. 2).

Part II

INFRARED AND RAMAN SPECTROSCOPY

6

Introduction to Vibrational Spectroscopy

6-1

Introduction

A rapid and simple method for obtaining preliminary information on the identity or structure of an organic molecule is to record an infrared (IR) absorption spectrum of the compound. The spectrum is a plot of the percent of IR radiation that passes through the sample vs. the wavelength (or wavenumber) of the radiation. An example is shown in Figure 6-1. The positions and relative strengths of the absorption peaks (often called bands) gives clues to the structure of the molecule, as indicated on the figure.

The instrument that produces the spectrum is known as an IR spectrophotometer (usually shortened to spectrometer). A typical bench-top instrument is shown in Figure 6-2. It is relatively simple to prepare a sample and record a spectrum such as that shown in Figure 6-1. Sample handling skills can be learned

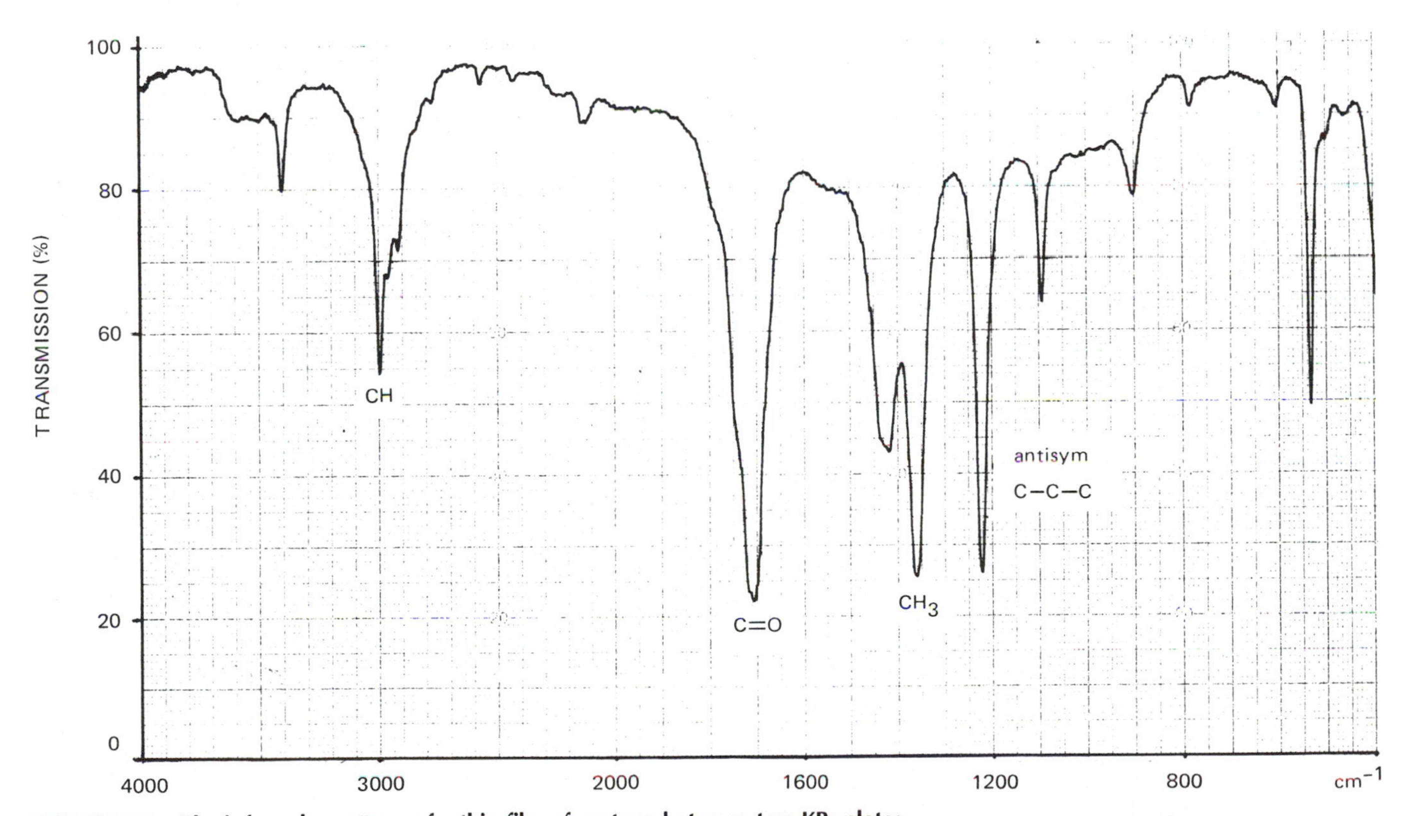

FIGURE 6-1 **The infrared spectrum of a thin film of acetone between two KBr plates.**

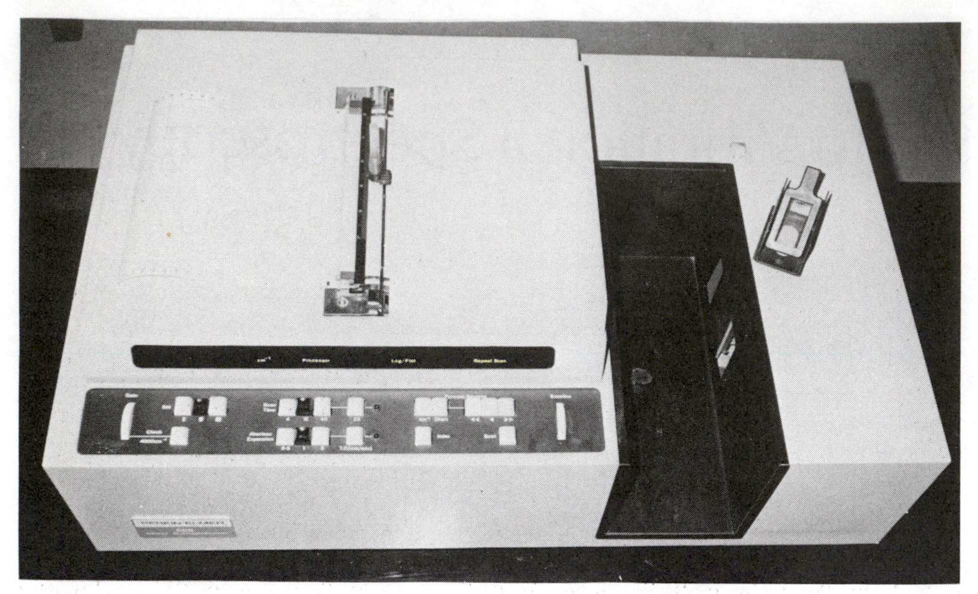

FIGURE 6-2 A simple benchtop infrared spectrophotometer, the Perkin-Elmer Model 598. [Courtesy of Perkin-Elmer (Canada) Ltd.]

quickly, and some tips on this procedure are given in Section 6-7. Recording the spectrum consists of placing the sample in the sample compartment, adjusting the recorder pen, setting the scan speed and resolution controls, and initiating the scan.

What an infrared spectrum actually is and how a molecule gives a spectrum are explained in Section 6-3. There are two types of IR spectrometer—dispersive and Fourier transform. How these instruments work is explained in Section 6-5.

Raman spectroscopy is a technique that provides information complementary to that obtained from IR spectroscopy. This technique is discussed in Sections 6-3 and 6-6.

Information on the structure of a molecule is obtained from a detailed study of those bands in the spectrum that are characteristic of certain functional groups. Some such bands are indicated on Figure 6-1. Characteristic group frequencies are discussed in detail in Chapter 7, and the interpretation of infrared and Raman spectra is the subject of Chapter 8.

6-2

Vibrations of Molecules

Infrared and Raman spectroscopy gives information on molecular structure through the frequencies of the normal modes of vibration of the molecule. A normal mode of vibration is one in which each atom executes a simple harmonic oscillation about its equilibrium position. All atoms move in phase with the same frequency while the center of gravity of the molecule does not move. A model of the molecule can be made using balls to represent the atoms and springs to represent the bonds. Vibrations of the model involve stretching and bending of the springs together with motions of the balls.

According to classical mechanics, the frequency of vibration ν (s^{-1}) of two balls of mass m (kg) connected by a spring with force constant k ($N\ m^{-1}$)[1] is given by eq. 6-1. The vibrational motions and frequencies of a

$$\nu = \frac{1}{\pi}\sqrt{\frac{k}{2m}} \tag{6-1}$$

structure containing several balls (atoms) of various masses connected by springs (bonds) with different force

[1]The force constant is a measure of the resistance to stretching of the spring. The force F needed to displace a mass m by a distance x is $F = kx$.

constants can also be studied using the methods of classical mechanics. The results of these calculations are very important because they form the basis for the interpretation of vibrational spectra.

There are $3N-6$ normal modes of vibration of a molecule, where N is the number of atoms. This number is increased by one (to $3N-5$) if the molecule is linear. The number of vibrations can be seen to be $3N-6$ if it is remembered that each atom has three degrees of motional freedom, which can be thought of as motions in the x, y, and z directions. Thus, N atoms have $3N$ independent motions. However, when the atoms are connected together in a molecule, the motions are no longer independent. Three motions become translations of the molecule, where all atoms move simultaneously in the x, y, or z directions. Another three are rotations, where all atoms rotate in phase about the x, y, and z axes. This leaves $3N-6$ motions, in which internuclear distances and bond angles change, but the center of gravity of the molecule does not move. These are the normal modes of vibration of the molecule.

It was stated above that the frequencies and forms of the normal modes of vibration of a ball and spring model can be calculated using methods of classical mechanics. These calculations also apply to molecules, if appropriate atomic masses, molecular dimensions, and force constants are used, and if certain rules of quantum mechanics are applied. This subject will be discussed further in Section 7-1.

6-3

Vibrational Spectra

An infrared spectrum is obtained when a sample *absorbs* radiation in the region of the electromagnetic spectrum known as the infrared. The expression *absorption band* is used to denote a feature that is observed in the spectrum. If the absorption band is quite narrow and sharp, the word *peak* is used. Thus, we have the more or less interchangeable expressions absorption band and absorption peak, or simply band, peak, and absorption.

In infrared absorption, energy is transferred from the incident radiation to the molecule, and a quantum mechanical transition occurs between two vibrational energy levels. The difference in energy (joules) between the two vibrational energy levels is directly related to the frequency (s^{-1}) of the electromagnetic radiation by eq. 6-2, in which h is Planck's constant (6.63×10^{-34} J s); the quantity of energy, $h\nu$ is known as a *photon*. A frequency of vibration of the molecule corresponds directly to the frequency of IR radiation absorbed.

$$E_2 - E_1 = h\nu \qquad (6\text{-}2)$$

The material for study will usually be in the form of a solid, liquid, or solution. However, gas or vapor phase spectra can also be obtained. Molecules in the gas phase can undergo changes in rotational energy at the same time as the vibrational transition, so that some rotational structure may be observed on the vibrational band.

A Raman spectrum is produced by a *scattering* process. Monochromatic incident radiation, from a laser operating in the visible region of the spectrum, is scattered by the sample. The scattered light is observed instrumentally in a direction at right angles to the incident radiation. Since early Raman spectra were recorded on photographic plates, the features appeared as *lines* and *bands* on the plates, and both of these words are used when discussing Raman spectra.

Raman spectra can be considered as resulting from inelastic collisions of photons with molecules. In an inelastic collision some energy is transferred either from the photon to the molecule or from the molecule to the photon, as illustrated in Figure 6-3. In the former case, the molecule will be left in a higher energy level (giving rise to the so-called *Stokes lines*). In the latter case, the molecule must *already be* in an excited state so that it can return to a lower state after giving up energy to the photon (giving rise to the so-called *anti-Stokes lines*). Since most molecules are in their ground vibrational state at normal temperatures, only the Stokes lines are important, and it is these that comprise the Raman spectrum of interest.

A typical Raman spectrum is shown in Figure 6-4. The zero on the abscissa corresponds to the frequency of the laser line (ν_L) used to "excite" the spectrum. The positions of the peaks correspond to differences or shifts between ν_L and the observed scattered frequencies (ν_{obs}). Since the frequencies are very large, it is customary to divide ($\nu_L - \nu_{obs}$) by c, the velocity of light (3×10^{10} cm s^{-1}) to obtain the wavenumber unit

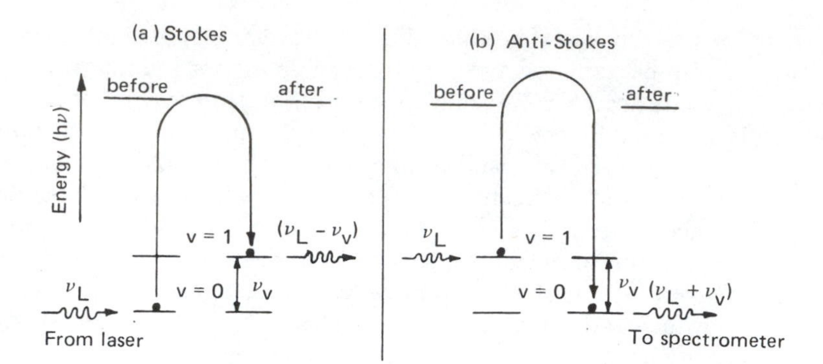

FIGURE 6-3 The mechanism of Raman scattering.

(cm^{-1}) (see Section 6-4). A Raman spectrum is then directly comparable to the IR spectrum of the same compound. For example, compare Figure 6-4 to Figure 6-1.

The goal of this section is to show the student how vibrational spectra can provide information on molecular structure. This process requires some knowledge of molecular vibrations, symmetry, and group frequencies, as well as factors affecting group frequencies. The infrared or Raman spectrum can also be used as a *fingerprint* for molecule identification. A computer searches a data bank to find a number of spectra that most closely match that of the unknown. The final identification is then made by consultation of an atlas of spectra. Further details of this method are given in Appendix II.A.

Vibrations of certain functional groups—OH, NH, CH_3, C=O, C_6H_5, etc.—always give rise to bands in the infrared and Raman spectra within well-defined frequency ranges regardless of the molecule containing the functional group. The exact position of the group frequency within the range gives further information on

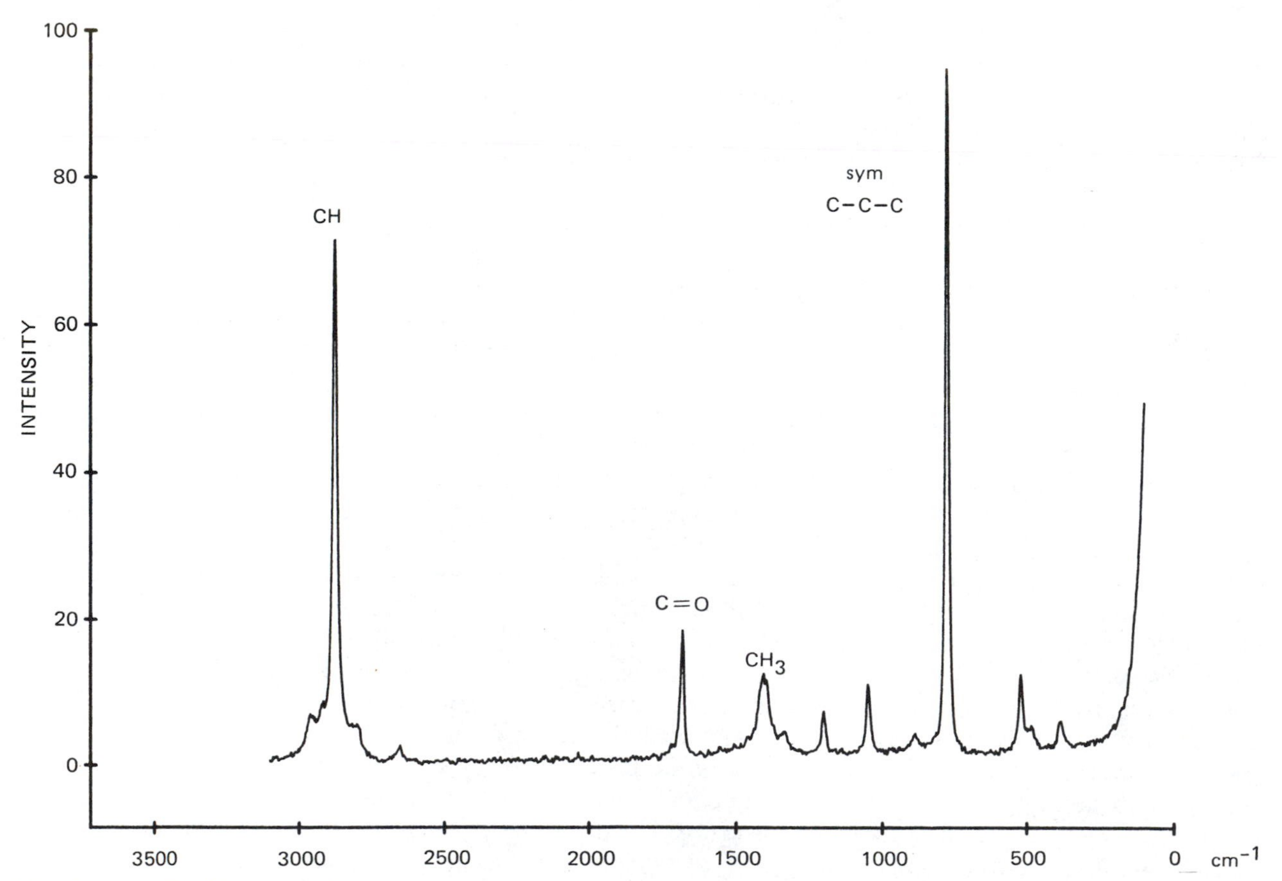

FIGURE 6-4 The Raman spectrum of acetone.

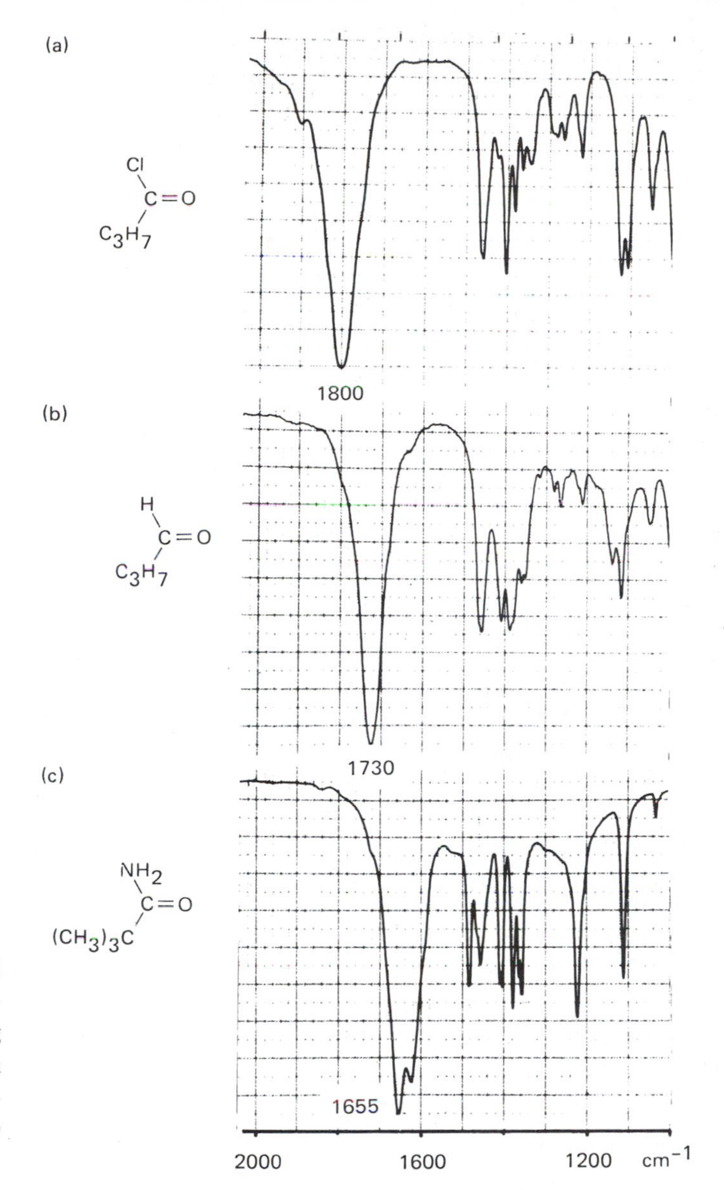

FIGURE 6-5 The carbonyl stretching bands of (a) butyryl chloride, (b) butyraldehyde, and (c) trimethylacetamide. (Note that the NH_2 bending band overlaps the C=O stretching band to give a doublet.) [Reproduced with permission of Aldrich Chemical Company, Inc., from *The Aldrich Library of FT–IR Spectra*.]

the environment of the functional group. A parallel can be drawn here with the NMR experiment, where the resonance frequency of a nucleus depends on its electronic environment. As an example, we might take the carbonyl stretching band of simple aliphatic aldehydes or ketones as the standard (1730 cm^{-1}). Carboxylic acid (monomers), acid chlorides, and esters have bands at higher frequencies, while amides and aromatic ketones have lower C=O stretching frequencies (Figure 6-5).

The observation of a band in the spectrum within an appropriate frequency range can indicate the presence of one *or more* different functional groups in the molecule because there is considerable overlap of the ranges of many functional groups. It is therefore necessary to examine other regions of the spectrum for confirmation of a particular group. Examples of this procedure are given in Chapter 8.

6-4

Units and Notation

As Figure 6-1 shows, a spectrum is recorded graphically with the wavelength or frequency as the abscissa and amount of absorption as the ordinate. There are several units used for both ordinate and abscissa scales. The unit of frequency ν is s^{-1} (vibrations per second). This is a very large number for molecular vibrations (of the order of 10^{13} s^{-1}) and is inconvenient. A more convenient unit, $\bar{\nu}$ (the wavenumber), is obtained (eq. 6-3) by

$$\bar{\nu} = \frac{\nu}{c} \tag{6-3}$$

dividing the frequency by c, the velocity of light. A vibration of frequency 3×10^{13} s^{-1} has a corresponding $\bar{\nu}$ of 1000 cm^{-1} as shown in eq. 6-4. Although the wavenumber is a frequency divided by a velocity, it is common

$$\frac{3 \times 10^{13}\ s^{-1}}{3 \times 10^{10}\ cm\ s^{-1}} = 10^3\ cm^{-1} \tag{6-4}$$

practice to refer to 1000 cm^{-1} as a "frequency" of 1000 cm^{-1} with the division by c understood. The three accepted ways of saying cm^{-1} are "centimeters to the minus one," "reciprocal centimeters," or "wavenumbers."

Infrared spectroscopists formerly used wavelength in micrometers (10^{-6} m) as the abscissa scale for their spectra. Wavelength (λ) is related to frequency (ν) or wavenumber ($\bar{\nu}$) by eq. 6-5. Other older wavelength units and their equivalent SI units are given in Table 6-1.

$$\frac{1}{\lambda} = \frac{\nu}{c} = \bar{\nu} \tag{6-5}$$

In many older texts and papers on infrared spectroscopy, the positions of the infrared absorption bands are given in micrometers (formerly called microns) and most of the earlier instruments produced a spectrum whose abscissa was linear in wavelength. Thus, many of the collections of reference spectra are also published in this format. This is unfortunate since all modern infrared instruments present the spectra with a linear wavenumber format. This problem will be addressed again in Chapter 8.

The positions of Raman lines cannot be expressed in units of wavelength because the lines are measured as *shifts* from the incident or exciting line. Hence the wavelength of a Raman line depends on the laser used. Since infrared and Raman spectra are used together to give information on molecular structure, it is convenient to use a common unit, the wavenumber (cm^{-1}). There is also strong support on theoretical grounds for using cm^{-1}, since this unit is related directly to energy ($E = h\nu = hc\bar{\nu}$).

Several units are used to measure the intensity of an infrared absorption peak. Percent *transmittance* (% T) is the most common, but *absorbance* (A) also will be encountered. Transmittance is the ratio of the radiant power or intensity (I) transmitted by a sample to the incident intensity (I_0) and can be expressed as equation 6-6, and percent transmittance as eq. 6-7. Absorbance is defined in the several ways given in eq. 6-8. In

$$T = \frac{I}{I_0} \tag{6-6}$$

$$\%\ T = 100\,\frac{I}{I_0} \tag{6-7}$$

$$A = \log_{10}\frac{I_0}{I} = \log_{10}\frac{1}{T} = \log_{10}\frac{100}{\%\ T} \tag{6-8}$$

solution spectra, the intensity of absorption can be related to the concentration and the pathlength by the Beer–Lambert law, eq. 6-9, in which C is the concentration in moles $liter^{-1}$ and l is the pathlength in cm. The

TABLE 6-1 Wavelength Units

Older Unit	SI Unit
1 micron (μ) = 10^{-4} cm	1 micrometer (μm) = 10^{-6} m
1 millimicron (mμ) = 10^{-7} cm	1 nanometer (nm) = 10^{-9} m
1 angstrom (Å) = 10^{-8} cm	1 angstrom (Å) = 10^{-10} m

$$A = \epsilon lC \quad (6\text{-}9)$$

constant ϵ is the *molar absorption coefficient*, with units liter mole^{-1} cm^{-1}. Thus ϵ is the absorbance produced by a solution of concentration 1.0 M in a cell of path length 1.0 cm. Other names for ϵ include *extinction coefficient* and *molar absorptivity*.

Intensities in Raman spectra are much less quantitative, since the height of a peak depends on many factors, such as the power of the laser, the wavelength, the slit width, the sensitivity of the photomultiplier detector, and the amplification system used. Thus, quantitative results can only be obtained if an *internal standard* is used to determine the amount of sample actually in the laser beam and giving rise to scattering.

The intensity of a Raman line is a *linear* function of concentration, whereas the intensity of an infrared absorption band is a *logarithmic* function of concentration. Thus, doubling the concentration of a solution should double the intensities of all Raman lines for identical instrumental settings, whereas the apparent effect on the infrared peak heights will depend on the peak. For example, a weak infrared band will appear to be affected much more than a strong absorption, since doubling the concentration will almost double the intensity of a weak band but change that of a strong band only about 10%. It is clear that caution must be exercised in discussing both infrared and Raman band intensities.

6-5

Infrared Spectroscopy: Dispersive and Fourier Transform

Infrared spectroscopy is a well-established technique, and commercial instruments have been available since the late 1940s. Over the years a very large number of infrared spectra have accumulated in the literature and collections of reference spectra are commercially available. This makes infrared spectroscopy a very useful tool for determination of molecular structure. Direct information about the presence of functional groups is immediately available from an infrared spectrum. Comparison of the infrared spectrum of an unknown material with a reference spectrum, or with the spectrum of a known compound, can provide absolute proof of the identity of the unknown substance.

For the present purposes the normal infrared range will be assumed to be 4000–400 cm^{-1}. However, many infrared spectrometers cover a somewhat wider range, overlapping the far infrared to 200 cm^{-1}. The region below 200 cm^{-1} is not readily accessible by infrared spectroscopy, but vibrational spectra can be obtained in this region by Raman spectroscopy (see Section 6-6).

Infrared spectra can be obtained by either dispersive or interferometric methods. Here an analogy can be drawn to NMR spectroscopy. In Section 2-11 it was seen that an NMR spectrum can be obtained by either continuous wave or Fourier transform methods. In the case of infrared spectra, dispersive instruments record the spectrum in the frequency domain, whereas interferometers record the spectrum in the time domain. The interferograms, as they are called, must be transformed to the frequency domain by means of a Fourier transformation.

An example of an infrared spectrum recorded on a dispersive infrared spectrometer is shown in Figure 6-6, and an FT–IR spectrum of the same sample can be seen in Figure 6-7. There is no fundamental difference between spectra recorded on dispersive and FT instruments. The latter is essential when a spectrum must be recorded rapidly. All the other advantages listed for the FT–NMR experiment also apply to FT–IR (see Section 6-5b).

6-5a Dispersive Infrared Spectrometers

An infrared spectrometer (or spectrophotometer, to give it its full name) consists of three basic parts: (1) a source of continuous infrared radiation, (2) a monochromator to disperse the radiation into its spectrum, and (3) a sensitive detector of infrared radiation. The sample is usually placed between the source and the monochromator, although in some instruments it can be placed after the monochromator.

The source of infrared radiation is usually a coil of wire with high resistance such as nichrome, a rod of partially conductive material such as silicon carbide, or rare earth oxides heated by passing an electrical current through it. Temperatures of about 1200° C (1500 K) give the optimum yield of energy in the infrared.

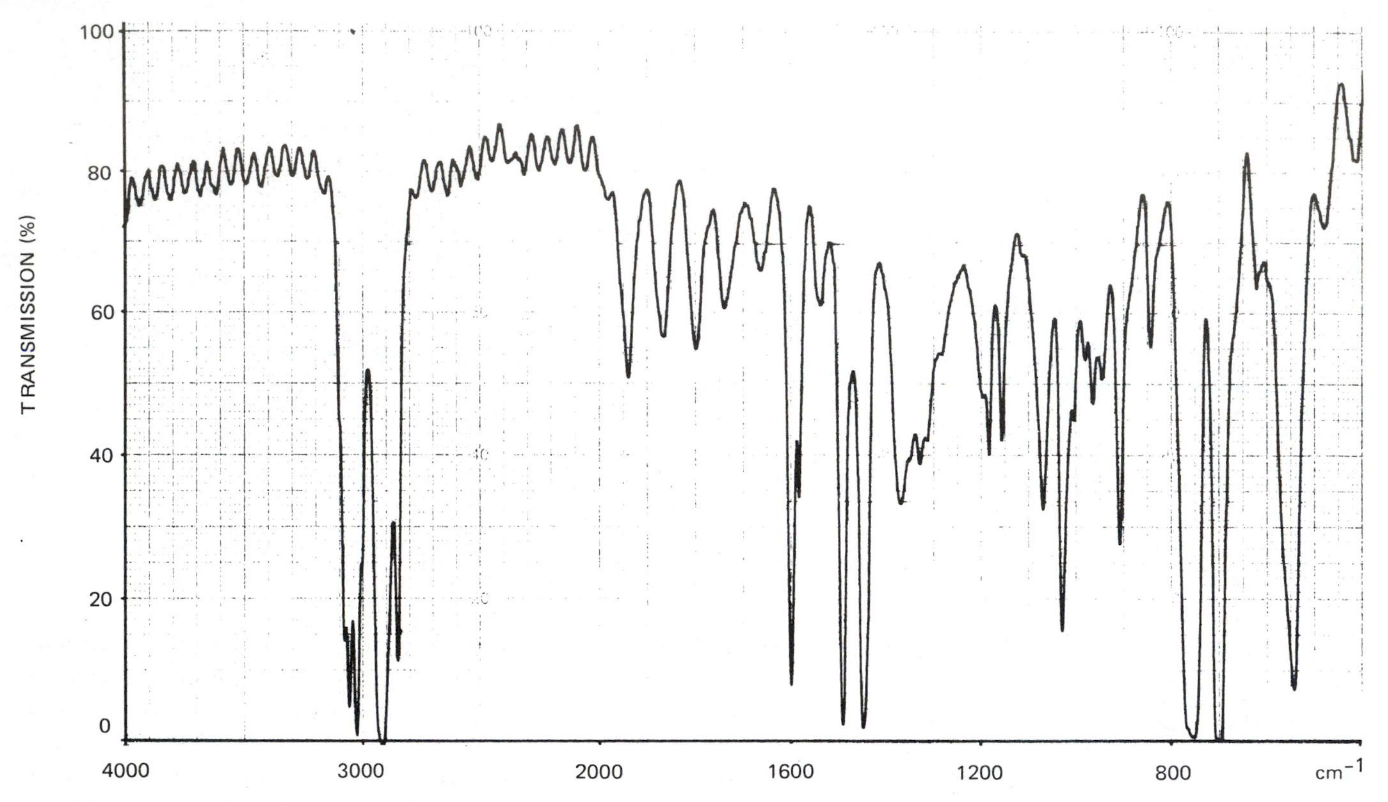

FIGURE 6-6 The infrared spectrum of a 0.05 mm film of polystyrene recorded on a dispersive infrared spectrometer.

The monochromator disperses the continuous radiation into its spectrum of monochromatic components. Prisms were used in earlier instruments, but today gratings are found in all dispersive spectrometers. The component frequencies are passed sequentially and continuously by a mechanical scanning device to the detector. In this way, the detector can sense which frequencies have been absorbed or partially absorbed by

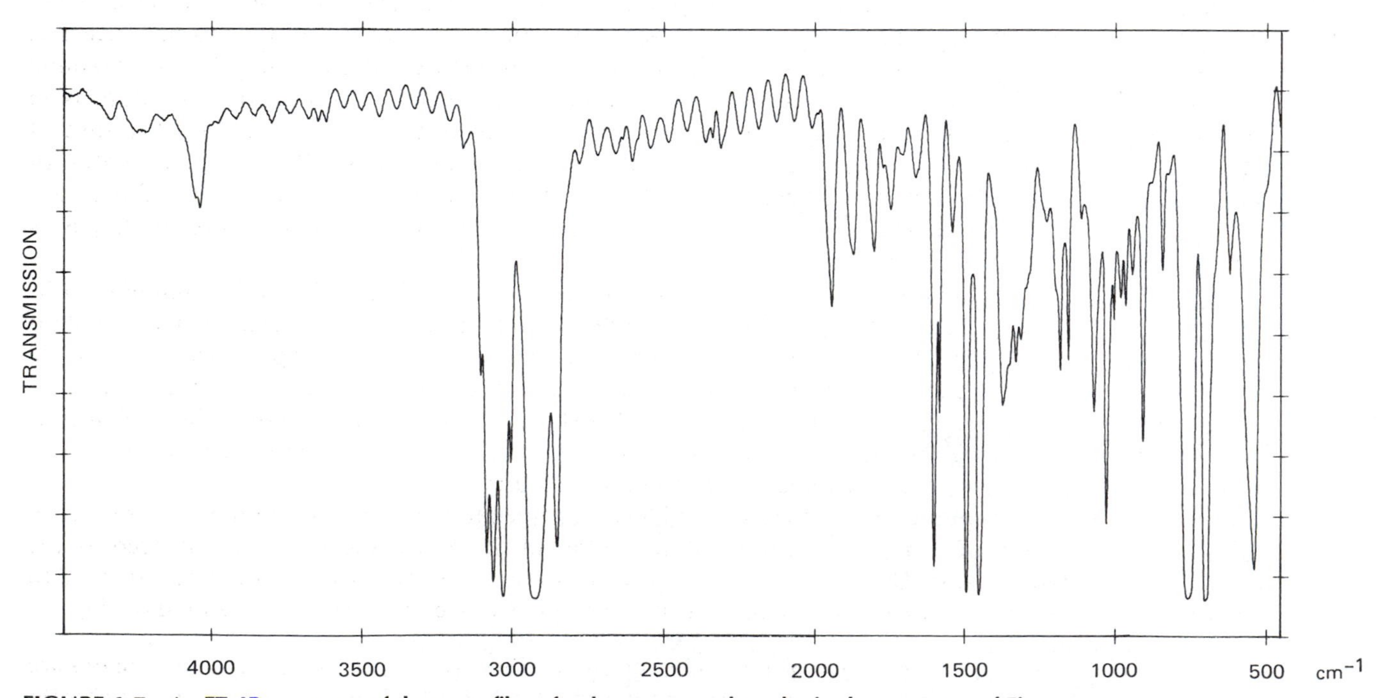

FIGURE 6-7 An FT-IR spectrum of the same film of polystyrene used to obtain the spectrum of Figure 6-6.

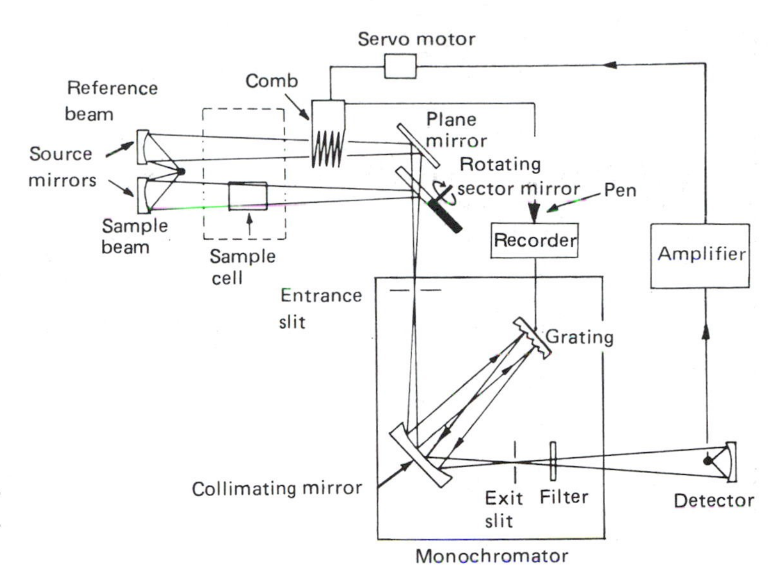

FIGURE 6-8 A schematic diagram of the operation of a double beam optical-null infrared spectrophotometer.

the sample and which frequencies have been unaffected. The radiation enters the monochromator through a slit and, after dispersion, leaves through another slit. The width of the entrance slit determines how much energy enters the monochromator, and the width of the exit slit determines the width of the narrow band of frequencies simultaneously reaching the detector.

The ability of the instrument to distinguish between absorptions at closely similar frequencies is known as the resolution. For most applications, a resolution of 4 cm^{-1} is adequate and is available from small grating instruments.

The detector is usually a very sensitive thermocouple maintained under vacuum. The radiation from the monochromator is focused by means of a mirror onto the thermocouple junction. The emf from the thermocouple is amplified and is used to produce the recorded spectrum.

A diagram of the path of the radiation through a typical double beam infrared spectrometer is shown in Figure 6-8. Radiation from the source is divided into two beams, a reference beam and a sample beam. The reference beam is needed to compensate for absorption by carbon dioxide and water vapor in the air in the instrument. The reference beam also compensates for the variations in source intensity, grating efficiency, and detector sensitivity over the spectral range covered. When solutions are studied, matched cells containing the solution and pure solvent may be placed in the sample and reference beams, respectively, to compensate for solvent absorptions. After passing through the sample compartment, the beams are transmitted alternately by means of a semicircular rotating mirror through an entrance slit into the monochromator. In the monochromator the radiation is dispersed into its constituent wavelengths by means of a diffraction *grating*, which can be rotated. At every position of the grating only a very narrow band of wavelengths passes through the exit slit to the detector.

When the beams are balanced, a steady dc signal is produced by the detector. When the sample absorbs energy, there will be a stronger signal from the reference beam. Thus, an alternating current will be produced by the detector. After amplification, this ac signal is used to drive a servomotor, which pushes a comb into the reference beam, and some energy is blocked off. When sufficient energy has been removed from the reference beam, the two beams will again be balanced and there will no longer be an ac signal produced by the detector. The movement of the comb is linked to a pen, which draws a trace usually on a moving chart. (In some instruments, the pen carriage moves while the chart remains stationary.)

The rate of movement of the chart (or the pen holder) is coupled to the rate of rotation of the grating, so that the spectrum is recorded as the grating rotates. Optical filters are automatically changed at the appropriate places in the scan. Often, two or more gratings mounted back to back, or on a turret, are used to cover the whole range of frequencies. The appropriate grating is moved into place as required during the course of the scan.

Another method of operation is known as *ratio recording*. In this case, the intensities of the two beams are alternately detected several times per second and compared electronically. The ratio of the intensities after amplification is displayed by a recorder. The more sophisticated ratio-recording instruments have a minicomputer built into the instrument, and the spectrum is stored digitally for future manipulations such as smoothing, baseline removal, spectrum comparison, and subtraction.

Two related parameters, scan speed and resolution, may be varied on an IR spectrometer. High resolution gives better separation of closely spaced peaks and more accurate intensity measurements for quantitative work. To achieve higher resolution, the monochromator slits must be narrowed and the spectrum must be scanned more slowly. Higher amplification is also needed when the slits are narrowed and this leads to more noise in the spectrum. Figure 6-9 shows an example of an IR spectrum recorded using two different sets of instrument settings for resolution.

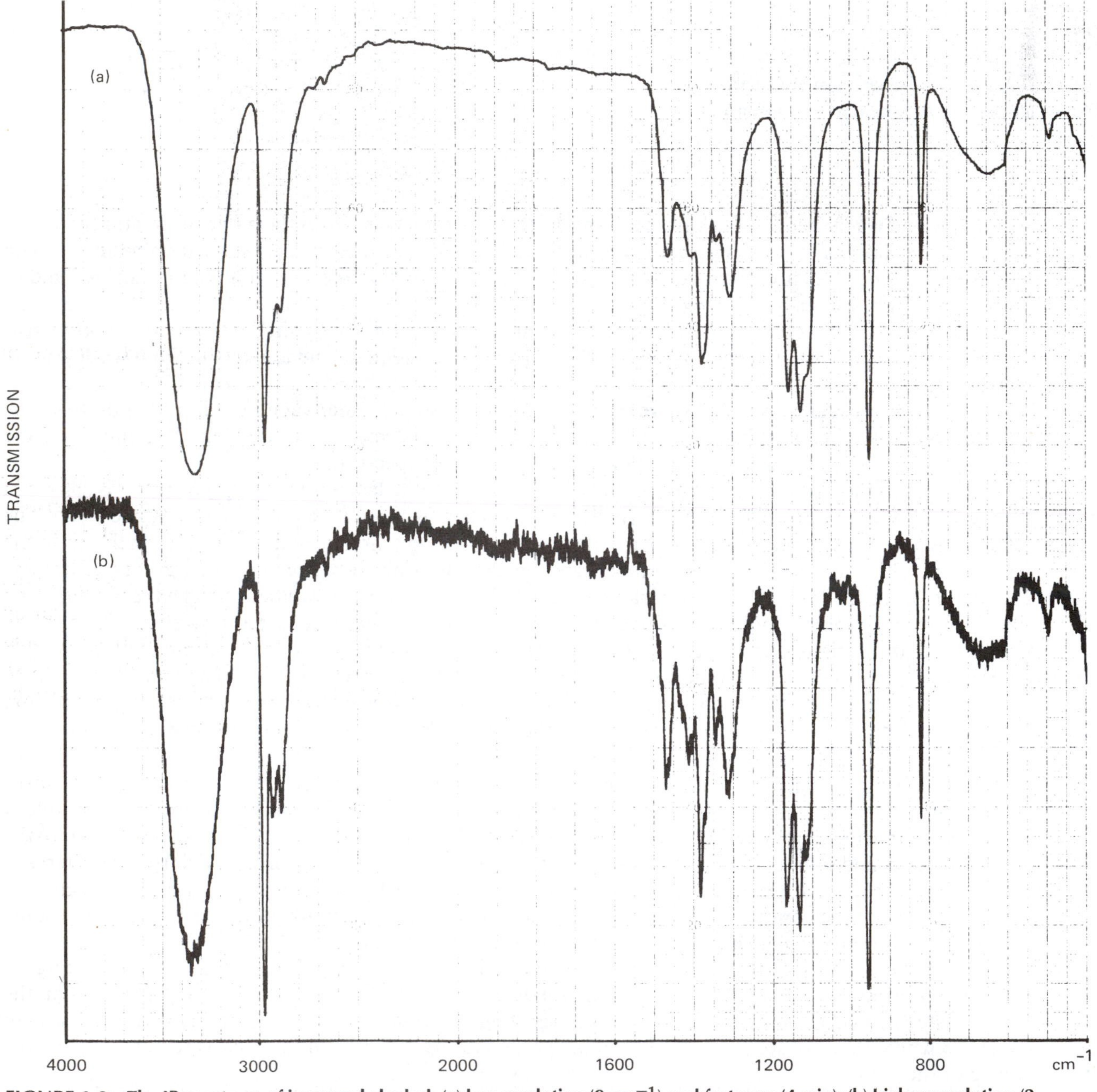

FIGURE 6-9 The IR spectrum of isopropyl alcohol: (a) low resolution (8 cm^{-1}) and fast scan (4 min); (b) higher resolution (2 cm^{-1}) and slow scan (24 min).

FIGURE 6-10 A Fourier transform infrared spectrophotometer, the Nicolet model 5DXB. [Courtesy of Nicolet Analytical Instruments.]

It is also possible to adjust the position of the baseline using either the 100% control or a reference beam attenuator.

6-5b Fourier Transform Infrared Spectrometers

An FT–IR spectrophotometer is shown in Figure 6-10. Important parts of the instrument are the computer and graphics terminal. Most FT–IR spectrometers are based on the *Michelson interferometer*. This instrument (Figure 6-11) consists of two plane mirrors, M1 and M2, mounted at 90° to each other and a semireflecting beam splitter (BS). One of the mirrors (M1) is fixed; the other (M2) can be moved very precisely and reproducibly through a distance (δ) of a few millimeters. The beam splitter transmits 50% of the incident radiation to one mirror and reflects 50% to the other. After reflection at M1, 50% of the radiation travels back through the BS and recombines with 50% of the radiation returned from mirror M2 and reflected by the BS. The optical path difference between the beams is known as the retardation x ($x = 2\delta$); unless $x = 0$, the recombined beams will interfere. With this arrangement 50% of the radiation returns to the source. The other 50% passes through the sample to a detector.

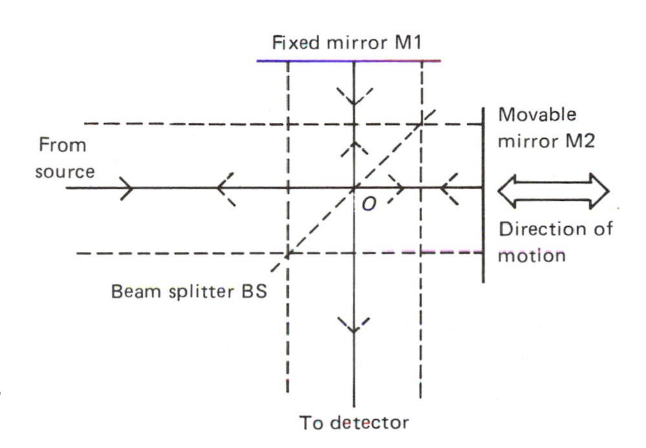

FIGURE 6-11 A schematic diagram of a Michelson interferometer.

Interferometers for the mid-IR region make use of the same sources as those used in dispersive instruments. However, thermocouples have a long response time and cannot be used as detectors for interferometers because the scan time is very short. One type of IR detector with response fast enough for an interferometer is the pyroelectric bolometer. The best pyroelectric material available at present is triglycine sulfate (TGS) or the deuterated compound (DTGS). The beam splitter is usually a very thin layer (typically 0.4 μm) of germanium deposited on an optically flat KBr plate. For the far-IR, the beam splitter is usually a thin film (typically 12.5 μm) of polyethylene terephthalate (Mylar). The thickness of the Mylar film controls the spectral range covered.

To understand how an interferometer can be used to measure an infrared spectrum, consider first a monochromatic beam of radiation of wavelength λ passing through the instrument. When $x=0$ or $n\lambda$ (n is an integer), the recombined beams will be exactly in phase, so the signal at the detector is a maximum. As mirror M2 is moved, the beams interfere and the signal falls to zero when $x=\lambda/2$. If mirror M2 continues to move at a constant velocity, the signal intensity $I(x)$ will vary according to a cosine function (eq. 6-10). In eq. 6-10 $I(\bar{\nu})$ is the intensity of the source at frequency $\bar{\nu}$ (cm^{-1}) and $\bar{\nu}=\lambda^{-1}$. A graph of $I(x)$ vs. x is known as an *interferogram*.

$$I(x) = 0.5\ I(\bar{\nu}) \cos 2\pi\bar{\nu}x \tag{6-10}$$

When a *continuous* source of infrared radiation is used, an infinite number of wavelengths will simultaneously pass through the interferometer and only when $x=0$ will all wavelengths be in phase. At any other position of mirror M2 a very complex interference pattern will result, giving rise to an interferogram like the one shown in Figure 6-12a. To obtain this interferogram, the mirror M2 was moved from a negative retardation through the zero path difference position to positive retardation. At $x = 0$, all wavelengths interfere constructively and produce the very strong central signal. The intensity $I(x)$ of the radiation reaching the detector at any other retardation is the sum of the intensities of all the interfering wavelengths at this mirror position.

A sample placed between the interferometer and the detector reduces the intensity of radiation at any frequency at which the sample absorbs. Thus the infrared absorption spectrum is contained in the resulting interferogram. Figure 6-12b shows the interferogram of Figure 6-12a modified by the absorption of a 0.05 mm thick polystyrene film placed between the interferometer and the detector. The interferograms are stored digitally by the computer as data files of $I(x)$ vs. x.

There is clearly a difference in the interferograms of Figure 6-12a and 6-12b, but in order to obtain quantitative information about the *absorption* from the interferograms we need the data in the form $I(\bar{\nu})$ vs. $\bar{\nu}$. To obtain this information, a mathematical procedure known as a *Fourier transformation* must be performed on both interferograms, as with FT–NMR. An FT–IR spectrum of the polystyrene sample used to

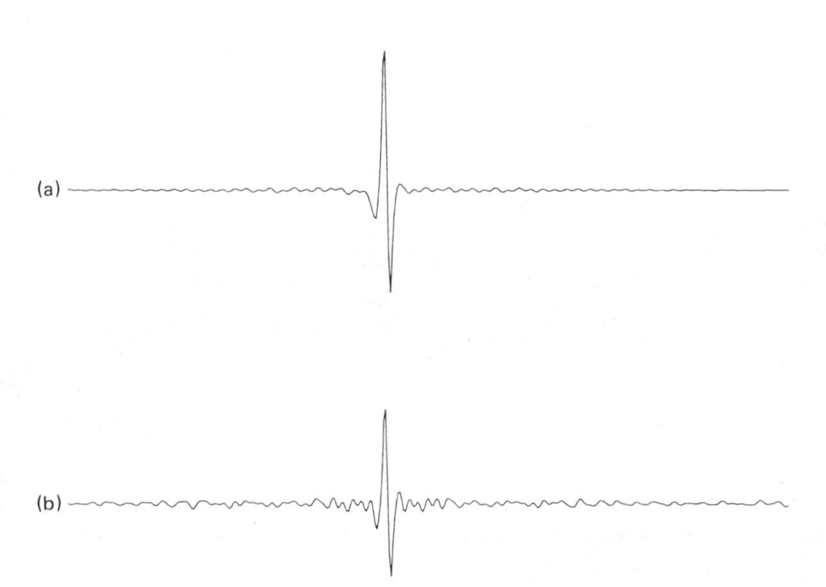

FIGURE 6-12 Interferograms from a continuous source: (a) with no sample; (b) with a 0.05 mm film of polystyrene placed between the interferometer and detector.

obtain the interferogram of Figure 6-12b is shown in Figure 6-7. The spectrum is virtually identical to that recorded by a dispersive IR spectrometer operating with the same resolution (compare with Figure 6-6).

The two main advantages of interferometers over dispersive spectrometers for infrared spectroscopy are speed and sensitivity. These advantages result from the increased energy throughput (Jacquinot's advantage) and higher signal-to-noise ratio (Fellgett's advantage) available from an interferometer. Limiting values for these advantages can be calculated (see Chapter 1 of reference 6-1). For mid-IR spectra (4000–400 cm^{-1}) at a resolution of 2 cm^{-1} the theoretical values are approximately 20 for Jacquinot's advantage and 40 for Fellgett's. However, sample size, detector sensitivity, and other instrumental factors can drastically reduce the magnitude of these advantages.

The resolution of an FT–IR spectrometer depends on the reciprocal of the retardation x. Hence, for a resolution of 2 cm^{-1} the moving mirror must move 0.25 cm. The Fourier transform is computed from the digitized values of the interferogram sampled at equal intervals (Δx) of retardation. The spectral range covered is inversely proportional to the sampling interval. For a spectral range of 4000–400 cm^{-1}, Δx is of the order of 10^{-4} cm. These sampling points must be separated by precisely equal intervals; otherwise noise will be introduced into the computed spectrum. This precision is possible when sampling points are triggered by the zero crossings of an interferogram generated by a helium–neon laser.

Co-addition of several scans results in an improved signal-to-noise ratio (S/N) because the signal increases directly with the number of scans (n), while the noise, because of its statistical nature, only increases with $\sqrt{n}$. Thus, in theory, 100 scans should improve the S/N by a factor of 10 over that of a single scan. In practice, instrumental restrictions and imperfections limit the improvement in S/N ultimately obtainable by co-addition of interferograms. It should also be noted that the spectrum cannot be plotted until all of the interferograms have been recorded and averaged, and the Fourier transformation carried out.

This section is not intended to be a comprehensive coverage of FT–IR spectroscopy. Consult references 6-1, 6-2, and 6-3 for further information on instrumentation and applications of this important technique. As far as organic structure determination is concerned, it makes little difference whether an IR spectrum is obtained from a dispersive or a Fourier transform instrument.

6-6

Raman Spectroscopy

Raman spectrometers such as the one shown in Figure 6-13 give information on vibrational frequencies in the range covered by both mid and far infrared spectrometers (10–4000 cm^{-1}). Samples in the form of liquids,

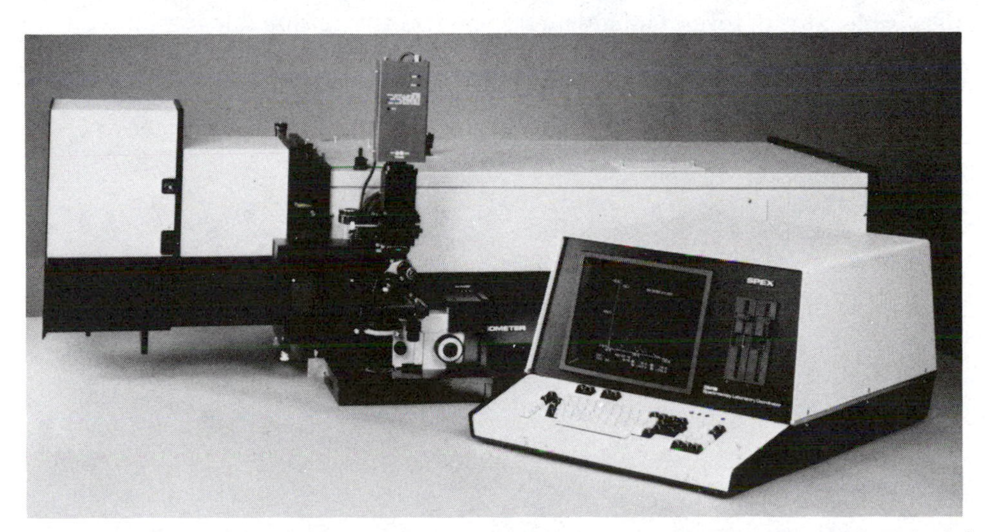

FIGURE 6-13 A Spex Industries Raman spectrometer. The sample compartment is at the left of the monochromator. The photograph also shows a microscope attachment. The desktop computer controls the instrument and displays the spectrum. [Courtesy of Spex Industries, Inc.]

solutions, powders, and single crystals can be handled by standard sampling techniques. However, gases, for which the Raman intensity is usually several orders of magnitude weaker, require a more sophisticated arrangement involving a higher power laser and multiple passing of the laser beam through the sample. Part of the low frequency end of the spectrum is always obscured by Rayleigh scattering of the exciting radiation. The width of the Rayleigh line, as it is called, increases from solids to pure liquids to solutions. An example of a Raman spectrum of a liquid organic compound appears in Figure 6-4.

Experienced users of Raman spectroscopy can obtain the same kind of structural information as the infrared spectroscopist. They use a somewhat different set of group frequencies, however, because vibrations that give rise to strong characteristic infrared absorption are often weak in the Raman spectrum. The converse is also true, as can be seen by comparing Figures 6-4 and 6-1. Raman spectra complement infrared spectra, and the two techniques used together provide a powerful tool in organic structure determination.

Raman spectroscopy has certain advantages over infrared. One is the simpler spectra usually observed in the Raman because of the absence of overtone or combination bands, which are an order of magnitude weaker in the Raman than in the infrared and are usually too weak to be observed. A second advantage is the wider choice of solvents for solution spectra—in particular, water can be used—and other solvents have more clear regions in the Raman than in the infrared. Information on the lower frequency region 50–200 cm^{-1}, corresponding to the far infrared, is easily obtained.

There are, however, certain disadvantages of Raman spectroscopy. One is the inherent weakness of Raman spectra, which can often be masked by the background and scattering from small particles in dirty samples. Fluorescence can sometimes be a nuisance, but this problem is virtually eliminated when the laser excitation is in the red. Also, it is often hard to get good spectra from solids unless they are crystalline. Another disadvantage of Raman spectroscopy is the high cost of the instruments. The cost of the lowest priced Raman instrument is in the price range of the best infrared grating spectrometers and of low cost FT–IR instruments.

6-6a Raman Spectrometers

Some of the instrumentation of a Raman spectrometer is similar to that of an infrared grating spectrometer. The Raman scattering is excited by the intense monochromatic radiation from a laser. The scattered light is focused by a lens onto the entrance slit of a monochromator, where it is dispersed into its spectrum. Several basic differences exist between Raman and infrared spectra. The infrared spectrum consists of a continuum with a few parts missing where the sample has absorbed radiation. The Raman spectrum, on the other hand, consists of mostly nothing, with a few narrow regions of radiation emitted from the sample.

The infrared continuum is relatively strong, but the Raman lines in the visible are inherently very weak. Infrared detectors, however, have low sensitivity, whereas very sensitive detectors (photomultipliers) can be obtained for radiation in the visible. Slits of several hundred micrometers give reasonable resolution in the

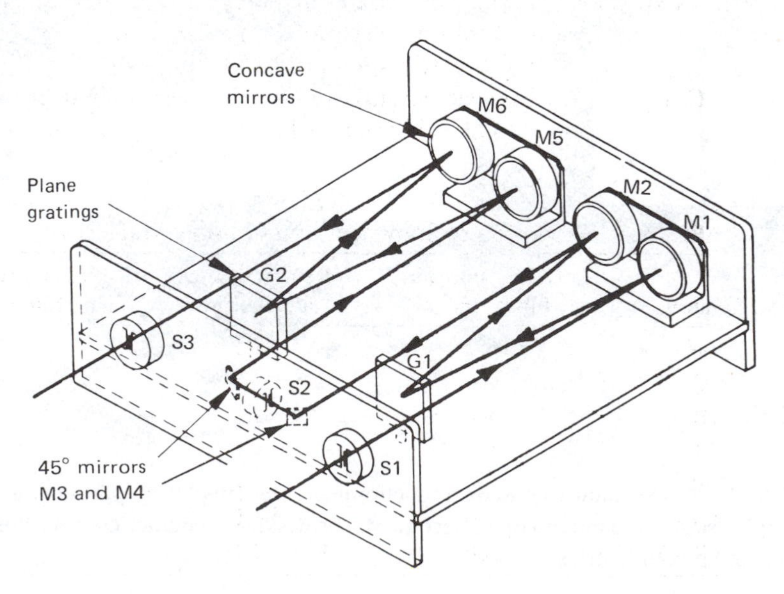

FIGURE 6-14 A schematic diagram of the Spex Industries Model 1401 monochromator.

infrared, but much smaller slits are needed in the visible. Another very important point is stray light, which if not effectively removed can produce such a high background that the Raman spectrum is lost. A Raman spectrum is usually recorded as variation of photomultiplier response with distance (in cm^{-1}) from the frequency of the exciting line.

In order to remove as much stray light as possible, a double, or even a triple, monochromator is used. A double monochromator, as the name implies, is simply two monochromators used in series, with a slit in between. During a scan the two gratings are rotated simultaneously by a common drive mechanism.

A typical arrangement of a double monochromator is shown in Figure 6-14. Light entering slit S1 is collimated by a concave mirror, M1, onto the first grating, G1, and then diffracted to M2. The radiation then passes through slit S2 into the identical second monochromator. A linear wavenumber counter can be set to zero at any excitation frequency. Thus, Raman frequencies can be read directly. The radiation is detected by a photomultiplier tube (PMT) mounted at the exit slit S3. The signal from the PMT is amplified and displayed on a recorder or a CRT screen. The Raman signal can also be interfaced through an analog-to-digital converter to a minicomputer, where the digitized spectrum can be stored for future manipulation. Further information on Raman spectrometers and instrumentation for Raman spectroscopy may be found in references 6-4, 6-5, and 6-6.

6-7

Sampling Methods for Infrared Transmission Spectra

The very simplest sampling techniques will often give quite satisfactory results. Sampling of pure liquids is relatively simple. Solutions and solid samples, on the other hand, require more care. In all transmission studies, except the pressed-pellet method, discs or plates transparent to infrared radiation are needed to support the sample. Gas cells and cryostats for low temperature work also need windows.

Several infrared transmitting materials are in common use. A summary of the more important of these is given in Table 6-2; most can be purchased as sawn crystal blanks ready for polishing.

NaCl. This is perhaps the most commonly used material. The rock salt region is the traditional range (2–16 μm) studied by earlier workers. This material is the least expensive and is easy to polish, but breaks easily and is only transparent as far as 650 cm^{-1}.

KCl. This is as cheap as NaCl, easy to polish, but fractures when subjected to stress. It is transparent to 500 cm^{-1}.

KBr. This is another traditional material. It costs about 50% more than NaCl or KCl. It is easy to polish but is fragile. KBr is transparent to 400 cm^{-1}.

CsI. This is a very useful material; it does not fracture under mechanical or thermal stress. However, it is very water soluble, very soft, and tricky to polish. Its cost is quite high (about ten times that of NaCl), but it is transparent to 150 cm^{-1}.

CaF_2. This is only very slightly soluble in water, so may be used for aqueous or D_2O solutions. The cost is about five times that of polished NaCl. CaF_2 windows should be purchased ready polished since this

TABLE 6-2 Properties of Some Infrared Window Materials

Material	IR Transmission Limit (cm^{-1})	Refractive Index at 4000 cm^{-1}	Solubility in Water (g/100 mL) at 20°C
NaCl	650	1.5	36
KCl	500	1.5	35
KBr	400	1.5	53
CsI	150	1.7	80
CaF_2	1100	1.4	0.002
BaF_2	850	1.5	0.1
KRS-5	200	2.4	.02

material is difficult to polish. Fortunately, CaF_2 is very hard and normally does not need repolishing. A disadvantage of CaF_2 is that it is only transparent to 1100 cm^{-1}.

BaF_2. This material is somewhat more soluble in water than is CaF_2, but has an extended transmission range to about 850 cm^{-1}. Both CaF_2 and BaF_2 fracture with thermal or mechanical shock.

KRS-5. This material is a mixed thallium bromide iodide compound. It forms bright red crystals, which are sparingly soluble in water, do not cleave, and transmit to 200 cm^{-1}. The main disadvantages of KRS-5 are the high price and the high refractive index, which may cause losses of transmitted energy by scattering. Also KRS-5 is toxic and may be attacked by some compounds in alkaline solution.

Further information on infrared optical materials can be found in Chapter 3 of reference 6-7.

6-7a Liquids and Solutions

Probably the easiest method to obtain a qualitative infrared spectrum of a liquid is to place one drop of the liquid onto a disc of NaCl, KBr, etc., cover the drop with a second disc, and mount the pair in a holder. Teflon spacers may be used to give various path lengths. Fixed-pathlength, sealed cells are also available. These usually have amalgamated silver or lead spacers. The cells are filled, emptied, or flushed by means of a syringe through conventional Luer ports. Teflon stoppers are used to close the ports. An example of a liquid cell is shown in Figure 6-15.

For the far infrared, polyethylene cells are available with various pathlengths. These become easily contaminated, but are of low enough cost to be disposable.

It is often convenient to record the infrared spectrum of a compound in solution. Unfortunately, some of the best solvents have very strong infrared absorption bands that obscure parts of the spectrum of the compound. Water, for example, absorbs strongly throughout the spectrum and is rarely used in routine infrared work. However, H_2O and D_2O can be useful solvents for infrared spectroscopy of compounds such as sugars, amino acids, and compounds of biochemical interest. Special window materials, such as CaF_2, BaF_2, or KRS-5, must be used. FT–IR instruments are especially useful for studying aqueous solutions because the capability of these instruments to signal-average over a large number of scans means that very low transmitted energy levels can be measured.

Weak to medium solvent absorption can be removed from the spectrum by using a pair of matched cells with the solvent in the reference beam and the solution in the sample beam. Since only regions in which the solvent does not absorb strongly can be used, a series of solvents will be needed for a complete spectrum. It should be noted that during the recording of a spectrum using matched cells, when a region in which the solvent absorbs strongly is scanned, essentially no energy passes in either the sample or reference beam. The instrument will be comparing "nothing" with "nothing," and the pen will be observed to be "dead" in such a region. FT–IR spectrometers and the more sophisticated dispersive infrared spectrometers can produce a spectrum from surprisingly little energy (as low as 0.1% T). However, for the low-priced instruments, the

FIGURE 6-15 A fixed-pathlength infrared cell.
[Courtesy of Spectra-Tech, Inc.]

TABLE 6-3 Useful Solvents for Infrared Solution Spectra

Solvent	Useful Regions (cm^{-1})	Typical Pathlength (mm)
CS_2	all except 2200–2100 and 1600–1400	0.5
CCl_4	all except 850–700	0.5
$CHCl_3$	all except 1250–1175 and below 820	0.25
$CHBr_3$	all above 700 except 1175–1100 and 3050–3000	0.5
C_2Cl_4	all except 950–750	0.5
Benzene	all above 750 except 3100–3000	0.1
CH_2Cl_2	all above 820 except 1300–1200	0.2
Acetone	2800–1850 and below 1100	0.1
Acetonitrile	all except 2300–2200 and 1600–1300	0.1
Cyclohexane	below 2600	0.1
N,*N*-Dimethylformamide	2750–1750 and below 1050	0.05
Diethyl ether	all except 3000–2700 and 1200–1050	0.05
Heptane and hexane	all except 3000–2800 and 1500–1400	0.2
Dimethyl sulfoxide	all except 1100–900	0.05

solvents used must be chosen so that no absorption band of the sample is overlapped by a strong absorption band of the solvent. A partial list of solvents is given in Table 6-3 with useful regions indicated.

6-7b Solids

Solid samples are handled either in the form of mulls or pressed discs. To make a mull, a small amount of the sample is ground in an agate or mullite mortar. Then a drop of a paraffin oil, usually Nujol, is added and the grinding continued. Nujol is available at very low cost in drugstores. The mixture should have the consistency of a thin paste. It is transferred to a window of NaCl, or KBr, etc., and a second window lowered onto it. A thin film is produced by gentle pressure accompanied by a slight rotational movement.

The two plates with the mull between are placed in a cell holder and the spectrum is recorded. There will be strong bands at 2900, 1470, and 1370 cm^{-1} and a weak band at 720 cm^{-1}. These are due to Nujol. If the Nujol bands are stronger than the peaks from the sample, then more sample and less Nujol must be ground. If the sample peaks are too strong, the two windows can be squeezed together or a small drop of Nujol can be added. The user will have to experiment with the mull to obtain the best results.

When the region near 2900 cm^{-1} is important, other mulling material must be used. The usual second compound is Fluorolube, a chlorofluorocarbon oil, which is available from chemical suppliers. This material is completely opaque below 1400 cm^{-1} in the pathlengths normally used in mulls. It also has a band at 1650 cm^{-1}. Another compound useful for mulls, when the 2900 cm^{-1} region is to be studied, is hexachlorobutadiene. This compound has no absorptions above 1650 cm^{-1}. It also has a useful "window" between 1500 and 1250 cm^{-1}.

For solid compounds that are insoluble in the usual solvents, a convenient sampling method is the pressed pellet technique. A few milligrams of the sample are ground together in an agate or mullite mortar with about 100 times the quantity of material (the matrix) that is transparent in the infrared. The usual material is KBr, although other compounds such as CsI, TlBr, and polyethylene are used in special circumstances (see below). The finely ground powder is introduced into a stainless steel die, usually 13 mm in diameter, which is then evacuated for a few minutes with a rotary vacuum pump to remove water vapor. The powder is then pressed in the die between polished stainless steel anvils at a pressure of about 30 tons per square inch, on the disc. Other devices such as the Mini-Press are available for making KBr pellets.

A well-made KBr pellet will have 80–90% transmittance in regions below 3000 cm^{-1} where the sample itself does not absorb. At the high frequency end of the spectrum the transmission will often be low because of scattering effects. The amount of scattering by mulls and pellets depends on the refractive indices of the sample and the matrix or mulling material. When the refractive indices are similar, large particles cause serious scattering at short wavelengths. At longer wavelengths, the particle size becomes less important. However, the particle size should be less than about 20 μm for good pellets. This size can usually be achieved by hand grinding in a hard mortar or by mechanical shaking of the sample in a tube containing stainless steel balls. When scattering reduces the transmission, a reference beam attenuator can be used to restore the baseline to near 100% T, but as the spectrum is scanned, less and less attenuation is needed, because the scattering becomes less serious at longer wavelengths.

Matrices other than KBr may be used. CsI is useful when spectra down to 180 cm^{-1} are required. One disadvantage of CsI is that it is a highly water-soluble material. Lower pressures are required for CsI pellet formation than for KBr. TlBr is used when materials of high refractive index are studied. It has a refractive index of 2.3 in the infrared region and is transparent out to 230 cm^{-1}. It has low water solubility and can be used in conditions of high humidity. Powdered polyethylene has been used for making pellets for far infrared spectra because, apart from a band at 80 cm^{-1}, its spectrum between 400 and 10 cm^{-1} is free of absorption.

In addition to the scattering problems discussed above, the pressed-disc method has other disadvantages. Changes may occur in the sample during the grinding and pressing process and the sample may react with absorbed water or even with the matrix material.

An excellent review of infrared sample handling techniques is given in Chapter 8 of reference 6-7.

6-8

Raman Sampling Methods

Sampling methods for Raman spectroscopy are usually simpler than for infrared. In principle, all that has to be done is to place the sample in the laser beam in front of the entrance slit of the monochromator. The sample itself then becomes the source of radiation passing into the monochromator. Focusing of the laser beam increases the Raman intensity but may also damage the sample. Sample damage can be prevented by rotating the sample rapidly in a special cell for liquids or holder for solids. Multi-passing of the laser beam also increases the Raman scattering from liquids, solutions, and gases.

The scattered light must be focused on the entrance slit of the monochromator by means by a lens. This lens is usually mounted on an optical bench together with the sample holder, a polarization analyser, and a polarization scrambler. The laser beam can impinge on the sample from below or from the side (known as 90° illumination), or in the direction of the slit and collection optics (known as 0° or 180° illumination), but in all cases it is simple enough to mount an interference filter, a half-wave plate, and a focusing lens in the laser beam before it reaches the sample.

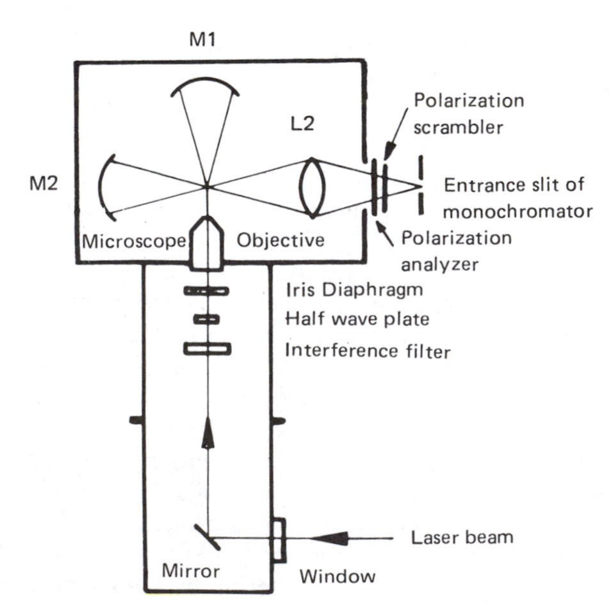

FIGURE 6-16 The arrangement of the optical components in a typical Raman spectrometer sample compartment.

All commercial instruments are sold complete with a sample compartment containing the items mentioned above. A schematic diagram of a typical sample compartment is shown in Figure 6-16. Starting from the laser itself, the components have the following uses. The interference filter is of a type that passes only a narrow band of wavelengths centered on the laser line. Nonlasing plasma lines are thus eliminated. However, it is often useful to remove this filter and use the scattered plasma lines for frequency calibration of the monochromator. The half-wave plate turns the plane of polarization through 90°. The iris diaphragm also reduces interference from the laser plasma. The microscope objective focuses the light to a spot in the sample.

The scattered light is focused by means of the lens (L2) onto the entrance slit of the monochromator. A spherical mirror (M1) causes double passing of the beam and a mirror (M2) reflects 180° Raman scattering back toward the monochromator. The polarization analyzer is usually a piece of polaroid, and the polarization scrambler is a quartz wedge that ensures that no polarized light gets into the monochromator. This component eliminates any effects produced by the polarizing properties of the monochromator itself in measurements of depolarization ratios (see Section 6-8c).

6-8a Liquids and Solutions

A small volume quartz cell (5 mm × 5 mm × 5 cm), such as supplied for UV absorption work, can be used as a Raman liquid sample container. This type of cell will have polished windows. The ends of the cell also should be polished. Although quartz is preferable because it has low fluorescence, cells made from ordinary Pyrex glass are usually satisfactory.

Raman spectroscopy is ideally suited for microsampling. The laser beam can be focused to a very small area, and samples contained in melting point capillaries yield virtually the same spectra as much larger samples. No special accessories are needed as in the case of infrared microsampling. It is important, however, to align the sample exactly with the slit of the monochromator. This operation may be done easily with the aid of a low-powered auxiliary laser. Figure 6-17 illustrates a typical arrangement.

The auxiliary laser is aligned with the slit, and the sample is adjusted so that it is exactly in line with both the exciting and the auxiliary laser beams.

The techniques for handling solutions are essentially the same as for pure liquids. However, care must be taken with the solvents. They should be clean, pure, and sometimes filtered and redistilled to avoid problems from fluorescence.

One of the advantages of Raman spectroscopy over infrared is in the area of solution spectra. As we saw in Section 6-7a most of the common solvents have extensive regions of absorption in the infrared. Although this is not the case for the Raman, there is still the drawback that the intensities obtained from dilute solutions cannot be increased by having longer path length. There is a lower limit to the useful concentrations that can be used, normally 1–5% by weight (about 0.1 M).

Of course, all solvents do have a Raman spectrum, but Raman lines are usually much narrower than infrared absorption bands. In addition, overtones and combinations are very much weaker in the Raman. Hence, there are many completely clear regions in the Raman spectra of all the usual solvents. Water is an excellent solvent for Raman solution studies, a fact that is extremely important in the examination of compounds of biological interest, since the natural environment is usually aqueous solution.

Common solvents are listed in Table 6-4. The most important Raman lines are given in the second column. The stronger lines extend for approximately 20–40 cm^{-1} on either side of the frequency quoted, and the weaker lines obscure a region of about 10–20 cm^{-1}. When Raman spectra of dilute solutions are recorded, the instrument is used at a high sensitivity. Under these conditions even the very weakest lines of the solvent may be observed. These lines are listed in the third column of Table 6-4. Of course, a spectrum of the solvent must

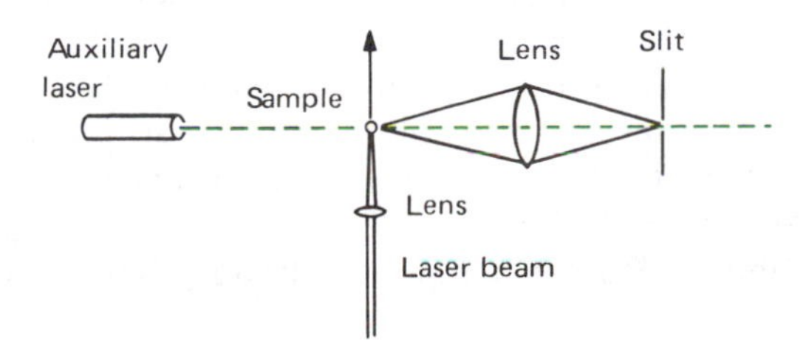

FIGURE 6-17 An arrangement for aligning a small sample with the monochromator slit.

TABLE 6-4 Some Solvents for Raman Spectroscopy

	Raman Frequencies[a] (rounded to nearest 5 cm^{-1})	
Solvent	The Most Important Lines	Other Weak Lines
Carbon disulfide	650 (vs), 795 (m)	400
Carbon tetrachloride	220 (s), 315 (s), 460 (vs), 760–790 (m)	1540
Chloroform	260 (s), 365 (vs), 670 (vs), 760 (m). 1220 (m), 3025 (w)	—
Methanol	1035 (m), 1460 (w), 2840 (m), 2900–2950 (w)	1110, 1165, 3405
Acetonitrile	380 (s), 920 (s), 1375 (w), 2250 (s), 2950 (s)	1440, 3000
Methylene chloride	285 (s), 700 (vs), 740 (w), 2985 (w)	1155, 1425, 3045
Nitromethane	480 (m), 655 (s), 920 (vs), 1370–1410 (m), 2965 (m)	610, 1105, 1560, 2765, 3040, 3060
Acetone	420 (m), 790 (vs), 1070 (m), 1225 (m), 1430 (m), 1710 (m), 2920 (s)	390, 490, 900, 1365, 2845, 2965
Benzene	605 (m), 990 (vs), 1180 (m), 1580–1610 (w), 3040–3070 (m)	405, 780, 825, 850, 2620, 2950
Cyclohexane	385 (w), 430 (w), 800 (s), 1030 (m), 1160 (w), 1265 (m), 1440–1470 (m), 2855 (m), 2920–2950 (m)	1345, 2630, 2670, 2700, 2905
Ethanol	880 (s), 1450–1490 (m)	430, 1050, 1095, 1275, 2875, 2930, 2975
Dimethyl formamide	320 (w), 360 (m), 410 (m), 660 (s), 870 (s), 1095 (m), 1405 (s), 1440 (m), 1660 (m)	1065, 2800, 2860, 2930
Dimethyl sulfoxide	300–350 (s), 385 (m), 650–710 (vs), 955 (w), 1045 (s), 1420 (m), 2915 (m), 3000 (w)	2885
Distilled water	3450 (m, br), 1650 (w)	—

[a]s = strong; m = medium; w = weak; v = very; br = broad.

be recorded in all cases. As far as possible, the same conditions should be used to observe the spectra of the solution and the solvent.

6-8b Solids

Polycrystalline powders are best handled in capillary tubes, or tamped into a hole in the end of a metal rod. Irregular pieces of solid materials may be glued to a support rod and held in the laser beam. Powders may be pressed into pellets and examined in the same way. Multipassing of the laser beam is not possible and high background scatter may be often observed, especially from amorphous materials. The interference filter described in Section 6-8 must be used with highly scattering samples to eliminate spurious laser lines from the spectrum.

A crystal 1 or 2 mm long can be mounted on a goniometer head in the same way that crystals are mounted for X-ray examination. Single crystals can be carefully positioned, so that the laser beam passes along one of the axes of the crystal. In this case, when a polarization analyzer is used, important information concerning the symmetry of the normal vibrations of the molecule can be obtained. A discussion of single crystal Raman spectroscopy is given in references 6-4 and 6-6 and Chapter 18 of reference 6-7.

Water-sensitive solid samples can be loaded into capillary tubes in a dry box and sealed. Larger bore tubes (NMR tubes) can be connected by a ground joint to a stopcock and evacuated on a vacuum line. The sensitive compound is transferred to the tube in the glove box and the stopcock rejoined. After evacuation the tube can be sealed.

For further details of Raman sampling methods, the reader is referred to Chapter 15 of reference 6-7.

6-8c Depolarization Measurements

Any Raman line can be classified as polarized or depolarized. The light from a laser is plane polarized and the Raman scattering is also partially polarized, so that when an analyzer is placed before the entrance slit of the monochromator there is a difference in intensity between the light transmitted through the analyzer in the parallel and perpendicular orientations. The ratio of the perpendicular intensity to the parallel is known as the depolarization ratio ρ. For laser-excited Raman spectra, ρ has a maximum value of 0.75. The depolarization ratio is usually measured by comparing peak heights. A band with ρ less than 0.75 is said to be polarized and a band with ρ exactly equal to 0.75 is said to be depolarized.

Depolarization measurements are useful for assigning frequencies to the totally symmetric vibrations of the molecule (polarized bands). Another use is in the separation of overlapping bands. When one band is due to a totally symmetric vibration, its intensity is often drastically reduced upon rotating the analyzer through 90°. For non-totally symmetric vibrations the reduction in intensity is much less, and the band due to the non-totally symmetric mode predominates.

To make depolarization measurements the spectrum is scanned twice with identical monochromator and amplifier settings. The only difference is the orientation of the analyzer (parallel and perpendicular). Small regions of the spectrum are scanned sequentially, peak heights (I) and baselines (I_o) are noted for each orientation, and depolarization ratios are calculated from eq. 6-11. It should be noted that depolarization ratios cannot be measured for powdered solid samples.

$$\rho = \frac{(I - I_o)_\perp}{(I - I_o)_{//}} \tag{6-11}$$

6-9

Special Techniques in Dispersive and Fourier Transform Infrared Spectroscopy

All the standard sampling methods used to obtain infrared transmission spectra can be used with FT–IR instruments. However, a number of special techniques have been developed specifically for FT–IR instruments and these will be outlined in this section together with some other techniques that can be used with either dispersive or FT spectrometers.

6-9a Gas Phase Infrared Spectroscopy

The gas or vapor phase spectrum of an organic molecule recorded under low resolution gives essentially the same information as the liquid or solution spectrum. In most cases, rotation of the molecules only causes some broadening of peaks. Gases are usually studied in a 10 cm glass cell with alkali halide windows at pressures of between 10 and 100 mm Hg. A vacuum line is needed to handle a gas or the vapor of a volatile liquid. Less volatile liquids or solids can be studied in the vapor phase if the cell is warmed with a heating tape until sufficient pressure of vapor is produced. In some cases this may be a useful alternative to other methods. If the vapor pressure or concentration of a gas is very low, the radiation can be reflected several times through the sample using special multiple-pass gas cells. These cells have effective pathlengths of up to 40 meters.

6-9b Microsampling

For infrared microsampling, special small volume cells are available. Very small KBr discs can also be prepared. Two types of microcell are in common use: the cavity cell and a miniaturized version of the standard liquid cell. The former consists of a small block of material such as NaCl, KBr, etc., with parallel polished faces and a microcavity drilled ultrasonically in the center. Various volumes are available from a fraction of a microliter up to 0.5 ml. Standard path lengths in the range 0.05 to 5 mm can be obtained. The second kind of microcell is usually purchased assembled and sealed with a fixed pathlength. Because of the small area exposed to the infrared beam, a beam condenser must be used with microcells.

Microgram quantities of sample can be pressed into KBr pellets using a special die. The pellets are pressed into openings centered in stainless steel discs. Various sizes of pellet from 0.5 mm up to 13 mm diameter can

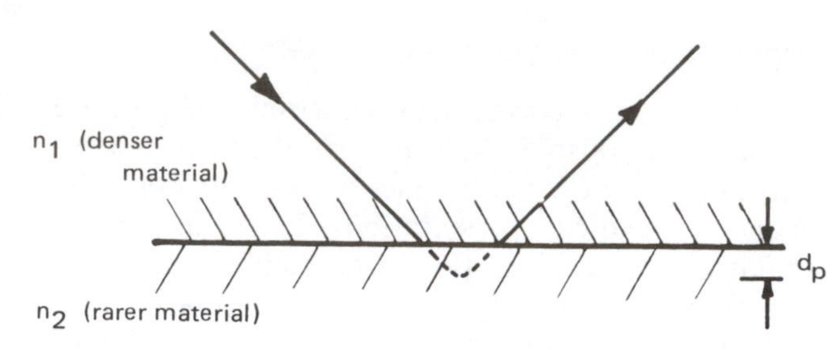

FIGURE 6-18 The path of a ray of light in total internal reflection. The ray penetrates a fraction of a wavelength (d) beyond the reflecting surface into the rarer medium of refractive index n_2.

be made. The 0.5 mm pellet requires about 1 mg of KBr mixed with 1 μg of sample. A standard KBr die could also be used with paper inserts, but the smallest practical diameter is somewhat larger.

6-9c Reflection Spectroscopy

When a ray of light strikes an interface between two *nonabsorbing* materials of different refractive index (n_1 and n_2), the light is partially transmitted and partially reflected. This property is used in the manufacture of beam splitters for FT–IR spectrometers. When light enters material 2 from 1 with n_1 less than n_2 (for example, air to glass), the result is *external reflection.* For external reflection the *reflectivity* can never be 100% and is usually much less. In the case when n_1 is greater than n_2, we have *internal reflection* and in this case the reflection is total when the angle of incidence is between the critical angle and 90° (grazing incidence). Both internal and external reflection can be used to obtain spectra.

Total internal reflection can be observed in a glass of water. When the inside of the glass is viewed through the water surface, it appears to be completely silvered and opaque. However, when the outside of the glass is touched with a finger, details of the ridges and whorls on the skin are clearly seen, but the silvered effect remains between these features. The total reflection is destroyed where the skin actually makes contact with the glass. This can be explained by a penetration of the electromagnetic field of the light into the rarer medium (smaller refractive index) by a fraction of a wavelength, as illustrated in Figure 6-18.

If the light is in the infrared region and if the rarer medium is a compound that absorbs infrared radiation, then the penetrating radiation field can interact by means of an absorbing mechanism to produce an attenuation of the total (internal) reflection (ATR). This interaction can be described in terms of an effective thickness, which corresponds to a sample thickness in a normal absorption process.

The technique of internal reflection spectroscopy consists of recording the wavelength dependence of the reflectivity, through the usual infrared region, of the interface between a sample compound and a material that has a higher refractive index but is transparent to the infrared radiation. The radiation approaches and leaves the interface through the denser medium. In the simplest experimental arrangement, the material of high refractive index is in the form of a prism, as shown in Figure 6-19. The ATR spectra recorded are very similar to those obtained by normal transmission techniques.

Multiple internal reflection (MIR) can be achieved using a crystal of the form shown in Figure 6-20. Multiple reflections can produce enhancement of the spectra of weakly absorbing samples. The angle of incidence is fixed in this method and more than one plate may be needed to study a variety of compounds. The thinner the plate, the

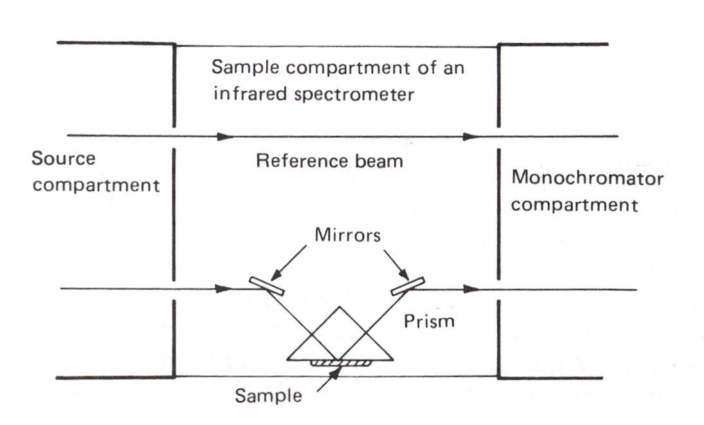

FIGURE 6-19 A diagram of a simple experimental arrangement for obtaining an internal reflection spectrum.

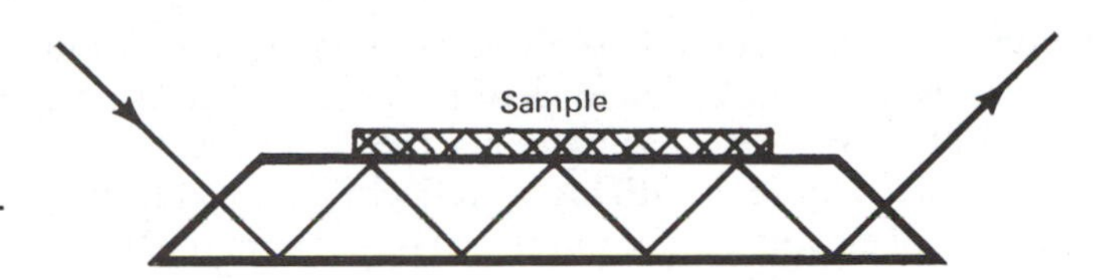

FIGURE 6-20 A multiple internal reflection element (IRE).

more internal reflections there will be. A typical plate is 2 mm thick, 5 cm long, and 2 cm high, with the ends cut at 45°. This plate gives about 25 reflections. The ends of the plates can be cut at various angles, with 30°, 45°, or 60° being the most commonly used. A large energy loss of 50% or more will occur in the sample beam so that an attenuator or a second internal reflection unit should be used in the reference beam. The infrared beam is directed through the IRE by means of a four mirror optical system.

The depth of penetration of the infrared radiation into the sample depends on the wavelength, the refractive indices of sample and crystal, and angle of incidence of the reflected ray at the interface. The most commonly used crystal material is KRS-5. Other materials such as Irtran-2 and germanium are also used. The properties of the first compound were discussed in Section 6-7. Germanium has a very high index of refraction (4.0 at 1000 cm^{-1}) and transmits to about 700 cm^{-1}.

Good spectra comparable to infrared transmission spectra often can be obtained using ATR methods, if proper choice is made of crystal material and angle of incidence. Differences will be observed between transmission and ATR spectra. In general, bands are shifted slightly to lower frequencies in ATR and the high frequency bands are weaker in ATR than in transmission. Although the refractive index of the internal reflection element must be greater than that of the sample, the best spectra are obtained when the difference in refractive indices is small. The method of ATR is extremely useful for difficult samples and for studying films or coatings.

For further details on the theory of ATR the reader is referred to the book by Harrick (ref. 6-8). A large number of applications of MIR are discussed in Chapter 14 of reference 6-7.

External reflection can be subdivided into two categories. Some of the reflected radiation from a sample travels along the theoretical path that would be followed if the surface were perfectly flat. This is called *specular reflection.* The rest of the reflected radiation is scattered in all directions. This is called *diffuse reflection.* Infrared spectra can be obtained from both kinds of reflection using dispersive spectrometers with reasonable sensitivity or FT–IR instruments. For specular reflection spectra, simple reflectance attachments are available from the manufacturers of spectrometer accessories. Alternatively, an MIR optical system can be adapted for specular reflection. Samples, which must be flat or well polished, are mounted in a holder that fits onto the pins on the MIR attachment.

Solid samples in the form of powders give mainly diffuse reflection, so the reflected radiation must be collected over a wide solid angle. Accessories that enable diffuse reflectance FT–IR spectra (DRIFTS) to be recorded are available from several manufacturers. The theory of diffuse reflectance spectrometry has been reviewed and many applications of DRIFTS discussed in reference 6-9.

6-9d Gas Chromatography and Infrared Spectroscopy (GC–IR)

Most of the methods discussed in this book apply to pure substances. However, in many practical situations the samples to be studied are in the form of mixtures. In larger scale preparations mixtures are separated into their components by standard methods such as distillation, crystallization, etc. For very small samples, some form of chromatography is invariably used. If the material is sufficiently volatile and relatively stable, gas chromatography (GC) is an ideal method for separation. Subsequent study by infrared (IR) can be conveniently carried out using the special technique known as (GC–IR).

There are two possibilities for GC–IR. Individual components in the effluent from the GC column can be collected and the IR spectra recorded later. This method is used with dispersive instruments. Alternatively, the rapid-scanning properties of FT–IR instruments can be used to record spectra of GC peaks as they are eluted from the gas chromatograph.

For the first method, GC effluents can be collected in small traps cooled with liquid nitrogen or dry ice and subsequently transferred to an infrared cell. Fraction collecting devices are available commercially. These usually collect the material directly on a thermoelectrically cooled MIR crystal. The ATR spectrum is then recorded. Another method involves trapping the sample onto a small quantity of KBr powder from which a pellet can be pressed. Chapter 13 of reference 6-7 gives details of several other GC–IR techniques.

In most GC–FT–IR measurements, the GC effluent is passed through a *light-pipe* gas cell, which consists of a long narrow gas cell with highly reflecting walls. Ideally, the volume of the cell should be slightly less than the volume of carrier gas containing a GC peak, so that the FT–IR spectrum of each individual GC peak can be recorded. It takes less than one second to acquire an interferogram, which when transformed will produce a spectrum with modest resolution (4–8 cm^{-1}). Only microgram amounts of sample mixtures need to be injected into the chromatograph. This technique is undoubtedly one of the most useful applications of FT–IR spectroscopy. Further information on GC–FT–IR is available in references 6-1 and 6-2.

REFERENCES/BIBLIOGRAPHY

6-1 J. R. Durig (Ed.), *Analytical Applications of FT–IR to Molecular and Biological Systems*, Reidel, Boston, 1980.
6-2 P. R. Griffiths, *Fourier Transform Infrared Spectrometry*, Wiley, New York, 1986.
6-3 A. E. Martin, *Infrared Interferometric Spectrometers*, vol. 8 of J. R. Durig (Ed.), *Vibrational Spectra and Structure*, Elsevier, Amsterdam, 1980.
6-4 T. R. Gilson and P. J. Hendra, *Laser Raman Spectroscopy*, Wiley–Interscience, London, 1970.
6-5 M. C. Tobin, *Laser Raman Spectroscropy*, Wiley–Interscience, New York, 1971.
6-6 D. A. Long, *Raman Spectroscopy*, McGraw-Hill, London, 1977.
6-7 R. G. J. Miller and B. C. Stace (Eds.), *Laboratory Methods in Infrared Spectroscopy*, 2nd ed., Heyden, London, 1972.
6-8 N. J. Harrick, *Internal Reflection Spectroscopy*, Interscience, New York, 1967 (available from Harrick Scientific Corporation).
6-9 P. R. Griffiths and M. P. Fuller, *Advan. Infrared Raman Spectrosc.*, **9**, 63, (1982).

7

Group Frequencies: Infrared and Raman

7-1

Introduction to Group Frequencies

The subject of group frequencies is essentially empirical in nature, but there is a sound theoretical basis for it. Infrared and Raman spectra of a large number of compounds containing a particular functional group—carbonyl, amino, phenyl, nitro, etc.—are found to have certain features that appear at more or less the same frequency for every compound containing the group. It is reasonable, then, to associate these features with the functional group, provided a sufficiently large number of different compounds containing the group have been studied. For example, the infrared spectrum of any compound that contains a C=O group has a strong band between 1800 and 1650 cm^{-1}. Compounds containing $—NH_2$ groups have two infrared bands between 3400 and 3300 cm^{-1}. The Raman spectrum of a compound containing the $C_6H_5—$ group has a strong polarized line near 1000 cm^{-1}, and nitro groups are characterized by infrared and Raman bands near 1550 and 1350 cm^{-1}. These are just four examples of the many characteristic frequencies of chemical groups observed in infrared or Raman spectra.

Various pairs of atoms joined by bonds in a molecule can be treated as diatomic molecules. This simple approach gives surprisingly good results when one of the atoms of the pair is a light atom, not bonded to any other atom, e.g., C—H and N—H in CH_3NH_2 or C—H and C=O in $(CH_3)_2CO$. The stretching frequencies (cm^{-1}; see Section 6-4 for a discussion of units) of these diatomic groups can be calculated from eq. 7-1, in

$$\nu\,(\mathrm{cm}^{-1}) = 130.3\sqrt{\frac{k}{\mu}} \qquad (7\text{-}1)$$

which k is the force constant (Nm^{-1}) and μ is the reduced mass, $m_1m_2/(m_1+m_2)$ in atomic mass units (amu). The numerical constant $130.3 = 1/2\pi c\sqrt{N} \times 10^{-1}$ (N is Avogadro's number, 6.02×10^{23}, and c is the velocity of light, 3.0×10^8 m s^{-1}). In older texts and papers, force constants have units of mdyn A^{-1} (10^5 dyn cm^{-1}) and c has units of cm s^{-1}. When these units are used, the numerical constant of eq. 7-1 becomes 1303.

Frequencies of numerous diatomic groups including C≡C and C=C can be calculated from eq. 7-1. A band characteristic of the group will be observed in the infrared or Raman spectrum in the predicted region, provided the vibrational frequency of the group is not close to that of another group in the molecule. Some diatomic group frequencies calculated from eq. 7-1 are given in Table 7-1. These are all examples of characteristic group frequencies and can be used to establish the presence of the functional group in the molecule.

It can be seen from Table 7-1 that the values of the force constants of double and triple bonds are approximately twice and three times those of single bonds, respectively. Carbon–carbon single bonds are included in Table 7-1, but C—C stretching does not usually give a well-defined group frequency. Most organic molecules contain several C—C single bonds and other groups that have vibrational frequencies in the same region as the C—C stretching mode. These vibrations interact with each other, and the simple

TABLE 7-1 Calculated Frequencies of Some Diatomic Groups

Group	Reduced Mass (amu)	Force Constant ($N\ m^{-1}$)	Frequency (cm^{-1})
O—H	0.94	700	3600
N—H	0.93	600	3300
C—H	0.92	500	3000
C—C	6.00	425	1100
C=C	6.00	960	1650
C=O	6.86	1200	1725
C≡C	6.00	1600	2100
C≡N	6.46	2100	2350

model (eq. 7-1) does not apply. Vibrational interactions can take several forms and are discussed in Section 7-2b.

When two or more identical groups are present in a molecule, there will be two or more similar frequencies in the spectrum, which may or may not be resolved. If the groups are attached to the same carbon atom or to two adjacent atoms the frequencies may be spread over a few hundred wavenumbers by strong interactions. The four CH groups in ethylene (C_2H_4) provide such an example. The four observed CH stretching frequencies are 3270, 3105, 3020, and 2990 cm^{-1}.

The explanation for these characteristic diatomic group frequencies lies in the approximately constant values of the stretching force constant of a group in different molecules. Polyatomic groups also have characteristic frequencies, which involve both stretching and bending vibrations or combinations of these. No simple relationship such as eq. 7-1 can be found for polyatomic groups, and the best way to establish whether or not a particular group such as $—CH_2$, $—CH_3$, $—NH_2$, or $—C_6H_5$ has characteristic bending in addition to stretching frequencies is to examine the vibrational spectra of a large number of compounds containing these groups.

Although many undergraduate laboratories do not have access to Raman spectrometers, some discussion of Raman group frequencies is necessary in a text on organic spectroscopy. Group frequencies in the Raman generally are not the same as in the infrared, and Raman spectra can provide important additional information on molecular structure and symmetry. An integrated discussion of infrared and Raman group frequencies is presented here to give the reader a more complete perspective on the subject.

7-2 Factors Affecting Group Frequencies

7-2a Symmetry

The vast majority of organic molecules have little or no symmetry. Nevertheless, some knowledge of symmetry can be of considerable help in understanding the factors that affect intensities of group frequencies.

Occasionally, a group frequency is not observed in the infrared spectrum. This is usually a consequence of symmetry. If a molecule possesses a *center of symmetry*, all vibrations that are symmetric with respect to that center are *inactive* in the infrared because they do not produce a change in the dipole moment. These symmetric modes, however, are always observed in the Raman spectrum because they give rise to a change in the polarizability of the molecule, as illustrated by the spectra of *trans*-dichloroethene (Figure 7-1). The C=C stretch is not observed in the infrared, but is seen at 1580 cm^{-1} in the Raman spectrum.

The C—Cl stretches give rise to two vibrational modes, a symmetric mode observed in the Raman spectrum at 840 cm^{-1} and an antisymmetric mode in the infrared at 895 cm^{-1}. Similarly, the two CH bending modes are observed at 1200 cm^{-1} (infrared) and 1270 cm^{-1} (Raman). This is an example of the infrared–Raman exclusion rule that holds when a molecule has a center of symmetry.

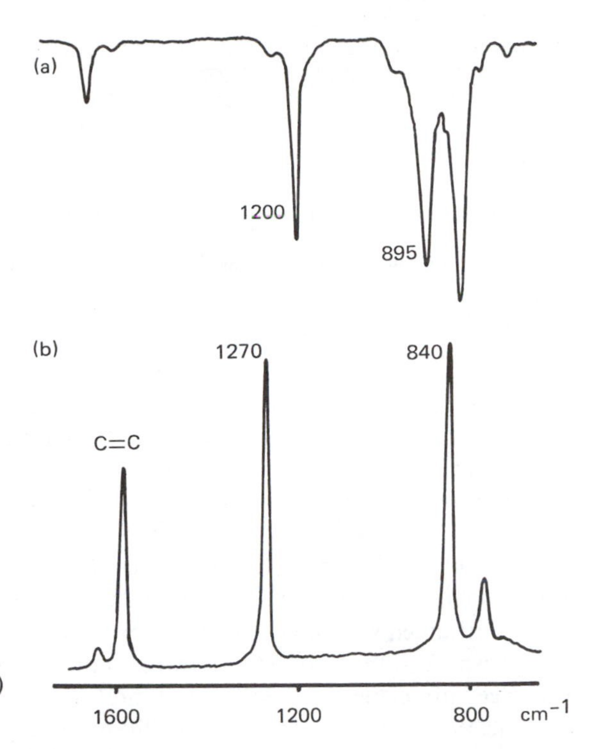

FIGURE 7-1 Portions of the (a) infrared and (b) Raman spectra of *trans*-dichloroethene.

Another example of the effect of a center of symmetry is given by the C≡C stretching mode. In methylacetylene ($CH_3C{\equiv}CH$) the vibration is both infrared and Raman active, and a strong infrared band is observed at 2150 cm^{-1}. On the other hand, in dimethylacetylene ($CH_3C{\equiv}CCH_3$), which has a center of symmetry, a strong C≡C stretching mode is found in the Raman spectrum, but no band is observed in the infrared near 2150 cm^{-1}. In larger, more complicated molecules, a local symmetry may exist for a homonuclear diatomic group such as C≡C, so that the infrared absorption from the group vibration may be weak or absent. In such cases a Raman spectrum can confirm the presence (or absence) of the functional group.

Molecules of high symmetry have simple infrared and Raman spectra. As an example consider the benzene molecule. It has 12 atoms and therefore has $3N-6=30$ normal modes of vibration. The first effect of the high symmetry is to make 10 pairs of these vibrations have identical frequencies (degenerate modes). This leaves 20 different normal frequencies. The second effect of the high symmetry is to reduce the number of modes for which there is a change in dipole moment (infrared active) or a change in polarizability (Raman active). In fact, the infrared spectrum of benzene contains only four fundamentals, whereas the Raman spectrum contains six.

When the symmetry of benzene is reduced, as in 1,3,5-trichlorobenzene, the number of infrared and Raman active modes increases, but there are still some degenerate modes and the spectra are relatively simple. When the symmetry is completely removed, as in 1-chloro-2-bromobenzene, all 30 normal modes are active in both infrared and Raman. However, because of the residual symmetry of the benzene ring, some of these vibrations, although allowed, appear only very weakly and are hard to distinguish from the weak bands of overtones and combinations.

Vibrations of the methyl group ($-CH_3$) in an unsymmetrical molecule can be described in terms of the local symmetry of the free group, which has a threefold axis and three planes of symmetry. A free methyl group would have $3N-6=6$ normal modes of vibration comprising symmetric and degenerate antisymmetric (with respect to the threefold axis) stretching and bending modes. When the methyl group is attached to a molecule, three new modes appear, a torsional mode and a degenerate pair of rocking vibrations. These motions would be rotation in the free methyl group. Thus, there are four regions of the spectrum where we expect to find methyl group vibrations. This conclusion is amply supported experimentally. The presence of the methyl group also contributes three skeletal modes to the vibrations of the molecule. These correspond to translations of the free methyl group.

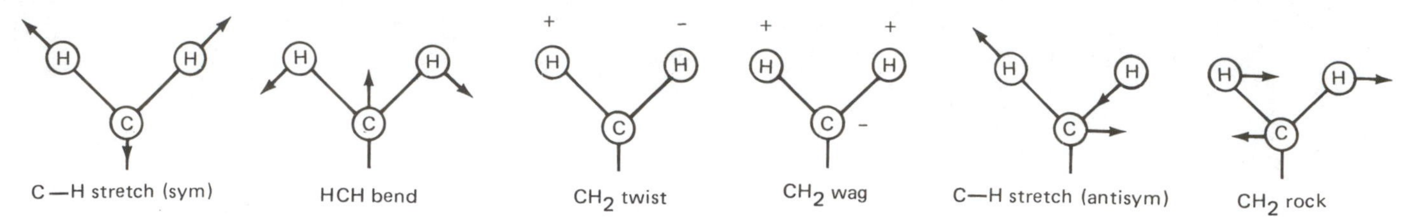

FIGURE 7-2 The vibrations of a CH_2 group. The arrows show the direction of motion of atoms in the plane of the CH_2 group, while the + and − signs denote motion above and below the plane, respectively.

When the methyl group is part of a molecule with lower symmetry, the degeneracies are removed, leading to the observation of doublets in some of the regions of the spectrum where the methyl group frequencies are expected. The torsional mode is actually inactive in the infrared because it produces no change in dipole moment. However, it may be allowed by the symmetry of the whole molecule and, in fact, methyl torsions are sometimes observed as weak bands in the far infrared.

The vibrations of a methylene group (CH_2) can also be described in terms of the local symmetry of the group that has a twofold axis and two planes of symmetry. Figure 7-2 shows the vibrations associated with a CH_2 group, when it is attached to a molecule. The free CH_2 group would have three modes, symmetric and antisymmetric (with respect to the twofold axis) stretching and the bending or scissors vibration. When the group is part of a larger molecule three additional modes described as twisting, wagging, and rocking are produced. Of these, the twisting mode produces no change in dipole moment and hence is not allowed in the infrared. However, it can give rise to a very weak band in the spectrum of an unsymmetrical molecule.

The terms twofold and threefold axis, plane of symmetry, and center of symmetry are examples of *symmetry elements*. The collection of all symmetry elements that a molecule possesses is known as a *point group* and provides a way of classifying the symmetry of the molecule. This, in turn, leads to an understanding of the symmetry of the normal vibrations of a molecule and to a prediction of the number of frequencies expected in the infrared and Raman spectra. The student is urged to consult references 7-12 and 7-13 for further information on the applications of symmetry and group theory to vibrational spectroscopy.

7-2b Mechanical Coupling of Vibrations

Two completely free identical diatomic molecules will, of course, vibrate with identical frequencies. When the two diatomic groups are part of a molecule, however, they can no longer vibrate independently of each other because the vibration of one group causes displacements of the other atoms in the molecule. These displacements are transmitted through the molecule and interact with the vibration of the second group. The resulting vibrations appear as in-phase and out-of-phase combinations of the two diatomic vibrations. When the groups are widely separated in the molecule, the coupling is very small and the two frequencies may not be resolved.

Consider the two C—H stretching modes in acetylene. These are observed at 3375 cm^{-1} in the Raman (in-phase) and 3280 cm^{-1} in the infrared (out-of-phase). In diacetylene, however, the two C—H stretching vibrations have closer frequencies, near 3330 and 3295 cm^{-1}. The vibrations of two *different* diatomic groups are not coupled unless the uncoupled frequencies are similar as the result of a combination of mass and force constant effects. For example, in thioamides and xanthates, the C=S group has a force constant of about 650 N m^{-1} and the reduced mass is 8.72 amu, so that the vibrational frequency calculated from eq. 7-1 is approximately 1120 cm^{-1}. The C—N and C—O groups have force constants of about 480 and 510 N m^{-1}, respectively, and the reduced masses are 6.46 and 6.86 amu. The calculated frequencies are both approximately 1120 cm^{-1}. Consequently, in any compound containing a C=S group adjacent to a C—O or a C—N group, there may be an interaction between the stretching vibrations of the groups. In compounds such as thioamides and xanthates, where the carbon atom is common to both groups, the coupling is large and the two vibrations interact with each other to produce two new frequencies, neither of which is in the expected region of the spectrum.

The way in which such mechanical coupling occurs can be illustrated for the case of two C=C groups

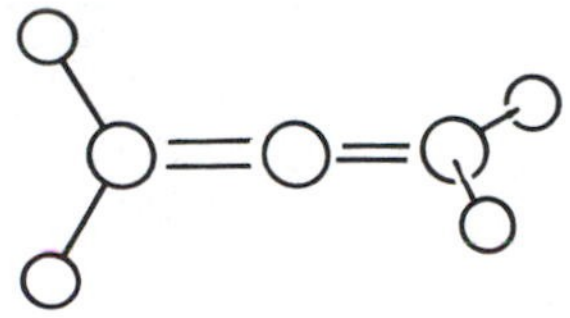

FIGURE 7-3 The allene molecule.

coupled through a common carbon atom, as in the allene molecule, $CH_2{=}C{=}CH_2$ shown in Figure 7-3. In the absence of strong coupling one might expect to observe a band in the infrared spectrum near 1600 cm^{-1} from the out-of-phase (unsymmetrical) vibrations of the C=C groups and a line in the Raman spectrum from the in-phase (symmetrical) modes at a similar frequency. For the 1,3-butadiene molecule ($CH_2{=}CH{-}CH{=}CH_2$), these bands are, in fact, observed (the observed frequencies being 1640 cm^{-1} in the infrared and 1600 cm^{-1} in the Raman). For allene, however, the observed frequencies are near 1960 and 1070 cm^{-1}. This result can be understood in terms of mechanical coupling of the two C=C group vibrations. When such coupling occurs, it is usually found that the higher frequency mode is the antisymmetric (out-of-phase) vibration and the lower frequency mode is the symmetrical (in-phase) vibration.

It is also possible for coupling to occur between dissimilar modes such as stretching and bending vibrations, when the frequencies of the vibrations are similar and the two groups involved are adjacent in the molecule. An example is found in secondary amides, in which the C—N stretching vibration is of the same frequency as the NH bending mode. Interaction of these two vibrations gives rise to two bands in the spectrum, one at a higher and one at a lower frequency than the uncoupled frequencies.

Singly bonded carbon atom chains are, of course, not linear, so that the simple model used for the allene molecule would have to be modified. In addition, we were able to ignore the bending of the C=C=C group in allene, which cannot couple with the stretching modes, because it takes place at right angles to the stretching vibrations. Mechanical coupling will always occur between C—C single bonds in an organic molecule, so that there is no simple C—C group stretching frequency. One can expect that there will always be several bands in the infrared and Raman spectra in the 1200–800 cm^{-1} range in compounds containing saturated carbon chains. Certain branched chain structures, such as the tertiary butyl group, $(CH_3)_3C{-}$, and the isopropyl group, $(CH_3)_2CH{-}$, do have characteristic group frequencies involving the coupled C—C stretching vibrations. These will be discussed further in later sections.

A special case of mechanical coupling, known as *Fermi resonance*, often occurs. This phenomenon, which results from coupling of a fundamental vibration with an overtone or combination, can shift group frequencies and introduce extra bands. For a polyatomic molecule there are $3N-6$ energy levels for which only one vibrational quantum number (v_i) is 1 when all the rest are zero. These are called the fundamental levels and a transition from the ground state to one of these levels is known as a *fundamental*. In addition, there are the levels for which one v_i is 2, 3, etc. (overtones) or for which more than one v_i is nonzero (combinations). There are therefore a very large number of vibrational energy levels, and it quite often happens that the energy of an overtone or combination level is very close to that of a fundamental. This situation is termed *accidental degeneracy*, and an interaction known as Fermi resonance can occur between these levels provided that the symmetries of the levels are the same. Since most organic molecules have no symmetry, all levels have the same symmetry and Fermi resonance effects occur frequently in vibrational spectra.

Normally, an overtone or combination band is very weak in comparison with a fundamental, because these transitions are not allowed. However, when Fermi resonance occurs there is a sharing of intensity and the overtone can be quite strong. The result is the same as that produced by two identical groups in the molecule. As an example, two peaks are observed in the carbonyl stretching band of benzoyl chloride, near 1760 and 1720 cm^{-1} (Figure 7-4). If this were an unknown compound, one might be tempted to suggest that there were two nonadjacent carbonyl groups in the molecule. However, the lower frequency band is due to the overtone of the CH out-of-plane bending mode at 865 cm^{-1}.

Numerous other well-characterized examples of Fermi resonance are known. The N—H stretching mode of the —CO—NH— group in polyamides (nylons), peptides, proteins, etc., appears as two bands near 3300 and 3050 cm^{-1}. The N—H stretching fundamental and the overtone of the N—H deformation mode near 1550 cm^{-1} combine through Fermi resonance to produce the two observed bands. The CH stretching region

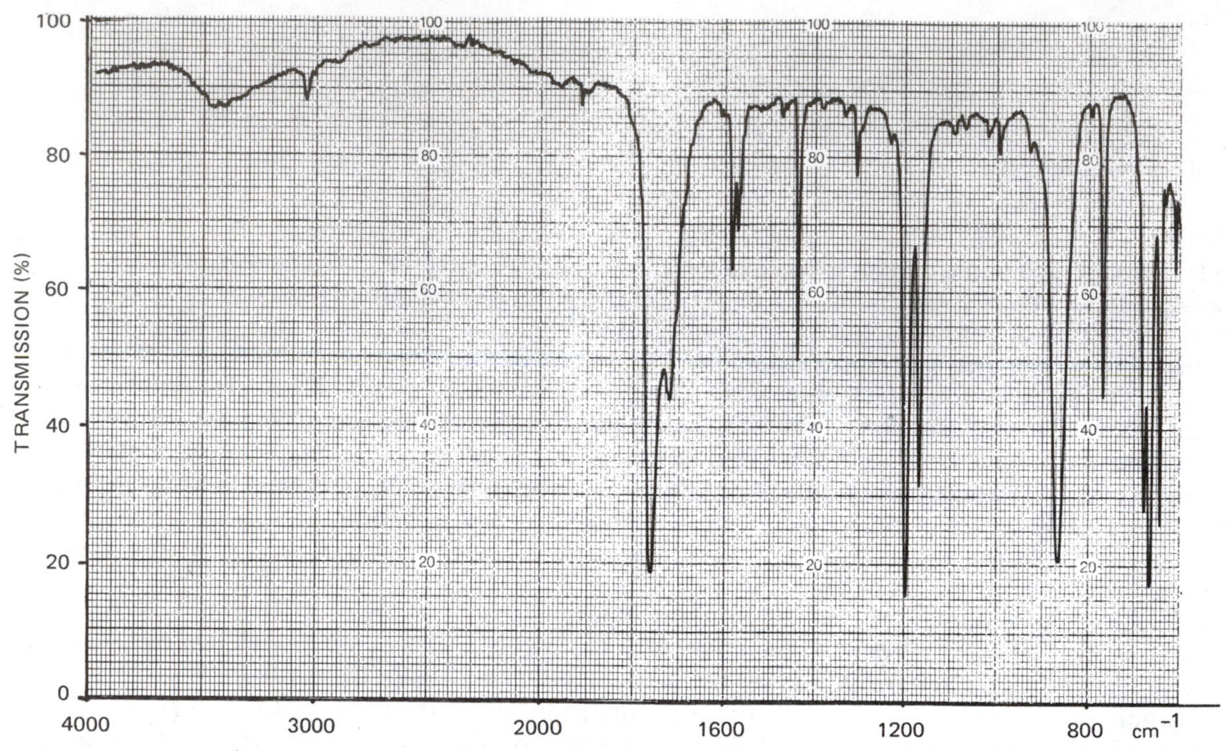

FIGURE 7-4 The infrared spectrum of benzoyl chloride, showing the Fermi doublet at 1760–1720 cm^{-1}.

of the —CHO group in aldehydes provides another example of Fermi resonance. Two bands are often observed near 2900 and 2700 cm^{-1} in the infrared spectra of aldehydes. This doubling is attributed to Fermi resonance between the overtone of the C—H deformation mode, which occurs near 1400 cm^{-1}, and the C—H stretching mode, which would also occur near 2800 cm^{-1} in the absence of Fermi resonance.

In many molecules, mechanical coupling of the group vibrations is so widespread that there are few, if any, frequencies assignable to functional groups. Many such examples are found in aliphatic fluorine compounds, in which the CF and CC stretching modes are coupled with each other and with FCF and CCF bending vibrations. The presence of fluorine can be deduced from several very strong infrared bands in the region between 1400 and 900 cm^{-1}. These vibrations give very weak Raman lines.

7-2c Chemical and Environmental Effects on Group Frequencies

Hydrogen bonding, electronic and steric effects, physical state, and solvent and temperature effects all contribute to the position, intensity, and appearance of the bands in the infrared and Raman spectra of a compound. Lowering the temperature usually makes the bands sharper, and better resolution can be achieved, especially in solids, at very low temperatures. However, there is a possibility of splittings due to crystal effects, which must be considered when examining the spectra of solids under moderately high resolution. Polar solvents can cause significant shifts of group frequencies through solvent–solute interactions, such as molecular association through hydrogen bonding.

Hydrogen bonding (written X—H···Y) occurs between the hydrogen atom of a donor X—H group such as OH or NH and an acceptor atom Y which is usually O or N. The main effects on the infrared and Raman spectra are broadening of bands in the spectra and shifts of group frequencies. X—H stretching frequencies are lowered by hydrogen bonding, and X—H bending frequencies are raised. Hydrogen bonding also affects the frequencies of the acceptor group, but the shifts are less than those of the X—H group. Inert solvents can reduce the extent of hydrogen bonding and even eliminate the effect in very dilute solutions. Figure 7-5 compares the infrared spectrum of glacial acetic acid with that of a 1% solution in CCl_4. The very broad band

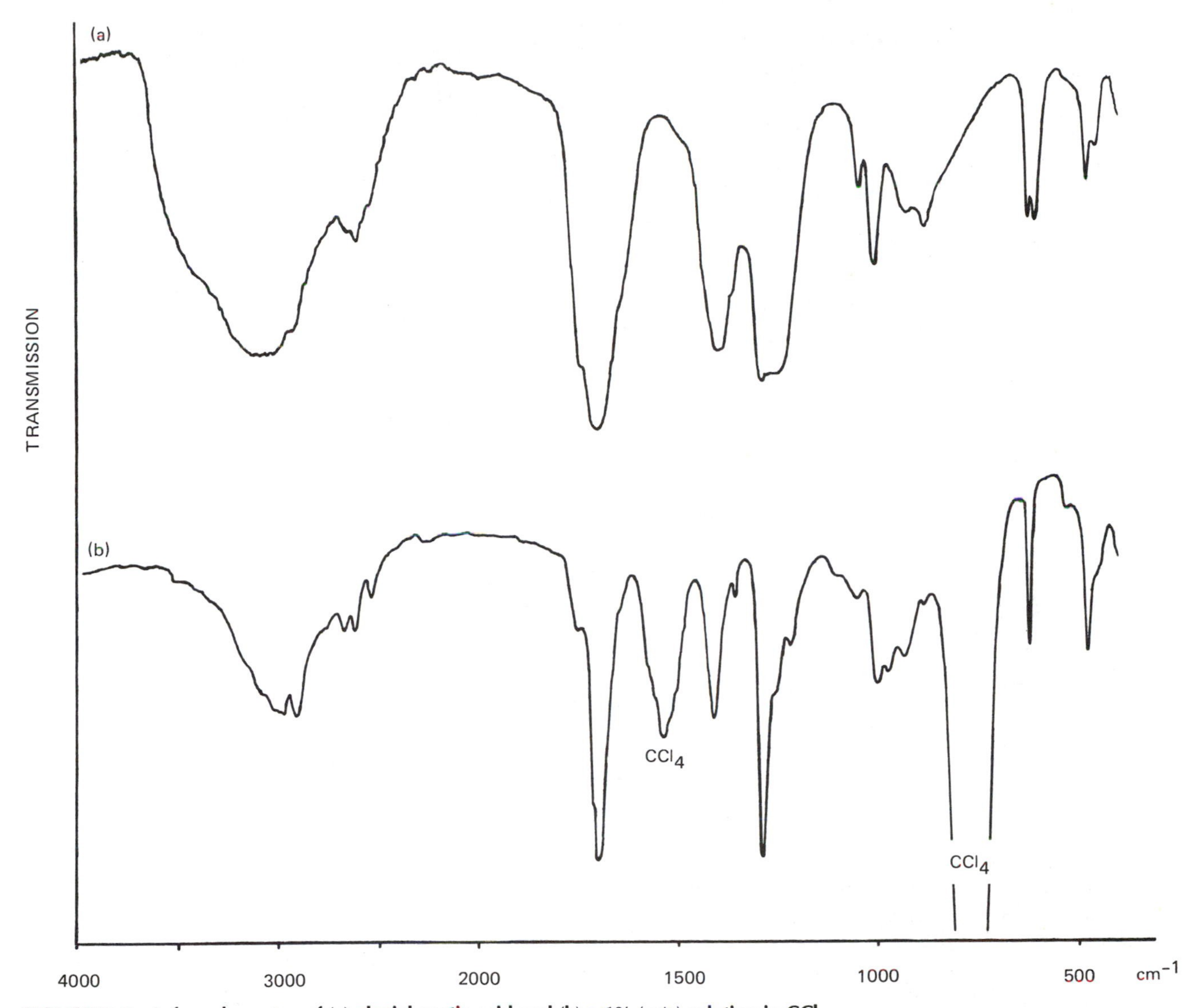

FIGURE 7-5 Infrared spectra of (a) glacial acetic acid and (b) a 1% (w/v) solution in CCl_4.

centered near 3100 cm^{-1} in the spectrum of the pure acid is due mainly to OH stretching of hydrogen-bonded polymers. In the solution spectra, it is seen that the bands are all sharper. However, hydrogen bonding persists even in dilute solution and the acid is present mainly as a cyclic dimer. Hydrogen bonding is discussed further in Section 8-16f.

Steric effects on group frequencies are quite interesting and useful for diagnostic purposes. As an example, consider the series of alicyclic ketones: cyclohexanone, cyclopentanone, and cyclobutanone. The observed carbonyl stretching frequencies are 1714, 1746, and 1783 cm^{-1}. This increase in frequency with increasing angle strain is generally observed for double bonds directly attached (exocyclic) to rings. Similar frequency changes are observed in the series of compounds, methylenecyclohexane, methylenecyclopentane, and methylenecyclobutane, in which a $C{=}CH_2$ group replaces the C=O group in the ring. The observed C=C stretching frequencies are 1649, 1656, and 1677 cm^{-1}, respectively. These shifts are illustrated in Figures 7-6 and 7-7.

When the double bond is a part of the ring, a decrease in the ring angle causes a *lowering* of the C=C stretching frequency. The observed frequencies for cyclohexene, cyclopentene, and cyclobutene are 1650, 1615, and 1565 cm^{-1}, respectively. Steric effects can give rise to rotational isomers, and certain group

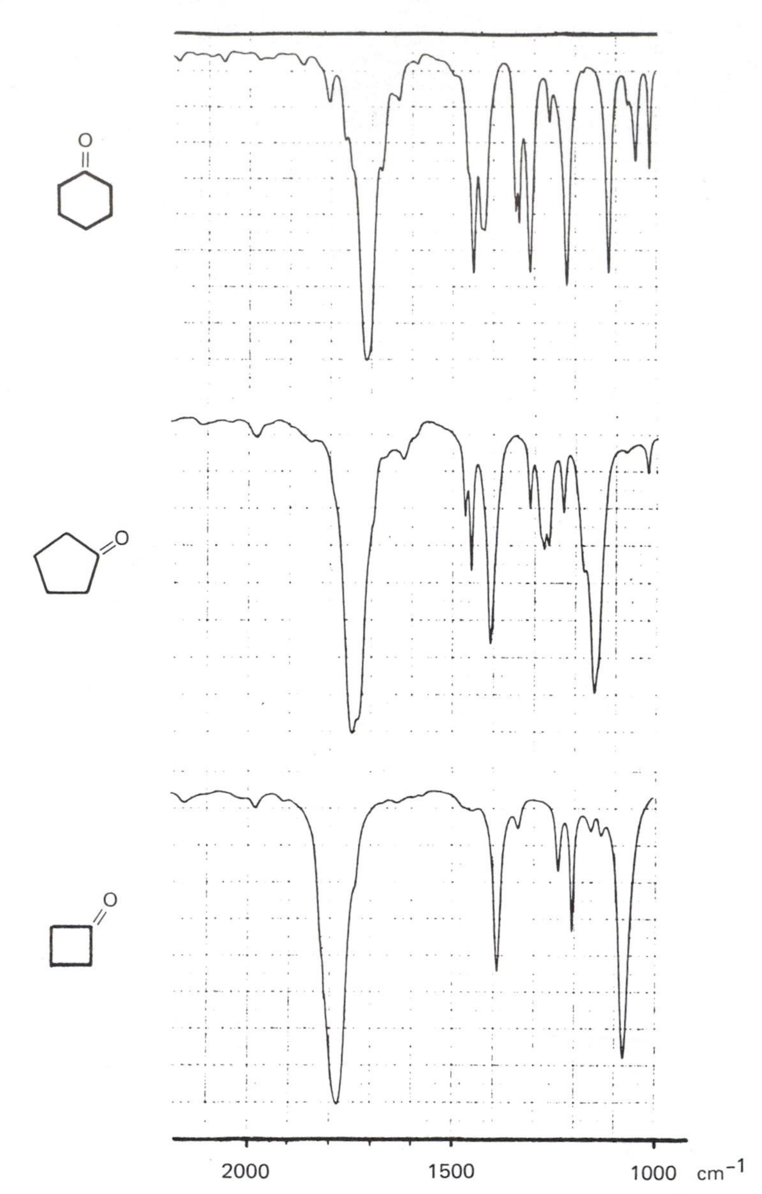

FIGURE 7-6 Portions of infrared spectra of three cyclic ketones showing the influence of increasing ring strain on the C=O stretching frequency. [Reproduced with permission of Aldrich Chemical Company, Inc., from *The Aldrich Library of FT–IR Spectra.*]

frequencies may be somewhat different in the different isomers. Two bands in the spectrum of 1,2-dichloroethane at 1290 cm^{-1} and 1235 cm^{-1} have been identified with trans and gauche forms of this compound. Substitution in the ortho position of phenols can change the position of the OH group vibration. It has been shown that the OH stretching frequency of cyclohexanols is slightly higher for axial isomers than for the equatorial isomers.

Effects from the change in the distribution of electrons in a molecule by a substituent atom or group can often be detected in the vibrational spectrum. There are several mechanisms such as inductive and resonance effects, which can be used to explain observed shifts and intensity changes in a qualitative way. These effects involve changes in electron distribution in a molecule and cause changes in the force constants that are, in turn, responsible for changes in group frequencies. Inductive and resonance effects have been used successfully to explain the shifts observed in C=O stretching frequencies produced by various substituent groups in compounds such as acid chlorides and amides. High C=O stretching frequencies are usually attributed to inductive effects and low frequencies arise when delocalized structures are possible. For example, in acid chlorides the C=O frequency is near 1800 cm^{-1}, which is high compared with a normal

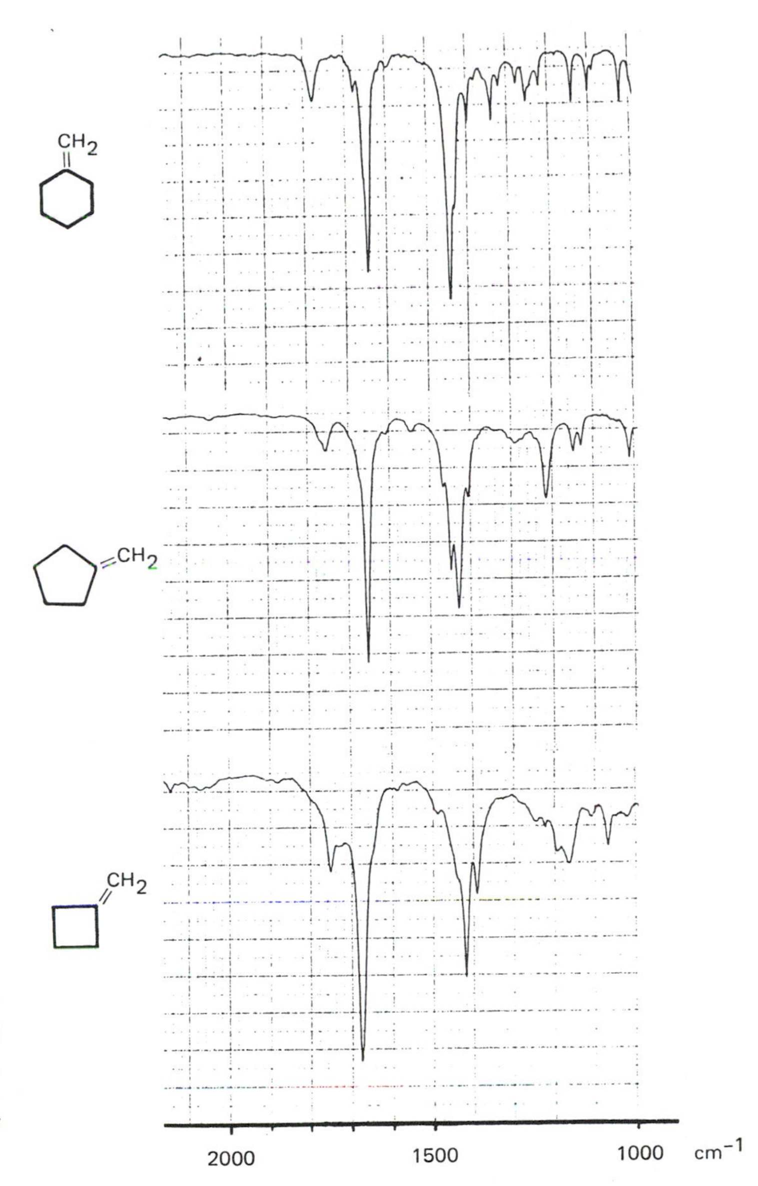

FIGURE 7-7 Portions of infrared spectra of three exocyclic alkenes showing the effect of increasing ring strain on the C=C stretching frequency. [Reproduced with permission of Aldrich Chemical Company, Inc., from *The Aldrich Library of FT–IR Spectra*.]

C=O frequency such as that observed for aldehydes or ketones (1730 cm^{-1}). On the other hand, in amides the carbonyl frequency is lower (near 1650 cm^{-1}). An example of this behavior was shown earlier in Figure 6-5. In acid chlorides the electronegative chlorine atom adjacent to the carbonyl group causes the increase in frequency, whereas in amides the delocalized electronic structure (**7-1** ⟷ **7-2**) lower the C=O stretching frequency.

$$-\overset{|}{\underset{|}{C}}-\overset{O}{\overset{\|}{C}}-NH_2 \quad \longleftrightarrow \quad -\overset{|}{\underset{|}{C}}-\overset{O^-}{\overset{|}{C}}=\overset{+}{N}H_2$$

7-1 **7-2**

Conjugation of double bonds tends to lower the double bond character and increase the bond order of the intervening single bond. For compounds in which a carbonyl group can be conjugated with an ethylenic double bond, the C=O stretching frequency is lowered by 20–30 cm^{-1}, as illustrated in Figure 7-8.

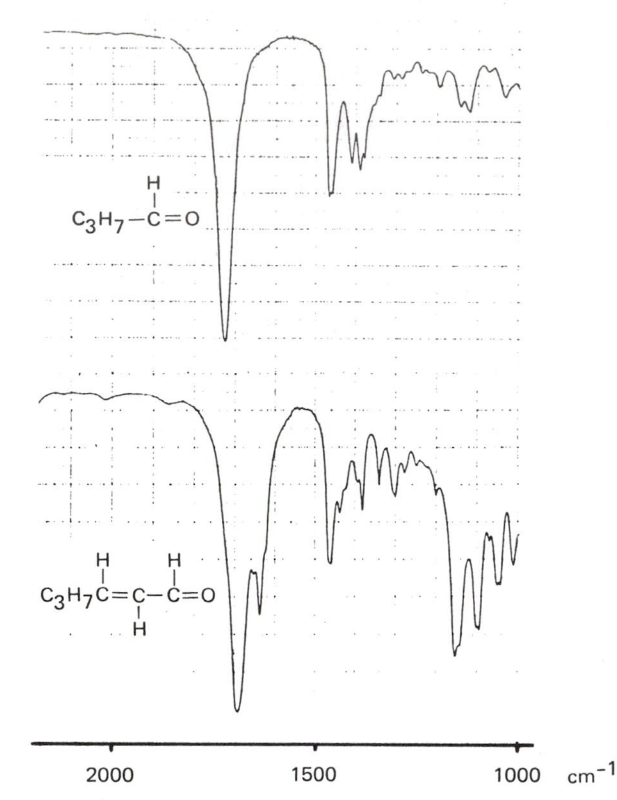

FIGURE 7-8 Portions of infrared spectra of two aliphatic aldehydes showing the influence of conjugation on the C=O stretching frequency. [Reproduced with permission of Aldrich Chemical Company, Inc., from *The Aldrich Library of FT–IR Spectra.*]

7-3

Infrared Group Frequencies

We shall define as a *good* group frequency one that falls within a fairly restricted range regardless of the compound in which the group is found.

Mechanical coupling, symmetry, or other effects discussed in the previous section may occasionally cause even a good group frequency to misbehave, so one should be aware of this possibility. Some of the vibrations of chemical groups giving rise to characteristic bands in the infrared spectrum between 4000 and 400 cm^{-1} are presented in this section. First, an alphabetical list of groups with frequency ranges and intensities is given in Table 7-2. Then, in Table 7-3, a list of frequency ranges from 4000 down to 400 cm^{-1} is presented with possible groups that could absorb within a given range. In the next chapter, several tables give further details of frequencies of the vibrations of certain groups in various types of compounds. Tables 7-2 and 7-3 are by no means comprehensive and in order to make full use of the group frequency method for structure determination, the references cited at the end of this chapter should be consulted.

TABLE 7-2 An Alphabetic Listing of Some Functional Groups and Classes of Compounds with Their Absorption Frequencies in the Infrared

Group or Class	Frequency Ranges (cm^{-1}) and Intensities[a]	Assignment and Remarks
Acid halides R—C(=O)—X		
aliphatic	1810–1790 (s)	C=O stretch; fluorides 50 cm^{-1} higher
	965–920 (m)	C—C stretch
	440–420 (s)	Cl—C=O in-plane deformation
aromatic	1785–1765 (s)	C=O stretch; also a weaker band (1750–1735 cm^{-1}) due to Fermi resonance
	890–850 (s)	C—C stretch (Ar—C) or C—Cl stretch
Alcohols		
primary —CH_2OH	3640–3630 (s)	OH stretch, dil CCl_4 soln
	1060–1030 (s)	C—OH stretch; lowered by unsaturation
secondary —CHROH	3630–3620 (s)	OH stretch, dil CCl_4 soln
	1120–1080 (s)	C—OH stretch; lower when R is a branched chain or cyclic
tertiary —CR_2OH	3620–3610 (s)	OH stretch, dil CCl_4 soln
	1160–1120 (s)	C—OH stretch; lower when R is branched
general —OH	3350–3250 (s)	OH stretch; broad band in pure solids or liquids
	1440–1260 (m–s, br)	C—OH in-plane bend
	700–600 (m–s, br)	C—OH out-of-plane deformation
Aldehydes R—C(=O)—H	2830–2810 (m) }	Fermi doublet; CH stretch with overtone of CH bend (see page 164)
	2740–2720 (m) }	
	1725–1695 (vs)	C=O stretch; slightly higher in CCl_4 soln
	1440–1320 (s)	H—C=O bend in aliphatic aldehydes
	695–635 (s)	C—C—CHO bend
	565–520 (s)	C—C=O bend
Alkenes		
monosubst —CH=CH_2	—	*See* Vinyl
disubst —CH=CH—	—	*See* Vinylene
>C=CH_2	—	*See* Vinylidene
trisubst >C=CH—	3050–3000 (w)	CH stretch
	1690–1655 (w–m)	C=C stretch
	850–790 (m)	CH out-of-plane bending
tetrasubst >C=C<	1690–1670 (w)	C=C stretch, may be absent for symmetrical compounds
Alkyl	2980–2850 (m)	CH stretch, several bands
	1470–1450 (m)	CH_2 deformation
	1400–1360 (m)	CH_3 deformation
	740–720 (w)	CH_2 rocking
Alkynes RC≡C—H	3300–3250 (m–s)	terminal ≡C—H stretch
	2250–2100 (w–m)	C≡C, frequency raised by conjugation
	680–580 (s)	—C≡CH bend

[a] s = strong; m = medium; w = weak; v = very; br = broad.

TABLE 7-2 (Continued)

Group or Class	Frequency Ranges (cm^{-1}) and Intensities[a]	Assignment and Remarks
Amides		
primary —$CONH_2$	3540–3520 (m)	NH_2 stretch (dil solns); bands shift to 3360–3340 and 3200–3180 in solid
	3400–3380 (m)	
	1680–1660 (vs)	C=O stretch (Amide I band)
	1650–1610 (m)	NH_2 deformation; sometimes appears as a shoulder (Amide II band)
	1420–1400 (m-s)	C—N stretch (Amide III band)
secondary —CONHR	3440–3420 (m)	NH stretch (dil soln); shifts to 3300–3280 in pure liquid or solid
	1680–1640 (vs)	C=O stretch (Amide I band)
	1560–1530 (vs)	NH bend (Amide II band)
	1310–1290 (m)	C—N stretch
	710–690 (m)	assignment uncertain
tertiary —$CONR_2$	1670–1640 (vs)	C=O stretch
general —$CONR_2$	630–570 (s)	N—C=O bend
	615–535 (s)	C=O out-of-plane bend
	520–430 (m-s)	C—C=O bend
Amines		
primary —NH_2	3460–3280 (m)	NH stretch; broad band, may have some structure
	2830–2810 (m)	CH stretch
	1650–1590 (s)	NH_2 deformation
secondary —NHR	3350–3300 (vw)	NH stretch
	1190–1130 (m)	C—N stretch
	740–700 (m)	NH deformation
	450–400 (w, br)	C—N—C bend
tertiary —NR_2	510–480 (s)	C—N—C bend
Amine hydrohalides $RNH_3^+X^-$, $R'\overset{+}{N}H_2R\ X^-$	2800–2300 (m-s)	NH_3^+ stretch, several peaks
	1600–1500 (m)	NH deformation (one or two bands)
Amino acids —C(NH_2)—COOH	3200–3000 (s)	H-bonded NH_2 and OH stretch; v broad band in solid state
(or —$C\overset{+}{N}H_3COO^-$)	1600–1590 (s)	COO^- antisym stretch
	1550–1480 (m-s)	—NH_3^+ deformation
	1425–1390 (w-m)	COO^- sym stretch
	560–500 (s)	COO^- rocking
Ammonium NH_4^+	3350–3050 (vs)	NH stretch; broad band
	1430–1390 (s)	NH_2 deformation; sharp peak
Anhydrides —CO—O—CO—	1850–1780 (variable)	antisym C=O stretch
	1770–1710 (m-s)	sym C=O stretch
	1220–1180 (vs)	C—O—C stretch (higher in cyclic anhydrides)
Aromatic compounds	3100–3000 (m)	CH stretch, several peaks
	2000–1660 (w)	overtone and combination bands
	1630–1430 (variable)	aromatic ring stretching (four bands)
	900–650 (s)	out-of-plane CH deformations (one or two bands depending on substitution)
	580–420 (m-s)	ring deformations (two bands)
Azides —$\overset{-}{N}=\overset{+}{N}\equiv N$	2160–2080 (s)	N≡N stretch

TABLE 7-2 (Continued)

Group or Class	Frequency Ranges (cm^{-1}) and Intensities[a]	Assignment and Remarks
Bromo —C—Br	650–500 (m)	C—Br stretch
t-Butyl $(CH_3)_3C$—	2980–2850 (m)	CH stretch; several bands
	1400–1370 (m) and 1380–1360 (s)	CH_3 deformations
Carbodiimides —N=C=N—	2150–2100 (vs)	N=C=N antisym stretch
Carbonyl >C=O	1870–1650 (vs, br)	C=O stretch
Carboxylic acids R—C(=O)OH	3550–3500 (s)	OH stretch (monomer, dil soln)
	3300–2400 (s, v br)	H-bonded OH stretch (solid & liq states)
	1800–1740 (s)	C=O stretch of monomer (dil soln)
	1710–1680 (vs)	C=O stretch of dimer (solid & liq states)
	960–910 (s)	C—OH deformation
	700–590 (s)	O—C=O bend
	550–465 (s)	C—C=O bend
Chloro —C—Cl	850–550 (m)	C—Cl stretch
Cycloalkanes	580–430 (s)	ring deformation
Diazonium salts —N≡N⁺	2300–2240 (s)	N≡N stretch
Esters R—C(=O)OR′	1765–1720 (vs)	C=O stretch
	1290–1180 (vs)	C—O—C antisym stretch
	645–575 (s)	O—C—O bend
Ethers —C—O—C—	1280–1220 (s)	C—O—C stretch in alkyl aryl ethers
	1140–1110 (vs)	C—O—C stretch in dialkyl ethers
	1275–1200 (vs)	C—O—C stretch in vinyl ethers
	1250–1170 (s)	C—O—C stretch in cyclic ethers
	1050–1000 (s)	R(alkyl)—C—O stretch in alkyl aryl ethers
Fluoroalkyl $—CF_3$, $—CF_2—$, etc.	1400–1000 (vs)	C—F stretch
Isocyanates —N=C=O	2280–2260 (vs)	N=C=O antisym stretch
Isothiocyanates —N=C=S	2140–2040 (vs, br)	C=N=S antisym stretch
Ketones R(R′)C=O	1725–1705 (vs)	C=O stretch in saturated aliphatic ketones
	1700–1650 (vs)	C=O stretch in aromatic ketones
	1705–1665 (s) and 1650–1580 (m)	C=O and C=C stretching in α,β-unsaturated ketones
Lactones R(CH_2—C(=O))(CH_2—O) (ring)	1850–1830 (s)	C=O stretch in β-lactones
	1780–1770 (s)	C=O stretch in γ-lactones
	1750–1730 (s)	C=O stretch in δ-lactones
Methyl $—CH_3$	2970–2850 (s)	CH stretch in $C—CH_3$ compounds
	2835–2815 (s)	CH stretch in methyl ethers ($O—CH_3$)
	2820–2780 (s)	CH stretch in $N—CH_3$ compounds
	1470–1440 (m)	CH_3 antisym deformation
	1390–1370 (m–s)	CH_3 sym deformation

TABLE 7-2 (Continued)

Group or Class	Frequency Ranges (cm^{-1}) and Intensities[a]	Assignment and Remarks
Methylene —CH_2—	2940–2920 (m) and 2860–2850 (m)	CH stretches in alkanes
	3090–3070 (m) and 3020–2980 (m)	CH stretches in alkenes
	1470–1450 (m)	CH_2 deformation
Naphthalenes	645–615 (m–s) and 545–520 (s)	in-plane ring bending
	490–465 (variable)	out-of-plane ring bending
Nitriles —C≡N	2260–2240 (w)	C≡N stretch in aliphatic nitriles
	2240–2220 (m)	C≡N stretch in aromatic nitriles
	580–530 (m–s)	C—C—CN bend
Nitro —NO_2	1570–1550 (vs) and 1380–1360 (vs)	NO_2 stretches in aliphatic nitro compounds
	1480–1460 (vs) and 1360–1320 (vs)	NO_2 stretches in aromatic nitro compounds
	920–830 (m)	C—N stretch
	650–600 (s)	NO_2 bend in aliphatic compounds
	580–520 (m)	NO_2 bend in aromatic compounds
	530–470 (m–s)	NO_2 rocking
Oximes =NOH	3600–3590 (vs)	OH stretch (dil soln)
	3260–3240 (vs)	OH stretch (solids)
	1680–1620 (w)	C=N stretch; strong in Raman
Phenols Ar—OH	720–600 (s, br)	O—H out-of-plane deformation
	450–375 (w)	C—OH deformation
Phenyl C_6H_5—	3100–3000 (w–m)	CH stretch
	2000–1700 (w)	four weak bands; overtones and combinations
	1625–1430 (m–s)	aromatic C=C stretches (four bands)
	1250–1025 (m–s)	CH in-plane bending (five bands)
	770–730 (vs)	CH out-of-plane bending
	710–690 (vs)	ring deformation
	560–420 (m–s)	ring deformation
Phosphates $(RO)_3P{=}O$		
R = alkyl	1285–1255 (vs)	P=O stretch
	1050–990 (vs)	P—O—C stretch
R = aryl	1315–1290 (vs)	P=O stretch
	1240–1190 (vs)	P—O—C stretch
Phosphines —PH_2, —PH	2410–2280 (m)	P—H stretch
	1100–1040 (w–m)	P—H deformation
	700–650 (m–s)	P—C stretch
Pyridyl —C_5H_4N	3080–3020 (m)	CH stretch
	1620–1580 (vs) and 1590–1560 (vs)	C=C and C=N stretches
	840–720 (s)	CH out-of-plane deformation (one or two bands, depending on substitution)
	635–605 (m–s)	in-plane ring bending
Silanes —SiH_3	2160–2110 (m)	SI—H stretch
—SiH_2—	950–800 (s)	Si—H deformation

TABLE 7-2 (Continued)

Group or Class	Frequency Ranges (cm^{-1}) and Intensities[a]	Assignment and Remarks
Silanes (fully substituted)	1280–1250 (m–s)	Si—C stretch
	1110–1050 (vs)	Si—O—C stretch (aliphatic)
	840–800 (m)	Si—O—C deformation
Sulfates R—O—SO_2—O—R	1140–1350 (s) and 1230–1150 (s)	S=O stretches in covalent sulfates
R—O—SO_3^- M^+	1260–1210 (vs)	S=O stretches in alkyl sulfate salts
(M = Na^+, K^+, etc.)	and 810–770 (s)	C—O—S stretch
Sulfides C—S—	710–570 (m)	C—S stretch
Sulfones —SO_2—	1360–1290 (vs)	SO_2 antisym stretch
	1170–1120 (vs)	SO_2 sym stretch
	610–545 (ms)	SO_2 scissor mode
Sulfonic acids —SO_2OH	1250–1150 (vs, br)	S=O stretch
Sulfoxides >S=O	1060–1030 (s, br)	S=O stretch
	610–545 (m–s)	SO_2 scissoring
Thiocyanates —S—C≡N	2175–2160 (m)	C≡N stretch
	650–600 (w)	S—CN stretch
	405–400 (s)	S—C≡N bend
Thiols —S—H	2590–2560 (w)	S—H stretch; strong in Raman
	700–550 (w)	C—S stretch; strong in Raman
Triazines $C_3N_3Y_3$	1600–1500 (vs)	ring stretching
1,3,5-trisubst	1380–1350 (vs)	ring stretching
	820–800 (s)	CH out-of-plane deformation
Vinyl —CH=CH_2	3095–3080 (m)	=CH_2 stretching
	and 3030–2980 (w–m)	=CH stretching
	1850–1800 (w–m)	overtone of CH_2 out-of-plane wagging
	1645–1615 (m–s)	C=C stretch
	1000–950 (s)	CH out-of-plane deformation
	950–900 (vs)	CH_2 out-of-plane wagging
Vinylene —CH=CH—	3040–3010 (m)	=CH_2 stretching
	1665–1635 (w–m)	C=C stretch (cis isomer)
	1675–1665 (w–m)	C=C stretch (trans isomer)
	980–955 (s)	CH out-of-plane deformation (cis isomer)
	730–665 (s)	CH out-of-plane deformation (trans isomer)
Vinylidene >C=CH_2	3095–3075 (m)	=CH_2 stretching
	1665–1620 (w–m)	C=C stretch
	895–885 (s)	CH_2 out-of-plane wagging

TABLE 7-3 A Numerical Listing of Wavenumber Ranges in Which Some Functional Groups and Classes of Compounds Absorb in the Infrared

Range (cm^{-1}) and Intensity[a]	Group and Class	Assignment and Remarks
3700–3600 (s)	—OH in alcohols and phenols	OH stretch (dil soln)
3520–3320 (m–s)	$—NH_2$ in aromatic amines, primary amines and amides	NH stretch (dil soln)
3420–3250 (s)	—OH in alcohols and phenols	OH stretch (solids & liquids)
3360–3340 (m)	$—NH_2$ in primary amides	NH_2 antisym stretch (solids)
3320–3250 (m)	—OH in oximes	O—H stretch
3300–3250 (m–s)	≡CH in acetylenes	≡CH—H stretch
3300–3280 (s)	—NH in secondary amides	NH stretch (solids); also in polypeptides and proteins
3200–3180 (s)	$—NH_2$ in primary amides	NH_2 sym stretch (solids)
3200–3000 (v br)	$—NH_3^+$ in amino acids	NH_3^+ antisym stretch
3100–2400 (v br)	—OH in carboxylic acids	H-bonded OH stretch
3100–3000 (m)	=CH in aromatic and unsaturated hydrocarbons	=C—H stretch
2990–2850 (m–s)	$—CH_3$ and $—CH_2—$ in aliphatic compounds	CH antisym and sym stretching
2850–2700 (m)	$—CH_3$ attached to O or N	CH stretching modes
2750–2650 (w–m)	—CHO in aldehydes	overtone of CH bending (Fermi resonance)
2750–2350 (br)	$—NH_3^+$ in amine hydrohalides	NH stretching modes
2720–2560 (m)	—OH in phosphorus oxyacids	associated OH stretching
2600–2540 (w)	—SH in alkyl mercaptans	S—H stretch; strong in Raman
2410–2280 (m)	—PH in phosphines	P—H stretch; sharp peak
2300–2230 (m)	N≡N in diazonium salts	N≡N stretch, aq soln
2285–2250 (s)	N=C=O in isocyanates	N=C=O antisym stretch
2260–2200 (m–s)	C≡N in nitriles	C≡N stretch
2260–2190 (w–m)	C≡C in alkynes (disubst)	C≡C stretch; strong in Raman
2190–2130 (m)	C≡N in thiocyanates	C≡N stretch
2175–2115 (s)	N≡C in isonitriles	N≡C stretch
2160–2080 (m)	$N=\overset{+}{N}=\overset{-}{N}$ in azides	N=N=N antisym stretch
2140–2100 (w–m)	C≡C in alkynes (monosubst)	C≡C stretch
2000–1650 (w)	substituted benzene rings	several bands from overtone and combination bands
1980–1950 (s)	C=C=C in allenes	C=C=C antisym stretch
1870–1650 (vs)	C=O in carbonyl compounds	C=O stretch
1870–1830 (s)	C=O in β-lactones	C=O stretch
1870–1790 (vs)	C=O in anhydrides	C=O antisym stretch; part of doublet
1820–1800 (s)	C=O in acid halides	C=O stretch; lower for aromatic acid halides
1780–1760 (s)	C=O in γ-lactones	C=O stretch
1765–1725 (vs)	C=O in anhydrides	C=O sym stretch; part of doublet
1760–1740 (vs)	C=O in α-keto esters	C=O stretch; enol form
1750–1730 (s)	C=O in δ-lactones	C=O stretch
1750–1740 (vs)	C=O in esters	C=O stretch; 20 cm^{-1} lower if unsaturated
1740–1720 (s)	C=O in aldehydes	C=O stretch; 30 cm^{-1} lower if unsaturated
1720–1700 (s)	C=O in ketones	C=O stretch; 20 cm^{-1} lower if unsaturated
1710–1690 (s)	C=O in carboxylic acids	C=O stretch; fairly broad
1690–1640 (s)	C=N in oximes	C=N stretch; also imines
1680–1620 (s)	C=O and NH_2 in primary amides	two bands from C=O stretch and NH_2 deformation
1680–1635 (s)	C=O in ureas	C=O stretch; broad band
1680–1630 (m–s)	C=C in alkenes, etc.	C=C stretch
1680–1630 (vs)	C=O in secondary amides	C=O stretch (Amide I band)

[a] v = very, s = strong, m = medium, w = weak, br = broad.

TABLE 7-3 (Continued)

Range (cm^{-1}) and Intensity[a]	Group and Class	Assignment and Remarks
1670–1640 (s–vs)	C=O in benzophenones	C=O stretch
1670–1650 (vs)	C=O in primary amides	C=O stretch (Amide I band)
1670–1630 (vs)	C=O in tertiary amides	C=O stretch
1655–1635 (vs)	C=O in β-ketone esters	C=O stretch; enol form
1650–1620 (w–m)	N—H in primary amides	NH deformation (Amide II band)
1650–1580 (m–s)	NH_2 in primary amines	NH_2 deformation
1640–1580 (s)	NH_3^+ in amino acids	NH_3 deformation
1640–1580 (vs)	C=O in β-diketones	C=O stretch; enol form
1620–1610 (s)	C=C in vinyl ethers	C=C stretch; doublet due to rotational isomerism
1615–1590 (m)	benzene ring in aromatic compounds	ring stretch; sharp peak
1615–1565 (s)	pyridine derivatives	ring stretch; doublet
1610–1580 (s)	NH_2 in amino acids	NH_2 deformation; broad band
1610–1560 (vs)	COO^- in carboxylic acid salts	—C(O)(O) – antisym stretch
1590–1580 (m)	NH_2 primary alkyl amide	NH_2 deformation (Amide II band)
1575–1545 (vs)	NO_2 in aliphatic nitro compounds	NO_2 antisym stretch
1565–1475 (vs)	NH in secondary amides	NH deformation (Amide II band)
1560–1510 (s)	triazine compounds	ring stretch; sharp band
1550–1490 (s)	NO_2 in aromatic nitro compounds	NO_2 antisym stretch
1530–1490 (s)	NH_3^+ in amino acids or hydrochlorides	NH_3^+ deformation
1530–1450 (m–s)	N=N—O in azoxy compounds	N=N—O antisym stretch
1515–1485 (m)	benzene ring in aromatic compounds	ring stretch, sharp band
1475–1450 (vs)	CH_2 in aliphatic compounds	CH_2 scissors vibration
1465–1440 (vs)	CH_3 in aliphatic compounds	CH_3 antisym deformation
1440–1400 (m)	OH in carboxylic acids	in-plane OH bending
1420–1400 (m)	C—N in primary amides	C—N stretch (Amide III band)
1400–1370 (m)	*t*-butyl group	CH_3 deformations (two bands)
1400–1310 (s)	COO^- group in carboxylic acid salts	C(O)(O) – sym stretch; broad band
1390–1360 (vs)	SO_2 in sulfonyl chlorides	SO_2 antisym stretch
1380 1370 (s)	CH_3 in aliphatic compounds	CH_3 sym deformation
1380–1360 (m)	isopropyl group	CH_3 deformations (two bands)
1375–1350 (s)	NO_2 in aliphatic nitro compounds	NO_2 sym stretch
1360–1335 (vs)	SO_2 in sulfonamides	SO_2 antisym stretch
1360–1320 (vs)	NO_2 in aromatic nitro compounds	NO_2 sym stretch
1350–1280 (m–s)	N=N—O in azoxy compounds	N=N—O sym stretch
1335–1295 (vs)	SO_2 in sulfones	SO_2 antisym stretch
1330–1310 (m–s)	CF_3 attached to a benzene ring	CF_3 antisym stretch
1300–1200 (vs)	$\overset{+}{N}—\overset{-}{O}$ in pyridine *N*-oxides	N—O stretch
1300–1175 (vs)	P=O in phosphorus oxyacids and phosphates	P=O stretch
1300–1000 (vs)	C—F in aliphatic fluoro compounds	C—F stretch
1285–1240 (vs)	Ar—O in alkyl aryl ethers	C—O stretch
1280–1250 (vs)	Si—CH_3 in silanes	CH_3 sym deformation
1280–1240 (m–s)	C—C(O) in epoxides	C—O stretch
1280–1180 (s)	C—N in aromatic amines	C—N stretch
1280–1150 (vs)	C—O—C in esters, lactones	C—O—C antisym stretch

TABLE 7-3 (Continued)

Range (cm^{-1}) and Intensity[a]	Group and Class	Assignment and Remarks
1255–1240 (m)	*t*-butyl in hydrocarbons	skeletal vibration; second band near 1200 cm^{-1}
1245–1155 (vs)	SO_3H in sulfonic acids	S=O stretch
1240–1070 (s–vs)	C—O—C in ethers	C—O—C stretch; also in esters
1230–1100 (s)	C—C—N in amines	C—C—N bending
1225–1200 (s)	C—O—C in vinyl ethers	C—O—C antisym stretch
1200–1165 (s)	SO_2Cl in sulfonyl chlorides	SO_2 sym stretch
1200–1015 (vs)	C—OH in alcohols	C—O stretch
1170–1145 (s)	SO_2NH_2 in sulfonamides	SO_2 sym stretch
1170–1140 (s)	SO_2— in sulfones	SO_2 sym stretch
1160–1100 (m)	C=S in thiocarbonyl compounds	C=S stretch; strong in Raman
1150–1070 (vs)	C—O—C in aliphatic ethers	C—O—C antisym stretch
1120–1080 (s)	C—O—H in secondary or tertiary alcohols	C—O stretch
1120–1030 (s)	C—NH_2 in primary aliphatic amines	C—N stretch
1100–1000 (vs)	Si—O—Si in siloxanes	Si—O—Si antisym stretch
1080–1040 (s)	SO_3H in sulfonic acids	SO_3 sym stretch
1065–1015 (s)	CH—O—H in cyclic alcohols	C—O stretch
1060–1025 (vs)	CH_2—O—H in primary alcohols	C—O stretch
1060–1045 (vs)	S=O in alkyl sulfoxides	S=O stretch
1055–915 (vs)	P—O—C in organophosphorus compounds	P—O—C antisym stretch
1030–950 (w)	carbon ring in cyclic compounds	ring breathing mode; strong in Raman
1000–950 (s)	CH=CH_2 in vinyl compounds	=CH out-of-plane deformation
980–960 (vs)	CH=CH— in trans disubstituted alkenes	=CH out-of-plane deformation
950–900 (vs)	CH=CH_2 in vinyl compounds	CH_2 out-of-plane wag
900–865 (vs)	CH_2=C(R)(R′) in vinylidenes	CH_2 out-of-plane wag
890–805 (vs)	1,2,4-trisubst benzenes	CH out-of-plane deformation (two bands)
860–760 (vs, br)	R—NH_2 primary amines	NH_2 wag
860–720 (vs)	Si—C in organosilicon compounds	Si—C stretch
850–830 (vs)	1,3,5-trisubst benzenes	CH out-of-plane deformation
850–810 (vs)	Si—CH_3 in silanes	Si—CH_3 rocking
850–790 (m)	CH=C(R)(R′) in trisubst alkenes	CH out-of-plane deformation
850–550 (m)	C—Cl in chloro compounds	C—Cl stretch
830–810 (vs)	*p*-disubst benzenes	CH out-of-plane deformation
825–805 (vs)	1,2,4-trisubst benzenes	CH out-of-plane deformation
820–800 (s)	triazines	CH out-of-plane deformation
815–810 (s)	CH=CH_2 in vinyl ethers	CH_2 out-of-plane wag
810–790 (vs)	1,2,3,4-tetrasubst benzenes	CH out-of-plane deformation
800–690 (vs)	*m*-disubst benzenes	CH out-of-plane deformation (two bands)
785–680 (vs)	1,2,3-trisubst benzenes	CH out-of-plane deformation (two bands)
775–650 (m)	C—S in sulfonyl chlorides	C—S stretch; strong in Raman
770–690 (vs)	monosubst benzenes	CH out-of-plane deformation (two bands)
760–740 (s)	*o*-disubst benzenes	CH out-of-plane deformation
760–510 (s)	C—Cl alkyl chlorides	C—Cl stretch
740–720 (w–m)	—$(CH_2)_n$— in hydrocarbons	CH_2 rocking in methylene chains; intensity depends on chain length
730–665 (s)	CH=CH in cis disubst alkenes	CH out-of-plane deformation
720–600 (s, br)	Ar—OH in phenols	OH out-of-plane deformation

TABLE 7-3 (Continued)

Range (cm^{-1}) and Intensity[a]	Group and Class	Assignment and Remarks
710–570 (m)	C—S in sulfides	C—S stretch; strong in Raman
700–590 (s)	O—C=O in carboxylic acids	O—C=O bending
695–635 (s)	C—C—CHO in aldehydes	C—C—CHO bending
680–620 (s)	C—OH in alcohols	C—O—H bending
680–580 (s)	C≡C—H in alkynes	C≡C—H bending
650–600 (w)	S—C≡N in thiocyanates	S—C stretch; strong in Raman
650–600 (s)	NO_2 in aliphatic nitro compounds	NO_2 deformation
650–500 (s)	Ar—CF_3 in aromatic trifluoro-methyl compounds	CF_3 deformation (two or three bands)
650–500 (s)	C—Br in bromo compounds	C—Br stretch
645–615 (m–s)	naphthalenes	in-plane ring deformation
645–575 (s)	O—C—O in esters	O—C—O bend
640–630 (s)	=CH_2 in vinyl compounds	=CH_2 twisting
635–605 (m–s)	pyridines	in-plane ring deformation
630–570 (s)	N—C=O in amides	N—C=O bend
630–565 (s)	C—CO—C in ketones	C—CO—C bend
615–535 (s)	C=O in amides	C=O out-of-plane bend
610–565 (vs)	SO_2 in sulfonyl chlorides	SO_2 deformation
610–545 (m–s)	SO_2 in sulfones	SO_2 scissoring
600–465 (s)	C—I in iodo compounds	C—I stretch
580–530 (m–s)	C—C—CN in nitriles	C—C—CN bend
580–520 (m)	NO_2 in aromatic nitro compounds	NO_2 deformation
580–430 (s)	ring in cycloalkanes	ring deformation
580–420 (m–s)	ring in benzene derivatives	in-plane and out-of-plane ring deformations (two bands)
570–530 (vs)	SO_2 in sulfonyl chlorides	SO_2 rocking
565–520 (s)	C—C=O in aldehydes	C—C=O bend
565–440 (w–m)	C_nH_{2n+1} in alkyl groups	chain deformation modes (two bands)
560–510 (s)	C—C=O in ketones	C—C=O bend
560–500 (s)	—C(O)(O) - in amino acids (carboxylate: C with two partial C–O bonds)	—C(O)(O) - rocking
555–545 (s)	=CH_2 in vinyl compounds	=CH_2 twisting
550–465 (s)	C—C=O in carboxylic acids	C—C=O bend
545–520 (s)	naphthalenes	in-plane ring deformation
530–470 (m–s)	NO_2 in nitro compounds	NO_2 rocking
520–430 (m–s)	C—O—C in ethers	C—O—C bend
510–400 (s)	C—N—C in amines	C—N—C bend
490–465 (variable)	naphthalenes	out-of-plane ring bending
440–420 (s)	Cl—C=O in acid chlorides	Cl—C=O in-plane deformation
405–400 (s)	S—C≡N in thiocyanates	S—C≡N bend

7-4

Raman Group Frequencies

Depending on the symmetry of a molecule, its vibrational frequencies may give rise to infrared absorption or Raman scattering or both. In the latter case, observed frequencies will be numerically the same in Raman and infrared, but the intensities will often be quite different. In most cases, information obtained from the Raman spectrum duplicates that obtained from the infrared, but in some cases the Raman spectrum provides additional information especially in the low frequency region where far infrared spectra may not be available.

Since infrared absorption depends on change of dipole moment, we expect polar bonds or groups to give strong infrared bands. On the other hand, a change in polarizability is necessary for Raman scattering so that bonds or groups with symmetrical charge distributions are expected to give rise to strong Raman lines. Some of the most important Raman group frequencies are given by C=C, N=N, C≡C, and S—S stretching modes.

A tabulation of some Raman group frequencies is given in Table 7-4. A bibliography of sources of Raman group frequencies is listed at the end of this chapter. The book by Dollish, Fateley, and Bentley (ref. 7-8) is particularly valuable. Weak Raman bands are included in Table 7-4 only if they are characteristic of a group.

TABLE 7-4 Characteristic Frequencies of Functional Groups in the Raman Spectra of Complex Molecules

Group or Class	Frequency Ranges (cm^{-1}) and Intensities[a]	Assignment and Remarks
Acetylenes ≡CH	3340–3270 (s)	CH stretch
(alkynes) R—C≡C—R	2300–2190 (s)	C≡C stretch in disubst acetylenes, sometimes two bands (Fermi doublet)
R—C≡CH	2140–2100 (s)	C≡C stretch in monoalkyl acetylenes
	650–600 (m)	C—C≡CH deformation
Acid chlorides R—C(=O)Cl	1800–1790 (s)	C=O stretch
Alcohols R—OH	3400–3300 (vw)	OH stretch; broad band
	1450–1350 (m)	OH in-plane bend
	1150–1050 (m–s)	C—O antisym stretch
	970–800 (s)	C—C—O sym stretch
Aldehydes R—C(=O)H	1730–1700 (m)	C=O stretch
n-Alkanes (general)	2980–2800 (vs)	CH stretch
	1475–1450 (s)	CH_3 antisym deformation
	1350–1300 (m–s)	CH_2 bend
	340–230 (s)	—C—C—C— bend
Alkenes (general)	3090–3010 (m)	CH stretch
	1675–1600 (m–s)	C=C stretch, stronger than IR
	1450–1200 (vs)	CH in-plane deformation
cis Alkenes R′CH=CHR	590–570 (m)	
	420–400 (m)	skeletal deformations
	310–290 (m)	
trans Alkenes R′CH=CHR	500–480 (m)	skeletal deformations
	220–200 (m)	
Terminal alkenes $RCH=CH_2$	500–480 (m)	
$RR'C=CH_2$	440–390 (m)	skeletal deformations
	270–250 (m)	

[a] v = very, s = strong, m = medium, w = weak.

TABLE 7-4 (Continued)

Group or Class	Frequency Ranges (cm^{-1}) and Intensities[a]	Assignment and Remarks
Allenes C=C=C	2000–1960 (s)	—C=C=C— antisym stretch
	1080–1060 (vs)	—C=C=C— sym stretch
Amides		
primary —$CONH_2$	3540–3520 (w)	NH_2 antisym stretch (dil soln)
	3400–3380 (w)	NH_2 sym stretch (dil soln)
	1680–1660 (m)	C=O stretch (amide I band)
	1420–1400 (s)	C—N stretch (Amide III band)
secondary —CONHR	3440–3420 (s)	NH stretch (dil soln)
	1680–1640 (w)	Amide I band
	1310–1280 (s)	Amide III band
tertiary —$CONR_2$	1670–1640 (m)	Amide I band
Amines, aliphatic		
primary RNH_2	3550–3330 (m)	NH_2 antisym stretch
	3450–3250 (m)	NH_2 sym stretch
	1090–1070 (m)	C—N stretch
secondary R′NHR	3350–3300 (w)	NH stretch
	1190–1130 (m)	C—N stretch
Amines, aromatic	1380–1250 (s)	C—N stretch
Amino acids —CNH_2COOH or ($CNH_3^+COO^-$)	1600–1590 (w)	OCO antisym stretch
	1400–1350 (vs)	OCO sym stretch
	900–850 (vs)	C—C—N sym stretch
Anhydrides —CO\O/—CO	1850–1780 (w–m)	C=O antisym stretch
	1770–1710 (m)	C=O sym stretch
Aromatic compounds	3070–3020 (s)	CH stretch
	1620–1580 (m–s)	C=C stretch; may be weak in IR
	1045–1015 (m)	CH in-plane bend
	1010–990 (vs)	ring breathing (absent in *o*- and *p*-disubst compounds)
	900–650 (m)	CH out-of-plane deformation (one or two bands)
Azides —N=$\overset{+}{N}$=$\overset{-}{N}$	2170–2080 (s)	NNN antisym stretch
	1345–1175 (s)	NNN sym stretch
Azo —N=N—	1580–1570 (vs)	nonconjugated compounds
	1420–1410 (vs)	conjugated to aromatic ring
	1060–1030 (vs)	—C—N—stretch in aromatic azo compounds
Benzenes		
monosubst	630–610 (s)	CH out-of-plane deformation
1,2-disubst and 1,2,4-trisubst	750–700 (s)	CH out-of-plane deformation
1,3-disubst	750–700 (s)	CH out-of plane deformation
	480–450 (m)	out-of-plane ring deformation
1,2,3-trisubst	655–645 (s)	CH out-of-plane deformation
1,3,5-trisubst	570–550 (s)	CH out-of-plane deformation
Bromo C—Br	650–490 (vs)	C—Br stretch
	310–270 (s)	C—C—Br bend
t-Butyl	1250–1200 (m–s)	CH_3 deformation (two bands)
	940–920 (s)	CH_3 rocking

TABLE 7-4 (Continued)

Group or Class	Frequency Ranges (cm^{-1}) and Intensities[a]	Assignment and Remarks
Carbonyl C=O	1870–1650 (w–s)	C=O stretch; weaker than in IR
Carboxylic acids	1680–1640 (s)	C=O sym stretch of dimer
Chloro C—Cl	850–650 (s)	C—Cl stretch
	340–290 (s)	C—C—Cl bend
Cumulenes	2070–2030 (vs)	C=C=C=C sym stretch
Cyanamides	1150–1140 (vs)	—N=C=N— sym stretch
Cyclobutanes	1000–960 (vs)	ring breathing
	700–680 (s)	ring deformation
	180–150 (s)	ring puckering
Cyclohexanes	1460–1440 (s)	CH_2 scissoring
	825–815 (s)	ring vibration (boat)
	810–795 (s)	ring vibration (chair)
Cyclopentanes	1450–1430 (s)	CH_2 scissoring
	900–880 (s)	ring breathing
Cyclopropanes	1210–1180 (s)	ring breathing
	830–810 (s)	ring deformation
Disulfides C—S—S—C	550–430 (vs)	S—S stretch
Epoxides	1280–1260 (s)	sym ring stretch
Esters R′COOR	1100–1025 (s)	C—O—C sym stretch
Esters		
cyclic	1040–1010 (s)	ring stretching, 4-membered ring
	920–900 (s)	ring stretching, 5-membered ring
	820–800 (s)	ring stretching, 6-membered ring
aliphatic satd	1140–1110 (m)	C—O—C stretch
aliphatic unsatd —C=C—O—C	1275–1200 (m)	C—O—C antisym stretch
	1075–1020 (s)	C—O—C sym stretch
aromatic	1310–1210 (m) and 1050–1010 (m)	C—O—C stretches
Hg—C (organomercury compounds)	570–510 (vvs)	C—Hg stretch
Isocyanates —N=C=O	1440–1400 (vs)	N=C=O sym stretch
Isopropyl $(CH_3)_2CH$—	1180–1160 (m)	CH_3 rocking
	835–795 (ms)	C—C stretching
Ketenes C=C=O	2060–2040 (vs)	C=C=O stretch
Ketones R(R′)C=O	1725–1705 (m)	C=O stretch in satd compounds
	1700–1650 (m)	C=O stretch in aromatic compounds
	1750–1705 (m)	C=O stretch in alicyclic ketones
Lactones R<(CH_2—C(=O)—O—CH_2)	1850–1730 (s)	C=O stretch
Mercaptans C—SH	850–820 (vs)	S—H in-plane deformation
	700–600 (vs)	C—S stretch; weak in IR

TABLE 7-4 (Continued)

Group or Class	Frequency Ranges (cm^{-1}) and Intensities[a]	Assignment and Remarks
Methyl $—CH_3$	2980–2800 (vs)	CH stretch
	1470–1460 (s)	CH_3 deformation
Methylene $=CH_2$ or $—CH_2—$	3090–3070 (s)	$=CH_2$ antisym stretch
	3020–2980 (s)	$=CH_2$ sym stretch
	2940–2920 (s)	$—CH_2—$ antisym stretch
	2860–2850 (s)	$—CH_2—$ sym stretch
	1350–1150 (m–s)	CH_2 wag and twist; weak or absent in IR
Nitrates $R—ONO_2$	1285–1260 (vs)	NO_2 sym stretch
Nitriles $—C—C{\equiv}N$	2260–2240 (s)	$C{\equiv}N$ stretch, nonconjugated
	2230–2220 (s)	$C{\equiv}N$ stretch, conjugated
	1080–1025 (s–vs)	C—C—C stretch
	840–800 (s–vs)	C—C—CN sym stretch
	380–280 (s–vs)	$C—C{\equiv}N$ bend
Nitrites —ONO	1660–1620 (s)	N=O stretch in alkyl nitrites
Nitro $—NO_2$	1570–1550 (w)	NO_2 antisym stretch
	1380–1360 (s)	NO_2 sym stretch
	920–830 (s)	C—N stretch
	650–520 (m)	NO_2 bend
Organoarsenic compounds	570–550 (vs)	C—As stretch
	240–220 (vs)	AsC_2 deformation
Organosilicon compounds	1300–1200 (s)	Si—C stretch
Oximes	1680–1620 (vs)	C=N stretch; may not be seen in IR
Peroxides—C—O—O—C—	900–850 (variable)	O—O stretch; weak in IR
Phosphines	2350–2240 (m)	P—H stretch
Pyridines	1620–1560 (m)	ring stretching
	1020–980 (vs)	ring breathing
Pyrroles	3450–3350 (s)	NH stretch
	1420–1360 (vs)	ring stretching
Sulfides C—S	705–570 (s)	C—S stretch
Sulfones $—SO_2—$	1360–1290 (m)	SO_2 antisym stretch
	1170–1120 (s)	SO_2 sym stretch
	610–545 (s)	SO_2 scissoring
Sulfonamides $—SO_2NH_2$	1155–1135 (vs)	SO_2 stretch
Sulfoxides >SO	1050–1010 (s)	S=O stretch
Sulfonyl chlorides R—OSOCl	1230–1200 (m)	S=O stretch
Thiocyanates $—S—C{\equiv}N$	650–600 (s)	S—CN stretch
Thiols RSH	2590–2560 (vs)	S—H stretch
	700–550 (vs)	C—S stretch
	340–320 (vs)	C—S—H out-of-plane bend
Thiophenes	740–680 (vs)	C—S—C stretch
	570–430 (s)	ring deformation
Xanthates —O—C(=S)—S—	670–620 (vs)	C=S stretch; not seen in IR
	480–450 (vs)	C—S stretch

REFERENCES/BIBLIOGRAPHY

Infrared Group Frequencies

7-1 G. Socrates, *Infrared Characteristic Group Frequencies*, Wiley, Chichester (UK), 1980.
7-2 M. St. C. Flett, *Characteristic Frequencies of Chemical Groups in the Infrared*, Elsevier, Amsterdam, 1963.
7-3 L. J. Bellamy, *The Infrared Spectra of Complex Molecules,* 2nd ed., Methuen, London, 1958.
7-4 L. J. Bellamy, *Advances in Infrared Group Frequencies*, 2nd ed., Chapman & Hall, London, 1980.
7-5 F. F. Bentley, L. D. Smithson, and A. L. Rozek, *Infrared Spectra and Characteristic Frequencies 700–300 cm*$^{-1}$, Interscience, New York, 1968.
7-6 H. A. Szymanski and R. F. Erickson, *Infrared Band Handbook* (2 vols.), IFI/Plenum, New York, 1970.
7-7 D. Dolphin and A. Wick, *Tabulation of Infrared Spectra Data*, Wiley, New York, 1977.

Raman Group Frequencies

7-8 F. E. Dollish, W. G. Fateley, and F. F. Bentley, *Characteristic Raman Frequencies of Organic Compounds*, Wiley–Interscience, New York, 1974.
7-9 M. C. Tobin, *Laser Raman Spectroscopy*, Wiley–Interscience, New York, 1971.
7-10 H. A. Szymanski, *Correlation of Infrared and Raman Spectra of Organic Compounds*, Hertillon Press, Cambridge Springs, PA, 1969.
7-11 H. J. Sloane, "The Use of Group Frequencies for Structural Analysis in the Raman Compared to the Infrared," in *Polymer Characterization*, C. D. Craver, Ed., Plenum, New York, 1971, pp. 15–36.

Additional Reading

7-12 F. A. Cotton, *Chemical Applications of Group Theory*, 2nd ed., Wiley–Interscience, New York, 1971.
7-13 J. B. Lambert, H. F. Shurvell, L. Verbit, R. G. Cooks, and G. H. Stout, *Organic Structural Analysis*, Macmillan, New York, 1976.
7-14 B. P. Straughan and S. Walker, *Spectroscopy*, vol. 2, Chapman & Hall, London, 1976.

8

Structure Determination

8-1

Introduction

The discussion in this chapter falls into two categories. First, it will be shown how the structure of an unknown compound can be deduced from its infrared and Raman spectra, with the help of other information such as elemental analysis and molecular weight. In most cases, the NMR, UV, and mass spectra will also be needed to confirm or assist in the determination of the structure. Second, we shall look at some examples of ways in which infrared and Raman spectra can give details of atomic arrangements in compounds of known gross structure. Information can be obtained on molecular conformations; structural, geometrical, and rotational isomerism (but *not* optical isomerism); tautomerism; hydrogen bonding; and various other structural features.

It is assumed in what follows that the student is able to obtain infrared spectra from a low or medium resolution spectrometer and, when necessary, to obtain Raman spectra to help in the investigations. The first and most important thing to do is to ensure that the sample is reasonably pure, since even 5% of an impurity could give rise to spurious peaks in the spectra. Gas chromatography usually gives an indication of the purity of a compound. As an example, a small amount of diethyl ether, present as an impurity in *tert*-butylacetylene, gives a band from the C—O—C stretching mode of the ether at 1120 cm^{-1} in the infrared spectrum. This is normally the strongest band in the infrared spectra of ethers, so that it has considerable intensity even when only a few percent are present as an impurity (Figure 8-1).

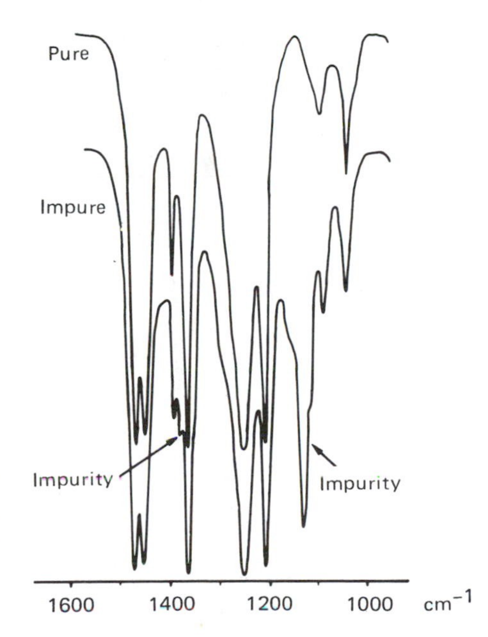

FIGURE 8-1 Part of the infrared spectra of thick films of two samples of *tert*-butylacetylene. The presence of an impurity is clearly indicated by the strong band at 1120 cm^{-1} and the peak at 1380 cm^{-1}.

A preliminary infrared spectrum of a solid is recorded using mulls (Nujol and Fluorolube) or a KBr pellet; that of a liquid as a film between two NaCl or KBr plates. When hydrogen bonding is evident from very broad bands, especially between 3500 and 2500 cm^{-1}, spectra should also be obtained from dilute solutions in CCl_4 or from some other suitable solvent (see Section 6-7a). For a detailed study, several spectra may be needed from solutions of various concentrations, from mulls or pellets containing various relative amounts of sample, or from liquid films of various thicknesses. The variation of concentrations, etc., should cover at least an order of magnitude, so that weak features are brought out as well as the very strong absorptions.

When Raman spectra of pure liquids or solids are recorded, the problems of pathlength or concentration do not arise. However, when Raman spectra of solutions are recorded, the highest concentrations possible should be used because the spectra are inherently weak and one does not have the option of increasing the pathlength.

An examination of the infrared spectrum indicates the various X—H bonds present in the molecule and the presence or absence of triple or double bonds, carbonyl groups, aromatic rings, and other functional groups. This information, together with other data such as elemental analysis, molecular weight, Raman, UV, mass, and NMR spectra, will suggest various possible structures. The infrared and Raman spectra can then be searched in detail for bands due to the various groups in the postulated structures. Comparison with published infrared spectra of compounds having the suggested or a similar structure should be made. (See the list of sources of published spectra given in Appendix II.A.) Also, a computer search of a data bank of spectra might be made (see Appendix II.B for details).

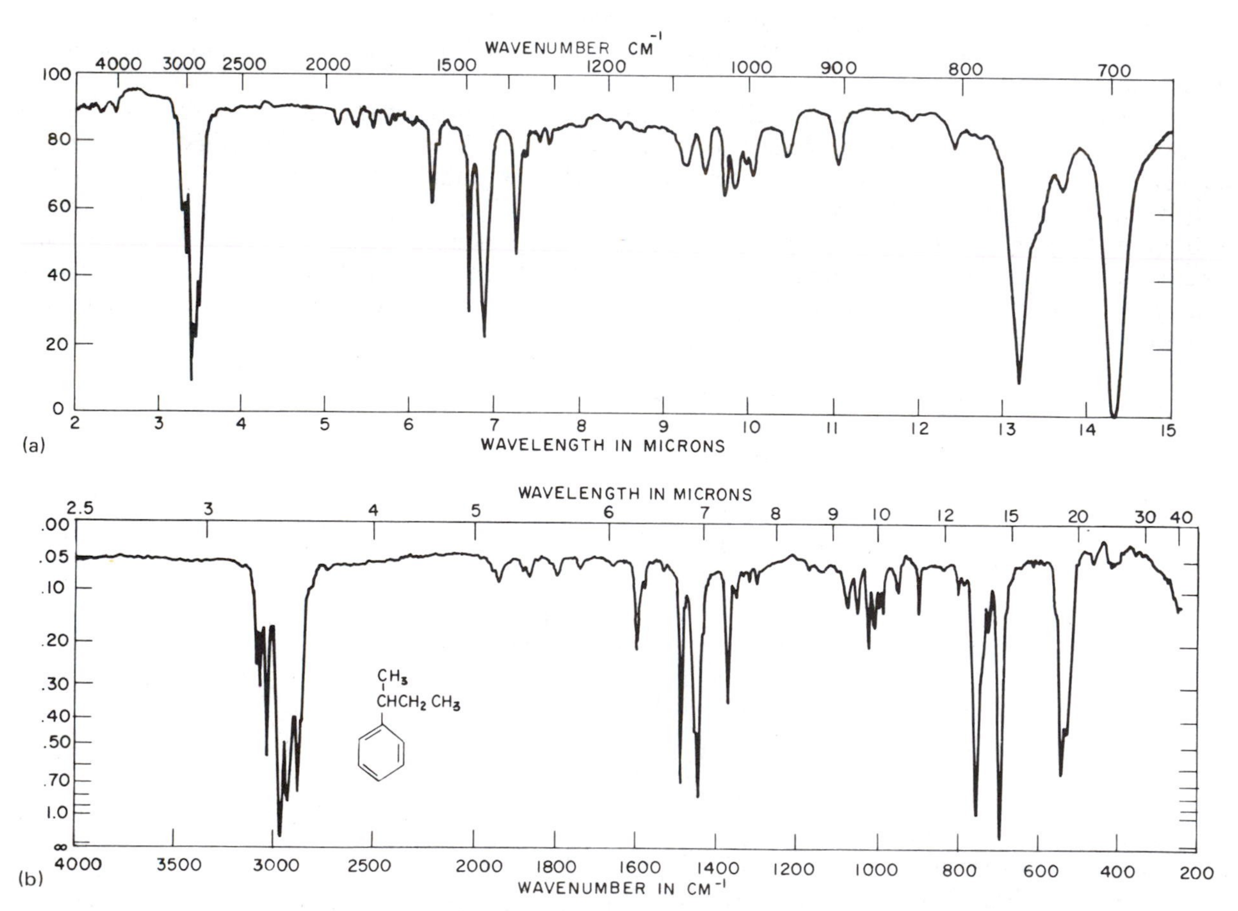

FIGURE 8-2 A comparison of (a) prism and (b) grating spectra of *sec*-butylbenzene. [© Sadtler Research Laboratories, Division of Bio-Rad Laboratories, Inc. (1967).]

8-2

A Note on Comparison of Prism and Grating Spectra

The appearance of spectra linear in wavenumber, obtained using grating or FT–IR spectrometers, is considerably different from spectra linear in wavelength recorded with prism instruments. Unfortunately, most of the earlier published work on infrared group frequencies was illustrated by spectra obtained using NaCl prism instruments, and many of the published collections present the spectra in the linear wavelength format. The differences are illustrated in Figure 8-2, which shows the infrared spectrum of *sec*-butylbenzene recorded with both linear wavelength and linear wavenumber formats. The most striking features are that the center part of the spectrum is displaced to the right and the low frequency bands are much narrower in the grating spectrum. Closer examination shows that the actual appearance of the central part of the spectrum is essentially the same in both spectra. However, the high frequency region near 3000 cm^{-1} is better resolved and more spread out in the grating spectrum. These features must be borne in mind when making visual comparisons with literature spectra. Note also the scale change at 2000 cm^{-1} (5 μm). This is a standard feature on all infrared spectrometers.

FT–IR spectra are always presented in the linear wavenumber format, so they are usually indistinguishable from grating spectra. Most of the spectra used as illustrations in this chapter are grating or FT–IR spectra.

8-3

Preliminary Analysis

8-3a Introduction

The infrared spectrum is arbitrarily divided into several regions (shown in Figure 8-3). The presence of bands in these regions gives immediate information. The *absence* of bands in these regions is also important information, since many groups can be excluded from further analysis, provided the factors discussed in Section 7-2 have been considered. A list of these regions together with absorbing groups and possible types of compounds is given in Table 8-1. This table is, of course, a condensation of Table 7-3. A similar table could be drawn up for Raman spectra.

Certain types of compounds give strong, broad absorptions, which are very prominent in the infrared spectrum. The hydrogen-bonded OH stretching bands of alcohols, phenols, and carboxylic acids are easily recognized at the high frequency end of the spectrum. The stretching of the NH_3^+ group in amino acids gives a very broad asymmetric band, which extends over several hundred wavenumbers. Broad bands associated

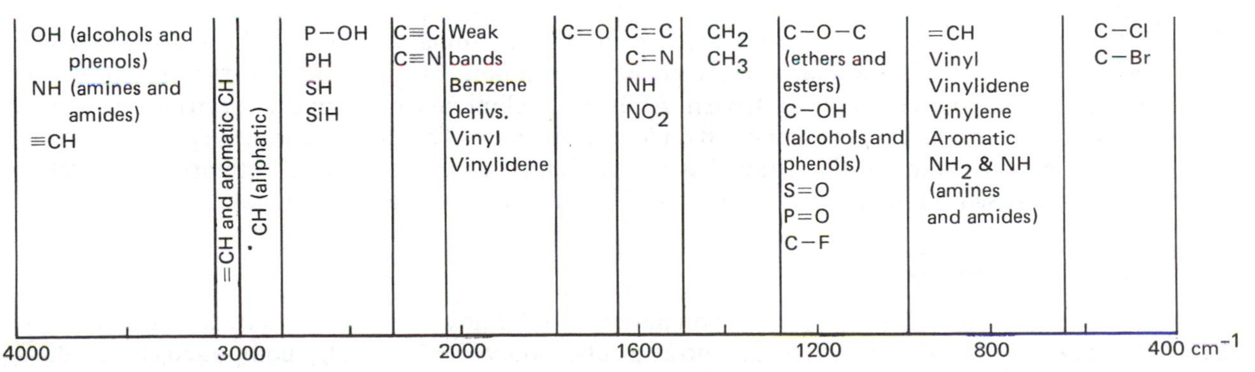

FIGURE 8-3 Regions of the infrared spectrum for preliminary analysis.

TABLE 8-1 Regions of the Infrared Spectrum for Preliminary Analysis

Region	Group	Possible Compounds Present (or Absent)
3700–3100	—OH	alcohols, aldehydes, carboxylic acids
	—NH	amides, amines
	≡C—H	acetylenes
3100–3000	=CH	aromatic compounds
	$=CH_2$ or —CH=CH—	alkenes or unsaturated rings
3000–2800	—CH, $—CH_2—$, $—CH_3$	aliphatic groups
2800–2600	—CHO	aldehydes (Fermi doublet)
2700–2400	—POH	phosphorus compounds
	—SH	mercaptans and thiols
	—PH	phosphines
2400–2000	—C≡N	nitriles
	$—N=\overset{+}{N}=\overset{-}{N}$	azides
	—C≡C—	acetylenes[a]
1870–1650	C=O	acid halides, aldehydes, amides, amino acids, anhydrides, carboxylic acids, esters, ketones, lactams, lactones, quinones
1650–1550	C=C, C=N, NH	unsaturated aliphatics,[a] aromatics, unsaturated heterocycles, amides, amines, amino acids
1550–1300	NO_2	nitro compounds
	CH_3 and CH_2	alkanes, alkenes, etc.
1300–1000	C—O—C and C—OH	ethers, alcohols, sugars
	S=O, P=O, C—F	sulfur, phosphorus, and fluorine compounds
1100–800	Si—O and P—O	organosilicon and phosphorus compounds
1000–650	=C—H	alkenes and aromatic compounds
	—NH	aliphatic amines
800–400	C—halogen	halogen compounds
	aromatic rings	aromatic compounds

[a]Band may be absent owing to symmetry (see Section 7-2a).

with bending of NH_2 or NH groups of primary or secondary amines are found at the low frequency end of the spectrum. Amides also give a broad band in this region. Examples of these characteristic absorptions are illustrated in Figure 8-4. These and other broad bands are listed in Table 8-2. It is well worth looking through one of the published collections of infrared spectra (see Appendix II.A) to familiarize oneself with these bands. The Aldrich Libraries of Infrared Spectra and FT–IR Spectra are particularly valuable for this purpose, since the spectra are arranged by class, with four or eight on a page.

In the spectra of certain compounds there are weak but characteristic bands that are known to be due to overtones or combinations. Some of these bands are shown in Figure 8-5 and listed in Table 8-3 with assignments.

In the following three sections, suggestions are given for the preliminary analysis of the infrared spectra of hydrocarbons or the hydrocarbon parts of molecules; of compounds containing carbon, hydrogen, and oxygen; and of compounds containing nitrogen. After this preliminary study of the spectrum, the analyst should have some idea of the kind of compound under investigation. The spectrum is then examined in more detail. In later sections, some of the detailed structural information that can be obtained from such an examination is discussed. In many cases, Raman spectra provide important additional information.

8-3b Hydrocarbons or Hydrocarbon Residues

The nature of a hydrocarbon or the hydrocarbon part of a molecule can be identified by first looking in the region between 3100 and 2800 cm^{-1}. If there is no absorption above 3000 cm^{-1}, the compound is aliphatic or alicyclic, with no ethylenic or aromatic structure. Cyclopropanes, which absorb above 3000 cm^{-1}, are an exception. If the absorption is entirely above 3000 cm^{-1}, the compound is probably aromatic or contains

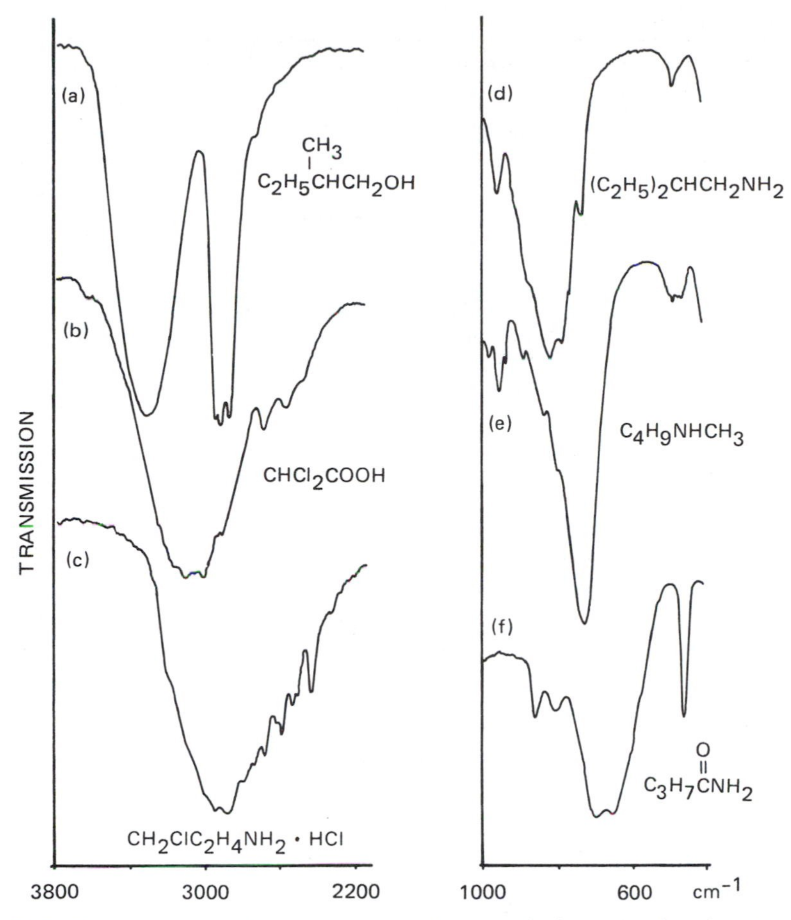

FIGURE 8-4 Some characteristic broad infrared absorption bands: (a) an alcohol, (b) a carboxylic acid, (c) an amine hydrochloride (Nujol mull), (d) a primary amine, (e) a secondary amine, (f) an amide. Note the different scales for the high and low frequency regions.

ethylenic $=CH$ or $=CH_2$ groups. Absorption both above and below 3000 cm^{-1} indicates both saturated and unsaturated or cyclic hydrocarbon moieties. (See Figure 8-2 for an example.)

Next, the region between 1000 and 650 cm^{-1} should be examined. Strong bands in this region suggest alkenes or aromatic structures. A small band near 720 cm^{-1} is indicative of a linear chain containing four or more CH_2 groups. The region between 1460 and 1350 cm^{-1} reveals the presence of methylene and methyl groups.

8-3c Compounds Containing Oxygen

If there is an oxygen atom present in the molecule, one should look in three regions for bands due to the oxygen-containing functional group. A strong broad band between 3500 and 3200 cm^{-1} may be from the hydrogen-bonded OH stretching mode of an alcohol or a phenol. Be aware that water also absorbs in this region. If hydrogen bonding is absent, the OH stretching band will be sharp and at higher frequencies (3650–3600 cm^{-1}). Carboxylic acids give very broad OH stretching bands between 3200 and 2700 cm^{-1} under the CH stretching bands.

One should next look for a very strong band between 1850 and 1650 cm^{-1} due to the $C{=}O$ stretching of a carbonyl group. The third region is between 1300 and 1000 cm^{-1}, where bands due to C—OH stretching of alcohols and carboxylic acids and C—O—C stretching modes of ethers and esters are observed. Ethers and esters have symmetric and antisymmetric C—O—C stretching modes, but the symmetric mode is usually weak in the infrared spectra of ethers and occurs below 1000 cm^{-1}. However, both symmetric and

TABLE 8-2 Characteristic Broad Absorption Bands

Range or Band Center	Possible Compounds	Assignment and Remarks
3600–3200 (vs)	alcohols, phenols, oximes	OH stretch (hydrogen-bonded)
3400–3000 (vs)	primary amides	NH_2 stretch; usually a doublet
3400–2400 (s)	carboxylic acids and other compounds with —OH groups	H-bonded OH stretch of dimers and polymers
3200–2400 (vs)	amino acids (zwitterion), amine hydrohalides	NH_3^+ stretching; a very broad asymmetric band
3000–2800 (vs)	hydrocarbons, all compounds containing CH_3 and CH_2 groups	CH stretch bands due to Nujol obscure this region when spectra are obtained from Nujol mulls
1700–1250 (s)	amino acids	C=O stretch; a broad region of absorption with much structure
1650–1500 (vs)	salts of carboxylic acids	$-C(\cdots O)(\cdots O)^-$ antisymmetric stretch
ca 1250 (vs)	perfluoro compounds	CF stretches; may cover the whole region from 1400–1100 cm^{-1} with several bands
ca 1200 (vs)	esters	C—O—C stretch; ester linkage (not always broad) with much structure
ca 1200 (vs)	phenols	C—OH stretch
ca 1150 (vs)	sulfonic acids	S=O stretch; with structure
1150–950 (vs)	sugars	with structure
ca 1100 (vs)	ethers	C—O—C stretch
1100–1000 (s)	alcohols	C—OH stretch
ca 1050 (vs)	anhydrides	not always reliable
1050–950 (vs)	phosphites and phosphates	P=O stretch; often two bands
ca 920 (ms)	carboxylic acids	H-bonded C—OH deformation
ca 830 (vs)	primary aliphatic amines	may cover the region 1000–700 cm^{-1}
ca 730 (s)	secondary aliphatic amines	may cover the region 850–650 cm^{-1}
ca 650 (s)	amides	may cover the region 750–550 cm^{-1}
800–500 (w)	alcohols	a weak broad band

antisymmetric C—O—C stretching modes of esters give strong bands between 1300 and 1000 cm^{-1}. Some examples of C—O stretching bands are shown in Figure 8-6. The presence or absence of the bands due to OH, C=O, and C—O stretching gives a good indication of the type of oxygen-containing compound. Infrared spectra of aldehydes and ketones show only the C=O band, while spectra of ethers have only the C—O band. Esters have bands due to both C=O and C—O, but no OH, while alcohols will have OH and C—O bands, but no C=O. Spectra of carboxylic acids contain bands in all three regions.

8-3d Compounds Containing Nitrogen

One or two bands in the region 3500–3300 cm^{-1} indicate primary or secondary amines or amides. These bands may be confused with OH stretching bands in the infrared, but are more easily identified in the Raman spectrum, where the OH stretching band is very weak. Amides may be identified by a strong doublet in the infrared spectrum centered near 1640 cm^{-1}. The Raman spectrum of an amide has bands near 1650 and 1400 cm^{-1}. These bands are associated with the stretching of the C=O group and bending of the NH_2 group of the amide. A sharp band near 2200 cm^{-1} in either infrared or Raman spectra is characteristic of a nitrile.

When both nitrogen and oxygen are present, but no NH groups are indicated, two very strong bands near 1560 and 1370 cm^{-1} provide evidence of the presence of nitro groups. An extremely broad band with some structure centered near 3000 cm^{-1} and extending as low as 2200 cm^{-1} is indicative of an amino acid or an amine hydrohalide, while a broad band below 1000 cm^{-1} suggests an amine or an amide (see Figure 8-4 for examples of these broad bands).

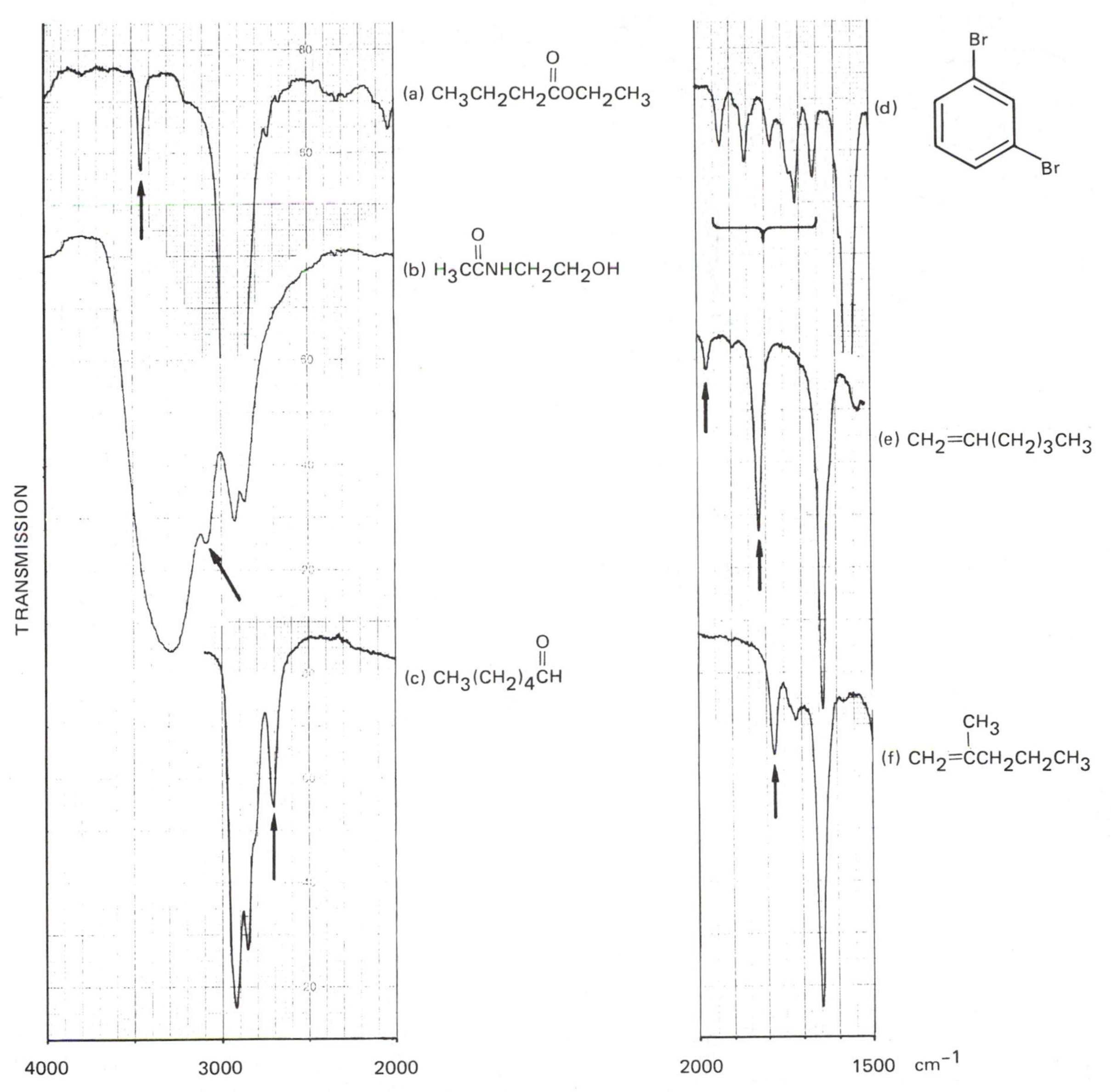

FIGURE 8-5 Some characteristic weak infrared bands from overtones and combinations: (a) an ester, (b) a secondary amide, (c) an aldehyde, (d) a substituted benzene, (e) a vinyl compound, (f) a vinylidene compound.

TABLE 8-3 Some Characteristic Overtone or Combination Bands

Range (cm^{-1})	Classes of Compounds	Assignment and Remarks
ca 3450	esters	overtone of C=O stretch
3100–3060	secondary amides	overtone of NH deformation
ca 2700	aldehydes	part of a Fermi doublet at 2800–2700
2200–2000	amino acids and amine hydrohalides	combination of NH_3^+ torsion and NH_3^+ antisym deformation
2000–1650	aromatic compounds	overtones and combinations of CH out-of-plane deformations
1990–1960 and 1830–1800	vinyl compounds	overtones of CH and CH_2 out-of-plane deformations; high frequency band is stronger
1800–1780	vinylidine compounds	overtone of CH_2 wag

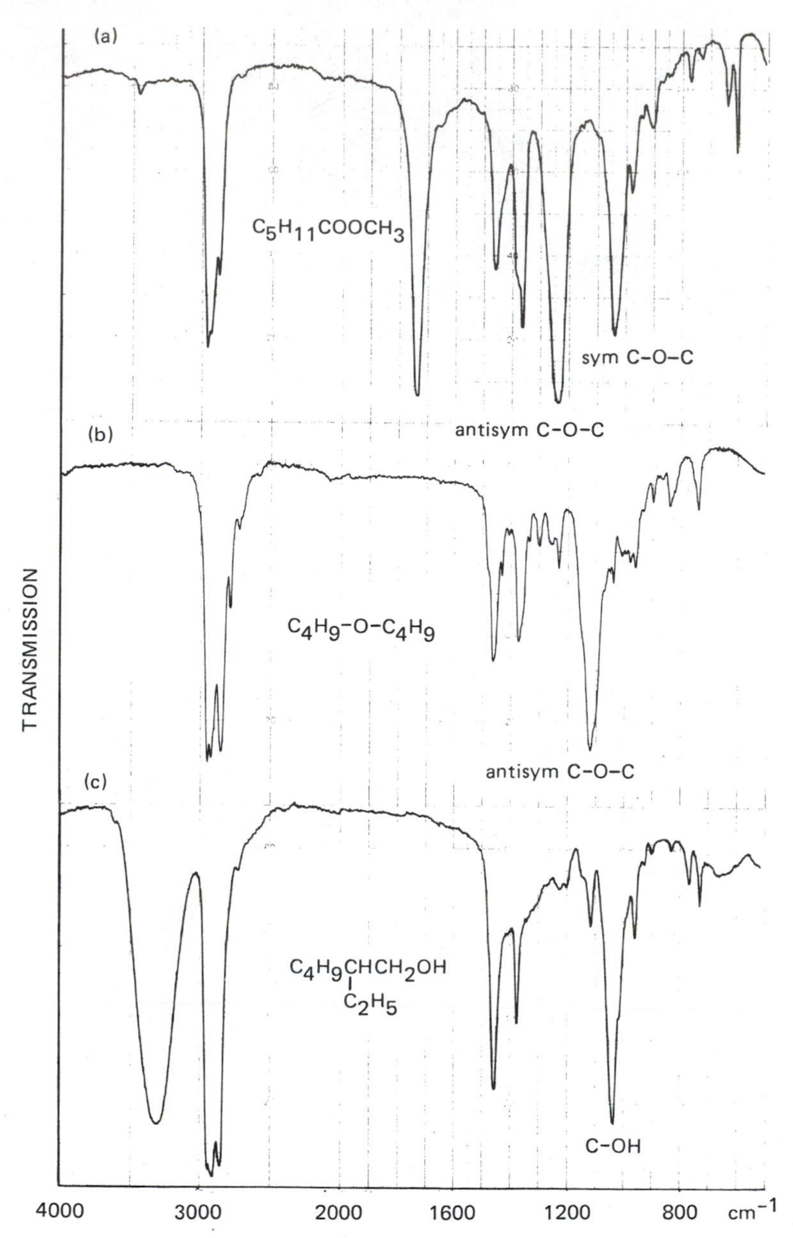

FIGURE 8-6 The infrared spectra of (a) an ester, (b) an ester, and (c) an alcohol, showing the C—O stretching bands between 1300 and 1000 cm^{-1}.

8-4

The CH Stretching Region (3340–2700 cm^{-1})

8-4a Introduction

Details of CH stretching frequencies in various compounds are summarized in Table 8-4. As indicated in Section 8-3b, this region is usually the first that one examines, since certain structural features are immediately revealed by the position of the CH stretching bands. A band at the high frequency end of the above range indicates the presence of an acetylenic hydrogen atom, while a band near 2710 cm^{-1} usually means that there is an aldehyde group in the molecule. It should be noted that, when spectra are recorded from Nujol mulls, the CH stretching region is obscured by strong Nujol bands. Alternative mulling materials such as Fluorolube or hexachlorobutadiene can be used for this region (see Section 6-7b).

TABLE 8-4 C—H Stretching Frequencies

Range (cm^{-1}) and Intensity[a]	Group or Class	Assignment and Remarks
3340–3270 (w–m)	≡CH terminal alkynes	≡CH stretch; sharp band
3100–3000 (w–m)	aromatic compounds	CH stretch; may be several bands, often weak
3100–3070 (m)	cyclopropanes (CH_2 group)	CH_2 antisym stretch
3095–3075 (m–s)	=CH_2 vinyl or vinylidine	=CH_2 stretch
3080–3040 (m)	epoxides (CH_2 group)	CH_2 stretch; sharp band
3035–2995 (m)	cyclopropanes (CH_2 group)	CH_2 sym stretch
3030–3000 (m–s)	=CHR aliphatic compounds	=CH stretch
3000–2970 (m)	cyclobutanes (CH_2 groups)	CH_2 antisym stretch
2970–2950 (vs)	—CH_3 group (in alkyl groups)	CH_3 antisym stretch
2960–2950 (vs)	cyclopentanes (CH_2 groups)	CH_2 antisym stretch
2940–2915 (vs)	—CH_2— (in alkyl groups)	CH_2 antisym stretch
2925–2875 (vs)	cyclobutanes (CH_2 groups)	CH_2 sym stretch
2885–2860 (vs)	—CH_3 group (in alkyl groups)	CH_3 sym stretch
2875–2855 (vs)	cyclobutanes (CHR groups) and cyclopentanes (CH_2 groups)	CH(R) stretch; CH_2 sym stretch
2870–2840 (vs)	—CH_2— group	CH_2 sym stretch
2850–2820 (m) and 2750–2720 (m)	aromatic aldehydes	Fermi doublet
2840–2815 (m–s)	methoxy	CH_3 stretch
2830–2810 (m) and 2725–2700 (m)	aliphatic aldehydes	Fermi doublet

[a] s = strong; m = medium; w = weak; v = very.

8-4b Acetylenes

If a sharp band is observed near 3300 cm^{-1} in the infrared spectrum, the presence of a terminal ≡CH group is suspected. Confirmation of this observation can be made by a small sharp peak in the infrared or a strong band in the Raman spectrum near 2100 cm^{-1} due to the C≡C stretch (see Section 8-9c). An example is given in Figure 8-7.

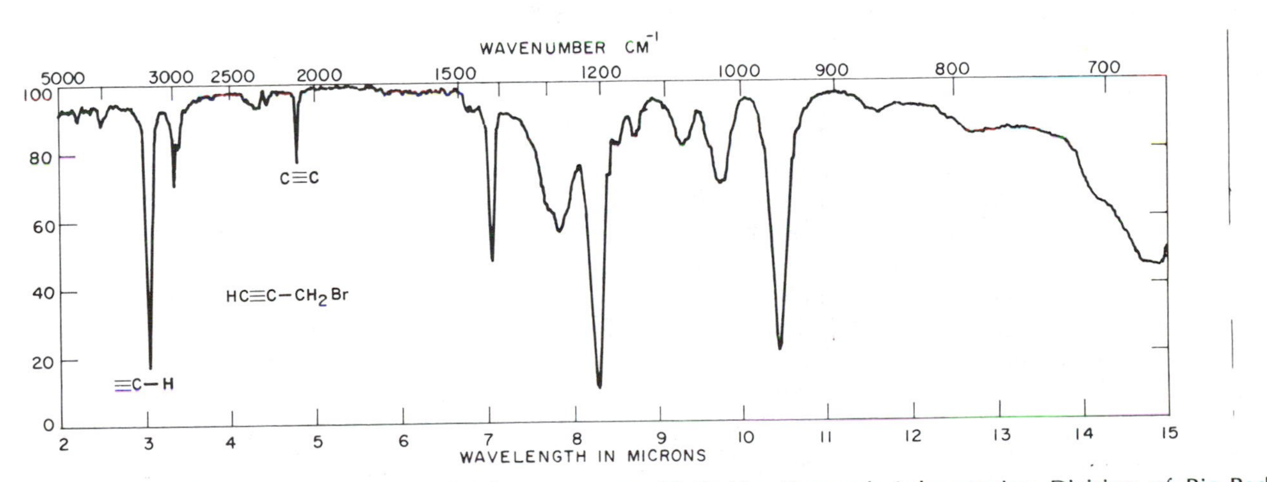

FIGURE 8-7 The infrared spectrum of 3-bromopropyne. [© Sadtler Research Laboratories, Division of Bio-Rad Laboratories, Inc. (1965).]

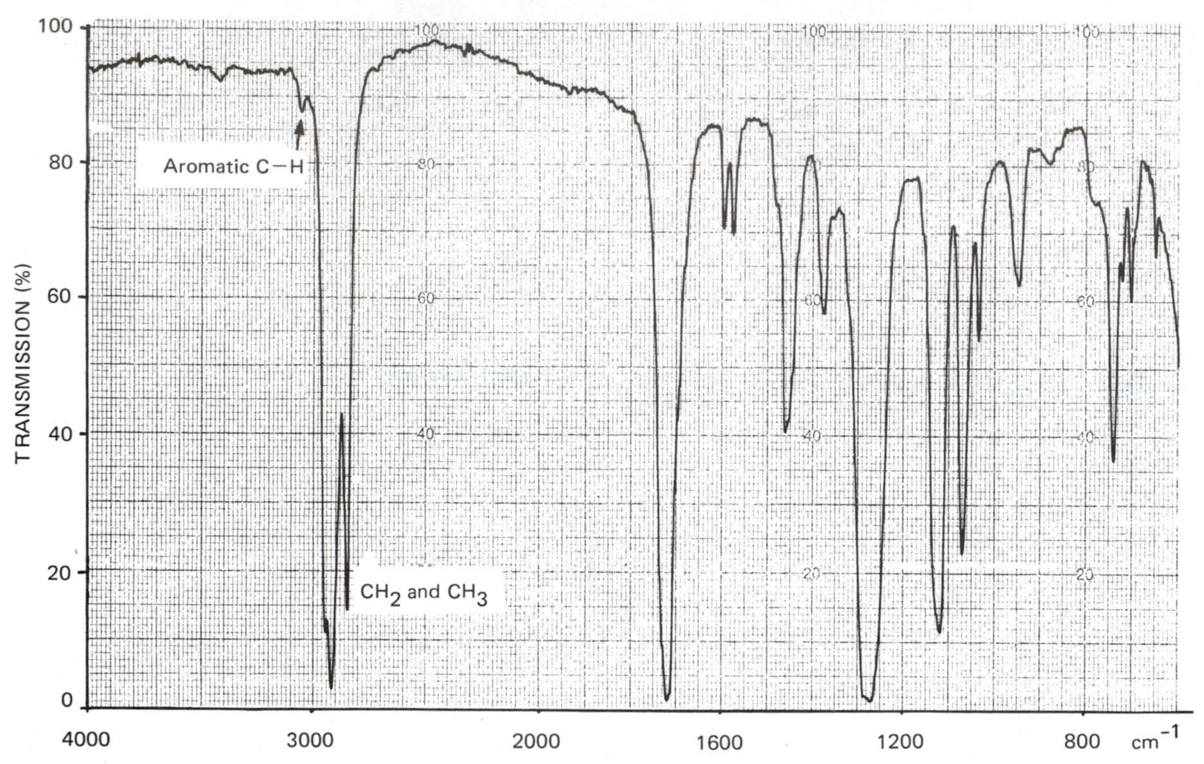

FIGURE 8-8 The infrared spectrum of dioctyl phthalate.

8-4c Aromatic Compounds

Aromatic compounds have one or more sharp peaks of weak or medium intensity between 3100 and 3000 cm^{-1}. A word of warning here: these bands may appear only as shoulders on a very strong CH_3 or CH_2 stretching band. This problem is illustrated by the spectrum of dioctyl phthalate shown in Figure 8-8.

8-4d Unsaturated Nonaromatic Compounds

Unsaturated compounds and small aliphatic ring compounds show absorption due to CH stretching in the 3100–3000 cm^{-1} region. Compounds containing the vinylidine ($=CH_2$) group absorb near 3080 cm^{-1}. Di- and trisubstituted ethylenes absorb at lower frequencies, nearer to 3000 cm^{-1}, and the band may be overlapped by the stronger CH_3 or CH_2 absorption (see Section 8-4e). Cyclopropane derivatives have a band between 3100 and 3070 cm^{-1}, while epoxides absorb between 3060 and 3040 cm^{-1}.

8-4e Saturated Groups: CH_3, CH_2, and CH

Saturated compounds can have methyl, methylene, or methine groups, each of which has characteristic CH stretching frequencies. CH_3 groups absorb near 2960 and 2870 cm^{-1} and the CH_2 bands are at 2930 and 2850 cm^{-1}. In many cases, only one band in the 2870–2850 cm^{-1} region can be resolved when both CH_3 and CH_2 groups are present in the molecule. Figure 8-9 shows the CH stretching region of three normal saturated molecules. As the carbon chain becomes longer, the CH_2 group bands increase in intensity relative to the CH_3 group absorptions. The doublet at 2960 cm^{-1} is the antisymmetric CH_3 stretching mode, which would be degenerate in a free CH_3 group or in a molecule in which the symmetry of the group was maintained, for example, CH_3Cl. The degeneracy is removed in the saturated molecules, and consequently two individual antisymmetric CH_3 stretching bands are observed. The methine CH stretch can only be observed (as a weak peak near 2885 cm^{-1}) when CH_3 and CH_2 groups are absent. The methoxy group CH_3O— has a characteristic sharp band of medium intensity near 2830 cm^{-1} separate from other CH stretching bands.

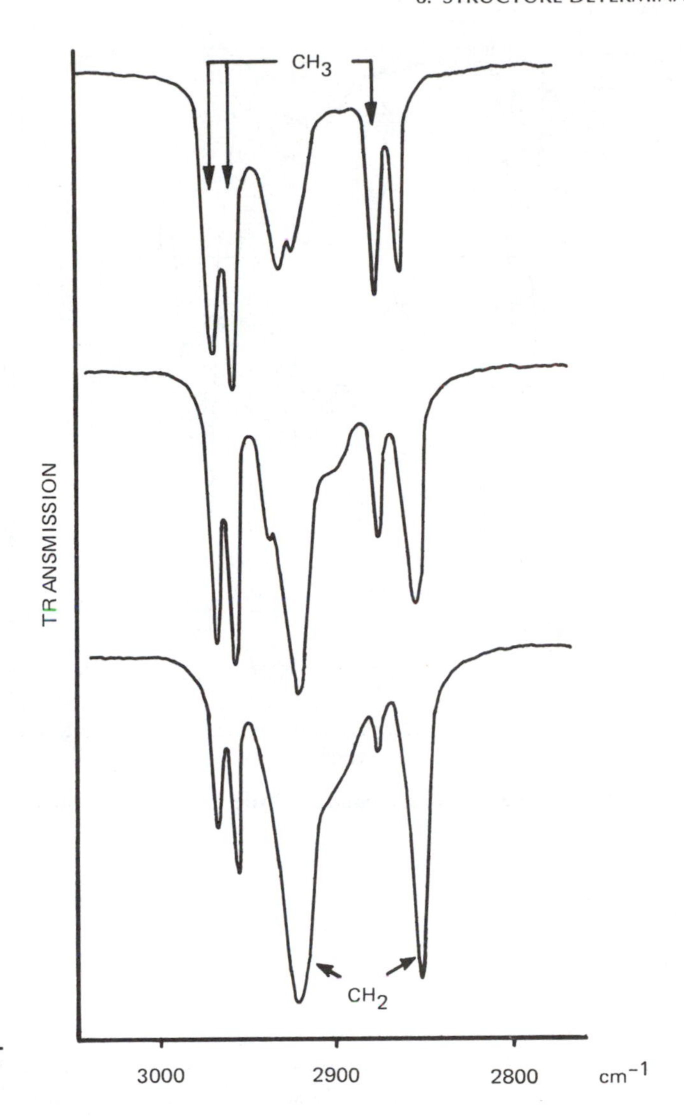

FIGURE 8-9 The C—H stretching regions of n-C_4H_{10}, n-C_8H_{18}, and n-$C_{20}H_{42}$.

8-4f Aldehydes

The CH stretching mode of the aldehyde group appears as a Fermi doublet near 2820 and 2710 cm^{-1} (see Section 7-2b). The 2710 cm^{-1} absorption is very useful and characteristic of aldehydes, but the higher frequency component often appears as a shoulder on the 2850 cm^{-1} CH_2 stretching band. In the spectrum of a purely aromatic aldehyde there is no overlap of other CH stretching bands and the doublet is clearly resolved. An example is given in Figure 8-10. In ortho-substituted aromatic aldehydes the frequencies of the doublet are about 40 cm^{-1} higher than usual with the high frequency component stronger and broader (Figure 8-10).

8-5

The Carbonyl Stretching Region (1850–1650 cm^{-1})

8-5a General

A very important region of the spectrum for structural analysis is the carbonyl stretching region. If we were to include metal carbonyls and salts of carboxylic acids, the region of the C═O stretching mode would extend from 2200 down to 1350 cm^{-1}. However, most organic compounds containing the C═O group show very strong infrared absorption in the range 1850–1650 cm^{-1}. The actual position of the peak (or peaks) within this range is

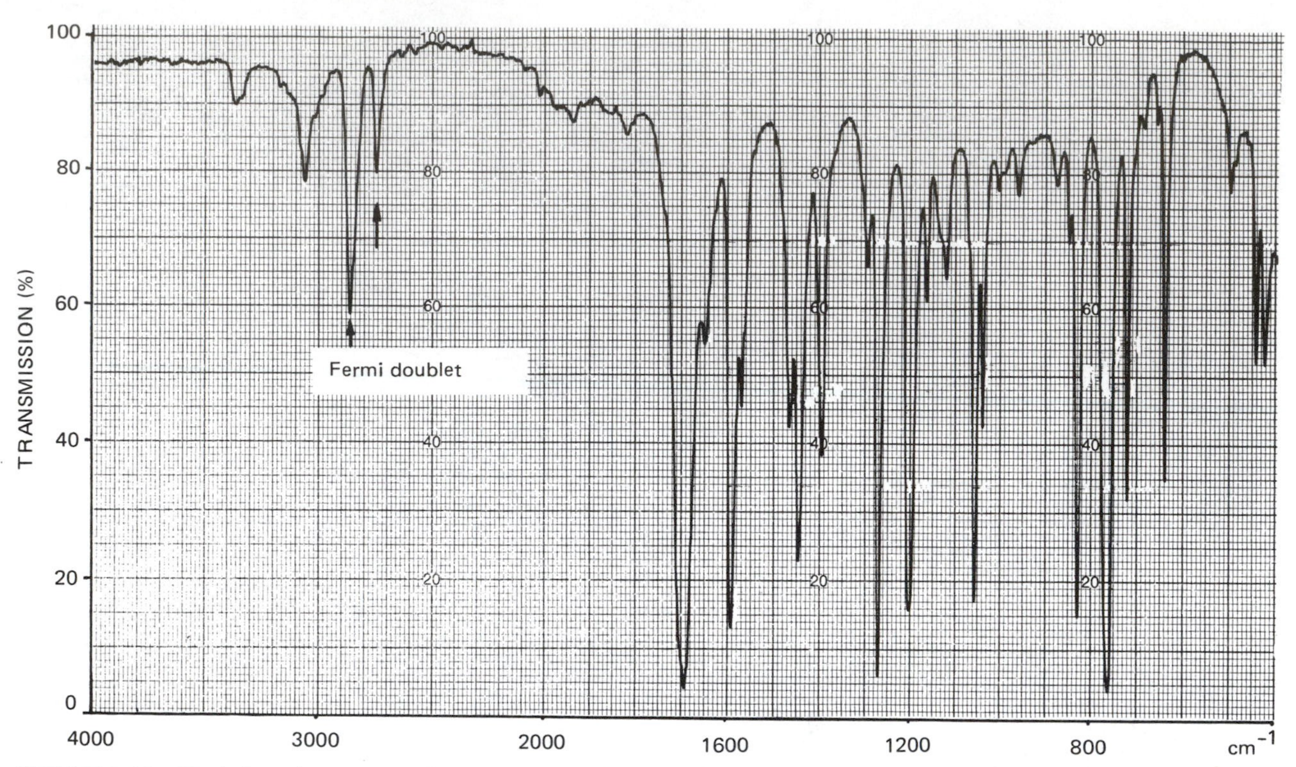

FIGURE 8-10 The infrared spectrum of an aromatic aldehyde, 2-chlorobenzaldehyde, showing the Fermi doublet at 2750 and 2870 cm^{-1}.

characteristic of the type of compound. At the upper end of the range are found bands due to anhydrides (two bands) and four-membered ring lactones (β lactones), while amides and substituted ureas absorb at the lower end of the range. Table 8-5 summarizes the ranges in which compounds with a single carbonyl group absorb. Table 8-6 lists compounds that have two interacting carbonyl groups. In these compounds, out-of-phase and in-phase (antisymmetric and symmetric) C=O stretching vibrations can occur (see Section 7-2b). Usually, the out-of-phase mode is found at higher frequencies than the in-phase mode.

8-5b Compounds Containing a Single C=O Group

The type of functional group usually cannot be identified from the C=O stretching band alone because there may be several carbonyl-containing functional groups that absorb within a given frequency range. However, an initial separation into possible compounds can be achieved with Table 8-5. For example, a single carbonyl peak in the region 1750–1700 cm^{-1} could indicate an ester, an aldehyde, a ketone (including cyclic ketones), a large ring lactone, a urethane derivative, an α-halo ketone, or an α-halo carboxylic acid. The presence of the halogen could be checked from the elemental analysis, an aldehyde would be identified by a peak near 2700 cm^{-1} (see Section 8-4f), and esters and lactones give a strong band near 1200 cm^{-1}, which is often quite broad in esters (see Section 8-10e). Urethanes would have an NH stretching band.

8-5c Compounds Containing Two C=O Groups

Compounds with two coupled carbonyl groups are smaller in number, but again ambiguities can exist and other parts of the infrared spectrum must be analyzed in conjunction with the positions of the carbonyl bands in order for a particular structural grouping to be identified. An example of this type of problem is provided by the spectrum shown in Figure 8-11. An elemental analysis indicates that only C, H, and O are present and the molecular formula is $C_5H_8O_3$. Looking at the CH stretching region (Table 8-4), we conclude that there is no unsaturated group present. This conclusion is confirmed by the absence of absorption between 1670 and

TABLE 8-5 Carbonyl Stretching Frequencies for Compounds Having One Carbonyl Group

Range (cm^{-1}) and Intensity	Classes of Compounds	Remarks
1840–1820 (vs)	β lactones	4-membered ring
1810–1790 (vs)	acid chlorides	saturated aliphatic compounds
1800–1750 (vs)	aromatic and unsaturated esters	the C=C stretch is higher than normal (1700–1650 cm^{-1})
1800–1740 (s)	carboxylic acid monomer	only observed in dil soln
1790–1740 (vs)	γ lactones	5-membered ring
1790–1760 (vs)	aromatic or unsaturated acid chlorides	second weaker combination band near 1740 cm^{-1} (Fermi resonance)
1780–1700 (s)	lactams	position depends on ring size
1770–1745 (vs)	α-halo esters	higher frequency due to electronegative halogen
1750–1740 (vs)	cyclopentanones	unconjugated structure
1750–1730 (vs)	esters and δ lactones	aliphatic compounds
1750–1700 (s)	urethanes	R—O—(C=O)—NHR compounds
1745–1730 (vs)	α-halo ketones	noncyclic compounds
1740–1720 (vs)	aldehydes	aliphatic compounds
1740–1720 (vs)	α-halo carboxylic acids	20 cm^{-1} higher frequency if halogen is fluorine
1730–1705 (vs)	aryl and α, β-unsaturated aliphatic esters	conjugated carbonyl group
1730–1700 (vs)	ketones	aliphatic and large ring alicyclic
1720–1680 (vs)	aromatic aldehydes	also α, β-unsaturated aliphatic aldehydes
1720–1680 (vs)	carboxylic acid dimer	broader band
1710–1640 (vs)	thiol esters	lower than normal esters
1700–1680 (vs)	aromatic ketones	position affected by substituents on ring
1700–1680 (vs)	aromatic carboxylic acids	dimer band
1700–1670 (s)	primary and secondary amides	in dil soln
1700–1650 (vs)	conjugated ketones	check C=C stretch region
1690–1660 (vs)	quinones	position affected by substituents on ring
1680–1630 (vs)	amides (solid state)	note second peak due to NH def near 1625 cm^{-1}
1670–1660 (s)	diaryl ketones	position affected by substituents on ring
1670–1640 (s)	ureas	second peak due to NH def near 1590 cm^{-1}
1670–1630 (vs)	ortho-OH or $—NH_2$ aromatic ketones	frequency lowered by chelation with ortho group

TABLE 8-6 Carbonyl Stretching Frequencies for Compounds Having Two Interacting Carbonyl Groups

Range (cm^{-1}) and Intensity[a]	Class of Compounds	Remarks
1870–1840 (m–s) 1800–1770 (vs)	cyclic anhydrides	low frequency band is stronger
1825–1815 (vs) 1755–1745 (s)	normal anhydrides	high frequency band is stronger
1780–1760 (m) 1720–1700 (vs)	imides	low frequency band is broad; high frequency band may be obscured
1760–1740 (vs)	α-keto esters	usually only one band
1740–1730 (vs)	β-keto esters (keto form)	may be a doublet due to two C=O groups
1660–1640 (vs)	β-keto esters (enol form)	may be a doublet due to a C=O and a C=C group
1710–1690 (vs) 1640–1540 (vs)	diketones	high frequency band due to keto form; low frequency band due to enol form
1690–1660 (vs)	quinones	frequency depends on substituents
1650–1550 (vs) 1440–1350 (s)	carboxylic acid salts	two broad bands due to antisym and sym —C(=O)O⁻ stretches

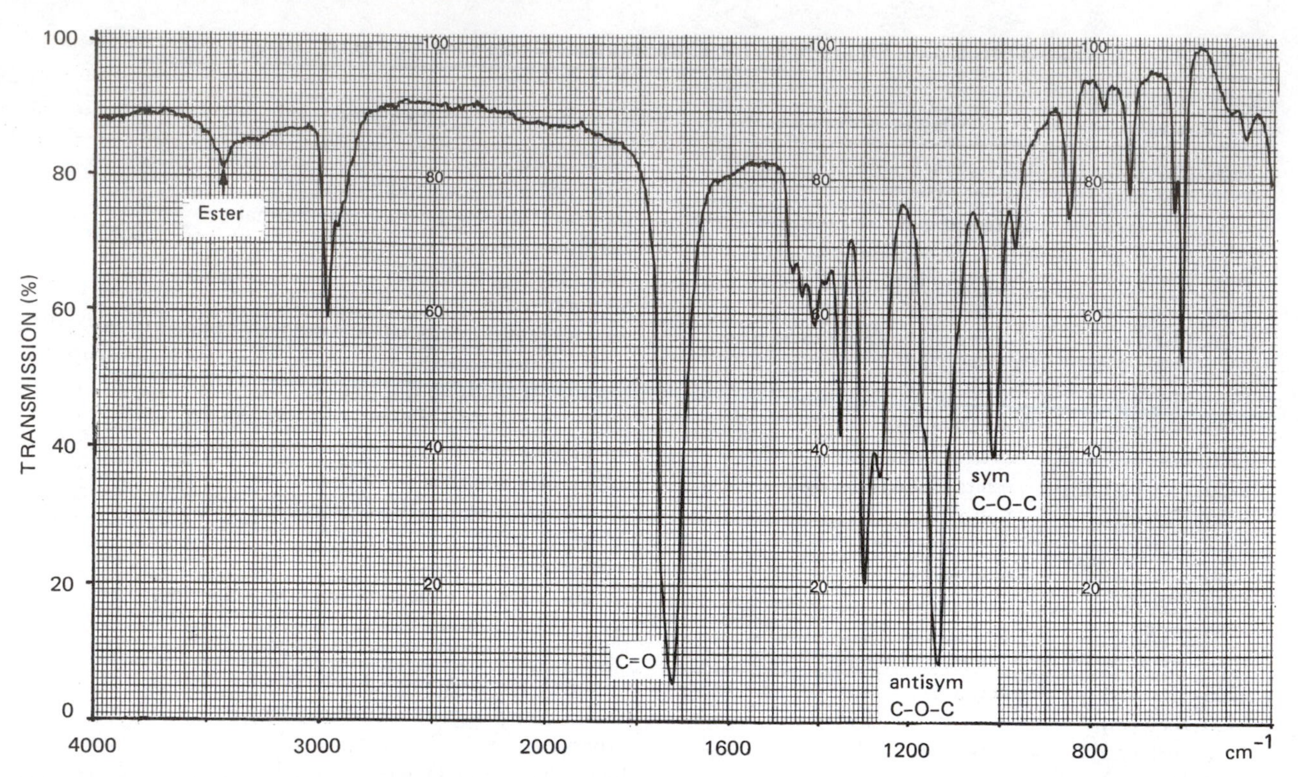

FIGURE 8-11 The infrared spectrum of an α-keto ester, ethyl pyruvate.

1540 cm^{-1}. A CH_3 group or groups (2980 and 1355 cm^{-1}) and possibly a CH_2 group (2930 and 1415 cm^{-1}) are present. Now turning to the carbonyl stretching region, we note from Tables 8-5 and 8-6 that the compound could be a cyclic ketone, an ester, a lactone, an aldehyde, or an α-keto ester. We can eliminate aldehyde (no band near 2700 cm^{-1}) and lactones or cyclic ketones (not possible with 3 oxygens and a methyl group). The bands at 1135 and 1020 cm^{-1} suggest an ester linkage. This assignment is supported by the weak band at 3440 cm^{-1} (see Table 8-3). The compound is then most likely to be an α-keto ester. Two possible structures are **8-1** and **8-2**. Structure **8-2** is more likely because an O—CH_3 group would give a sharp band near 2830 cm^{-1} (see Section 8-4e).

$$C_2H_5-\overset{\overset{O}{\|}}{C}-\overset{\overset{O}{\|}}{C}-O-CH_3$$

8-1

$$CH_3-\overset{\overset{O}{\|}}{C}-\overset{\overset{O}{\|}}{C}-O-C_2H_5$$

8-2

8-6

Aromatic Compounds

8-6a General

Bands characteristic of aromatic compounds can be found in five regions of the infrared spectrum. These are 3100–3000 cm^{-1} (CH stretching), 1650–1430 cm^{-1} (C=C stretching), 1275–1000 cm^{-1} (in-plane CH deformation), 900–690 cm^{-1} (out-of-plane CH deformation), and 2000–1700 cm^{-1} (overtones and combinations). Examples of infrared spectra of three aromatic compounds can be seen in Figures 8-2, 8-8, and 8-10. In these spectra, there are bands in all five regions. However, the intensities of the bands vary widely.

The intensities of the CH stretching bands range from medium, as in Figure 8-2, to weak, as in Figure 8-8. Occasionally, these bands are seen only as shoulders on a strong aliphatic CH stretching band. There are usually sharp bands near 1600, 1500, and 1430 cm^{-1} in benzene derivatives. The 1600 cm^{-1} absorption can be hidden by a strong C=O stretching band in some carbonyl compounds, or an NH deformation band in amines. There are also several sharp bands between 1275 and 1000 cm^{-1} in aromatic compounds. In the 900–690 cm^{-1} region, one or two strong bands are observed. These bands, together with the pattern of weak bands between 2000 and 1700 cm^{-1}, give an indication of the order of substitution on the benzene ring.

8-6b Substituted Aromatic Compounds

The monosubstitution pattern is easily recognized by two very strong bands near 750 and 700 cm^{-1} due to out-of-plane CH bending. These are very prominent in the spectrum of Figure 8-2, which will now be recognized as a monosubstituted benzene derivative. Other examples can be seen in Figures 6-6 and 8-20b. In addition, monosubstitution is indicated by four weak but clear absorptions between 2000 and 1700 cm^{-1}. Again, Figure 8-2 illustrates this pattern. The patterns of bands in these two regions can also give an indication of the positions of substitution in di- and trisubstituted benzene rings. These regions are less useful for highly substituted derivatives. The characteristic bands of mono-, di-, and trisubstituted aromatic compounds are summarized in Table 8-7. Examples of the three disubstitution patterns are shown in Figure 8-12.

TABLE 8-7 Absorption Bands Characteristic of Substitution in Benzene Rings

Type of Substitution	Range (cm^{-1}) and Intensity	Remarks
monosubstitution	770–730 (vs) and 710–690 (s)	two bands; very characteristic of monosubstitution
	2000–1700 (w)	four weak but prominent bands; also very characteristic of monosubstitution
ortho disubstitution	770–730 (vs)	a single strong band
	1950–1650 (vw)	several bands, the two most prominent near 1900 and 1800 cm^{-1}
meta disubstitution	810–750 (vs) and 725–680 (s)	two bands similar to monosubstitution, but the higher frequency band is 50 cm^{-1} higher, and the lower is more variable in position and intensity than in monosubstituted benzenes
	1930–1740 (vw)	three weak but prominent bands; the lowest frequency band may be broader with some structure
para disubstitution	860–800 (vs)	a single strong band, similar to ortho disubstitution, but 50 cm^{-1} higher in frequency
	1900–1750 (w)	two bands; the higher frequency one is usually stronger
1,3,5-trisubst	865–810(s) and 765–730 (s)	two bands with wider separation than mono or *m*-disubstitution
	1800–1700 (w)	one fairly broad band with a much weaker one near 1900 cm^{-1}
1,2,3-trisubst	780–760 (s) and 745–705 (s)	two bands, similar to *m*-disubst, but closer in frequency
	2000–1700 (w)	similar to monosubst, but only three bands
1,2,4-trisubst	885–870 (s) and 825–805 (s)	two bands at higher frequencies than mono-, *m*-di-, and the other trisubst compounds
	1900–1700 (w)	two prominent bands near 1880 and 1740 cm^{-1} with a much weaker one between

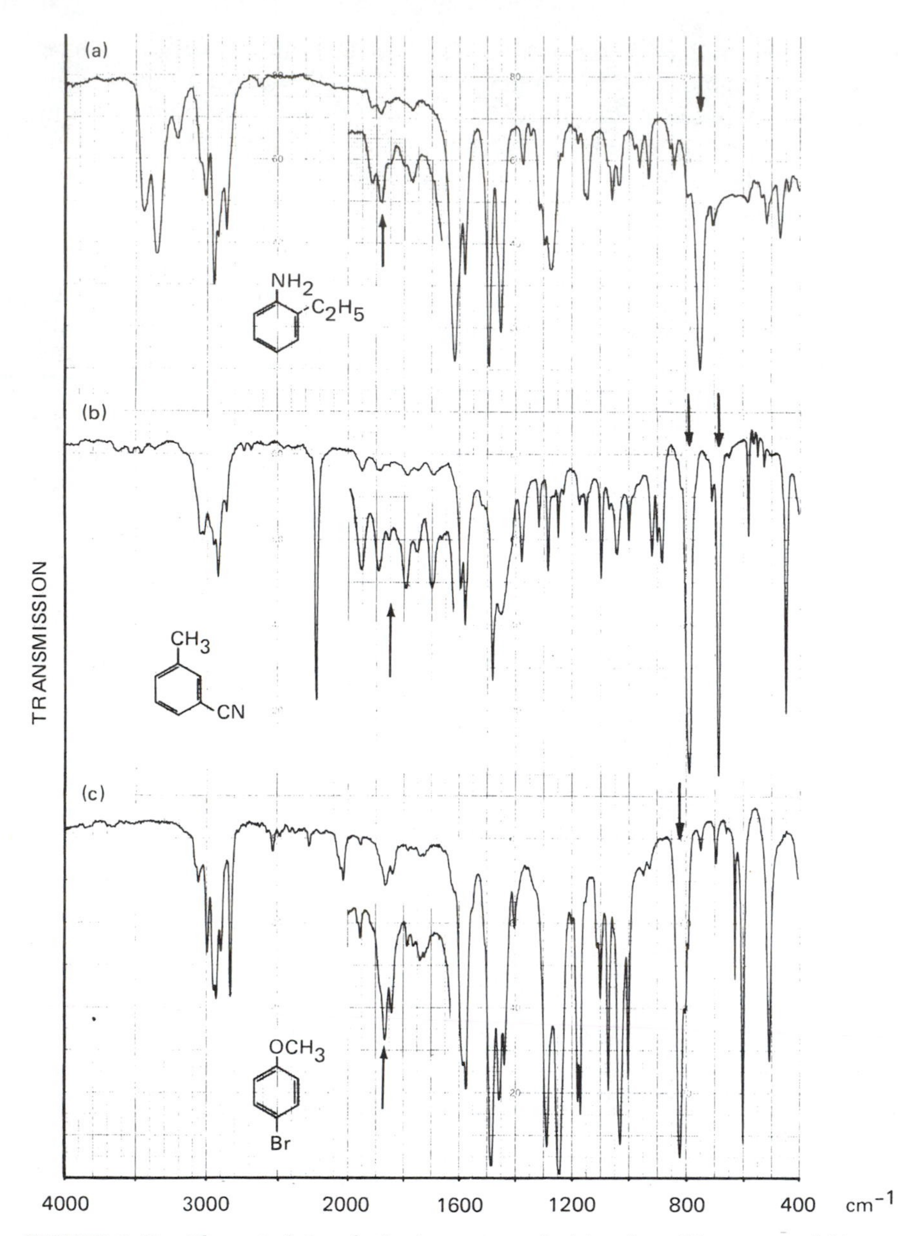

FIGURE 8-12 Characteristic substitution patterns in (a) ortho-, (b) meta-, and (c) para-disubstituted benzene derivatives.

8-6c Raman Spectra of Aromatic Compounds

The Raman spectra of aromatic compounds also contain several characteristic bands, which can be useful in cases for which the infrared spectrum gives ambiguous results. A prominent band in benzene derivatives is found near 1000 cm^{-1}. This band is due to a ring-breathing vibration. Two examples are shown in Figure 8-13. All aromatic compounds show a Raman line near 1600 cm^{-1}, which is usually seen in the infrared but may be weak or obscured. Monosubstituted benzene rings also have a Raman line of medium intensity between 1030 and 1010 cm^{-1} due to in-plane CH bending, and a weaker line between 625 and 605 cm^{-1} due to an in-plane ring bending mode.

The infrared and Raman spectra of nitrobenzene are compared in Figure 8-14.

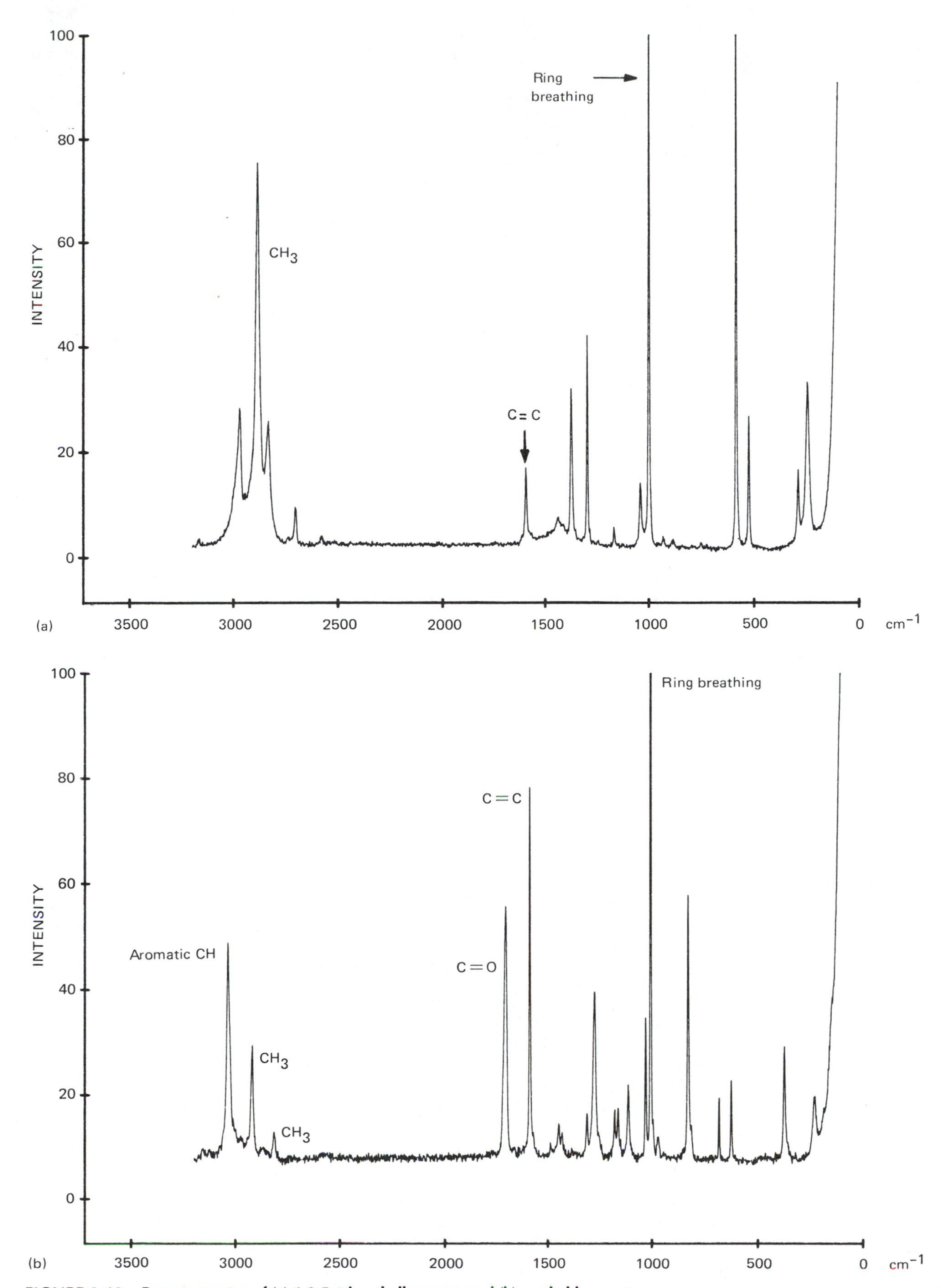

FIGURE 8-13 Raman spectra of (a) 1,3,5-trimethylbenzene and (b) methyl benzoate.

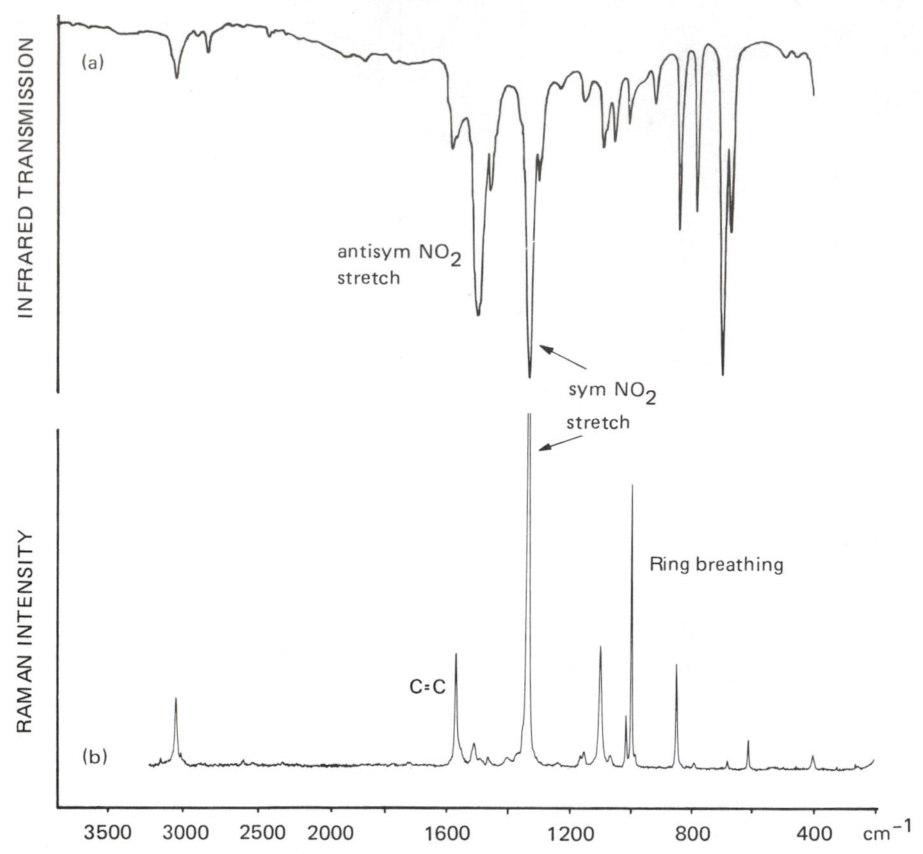

FIGURE 8-14 **(a) Infrared and (b) Raman spectra of nitrobenzene.**

8-7

Compounds Containing Methyl Groups

8-7a General

In Section 7-2a the vibrations of a methyl group were discussed in terms of the symmetry of the group. These vibrations, stretching, bending (deformation), rocking, and torsion, give rise to infrared absorption and Raman scattering in four different regions of the spectrum. The frequencies of CH stretching vibrations of methyl groups in various environments have been discussed in Section 8-4. Antisymmetric deformation of the HCH angles of a CH_3 group gives rise to very strong infrared absorption and Raman scattering in the 1470–1440 cm^{-1} region. Bending of methylene (—CH_2—) groups also gives rise to a band in the same region. The symmetric CH_3 deformation gives a strong, sharp band between 1380 and 1360 cm^{-1}. This band appears as a doublet when more than one CH_3 group is attached to the same carbon atom and gives a good indication of the presence of isopropyl or *tert*-butyl groups (see Section 8-7b). When the methyl group is attached to an atom other than carbon, there is a significant shift in the symmetric CH_3 deformation frequency. Table 8-8 lists some typical frequency ranges.

The CH_3 rocking vibrations are usually coupled with skeletal modes and may be found anywhere between 1250 and 800 cm^{-1}. Medium to strong bands in both infrared and Raman spectra may be observed, but these are of little use for structure determination. The methyl torsion vibration has a frequency between 250 and 100 cm^{-1}, but is often not observed in either infrared or Raman spectra. In cases where torsional frequencies can be observed or estimated, information on rotational isomerism, conformation, and barriers to internal rotation can be obtained.

TABLE 8-8 Frequencies of the Symmetric CH_3 Deformation in Various Compounds

Compounds	Group	Range (cm^{-1})
esters, ethers, etc.	$O—CH_3$	1460–1430
amines, amides	$N—CH_3$	1440–1410
hydrocarbons	$C—CH_3$	1380–1360
sulfoxides, thioethers, etc.	$S—CH_3$	1330–1290
phosphines	$P—CH_3$	1310–1280
silanes	$Si—CH_3$	1280–1250
organomercury compounds	$Hg—CH_3$	1210–1190

8-7b Isopropyl and *tert*-Butyl Groups

Isopropyl and tertiary butyl groups give characteristic doublets in the symmetric CH_3 deformation region of the infrared spectrum. The isopropyl group gives a strong doublet at 1385/1370 cm^{-1}, while the *tert*-butyl group gives a strong band at 1370 cm^{-1} with a weaker peak at 1395 cm^{-1}. Examples of these doublets are seen in Figure 8-15. Other examples of spectra of compounds containing *tert*-butyl and isopropyl groups can be seen in Figures 8-1 and 8-31, respectively.

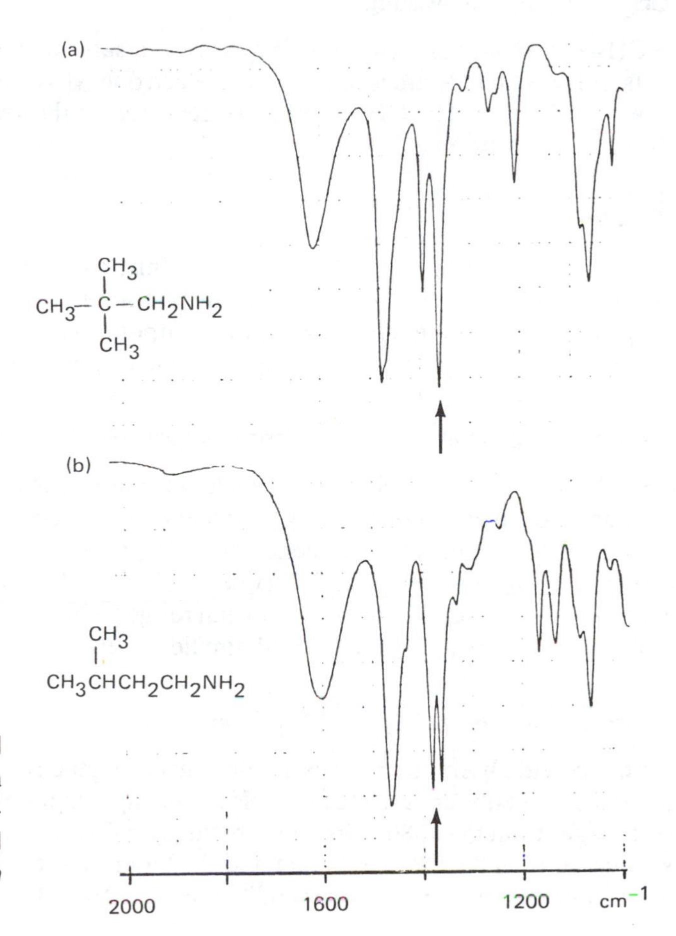

FIGURE 8-15 Examples of the doublets observed in the symmetric CH_3 stretching region for (a) a *tert*-butyl group and (b) an isopropyl group. [Reproduced with permission of Aldrich Chemical Company, from C. J. Pouchert (Ed.), *The Aldrich Library of FT–IR Spectra*.]

8-8

Compounds Containing Methylene Groups

8-8a Introduction

There are two kinds of methylene groups, the $—CH_2—$ group in a saturated chain and the terminal $=CH_2$ group in vinyl, allyl, or vinylidene compounds. Diagrams of stretching, bending, wagging, twisting, and rocking motions of a CH_2 group were given in Figure 7-2. The CH stretching vibrations were covered in Section 8-4. However, bending and wagging and rocking modes also give rise to important group frequencies.

8-8b CH_2 Bending (Scissoring)

The bending (sometimes called scissoring) motion of saturated $—CH_2—$ groups gives a band of medium to strong intensity between 1480 and 1440 cm^{-1}. When the $—CH_2—$ group is adjacent to a carbonyl or nitro group, the frequency is lowered to 1430–1420 cm^{-1}. A vinyl $=CH_2$ group gives a band of medium intensity between 1420 and 1410 cm^{-1}. This band is sometimes assigned as an in-plane deformation, since the two hydrogen atoms are in the same plane as the C=C group.

8-8c CH_2 Wagging and Twisting

The $—CH_2—$ wagging and twisting frequencies in saturated groups are observed between 1350 and 1150 cm^{-1}. The infrared bands are weak unless an electronegative atom such as a halogen or sulfur is attached to the same carbon atom. The CH_2 twisting modes occur at the lower end of the frequency range and give very weak infrared absorption.

8-8d CH_2 Rocking

A small band is observed near 725 cm^{-1} in the infrared spectrum when there are four or more $—CH_2—$ groups in a chain. The intensity increases with increasing chain length. This band is assigned to the rocking of the CH_2 groups in the chain. However, many compounds have bands in this region, so the CH_2 rocking band is only useful for aliphatic molecules. An example of this band can be seen in Figure 8-20a.

8-8e CH_2 Wagging in Vinyl and Vinylidene Compounds

The CH_2 wagging modes in vinyl and vinylidene compounds are found at much lower frequencies than in saturated groups. In vinyl compounds, a strong band is observed in the infrared between 910 and 900 cm^{-1}. For vinylidene compounds the frequency range is 10 cm^{-1} lower. The overtone of the CH_2 wag often can be clearly seen as a band of medium intensity near 1820 cm^{-1} for vinyl and 1780 cm^{-1} for vinylidene compounds. These frequencies are raised above the normal range by halogens or other functional groups on the α carbon atom. More will be said about vinyl and vinylidene groups in Section 8-9d.

8-8f Relative Numbers of CH_2 and CH_3 Groups

One further useful observation can be made concerning the H—C—H deformation bands in saturated parts of a molecule. When there are more $—CH_2—$ groups than CH_3 groups present, the 1480–1440 cm^{-1} band will be stronger than the 1380–1360 cm^{-1} band (sym CH_3 deformation). The relative intensities of these two bands, coupled with the 725 cm^{-1} band and the bands in the CH stretching region (see Figure 8-9), can give information on the relative numbers of $—CH_2—$ and CH_3 groups as well as the saturated carbon chain length.

8-9

Unsaturated Compounds

8-9a The C=C Stretching Mode

Stretching of a C=C bond usually gives rise to infrared and Raman bands in the region 1690–1560 cm^{-1}. The band is often weak in the infrared and sometimes is not observed at all in symmetrical molecules (see Figures 8-16a and 7-1 for examples). Weak absorption due to overtones or combinations may occur in this region and

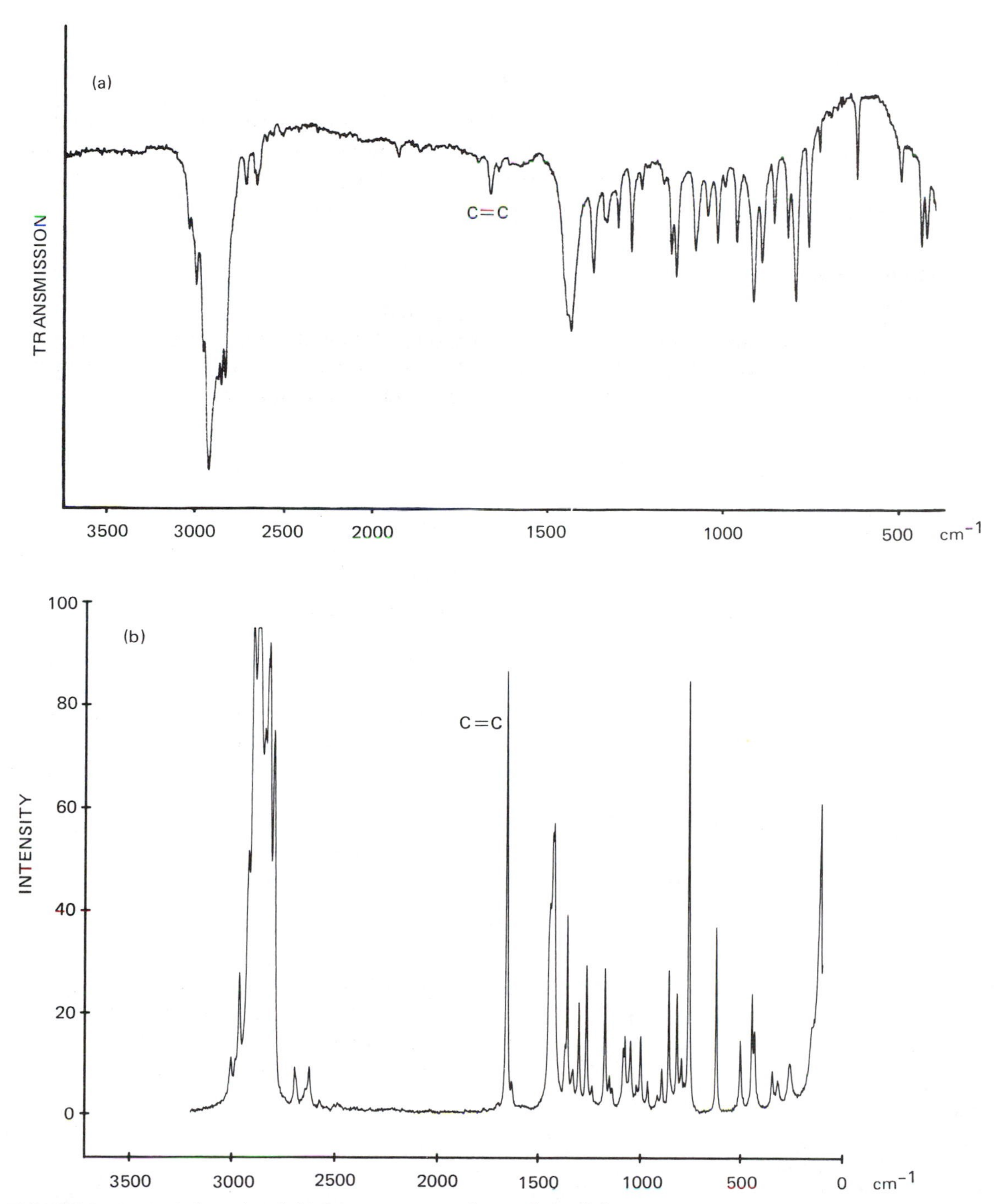

FIGURE 8-16 (a) Infrared and (b) Raman spectra of 1-methylcyclohexene.

could be mistaken for a C=C stretching mode. In such cases a Raman spectrum will confirm the presence or absence of a C=C bond because this group always gives a strong Raman line in the 1690–1560 cm^{-1}1 region (see Figures 7-1 and 8-16b).

The exact frequency of the C=C stretching mode gives some additional information on the environment of the double bond. Tri- or tetraalkyl-substituted groups and trans disubstituted alkenes have frequencies in the 1690–1660 cm^{-1} range. These bands are weak or absent in the infrared, but strong in the Raman. Vinyl and vinylidene compounds as well as cis alkenes absorb between 1660 and 1630 cm^{-1}. Substitution by halogens may shift the C=C stretching band out of the usual range. Fluorinated alkenes have very high C=C stretching frequencies (1800–1730 cm^{-1}). Chlorine and other heavy substituents, on the other hand, usually lower the frequency.

8-9b Cyclic Compounds

The C=C stretching frequencies of cyclic unsaturated compounds depend on ring size and substitution. Cyclobutene, for example, has its C=C stretching mode at 1565 cm^{-1}, whereas cyclopentene, cyclohexene, and cycloheptene have C=C stretching frequencies of 1610, 1645, and 1655 cm^{-1}, respectively. These frequencies are increased by substitution. An example can be seen in Figure 8-16a, in which the weak C=C stretching band of 1-methylcyclohexene is found at 1670 cm^{-1}.

8-9c The C≡C Stretching Mode

A small sharp peak due to C≡C stretching is observed in the infrared spectra of terminal acetylenes near 2100 cm^{-1} (see Figure 8-7). In substituted acetylenes the band is shifted by 100–150 cm^{-1} to higher frequencies. If the substitution is symmetric, no band is observed in the infrared because there is no change in dipole moment during the C≡C stretching vibration. Even when substitution is unsymmetric, the band may

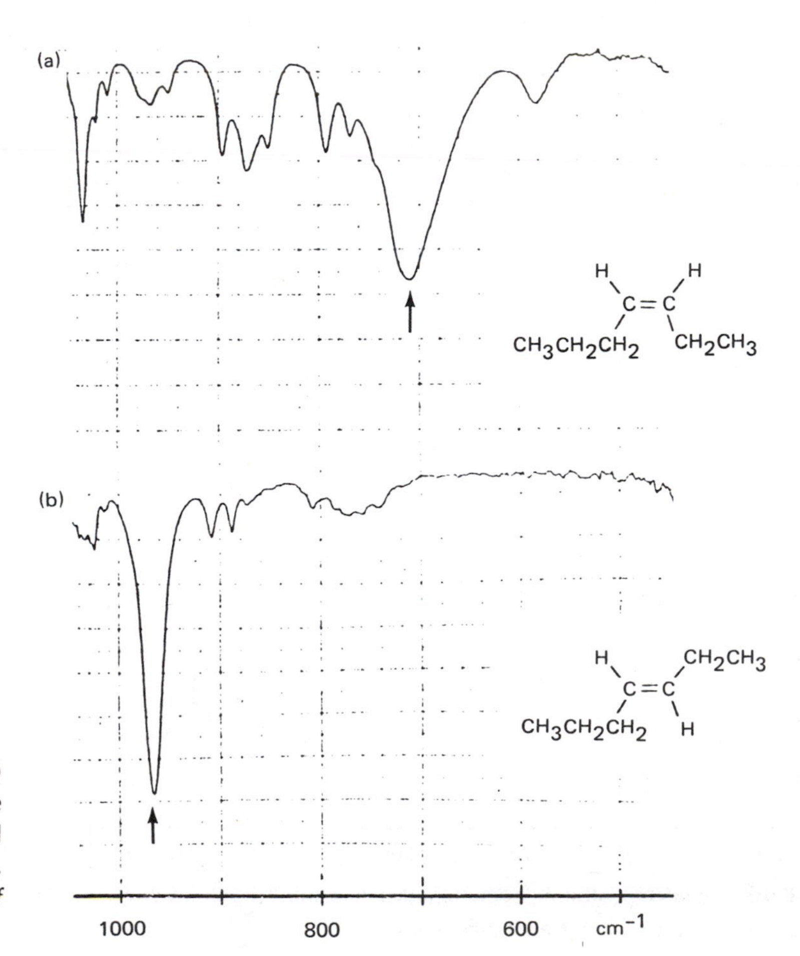

FIGURE 8-17 Portions of infrared spectra of (a) cis- and (b) trans-disubstituted alkenes showing the out-of-plane CH bending bands. [Reproduced with permission of Aldrich Chemical Company, from C. J. Pouchert (Ed.), *The Aldrich Library of FT–IR Spectra, 1985.]*

be very weak in the infrared and could be missed. In such cases a Raman spectrum is very valuable, since the C≡C stretching mode always gives a strong line.

8-9d =CH and =CH_2 Bending Modes

The CH and CH_2 wagging or out-of-plane bending modes are very important for structure identification in unsaturated compounds. They occur in the region between 1000 and 650 cm^{-1}. The trans CH bending of a vinyl group gives rise to a strong infrared band between 1000 and 980 cm^{-1}, while a trans disubstituted alkene is characterized by a very strong band in the 980–950 cm^{-1} frequency range. Electronegative substituents tend to lower this frequency. The cis disubstituted alkenes give a medium to strong, but less reliable band between 750 and 650 cm^{-1}. Infrared spectra of trans and cis disubstituted alkenes are shown in Figure 8-17.

The CH_2 out-of-plane wagging vibration of vinyl and vinylidene compounds gives a strong band between 910 and 890 cm^{-1}. This band coupled with the trans CH bending mode gives a very characteristic doublet (1000 and 900 cm^{-1}) and distinguishes the vinyl group from the vinylidene group (900 cm^{-1} only). Examples of infrared spectra of vinyl and vinylidene compounds are seen in Figure 8-18. The overtones of the out-of-plane CH bend and the CH_2 wagging modes give weak but characteristic bands near 1950 and 1800 cm^{-1}, respectively. These are clearly seen in Figure 8-18 and provide a useful confirmation of the structural grouping.

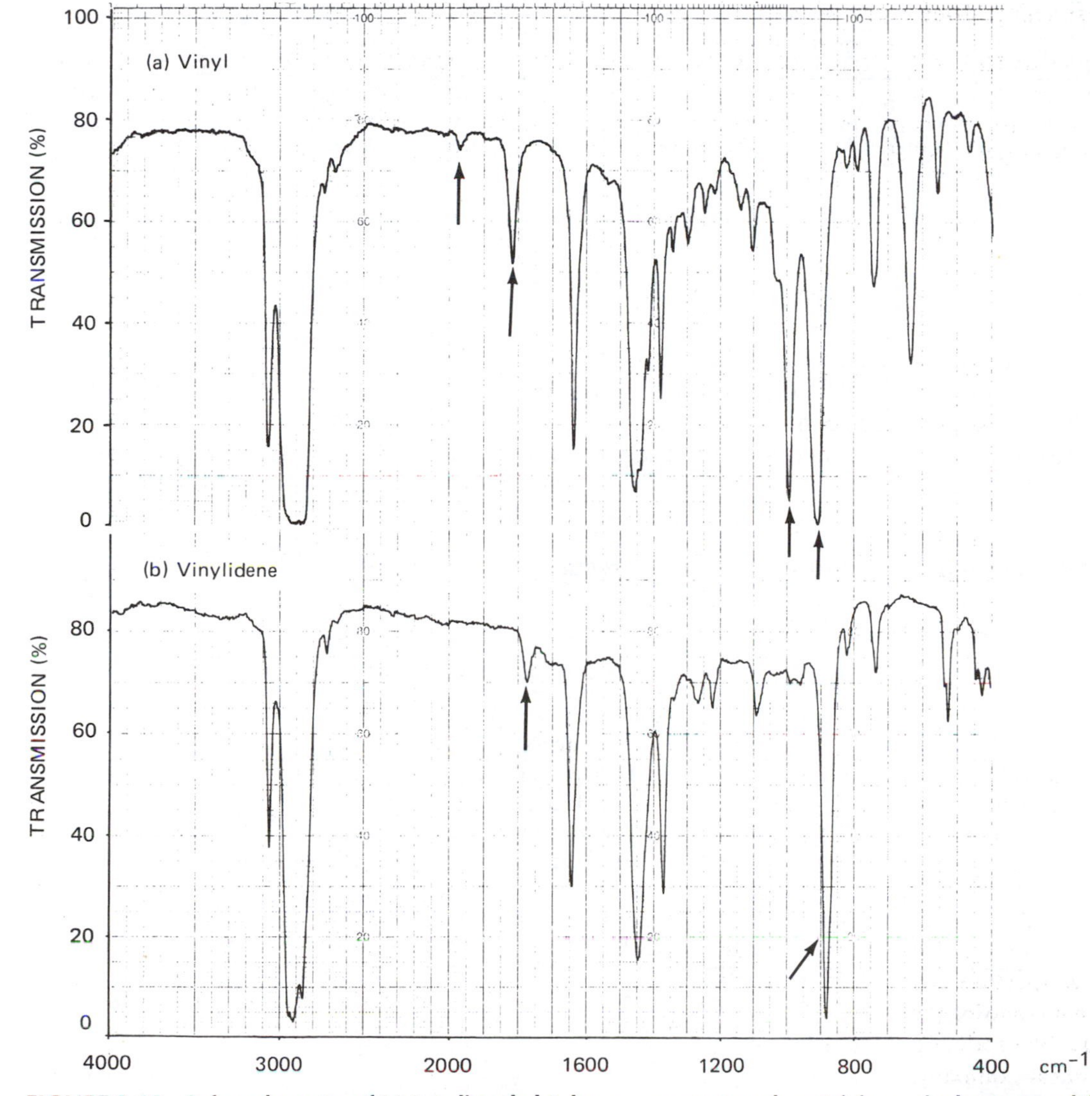

FIGURE 8-18 Infrared spectra of (a) 3,4-dimethyl-1-hexene, a compound containing a vinyl group, and (b) 2,3-dimethyl-1-pentene, a compound containing a vinylidene group.

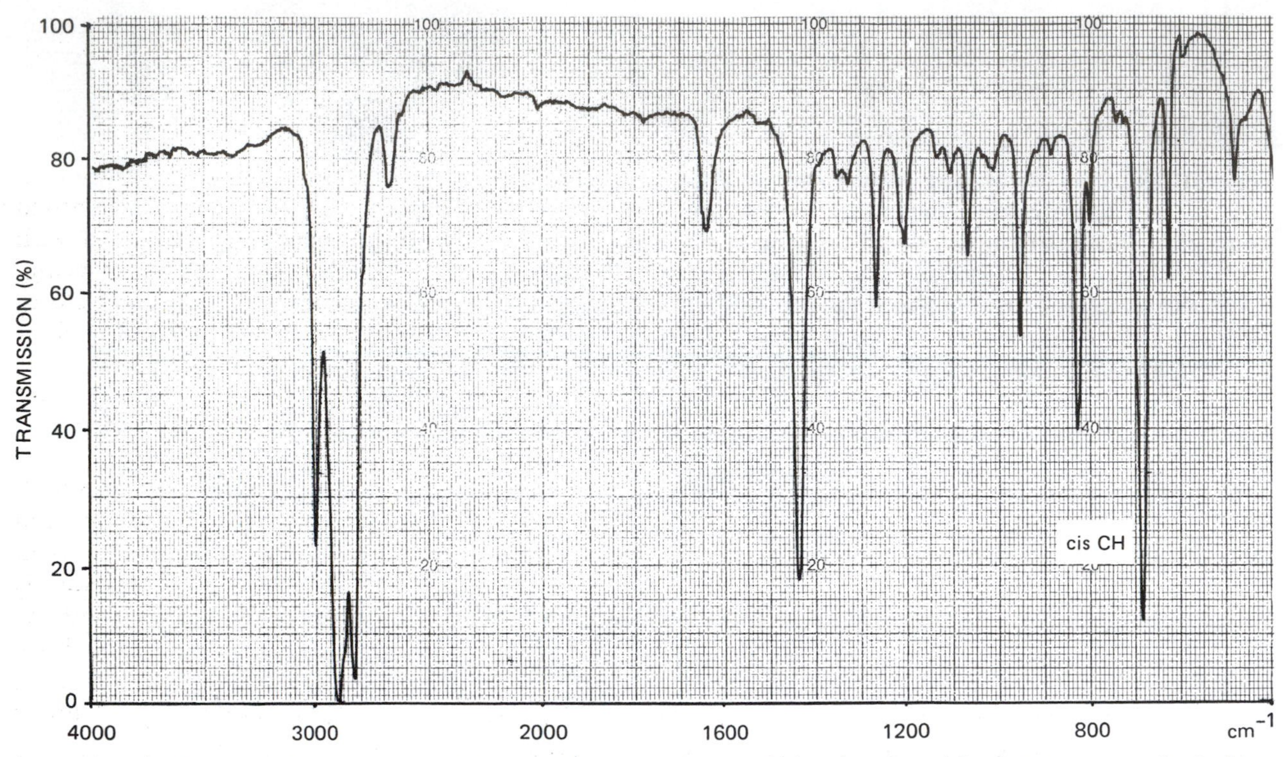

FIGURE 8-19 The infrared spectrum of cycloheptene illustrating the out-of-plane bending of the cis CH groups at the double bond (690 cm^{-1}).

Cyclic alkenes usually have a strong band between 750 and 650 cm^{-1} due to out-of-plane bending of the two CH groups in a cis arrangement. An example is shown in Figure 8-19. However, when one of these hydrogens is substituted by another group such as methyl, the band between 750 and 650 cm^{-1} is absent (Figure 8-16). There are always several bands of medium intensity between 1200 and 800 cm^{-1} in the infrared spectra of cycloalkyl and cycloalkene compounds due to —CH_2— rocking modes.

A summary of the out-of-plane CH bending or wagging modes in alkenic compounds is given in Table 8-9.

TABLE 8-9 CH_2 Wagging and CH Out-of-Plane Bending Frequencies of Alkenes in the Infrared

Range (cm^{-1}) and Intensity	Group or Class	Assignment and Remarks
1000–980 (s)	vinyl group (—CH=CH_2)	trans CH=CH bending, lower in vinyl ethers
980–950 (vs)	trans disubstituted alkenes (vinylenes)	CH=CH bending, frequency lowered by halogens
920–900 (s)	vinyl group (—CH=CH_2)	CH_2 out-of-plane wagging, higher in vinyl ketones, but lower in vinyl ethers
900–880 (s)	vinylidene group (C=CH_2)	terminal =CH_2 out-of-plane wagging
750–650 (m–s)	cis disubstituted and cyclic alkenes	cis CH=CH bending

8-10

Compounds Containing Oxygen

8-10a General

Carboxylic acids and anhydrides, alcohols, phenols, sugars, and carbohydrates all give strong, often broad infrared absorption bands somewhere between 1400 and 900 cm^{-1}. These bands are associated with stretching of the C—O—C or C—OH bonds or bending of the C—O—H group. The position and multiplicity of the absorption together with evidence from other regions of the spectrum can help to distinguish the particular functional group. The usual frequency ranges for these groups in various compounds are summarized in Table 8-10.

8-10b Ethers

The simplest structure with the C—O—C link is the ether group. Aliphatic ethers absorb near 1100 cm^{-1}, while alkyl aryl ethers have a very strong band between 1280 and 1220 cm^{-1} and another strong band between 1050 and 1000 cm^{-1}. In vinyl ethers the C—O—C stretching mode is found near 1200 cm^{-1}. Vinyl ethers can be further distinguished by a very strong C=C stretching band and the out-of-plane CH bending and CH_2 wagging bands, which are observed near 960 and 820 cm^{-1}, respectively. These frequencies are below the usual ranges discussed in Section 8-9d. Examples of these three types of ether are compared in Figure 8-20. Cyclic saturated ethers such as tetrahydrofuran have a strong antisymmetric C—O—C stretching band in the range 1250–1150 cm^{-1}, while unsaturated cyclic ethers have their C—O—C stretching modes at lower frequencies.

TABLE 8-10 C—O—C and C—O—H Group Vibrations

Range (cm^{-1}) and Intensity	Group or Class	Assignment and Remarks
1440–1400 (m)	aliphatic carboxylic acids	C—O—H deformation; may be obscured by CH_3 and CH_2 deformation bands
1430–1280 (m)	alcohols	C—O—H deformation; broad band
1390–1310 (m–s)	phenols	C—O—H deformation
1340–1160 (vs)	phenols	C—O stretch; broad band with structure
1310–1250 (vs)	aromatic esters	C—O—C antisym stretch
1300–1200 (s)	aromatic carboxylic acids	C—O stretch
1300–1100 (vs)	aliphatic esters	C—O—C antisym stretch
1160–1000 (s)		C—O—C sym stretch
1280–1220 (vs)	alkyl aryl ethers	aryl C—O stretch; a second band near 1030 cm^{-1}
1270–1200 (s)	vinyl ethers	C—O—C stretch; a second band near 1050 cm^{-1}
1265–1245 (vs)	acetate esters	C—O—C antisym stretch
1250–900 (s)	cyclic ethers	C—O—C stretch, position varies with compound
1230–1000 (s)	alcohols	C—O stretch; see below for more specific frequency ranges
1200–1180 (vs)	formate and propionate esters	C—O—C stretch
1180–1150 (m)	alkyl-substituted phenols	C—O stretch
1150–1050 (vs)	aliphatic ethers	C—O—C stretch; usually centered near 1100 cm^{-1}
1150–1130 (s)	tertiary alcohols	C—O stretch; lowered by chain branching or adjacent unsaturated groups
1110–1090 (s)	secondary alcohols	C—O stretch; lowered 10–20 cm^{-1} by chain branching
1060–1040 (s–vs)	primary alcohols	C—O stretch; often fairly broad
1060–1020 (s)	saturated cyclic alcohols	C—O stretch; not cyclopropanol or cyclobutanol
1050–1000 (s)	alkyl aryl ethers	alkyl C—O stretch
960–900 (m–s)	carboxylic acids	C—O—H deformation of dimer

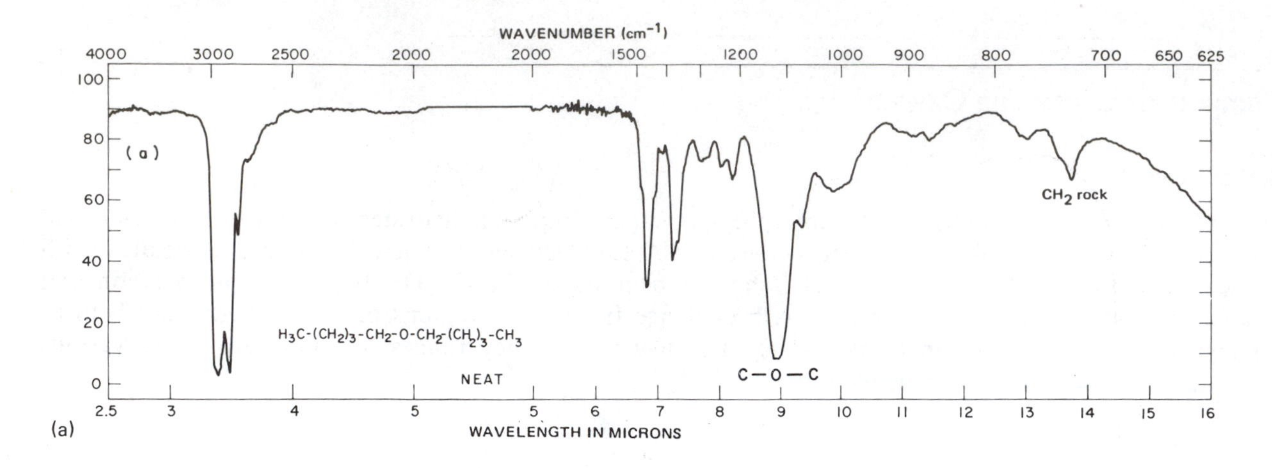

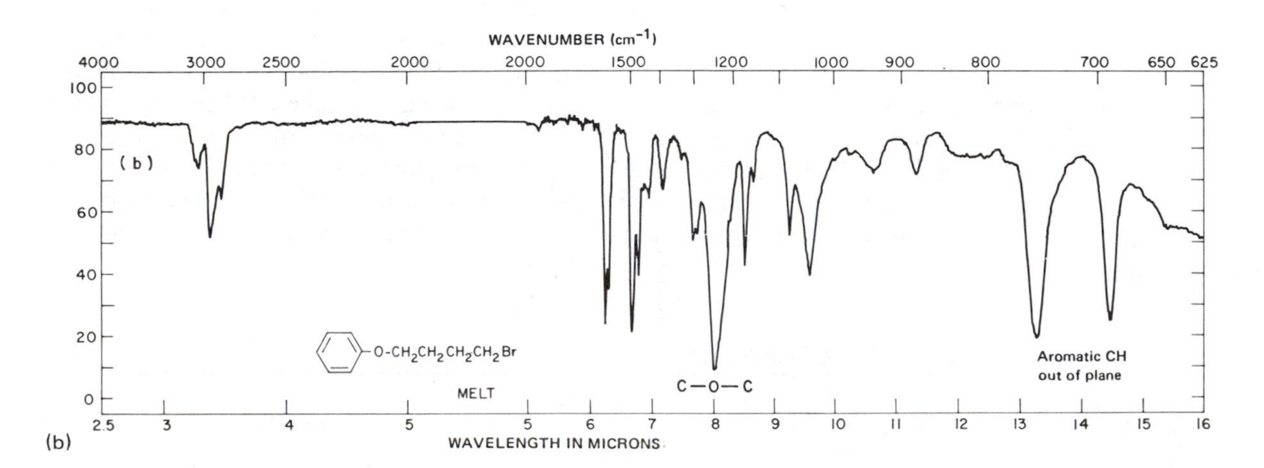

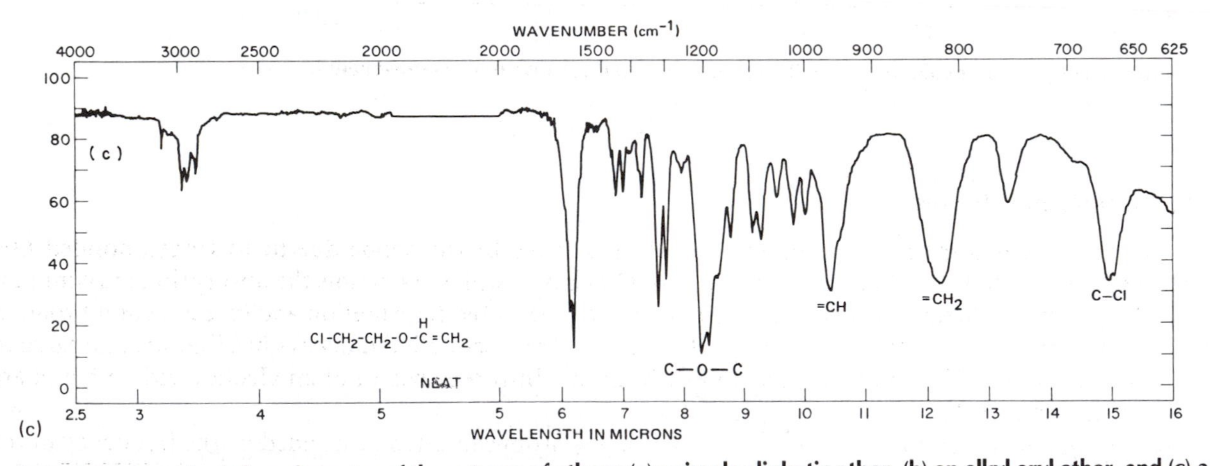

FIGURE 8-20 The infrared spectra of three types of ethers: (a) a simple aliphatic ether, (b) an alkyl aryl ether, and (c) a vinyl ether. [Reproduced with permission of Aldrich Chemical Company, from C. J. Pouchert (Ed.), *The Aldrich Library of Infrared Spectra*.]

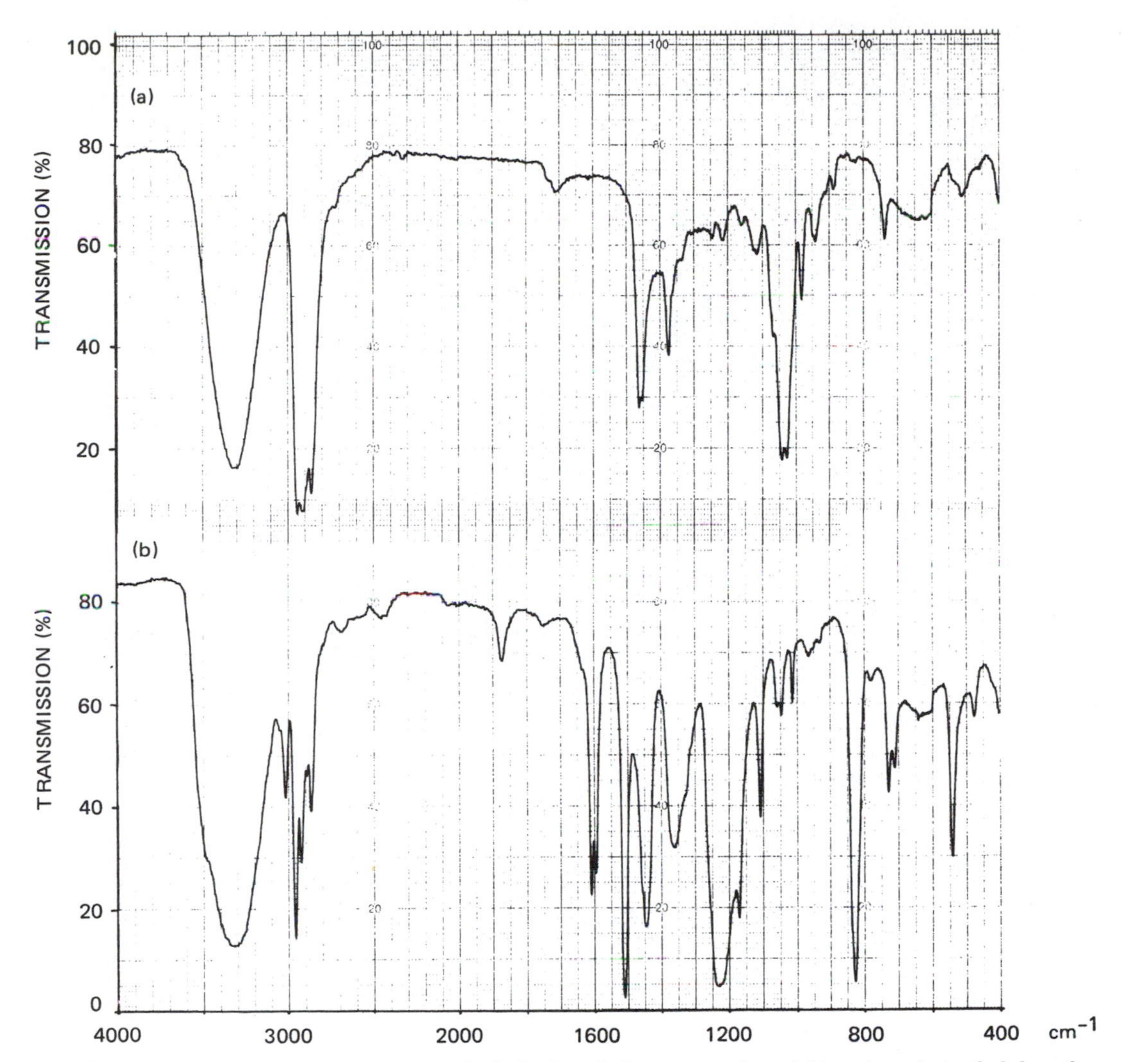

FIGURE 8-21 Infrared spectra of (a) an alcohol, 2-methyl-1-pentanol, and (b) a phenol, 4-ethylphenol.

8-10c Alcohols and Phenols

Alcohols and phenols in the pure liquid or solid state have broad bands due to hydrogen-bonded OH stretching. For alcohols this band is centered near 3300 cm^{-1}, while in phenols the absorption maximum is 50–100 cm^{-1} lower. Phenols absorb near 1350 cm^{-1} due to the OH deformation and give a second broader, stronger band due to C—OH stretching near 1200 cm^{-1}. This second band always has fine structure due to underlying aromatic CH in-plane deformation vibrations. Infrared spectra of an alcohol and a phenol are compared in Figure 8-21.

In simple alcohols, substitution on the C—OH carbon atom can often be decided by the frequency of the C—OH stretching band (see Table 8-10). Sugars and carbohydrates give very broad absorption centered in the 1150–950 cm^{-1} region. The infrared spectrum of a sugar is shown in Figure 8-22.

8-10d Carboxylic Acids and Anhydrides

Carboxylic acids usually exist as dimers except in dilute solution. The carbonyl stretching band of the dimer is found near 1700 cm^{-1}, while in the monomer spectrum the band is located at higher frequencies (1800–1740 cm^{-1}). In addition to the very broad OH stretching band mentioned in Section 8-3a, there are three vibrations associated with the C—OH group in carboxylic acids: a band of medium intensity near 1430 cm^{-1},

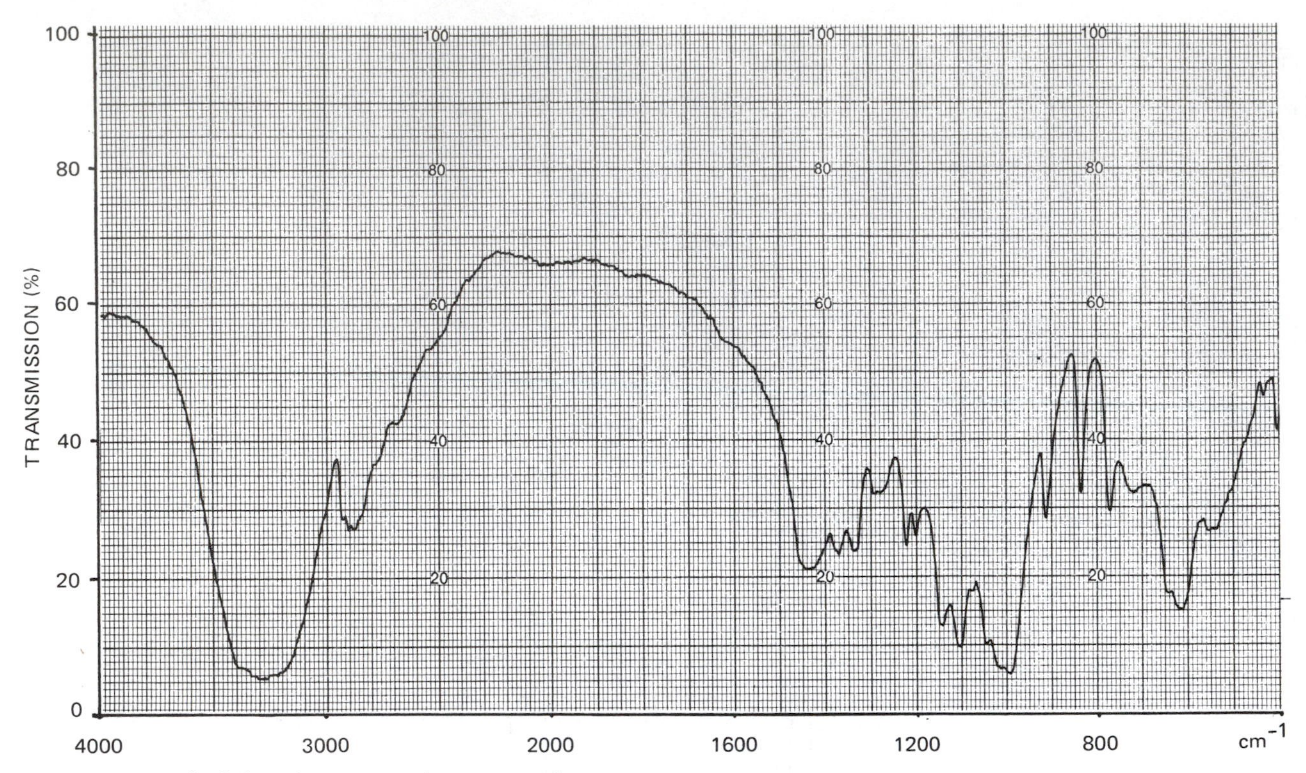

FIGURE 8-22 The infrared spectrum of a sugar, D-glucose.

a stronger band near 1240 cm^{-1}, and another band of medium intensity near 930 cm^{-1}. The presence of an anhydride is detected by the characteristic absorption in the C=O stretching region. This consists of a doublet with one band at unusually high frequency (see Section 8-5c). The C—O—C stretch gives rise to a band near 1190 cm^{-1} in open chain anhydrides and at higher frequencies in cyclic structures.

8-10e Esters

The antisymmetric C—O—C stretching mode in esters gives rise to a very strong and often quite broad band near 1200 cm^{-1}. The actual frequency of the maximum of this band can vary from 1290 cm^{-1} in benzoates down to 1100 cm^{-1} in aliphatic esters. There may be structure on this band due to CH deformation vibrations that absorb in the same region. The band may be even stronger than the C=O stretching band near 1750 cm^{-1}. The symmetric C—O—C stretch also gives a strong band at lower frequencies between 1160 and 1000 cm^{-1} in aliphatic esters.

8-11

Compounds Containing Nitrogen

8-11a General

The presence of primary or secondary amines and amides can be detected by absorption due to stretching of NH_2 or NH groups between 3350 and 3200 cm^{-1}. Tertiary amines and amides on the other hand are more difficult to identify. Nitriles and nitro compounds also give characteristic infrared absorption bands. Isocyanates and carbodiimides have very strong infrared bands near 2260 and 2140 cm^{-1}, respectively, where very few absorptions due to other groupings occur. Oximes, imines, and azo compounds give weak infrared bands in the 1700–1600 cm^{-1} region due to the —C=N—, or —N=N— group. A Raman spectrum is useful in these cases. Some characteristic group frequencies of nonheterocyclic nitrogen-containing compounds are listed in Table 8-11. Much of the information in this table is taken from Tables 7-2 and 7-3 but it is useful to collect the data on nitrogen-containing groups in a single table.

TABLE 8-11 Details of Infrared Frequencies of Some Nitrogen-Containing Groups

Group and Class	Range (cm^{-1}) and Intensity	Assignment and Remarks
The $-NH_2$ Group		
Primary amides (dil soln)	3530–3520 (s)	NH_2 antisym stretch
	3400–3390 (s)	NH_2 sym stretch
(solid state)	3360–3340 (vs)	NH_2 antisym stretch
	3190–3170 (vs)	NH_2 sym stretch
	1680–1660 (vs)	C=O stretch; (Amide I band)
	1650–1620 (m)	NH_2 deformation (Amide II band)
Primary amines (dil soln)	3550–3350 (m)	NH_2 antisym stretch
	3450–3250 (m)	NH_2 sym stretch
(condensed phase)	3450–3250 (m)	NH_2 stretching; br with structure
	1650–1590 (s)	NH_2 deformation
	850–750 (s)	NH_2 wagging; br
The $-NH-$ Group		
Secondary amides	3450–3400 (m)	NH stretch; dil soln
	3300–3250 (m)	NH stretch; solid state
	3100–3060 (w)	overtone band
	1680–1640 (vs)	C=O stretch (Amide I band)
	1560–1530 (vs)	coupled NH deformation and C—N stretch (Amide II band)
	750–650 (s)	NH wag; br
Secondary amines	3500–3300 (m)	NH stretch
	750–650 (s, br)	NH wag
The C≡N Group		
Nitriles		
saturated aliphatic	2260–2240 (w)	C≡N stretch; strong in Raman
unsaturated aliphatic adjacent to a double bond	2230–2220 (m)	C≡N stretch; doublet when the adjacent double bond is disubstituted
aromatic	2240–2220 (variable)	C≡N stretch; stronger than in saturated aliphatic nitriles
Isonitriles		
alkyl	2180–2150 (w–m)	—N≡C stretch; strong in Raman
aryl	2130–2110 (w–m)	—N≡C stretch; strong in Raman
The C=N Group		
Oximes	1690–1620 (w–m)	C=NOH stretch
Pyridines	1615–1565 (s)	two bands, due to C=C and C=N stretch in ring
The C—N Group		
Amines and amides		
primary aliphatic	1140–1070 (m)	C—C—N antisym stretch
secondary aliphatic	1190–1130 (m–s)	C—N—C antisym stretch
primary aromatic	1330–1260 (s)	phenyl—N stretch
secondary aromatic	1340–1250 (s)	phenyl—N stretch
The NO_2 Group		
Aliphatic nitro compounds	1560–1530 (vs)	NO_2 antisym stretch
	1390–1370 (m–s)	NO_2 sym stretch
Aromatic nitro compounds	1540–1500 (vs)	NO_2 antisym stretch
	1370–1330 (s–vs)	NO_2 sym stretch
Nitrates R—O—NO_2	1660–1620 (vs)	NO_2 antisym stretch
	1300–1270 (s)	NO_2 sym stretch
	710–690 (s)	NO_2 deformation

TABLE 8-11 (continued)

Group and Class	Range (cm^{-1}) and Intensity	Assignment and Remarks
The NO Groups ($N{=}O$, $N{-}O$, $\overset{+}{N}{-}\overset{-}{O}$)		
Nitrites R—ONO	1680–1650 (vs)	N=O stretch; often a weaker band is seen between 1630 and 1600 cm^{-1}
Oximes	965–930 (s)	N—O stretch
Nitrates R—O—NO_2	870–840 (s)	N—O stretch
N-Oxides		
aromatic	1300–1200 (vs)	$\overset{+}{N}{-}\overset{-}{O}$ stretch
aliphatic	970–950 (vs)	$\overset{+}{N}{-}\overset{-}{O}$ stretch
The N≡N Group		
Azides	2120–2160 (variable)	N≡N stretch; strong in Raman
The N=N Group		
Azo compounds	1450–1400 (vw)	N=N stretch; strong in Raman

8-11b Amino Acids, Amines, and Amine Hydrohalides

Three classes of nitrogen-containing compounds (amino acids, amines, and amine hydrohalides) give rise to very characteristic broad absorption bands. Perhaps the most striking of these are found in the infrared spectra of amino acids, which contain an extremely broad band centered near 3000 cm^{-1} and often extending as low as 2200 cm^{-1}, with some structure. The infrared spectrum of an amino acid is shown in Figure 8-23. Amine hydrohalides (ammonium halides) give a similar, very broad band with structure on the low frequency side. The center of the band tends to be lower than in amino acids, especially in the case of tertiary amine hydrohalides, the band center for which may be as low as 2500 cm^{-1}. In fact, this band gives a very useful indication of the presence of a tertiary amine, and an example is shown in Figure 8-24. Both amino acids and primary amine hydrohalides have a weak but characteristic band between 2200 and 2000 cm^{-1} (see Figure 8-23), which is believed to be a combination of the $-NH_3^+$ deformation near 1600 cm^{-1} and the $-NH_3^+$ torsion near 500 cm^{-1}.

Primary amines have a fairly broad band in their infrared spectra centered near 830 cm^{-1}, whereas the frequency for secondary amines is about 100 cm^{-1} lower (see Figure 8-4d and e). This band is not present in the spectra of tertiary amines or amine hydrohalides.

8-11c Anilines

In anilines the characteristic broad band shown by aliphatic amines in the 830–730 cm^{-1} region is not present, so that the out-of-plane CH deformations of the benzene ring can be observed. These bands permit the ring substitution pattern to be determined. Of course, when an aliphatic amine is joined to a benzene ring through a carbon chain, both the characteristic amine band and the CH deformation pattern will be present.

In Figure 8-25 the infrared spectrum of an aniline derivative is shown. Figure 8-26 shows the spectrum of an aliphatic amine joined to a benzene ring. The presence of the benzene ring is identified in both compounds by CH stretching bands between 3100 and 3000 cm^{-1}, out-of-plane CH bending bands between 850 and 700 cm^{-1}, and bands diagnostic of the substitution patterns between 2000 and 1700 cm^{-1}. In the out-of-plane bending region, the single band at 825 cm^{-1} in Figure 8-25 indicates the presence of a para-disubstituted benzene ring, while the doublet at 740 and 700 cm^{-1} in Figure 8-26 indicates monosubstitution (see Section 8-6).

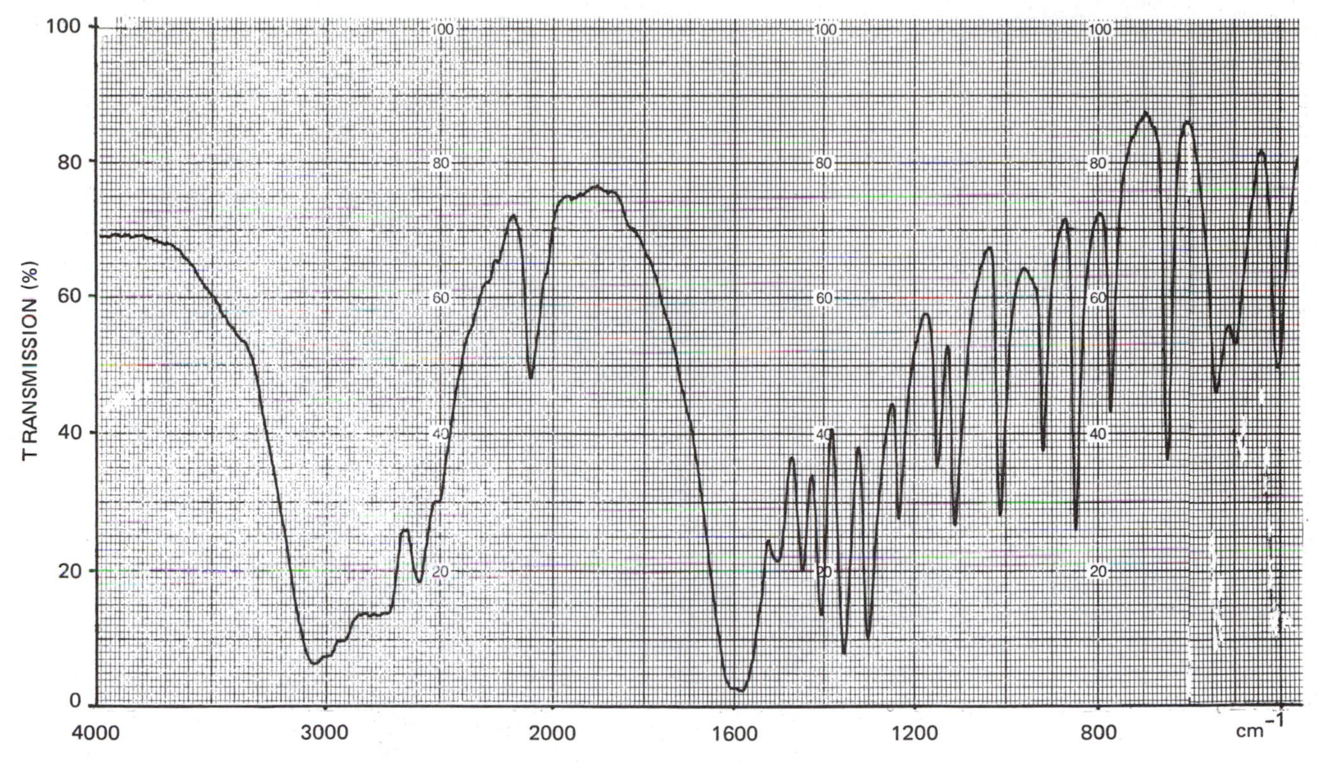

FIGURE 8-23 The infrared spectrum of an amino acid, L-alanine.

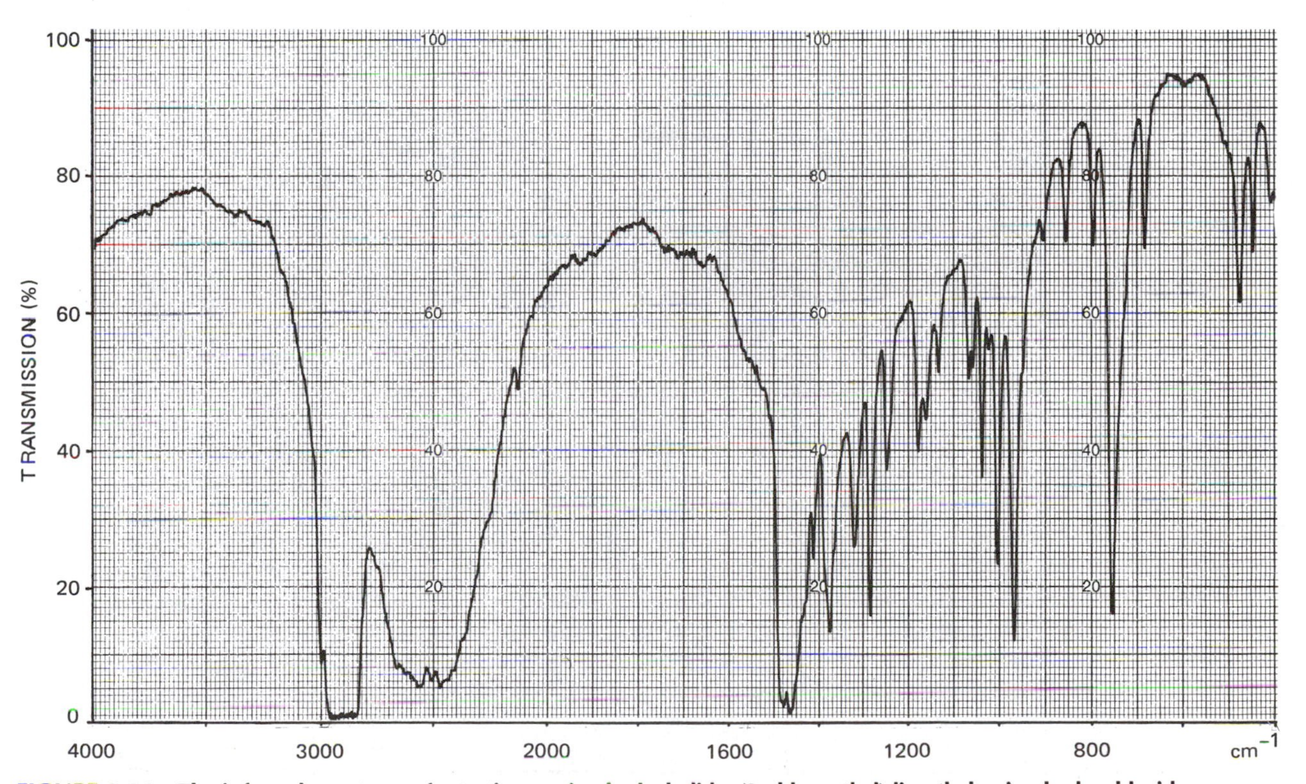

FIGURE 8-24 The infrared spectrum of a tertiary amine hydrohalide, (2-chloroethyl)dimethylamine hydrochloride.

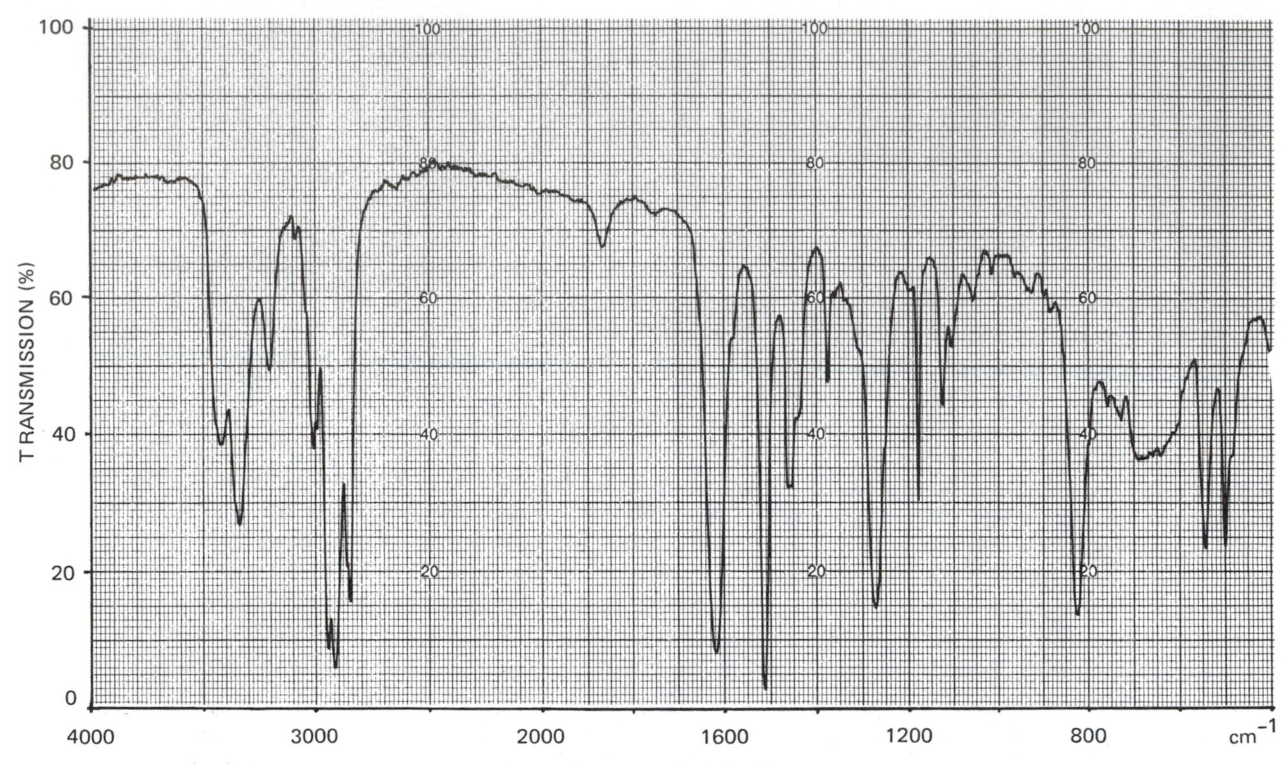

FIGURE 8-25 The infrared spectrum of an aniline derivative, *p*-butylaniline.

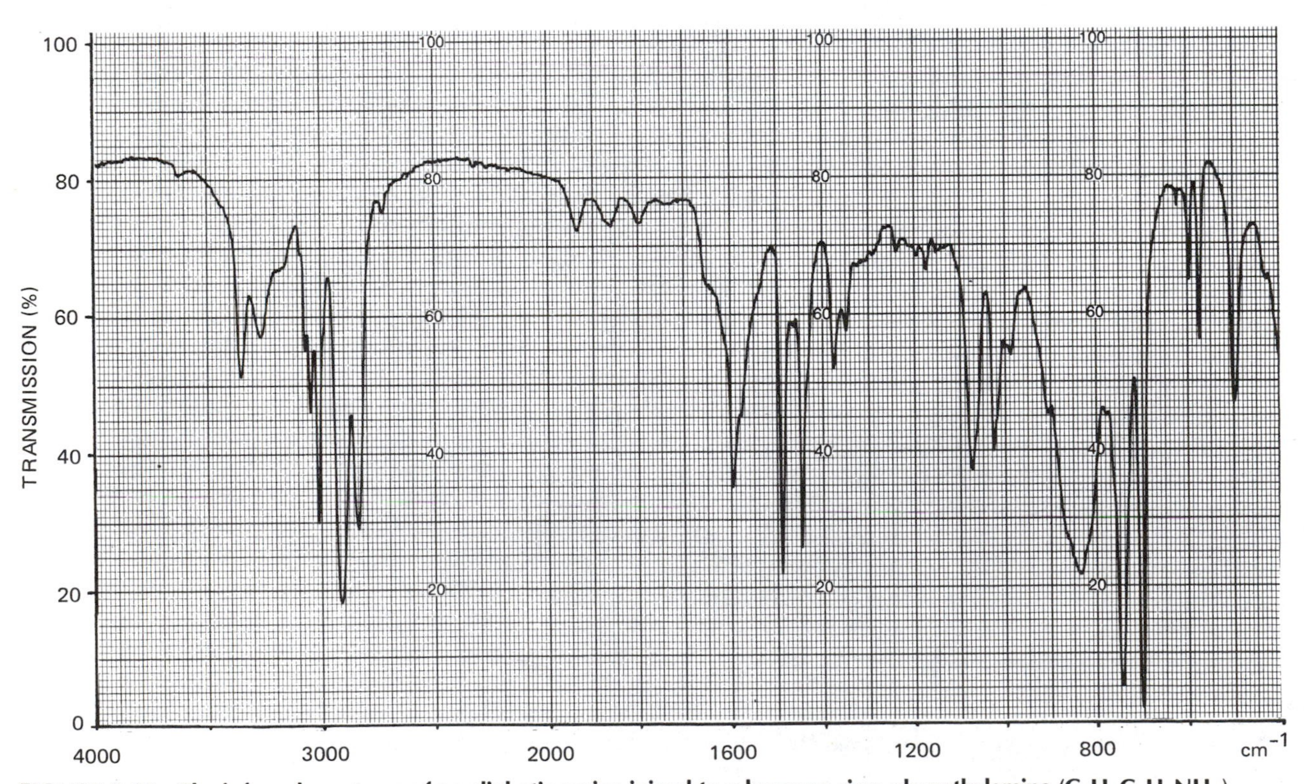

FIGURE 8-26 The infrared spectrum of an aliphatic amine joined to a benzene ring, phenethylamine ($C_6H_5C_2H_5NH_2$).

8-11d Nitriles

Saturated nitriles absorb weakly in the infrared near 2250 cm^{-1}. The band is strong in the Raman spectrum. Unsaturated or aromatic nitriles for which the double bond or ring is adjacent to the C≡N group absorb more strongly in the infrared than saturated compounds, and the band occurs at somewhat lower frequencies near 2230 cm^{-1}.

8-11e Nitro Compounds

Nitro compounds have two very strong absorption bands due to symmetric and antisymmetric NO_2 stretching (see Figure 8-14). In aliphatic compounds, the frequencies are near 1550 and 1380 cm^{-1}, whereas in aromatic compounds the bands are observed near 1520 and 1350 cm^{-1}. These frequencies are somewhat sensitive to nearby substituents. In particular, the 1350 cm^{-1} band in aromatic nitro compounds is intensified by electron-donating substituents in the ring. The out-of-plane CH bending patterns of ortho-, meta-, and para-disubstituted benzene rings are often perturbed in nitro compounds. Other compounds containing N—O bonds have strong characteristic infrared absorption bands (see Table 8-11).

8-11f Amides

Secondary amides (*N*-monosubstituted amides) usually have their NH and O groups trans to each other, as in **8-3**. The carbonyl stretching mode gives rise to a very strong infrared band between 1680 and 1640 cm^{-1}.

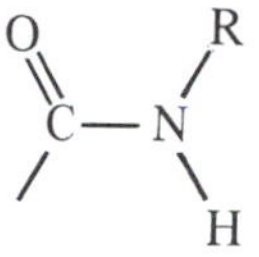

```
  O       R
   \\    /
    C—N
   /    \
          H
```

8-3

This band is known as the Amide I band. A second, very strong absorption that occurs between 1560 and 1530 cm^{-1} is known as the Amide II band. It is believed to be due to coupling of the NH bending and C—N stretching vibrations. The trans amide linkage **8-3** also gives rise to absorption between 1300 and 1250 cm^{-1} and to a broad band centered near 700 cm^{-1}. Occasionally, the amide linkage is cis in cyclic compounds such as lactams. In such cases, a strong NH stretching band is seen near 3200 cm^{-1} and a weaker combination band near 3100 cm^{-1} involving simultaneous excitation of C=O stretching and NH bending. The Amide II band is absent but a cis NH bending mode absorbs between 1500 and 1450 cm^{-1}. This band may be confused with the CH_2 or antisymmetric CH_3 deformation bands.

8-11g Oximes

Solid oximes exhibit a very strong broad absorption due to the hydrogen-bonded NOH group centered between 3250 and 3150 cm^{-1}. This band is observed between 3650 and 3500 cm^{-1} in CCl_4 solution. A weak band due to C=N stretching may be observed between 1685 and 1650 cm^{-1} in aliphatic oximes and about 30 cm^{-1} lower in aromatic oximes. This group vibration usually gives a strong band in the Raman spectrum. There is also a strong absorption between 965 and 930 cm^{-1} due to the N—O stretch. An infrared spectrum of an oxime illustrating these features is shown in Figure 8-27.

8-12

Compounds Containing Phosphorus and Sulfur

8-12a General

The presence of phosphorus in organic compounds can be detected by the infrared absorption bands arising from the P—H, P—OH, P—O—C, P=O, and P=S groups. A phosphorus atom directly attached to an aromatic ring is also well characterized. The usual frequencies of these groups in various compounds are

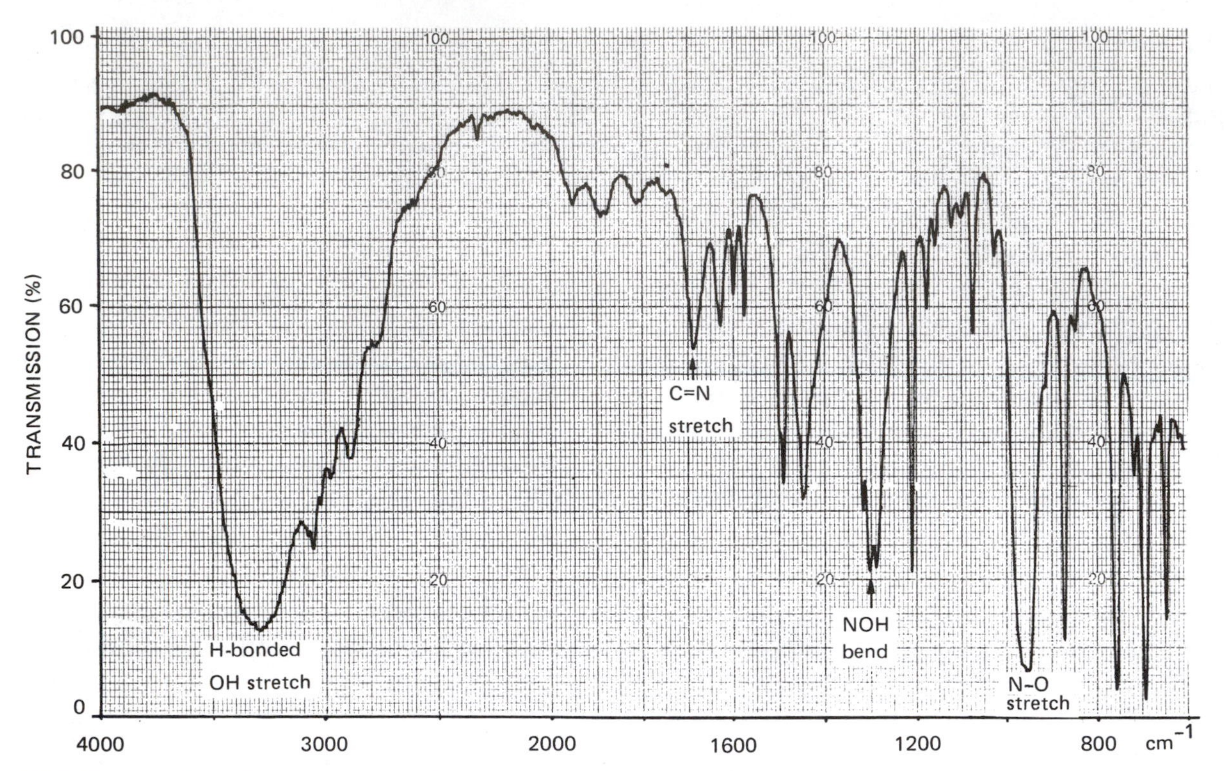

FIGURE 8-27 The infrared spectrum of an oxime, benzaldehyde oxime.

listed in Table 8-12. Most of these groups absorb strongly or very strongly in the infrared with the exception of P═S. The Raman spectrum is valuable for detecting this group, which has a frequency between 700 and 600 cm^{-1}. There is no characteristic P—C group frequency in aliphatic compounds.

The SO_2 and SO groups give rise to very strong infrared bands in various compounds between 1400 and 1000 cm^{-1}. Other bonds involving sulfur, such as C—S, S—S, and S—H, give very weak infrared absorption, so that a Raman spectrum is needed to identify these groups. Characteristic frequencies of some sulfur-containing groups are also listed in Table 8-12. The C═S group has been omitted from the table because the C═S stretching vibration is invariably coupled with vibrations of other groups in the molecule. Frequencies in the 1400–850 cm^{-1} range have been assigned to this group with thioamides at the low frequency end of the range. The infrared bands involving C═S groups are usually weak.

8-12b Phosphorus Acids and Esters

Several phosphorus acids have P—OH groups that give one or two broad bands of medium intensity between 2700 and 2100 cm^{-1}. Esters and acid salts that have P—OH groups also absorb in this region. The presence of a PH group is indicated by a small, sharp band near 2400 cm^{-1}. In ethoxy and methoxy phosphorus compounds, as well as other aliphatic compounds with a P—O—C linkage, a very strong and quite broad infrared band is observed between 1050 and 950 cm^{-1}. The presence of a P═O bond is indicated by a strong band close to 1250 cm^{-1}. An example of a spectrum of an aliphatic compound containing P—H, P═O, and P—O—C groups is given in Figure 8-28.

8-12c Aromatic Phosphorus Compounds

Aromatic phosphorus compounds have several characteristic group frequencies. A fairly strong, sharp infrared peak is observed near 1440 cm^{-1} in compounds in which a phosphorus atom is attached directly to a benzene ring (Figure 8-29). A quaternary phosphorus atom attached to a benzene ring has a characteristic strong, sharp band near 1100 cm^{-1}. The P—O group attached to an aromatic ring gives rise to two strong bands between 1250 and 1160 cm^{-1} and between 1050 and 870 cm^{-1} due to stretching of the Ar—P—O

TABLE 8-12 Characteristic Infrared Frequencies of Groups Containing Phosphorus or Sulfur

Group and Class	Range (cm^{-1}) and Intensity	Assignment and Remarks
The P—H Group		
Phosphorus acids and esters	2425–2325 (m)	P—H stretch
Phosphines	2320–2270 (m)	P—H stretch; sharp band
	1090–1080 (m)	PH_2 deformation
	990–910 (m–s)	P—H wag
The P—OH Group		
Phosphoric or phosphorus acids, esters and salts	2700–2100 (w)	OH stretch; one or two broad and often weak bands
	1040–920 (s)	P—OH stretch
The P—O—C Group		
Aliphatic compounds	1050–950 (vs)	antisym P—O—C stretch
	830–750 (s)	sym P—O—C stretch (methoxy and ethoxy phosphorus compounds only)
Aromatic compounds	1250–1160 (vs)	aromatic C—O stretch
	1050–870 (vs)	P—O stretch
The P—C Group		
Aromatic compounds	1450–1430 (s)	P joined directly to a ring; sharp band
Quaternary aromatic	1110–1090 (s)	P^+ joined directly to a ring; sharp band
The P=O Group		
Aliphatic compounds R—O—P(=O)—	1260–1240 (s)	strong, sharp band
Aromatic compounds Ar—O—P(=O)—	1350–1300 (s)	lower frequency (1250–1180 cm^{-1}) when OH group is attached to the P atom
Phosphine oxides	1200–1140 (s)	P=O stretch
The S—H Group		
Thiols (mercaptans)	2580–2500 (w)	S—H stretch; strong in Raman
The C—S Group	720–600 (w)	C—S stretch; strong in Raman
The S—S Group		
Disulfides	550–450 (vw or absent)	S—S stretch; strong in Raman
The S=O Group		
Sulfoxides	1060–1020 (vs)	S=O stretch
Dialkyl sulfites	1220–1190 (vs)	S=O stretch
The SO_2 Group		
Sulfones, sulfonamides, sulfonic acids, sulfonates, and sulfonyl chlorides	1390–1290 (vs)	SO_2 antisym stretch
	1190–1120 (vs)	SO_2 sym stretch
Dialkyl sulfates and sulfonyl fluorides	1420–1390 (vs)	SO_2 antisym stretch
	1220–1190 (vs)	SO_2 sym stretch
The S—O—C Group		
Dialkyl sulfites	1050–850 (vs)	S—O—C stretching (two bands)
Sulfates	1050–770 (vs)	two or more bands

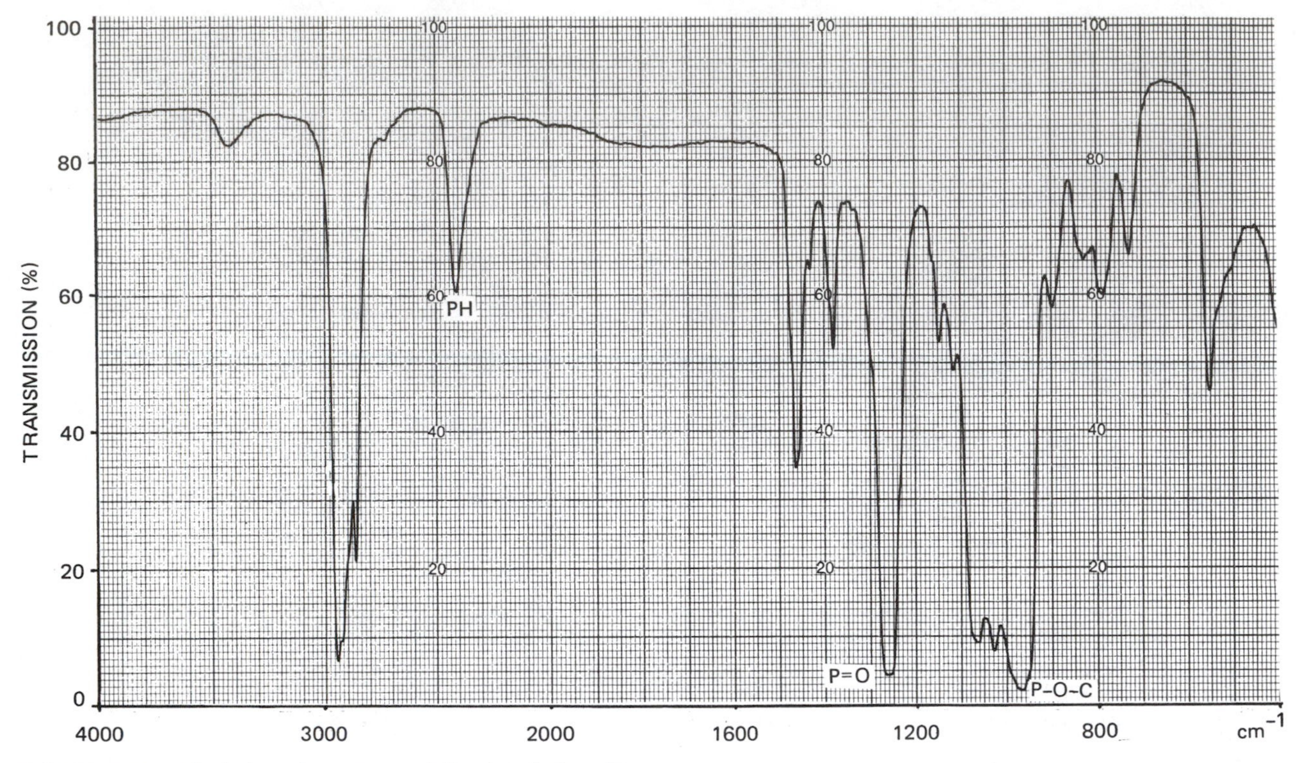

FIGURE 8-28 The infrared spectrum of di-*n*-butyl phosphite.

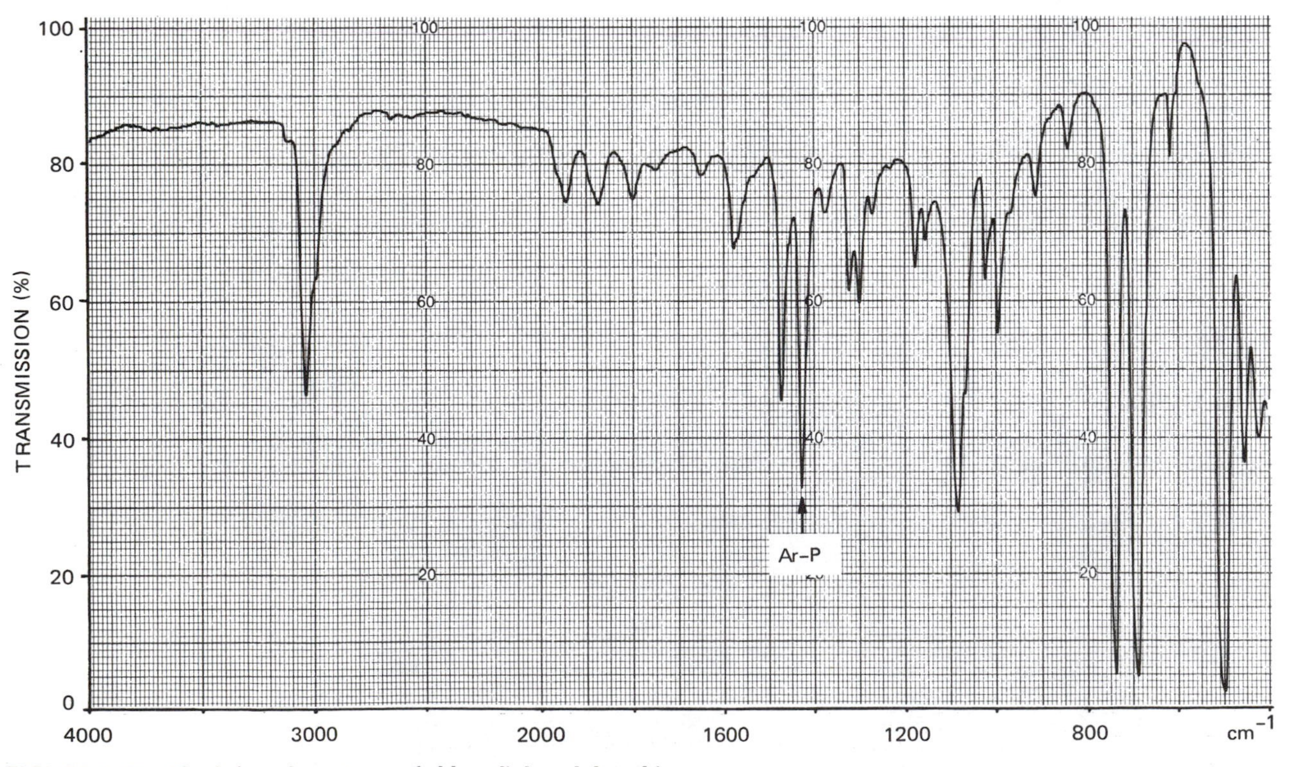

FIGURE 8-29 The infrared spectrum of chlorodiphenylphosphine.

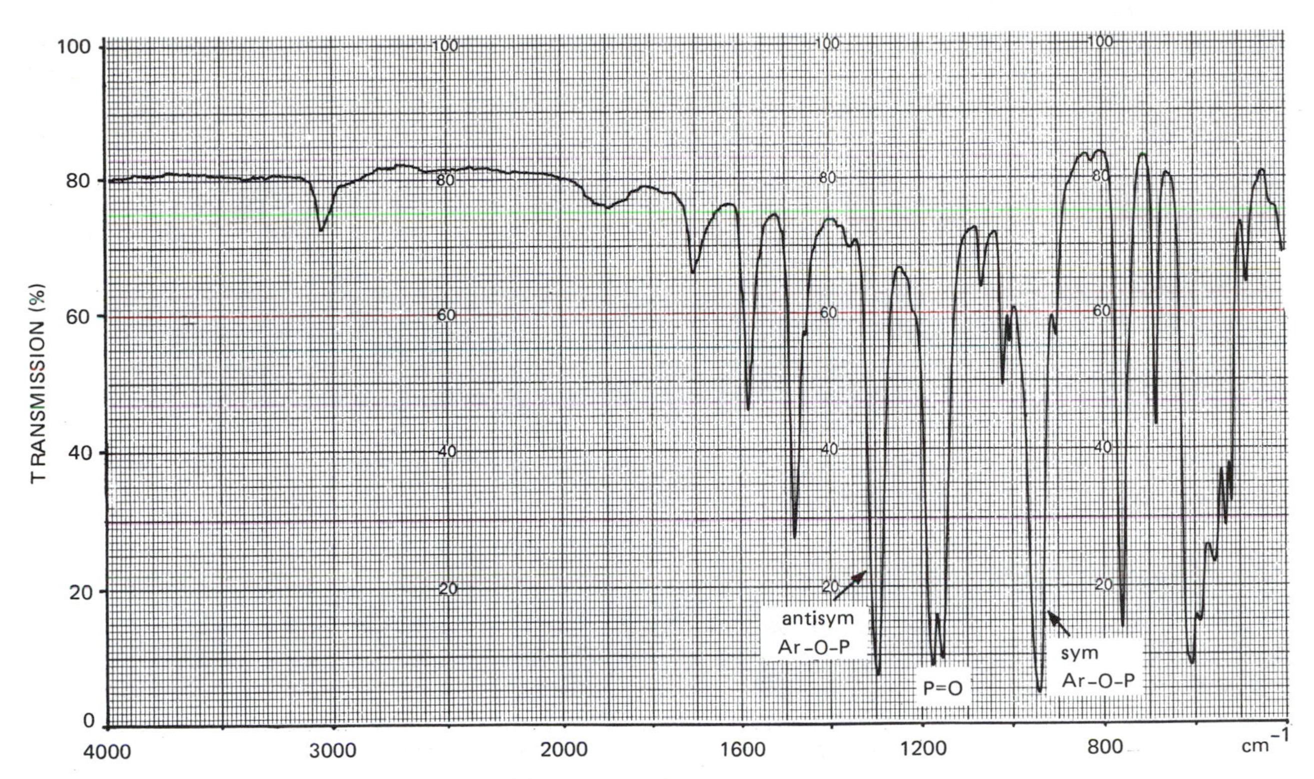

FIGURE 8-30 The infrared spectrum of phenyl dichlorophosphate.

linkage. When the P—O group is attached to a ring through the oxygen atom, the Ar—O—P group again gives two strong bands, but the higher frequency of these is found between 1350 and 1250 cm^{-1}, as illustrated in Figure 8-30.

8-12d Compounds Containing C—S, S—S, and S—H Groups

Raman spectra of compounds containing C—S and S—S bonds contain very strong lines due to these groups between 700 and 600 and near 500 cm^{-1}, respectively. These group frequencies, especially S—S, are either absent or appear only very weakly in the infrared. The S—H stretching band near 2500 cm^{-1} is normally quite weak in the infrared but shows a high intensity in the Raman spectrum. The spectrum of a simple aliphatic thiol is shown in Figure 8-31, where the S—H stretch can be seen at 2530 cm^{-1} and the C—S

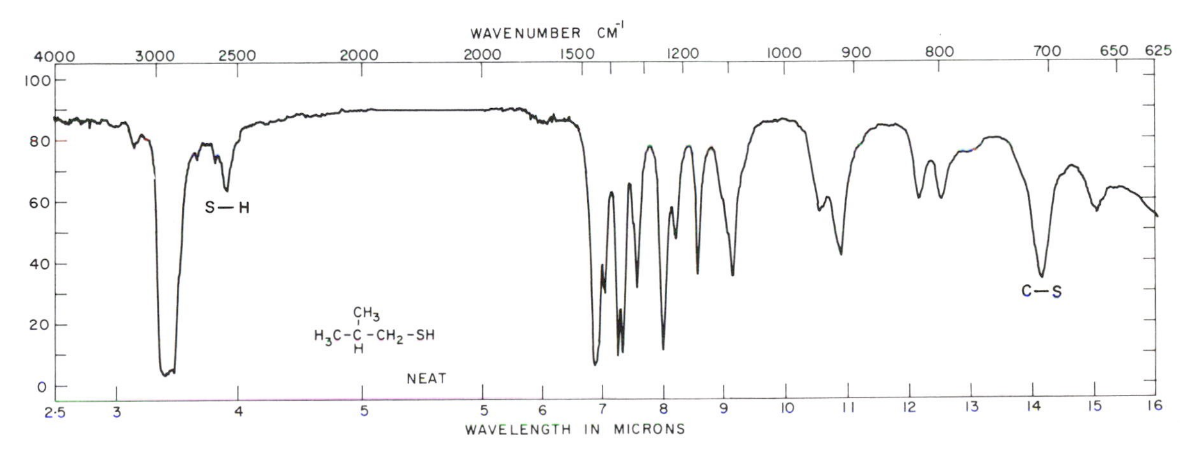

FIGURE 8-31 The infrared spectrum of 2-methyl-1-propanethiol (isobutyl mercaptan). [Reproduced with permission of Aldrich Chemical Company, from C. J. Pouchert (Ed.), *The Aldrich Library of Infrared Spectra*.]

stretch near 700 cm^{-1}. The compound contains an isopropyl group and the symmetric deformations of the two methyl groups give rise to a strong doublet centered at 1375 cm^{-1} (see Section 8-7b).

8-13

Heterocyclic Compounds

8-13a General

Heterocyclic compounds containing nitrogen, oxygen, or sulfur may exhibit three kinds of group frequencies: those involving CH or NH vibrations, those involving motion of the ring, and those due to the group frequencies of substituents on the ring. The identification of a heterocyclic compound from its infrared and Raman spectra is a difficult task and beyond the scope of an introductory treatment. However, characteristic frequencies for some heterocyclic compounds are collected in Table 8-13, and the identification of a few types of compound are discussed in this section.

8-13b Aromatic Heterocycles

A classification of heterocyclic compounds into aromatic or nonaromatic can be made. Hydrogen atoms attached to carbon atoms in an aromatic ring give rise to CH stretching modes in the usual 3100–3000 cm^{-1} region, or a little higher in furans, pyrroles, and some other compounds. Characteristic ring stretching modes, similar to benzene, are observed, and the out-of-plane CH deformation vibrations give rise to strong infrared bands in the 1000–650 cm^{-1} region. In some cases, these patterns are characteristic of the type of substitution in the heterocyclic ring. Some examples include furans, indoles, pyridines, pyrimidines, and quinolines. The in-plane CH bending modes also give several bands in the 1300–1000 cm^{-1} region for aromatic heterocyclic compounds. CH vibrations in benzene derivatives and analogous modes in related heterocyclic compounds can be correlated and may be useful in structure determination.

In aromatic heterocyclic compounds involving nitrogen, the coupled C=C and C=N stretching modes give rise to several characteristic vibrations. These are similar in frequency to their counterparts in the corresponding nonheterocyclic compounds. Ring stretching modes are found in the 1600–1300 cm^{-1} region. Other skeletal ring modes include ring breathing modes near 1000 cm^{-1}, which give very strong Raman bands, in-plane ring deformations between 700 and 600 cm^{-1}, and out-of-plane ring deformation modes, which may be observed between 700 and 300 cm^{-1}.

Nonaromatic heterocyclic compounds usually have one or more CH_2 groups present. The stretching and deformation (scissoring) modes give rise to bands in the usual regions (see Section 8-8). However, the wagging, twisting, and rocking modes often interact with skeletal ring modes and may be observed over a wide range of frequencies.

8-13c NH Stretching Bands

Spectra of heterocyclic nitrogen compounds may contain bands due to a secondary or tertiary amine group. Pyrroles, indoles, and carbazoles in nonpolar solvents have their NH stretching vibration between 3500 and 3450 cm^{-1}, and the band is very strong in the infrared. In saturated heterocyclics, such as pyrrolidines and piperidines, the band is at lower frequencies. Azoles have a very broad hydrogen-bonded NH stretching band between 3300 and 2500 cm^{-1}. This band might be confused with the broad OH stretching band of carboxylic acids.

TABLE 8-13 Characteristic Infrared Frequencies for Some Heterocyclic Compounds

Classes of Compounds	Range (cm^{-1}) and Intensity	Assignment and Remarks
Azoles (imidazoles, isoxazoles, oxazoles, pyrazoles, triazoles, tetrazoles)	3300–2500 (s, br)	H-bonded NH stretch; resembles carboxylic acids
	1650–1380 (m–s)	three ring-stretching bands
	1040–980 (s)	ring breathing
Carbazoles	3490–3470 (vs)	NH stretch (dil soln, nonpolar solvents)
1-4-Dioxanes	1460–1440 (vs)	CH_2 deformation
	1400–1150 (s)	CH_2 twist and wag
	1130–1000 (m)	ring mode; strong in Raman
	850–830 (w)	very strong in Raman
Furans	3140–3120 (m)	CH stretch; higher than most aromatics
	1600–1400 (m–s)	ring stretching (three bands)
	770–720 (vs)	band becomes weaker as number of substituents increases
Indoles	3470–3450 (vs)	NH stretch
	1600–1500 (m–s)	two bands
	900–660 (vs)	substitution patterns due to both 6- and 5-membered rings
Pyridines (general)	3080–3020 (w–m)	CH stretch; several bands
	2080–1670 (w)	combination bands
	1615–1565 (s)	two bands due to C=C and C=N stretch in ring
	1030–990 (s)	ring-breathing
2-subst	780–740 (s)	CH out-of-plane deformation
	630–605 (m–s)	in-plane ring deformation
	420–400 (s)	out-of-plane ring deformation
3-subst	820–770 (s)	CH out-of-plane deformation
	730–690 (s)	ring deformation
	635–610 (m–s)	in-plane ring deformation
	420–380 (s)	out-of-plane ring deformation
4-subst	850–790 (s)	CH out-of-plane deformation
disubst	830–810 (s)	two bands due to CH out-of-plane deformations
	740–720 (s)	
trisubst	730–720 (s)	CH out-of-plane deformation
Pyrimidines	1590–1370 (m–s)	ring stretching; four bands
	685–660 (m–vs)	ring deformation
Pyrroles	3480–3430 (vs)	NH stretch; often a sharp band
	3130–3120 (w)	CH stretch; higher than normal
	1560–1390 (variable)	ring stretch; usually three bands
	770–720 (s, br)	CH out-of-plane deformation
Thiophenes	1590–1350 (m–vs)	several bands due to ring stretching modes
	810–680 (vs)	CH out-of-plane deformation; lower than in pyrroles and furans
Triazines	1560–1520 (vs)	two bands due to ring stretching modes
	1420–1400 (s)	
	820–740 (s)	out-of-plane ring deformation

8-14

Compounds Containing Halogens

8-14a General

A halogen atom attached to a carbon atom adjacent to a functional group often causes a significant shift in the group frequency. Some examples are listed in Table 8-14. Fluorine is particularly important in this regard,

TABLE 8-14 The Effect of Halogen Substituents on Some Group Frequencies

Group or Class	Range (cm^{-1})	Assignment and Remarks
Fluorine		
Fluorocarbons —FCH— and CF_2H	3010–2990	CH stretch; higher frequency than normal
$F_2C{=}C(/)(\backslash F)$	1870–1800	C=C stretch; much higher frequency than normal
$F_2C{=}C(/)(\backslash)$	1760–1730	C=C stretch; much higher frequency than normal
Acid fluorides F—C=O	1900–1820	C=O stretch; extremely high carbonyl group frequency
Ketones $—CF_2COCH_2$ and $—CF_2COCF_2$	1800–1770	C=O stretch; the normal range for ketones is 1730–1700 cm^{-1}
Carboxylic acids $—CF_2—COOH$	1780–1740	C=O stretch; the normal range for carboxylic acids (dimers) is 1720–1680 cm^{-1}
Nitriles $—CF_2—C{\equiv}N$	2280–2260	C≡N stretch; 20 cm^{-1} higher than normal
Amides $—CF_2—CONH_2$	1730–1700	C=O stretch; 30 cm^{-1} higher than normal
Chlorine, Bromine, and Iodine		
CH_2Cl	1300–1240	CH_2 wag; strong IR band
CH_2Br	1240–1190	CH_2 wag; strong IR band
CH_2I	1190–1150	CH_2 wag; strong IR band
Acid chlorides	1810–1790	C=O stretch
α-Halo esters	1770–1745	C=O stretch; 20 cm^{-1} higher than the normal range
Noncyclic halo ketones	1745–1730	C=O stretch; 15 cm^{-1} higher than the normal range
α-Halo carboxylic acids	1740–1720	C=O stretch; the normal range is 1720–1680 cm^{-1}
Chloroformates	1800–1760	C=O stretch; near 1720 cm^{-1} in formate esters
α-Chloro aldehydes	1770–1730	higher than normal aldehydes

and special care must be exercised in conclusions drawn from infrared and Raman spectra when this element is present. Carbon–fluorine stretching bands are very strong in the infrared, usually between 1350 and 1100 cm^{-1}, but they are often weak in the Raman. Other functional groups that absorb in this region of the spectrum may be hidden by the CF stretching band. These can often be detected in the Raman spectrum. It should be mentioned that there are many known cases of symmetrical C—F stretching modes at frequencies much lower than the usual 1350–1100 cm^{-1}. The usual regions for the C—X stretching and bending vibrations have been given previously in Tables 7-2, 7-3, and 7-4.

8-14b CH_2X Groups

The CH_2 wagging mode in compounds with a CH_2X group gives rise to a strong band whose frequency depends on X. When X is Cl, the range is 1300–1250 cm^{-1}. For Br, the band is near 1230 cm^{-1}, and for I, a still lower frequency near 1170 cm^{-1} is observed. Halogen atoms attached to aromatic rings are involved in certain vibrations that are sensitive to the mass of the halogen atom. One of the benzene ring vibrations that involves motion of the substituent atom gives rise to bands between 1250 and 1100 cm^{-1} when the substituent is fluorine, between 1100 and 1040 cm^{-1} for chlorine, and between 1070 and 1020 cm^{-1} for bromine.

8-14c Haloalkyl Groups

In haloalkyl groups, the presence of more than one halogen atom on a single carbon atom shifts the C—X stretching frequency to the high wavenumber end of the range. The CCl_3 group antisymmetric stretching frequency is found in the 830–700 cm^{-1} range.

8-15

Boron, Silicon, and Organometallic Compounds

8-15a General

Boron–carbon and silicon–carbon stretching modes are not usually identifiable, since they are coupled with other skeletal modes. However, the C—B—C antisymmetric stretching mode in phenylboron compounds gives a strong infrared band between 1280 and 1250 cm^{-1}, and a silicon atom attached to an aromatic ring gives two very strong bands near 1430 and 1110 cm^{-1}. Metal–carbon stretching frequencies are found between 600 and 400 cm^{-1} with the lighter metals at the high frequency end of the range, as expected from eq. 7-1. These bonds usually give rise to very strong Raman lines.

The B—O and B—N bonds in organoboron compounds give very strong infrared bands between 1430 and 1330 cm^{-1}. The Si—O—C vibration gives a very strong infrared absorption, which is often quite broad in the 1100–1050 cm^{-1} range. Some characteristic frequencies for boron and silicon compounds are listed in Table 8-15. The B—CH_3 and Si—CH_3 symmetric CH_3 deformation modes occur at 1330–1280 cm^{-1} and 1280–1250 cm^{-1}, respectively. The CH_3 deformations in metal–CH_3 groups give rise to bands between 1210 and 1180 cm^{-1} in organomercury and organotin compounds and between 1170–1150 in organolead compounds.

The infrared spectra of aromatic organometallic compounds usually contain a fairly strong, sharp band near 1430 cm^{-1} due to a benzene ring vibration. This band has been observed for compounds in which As, Sb, Sn, Pb, B, Si, and P atoms are attached directly to the ring.

8-15b Organomercury Compounds

The mercury–carbon bond in aliphatic organomercury compounds can be characterized by a very strong Raman band between 550 and 500 cm^{-1}. For aromatic compounds, a band between 250 and 200 cm^{-1} is assigned to the phenyl—Hg stretch. These bands are so strong that they can be seen in Raman spectra of dilute aqueous solutions.

TABLE 8-15 Some Infrared Group Frequencies in Boron and Silicon Compounds

	Group	Range (cm^{-1}) and Intensity	Assignment and Remarks
Boron	—BOH	3300–3200 (s)	broad band due to H-bonded OH stretch
	—BH and —BH_2	2650–2350 (s)	doublet for —BH_2 stretch
		1200–1150 (ms)	—BH_2 deformation or B—H bend
		980–920 (m)	—BH_2 wag
	—B(—)—Ar	ca 1430 (m–s)	benzene ring vibration
	B—N	1460–1330 (vs)	B—N stretch; borazines and aminoboranes
	B—O	1380–1310 (vvs)	B—O stretch: boronates, boronic acids
	C—B—C	1280–1250 (vs)	C—B—C antisym stretch
Silicon	—SiOH	3700–3200 (s)	OH stretch, similar to alcohols
		900–820 (s)	Si—O stretch
	—SiH, —SiH_2, and —SiH_3	2150–2100 (m)	Si—H stretch
		950–800 (s)	Si—H deformation and wag
	Si—Ar	ca 1430 (m–s)	ring mode
		1100 (vs)	ring mode
	Si—O—C (aliphatic)	1100–1050 (vvs)	Si—O—C antisym stretch
	Si—O—Ar	970–920 (vs)	Si—O stretch
	Si—O—Si	1100–1000 (s)	Si—O—Si antisym stretch

8-16

Some Detailed Structural Studies

8-16a Introduction

In this section, some applications of vibrational spectroscopy in detailed structural and conformational studies will be outlined. Structural, geometrical, and rotational isomers and tautomers can be observed in infrared and Raman spectra. Equilibria involved in these systems can be studied by varying concentrations and temperature. Monomer–dimer equilibria can be studied for molecules such as carboxylic acids and phenols that self-associate to form hydrogen-bonded species. Vibrational spectroscopy is a valuable tool in studies of both inter- and intramolecular hydrogen bonding.

8-16b Structural Isomerism

Structural isomers often differ in the functional groups present, and their vibrational spectra then differ considerably. Some examples are the α-amino acid alanine, the ester ethyl carbamate (urethane), and the nitro compound 1-nitropropane. All three compounds have the empirical formula $C_3H_7O_2N$. The infrared survey spectra of these three compounds are shown in Figure 8-32.

A good example of the importance of vibrational spectra in differentiating between structural isomers is found in ortho-, meta-, and para-disubstituted benzenes. The CH out-of-plane deformation patterns in the 850–700 cm^{-1} region, which were discussed in Section 8-6, are different for each isomer. Substituted pyridines, pyrimidines, and other heterocyclic compounds provide further examples of structural isomers that can be distinguished by their vibrational spectra.

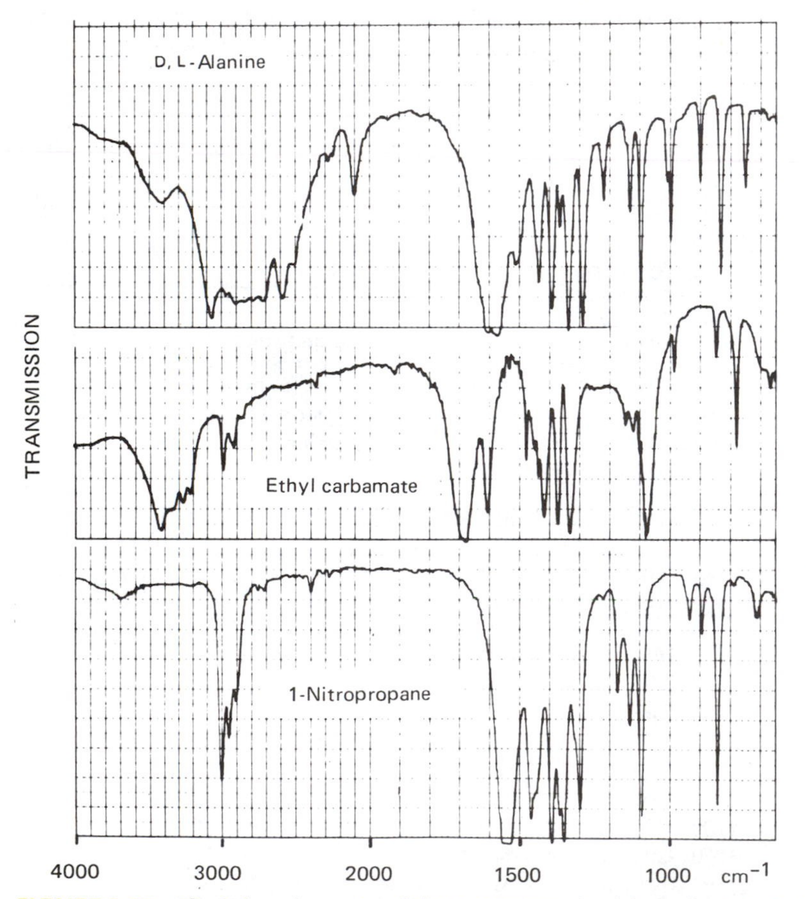

FIGURE 8-32 The infrared spectra of three compounds of formula $C_3H_7O_2N$.

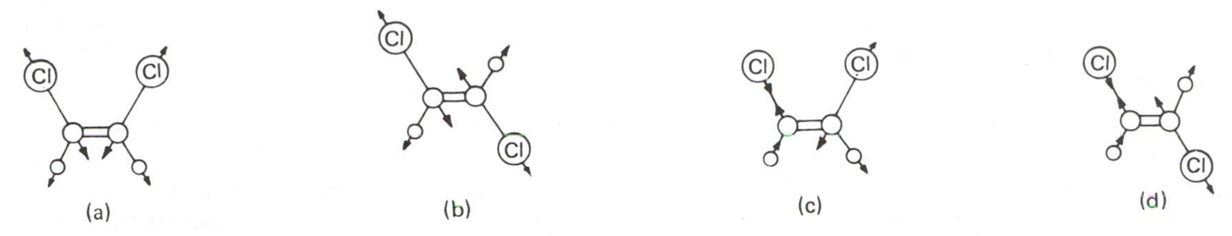

FIGURE 8-33 The C—Cl stretching modes, of *cis*- and *trans*-dichloroethene (the plane of the molecule is assumed to be the *xz* plane): (a) cis (symmetric): the dipole moment changes (observed in both IR and Raman at 710 cm^{-1}); (b) trans (symmetric): no change in dipole moment (not observed in IR but observed in Raman at 840 cm^{-1}-see Figure 7-1); (c) cis (antisymmetric): the dipole moment changes (observed in both IR and Raman at 840 cm^{-1}); (d) trans (antisymmetric): the dipole moment changes, but no change in polarizability (observed only in IR at 895 cm^{-1}-see Figure 7-1).

8-16c Geometrical (cis-trans) Isomers

Infrared/Raman spectroscopy is useful in distinguishing between cis and trans isomers. Absorption of infrared radiation by a molecule can only occur if there is a change in dipole moment accompanying a vibration. For cis isomers a dipole moment change occurs for most of the normal vibrations. However, trans isomers usually have higher symmetry, which leads to a zero or very small dipole moment change for some vibrations, so that they are not observed in the infrared spectrum. This situation is illustrated in Figure 8-33 for the C—Cl stretching vibrations of *cis*- and *trans*-dichloroethene.

It is seen that in case (b) the symmetric vibration involving stretching of the C—Cl bonds produces no change in dipole moment. If the chlorine atoms were to be replaced by similar, but not identical groups, then a small dipole moment change would be produced during the vibration. However, this oscillating dipole might be too weak to give rise to an observable infrared absorption. Thus, we can conclude that trans compounds often have simpler infrared spectra than the cis isomers.

For a vibration to be active in the Raman effect there must be a change in *polarizability* during the vibration. This is not so easily visualized, but highly symmetrical modes *always* produce a change in polarizability, while less symmetrical vibrations sometimes give rise to no change and are therefore not seen in the Raman spectrum.

For the C—Cl stretching modes of the two dichloroethylenes of Figure 8-33, both vibrations of the cis isomer are seen in the Raman spectrum. However, only the symmetric mode of the trans compound is Raman active. Once again, we conclude that trans isomers have *simpler* spectra than the cis compounds. We also note that for the trans compound some vibrations not seen in the infrared spectrum are Raman active (Figure 8-33b) and, conversely, certain frequencies not seen in the Raman spectrum are infrared active (Figure 8-33d). Trans-substituted alkenes are characterized by a very strong IR band near 970 cm^{-1} due to the wagging vibration. For the cis isomer a medium to strong band is observed between 730 and 650 cm^{-1} for this mode. Another point that is helpful in determining which structural isomer is present is the observation that for a trans isomer the antisymmetric C—X stretching mode in an XC=CX structure is observed at frequencies 20–40 cm^{-1} higher than for the corresponding cis isomer. Many such examples are found in the spectra of long chain acids, alcohols, and esters. A similar observation has been made concerning the C=C stretch in unsaturated ketones and unsaturated hydrocarbons. Again, the trans isomer has a slightly higher frequency (5–10 cm^{-1}). However, the infrared absorption may be very weak for the trans compound for the reasons noted above. The lower frequency in the cis compound is probably due to the lower symmetry (or pseudo symmetry), which gives rise to greater coupling of —C—C= or —C=C— vibrations with other lower frequency skeletal modes of the molecule.

8-16d Rotational Isomerism

In open chain compounds, the barriers to internal rotation about one or more carbon–carbon single bonds may be too high for rapid interconversion between different configurations. In such cases, two or more different isomers can exist and their presence may be detected in their infrared or Raman spectra. The restriction of rotation about double bonds can be thought of as an extension of the above concept. In this case, very high barriers are involved and cis and trans compounds result. The axial–equatorial conformations of cyclohexane and cyclopentane derivatives are examples of another kind of conformational (rotational) isomerism.

In noncyclic structures, rotation about a single bond can produce an infinite number of configurations. Some of these are energetically favored (energy minima). The simplest examples are the substituted ethanes, CH_2XCH_2Y, for which there are several preferred staggered conformations. Two stable conformers, **8-4** and **8-5**, are illustrated below together with the unstable (in this case) eclipsed form **8-6**.

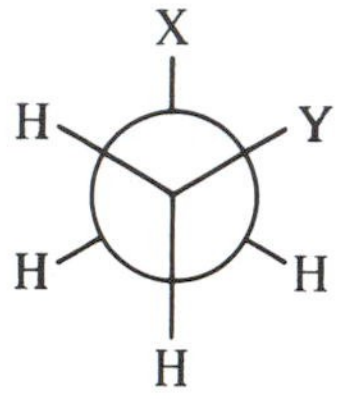

8-4
staggered
(gauche)

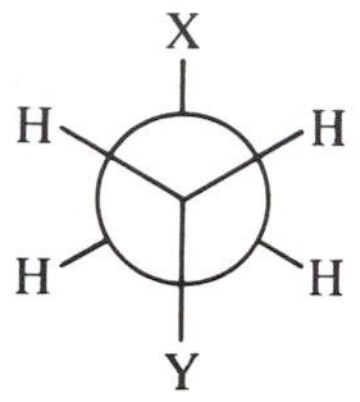

8-5
staggered
(trans)

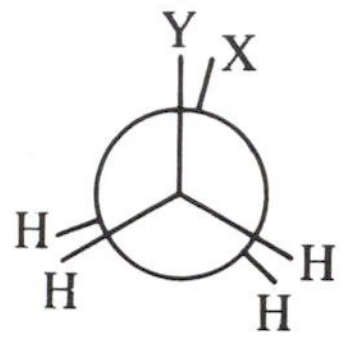

8-6
eclipsed
(cis)

When there is a stabilizing interaction, the eclipsed form may be one of the stable conformations. Many such examples can be found in α-halo ketones, esters, acid halides, and amides. In these compounds, the halogen atom is believed to be either cis (eclipsed) or gauche (staggered) with respect to the carbonyl group. Two C=O stretching bands are observed in such cases. One is at higher frequencies, due to the eclipsed interaction between the halogen atom and the C=O group. The other is found at the normal frequency. In α-halo ketones, substitution of a second halogen on the other side of the carbonyl group leads to three preferred isomers, eclipsed–eclipsed, eclipsed–gauche, and gauche–gauche. Three C=O stretching frequencies can be observed in such cases.

For α-halo carboxylic acids, multiple carbonyl bands are not usually observed because of complications from hydrogen bonding. An exception is found in FCH_2COOH, for which five bands can be resolved. These arise from the five possible rotational isomers of monomer and dimer.

In cyclic compounds, the possibility of axial and equatorial conformations exists. For example, in α-chloro substituted cyclopentanones or cyclohexanones, two distinct carbonyl stretching frequencies can be observed. One band is found near 1745 cm^{-1}, due to the equatorial conformation **8-7**, in which interaction between the Cl and C=O groups can occur. A second band near 1725 cm^{-1} is attributed to the axial isomer **8-8**, in which interaction is minimized. The relative proportions of axial and equatorial forms change with

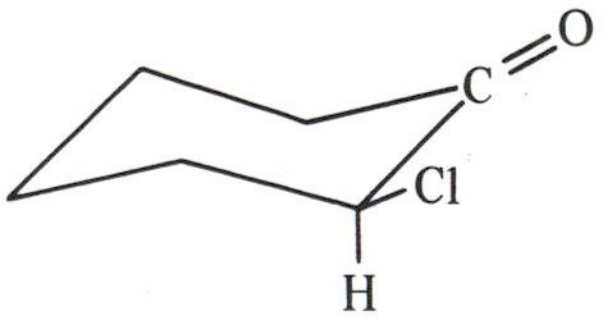

8-7
equatorial
$\nu_{CO} = 1745\ cm^{-1}$

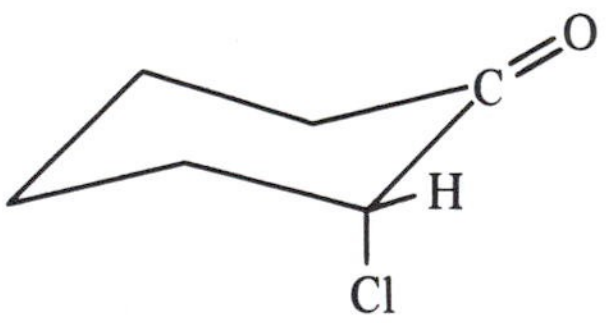

8-8
axial
$\nu_{CO} = 1725\ cm^{-1}$

phase, temperature, and solvent, and such changes can be readily followed in the vibrational spectra. In cyclohexanols, the equatorial C—OH stretching frequency is 1050–1030 cm^{-1}, while in the axial conformation the frequency is 10–30 cm^{-1} lower.

Ortho-halogenated benzoic acids also show two carbonyl stretching frequencies, due to the two rotational isomers, **8-9** and **8-10**, which could be described as cis and trans with respect to the halogen and C=O groups.

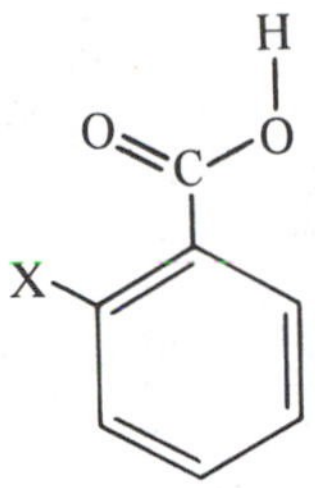

8-9
cis

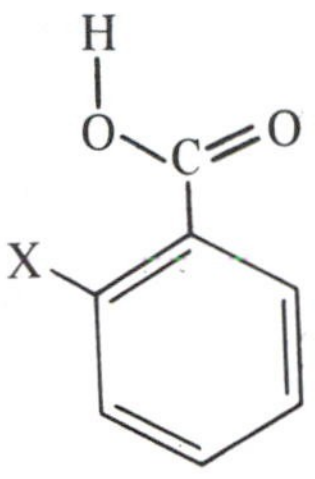

8-10
trans

Vinyl ethers show a doublet for the C=C stretching mode at 1640–1620 and 1620–1610 cm^{-1}. These bands correspond to rational isomers about the C—O bond. The two bonds show variations in intensity with temperature. The CH_2 deformation band is also found to be a doublet that is due to the two different rotational isomers.

8-16e Tautomerism

Numerous examples of tautomerism can be found in the literature. Infrared spectroscopy offers a useful means of distinguishing between possible tautomeric structures. A simple example is found in β-keto esters or β-diketones. The keto form has two C=O groups, which have separate stretching frequencies, and a

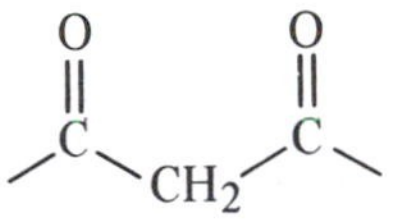

keto

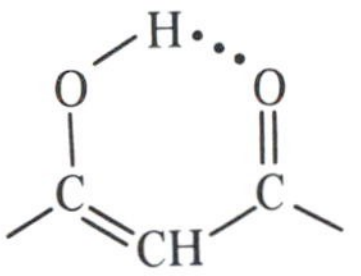

enol

doublet is often observed in the usual ketone carbonyl stretching region, near 1730 cm^{-1}. The enol form, on the other hand, has only one carbonyl group, the frequency of which is lowered by hydrogen bonding and conjugation by 80–100 cm^{-1}. This structure also has an alkenic double bond that should give a band between 1650 and 1600 cm^{-1}. The C=O and C=C peaks may then appear as a doublet. An example of tautomerism is shown in Figure 8-34. The compound ethyl methylacetoacetate clearly shows both keto and enol forms.

8-16f Hydrogen Bonding

Hydrogen bonding manifests itself in very broad OH and NH stretching bands at frequencies considerably lower than normal. Changes in the intensity of these bands can be brought about by changes in temperature and concentration. In solutions of carboxylic acids in an inert solvent such as CCl_4, the presence of monomer, dimer, and polymeric species can be identified in the carbonyl stretching region (ref. 8-1).

In addition to the hydrogen-bonded OH and NH stretching bands (ν_{OH} or ν_{NH}) between 3500 and 2500 cm^{-1}, the R—OH or R—NH bending modes can also be observed between 1700 and 1000 cm^{-1}. The torsional motion of the R—OH or R—NH bonds gives rise to absorption between 900 and 300 cm^{-1}. Stretching of the hydrogen bond itself has been observed in the far infrared in many cases between 200 and 50 cm^{-1}, and bending of the hydrogen bond occurs at very low frequencies, usually below 50 cm^{-1}.

Both intra- and intermolecular hydrogen bonding can occur between OH groups in alcohols or phenols and halogen atoms. In 2-chloroethanol, for example, an intramolecular hydrogen bond stabilizes the gauche rotational isomer (structure **8-4**, Section 8-16d). The free ν_{OH} in the trans conformation absorbs at 3623 cm^{-1}, whereas for the hydrogen-bonded isomer the frequency is 3597 cm^{-1}. Halophenols also show two ν_{OH} bands separated by 50–100 cm^{-1} due to bonded and nonbonded conformations. Infrared spectra indicate that intermolecular OH···halogen bonding occurs between alkyl halides and phenols or alcohols.

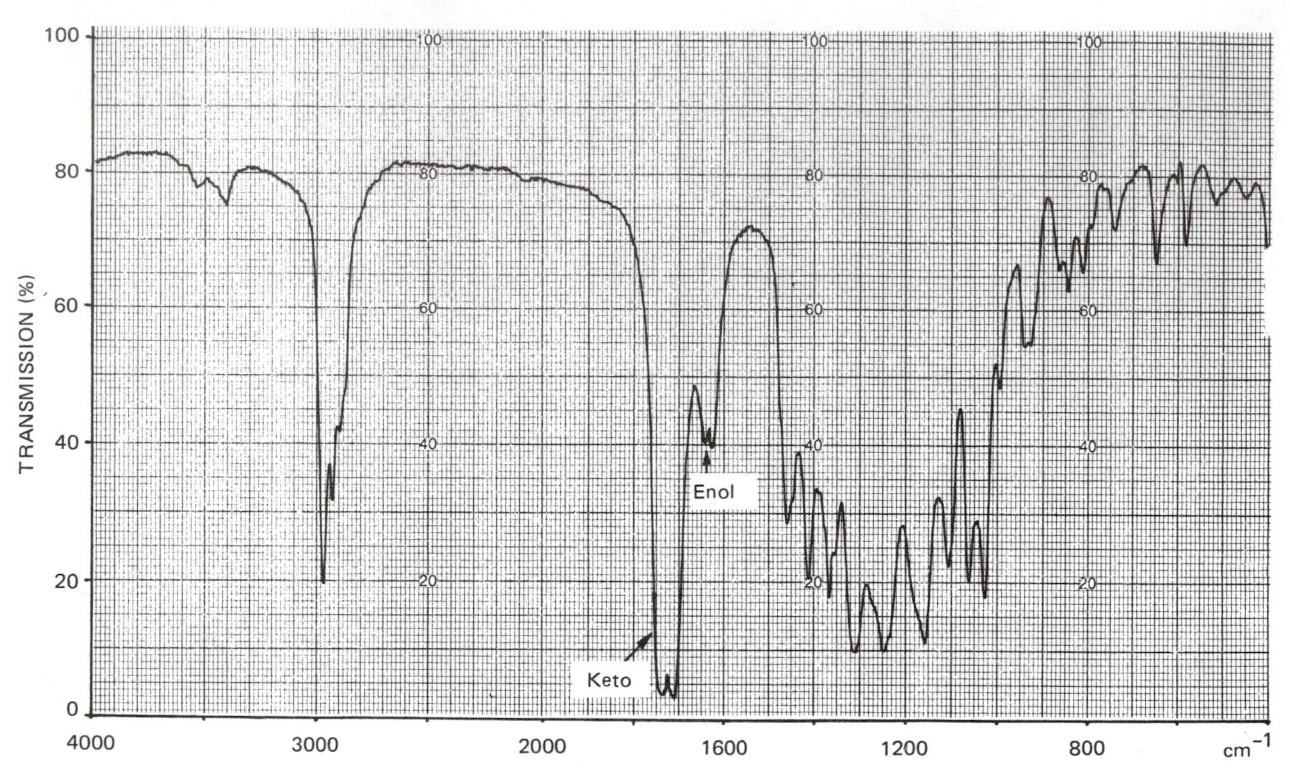

FIGURE 8-34 The infrared spectrum of a β-keto ester, ethyl methylacetoacetate (ethyl 3-oxopentanoate).

In this section only a very brief mention has been made of some of the cases in which hydrogen bonding is found. The reader is referred to the books by Bellamy (ref. 8-2) and Pimentel and McClellan (ref. 8-3) for further discussion and references to the literature of the subject.

PROBLEMS

Deducing the structure of a compound from its infrared and Raman spectra is not easy; in fact, for large complicated molecules it is not possible. The best way to learn how to obtain structural information from spectra is by practice. A list of sources of interpreted spectra and problems is given in references 8-4 through 8-10.

It is useful to have a checklist of questions giving basic information on the structure. For example, which elements are present, what is the molecular formula, what other information is available from other instrumental or chemical methods? Then, from an infrared survey spectrum, further clues can be gathered. Again, a checklist is useful. Are there any broad absorption bands (see Table 8-2)? Is there an aromatic ring present (see Section 8-6)? What kind of X—H bonds are present? Is there a carbonyl group in the molecule? Then a systematic analysis of the bands in the infrared spectrum can be made, first using Table 8-1, then Table 8-3. Tentative assignments can be made to each band, cross-checking where possible in other regions of the spectrum and referring to Tables 7-2 and 8-3 through 8-15. A Raman spectrum may be useful in some cases to confirm or eliminate certain assignments (Table 7-4).

This method will now be illustrated with some examples. The spectra were all recorded, linear in wavenumber, on a small grating instrument. References are given to corresponding spectra in the *Aldrich Library of Infrared Spectra* (see Appendix II.A) so that the reader can compare the example spectra with the spectra recorded in the linear wavelength format (see Section 8-2).

8-1 The compounds represented in the following spectra contain only carbon and hydrogen. Try to identify the type of compound and suggest possible structures.

a. A volatile liquid hydrocarbon. The molecular weight determined by mass spectrometry is 84.

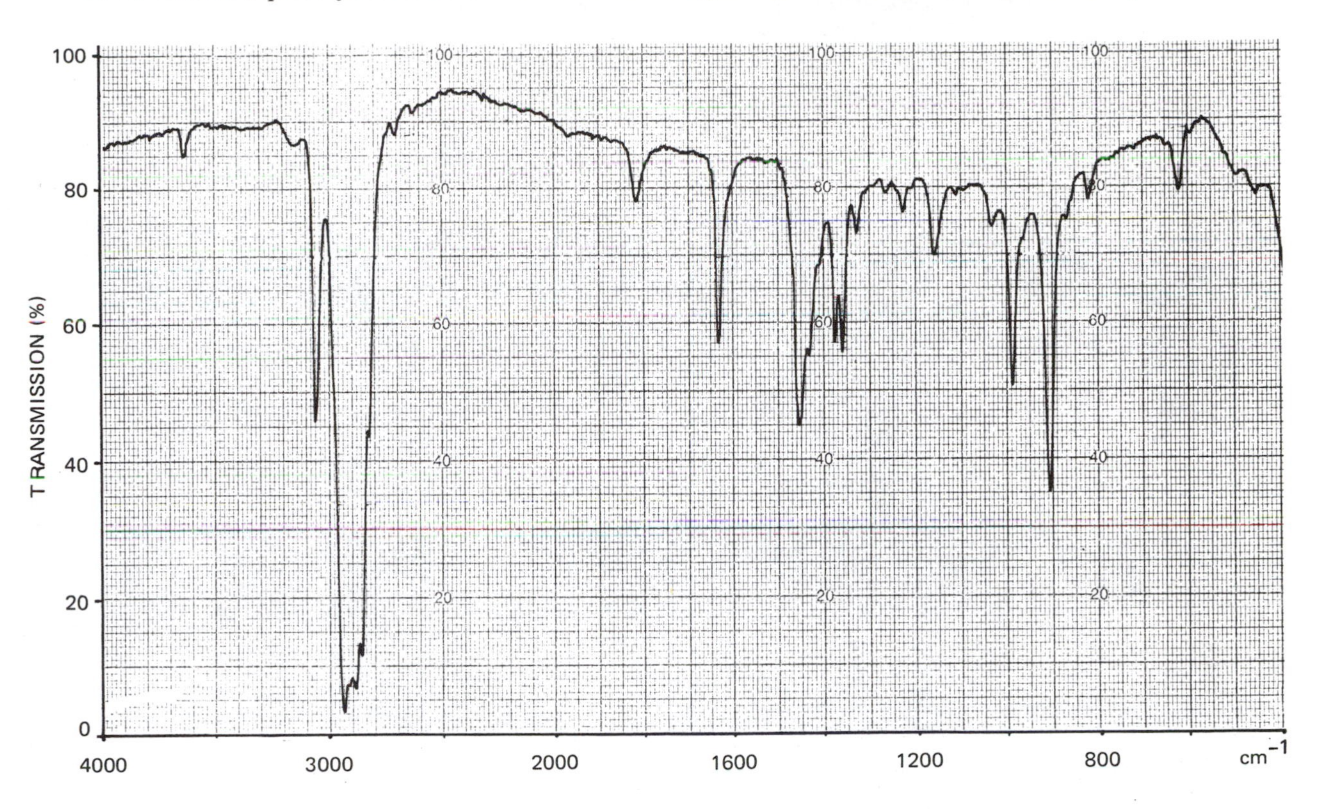

b. A liquid hydrocarbon of molecular formula C_7H_{14}.

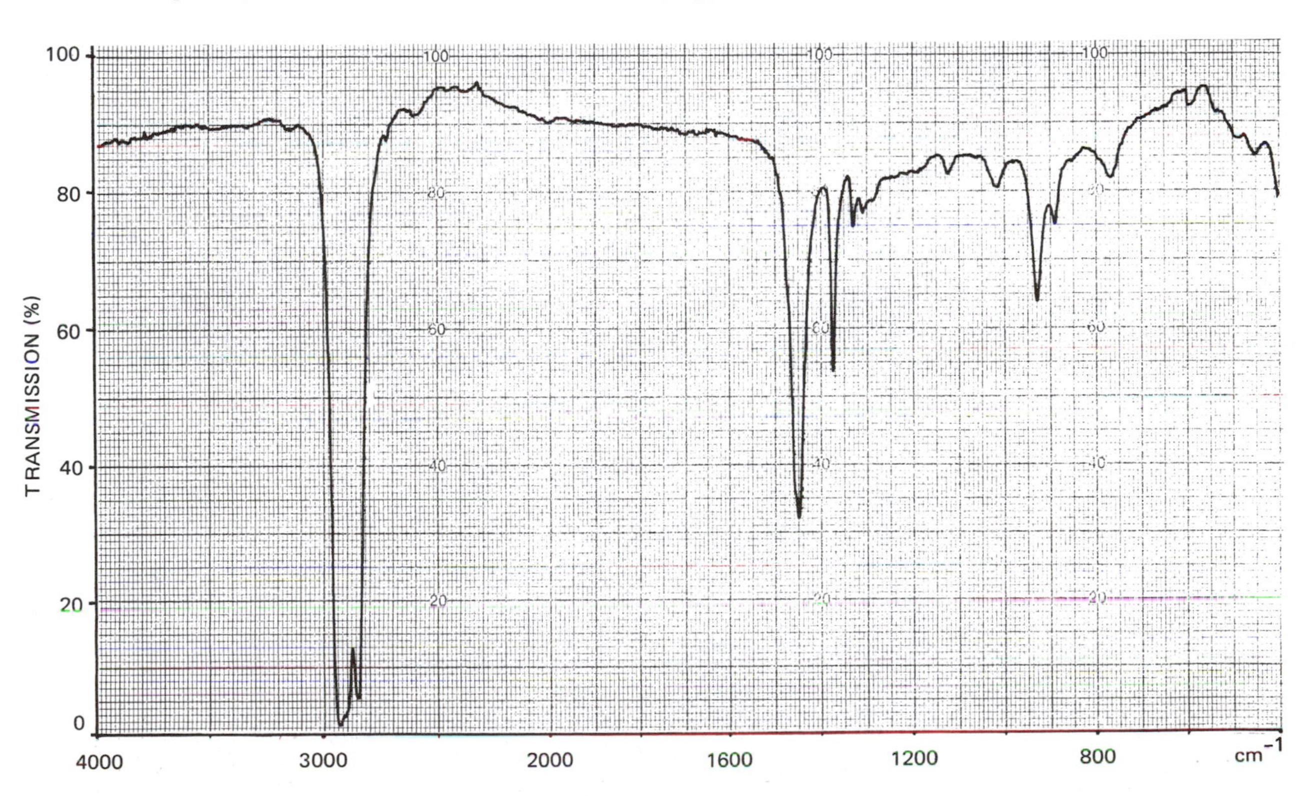

c. A liquid hydrocarbon of boiling point 159° C and molecular weight 120.

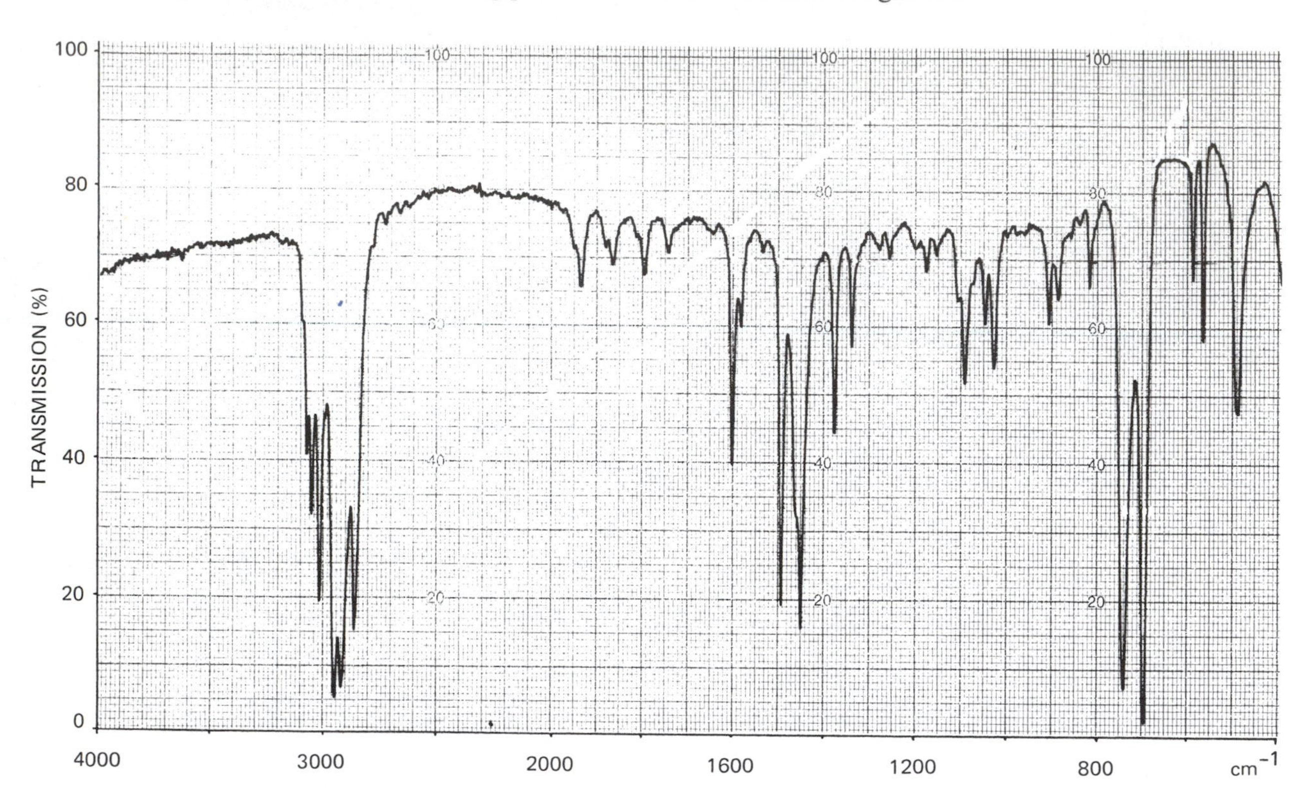

d. A liquid hydrocarbon of molecular weight 104.

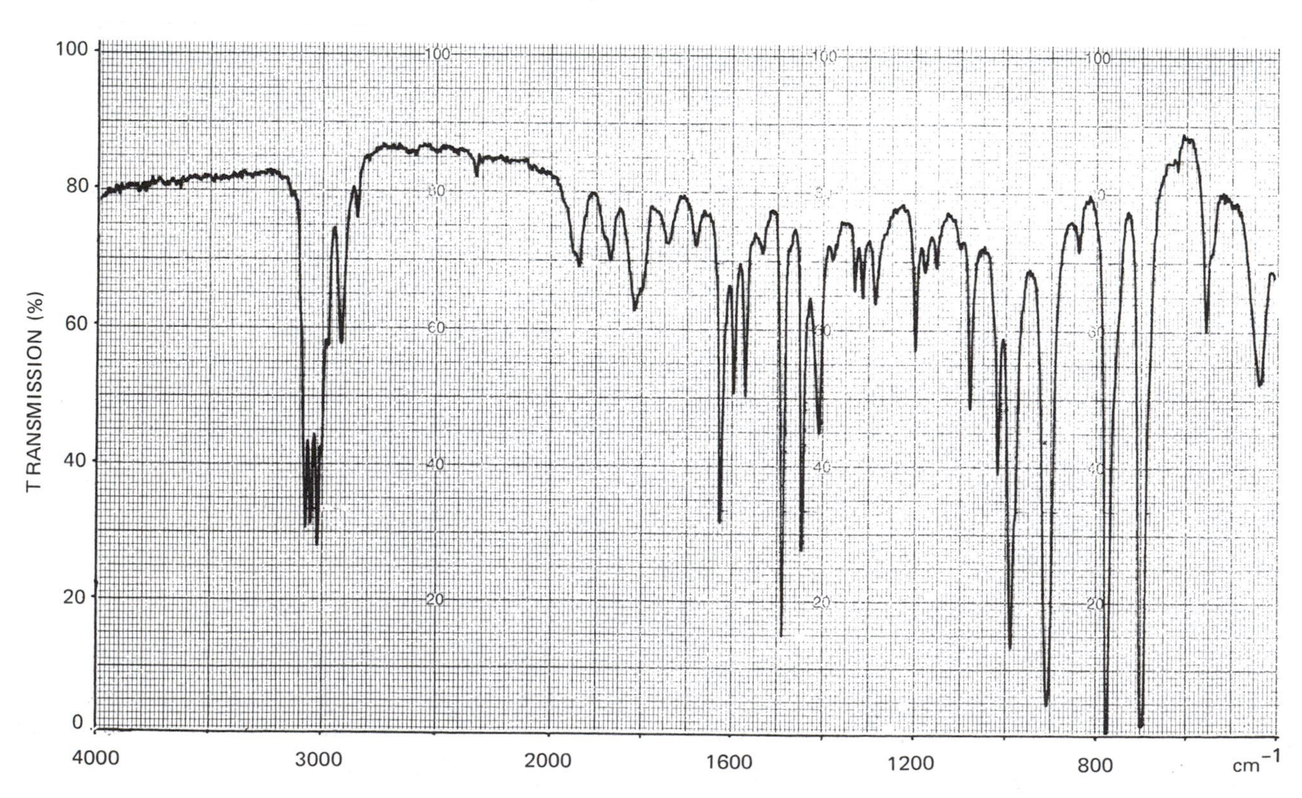

8-2 The following spectra are of compounds containing only C, H, and O. Each compound contains only one kind of functional group involving oxygen. Consult Sections 8-5 and 8-10, as well as the sections on carbon–hydrogen vibrations to deduce the structures.

a. A high-boiling (206° C) liquid of molecular weight 138.

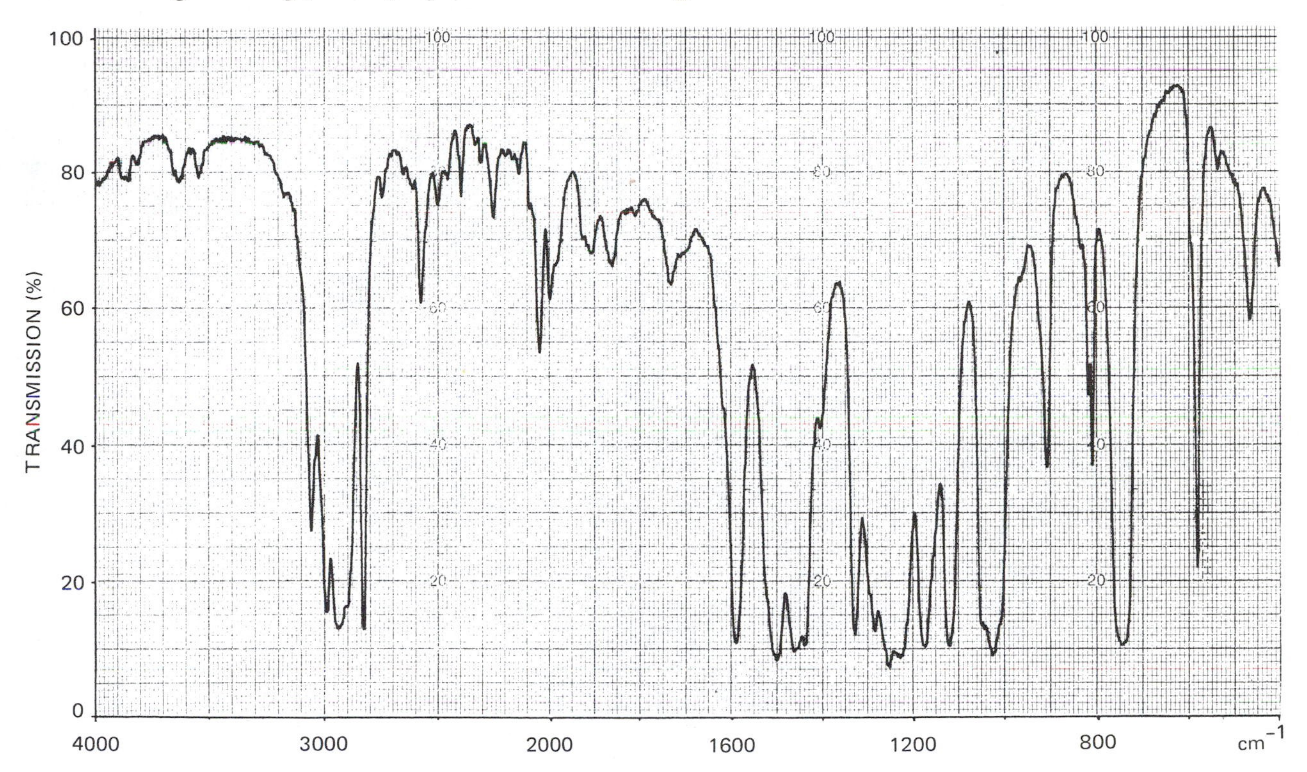

b. A liquid compound of molecular weight 100.

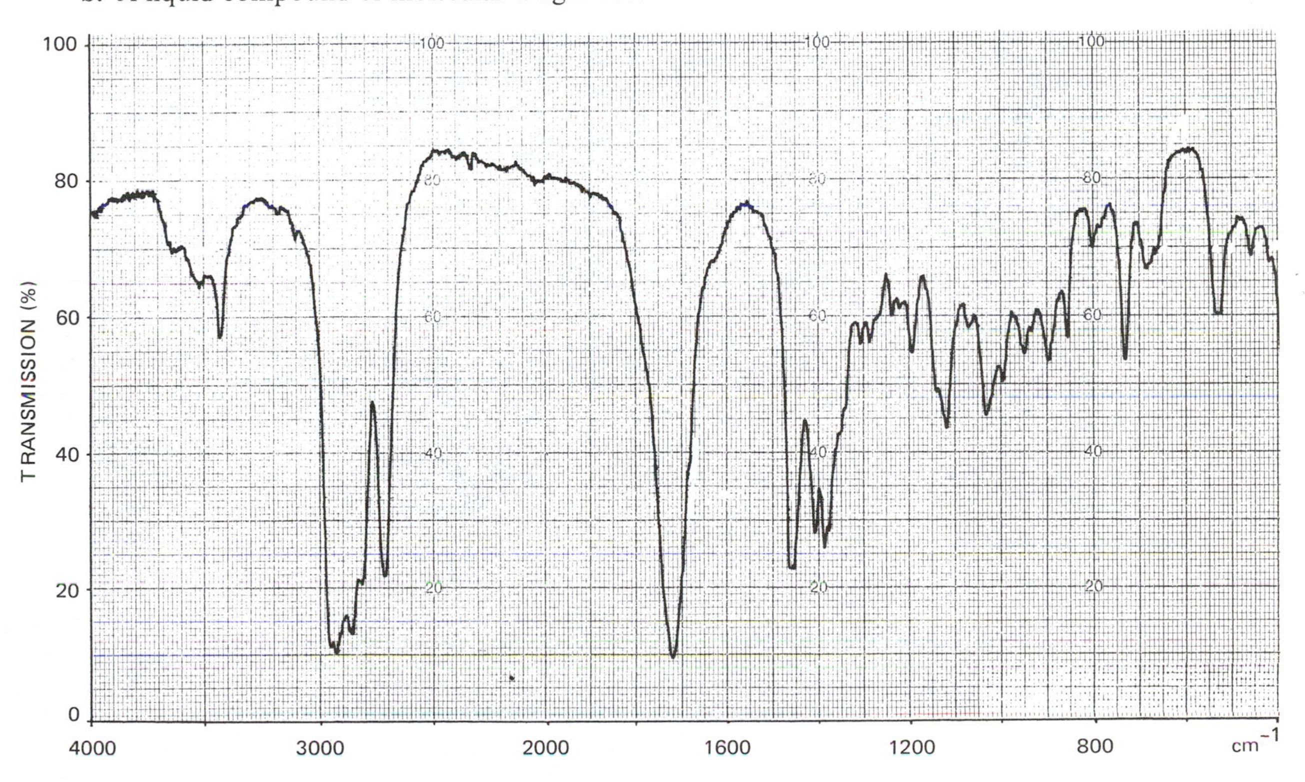

c. A liquid compound of molecular weight 74.

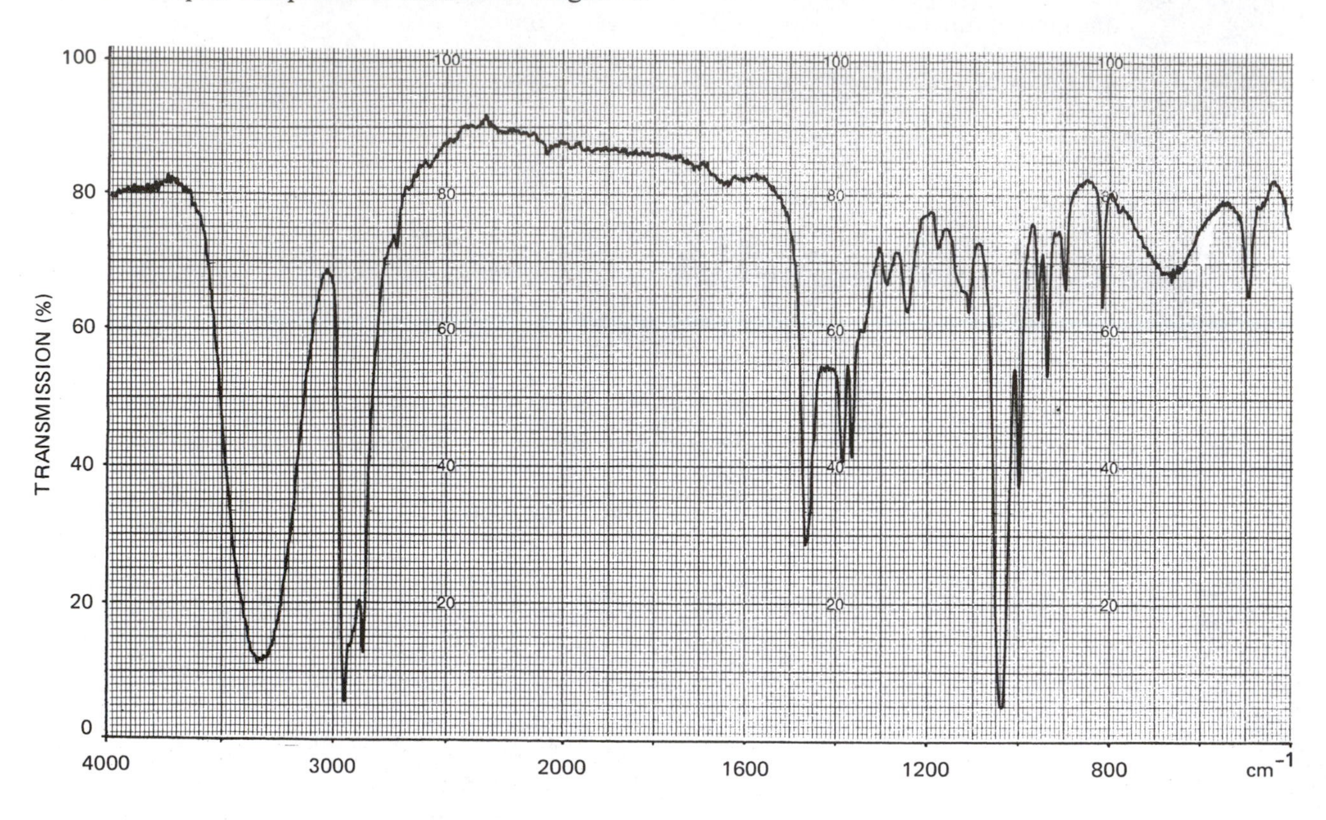

d. A low-melting (43° C) solid of molecular weight 122.

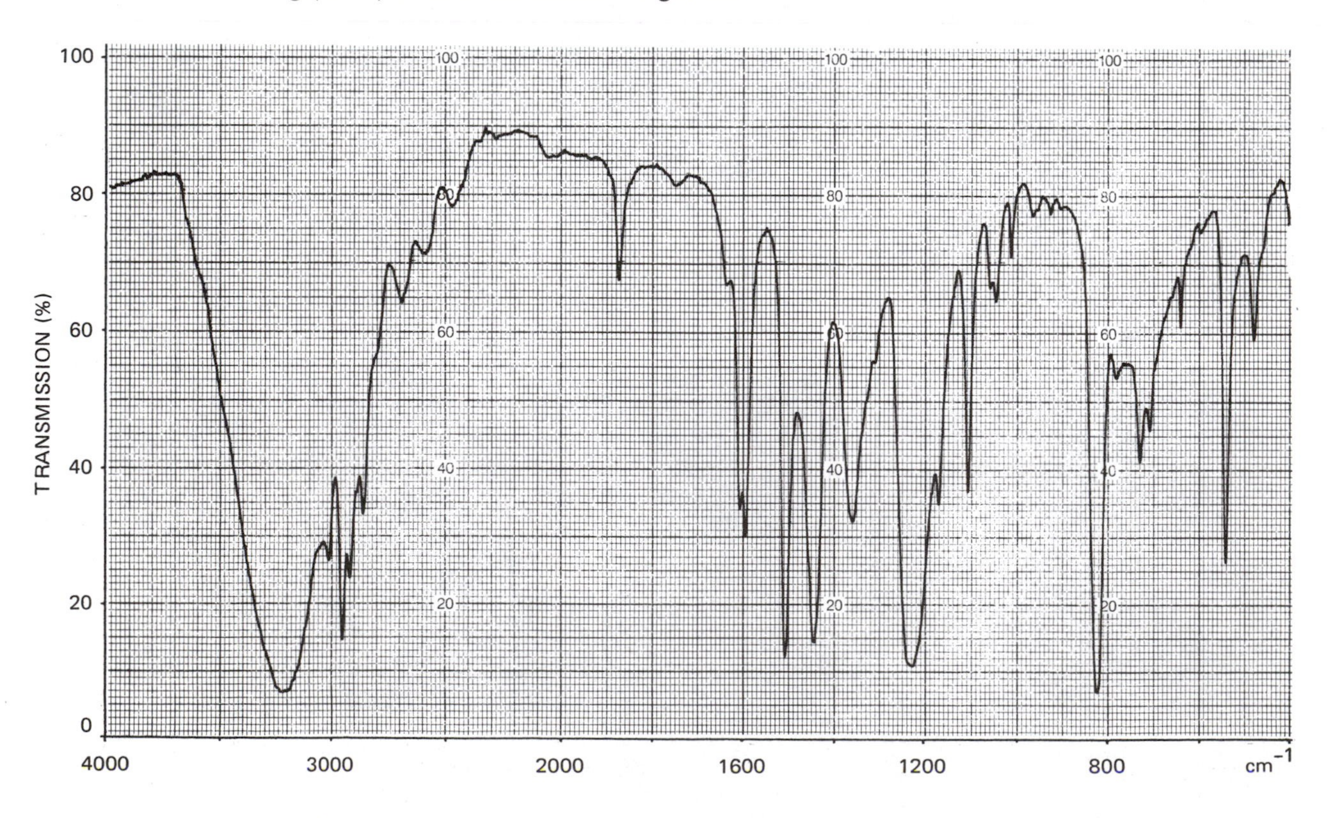

e. A liquid with boiling point 155° C and molecular formula $C_6H_{10}O$.

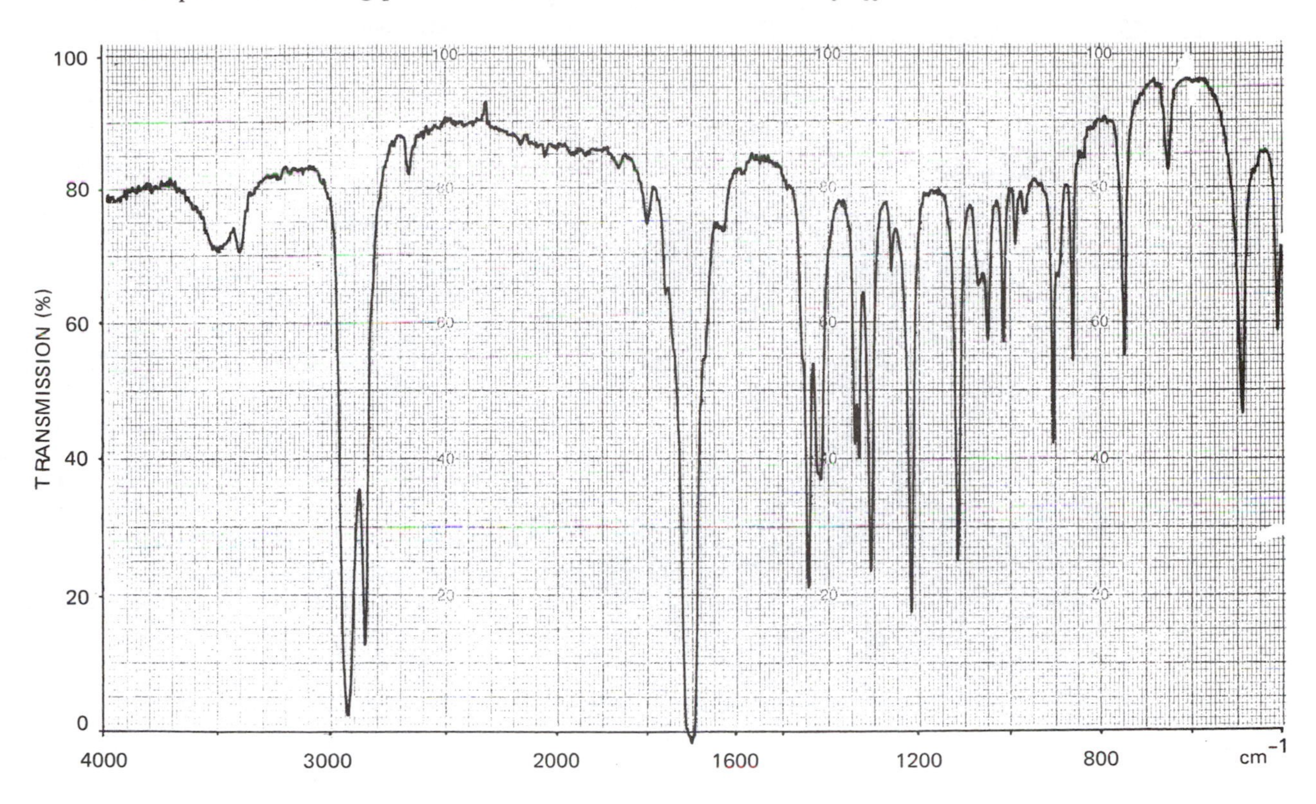

f. A low-melting (50–52° C) solid of molecular weight 164.

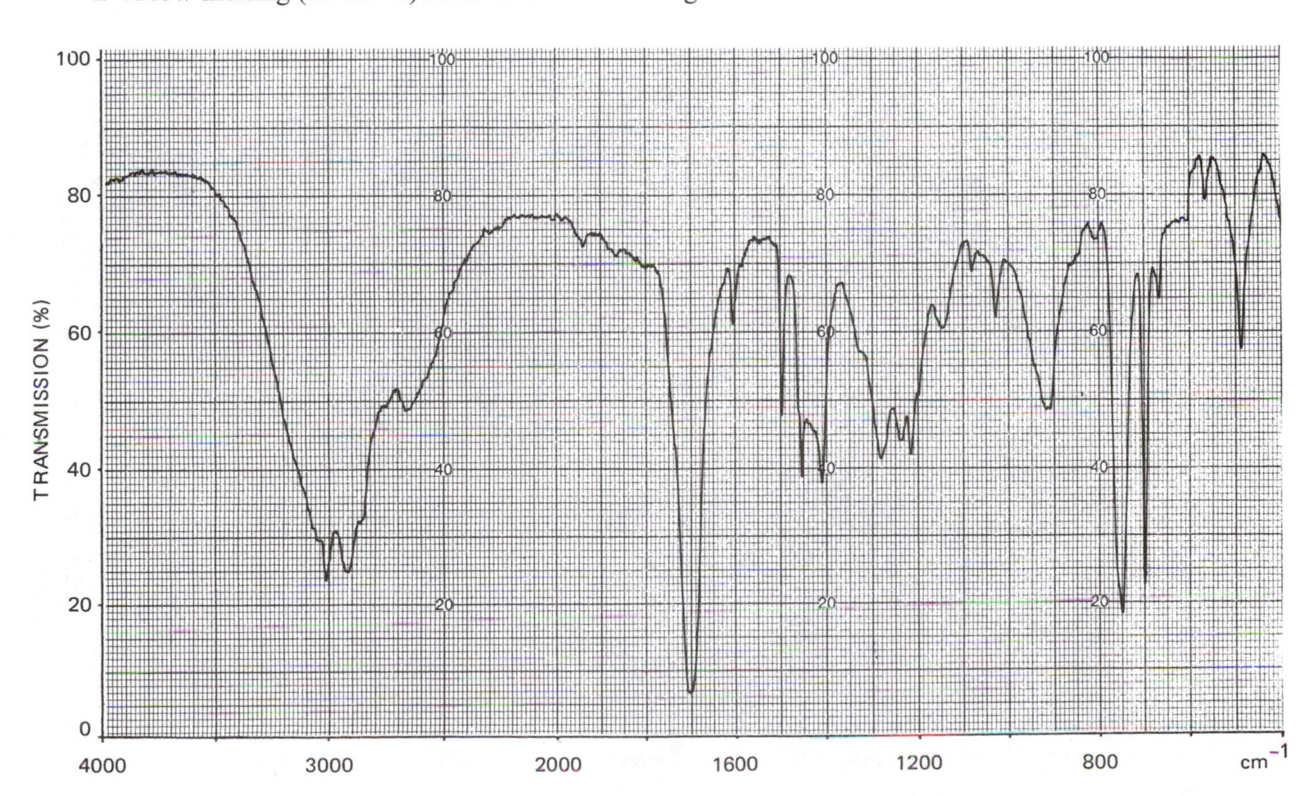

g. An aromatic compound containing only C, H, and O.

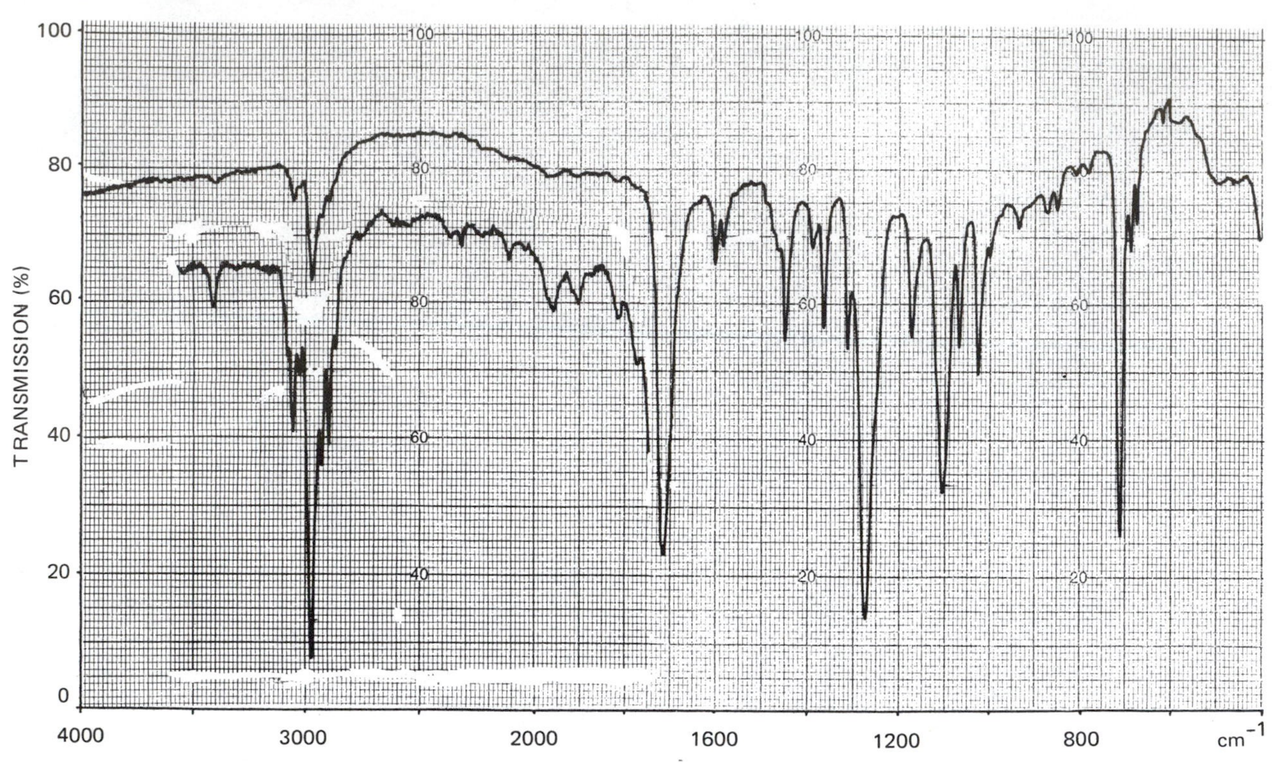

8-3 Compounds containing only C, H, and N yielded the following infrared spectra. In each compound there is only one kind of nitrogen-containing functional group. Consult Section 8-11 as well as the sections on carbon–hydrogen vibrations to deduce the structures.

a. A compound of molecular weight 103.

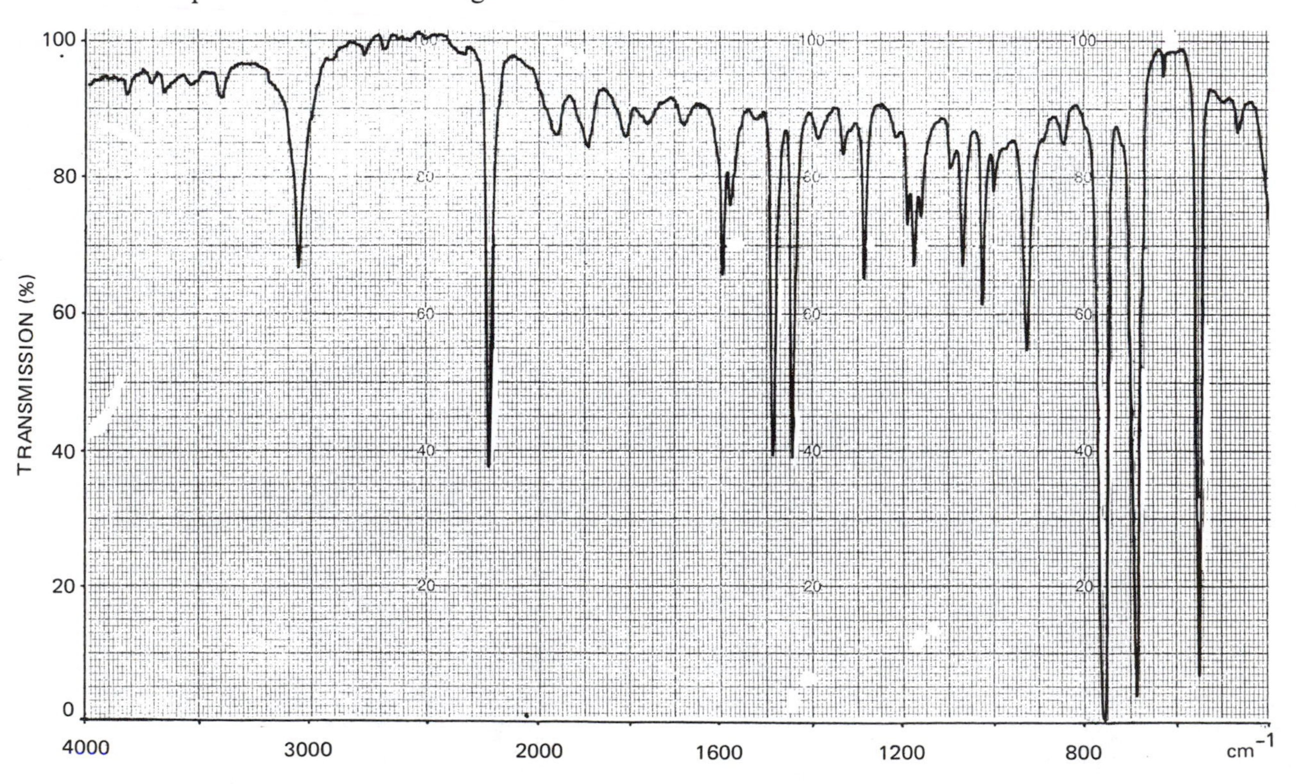

b. A liquid compound of molecular weight 101.

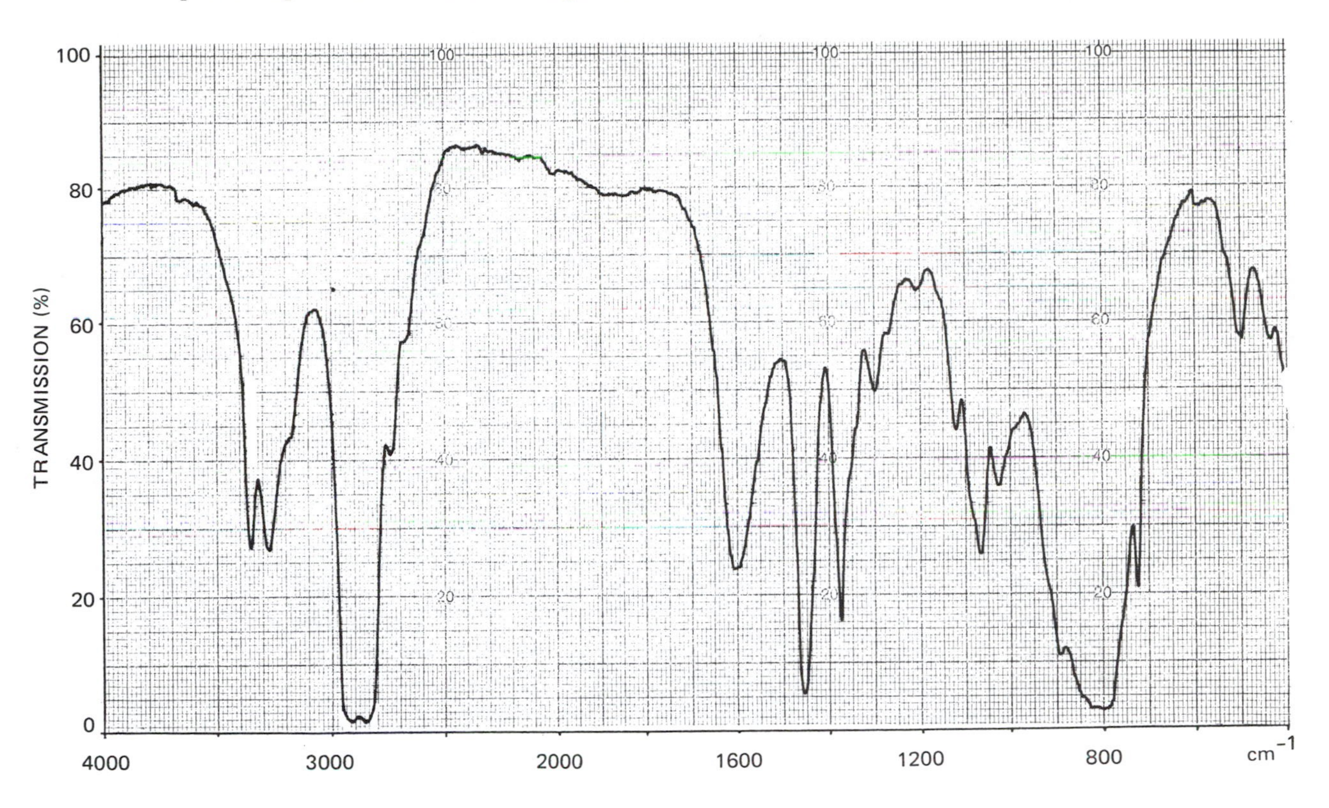

c. A compound with molecular formula $C_{10}H_{15}N$. Try first to deduce the functional groups present.

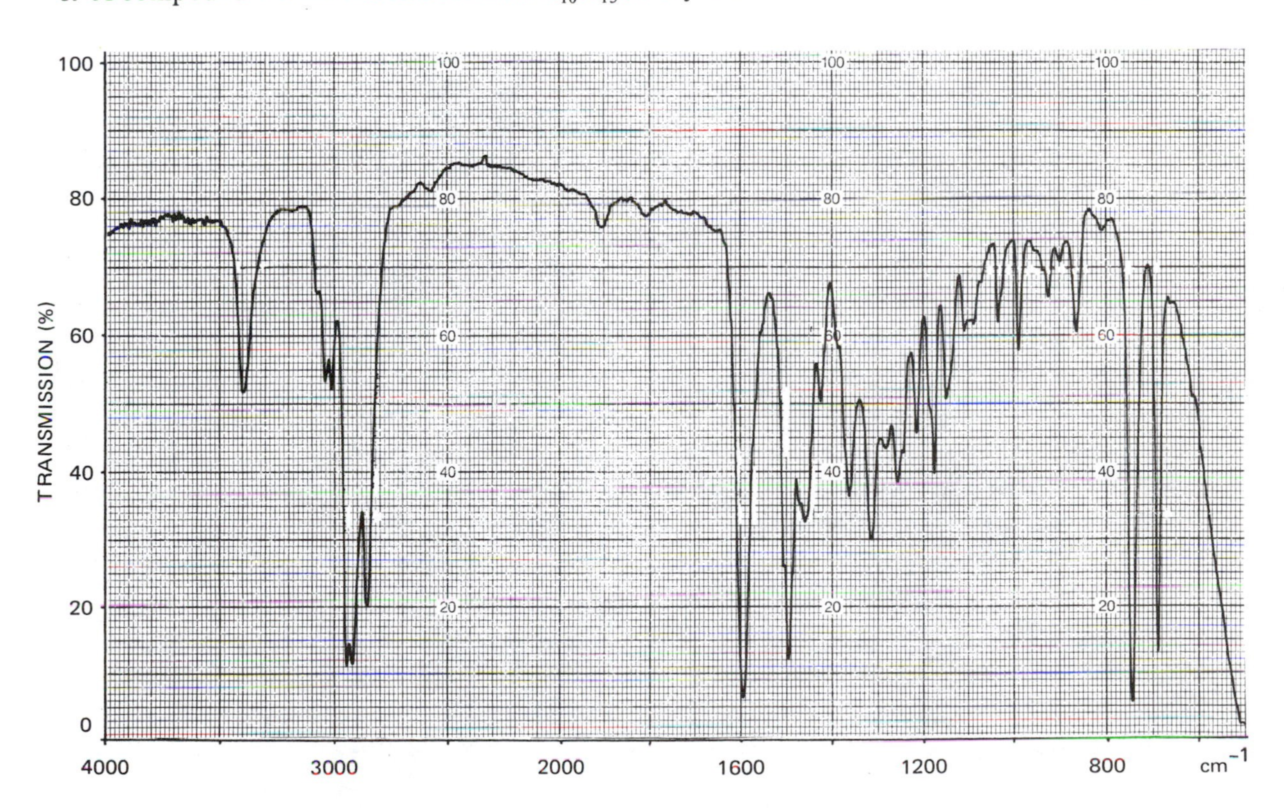

d. A compound of molecular weight 79.

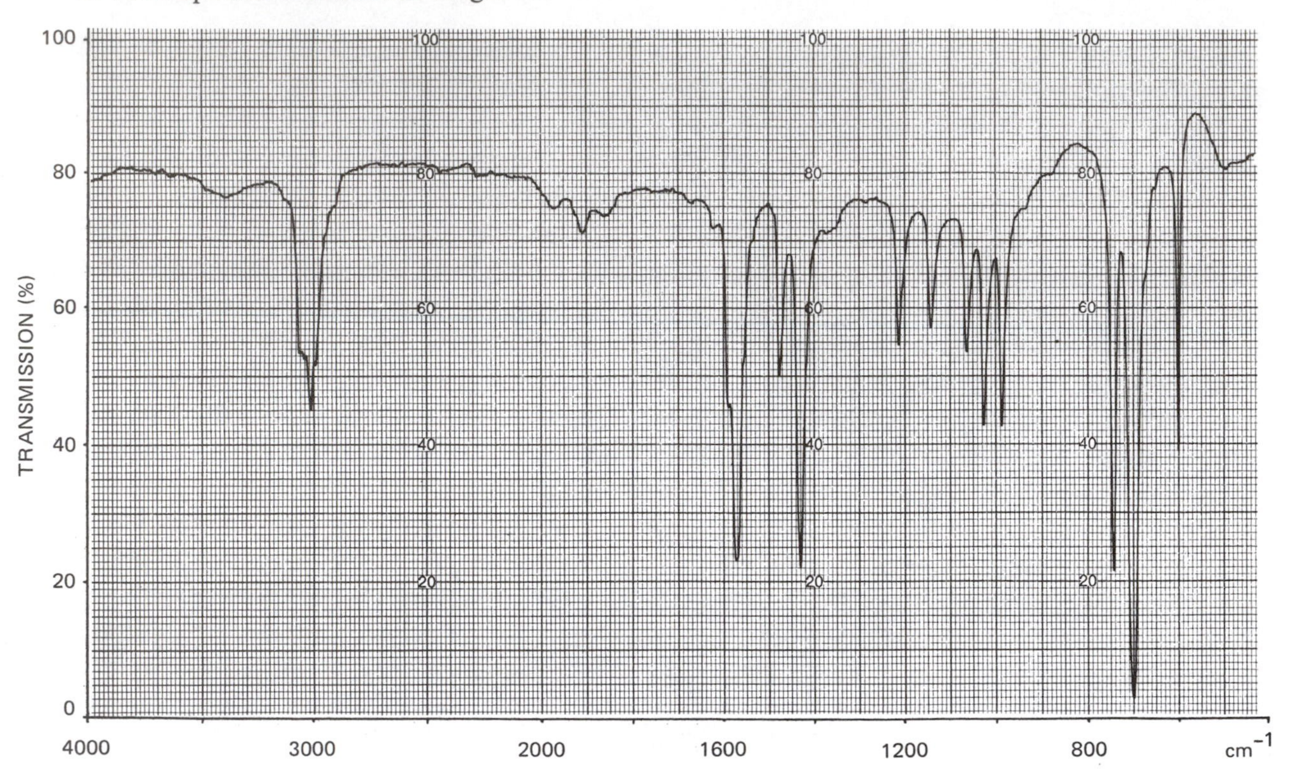

8-4 The spectra that follow are of compounds containing both oxygen and nitrogen in addition to carbon and hydrogen. Suggest a structure for each compound.

a. A KBr pellet of a compound of molecular formula $C_3H_7O_3N$.

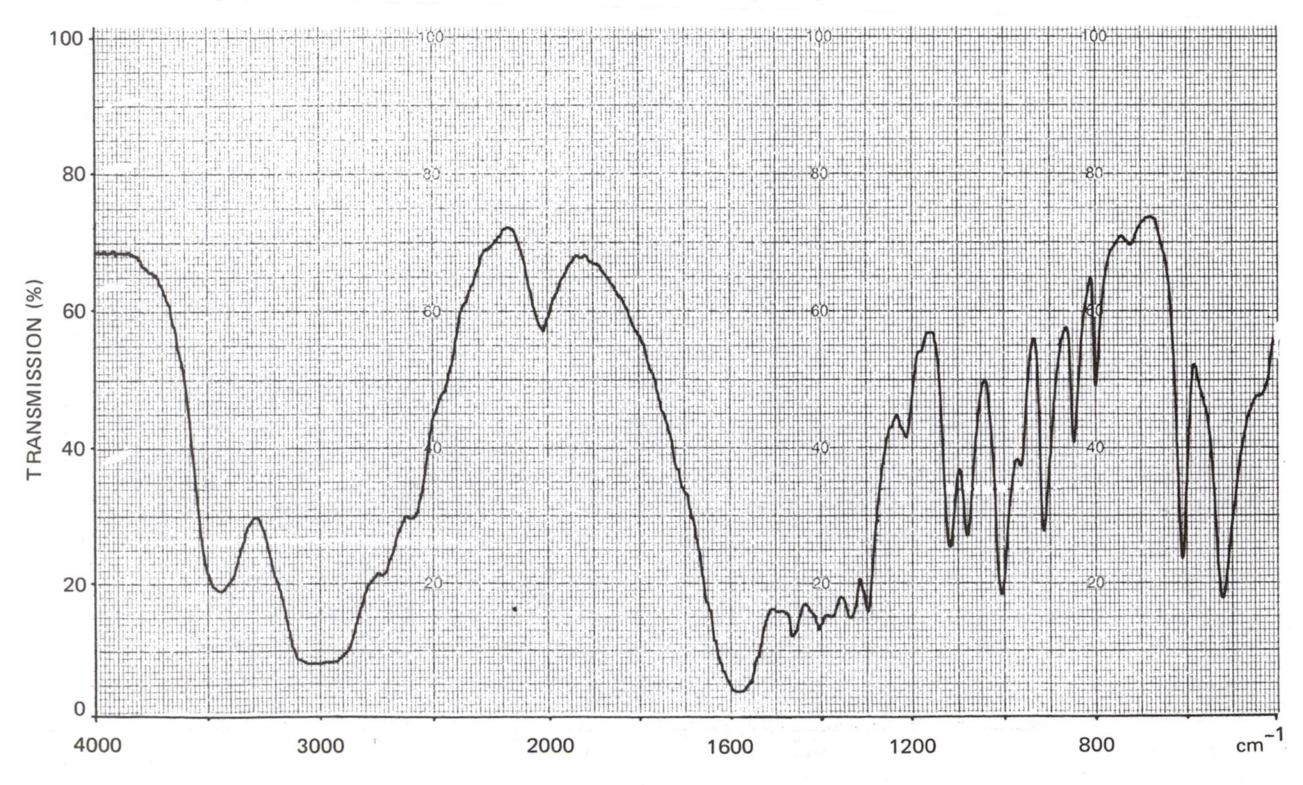

b. A thin film of a compound of molecular weight 153.

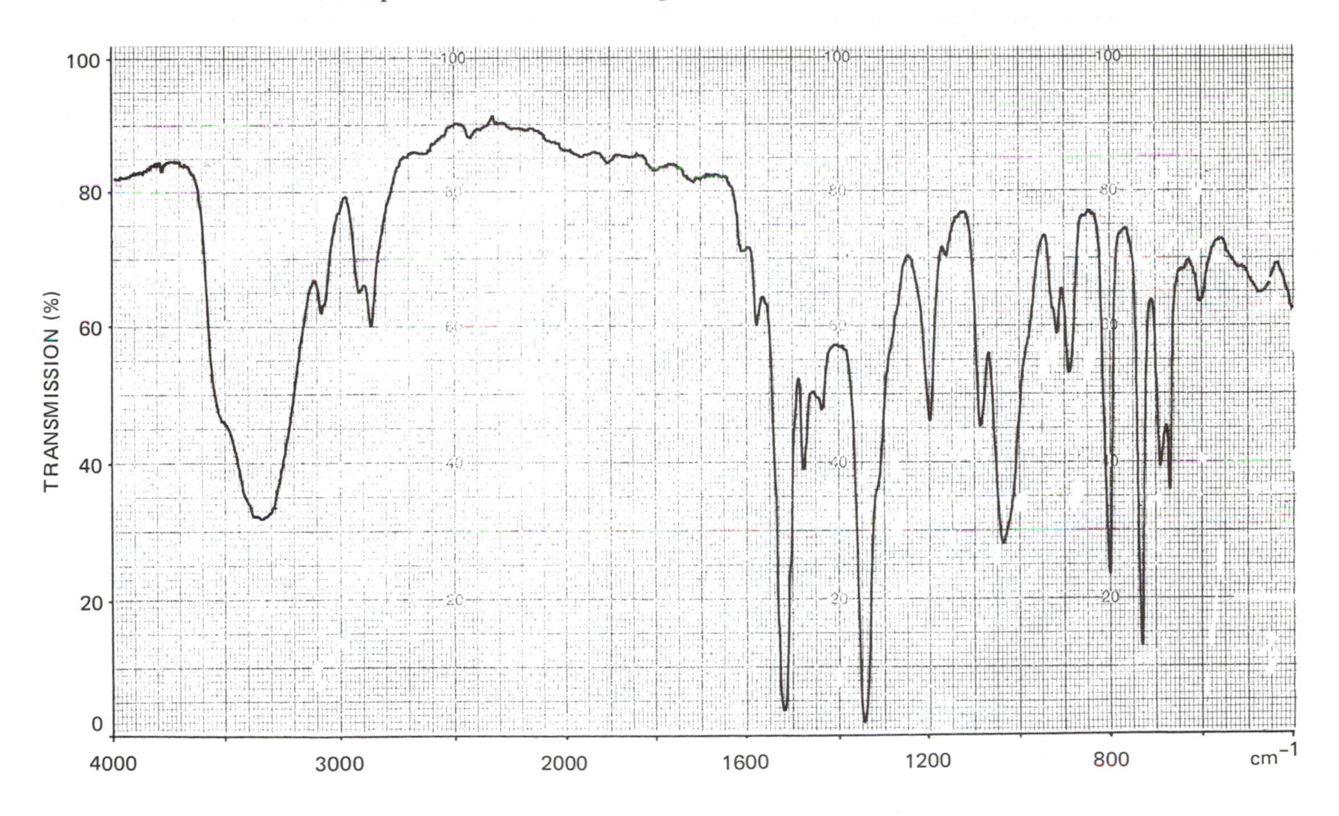

c. A Nujol mull of a compound of molecular weight 87.

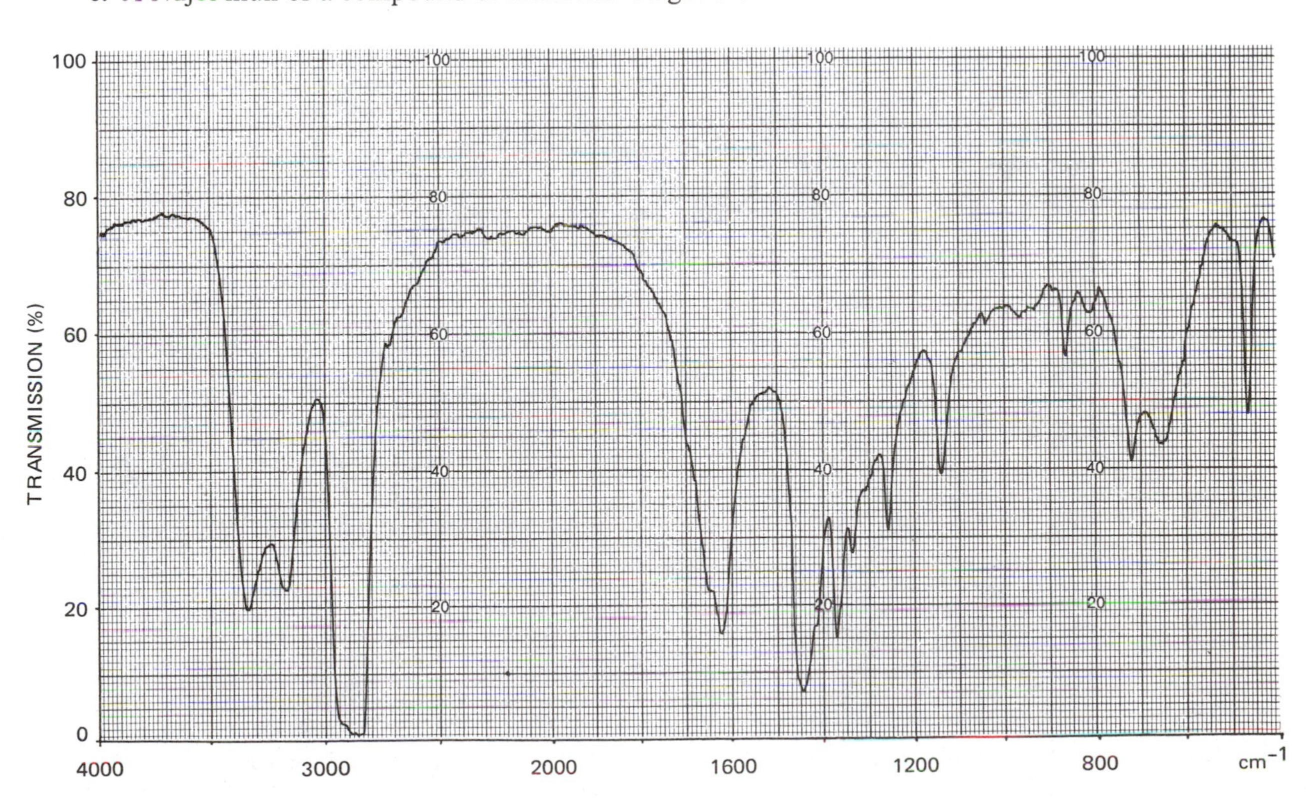

8-5 The spectrum below is of a compound that contains sulfur and oxygen in addition to carbon and hydrogen. The compound is a liquid (bp 158–160°C) of molecular weight 138. Suggest possible structures.

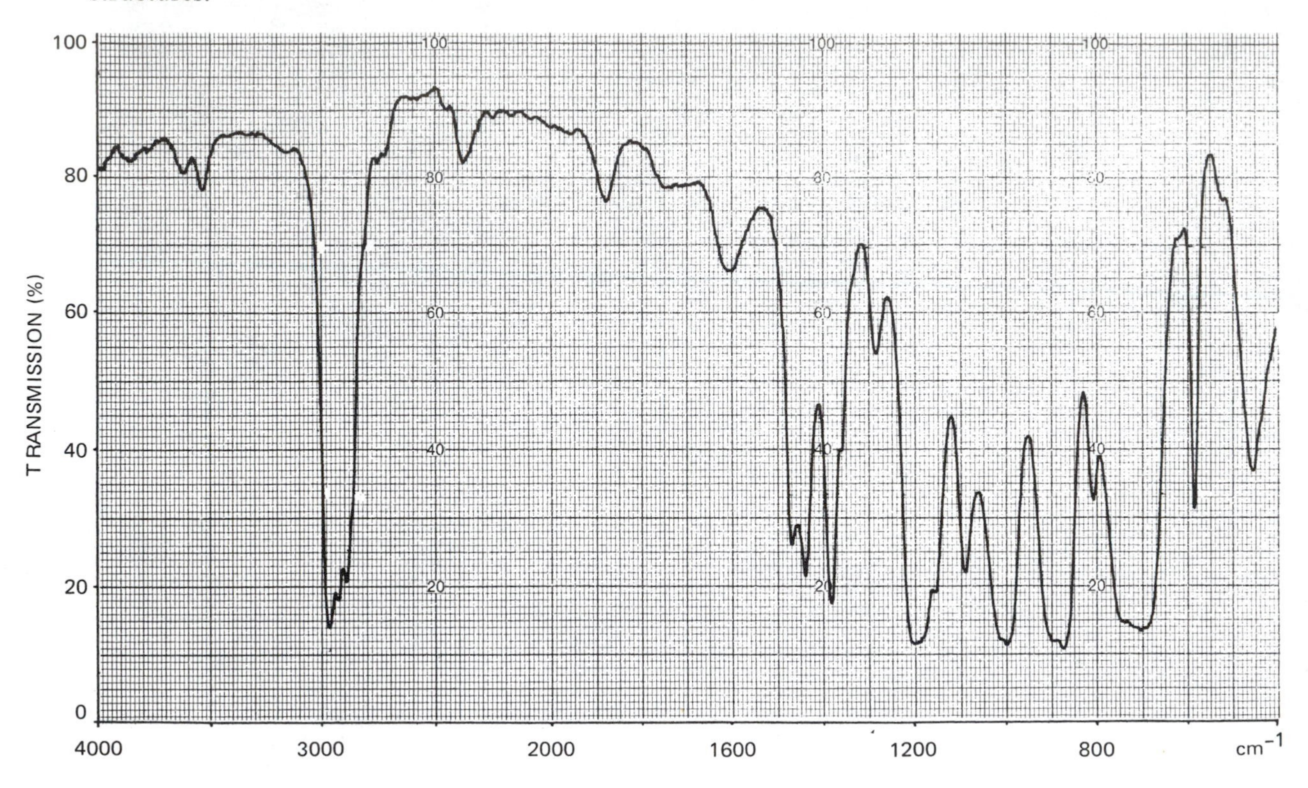

8-6 This problem illustrates that both infrared and Raman spectra may be needed to identify a structure. The compound, of molecular weight 71, contains oxygen, nitrogen, carbon, and hydrogen.

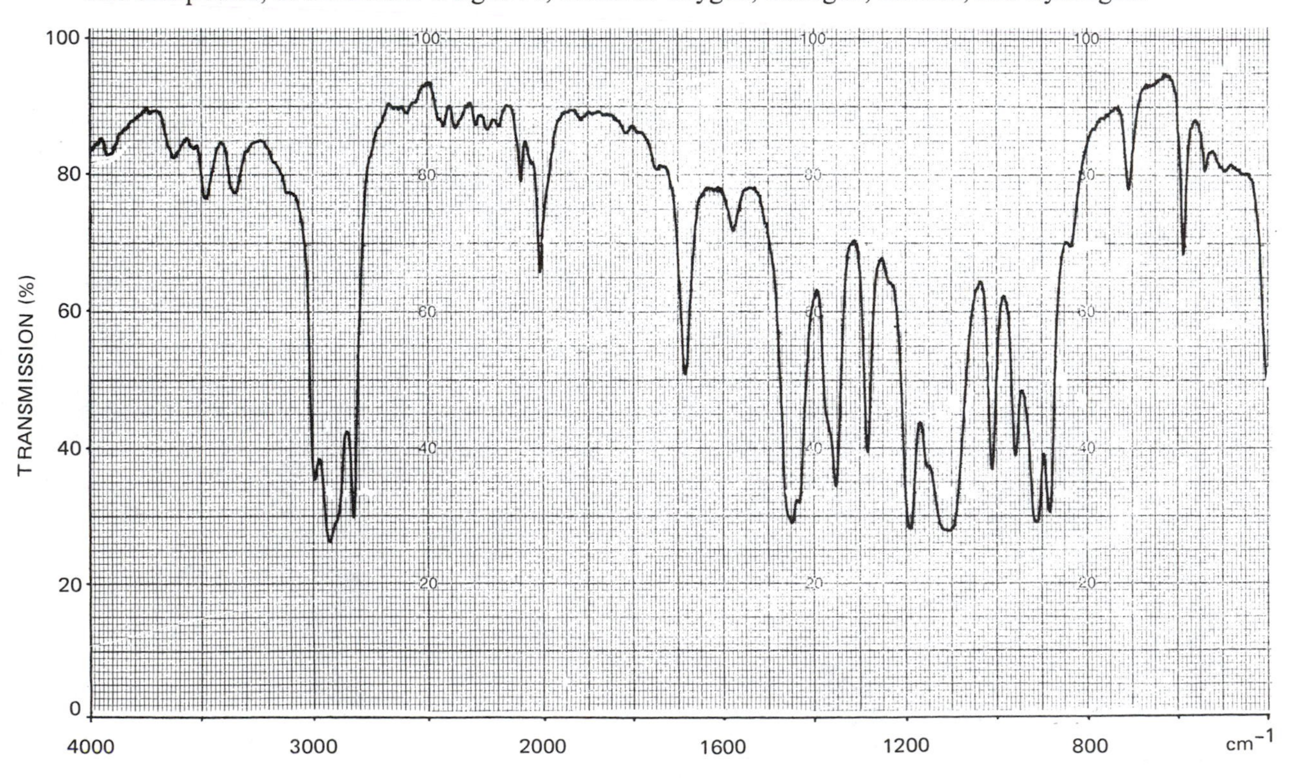

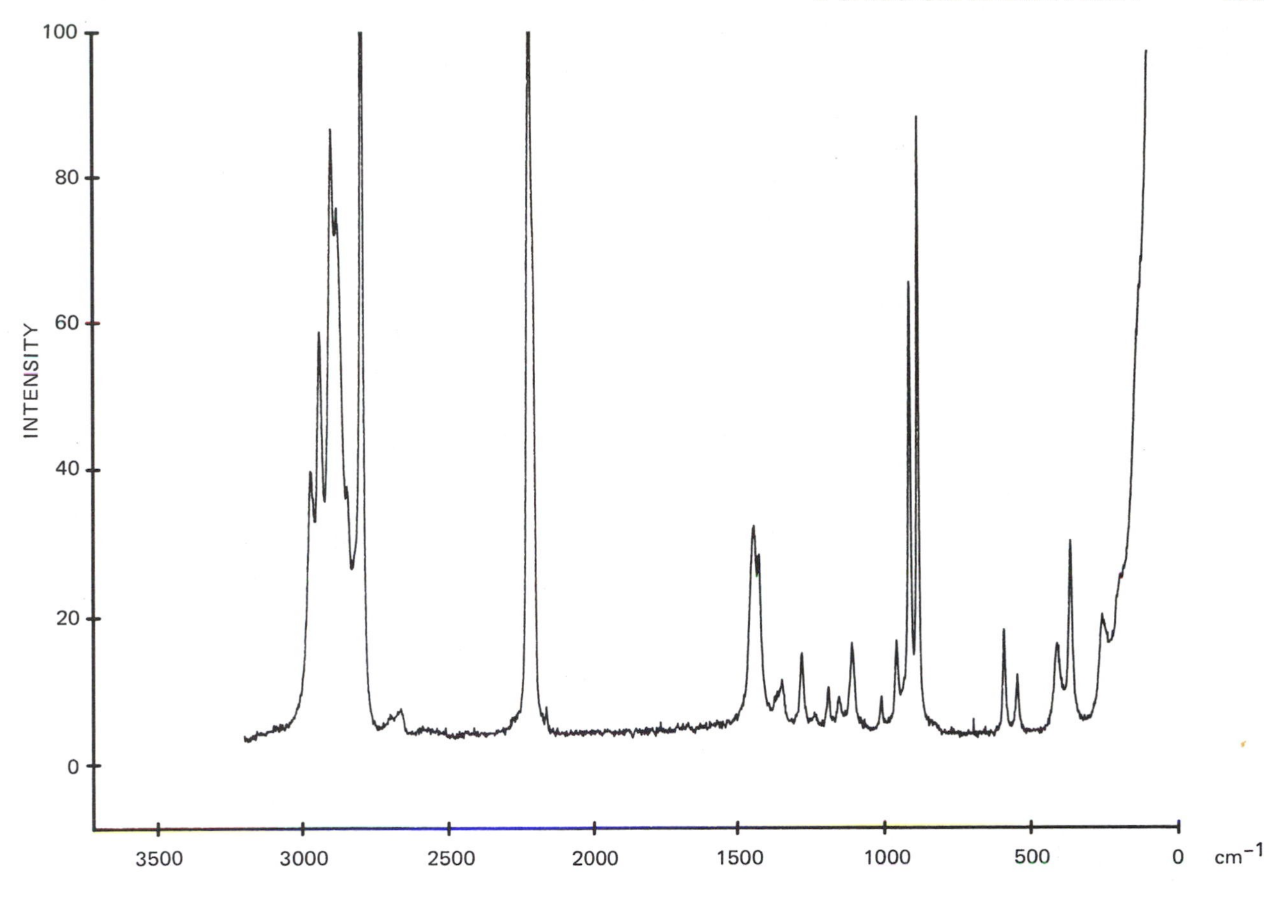

REFERENCES/BIBLIOGRAPHY

8-1 J. T. Bulmer and H. F. Shurvell, *J. Phys. Chem.,* 77, 256 (1973); *Can. J. Chem.*, **53**, 1251 (1975).

8-2 L. J. Bellamy, *Advances in Infrared Group Frequencies*, Methuen, London, 1968.

8-3 G. C. Pimentel and A. L. McClellan, *The Hydrogen Bond*, Freeman, San Francisco, 1960.

Sources of Interpreted Spectra and Problems

8-4 K. Nakanishi and P. H. Solomon, *Infrared Absorption Spectroscopy*, 2nd ed., Holden Day, San Francisco, 1977. The book contains 100 problems with detailed solutions.

8-5 H. A. Szymanski, *Interpreted Infrared Spectra*, 3 vols., Plenum, New York, 1964, 1966, 1967.

8-6 T. Cairns, *Spectroscopic Problems in Organic Chemistry*, Heyden, London, 1964.

8-7 A. J. Baker, T. Cairns, G. Eglinton, and F. J. Preston, *More Spectroscopic Problems in Organic Chemistry*, Heyden, London, 1967.

8-8 D. Steele, *The Interpretation of Vibrational Spectra*, Chapman and Hall, London, 1971. This book contains 26 infrared (and other) spectra of organic molecules with interpretation.

8-9 R. K. Smalley and B. J. Wakefield, "Infrared Spectroscopic Problems and Answers," in *An Introduction to Spectroscopic Methods for the Identification of Organic Compounds*, vol. 1, *Nuclear Magnetic Resonance and Infrared Spectroscopy*, F. Scheinmann (Ed.), Pergamon, Oxford, 1970. The chapter contains 14 problems followed by detailed answers.

8-10 R. T. Conley, *Infrared Spectroscopy*, 2nd ed., Allyn and Bacon, Boston, 1972, chs. 9 and 10.

Suggested Additional Reading

8-11 N. L. Alpert, W. E. Keiser, and H. A. Szymanski, *IR: Theory and Practice of Infrared Spectroscopy*, 2nd ed., Plenum, New York, 1970, ch. 5.

8-12 H. A. Szymanski, *A Systematic Approach to the Interpretation of Infrared Spectra*, Hertillon, Buffalo, NY, 1967.

8-13 R. R. Hill and D. A. E. Rendell, *The Interpretation of Infrared Spectra, A Programmed Introduction,* Heyden, London, 1975.

Appendix

II.A Using the Literature on Vibrational Spectra

We have seen that vibrational spectra are extremely useful for checking for the presence of a functional group in a compound. Identification of a compound can be obtained by direct comparison of its infrared spectrum with the spectra of known compounds. To accomplish this task, collections of spectra or references to spectra in the literature are needed. Published spectra, either in the literature or in collections, are currently available for about 200,000 compounds.

In the next two sections, sources of collections of spectra, sources of literature, and references to spectra are listed.

II.A-1 Collections of Spectra

In addition to the very extensive *Sadtler Standard Infrared Spectra* collection and other large collections of infrared spectra, there are several smaller collections of infrared, FT–IR and Raman spectra. Perhaps the most useful of these are *The Aldrich Library of Infrared Spectra* and *The Coblentz Society Desk Book of Infrared Spectra.*

II.A-1a Infrared Spectra

1. *Sadtler Standard Infrared Spectra*, Sadtler Research Laboratories, Inc., 3314 Spring Garden Street, Philadelphia, PA 19104. The main collection consists of over 50,000 prism or small grating spectra in loose leaf volumes, containing 1000 spectra per volume. Approximately 2000 new spectra are added each year. The format of these spectra is linear in wavelength (2–15 μm). The index to this collection consists of four sections: Chemical Classes, Alphabetical, Molecular Formula, and Numerical.
2. *The Aldrich Library of Infrared Spectra*, C. J. Pouchert (Ed.), Aldrich Chemical Company, Inc., P.O. Box 355, Milwaukee, WI 53201. A collection of 12,000 spectra in one volume. The format is linear in wavelength (2.5–16 μm) and there are 8 spectra to a page. There are alphabetical and molecular formula indices.
3. *Documentation of Molecular Spectra (DMS)*, Butterworth and Co. Ltd., London WC2, England. A collection of data cards with spectra, frequencies, and structural information. The spectra are presented in a linear wavenumber format. There are about 23,000 spectra in the collection.
4. *Selected Infrared Spectral Data*, American Petroleum Institute (API), Research Project 44, Department of Chemistry, Texas A&M University, College Station, TX 77843. A large collection of spectra that is continually updated. The presentation is usually linear in wavelength on the older entries in the collection, but linear in wavenumber on more recent spectra. This is the most extensive collection of spectra of petroleum hydrocarbons on high purity compounds. Also included are N and S compounds found in petroleum.
5. *Infrared Data Committee of Japan*, Sanyo Shuppan Boeki Co., Hoyu Bldg., 8, 2-Chome, Takara-cho, Chuo-ku, Tokyo, Japan. Approximately 14,000 spectra are available, the most recent ones being grating

spectra. These spectra are printed on edge punched cards, with structural and other data; similar to the DMS Collection.

6. *Coblentz Society Spectra*, P.O. Box 9952, Kirkwood, MO 63122. 10,000 spectra in volumes of 1000 spectra each, in notebook format and as 16 mm microfilm. Also *The Coblentz Society Desk Book of Infrared Spectra*, Clara Craver, 870 grating spectra grouped by chemical classes, with text; designed as a reference and teaching aid.
7. *Infrared Spectra of Selected Chemical Compounds*, R. Mecke and F. Langenbucher, Heyden & Son Ltd., Spectrum House, Alderton Crescent, London NW4, England. 1800 spectra in eight volumes selected as "most useful" compounds. The spectra are linear in wavelength and band tables are given in wavenumbers with intensities.

II.A-1b FT–IR Spectra

1. *The Aldrich Library of FT–IR Spectra*, C. J. Pouchert (Ed.), Aldrich Chemical Company, Inc., P.O. Box 355, Milwaukee, WI 53201. A two volume set containing 10,780 FT–IR spectra arranged 4 to a page, linear in wavenumber, with alphabetic and molecular formula indices.

II.A-1c Raman Spectra

1. *Sadtler Standard Raman Spectra*, Sadtler Research Laboratories, Inc., 3314 Spring Garden Street, Philadelphia, PA 19104. A collection of 4000 Raman spectra. Both parallel and perpendicular polarized spectra are presented together with the corresponding infrared spectrum, linear in wavenumber.
2. *Selected Raman Spectra Data*, American Petroleum Institute (API), Research Project 44. This compilation is produced in the same format as the API infrared spectra (Section II.A1-1a, No. 4). There are 500 Raman spectra obtained using mercury lamp excitation and 200 laser-excited spectra.
3. *Thermodynamic Research Center Data Project (TRC)*, Chemistry Department, Texas A&M University, College Station, TX, 77843. A subscription publication, formerly the Manufacturing Chemists' Association, similar to API described above, but with the emphasis on spectra of petrochemicals and other major industrial chemicals. 1179 infrared spectra and 113 Raman (30 laser-excited) spectra were published.
4. *Characteristic Raman Frequencies of Organic Compounds*, F. E. Dollish, W. G. Fateley, and F. F. Bentley, Wiley–Interscience, New York, 1973. This work includes 108 representative Raman spectra.
5. *Ramanspektren*, K. F. W. Kohlrausch, Heyden and Sons Ltd., London, 1972, reprinted from the original German edition. This work contains data on Raman spectra obtained using mercury arc excitation.

II.A-2

Sources of References to Published Spectra

In order to find a spectrum in a collection, or in the original literature, it is necessary to have a reference. The following list of indices are very useful. The ultimate source is, of course, the Chemical Abstracts Index. One should first look for the reference to a spectrum of a compound in the 5 or 10 year cumulative indices under Spectra, Infrared (or Raman). Then the abstract is looked up. The abstract gives the reference to the paper in which the spectrum was published. However, this can be a tedious procedure leading to many papers in which the complete spectrum may not be included.

1. *American Society for Testing and Materials (ASTM)*, distributed by Sadtler Research Laboratories, Inc., 3314 Spring Garden Street, Philadelphia, PA 19104. This source contains comprehensive indices for the infrared spectra in all of the general collections listed above in II.A-1a, plus infrared spectra abstracted from technical journals through 1972. There is a molecular formula list and a serial number list, each with names and references to published infrared spectra. There is also an alphabetical list of compound names, formulas, and references.
2. *Atlas of Spectral Data & Physical Constants for Organic Compounds*, 2nd ed., J. G. Grasselli and W. M.

Ritchey (Eds.), CRC Press Inc., 2000 Corporate Boulevard NW, Boca Raton, FL 33431, 1975. This index contains coded infrared spectra for 22,000 compounds. It lists strong bands in the infrared and includes Raman, UV, NMR, and mass spectral data when available.

3. *Infrared Absorption Spectra*, H. M. Hershenson, Indices for 1945–1957 and 1958–1962, Academic Press, New York, 1959 and 1964. A total of 36,000 references to ir absorption spectra. The indices are alphabetic and references are made to 66 journals and one collection of spectra.
4. *Current Literature Lists of IR, Raman, and Microwave Spectra*, Butterworth and Co. Ltd., London, 1967 onward. A list of references issued by the publishers of the DMS collection (II.A-1a, No. 3).

Appendix

II.B Computer Search Programs for Infrared Spectra

There are many search programs in use in the United States, Canada, Europe, and Japan. These are usually available through remote terminals in industrial or university libraries and laboratories. Manufacturers of FT-IR spectrometers often supply their own search program with a data base of spectra. This method is possible because an FT-IR spectrometer always comes with its own computer. A partial list of search programs available in North America is given in the table below.

II.B-1

Computer Search Programs

The use of a computer search program to identify a compound from its infrared spectrum depends on comparison of the spectral pattern of an unknown compound with the patterns of all spectra in a data base, which may contain as many as 100,000 spectra. Thus, the unknown spectrum must be coded into a digital form in exactly the same format that the spectra in the data base are coded.

Most computer search programs use only the central region of the infrared spectrum (the fingerprint region) 1850–650 cm^{-1}. The prominent bands in this region are coded, together with other information including which elements are present and which are absent and any functional groups known to be present. Regions where no absorption bands occur are also coded.

The search program produces a list of a small number of compounds (usually 10–20) with patterns that most closely match the coded pattern that the analyst has entered for the unknown. Each entry in the list is

Computer Search Programs for Infrared Spectra

Program	Source
1. FIRST 1 FIRST 2	D.N.A. Systems Inc. 2415 West Stewart Ave., Flint, MI 48504
2. IRIS	Sadtler Research Laboratories, Division of Bio-Rad Labs, 3316 Spring Garden Street, Philadelphia, PA 19104
3. ISIS	Triangle Universities Computation Center.
4. SPIR 2	National Research Council of Canada, Ottawa, Ontario, K1A OR6
5. SEARCH/PSU[a]	Perkin-Elmer Corp., Main Avenue, Norwalk, CT 06856
6. NICOS Search Algorithms	Nicolet Analytical Instruments, 5225 Verona Road Madison, WI 53711

[a]PSU is an additional feature of the Perkin-Elmer system that lists Probable Structural Units following the list of compounds found in the search.

accompanied by a reference to the original source of the spectrum and a correlation coefficient (or score) between 0 and 100, which indicates the closeness of the match. A score of 100 means an exact fit with the pattern coded for the unknown, while 0 means no coincidences at all between the coded spectrum and any spectrum in the data base.

It is often possible to eliminate several of the compounds listed in the computer printout, on the basis of other knowledge of the sample. For example, the elemental analysis, mass spectrum, and NMR or UV–visible spectrum may enable the analyst to exclude certain compounds or functional groups. The analyst must then look up the spectra of the remaining compounds, either in the original references, or in a collection of infrared spectra (see Appendix II.A). Final identification can then be made.

As an example of the use of a typical search program, we consider the spectrum shown in Figure II.B-1. The program used is SPIR 2, which is based on FIRST-1 written by D. S. Erley.

The spectrum has bands at *1742, 1437, *1369, *1242, 1048, and 848 cm^{-1}. (The asterisks (*) indicate that special importance is attached to these bands.) These data are entered, followed by entry of the regions in which no bands are found: 1818–1785, 1695–1470, 1333–1316, 1205–1075, 1020–990, 970–860, and 820–667 cm^{-1}. The SPIR program can also make use of other information such as functional groups known to be present or absent. In our example, a carbonyl group is assumed to be present, but nothing is known about the presence of elements other than C, H, and O.

The program then compares the coded information on the unknown with the spectra in the data base and produces the output shown in Figure II.B-2. The first four entries are methyl acetate with scores ranging from 81 to 92. A comparison of the spectrum of Figure II.B-1 with that of methyl acetate, spectrum 359E in the Aldrich Library (see Section II.A-1a, No. 2), confirms the identity of the unknown.

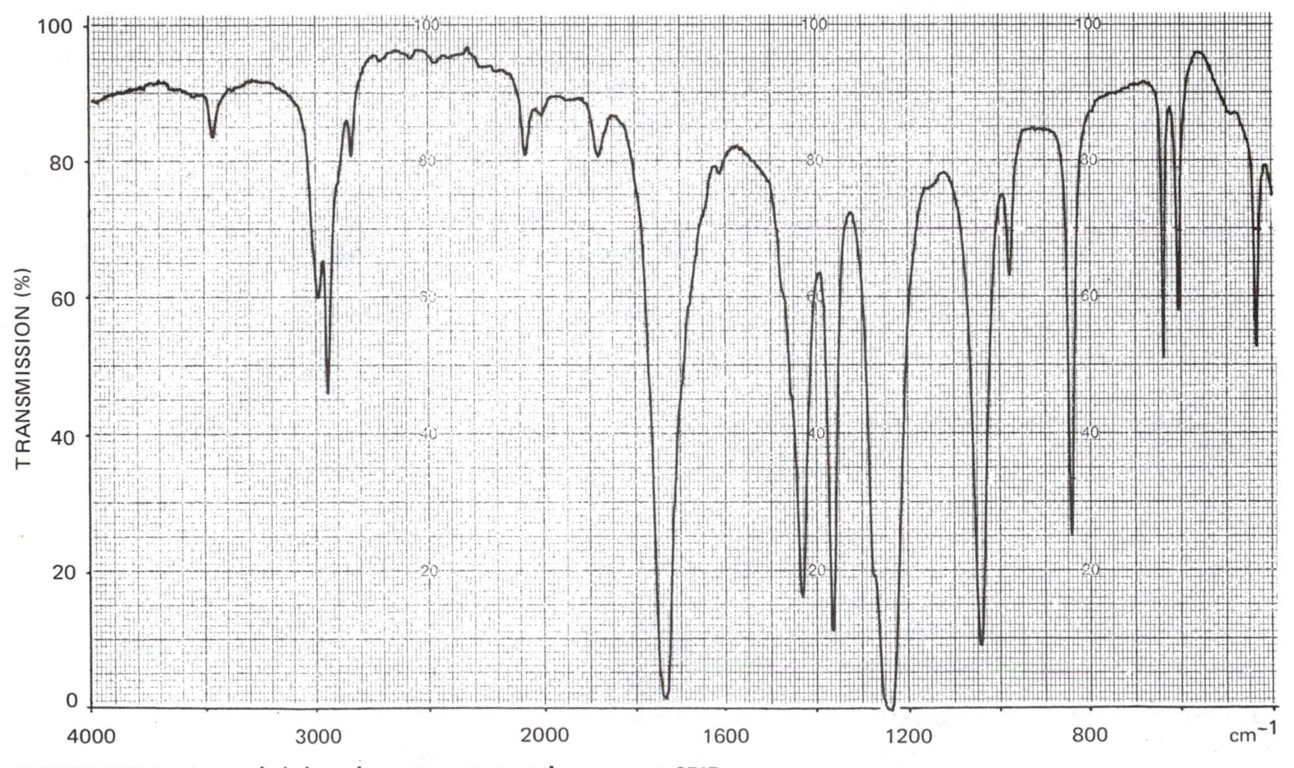

FIGURE II.B-1 A sample infrared spectrum to test the program SPIR.

```
SEARCH RESULTS FOR: TEST RUN
SEARCH FILES       : SPIR MASTER FILE (ASTM)
DATE/TIME          : JAN.  07 1985 (15.09.18)


SERIAL #        COMPOUND NAME                                     PROPERTIES PRESENT
  WT            & CHEMICAL FORMULA                                & BANDS PRESENT (WAVENUMBERS)
------------------------------------------------------------------------------------------------
603JA      METHYL ACETATE                                        LIQ,O,ACYC,C=O
   +92       C3 H6 O2                                            1754(14) 1428(9) 1370(9)
                                                                 1234(7) 1053(5) 840(3)


L695CA     METHYL ACETATE                                        LIQ,O,ACYC,C=O
   +84       C3 H6 O2                                            1754(14) 1449(10) 1370(9)
                                                                 1234(7) 1053(5) 840(3)


8460EA     ACETATE, METHYL-                                      LIQ,SLN,O,ACYC,C=O
   +84       C3 H6 O2                                            1754(14) 1428(9) 1370(9)
                                                                 1234(7) 1042(5) 840(3)


204HA      ACETIC ACID, METHYL ESTER                             LIQ,O,AROM,C=O
   +81       C3 H6 O2                                            1754(14) 1449(10) 1370(9)
                                                                 1250(7) 1053(5) 980(4) 847(3)


10743JA    DECANE, 1,2-DIACETOXY-                                LIQ,O,ACYC,C=O
   +80       C14 H26 O4                                          1754(14) 1449(10) 1370(9)
                                                                 1234(7) 1053(5)


More? (press (CR) for more or 'n' to stop)
:
353JA      ETHYL ACETATE                                         LIQ,O,ACYC,C=O
   +80       C4 H8 O2                                            1754(14) 1449(10) 1370(9)
                                                                 1234(7) 1053(5)


13225EA    ACETIN, MONO-                                         SOL,O,ACYC,C=O,OH
   +78       C5 H10 O4                                           1754(14) 1370(9) 1234(7)
                                                                 1053(5) 980(4) 934(3) 847(3)


3278EA     METHYL ACETATE                                        LIQ,SLN,O,ACYC,C=O
   +77       C3 H6 O2                                            1754(14) 1428(9) 1370(9)
                                                                 1250(7) 1042(5) 840(3)


2228CA     ACETIC ACID,METHYL ESTER                              LIQ,O,ACYC,C=O
   +77       C3 H6 O2                                            1754(14) 1449(10) 1370(9)
                                                                 1250(7) 1053(5) 840(3)


5407EA     MANNOPYRANOSIDE, METHYL TETRA-O-                      SOL,SLN,O,ACYC,HET,C=O,COC
   +75     ACETYL-BETA-                                   -DI-   1754(14) 1370(9) 1234(7)
             C15 H22 O10                                         1053(5)

***  SPIR SEARCH COMPLETE


---------------------------------------------
Choose the next activity  : QUIT


***  SPIR SIGNING OFF    15:11:57
```

FIGURE II.B-2 The computer output from the program SPIR applied to the spectrum of Figure II.B-1.

Part III

ELECTRONIC ABSORPTION SPECTROSCOPY: ULTRAVIOLET–VISIBLE AND CHIROPTICAL

9

Characteristics of Electronic Spectra

9-1

Electronic Transitions and Chromophores

All organic compounds absorb light in the ultraviolet (UV) region of the electromagnetic spectrum, and some absorb it in the visible region as well. Absorption of UV or visible light by molecules occurs only when the energy of incident radiation is the same as that of a possible electronic transition in the molecules involved (quantization of energy). Such absorption of energy is termed *electronic excitation* and is typically associated with the transition from the molecular ground state to a higher energy electronic excited state. These excitations are usually viewed, approximately, in terms of the promotion of a single electron from an occupied to an unoccupied molecular orbital (Figure 9-1). In a given molecule many different electronic transitions are possible, and those of greatest importance in organic chemistry usually involve the promotion of an electron from the highest occupied bonding or nonbonding molecular orbital (HOMO) to the lowest unoccupied molecular orbital (LUMO). The electronically excited state thus formed may decay unimolecularly back to the ground state *photophysically* by emitting energy of fluorescence (from an excited singlet state) or phosphorescence (from an excited triplet state). Alternatively, it may decay *photochemically* to a different ground state (thus a different structure).

The *UV–vis spectrum* typically represents the absorption of light as a plot (Figure 9-2) of energy (as wavelength, λ, in nanometers, from $E = hc/\lambda$) versus the intensity of absorption (as absorbance, A, or molar extinction coefficient, ϵ, where ϵ is a rough measure of the transition probability).

A difference spectrum for absorption may be obtained if the excitation light is decomposed into its left (L) and right (R) circularly polarized components and the *difference* in absorption, $\Delta\epsilon = (\epsilon_L - \epsilon_R)$, of each component is recorded at each λ of the absorption band (Figure 9-3). This type of spectrum with $\Delta\epsilon_\lambda$ either <0 or >0 is called a circular dichroism (CD) spectrum and is found with chiral molecules. For achiral molecules and racemic mixtures $\epsilon_L = \epsilon_R$, and in these cases $\Delta\epsilon_\lambda = 0$ for each electronic transition in an isotropic medium.

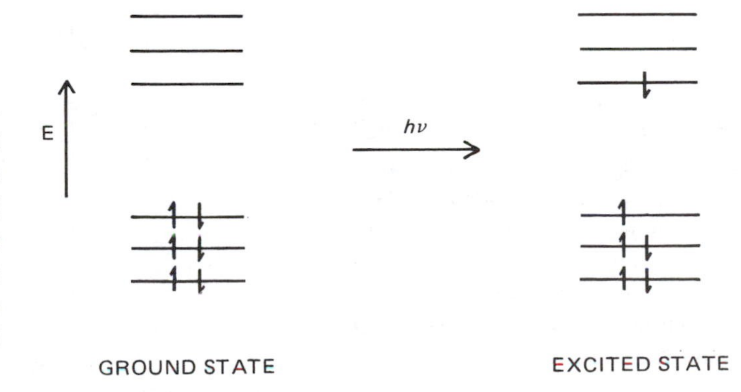

FIGURE 9-1 Idealized representation on a potential energy scale of occupied and unoccupied molecular orbitals in the electronic ground state (left) and electronic configuration of an excited state arising by promotion of an electron from the highest occupied molecular orbital to the lowest unoccupied molecular orbital (right). The electrons and their relative spin orientations are represented by small arrows.

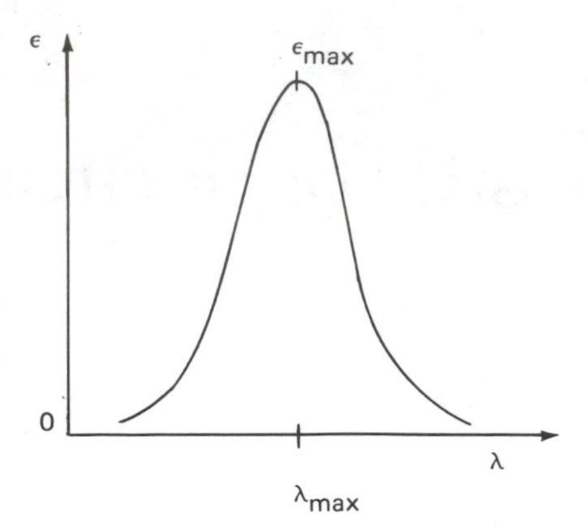

FIGURE 9-2 Absorption of UV–vis light, through an electronic transition, recorded as molar extinction coefficient, ε, (or absorbance, *A*) versus wavelength, λ, in nanometers. Typically, UV–vis spectral data are reported as the maximum value of ε, ϵ_{max}, at the corresponding wavelength, λ_{max}.

The energy states for the carbonyl chromophore of, e.g., acetone, are shown in Figure 9-4. Note that both singlet (S) and triplet (T) states are possible. Selection rules permit S→S and T→T processes but not S→T and T→S. Ground states are usually singlets; thus, most excitations are to singlet excited states. Triplet excited states are usually formed from singlet excited states by intersystem crossing rather than by direct excitation.

One can thus measure electronic absorption and emission from molecules. In this chapter we will be concerned mainly with absorption processes.

UV–vis (called ordinary absorption) spectroscopy and circular dichroism (called chiroptical absorption) spectroscopy arise from the same photophysical process: typically the promotion of an electron from a ground state orbital to an excited state orbital. As such they are intimately related. Both types of spectroscopy reveal (1) the energy (usually reported as a wavelength, λ_{max}, in organic chemistry) associated with the transition from the ground state to an electronically excited state and (2) the probability or intensity of the transition (approximated by ϵ_{max} in UV–vis and $\Delta\epsilon_{max}$ in CD but in each case better represented by the integrated areas under the curves). In CD spectroscopy $\Delta\epsilon_{max}$, unlike the UV–vis ϵ_{max}, is a signed quantity, and the sign (+ or −) is related to the absolute stereochemistry of the molecule being measured. Although electronic transitions arise between, e.g., ground and excited states of the *entire* molecule, most of the action can usually be assigned to parts of the molecule where electrons are loosely bound (*chromophores*). One speaks thus of an electronic transition in a chromophore, which in organic molecules is typically a functional group, e.g., carbonyl, carbon–carbon double bond, or aromatic ring. Representative chromophores are

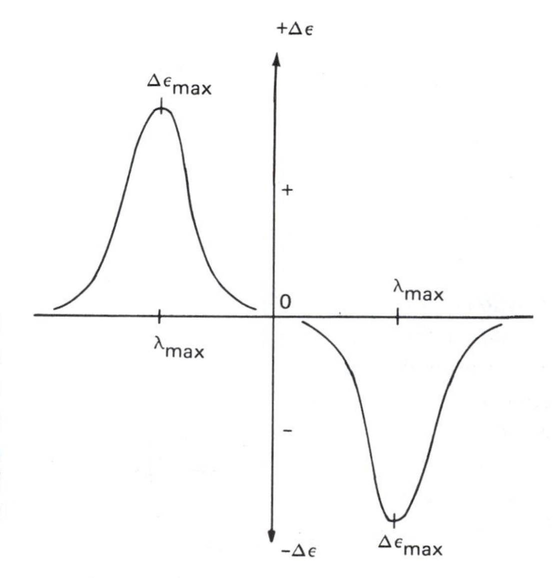

FIGURE 9-3 Circular dichroism (CD) recorded through an electronic transition, e.g., that of Figure 9-2, as (vertical axis) Δε versus (horizontal axis) wavelength, λ, in nanometers. Note that Δε values may be (+) or (−). The left half of the figure corresponds to a (+) CD Cotton effect, the right half to a (−) CD Cotton effect. These two curves might correspond to the CD spectra of enantiomers.

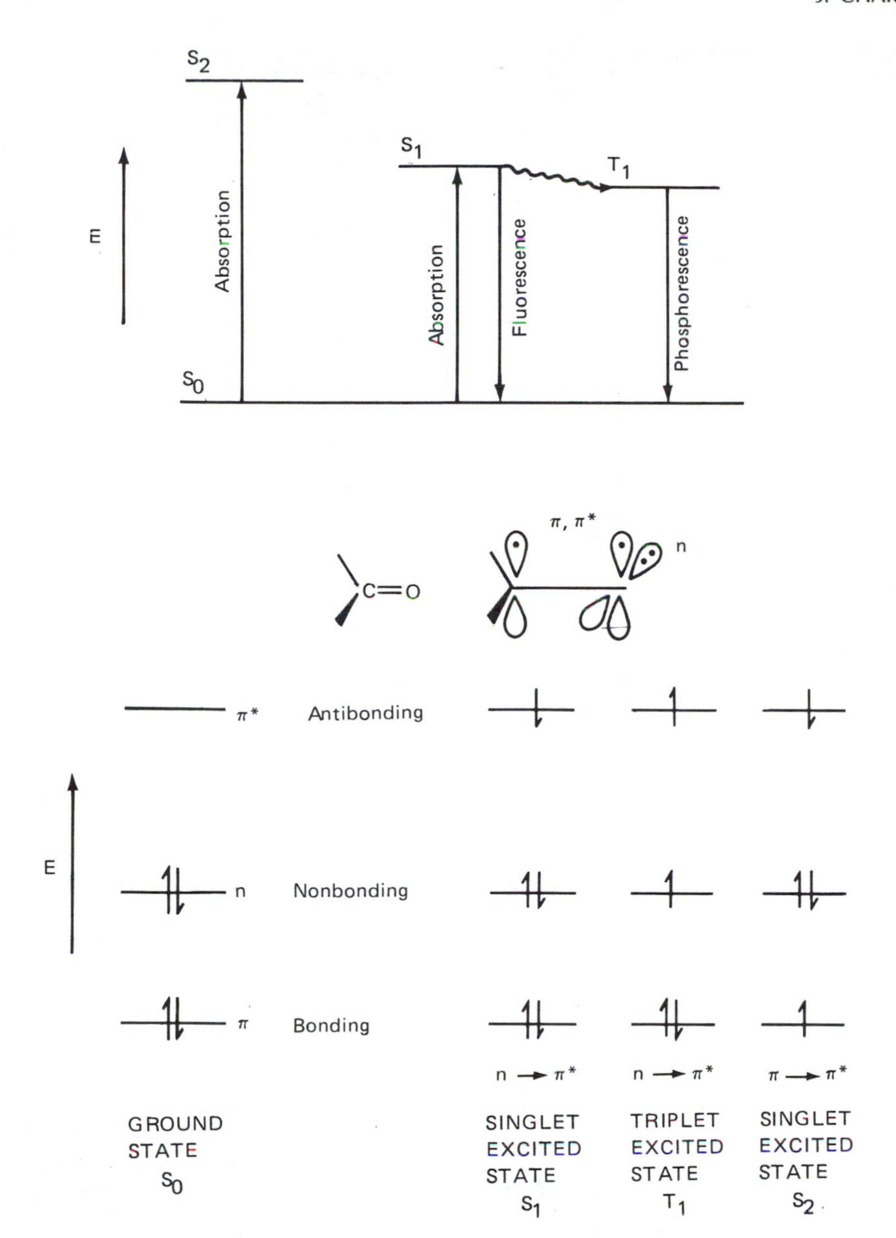

FIGURE 9-4 (Upper) Energy diagram for electronic excitation and decay processes in the (ketone) carbonyl chromophore. (Lower) Diagram of selected electronic molecular orbital energies for the isolated carbonyl chromophore of, e.g., acetone or di-*tert*-butyl ketone, showing the ground state and excited state configurations. Single states (S) all have electron spins paired, triplet states (T) have an electron spin inverted. Note that the n orbital containing two electrons is orthogonal to the π and π^* orbitals.

shown in Table 9-1, along with λ_{max} and ϵ_{max}. And a representative UV spectrum for the lowest energy electronic transition of di-*tert*-butyl ketone is shown in Figure 9-5.

Molecules continually vibrate and rotate about the axes between atoms. Since these vibrational and rotational motions are quantized, absorption of light also causes transitions to higher vibrational and rotational energy levels. Thus all electronic transitions are necessarily accompanied by corresponding transitions in the vibrational and rotational energy levels of the molecule. The total excitation energy ΔE is a sum of three terms, electronic, vibrational, and rotational (eq. 9-1). The energy changes involved in the three

$$\Delta E = \Delta E_{elec} + \Delta E_{vib} + \Delta E_{rot} \tag{9-1}$$

TABLE 9-1 Electronic Absorption Data for Isolated Chromophores

Chromophore	Example	Solvent	λ_{max} (nm)	ϵ (liter mol^{-1} cm^{-1})
C=C	1-hexene	heptane	180	12,500
—C≡C—	1-butyne	vapor	172	4,500
C=O	acetaldehyde	vapor	289	12.5
			182	10,000
	acetone	cyclohexane	275	22
			190	1,000
	camphor	hexane	295	14
—COOH	acetic acid	ethanol	204	41
—COCl	acetyl chloride	heptane	240	34
—COOR	ethyl acetate	water	204	60
$—CONH_2$	acetamide	methanol	205	160
$—NO_2$	nitromethane	hexane	279	15.8
			202	4,400
$=\overset{+}{N}=\overset{-}{N}$	diazomethane	diethyl ether	417	7
—N=N—	*trans*-azomethane	water	343	25
>C=N—	$C_2H_5CH=NC_4H_9$	isooctane	238	200
	benzene	water	254	205
			203.5	7,400
	toluene	water	261	225
			206.5	7,000

From J. B. Lambert, H. F. Shurvell, L. Verbit, R. G. Cooks, and G. H. Stout, *Organic Structural Analysis*, Macmillan, New York, 1976.

terms of eq. 9-1 decrease on going from left to right as written. Rotational energy changes in a molecule may be observed as pure transitions only in the microwave spectral region. Changes in vibrational levels require more energy and are observed in the infrared region (Part II). In this Part we focus on the chemical information obtainable by the measurement of the relatively much more energetic electronic transitions in the ultraviolet and visible regions.

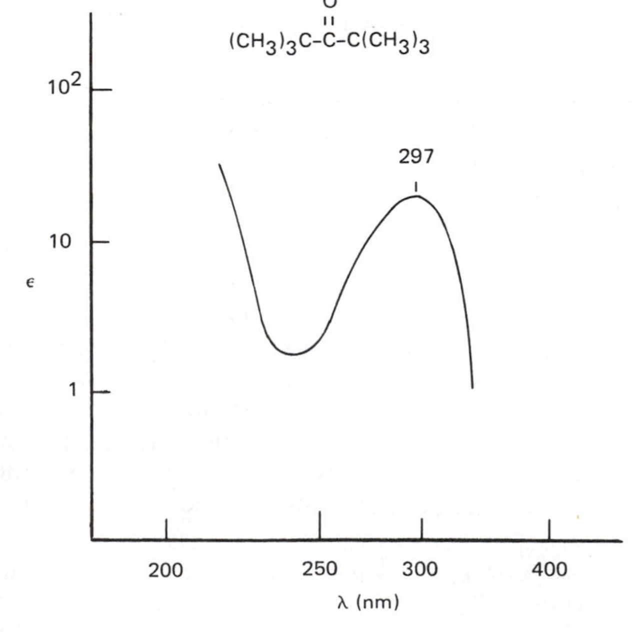

FIGURE 9-5 The UV spectrum of di-*tert*-butyl ketone, 0.025 M in cyclohexane. The lowest energy electronic transition is centered near 297 nm with an extinction coefficient, ϵ_{max}, of ~20. This absorption corresponds to the n→π* transition (Figure 9-1). The chromophore of di-*tert*-butyl ketone is the carbonyl, and there are other, higher energy (shorter λ) and more intense (greater ε) electronic transitions, e.g., π→π* (Figure 9-1), associated with the ketone carbonyl chromophore, as noted by the increasing ε values as λ decreases.

9-2

The Shape of Absorption Curves; The Franck-Condon Principle

Absorption of UV–vis light is typically recorded as broad absorption maxima (e.g., Figure 9-5) and not single, sharp lines representing the absorption in an extremely narrow energy range. The absorption curves are broadened because the electronic levels have vibrational levels on them.

For simplicity, let us look at the ground and excited electronic states of a diatomic molecule. The case for a polyatomic molecule is similar but more difficult to visualize since it requires the superposition of many two-dimensional potential energy surfaces.

In the more common case, the bond strength in the excited electronic state is less than that in the ground state, and equilibrium internuclear distance is longer than in the ground state, giving the potential curves shown in Figure 9-6.

If the vibrational frequency is fairly high, essentially all the molecules exist in their ground vibrational state. Excitation can occur to any of the excited state vibrational levels so that the absorption due to the electronic transition consists, in theory, of a large number of lines. (Excitation also occurs to various excited state rotational levels, but the rotational fine structure is almost never resolved. It contributes only a bandwidth to each vibrational subband.) In practice, for most organic molecules the lines overlap so that a continuous band is observed. Hence the shape of an absorption band may be considered to be determined by the spacing of the vibrational levels and by the distribution of the total band intensity over the vibrational subbands. The intensity distribution is determined by the *Franck–Condon principle*, which states that *nuclear motion may be considered negligible during the time required for an electronic excitation.* For example, the time required for an electron to circle a hydrogen nucleus can be calculated from Bohr's model to be about 10^{-16} s, whereas a typical molecular vibration is about 10^{-13} s, about a thousand times longer. Another statement of the Franck–Condon principle based on classical mechanics is that the most probable vibrational component of an electronic transition is one that involves no change in the position of the nuclei, a so-called *vertical transition* represented by the vertical arrow in Figure 9-6. The most probable transition is to the excited $\nu = 3$ state. This state has a maximum at the same internuclear distance (r) as that corresponding to the starting point of the transition. Figure 9-7 shows the vibrational–electronic (vibronic) spectrum corresponding to Figure 9-6, with the 0–3 band (from $\nu = 0$ in the ground state to $\nu = 3$ in the excited state) the most intense one. Note that the other transitions, including the 0–0 band, have significant probabilities. This result is not necessarily from *nonvertical* transitions but may be from the fact that even in the ground electronic state (zeroth vibrational level), the internuclear distance is described by a probability distribution (Figure 9-6). Therefore, transitions may originate over a range of r values so that more than one band originating from $\nu = 0$ may be observed.

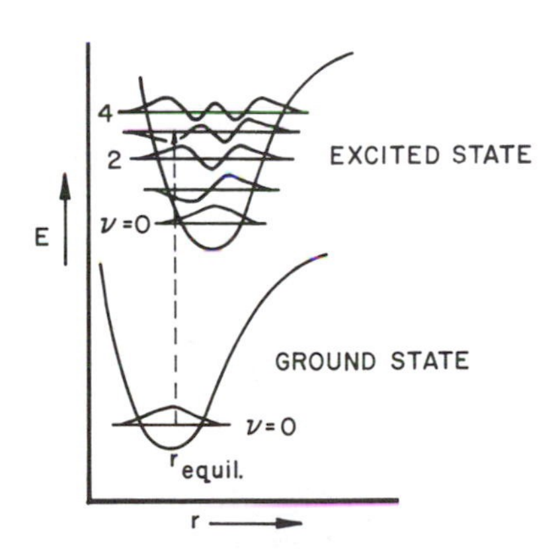

FIGURE 9-6 Potential energy diagram for a diatomic molecule illustrating Franck-Condon excitation. The equilibrium separation is longer in the excited than in the ground state.

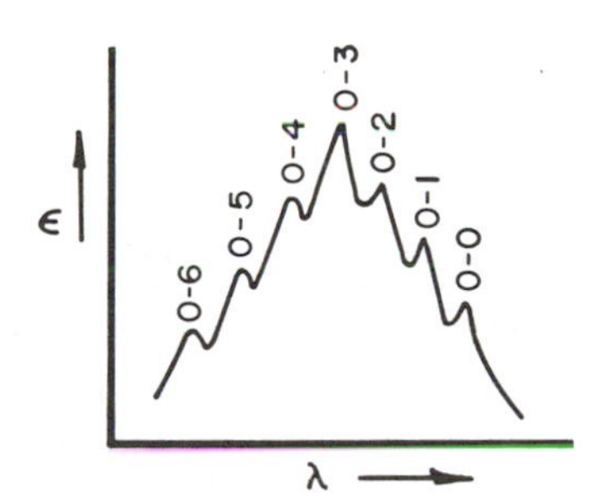

FIGURE 9-7 Intensity distribution among vibronic bands as determined by the *Franck-Condon principle.*

Sometimes, on raising the temperature, the vibrational structure of a band is lost. This band broadening is due to the population of several ground vibrational states at higher temperature so that a larger number of possible vibrational transitions can occur upon electronic excitation. Featureless or broad bands are also observed at ambient temperatures, usually in solution spectra where solute–solvent vibrational interactions become important.

9-3

Measuring and Reporting Spectral Data

As indicated earlier, in organic chemistry, electronic transitions are usually characterized by a wavelength (λ_{max}) and by an intensity (ϵ_{max} in UV–vis or $\Delta\epsilon_{max}$ in CD).

9-3a Wavelength (λ_{max})

Electromagnetic radiation may be described by the wavelength λ between its waves, by the frequency ν (s^{-1}), or by the wavenumber $\bar{\nu}$ (cm^{-1}). As mentioned in Part II, these quantities are related to each other by eq. 9-2,

$$c = \lambda\nu \tag{9-2}$$

in which c is the velocity of light ($3 \times 10^8 m\ s^{-1}$), and by eq. 9-3. Commonly used wavelength units in ultra-

$$\bar{\nu} = \frac{1}{\lambda} \tag{9-3}$$

violet and visible regions are either nanometers (nm; 1 nm = 10^{-9} m) or Ångstrom units (Å; 1 Å = 10^{-10} m).

According to the Planck equation (9-4), frequency is directly proportional to energy. However, it is

$$E = h\nu \tag{9-4}$$

common practice to use units of either wavelength or wavenumber. Table 9-2 lists the units commonly used for λ, ν, and $\bar{\nu}$, and Table 9-3 gives some useful conversion factors. Equation 9-5 is convenient for calcula-

$$E = \frac{hc}{\lambda} = \frac{28{,}636}{\lambda}\ \text{kcal mol}^{-1} = \frac{119{,}809}{\lambda}\ \text{kJ mol}^{-1} \quad (\text{for } \lambda \text{ in nm}) \tag{9-5}$$

tion of energies in the familiar units of kcal mol^{-1}. Hence, light of 300 nm wavelength corresponds to an energy of 95.4 kcal mol^{-1} or 399 kJ mol^{-1}, depending on the units of h. Combination of eqs. 9-2, 9-3, and 9-4 leads to eqs. 9-6 and 9-7.

$$E = hc\bar{\nu} \tag{9-6}$$

TABLE 9-2 Definitions of Terms and Equations

Quantity	Equation	Unit	Dimensions
Wavelength, λ	—	nanometer, nm Ångstrom, Å	1 nm = 10^{-9} m 1 Å = 10^{-10} m
Wavenumber, $\bar{\nu}$	$\nu = \frac{1}{\lambda}$	reciprocal cm, cm^{-1}	the wavenumber is the reciprocal of the wavelength in cm
Frequency, ν	$\nu = \frac{c}{\lambda}$	hertz, Hz, or s^{-1}	cycles per second
Energy	$E = h\nu = \frac{hc}{\lambda} = hc\bar{\nu}$	depends on the units of h	—

TABLE 9-3 Useful Conversion Factors

cm^{-1}	Hz	kcal mol^{-1}	kJ mol^{-1}
1	3.00×10^{10}	2.86×10^{-3}	1.20×10^{-2}
3.33×10^{-11}	1	9.53×10^{-14}	3.99×10^{-13}
3.50×10^{2}	1.05×10^{13}	1	4.18

$$E = 28.635 \times 10^{-4} \times \bar{\nu} \text{ kcal mol}^{-1}$$
$$= 119.8 \times 10^{-4} \times \bar{\nu} \text{ kJ mol}^{-1} \quad (\text{for } \bar{\nu} \text{ in cm}^{-1}) \tag{9-7}$$

Wavenumbers are thus directly proportional to energy so that a given number of reciprocal centimeters (cm^{-1}) represents the same energy anywhere in the electromagnetic spectrum. For example, a shift of λ_{max} of 700 cm^{-1} anywhere in the spectrum corresponds to 1.95 kcal mol^{-1}. On the other hand, wavelength is inversely proportional to energy and thus the relationship is not linear. As an example, an energy change of 1.95 kcal mol^{-1} at 200 nm corresponds to a shift of 2.7 nm, but the same energy change at 800 nm corresponds to a shift of approximately 4.4 nm.

At the lower end of the visible spectrum, below 400 nm, is the UV region. It is convenient to divide the UV into two parts, the near UV, 190–400 nm (53,000–25,000 cm^{-1}), and the far or vacuum UV, below 190 nm (>53,000 cm^{-1}). This seemingly arbitrary division is due mainly to the fact that atmospheric oxygen begins to absorb around 190 nm. Oxygen must be removed from the spectrophotometer, either by using a vacuum instrument or by vigorous purging with nitrogen. The UV–vis spectroscopic regions and some associated energies are given in Figure 9-8.

9-3b ϵ_{max}, $\Delta\epsilon_{max}$, and the Beer-Bouger-Lambert Law

The laws of Lambert, Bouger, and Beer state that at a given wavelength the proportion of light absorbed by a transparent medium is independent of the intensity of the incident light and is proportional to the number of absorbing molecules through which the light passes. According to eq. 9-8, I_0 is the intensity of incident light,

$$I = I_0 10^{-alc} \quad \text{or} \quad \log \frac{I_0}{I} = alc \tag{9-8}$$

I the intensity of transmitted light, a the absorptivity, l the pathlength (cm), and c the concentration. Since the absorbance A is the quantity actually measured, eq. 9-8 is rewritten as eq. 9-9. When concentration is in units

$$\log \frac{I_0}{I} = A = alc \tag{9-9}$$

of mol $liter^{-1}$, the molar absorption coefficient (molar a) is denoted by ϵ, the molar extinction coefficient or

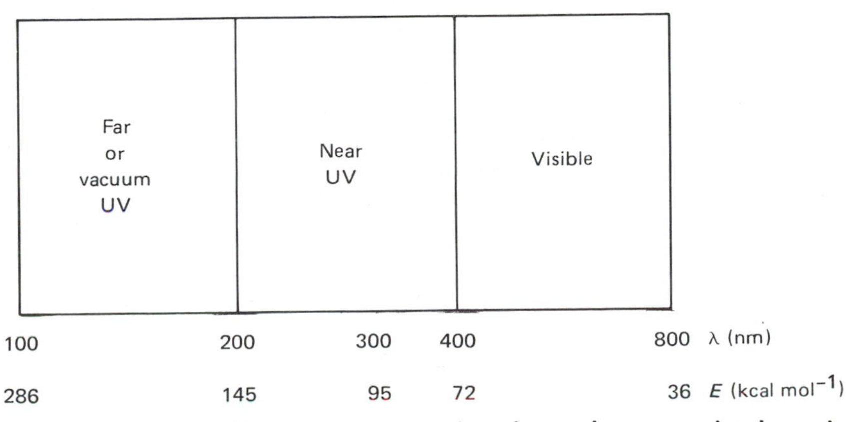

FIGURE 9-8 Ultraviolet-visible spectroscopic regions and some associated energies.

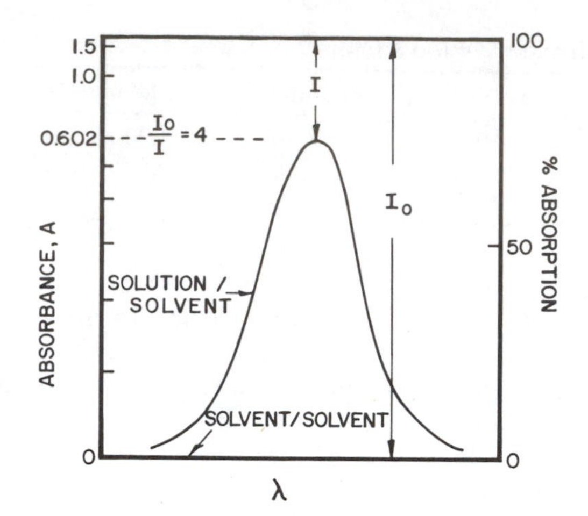

FIGURE 9-9 Measurement of solute absorbance *A* by a double-beam spectrophotometer.

molar absorptivity (eq. 9-10). The units of ϵ are $cm^2 \, mol^{-1}$ (liter $mol^{-1} \, cm^{-1}$) but are usually omitted. (For a derivation of Beer's law see ref. 9-2.)

$$A = \epsilon c l \tag{9-10}$$

In practice, the quantities actually measured are the relative intensities of the light beams transmitted by a reference cell containing pure solvent and by an identical cell containing the solution. When the intensities are taken as I_0 and I, respectively, the resulting absorption is that of the dissolved solute only (Figure 9-9). One also can see from Figure 9-9 that ϵ is different at different λ, and one refers to ϵ_{max} at λ_{max}.

When the incident light has handedness, as in left and right circularly polarized light, one may measure the absorbance (and hence ϵ) for left and right circularly polarized light. Subtraction ($\epsilon_L - \epsilon_R$) gives $\Delta\epsilon$, which is of course also dependent on λ, and one refers to $\Delta\epsilon_{max}$ at λ_{max}. When $\epsilon_L - \epsilon_R = 0$ at all λ, the sample is a racemic or the molecule is achiral, but chiral molecules always have $\Delta\epsilon \neq 0$ in the region of an electronic transition.

It is important to note that the actual intensity of absorption is related to the *integrated area* under the particular absorption band: for UV–vis this quantity is the oscillator strength (f); for CD it is the rotatory strength (R). Although f is a weighted sum of ϵ at each λ, and R a weighted sum of $\Delta\epsilon$ at each λ in the absorption band, unfortunately neither ϵ_{max} nor $\Delta\epsilon_{max}$ is directly related to f or R. Nonetheless, ϵ_{max} and $\Delta\epsilon_{max}$ serve as very useful approximations of intensity.

9-3c Solvents

The methods and procedures for UV–vis (ordinary or isotropic absorption) and CD (chiroptical) measurements are virtually the same. Most measurements are carried out on fairly dilute solutions (10^{-2} to 10^{-6} M) of the sample in an appropriate solvent. Such a solvent should not interact with the solute, should not absorb in the spectral region of interest, and, in the case of chiroptical measurements, should not be optically active—unless one is looking for solvent-induced CD. It is well to point out that, in both ordinary absorption and chiroptical techniques, measurements may be made on pure liquids, gases, and solids.

An important difference between UV–vis spectrophotometers and CD instruments is that the former are often double-beam instruments, whereas the latter are invariably single-beam instruments. Hence chiroptical measurements involve the examination of a solution of the desired compound followed by rescanning the spectrum with all parameters held the same and using pure solvent in the sample cell to obtain the baseline.

Some useful solvents and their short wavelength cutoff limits are given in Table 9-4. Note the significant advantage in UV penetration to be gained by the use of short pathlength cells (1 mm or less).

Commonly used polar solvents are 95% ethanol, water, and methanol. Aliphatic hydrocarbons (hexane, heptane, cyclohexane, etc.) are examples of nonpolar solvents that allow good UV penetration (Table 9-4) and have boiling points high enough so that solvent evaporation does not become a problem. However, they

TABLE 9-4 Short Wavelength Cutoff Limits of Various Solvents

Solvent	Cutoff Point, λ (nm)[a]		Boiling Point (°C)
	10 mm Cell	0.1 mm Cell	
Acetonitrile	190	180	81.6
2,2,2-Trifluorethanol	190	170	79
Pentane	190	170	36.1
2-Methylbutane	192	170	28
Hexane	195	173	68.8
Heptane	197	173	98.4
2,2,4-Trimethylpentane (isooctane)	197	180	99.2
Cyclopentane	198	173	49.3
Ethanol (95%)	204	187	78.1
Water	205	172	100.0
Cyclohexane	205	180	80.8
2-Propanol	205	187	82.4
Methanol	205	186	64.7
Methylcyclohexane	209	180	100.8
Dibutyl ether	210	195	142
EPA[b]	212	190	—
Diethyl ether	215	197	34.6
1,4-Dioxane	215	205	101.4
Bis(2-methoxyethyl) ether (glyme)	220	199	162
1,1,2-Trichlorotrifluorethane	231	220	47.6
Dichloromethane	232	220	41.6
Chloroform	245	235	62
Carbon tetrachloride	265	255	76.9
N,N-Dimethylformamide	270	258	153
Benzene	280	265	80.1
Toluene	285	268	110.8
Tetrachloroethylene	290	278	121.2
Pyridine	305	292	116
Acetone	330	325	56
Nitromethane	380	360	101.2
Carbon disulfide	380	360	46.5

From J. B. Lambert, H. F. Shurvell, L. Verbit, R. G. Cooks, and G. H. Stout, *Organic Structural Analysis*, Macmillan, New York, 1976.

[a]The cutoff point is taken as the wavelength at which the absorbance in the indicated cell is about one.

[b]5/5/2 by volume mixture of ethyl ether, isopentane, and ethanol.

must be rigorously purified since these hydrocarbons may contain alkenic impurities or traces of aromatic compounds. It has been observed that fluoroalkanes have enhanced transparency relative to the alkanes, and a similar finding has been made for the fluorinated alcohols such as 2,2,2-trifluoroethanol.

Organic cyanides such as acetonitrile and propionitrile are polar, nonhydroxylic solvents with excellent spectral transparency. A widely used polar, nonhydroxylic solvent is 1,4-dioxane, transparent to about 205 nm. A highly transparent liquid for use in special cases is anhydrous sulfuric acid.

Several mixed solvents have found use in spectroscopic studies at very low temperatures, usually down to liquid nitrogen temperatures, about −190°C. These solvent systems do not crystallize when cooled but instead become viscous and glassy. Low temperature solvents include (1) EPA, a 5/5/2 by volume mixture of diethyl ether, isopentane, and ethanol, (2) methanol and glycerol, 9/1 v/v, (3) tetrahydrofuran and diglyme, 4/1 v/v, and (4) methylcyclohexane and isopentane, 1/3 v/v. A table of the percent degree of contraction of these solvents over the range 25 to −190°C has been published (ref. 9-3).

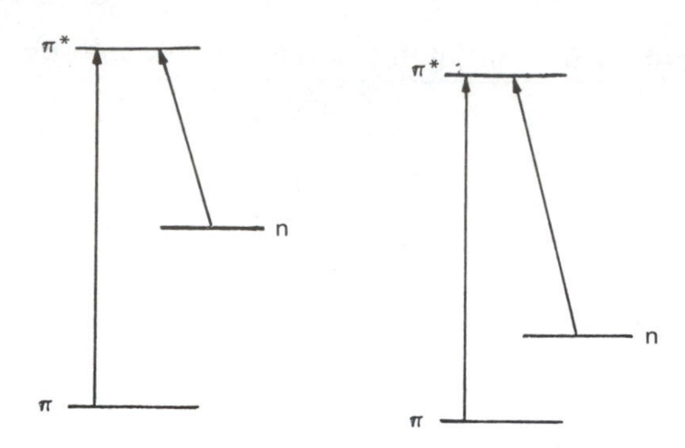

FIGURE 9-10 Influence of solvent on the orbitals involved in $\pi \rightarrow \pi^*$ and $n \rightarrow \pi^*$ electronic transitions.

9-3d Solvent Effects and λ_{max} Shifts

Moving an electron from the ground state to an excited state configuration typically leads to an excited state that is more polar than the ground state and more sensitive to solvation effects. For $\pi \rightarrow \pi^*$ excited states, dipole–dipole interactions and hydrogen bonding with solvent molecules tend to lower the energy of the excited state more than the ground state with the result that the λ_{max} *increases* (red shift) about 10 nm in going from a fairly noninteractive solvent like hexane to methanol (Figure 9-10). For $n \rightarrow \pi^*$ excited states, both the ground and the excited states are lowered in energy by dipole–dipole and hydrogen-bonding interaction with solvent. In hydrogen-bonding solvents, the ground state n electrons coordinate with the solvent more strongly than excited state n electrons, with the result that the λ_{max} *decreases* about 15 nm in going from hexane to methanol solvent (Figure 9-10). Solvent effects on $\pi \rightarrow \pi^*$ and $n \rightarrow \pi^*$ transitions are summarized for the example of mesityl oxide in Table 9-5.

9-3e Sample Preparation

Quantitative analytical techniques are applied to sample preparation. Volumetric glassware and cells must be clean and dry. Solid samples should be dried to constant weight in a desiccator in order to remove adhering water or solvent. A typical 1–10 mm cell holds anywhere from 0.2 to 3 ml of solution so that an appropriate amount of stock sample solution should be prepared. Since the measurement is nondestructive, the sample may be recovered by evaporation of the solvent. If the amount of sample available permits, 10–25 ml of solution is a convenient size to prepare. The amount of sample required for this volume is sufficient to minimize errors associated with weighing small quantities. Naturally, one must use an analytical balance

TABLE 9-5 Influence of Solvent on the UV λ_{max} and ϵ_{max} of the $n \rightarrow \pi^*$ and $\pi \rightarrow \pi^*$ Excitations of 4-Methyl-3-penten-2-one (Mesityl Oxide)

$$(CH_3)_2C{=}CH{-}\overset{\overset{\displaystyle O}{\|}}{C}{-}CH_3$$

	$\pi \rightarrow \pi^*$ Transition		$n \rightarrow \pi^*$ Transition	
Solvent	λ_{max} (nm)	ϵ_{max} (liter mol^{-1} cm^{-1})	λ_{max} (nm)	ϵ_{max} (liter mol^{-1} cm^{-1})
Hexane	229.5	12,600	327	40
Diethyl ether	230	12,600	326	40
Ethanol	237	12,600	325	90
Methanol	238	10,700	312	55
Water	244.5	10,000	305	60

From G. J. Brealey and M. Kasha, *J. Am. Chem. Soc.*, 77, 4462 (1955).

capable of weighing directly to at least 0.1 mg. As most compounds being measured probably will not have been run previously by the operator, an initial sample concentration should be approximately 0.05% (about 0.5 mg ml^{-1}) for small molecules and about 0.005% for polypeptides and large macromolecules. A peak absorbance in the range 0.7–1.2 absorbance units is desirable for most instruments, since it gives a good pen deflection and the electronics are usually most sensitive in this range.

9-4

Types of Electronic Transitions

9-4a Assignments (ref. 9-5)

The wavelength of an electronic transition depends on the energy difference between the ground state and the excited state. It is a useful *approximation* to consider the wavelength of an electronic transition to be determined by the energy difference between the molecular orbital originally occupied by the electron and the higher orbital to which it is excited. Saturated hydrocarbons contain only strongly bound σ electrons. Their excitation to antibonding σ^* orbitals ($\sigma \rightarrow \sigma^*$) or to molecular Rydberg orbitals requires relatively large energies, corresponding to absorption in the far UV region. One exception is cyclopropane, which has λ_{max} at 190 nm. Contrast this cycloalkane to propane, which has λ_{max} about 135 nm.

Electronic transitions commonly observed in the readily accessible UV (above ~190 nm) and visible regions have been grouped into several main classes (Figure 9-11).[1]

$n \rightarrow \pi^*$ Transitions. These transitions can be considered to involve the excitation of an electron in a nonbonding atomic orbital, i.e., unshared electrons on O, N, S, or halogen atom, to an antibonding π^* orbital associated with an unsaturated center in the molecule. The transitions occur with compounds possessing double bonds involving heteroatoms, e.g., C=O, C=S, N=O. A familiar example is the low intensity absorption in the 285–300 nm region of saturated aldehydes and ketones (Figure 9-4).

$\pi \rightarrow \pi^*$ Transitions. Molecules that contain double or triple bonds or aromatic rings can undergo transitions in which a π electron is excited to an antibonding π^* orbital. Although ethylene itself does not absorb strongly above about 185 nm, conjugated π electron systems are generally of lower energy and absorb in the accessible spectral region. An important application of UV–vis spectroscopy is to define the presence, nature, and extent of conjugation. Increasing conjugation generally moves the absorption to longer wavelengths and finally into the visible region; this principle is illustrated in Table 9-6.

$n \rightarrow \sigma^*$ Transitions. These transitions, which are of less importance than the first two classes, involve excitation of an electron from a nonbonding orbital to an antibonding σ^* orbital. Since *n* electrons do not form bonds, there are no antibonding orbitals associated with them. Some examples of $n \rightarrow \sigma^*$ transitions are

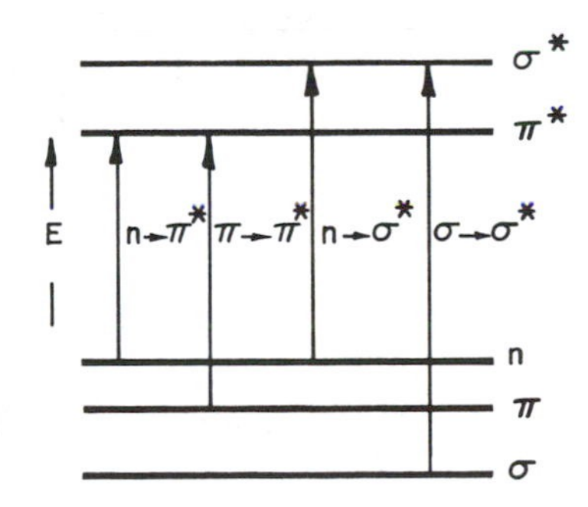

FIGURE 9-11 Relative electronic orbital energies and selected transitions in order of increasing energy.

[1]In addition to that given here, several other systems of classification exist. For example, the designation N–V is used to describe transitions from a bonding to an antibonding orbital ($\sigma \rightarrow \sigma^*$, $\pi \rightarrow \pi^*$). The term N–Q designates transitions from a nonbonding atomic orbital to a higher energy molecular orbital ($n \rightarrow \sigma^*$, $n \rightarrow \pi^*$). Burawoy has termed $\pi \rightarrow \pi^*$ transitions K-bands (from the German *Konjugation*) and $n \rightarrow \pi^*$ transitions R-bands (from an early theory that the excited state was a radical). Numerous terms also exist in the literature for the classification of ground and excited states on the basis of symmetry (ref. 9-5).

TABLE 9-6 Effect of Extended Conjugation in Alkenes on Position of Maximum Absorption

n in $H(CH{=}CH)_nH$	λ_{max} (nm)	ϵ_{max} (liter mol^{-1} cm^{-1})	Color
1	162	10,000	colorless
2	217	21,000	colorless
3	258	35,000	colorless
4	296	52,000	colorless
5	335	118,000	pale yellow
8	415	210,000	orange
11	470	185,000	red
15	547[a]	150,000	violet

From J. B. Lambert, H. F. Shurvell, L. Verbit, R. G. Cooks, and G. H. Stout, *Organic Structural Analysis*, Macmillan, New York, 1976.

[a]Not a maximum.

CH_3OH (vapor), λ_{max} 183 nm, ϵ 150; trimethylamine (vapor), λ_{max} 227 nm, ϵ 900; and CH_3I (hexane), λ_{max} 258 nm, ϵ 380.

Rydberg transitions are mainly to highly excited states. For most organic molecules, they occur at wavelengths below about 200 nm. A Rydberg transition is often part of a series that terminates at a limit representing the ionization potential of the molecule.

Groups that give rise to electronic absorption are known as *chromophores* (color bearer, from an early theory of color). The term *auxochrome* (color increaser) is used for substituents containing unshared electrons (OH, NH, SH, halogens, etc.). When attached to π electron chromophores, auxochromes generally move the absorption maximum to longer wavelengths (lower energies). Such a movement is described as a *bathochromic* or *red shift*. The term *hypsochromic* denotes a shift to shorter wavelength (*blue shift*). Increased conjugation usually results in increased intensity termed *hyperchromism*. A decrease in intensity of an absorption band is termed *hypochromism*. These terms are summarized in Figure 9-12.

9-4b Allowed and Forbidden Transitions

Electronic transitions may be classed as intense or weak according to the magnitude of, e.g., ϵ_{max}. These correspond to *allowed* or *forbidden* transitions. Allowed transitions are those for which (a) there is no change in the orientation of electron spin, (b) the change in angular momentum is 0 or ± 1, and (c) the product of the electric dipole vector and the group theory representations of the two states is totally symmetric.

The first rule is the *spin selection rule* and may be stated as follows: transitions between states of different spin multiplicities are invariably forbidden since electrons cannot undergo spin inversion except for spin–orbit and spin–spin interactions. The second rule usually presents no problem since most states are within one unit of angular momentum of each other. The last rule is the *symmetry selection rule*. If the direct product of the representations to which the initial and final state functions belong is different from all the representations to which the coordinate axes belong, the transition moment of that transition is zero. A good

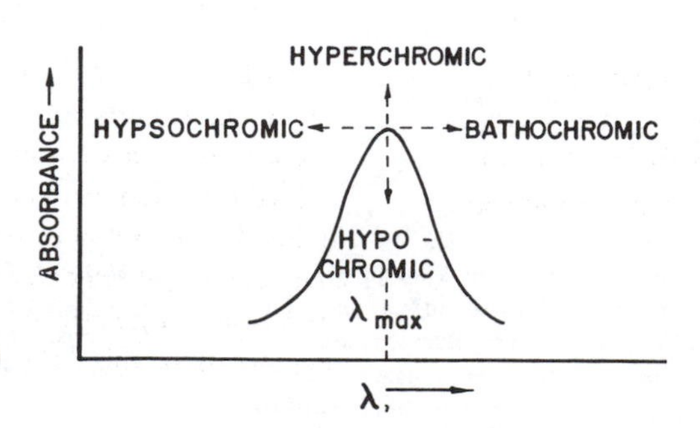

FIGURE 9-12 Terminology of shifts in the position of an absorption band.

example of this rule is the n→π^* transition of saturated alkyl ketones, where a carbonyl n electron is promoted to an *orthogonal* π^* orbital (90° movement of charge), e.g., Figure 9-4. Such a transition is said to be symmetry forbidden. For most organic molecules, such forbidden transitions are usually observable but of weak intensity. They arise because the intensity of the electronic absorption band really depends on the average of the electronic transition moments over all the nuclear orientations of the vibrating molecule and this average is not necessarily zero. When the symmetry of a molecule is periodically changed by some vibration that is not totally symmetric, the symmetry of the electronic wavefunctions is also periodically changed since the electrons adapt instantaneously to the motion of the nuclei. Hence a symmetry forbidden transition may become allowed. The intensity of a transition that is symmetry forbidden but has become vibrationally allowed is much less than that of an ordinarily allowed transition. Such vibrational contributions are temperature dependent.

9-5

Chiroptical Methods: Circular Dichroism and Optical Rotatory Dispersion

9-5a The Nature of Polarized Light

The light beam used in UV–vis spectroscopy is essentially unpolarized. Use of linearly polarized light (sometimes less rigorously referred to as plane-polarized light) to investigate optically active (chiral) molecules is a powerful technique for obtaining structural and stereochemical information.

Figure 9-13 considers light in the context of a wave phenomenon caused by transverse vibrations of the electric field vector (vertical arrows). There is an associated magnetic field vector perpendicular to the oscillating electric field vector, but we can ignore it for purposes of the present discussion. Note that the electric field vector vibrates perpendicular to the direction of travel of the light wave. Now there are an infinite number of planes that we can pass through the line *OX* in Figure 9-13. Ordinary light consists of different wavelengths vibrating in many different planes.

If we could place ourselves at the point *X* and look toward *O*, we would see the cross section of the light wave depicted in Figure 9-14, a schematic representation of unpolarized and linearly polarized light. The radial electric field vectors (the arrows) are meant to indicate that no single direction predominates in completely unpolarized light.

Even if we were to use polarized light of a single wavelength, it would still consist of waves vibrating in many planes at right angles to the direction of propagation. Since several directions of propagation are possible within a plane, it is correct to refer to light traveling in a specified direction as linearly polarized rather than plane polarized.

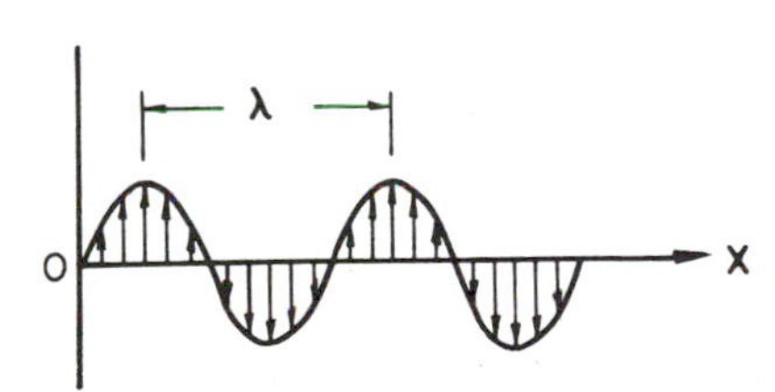

FIGURE 9-13 Wave motion propagated in the *x* direction by transverse vibration; λ is the wavelength. The arrows denote the electric field vector at a given instant as the light wave progresses along the *x* axis.

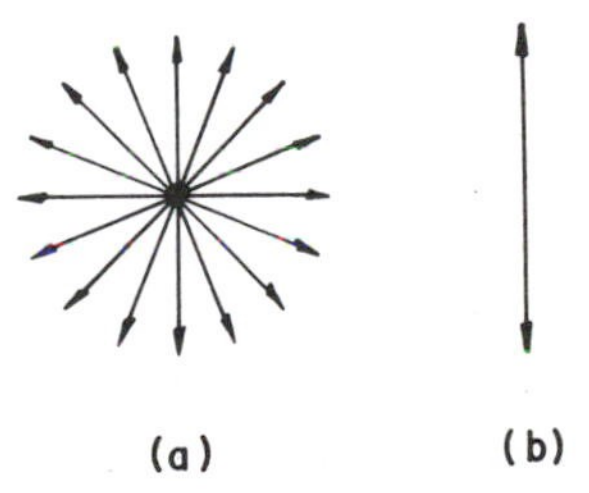

FIGURE 9-14 Schematic representation of unpolarized and linearly polarized light. (a) Cross section of a narrow beam of ordinary light traveling directly toward the observer. Vibration of the light may be in any direction that is perpendicular to the direction of travel, as indicated by the numerous arrows. (b) A beam of polarized light has vibration in only one direction. This direction is the plane of polarization.

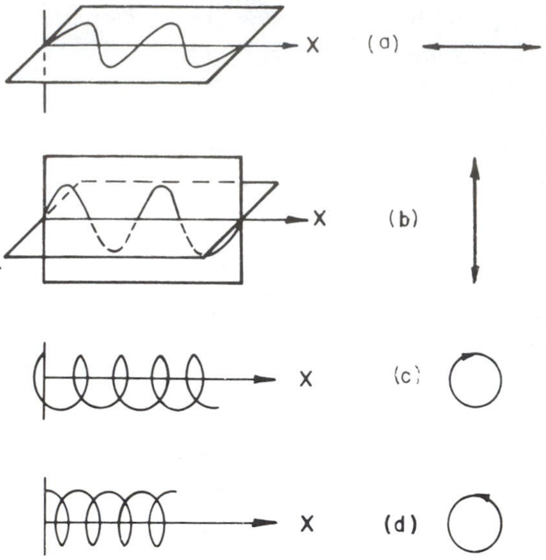

FIGURE 9-15 Linearly and circularly polarized radiation. (Left) The light wave as a function of time. (Right) Cross-section of the light wave. Light polarized (a) linearly and horizontally, (b) linearly and vertically, (c) right circularly, (d) left circularly. Light having a right-handed helical pattern is termed right circularly polarized. The cross-sectional clockwise rotation of the electric field vector is obtained as the helix is moved forward without rotation through a perpendicular plane.

In addition to linearly polarized light, other kinds of light exist. Figure 9-15 shows light beams that are polarized (a) linearly and horizontally, (b) linearly and vertically, (c) right circularly, and (d) left circularly.

In the case of circularly polarized light, the transverse vibrations trace out a helix as a function of time. The helix may be either right-handed (Figure 9-15c) or left-handed (Figure 9-15d). Viewed in cross section, i.e., as if an observer were situated on the x axis looking toward the light source, the transverse vibrations trace out a circle. Light whose electric field vector traces out a right-handed helical pattern is termed *right circularly polarized light*. The cross-sectional appearance of clockwise rotation of the electric field vector is obtained by pushing the helix forward through a perpendicular plane without rotating it. In other words, the helix is moved forward, but it is not turned like a mechanical screw.

Another type of polarized light, which we have not pictured, resembles a flattened helix and has a cross section that is an ellipse. Elliptically polarized light may also be right- or left-handed.

The French physicist Biot discovered early in the nineteenth century that certain naturally occurring organic compounds possessed the unusual property of rotating the plane of polarization of a linearly polarized incident light beam. A few years later, in 1817, Biot and his countryman Fresnel independently found that the extent of optical rotation of a compound increases as one used light of increasingly shorter wavelength for the measurement. The change in optical rotation with wavelength is termed *optical rotatory dispersion* (ORD).

Thirty years later, Haidenger reported his observations on the differences in the absorption of the left- and right-handed components of circularly polarized light by crystals of amethyst quartz. Such differential absorption of left- and right-handed circularly polarized light is termed *circular dichroism* (CD).

Since both CD and ORD involve optical measurements on chiral molecules, they have been termed chiroptical methods.

9-5b Optical Rotation

As a useful model for conceptualizing the rotation of linearly polarized light, consider the light as composed of two oppositely rotating coherent beams of circularly polarized light. The linearly polarized light is then the vector sum of the left and right circularly rotating components as shown in Figure 9-16. The vector sums are indicated at points A to E with the resultant vectors having the properties of a linearly polarized light wave.

Fresnel, in 1825, postulated that when the circularly polarized light beams pass through an optically active medium, which may be a solid, liquid, or gas, the refractive index (n) for one circularly polarized component is different from that for the other. The medium is said to be *circularly birefringent* and to have the property given by eq. 9-14. Differences in refractive indices correspond to differences in light velocities. Consequently,

$$n_L - n_R \neq 0 \tag{9-14}$$

one of the two circularly polarized components of the linearly polarized light becomes retarded with respect

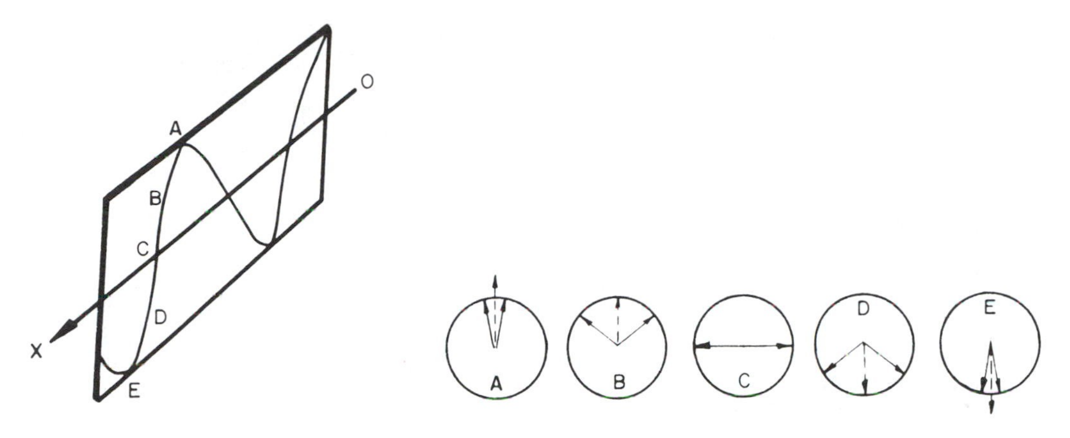

FIGURE 9-16 A representation of linearly polarized radiation as the vector sum (dotted arrows) of two oppositely rotating beams of circularly polarized radiation. [From J. D. Roberts and M. C. Caserio, *Basic Principles of Organic Chemistry*, 2nd ed., 1977 copyright © 1964 by W. A. Benjamin, Inc., Menlo Park, CA.]

to the other. Upon emerging from the optically active medium, the two components are no longer in phase and the resultant vector has been rotated by the angle α to the original plane of polarization (Figure 9-17).

In the region of an absorption band, the two circularly polarized components, in addition to suffering a differential retardation because of the circular birefringence of the medium, also are absorbed to different extents. In other words, the optically active medium has an unequal molar absorption coefficient ϵ for left and right circularly polarized light. This difference in molar absorptivity (eq. 9-15) is termed *circular dichroism.*

$$\Delta\epsilon = \epsilon_L - \epsilon_R \neq 0 \tag{9-15}$$

Upon emerging from the optically active medium, the two circularly polarized components are not only out of phase but also of unequal amplitude. The resultant vector no longer oscillates along a single line but now traces out an ellipse, as shown in Figure 9-18. The linearly polarized light beam has been converted to elliptically polarized light by the unequal absorption of its two circularly polarized components.

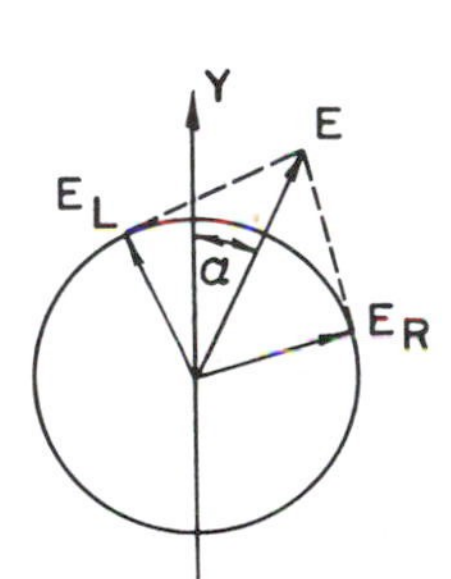

FIGURE 9-17 Rotation of the plane of polarized light as the result of a change in the velocity of E_R relative to E_L.

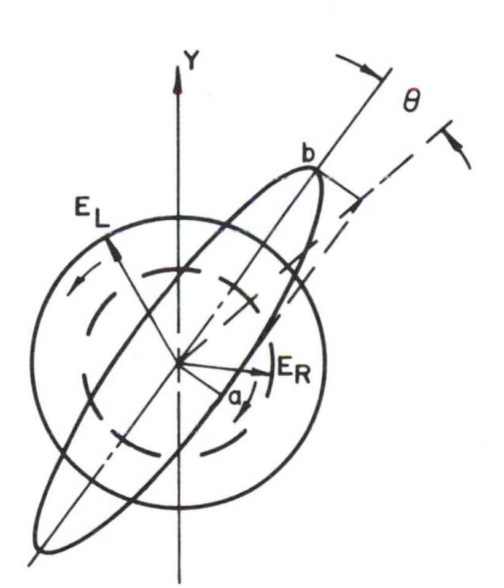

FIGURE 9-18 Elliptically polarized light caused by the unequal speed and unequal absorption of left and right circularly polarized light by a chiral medium. The tangent of the ratio of the minor axis *a* to the major axis *b* is θ, the angle of ellipticity. The major axis of the ellipse forms the angle of rotational, α, to the original plane of polarization, the *y* plane.

Note that ORD involves measurement of a rotation, whereas CD involves an absorption measurement, namely the differential absorption of left- and right-handed circularly polarized radiation. Hence CD occurs only in the vicinity of an absorption band, whereas ORD is theoretically finite everywhere.

9-5c CD and ORD Quantities

The angle of rotation α, in degrees per decimeter, is given by eq. 9-16. We may note that for an observed optical

$$\alpha = \frac{1800}{\lambda\ (\mathrm{cm})} (n_L - n_R) \tag{9-16}$$

rotation of 1° at 360 nm in a 1 decimeter (10 cm) cell, $n_L - n_R$ is 2×10^{-8}. Typical indices of refraction are of the order of unity. Hence the difference in refractive indices is extremely small, of the order of one millionth of 1%.

Spectropolarimeters, as polarimeters able to make measurements at a variety of wavelengths are termed, record the angle of rotation α as a function of wavelength. Eq. 9-17 is used to calculate the specific rotation

$$[\alpha]^t_\lambda = \frac{\alpha}{cl} = \frac{\text{observed rotation (degrees)}}{\text{concn (g ml}^{-1}) \times \text{length of sample tube (dm} = 0.1 \text{ m)}} \tag{9-17}$$

$[\alpha]$. For ORD work, it is more common to use the molar rotation $[\Phi]$, which is simply the specific rotation multiplied by the molecular weight M over 100 (eq. 9-18). The units of $[\Phi]$ are degrees cm^2 dmol^{-1}.

$$[\Phi] = \frac{[\alpha]\, M}{100} \tag{9-18}$$

Just as $(n_L - n_R)$ is small in magnitude compared with the mean index of refraction, so the difference $(\epsilon_L - \epsilon_R)$ between the absorption coefficients for left and right circularly polarized light is of the order of 10^{-2} to 10^{-3}. Most CD instruments measure the differential absorbance, $\Delta A = A_L - A_R$. This quantity is related to the difference in molar absorption coefficients, $\Delta\epsilon = (\epsilon_L - \epsilon_R)$, by eq. 9-19, in which c is in mol liter^{-1} and l is

$$\Delta A = \Delta\epsilon c l \tag{9-19}$$

the pathlength in cm. By analogy with the molar rotation, the molar ellipticity $[\theta]$ is defined by eq. 9-20, in which $[\theta]$ has the units degrees cm^2 dmol^{-1}.

$$[\theta] = 3300 \Delta\epsilon \tag{9-20}$$

9-5d CD and ORD Cotton Effects

The shape and appearance of a CD curve closely resembles that of the isotropic (UV–vis) absorption curve of the electronic transition to which it corresponds. Unlike ordinary absorption curves, e.g., Figure 9-5, however, CD curves may be positive or negative, as in Figure 9-19. A CD is a plot of $\Delta\epsilon$ (or $[\theta]$) vs. wavelength. It is a *difference* spectrum representing the difference in absorption of left and right circularly polarized light, hence the *signed* nature of the curve. Each CD curve for each electronic transition also

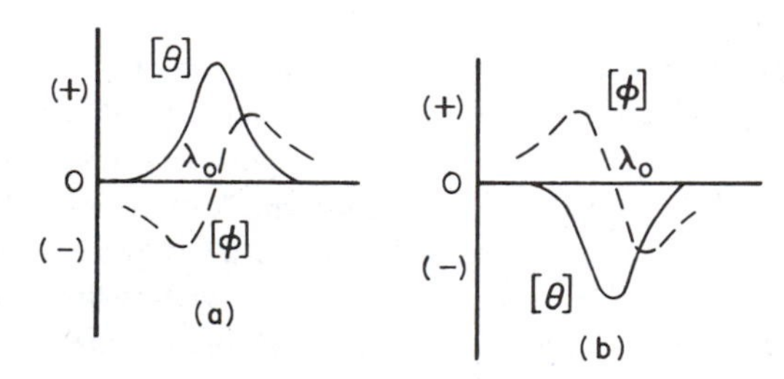

FIGURE 9-19 Positive (a) and negative (b) circular dichroism Cotton effects of an isolated absorption band with their corresponding optical rotatory dispersion curves (dashed lines).

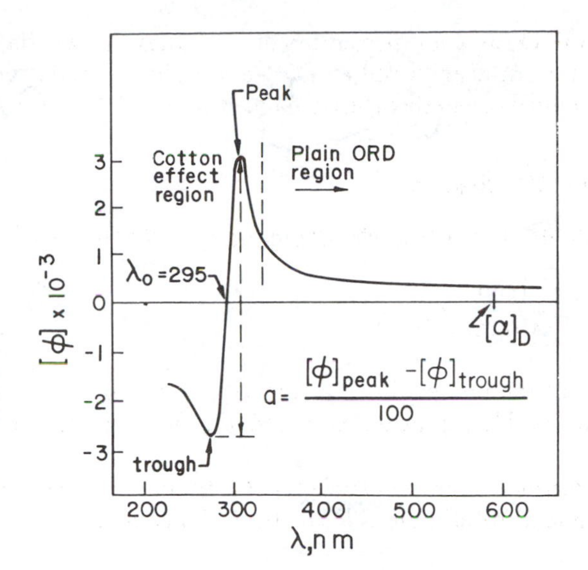

FIGURE 9-20 Optical rotatory dispersion curve of a typical saturated ketone. The amplitude of the Cotton effect *a* is defined as shown. The crossover point from positive to negative rotational values, λ_0 = 295 nm, corresponds closely to the UV absorption maximum of the ketone.

represents a positive or negative *Cotton effect* for the transition. For every CD Cotton effect there exists a corresponding ORD Cotton effect of the same sign (Figure 9-19). (The *S*-shaped ORD curve is known as a Cotton effect, in honor of the French physicist Aimé Cotton, who observed both ORD and CD phenomena beginning in 1896.)

Cotton effects are also seen in ORD spectra because, in the region of an absorption band, left and right circularly polarized light is not only absorbed to different extents but also propagated with different velocities. ORD curves, especially those of close-lying electronic transitions, are generally more difficult to read than are the bell-shaped CD curves, but information can be derived beyond the immediate region (λ) of the electronic transitions. In ORD curves, the angle of rotation α of a chiral compound is plotted against wavelength. A typical ORD curve for a dextrorotatory ketone having an absorption maximum at 295 nm is shown in Figure 9-20. The part of the ORD curve above about 325 nm, labeled plain ORD region, is characteristic of compounds that have no optically active absorption bands in the spectral region being measured. The smoothly rising part of the curve is referred to as a plain dispersion curve; in the example shown it is a plain, *positive* dispersion curve. A plain *negative* curve would be one whose rotational values fall or become increasingly more negative on going toward shorter wavelength. The so-named anomalous part of the ORD curve falls in the region of an electronic transition and corresponds to the Cotton effect.

As measurements of optical rotation in Figure 9-20 are made to shorter wavelengths, the rotation increases. It is found to increase rapidly as the absorption maximum is approached. Somewhat before this maximum, rotational values reach a maximum (termed a *peak*), then drop drastically, *going through zero rotation*, until another inflection point (termed a *trough*) is reached. The rotation will then tend to increase again. In the ideal case for which the molecule possesses no other absorption bands near the one measured, λ_0 will closely correspond to λ_{max}, the maximum of the absorption band.

The vertical distance between the peak and trough divided by 100 is termed the amplitude α (eq. 9-21). If

$$\alpha = \frac{[\Phi]_1 - [\Phi]_2}{100} \tag{9-21}$$

the peak precedes the trough on measuring from longer to shorter wavelength, the Cotton effect is termed positive. Conversely, if a trough precedes a peak, it is a negative Cotton effect.

An entire CD spectrum may be transformed into an ORD spectrum, and vice versa, by means of a Kronig–Kramers transform. In a simpler example, the Kronig–Kramers relations lead to an expression, derived from the n→π^* transition of saturated ketones, which relates the amplitude α of an ORD Cotton effect (eq. 9-22) to $\Delta\epsilon$ of the corresponding CD peak (eq. 9-22).

$$\alpha = 40.28\Delta\epsilon_{max} = (1.22 \times 10^{-2})[\theta]_{max} \tag{9-22}$$

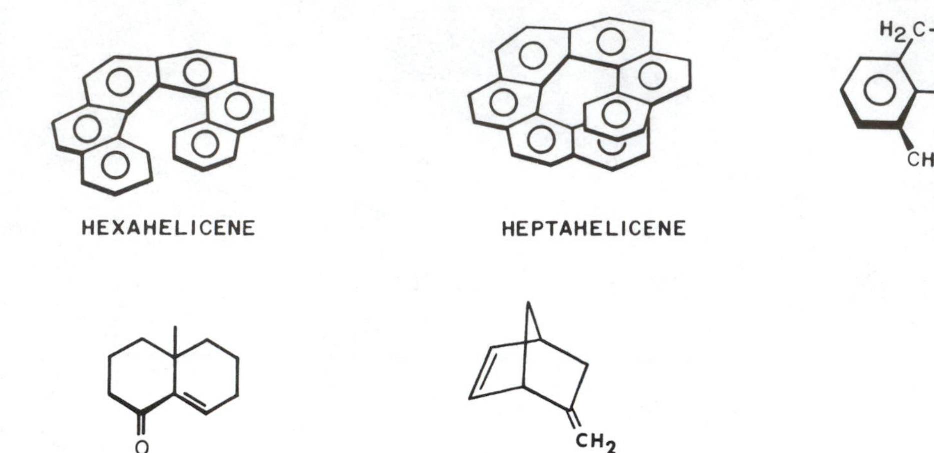

FIGURE 9-21 Some inherently chiral chromophores.

Because they are easy to read and yield quantitative data, CD spectra are usually favored over ORD spectra. Nowadays, most chiroptical data are measured in the CD form, but in either form the most meaningful structural information is derived from the sign and magnitude of the Cotton effect.

9-5e Optically Active Chromophores

An optically active chromophore is one that gives rise to a Cotton effect in the CD or ORD spectrum of the compound. All chromophores in a chiral molecule should exhibit Cotton effects in the region of their electronic absorption bands. However, rotatory strength falls off with distance from a chiral center so that a distant chromophore may have a Cotton effect too weak to measure.

It is useful to divide optically active chromophores into two broad classes: (1) the inherently chiral chromophore and (2) the inherently achiral but chirally perturbed chromophore. (The original classification of Moscowitz (ref. 9-8) used the terms dissymmetric and symmetric for chiral and achiral.) The optical activity of compounds belonging to the first class is inherent in the geometry of the chromophore. Some examples of inherently chiral chromophores are given in Figure 9-21. Inherently achiral but chirally perturbed chromophores include most of the common functional groups; some examples of this class are given in Figure 9-22. Inherently chiral chromophores are characterized by large rotational strengths, on the order of 10^{-38} cgs unit, whereas the achiral but chirally perturbed chromophores are characterized by rather weak Cotton effects. Their rotatory strength is generally less than 10^{-41} cgs unit.

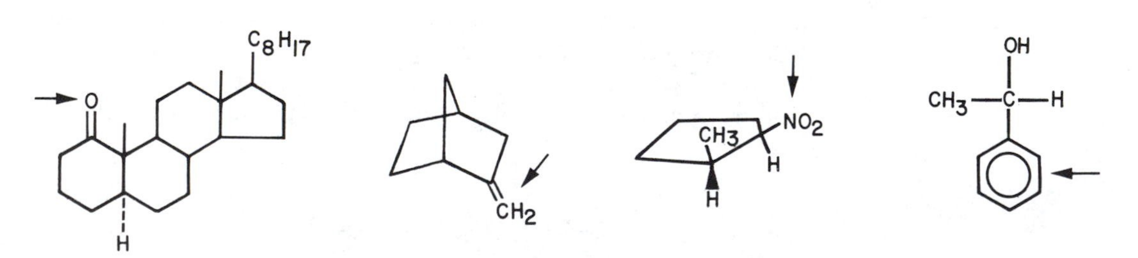

FIGURE 9-22 Some inherently achiral but chirally perturbed chromophores.

PROBLEMS

9-1 Using the data in Table 9-1, calculate the electronic transition energies (in kcal mol^{-1}) for the $n \rightarrow \pi^*$ and $\pi \rightarrow \pi^*$ excitations of acetaldehyde.

9-2 Calculate the molar extinction coefficient of di-*tert*-butyl ketone at 280 nm and 320 nm from the data of Figure 9-4.

9-3 A compound $C_5H_8O_2$ has a UV λ_{max} at 270 nm ($\epsilon = 32$) in methanol; in hexane, the band shifts to 290 nm ($\epsilon_{max} = 40$). What functional groups could be present? Draw a possible structure.

9-4 Calculate the absorbance (A) of a 0.005 M solution of cyclohexanone in isooctane in a 10 cm pathlength quartz cuvette at $\lambda = 280$ nm.

9-5 A compound, 0.0002 M in methanol, shows $\lambda_{max} = 235$ nm with an absorbance (A) of 1.05 when measured in an 0.5 cm pathlength quartz cuvette. Calculate its ϵ_{max} at 235 nm.

9-6 The CD spectrum of a chiral compound, 0.0005 M in isooctane, shows $\lambda_{max} = 285$ nm with $\Delta A = -0.0003$ when measured in a 1.0 cm pathlength quartz cuvette. Calculate $\Delta\epsilon_{max}$ at 285 nm.

9-7 The amplitude of the ORD $n \rightarrow \pi^*$ cotton effect of 3-methyl-cyclohexanone was determined to be $[a] = -25°$. Calculate $\Delta\epsilon_{max}$.

9-8 Calculate the absorbance (A) of a 0.005 M solution of acetone in isooctane in a 0.1 cm pathlength quartz cuvette at $\lambda = 280$ nm.

9-9 A compound, 0.0002 M in ethanol, shows $\lambda_{max} = 235$ nm with an absorbance (A) of 1.05 when measured in a 0.5 cm pathlength quartz cuvette. Calculate its ϵ_{max}.

9-10 At pH 13, the absorbance of a particular phenol solution is 1.5 at 400 nm and 0.0 at 270 nm. At pH 4, the values for a solution of the same concentrations are 0.0 and 1.0 at these two wavelengths, respectively. At pH 9, the values are 0.9 and 0.4, respectively.

- **a.** Explain the spectral change.
- **b.** Calculate the pK_a of the phenol (negative log of the dissociation constant).
- **c.** If the concentration used were 18.8 mg of phenol in 20 ml of solvent, for an $\epsilon_{270} = 100$ liter mol^{-1} cm^{-1} and a cell length of 1 cm, calculate the molecular weight of the phenol.
- **d.** Draw a possible structure.

9-11 At pH 13 the absorbance of a certain phenol is 1.5 at 430 nm and 0.0 at 290 nm. At pH 4, the values for a solution of the same concentration are 0.0 and 0.5, respectively. At pH 8, the values are 0.6 and 0.3, respectively.

- **a.** Explain the spectral changes.
- **b.** Calculate the pK_a of the phenol.
- **c.** If the concentration used was 10.8 mg of phenol in 20 ml of solvent, for an $\epsilon = 300$ liter mol^{-1} at 290 nm and a cell length of 1 cm, calculate the molecular weight of the phenol.
- **d.** Draw the possible structure(s) for the substance that absorbs light at 290 nm.

9-12 A 250 mg sample containing a colored component X is dissolved and diluted to 250 ml. The absorbance of an aliquot of this solution, measured at 500 nm in a 1.00 cm cell, is 0.900. Pure X (10.0 mg) is dissolved in 1 liter of the same solvent, and the absorbance measured in a 0.100 cm cell at the same wavelength is 0.300. What is the percent of X in the first sample?

9-13 The molar absorptivity of benzoic acid (mol. wt. = 122.1) in ethanol at 273 nm is about 2000. If an absorbance not exceeding 1.35 is desired, what is the maximum allowable concentration in g $liter^{-1}$ that can be used in a 2.00 cm cell?

9-14
- **a.** Estimate the K_a of a weak acid from the data below. All of the various buffered solutions are *one millimolar* in sample, and all solutions were measured under the same conditions. The anion of the acid is the only substance that absorbs at the wavelength used. A is the absorbance.

pH:	4	5	6	7	8	9	10	11
A:	0.00	0.00	0.10	0.75	1.00	1.25	1.50	1.50

- **b.** What is the value of the molar extinction coefficient for the anion at this wavelength if a 1 cm cell is used?

9-15 Estimate the K_a of a weak acid from the data below. Samples (1 g) are dissolved in equal quantities of the various buffers, and all solutions are measured under the same conditions. The anion of the acid is the only substance that absorbs at the wavelength used.

pH:	4	5	6	7	8	9	10	11
A:	0.00	0.00	0.06	0.39	0.95	1.13	1.18	1.18

9-16 **a.** Estimate the K_a of a weak acid from the data below. All the various buffered solutions are 1.00 millimolar in sample, and all solutions were measured under the same conditions. The anion of the acid is the only substance that absorbs at the wavelength used. A is the absorbance.

pH:	4	5	6	7	8	9	10	11
A:	0.000	0.000	0.100	0.750	1.000	1.150	1.250	1.250

b. What is the value of the molar extinction coefficient for the anion at this wavelength if a 1.00 cm cell is used?

9-17 Substances X and Y, which are colorless, form the colored compound XY: $X + Y \rightleftharpoons XY$. When 2.00×10^{-3} mol of X is mixed with a large excess of Y and diluted to 1 liter, the solution has an absorbance that is twice as large as when 2.00×10^{-3} mol of X is mixed with 2.00×10^{-3} mol of Y and treated similarly. What is the equilibrium constant for the formation of XY?

9-18 Absorbances were measured for three solutions containing A and B separately and in a mixture, all in the same cell. Calculate the concentrations of A and B in the mixture.

	Absorbance	
	475 nm	**670 nm**
0.001 M A	0.90	0.20
0.01 M B	0.15	0.65
Mixture	1.65	1.65

REFERENCES/BIBLIOGRAPHY

9-1 J. B. Lambert, H. F. Shurvell, L. Verbit, R. G. Cooks, and G. H. Stout, *Organic Structural Analysis*, Macmillan, New York, 1976.

9-2 R. L. Pecsok, L. D. Shields, T. Cairns, and I. G. McWilliam, *Modern Methods of Chemical Analysis*, 2nd ed., Wiley, New York, 1976.

9-3 O. Korver and J. Bosma, *Anal. Chem.*, **43**, 1119 (1971).

9-4 G. J. Brealey and M. Kasha, *J. Am. Chem. Soc.*, **77**, 4462 (1955).

9-5 H. H. Jaffé and M. Orchin, *Theory and Applications of Ultraviolet Spectroscopy*, Wiley, New York, 1962.

9-6 C. Djerassi, *Optical Rotatory Dispersion*, McGraw-Hill, New York, 1960.

9-7 P. Crabbé, *Optical Rotatory Dispersion and Circular Dichroism in Organic Chemistry*, Holden-Day, San Francisco, 1965.

9-8 A. Moscowitz, *Tetrahedron*, **13**, 48 (1961).

10 Molecular Structure from Electronic Spectroscopy of Chromophores

10-1 Isolated Chromophores

10-1a The Ketone and Aldehyde Carbonyl

The longest wavelength transition in aliphatic aldehydes and ketones, the n→π^* band, is probably the best studied of any transition (refs. 10-1 and 10-2 and Figure 9-4). It is a low intensity ($\epsilon \sim$ 10–20) and rather broad band, occurring in the neighborhood of 270–300 nm. As noted in Section 9-3d, its position is quite solvent sensitive.

A second carbonyl band, attributed to a $\pi \rightarrow \pi^*$ transition, occurs near 190 nm in ketones and is considerably more intense than the n→π^* transition. This wavelength region is just beyond the range of most UV instruments so that only the beginning of the band is observed as so-called end absorption. Transitions at wavelengths shorter than about 190 nm are most likely due to $\sigma_{CO} \rightarrow \pi^*$ and Rydberg transitions.

The lowest energy transition involves the promotion of an electron from the nonbonding p orbital on oxygen to the antibonding π^* orbital associated with the entire carbonyl group. The transition is symmetry forbidden (see Section 9-4b, formaldehyde and symmetrically disubstituted ketones belong to the C_{2v} point group) and thus is of low intensity. The $\pi \rightarrow \pi^*$ transition, on the other hand, is allowed and relatively intense. The symmetries of the orbitals involved in the transitions are important. The bonding π orbital and the antibonding π^* orbital lie in the same plane, whereas the nonbonding orbital is in an orthogonal plane. Hence, promotion of an electron from the nonbonding orbital is not possible without a significant change in the geometry of the molecule. The weak intensity is due to nonsymmetrical vibrations that slightly deform the molecule and lower its symmetry, allowing the n→π^* transition to acquire a finite probability.

Cyclic ketones absorb at longer wavelength than the corresponding open-chain analogues. In addition, there is a variation in the position of the absorption band with ring size in nonpolar solvents, as illustrated in Table 10-1.

In addition to alkyl groups, other substituents α to the carbonyl group affect the position of the n→π^* transition. The presence of an α bromine in the cyclohexanone series causes a bathochromic shift of λ_{max} of about 23 nm when the bromine is axial but a 5 nm shift when it is equatorial. Equatorially substituted 2-chloro-4-*tert*-butylcyclohexanone has its n→π^* maximum at a slightly shorter wavelength than that of the parent ketone, whereas the axial chlorine isomer has a more intense absorption band at considerably longer wavelength (Table 10-1). The strong bathochromic and hyperchromic effect of an α-halo substituent also is observed in steroidal ketones, but a satisfactory explanation for the phenomenon has not yet been advanced. In general, n→π^* transitions are easily recognizable by their low intensities, by the spectral shifts caused by substitution, and by the sensitivity of the position of the band to solvent effects.

Shifts in the position of absorption bands on going from the vapor phase to solution or from one solvent to another are mainly caused by differences in the solvation energies of the solute in the ground and excited electronic states.

TABLE 10-1 Absorption Data for Aliphatic Aldehydes and Ketones

		n→π* Transition		π→π* Transition	
Compound	Solvent	λ_{max} (nm)	ϵ_{max} (liter mol^{-1} cm^{-1})	λ_{max} (nm)	ϵ_{max} (liter mol^{-1} cm^{-1})
Formaldehyde	vapor	304	18	175	18,000
	isopentane	310	5		
Acetaldehyde	vapor	289	12.5	182	10,000
Acetone	vapor	274	13.6	195	9,000
	cyclohexane	275	22	190	1,000
Butanone	isooctane	278	17		
2-Pentanone	hexane	278	15		
4-Methyl-2-pentanone	isooctane	283	20		
Cyclobutanone	isooctane	281	20		
Cyclopentanone	isooctane	300	18		
Cyclohexanone	isooctane	291	15		
Cycloheptanone	isooctane	292	17		
Cyclooctanone	isooctane	291	15		
Cyclononanone	isooctane	293	17		
Cyclodecanone	isooctane	288	15		
2-Chloro-4-*tert*-butyl-cyclohexanone					
equatorial Cl	isooctane	286	17		
axial Cl	isooctane	306	49		

The effect of solvent on the position of the n→π* absorption has served as an important diagnostic tool (Section 9-3d). Indeed, the fundamental role of the unshared electron pair in this transition can be demonstrated by the disappearance of the n→π* band in acid solution in which the unshared pair is protonated.

Changing from a nonpolar solvent to a polar one results in a significant hypsochromic shift in the position of the n→π* transition. It has been shown that hydroxylic solvents of comparable dielectric constant cause a larger blue shift than do nonhydroxylic, polar solvents. The larger shifts occasioned by hydroxylic solvents are attributable in part to greater hydrogen bonding to the carbonyl oxygen lone pairs than to the π* electrons, thus lowering the energy of the ground state relative to that of the excited state. An example of solvent effects on the n→π* band of acetone is shown in Figure 10-1, and of an α,β-unsaturated ketone in Table 9-5.

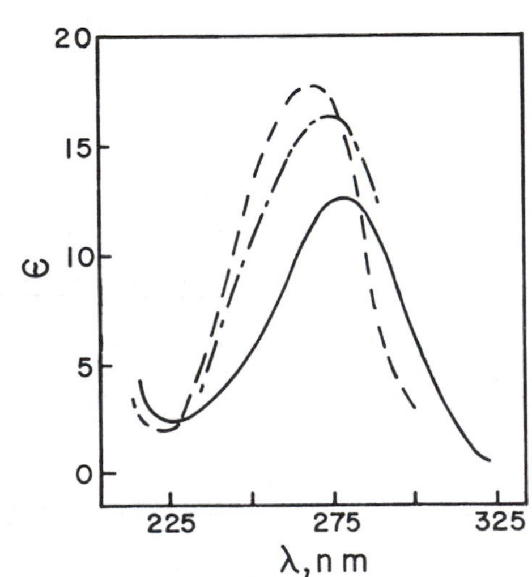

FIGURE 10-1 Solvent effects on the n→π* transition of acetone in hexane (———), 95% ethanol (—·—), and water (- - -).

10-1b The Octant Rule for Saturated Alkyl Ketones

The renaissance of the chiroptical techniques in the mid-1950s began with Djerassi's studies of the ketone n→π* transition (ref. 10-3). Optical rotatory dispersion was first applied to stereochemical and configurational investigations of compounds such as the 5α-cholestanones (**10-1, 10-2, and 10-3**). The n→π* carbonyl

C_8H_{17} C_8H_1 C_8H_{17}

O 1 3 5 H O H O H

10-1 **10-2** **10-3**

band is an example of an electric-dipole-forbidden, magnetic-dipole-allowed transition (see Section 9-4b) that is rendered optically active by its chiral environment. It has a low absorption intensity as measured by ϵ but a relatively strong rotational strength as measured by Δϵ, so that the *anisotropy ratio* Δϵ/ϵ of the carbonyl group is fairly large. In practice, this means that optically active ketones are generally favorable compounds to study, giving strong Cotton effects and good signal (Δϵ)/noise (ϵ) ratios.

The ORD curves of the isomeric 1-, 2-, and 3-keto steroids (**10-1, 10-2**, and **10-3**) are shown in Figure 10-2. While **10-2** and **10-3** exhibit positive Cotton effects, the 1-keto isomer (**10-1**) shows only a weak negative n→π* Cotton effect ($a=-25$) superimposed on a positive background rotation due to more intense Cotton effects at shorter wavelengths. The sign and magnitude of the Cotton effects are due to the chiral environment in the vicinity of the carbonyl chromophore. In principle, then, optically active chromophores in a molecule can be used as stereochemical probes.

Distinguishing between a 2- and a 3-keto steroid by means of their UV spectra is practically impossible; even the IR spectra of such compounds show only slight differences. However, ORD–CD allows a clear distinction. As seen in Figure 10-2, the magnitudes of the two positive Cotton effects are quite different: $a=+121$ for the 2-keto isomer (**10-2**) and +55 for the 3-isomer (**10-3**). Since these compounds are derived from natural products, they are optically pure and the amplitudes of their 290 nm Cotton effects can serve to differentiate them.

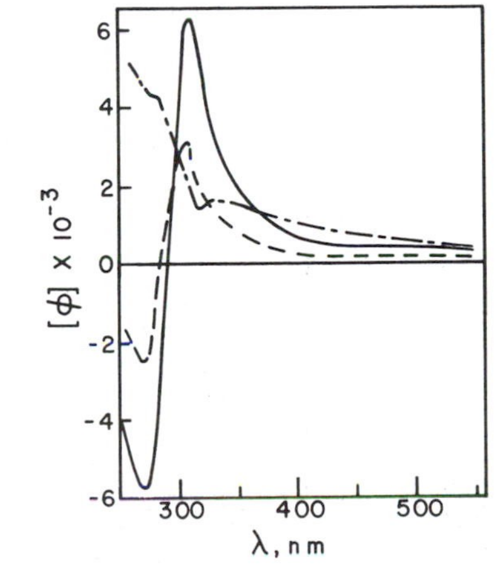

FIGURE 10-2 The ORD curves in methanol solution of 5α-cholestan-1-one (10-1) (— · —), -2-one (10-2) (———), and -3-one (10-3) (- - -). [Adapted with permission from C. Djerassi, *Optical Rotatory Dispersion*, McGraw-Hill Book Co., New York, 1960.]

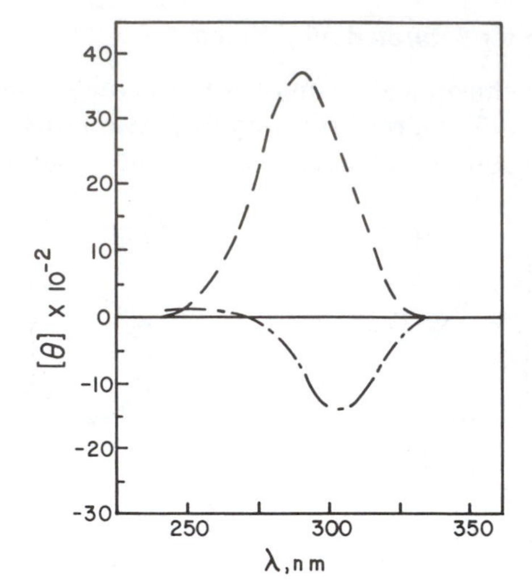

FIGURE 10-3 The CD spectra in methanol solution of 5-cholestan-1-one (10-1) (—·—) and -3-one (10-3) (- - -). [Reproduced with permission from C. Djerassi, H. Wolf, and E. Bunnenberg, *J. Am. Chem. Soc.*, **84**, 4552 (1962). Copyright 1962 American Chemical Society.]

The corresponding CD spectra of **10-1** and **10-3** are shown in Figure 10-3. The negative Cotton band of the 1-keto steroid (**10-1**) is seen much more clearly than by ORD, owing to the absence of background effects.

By 1961 a considerable amount of chiroptical data for the carbonyl chromophore had accumulated, and a semiempirical generalization, the *octant rule*, was proposed (references 10-4, 10-5, and 10-6). The rule allows the sign of the n→π^* Cotton effect to be deduced from the contributions of perturbing groups in each of eight sectors (octants) about the carbonyl group. Hence the octant rule allows one to decide upon not only the absolute configuration but also a likely conformation of the molecule. Conversely, configuration and conformation may be deduced from the sign of the Cotton effect.

In order to understand the octant rule, let us consider the symmetry properties of the orbitals that are involved in the n→π^* transition near 290 nm. As depicted in Figure 10-4(a), the n orbital on oxygen has a nodal plane that bisects the R—C—R′ bond angle and is perpendicular to the plane of the carbonyl group. For the π^* orbital, the plane containing the carbonyl group is a nodal plane. There is another nodal surface, not necessarily a plane, perpendicular to the C—O bond axis and intersecting it approximately midway between the C and O. The three nodal surfaces combine together to divide the entire space into the octants having the signs shown in Figure 10-5 for cyclohexanone.

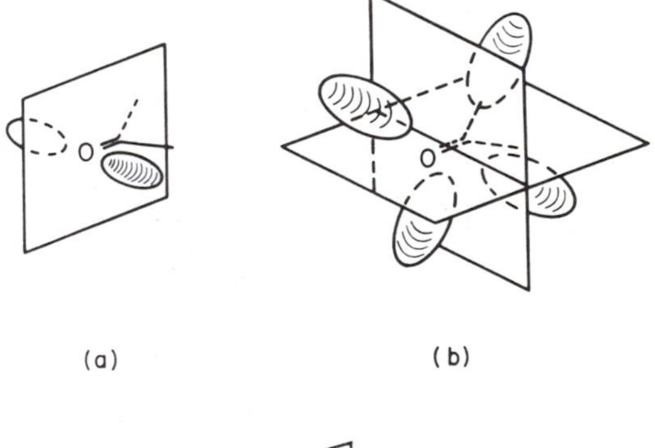

FIGURE 10-4 Nodal surfaces for a saturated ketone, R—CO—R′. (a) The nodal plane of the n orbital. It bisects the R—C—R′ angle and is perpendicular to the plane of the ketone. (b) The nodal surfaces of the π^* orbital. The plane of the carbonyl group is a nodal plane and there is another nodal surface, not necessarily a plane, perpendicular to the C-O axis and intersecting it between the carbon and oxygen atoms.

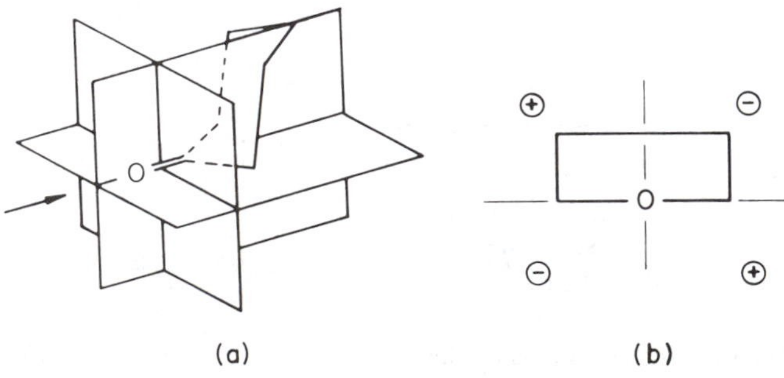

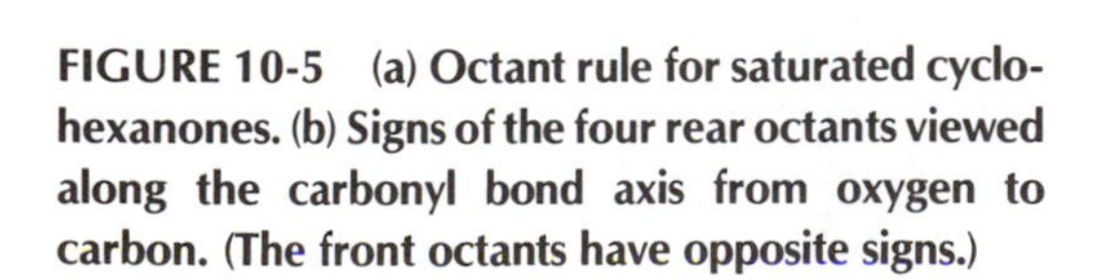

FIGURE 10-5 (a) Octant rule for saturated cyclohexanones. (b) Signs of the four rear octants viewed along the carbonyl bond axis from oxygen to carbon. (The front octants have opposite signs.)

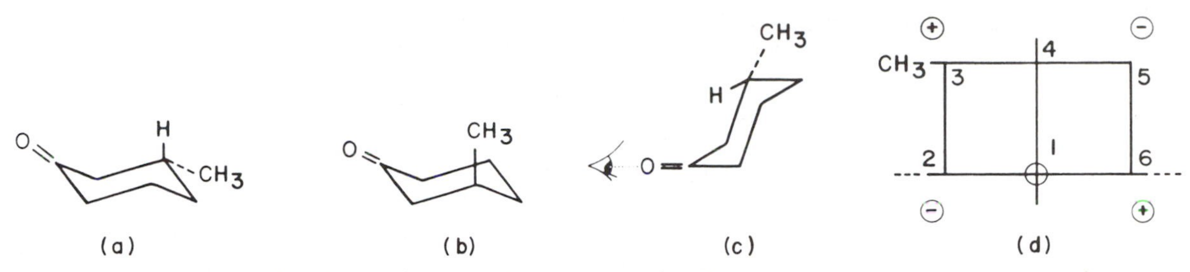

FIGURE 10-6 Equatorial and axial conformations of (+)-3-methylcyclohexanone and the octant rule projection of the equatorial conformer.

The octant rule states that substituents lying in the nodal planes make no contributions to the n→π* Cotton effect. Substituents within an octant contribute the sign of that octant to the overall sign of the Cotton effect. Since most substituents are usually on the same side of the nodal surface as the carbonyl carbon, the octant diagram is simplified by considering only the four rear octants. Relative intensities are determined qualitatively. For example, when both negative rear octants are occupied, the magnitude of the Cotton effect is enhanced.

A few substituents such as —F, $—\overset{+}{N}Me_3$, and cyclopropyl exhibit *antioctant* behavior, and several breakdowns of the octant rule for certain aliphatic ketones have been reported (refs. 10-5 and 10-7). Ketones having an axial α-chloro or α-bromo substituent display Cotton effects whose signs are determined only by the octant location of the halogen, even if most of the molecule occupies an oppositely signed octant; this is the *axial halo ketone rule* of Djerassi and Klyne (refs. 10-3 and 10-8).

As an illustration of the use of the octant rule, consider (+)-3-methylcyclohexanone, which exhibits a positive n→π* Cotton effect. We would predict that the chair form with the methyl group equatorial (Figure 10-6a) would be energetically favored over the conformer with the axial methyl group (Figure 10-6b). In the octant projection of (+)-3-methylcyclohexanone (Figure 10-6d), carbon atoms 2, 4, and 6 lie in nodal planes and thus make no contribution to the sign of the Cotton effect. The contribution of C-3 is canceled by that of C-5, which is equal but opposite. However, the methyl group at C-3 lies in a positive octant and is responsible for the sign of the positive Cotton effect. The absolute configuration of (+)-3-methylcyclohexanone is deduced to be *R*, since the enantiomeric molecule would lead to an octant prediction of a negative n→π* Cotton effect.

The octant rule is also of great value in the field of natural products. We saw in Figure 10-2 that the Cotton effect of cholestan-2-one (**10-2**) was about twice as large as that of the 3-isomer (**10-3**). The octant diagrams for the two steroids are shown in Figure 10-7. In the top octant diagram for **10-3**, the contributions of rings A and C are seen to cancel. The two angular methyl groups lie in the vertical nodal plane, so that only the methylene groups C-6 and C-7 (and C-15 and C-16, these two being very remote from the chromophore) contribute to the Cotton effect. According to the octant diagram, these contributions should be positive. In the case of cholestan-2-one (Figure 10-7b), the positive contributions are very strong and give rise to the relatively intense Cotton effect.

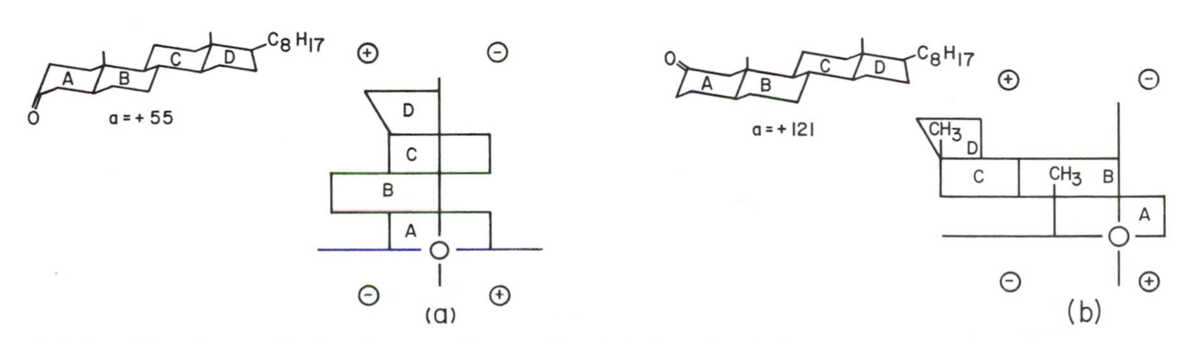

FIGURE 10-7 Octant diagrams for (a) cholestan-3-one and (b) cholestan-2-one.

TABLE 10-2 Absorption Data for Alkenes

Compound	Solvent	λ_{max} (nm)	ϵ_{max} (liter mol^{-1} cm^{-1})
Ethylene	vapor	162	10,000
cis-2-Butene	vapor	174	—
trans-2-Butene	vapor	178	13,000
1-Hexene	vapor	177	12,000
	hexane	179	—
Allyl alcohol	hexane	189	7,600
Cyclohexene	vapor	176	8,000
	cyclohexane	183.5	6,800
Cholest-4-ene	cyclohexane	193	10,000
1,5-Hexadiene	vapor	178	26,000
1,3-Butadiene	vapor	210	—
	hexane	217	21,000
1,3,5-Hexatriene	isooctane	268	43,000
1,3,5,7-Octatetraene	cyclohexane	304	—
1,3,5,7,9-Decapentaene	isooctane	334	121,000
1,3,5,7,9,11-Dodecahexaene	isooctane	364	138,000

A. J. Merer and R. S. Mulliken, *Chem. Rev.*, **63**, 639 (1969).

It is generally found that the contribution of a substituent to the total Cotton effect depends upon its octant for the sign and upon the distance from the chromophore for its magnitude, e.g., the more distant from the chromophore, the smaller the contribution.

10-1c Alkenes

Ethylene itself absorbs well outside of the generally accessible UV region with a broad absorption maximum at about 162 nm. The rather intense absorption ($\epsilon \sim 10{,}000$) is attributed to a $\pi \rightarrow \pi^*$ absorption maximum.

Geometrical isomers of disubstituted alkenes often can be differentiated by the longer wavelength absorption of the trans compounds. Solution spectra are displaced to longer wavelength than the corresponding vapor spectra. When two alkenic chromophores in the same molecule are insulated from each other by saturated carbons, their spectrum approximates the sum of the two chromophores, e.g., 1,5-hexadiene in Table 10-2. Conjugation of π electron systems of double bonds results in dramatic bathochromic shifts and increased intensities. Spectral data illustrating some of the above effects are given in Table 10-2.

The bathochromic shifts resulting from increased double bond conjugation may be illustrated in terms of Hückel molecular orbital theory (more sophisticated theories lead to the same result). The molecular orbitals and their relative energies for ethylene, 1,3-butadiene, and two higher conjugated homologues are depicted in Figure 10-8. Note that the energies of the highest occupied molecular orbitals (HOMO) increase, while those

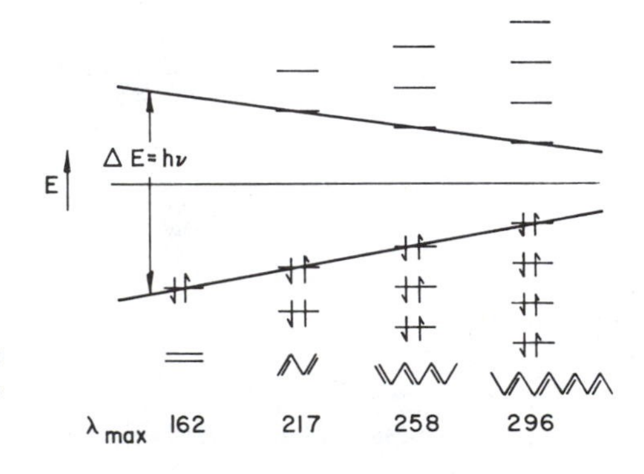

FIGURE 10-8 Schematic Hückel MO diagram illustrating the effect of conjugation on the $\pi \rightarrow \pi^*$ absorption maximum.

of the lowest unfilled molecular orbitals (LUMO) decrease with increasing conjugation. The observed transition involves promotion of an electron from the HOMO to the LUMO. One notes from Figure 10-8 that as the conjugated π system increases in length, the energy required for the transition becomes less, i.e., a bathochromic shift results.

The double bond, when rendered optically active by a chiral environment, gives rise to Cotton effects. While the UV spectra of the isomeric steroidal alkenes cholest-4-ene (**10-4**) and cholest-5-ene (**10-5**) are almost

C_8H_{17} C_8H_{17}

10-4 **10-5**

identical, the ORD curves are quite different. The data are presented in Figure 10-9. The shape of the ORD curve in the Cotton effect region is characteristic of the position of the double bond for various steroidal alkenes. Hence, once the ORD curves are known for these model compounds, they can be used to locate the position of double bonds in unknown compounds so long as the latter do not contain other chromophores that absorb in the same spectral region as the alkenic chromophore or interact with it.

The UV spectra of unconjugated alkenes have been shown to exhibit more than just the high intensity $\pi \rightarrow \pi^*$ transition in the accessible spectral region. Within the past decade, other bands have been observed in the vapor phase spectra of alkyl-substituted alkenes: higher energy Rydberg transitions and a weaker band somewhat above 200 nm, which has been assigned to a $\pi \rightarrow \sigma^*$ transition (refs. 10-6, 10-9, and 10-11). Studies by Yogev, Sagiv, and Mazur (ref. 10-10) based on linear dichroism (the measurement of $\epsilon_{\parallel}$ and $\epsilon_{\perp}$ of an oriented molecule) have shown that the broad UV band of **10-4** and **10-5** near 190 nm is actually composed of two transitions, a weaker one at about 203 nm (ϵ 4,000) and a more intense band at approximately 185 nm (ϵ 8,000). The first band is polarized at an angle of about 16.5° to the double bond axis, whereas the second transition is polarized along the direction of the double bond. The observed and resolved UV spectra, together with the CD curve of cholest-4-ene (**10-4**), are shown in Figure 10-10. This example illustrates the usefulness of CD in detecting transitions that are not resolved in the isotropic absorption spectrum. The CD spectrum of **10-4** reveals a negative Cotton effect associated with the longer wavelength transition and, overlapping it on the shorter wavelength side, a more intense positive Cotton effect whose maximum was not reached but which presumably is associated with the second transition. A caveat should be given here. Oppositely signed, overlapping Cotton effects will appear essentially as the algebraic sum of the component Cotton effects. Hence, depending upon their relative magnitudes and proximity to each other, the observed bands will be shifted in position and diminished in intensity relative to the isolated Cotton effects.

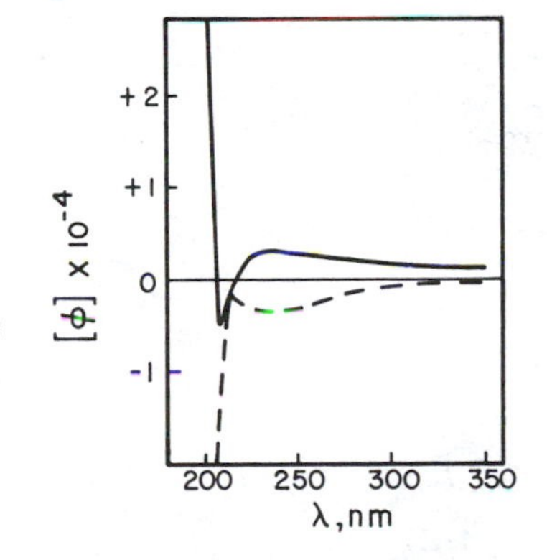

FIGURE 10-9 The ORD curves in cyclohexane of cholest-4-ene (10-4) (——) and cholest-5-ene (10-5) (- - -). [Reproduced with permission from A. Yogev, D. Amar, and Y. Mazur, *Chem. Commun.*, 339 (1967).]

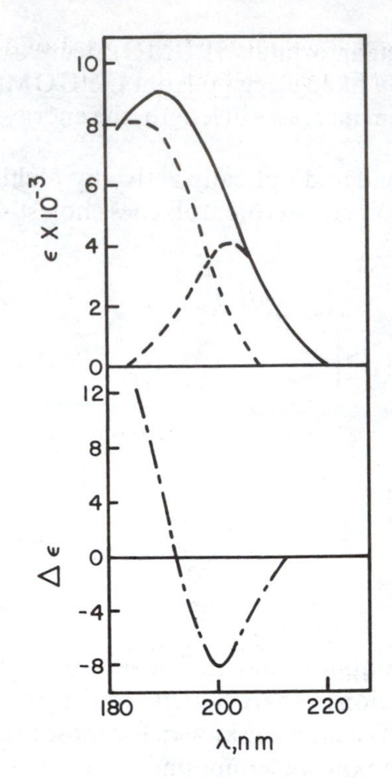

FIGURE 10-10 Isotropic absorption (——), resolved isotropic absorption (- - -), and CD (— · —) spectra of cholest-4-ene (10-4) in cyclohexane. [Reproduced with permission from A. Yogev, J. Sogiv, and Y. Mazur, *Chem. Commun.*, 411 (1972).]

10-2

Conjugated Chromophores

10-2a Dienes, Polyenes, and Woodward's Rules

Extensive studies of the UV spectra of alkenes, in particular of terpenes and steroids, led Woodward and then the Fiesers to formulate empirical rules for the prediction of the absorption maxima of various dienes and polyenes. These rules have proved quite useful in the solution of structural problems, particularly in the natural products field. The procedure involves beginning with a base absorption maximum for the parent chromophore and then incrementing it by values corresponding to each substituent attached to the parent π electron system. The values used in the Woodward–Fieser rules for diene absorption are given in Table 10-3.

EXAMPLES

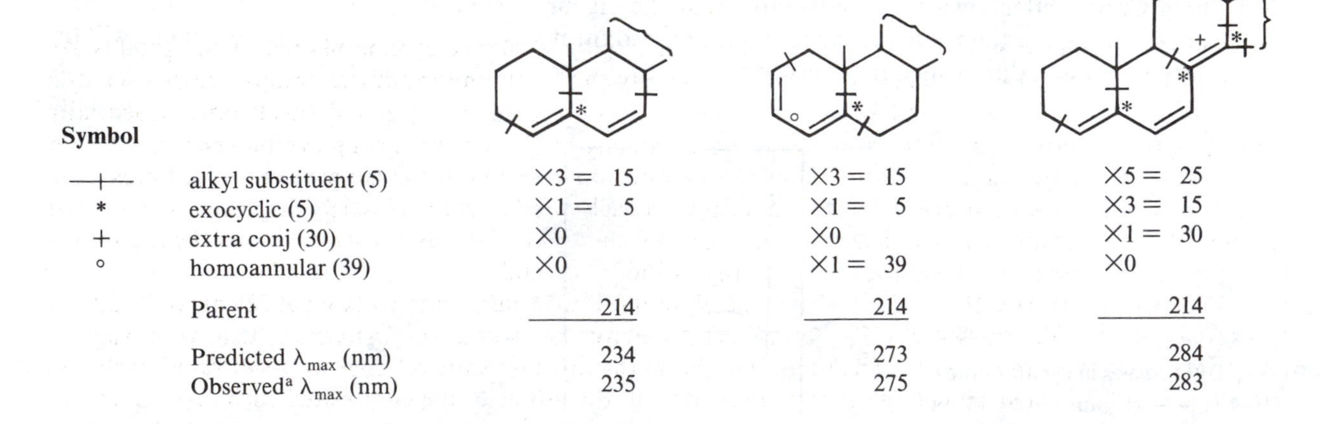

Symbol				
—+—	alkyl substituent (5)	×3 = 15	×3 = 15	×5 = 25
*	exocyclic (5)	×1 = 5	×1 = 5	×3 = 15
+	extra conj (30)	×0	×0	×1 = 30
°	homoannular (39)	×0	×1 = 39	×0
	Parent	214	214	214
	Predicted λ_{max} (nm)	234	273	284
	Observed[a] λ_{max} (nm)	235	275	283

[a]Data from A. I. Scott, *Interpretation of the Ultraviolet Spectra of Natural Products*, Pergamon, New York, 1964.

TABLE 10-3 Woodward–Fieser Rules for the Calculation of Absorption Maxima of Dienes and Polyenes (good to about ±3 nm)

	λ (nm)
Parent chromophore	214
Each alkyl substituent (at any position) add	5
Each exocyclic double bond add	5
*exocyclic to ring B only	
Each additional conjugated double bond (one end only) add	30
Each homoannular (rather than acyclic or heteroannular) add	39
°homoannular (same ring)	
Note: In cases for which both types of diene systems are present the longer wavelength one is chosen as the parent system.	
Do *not* count the double bond as a substituent, since this effect is included.	
Each polar group	
—O—acyl	0
—OR	6
—SR	30
—Cl, —Br	5
$—NR_2$	60
Solvent correction	0

R. B. Woodward, *J. Am. Chem. Soc.*, **63**, 1123 (1941); **64**, 72, 76 (1942); L. F. Fieser and M. Fieser, *Natural Products Related to Phenanthrene*, Reinhold, New York, 1949.

10-2b Conjugated Ketones and Woodward's Rules

Many organic molecules of interest contain both carbonyl and alkene chromophores. If the groups are separated by two or more sigma bonds, there is generally (but with important exceptions, see below) little electronic interaction and the effect of the two chromophores on the observed spectrum is essentially additive. Compounds in which the double bond is conjugated to a carbonyl group exhibit spectra in which both the alkenic $\pi \rightarrow \pi^*$ and the carbonyl $n \rightarrow \pi^*$ absorption maxima of the isolated chromophores have undergone bathochromic shifts of 15–45 nm, although each band is not necessarily displaced by an equal amount. Photoionization data indicate that the n orbital energy is relatively constant, so that the red shift is most probably caused by a lowering of the energy of the π^* orbital.

Crotonaldehyde ($CH_3CH{=}CH{-}CH{=}O$) in ethanol solution has an intense band at 220 nm ($\epsilon = 15{,}000$) and a weak band at 322 nm ($\epsilon = 28$). The low intensity and hypsochromic shift in hydroxylic solvents suggest that the 322 band is the $n \rightarrow \pi^*$ transition, and the bathochromic shift relative to a saturated carbonyl indicates that the excited π^* orbital now extends over all the atoms of the conjugated carbonyl group.

The solvent effect on the $\pi \rightarrow \pi^*$ transition is opposite to that on the $n \rightarrow \pi^*$ peak; the $\pi \rightarrow \pi^*$ absorption shifts to longer wavelength with increasing solvent polarity. The effect of solvent on the $n \rightarrow \pi^*$ and $\pi \rightarrow \pi^*$ transitions of mesityl oxide, $(CH_3)_2C{=}CH{-}CO{-}CH_3$, is illustrated in Table 9-5.

As the number of double bonds conjugated with the carbonyl increases, the $\pi \rightarrow \pi^*$ transition shifts to longer wavelength and its intensity increases, with the result that the much weaker $n \rightarrow \pi^*$ absorption appears as a shoulder or becomes completely obscured by the more intense, overlapping $\pi \rightarrow \pi^*$ band.

Alkyl substitution shifts the $\pi \rightarrow \pi^*$ and $n \rightarrow \pi^*$ maxima in opposite directions, the $\pi \rightarrow \pi^*$ being displaced to longer wavelength. Such effects of substitution on the position of the $\pi \rightarrow \pi^*$ transition can be predicted through the use of empirical rules similar to those already discussed for dienes and also first formulated by R. B. Woodward, then modified by the Fiesers. These rules, which have played an important role in assigning the structures of steroids and other natural products, are given in Table 10-4.

TABLE 10-4 Rules for the Calculation of the Position of $\pi \rightarrow \pi^*$ Absorption of Unsaturated Carbonyl Compounds

$-\overset{\delta}{C}=\overset{\gamma}{C}-\overset{\beta}{C}=\overset{\alpha}{C}-\underset{R}{\underset{\vert}{C}}=O$		λ (nm)
Parent α, β-unsaturated carbonyl compound (acyclic, six-membered, or larger ring ketone)		215
Each alkyl substituent: α add		10
β add		12
If other double bonds, for each γ, δ, etc., add		18
Each exocyclic carbon–carbon double bond add		5
Each extra conjugation add (Do not count double bond as substituent, as this effect is included.)		30
Each homoannular add		39
α,β bond in five-membered ring		−13
Aldehyde		−6
Each polar group		
—OH	α	35
	β	30
	δ	50
—O—Ac	α, β, or δ	6
—OR	α	35
	β	30
	γ	17
	δ	31
—SR	β	85
—Cl	α	15
	β	12
—Br	α	25
	β	30
—NR_2	β	95
Solvent correction		
Ethanol, methanol		0
Chloroform		1
Dioxane		5
Diethyl ether		7
Hexane, cyclohexane		11
Water		−8

R. B. Woodward, *J. Am. Chem. Soc.*, **63**, 1123 (1941); **64**, 72, 76 (1942); L. F. Fieser and M. Fieser, *Natural Products Related to Phenanthrene*, Reinhold, New York, 1949; A. I. Scott, *Interpretation of the UV Spectra of Natural Products*, Pergamon, New York, 1964.

EXAMPLES

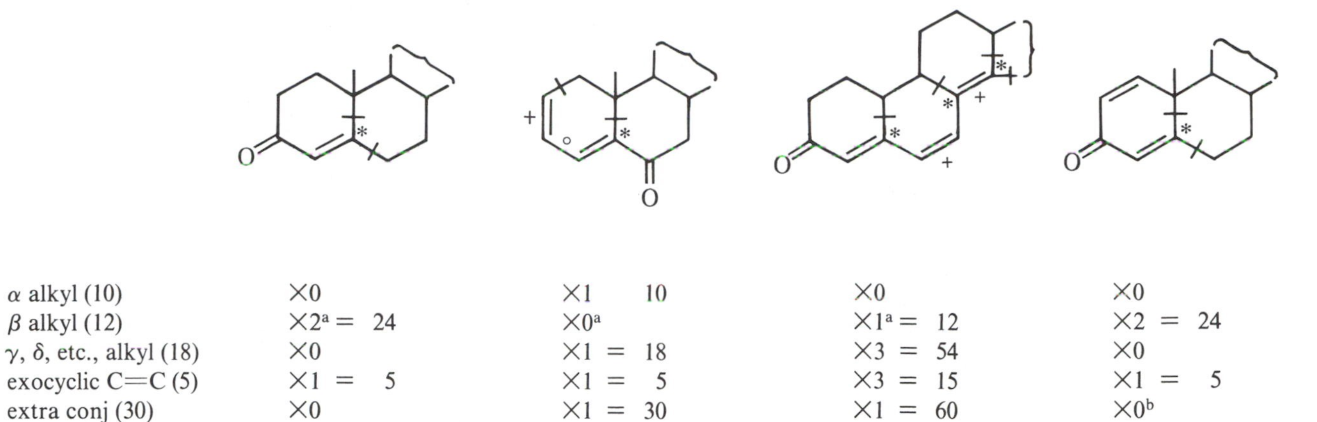

Symbol					
—+—	α alkyl (10)	×0	×1 10	×0	×0
—+—	β alkyl (12)	×2[a] = 24	×0[a]	×1[a] = 12	×2 = 24
—+—	γ, δ, etc., alkyl (18)	×0	×1 = 18	×3 = 54	×0
*	exocyclic C=C (5)	×1 = 5	×1 = 5	×3 = 15	×1 = 5
+	extra conj (30)	×0	×1 = 30	×1 = 60	×0[b]
°	homoannular (39)	×0	×1 = 39	×0	×0
	Parent	215	215	215	215
	Predicted λ_{max} (nm)	244	317	356	244
	Observed[c] λ_{max} (nm)	241	314	348	244

[a]Do *not* count the double bond as a substituent; this effect is included.
[b]Cross conjugated, count only the most substituted double bond.
[c]Data from A. I. Scott, *Interpretation of the Ultraviolet Spectra of Natural Products*, Pergamon, New York, 1964.

Both the $\pi \rightarrow \pi^*$ and $n \rightarrow \pi^*$ transitions of α,β-unsaturated aldehydes and ketones give rise to Cotton effects. In rigid systems such as the steroids, the electronic interactions of two *achiral* chromophores, the alkene and the carbonyl, can result in an inherently chiral α,β-unsaturated ketone chromophore. As a consequence, the octant rule, based on the concept of the inherently achiral carbonyl group, cannot be generally applied.

The $\pi \rightarrow \pi^*$ transition near 240 nm usually exhibits a Cotton effect of opposite sign to that of the $n \rightarrow \pi^*$ band. Recently several authors have reported a new Cotton effect of relatively high intensity near 215 nm, which overlaps the $\pi \rightarrow \pi^*$ transition and sometimes obscures its sign (ref. 10-12). Thus, caution should be used in the analysis of chiroptical spectra for α,β-unsaturated ketones.

Unsaturated carbonyl compounds such as β,γ-unsaturated ketones can also form an inherently chiral chromophoric system. Spectroscopically, in these compounds there is a chromophoric interaction, *coupling*, in which the forbidden $n \rightarrow \pi^*$ transition borrows intensity from the $\pi \rightarrow \pi^*$ band resulting in a substantial increase in the $n \rightarrow \pi^*$ Cotton effect near 310 nm.

A chirality rule, depicted in Figure 10-11, has been proposed (ref. 10-14) for this chromophore to correlate the sign of the $n \rightarrow \pi^*$ Cotton band. The plus and minus signs refer to the sign of the $n \rightarrow \pi^*$ Cotton effect. An application of this chirality rule for β,γ-unsaturated ketones (ref. 10-13) is given in Figure 10-12, which shows the UV and CD spectra of (+)-2-benzonorbornenone. The high rotational strength expected of such inherently chiral chromophores is reflected in the large molar ellipticity, $[\theta]_{307.5}$ +62,000 ($\Delta\epsilon$ +18.8). In order to have the orientation corresponding to that shown in the chirality rule (Figure 10-11), the structure shown on the UV–CD spectra would need to be turned over. The positive $n \rightarrow \pi^*$ Cotton effect centered at 307.5 nm then corresponds to the positive geometry in Figure 10-11. Note the three fingers and a shoulder of this Cotton effect, due to vibrational transitions within the $n \rightarrow \pi^*$ electronic transition. On the basis of the

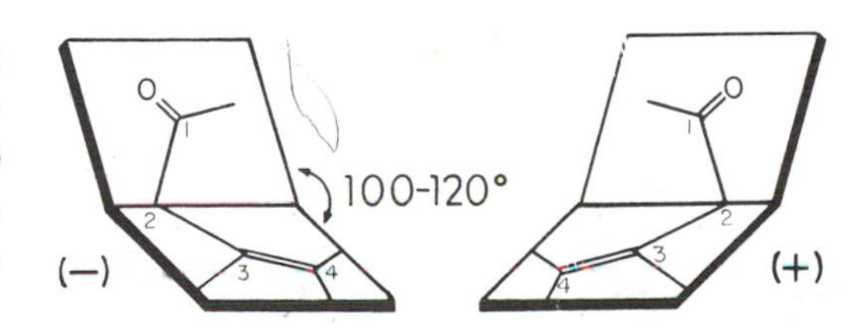

FIGURE 10-11 Correlation between chirality and sign of Cotton effect for inherently chiral β,γ-unsaturated ketones. The (+) and (−) signs refer to the $n \rightarrow \pi^*$ Cotton effect near 300 nm. [See A. Moscowitz, K. Mislow, M. A. W. Glass, and C. Djerassi, *J. Am. Chem. Soc.*, **84**, 1945 (1962).]

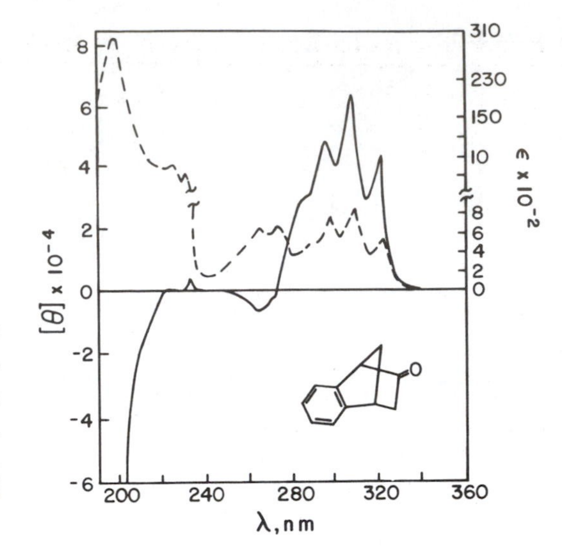

FIGURE 10-12 The UV (- - -) and CD (——) spectra in isooctane of (1*R*)-(+)-2-benzonorbornenone, a homoconjugated system. [Adapted with permission from D. J. Sandman, K. Mislow, W. P. Giddings, J. Dirlam, and G. C. Hanson, *J. Am. Chem. Soc.*, **90**, 4877 (1968). Copyright 1968 American Chemical Society.]

chiroptical correlation, which was in agreement with chemical evidence, Mislow and co-workers (ref. 10-15) were able to assign the 1*R* configuration to (+)-2-benzonorbornenone.

10-3

Aromatic Compounds

The benzene ring ranks with the carbonyl group and alkenes as one of the most widely studied chromophores. The spectrum of benzene above 180 nm consists of three well-defined absorption bands due to $\pi \rightarrow \pi^*$ transitions (Figure 10-13). An intense structureless band occurs at about 185 nm, with a somewhat weaker band (λ_{max} ~200) of poorly resolved vibrational structure overlapping the 185 nm absorption. The longest wavelength transition is a low intensity system centered near 255 nm and exhibiting characteristic vibrational structure. Data on the benzene absorption bands and some of the various nomenclature systems used to describe them are given in Table 10-5.

The benzene absorptions at 254 and 204, termed ^1L_b and ^1L_a in the Platt notation (ref. 10-16), are both forbidden, but the ^1L_a is able to "borrow" intensity from the allowed ^1B transition, which overlaps it at shorter wavelength. The different transition probabilities relate to configuration interaction because of the degeneracy of the highest occupied and lowest vacant orbitals in benzene. The superscript 1 indicates that the transition is to a singlet excited state. Benzene belongs to the D_{6h} point group, and the intensity of the symmetry-forbidden 254 nm ^1L_b transition should be zero. However, vibrational distortions from hexagonal symmetry result in a small net transition dipole moment and the observed low intensity.

The ^1L_b absorption, sometimes called the benzenoid band, is usually easily identifiable; it has about the

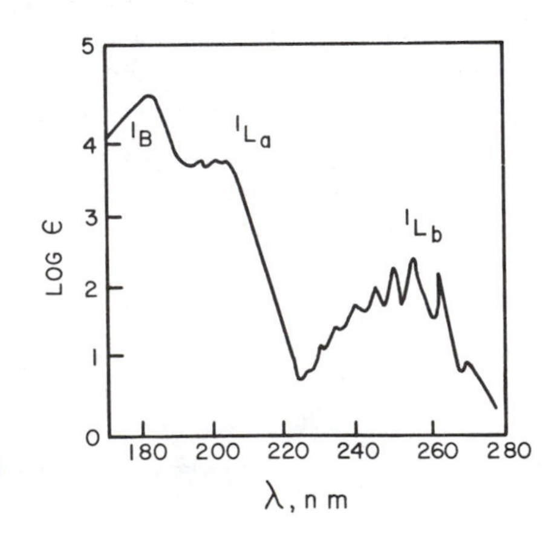

FIGURE 10-13 The UV spectrum of benzene in hexane.

TABLE 10-5 Notation Systems for Benzene Absorption Bands

λ_{max} (ϵ_{max}) in Hexane			Origin of the Spectral Notation
184 nm (68,000)	204 nm (8800)	254 nm (250)	
^1B	^1L_a	^1L_b	a
β	*p* (para)	α	b
$^1E_{2u}$	$^1B_{1u}$	$^1B_{2u}$	c
Second primary	primary	secondary	d
	K (conjugation)	*B* (benzenoid)	e

[a]Platt free-electron method notation: J. R. Platt, *J. Chem. Phys.*, **17**, 484 (1949).
[b]Empirical notation of Clar based on behavior of bands with temperature: E. P. Clar, *Aromatische Kohlenwasserstoffe*, Springer Verlag, Berlin, 1952.
[c]Molecular orbital approach based on the group theoretical notation of the transitions.
[d]Empirical notation: L. Doub and J. M. Vandenbelt, *J. Am. Chem. Soc.*, **69**, 2714 (1947); **71**, 2414 (1949).
[e]Early empirical notation: A Burawoy, *J. Chem. Soc.*, 1177 (1939); 20 (1941).

same intensity in benzene and its simple derivatives, ϵ 250–300. It also has similar well-defined vibrational structure with up to six vibrational bands, as in benezene itself. The vibrational structure is less evident in polar solvents and more sharply defined in vapor spectra or in nonpolar solvents.

Substitution of benzene by auxochromes (nonchromophoric groups, usually containing unshared electrons, e.g., —Cl, —NR_2, —OH), chromophores, or fused rings has different effects on the absorption spectrum. Because of their importance, we shall consider these effects in some detail.

Alkyl substitution shifts the benzene absorption to longer wavelengths and tends to reduce the amount of vibrational structure. Increases in band intensities are commonly observed. In general, substitution can perturb the benzene ring by both inductive and resonance effects. A methyl substituent causes the largest wavelength shift, hyperchromism, and the greatest change in vibrational intensities. The effect decreases as the methyl hydrogens are replaced by alkyl groups. This result is often cited as evidence for the importance of C—H hyperconjugation ($\sigma \rightarrow \pi$ electron interaction).

Data in Table 10-6 for the xylenes illustrate that bathochromic shifts caused by alkyl disubstitution are usually in the order para > meta > ortho. Alkylbenzenes, like alkyl-substituted alkenes, normally do not undergo any significant spectral changes on varying the solvent.

Introduction of polar substituents such as —NH_2, —OH, —OCH_3, —CHO, —COOH, and —NO_2 causes marked spectral changes. With these groups, the intensity of the ^1L_b band is enhanced. Much of the fine structure is lost in polar solvents, although it may be observed to some extent in nonpolar solvents. In addition, the ^1L_a band is shifted bathochromically; for example, in aniline, thiophenol, and benzoic acid it occurs in the 230 nm region (Table 10-6).

Substituents possessing nonbonding electrons can conjugate with the π system of the ring. Since the energy of the π^* state is lowered by delocalization over the entire conjugated system, the $n \rightarrow \pi^*$ absorption occurs at longer wavelength than in the corresponding unconjugated chromophoric substituent. For example, acetophenone, C_6H_5—CO—CH_3, exhibits an $n \rightarrow \pi^*$ absorption at 320 nm and the ^1L_b aromatic transition at 276 nm. Both bands are bathochromically shifted and considerably increased in intensity due in part to conjugation of the benzene π electrons with the π electrons of the carbonyl group.

The spectral changes found on conversion from phenol to phenoxide anion and from anilinium cation to aniline are of considerable interest and practical importance. In the case of aniline, the 280 nm band is most probably due to the benzenoid ^1L_b transition, red-shifted and enhanced by electron donation from the amino group to the ring. Resonance structures involving intramolecular charge transfer (e.g., **10-6**) make a sub-

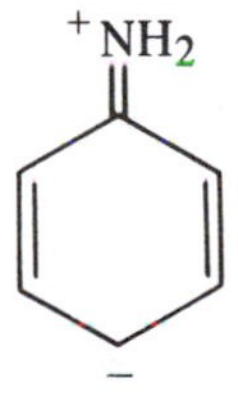

$^{+}NH_2$

10-6

TABLE 10-6 Absorption Data for Benzene and Derivatives[a]

Compound	Solvent	λ_{max} (nm)	ϵ_{max}	λ_{max} (nm)	ϵ_{max}	λ_{max} (nm)	ϵ_{max}	λ_{max} (nm)	ϵ_{max}
Benzene	hexane	184	68,000	204	8,800	254	250		
	water	180	55,000	203.5	7,000	254	205		
Toluene	hexane	189	55,000	208	7,900	262	260		
	water			206	7,000	261	225		
Ethylbenzene	ethanol[b]			208	7,800	260	220		
t-Butylbenzene	ethanol			207.5	7,800	257	170		
o-Xylene	25% methanol			210	8,300	262	300		
m-Xylene	25% methanol			212	7,300	264	300		
p-Xylene	ethanol			216	7,600	274	620		
1,3,5-Trimethylbenzene	ethanol			215	7,500	265	220		
Fluorobenzene	ethanol			204	6,200	254	900		
Chlorobenzene	ethanol			210	7,500	257	170		
Bromobenzene	ethanol			210	7,500	257	170		
Iodobenzene	ethanol			226	13,000	256	800		
	hexane			207	7,000	258	610	285 (sh)	180
Phenol	water			211	6,200	270	1,450		
Phenolate ion	aq NaOH			236	9,400	287	2,600		
Aniline	water			230	8,600	280	1,400		
	methanol			230	7,000	280	1,300		
Anilinium ion	aq acid			203	7,500	254	160		
N,N-Dimethylaniline	ethanol			251	14,000	299	2,100		
Thiophenol	hexane			236	10,000	269	700		
Anisole	water			217	6,400	269	1,500		
Benzonitrile	water			224	13,000	271	1,000		
Benzoic acid	water			230	10,000	270	800		
	ethanol			226	9,800	272	850		
Nitrobenzene	hexane			252	10,000	280 (sh)	1,000	330 (sh)	140
Benzaldehyde	hexane			242	14,000	280	1,400	328	55
	ethanol			240	16,000	280	1,700	328	20
Acetophenone	hexane			238	13,000	276	800	320	40
	ethanol			243	13,000	279	1,200	315	55
Styrene	hexane			248	15,000	282	740		
	ethanol			248	14,000	282	760		
Cinnamic acid									
cis-	hexane	200	31,000	215	17,000	280	25,000		
trans-	hexane	204	36,000	215	35,000	283	56,000		
	ethanol			215	19,000	268	20,000		
Stilbene									
cis-	ethanol			225	24,000	274	10,000		
trans-	heptane	202	24,000	228	16,000	294	28,000		
Phenylacetylene	hexane	202	44,000	248	17,000	hidden			
2,2′-Dimethylbiphenyl	hexane	198	43,000	228 (sh)	6,000	264	800		
Diphenylmethane	ethanol			220	10,000	262	500		

[a] If vibrational structure is present, λ_{max} refers to the subband of highest intensity.
[b] "Ethanol" should be taken to mean 95% ethanol.

stantial contribution to the ground electronic state, but the predominant contribution is to the excited state. In general, substituted benzenes for which this type of resonance form can be written have bathochromically shifted spectra relative to benzene and exhibit hyperchromism.

Conversion of aniline to the anilinium cation involves attachment of a proton to the nonbonding electron pair, removing it from conjugation with the π electrons of the ring (eq. 10-1). The absorption characteristics

$$C_6H_5\ddot{N}H_2 \xrightarrow{H_3O^+} C_6H_5\overset{+}{N}H_3 \qquad (10\text{-}1)$$

of this ion closely resemble those of benzene. The blue shift observed in the conversion of aniline to anilinium ion is typical of the spectral changes due to protonation of basic groups and can serve as a useful tool in structure elucidation.

Conversion of phenol to the phenolate anion makes an additional pair of nonbonding electrons available to the conjugated system, and both the wavelengths and the intensities of the absorption bands are increased (Table 10-6). Analogous to the information obtainable in the aniline–anilinium ion conversion, a suspected phenolic group may be determined by comparison of the UV spectrum of the compound in neutral and in alkaline (pH 13) solution.

The aniline–anilinium or phenol–phenolate conversion as a function of pH can demonstrate the presence of the two species in equilibrium by the appearance of an *isosbestic point* in the UV spectrum. If two substances, each of which obeys Beer's law, are in equilibrium, the spectra of all equilibrium mixtures at a constant total concentration intersect at a fixed wavelength. This point, termed the *isosbestic point*, is the wavelength at which the absorbances of the two species are equal. An example is shown in Figure 10-14 for 4-methoxy-2-nitrophenol.

In the UV spectra of the *cis*- and *trans*-cinnamic acids ($C_6H_5CH{=}CHCOOH$), the band at about 280 nm represents the ${}^1L_b$ benzenoid absorption displaced to longer wavelength and intensified by conjugation with the double bond and the carbonyl group of the carboxylic acid. The molar absorption coefficient of the trans isomer is more than twice that for the cis and is thought to be related to the longer chromophoric length in the trans compound. A similar relation is found in the *cis*- and *trans*-stilbenes. A generally applicable rule for many cis-trans isomer pairs is that the lowest energy $\pi \rightarrow \pi^*$ transition occurs at longer wavelength and is more intense for the trans isomer.

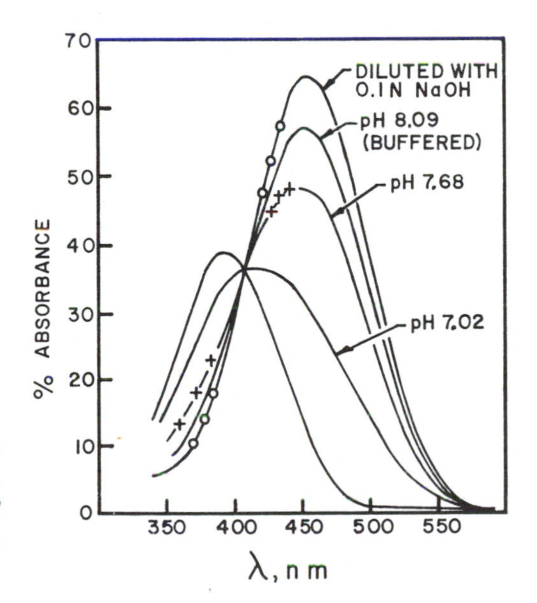

FIGURE 10-14 The spectra of 4-methoxy-2-nitrophenol as a function of pH. [Adapted with permission from H. H. Jaffé and M. Orchin, *Theory and Applications of Ultraviolet Spectroscopy*, John Wiley & Sons, Inc., New York, 1962, p. 562.]

The spectral data for biphenyl illustrate the effects of conjugation of adjacent benzene chromophores. The spectrum above 185 nm consists of two broad and intense bands at 202 nm (ϵ 44,000) and 248 nm (ϵ 17,000). The strong intensity of the 248 nm band indicates that it corresponds to the 205 nm ^1L_a band of benzene, shifted by conjugation between the two rings. The ^1L_b absorption is concealed beneath the broad envelope of the intense ^1L_a band.

The deviation from coplanarity in biphenyl is thought to be about 23°. Ortho substituents increase the deviation of the rings from coplanarity with concomitant loss of conjugation. This effect can be seen in the data for 2,2′-dimethylbiphenyl, which more resembles the sum of two independent alkylbenzene systems.

The presence of a saturated methylene group between two chromophores in a molecule results in almost complete loss of conjugation. This is well illustrated in the data for diphenylmethane (C_6H_5—CH_2—C_6H_5), for which the ^1L_b band at 262 nm has an ϵ of 500, almost exactly the sum of two isolated benzene rings.

Like the carbonyl group, a monosubstituted benzene ring is an example of an inherently achiral but chirally perturbed chromophore. Chiroptical studies of benzene chromophores are surprisingly recent. Prior to 1965 only a few measurements had been reported, and disagreement existed in the literature as to whether a benzene ring in a chiral molecule could exhibit optically active transitions. The confusion was in part caused by conflicting reports on compounds such as α-phenylethanol (**10-7**). Some workers had reported

CH_3
H—C—OH
C_6H_5

10-7

that this alcohol in its chiral form exhibits an ORD Cotton effect in the 260 nm region, while others measured the same compound and reported no Cotton effect, only a plain ORD curve. The CD of (*S*)-(−)-α-phenylethanol (Figure 10-15) clearly shows the presence of a positive Cotton effect containing vibrational fine structure in the 260 nm region as well as a more intense ^1L_a Cotton band near 210 nm. Both Cotton effects are due to aromatic transitions since only the benzene ring absorbs in the region above 200 nm. The 260 nm band is due to the symmetry-forbidden ^1L_b transition. Its relative weakness, $[\theta]_{max}$ +200, explains why previous workers sometimes had difficulty in discerning this Cotton effect superposed on a steeply falling background ORD curve. The background rotation is caused by the more intense negative Cotton effect near 210 nm, which also determines the sign of this compound in the visible region. Hence, difficulties in observing benzenoid Cotton effects were a question of instrument sensitivity rather than any intrinsic difference in the nature of the benzene ring as an optically active chromophore.

The benzenoid transitions are better resolved in CD spectroscopy. Thus, Salvadori and co-workers (ref. 10-17) reported the UV and CD spectra of a chiral aromatic hydrocarbon, (*S*)-(+)-2-phenyl-3,3-dimethylbutane (**10-8**), for which they were able to measure three aromatic Cotton effects in the region above

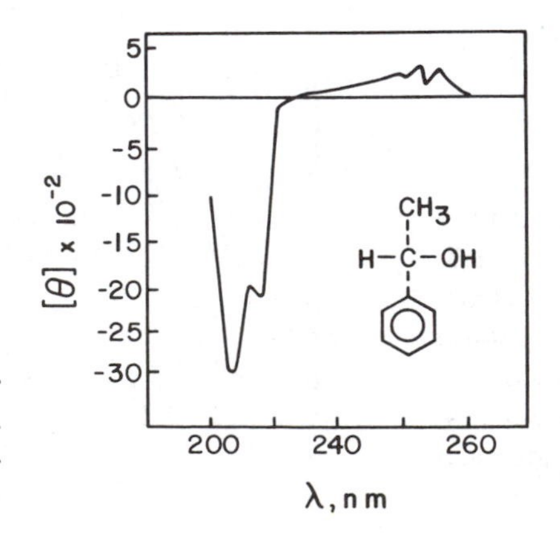

FIGURE 10-15 The CD in heptane of (*S*)-(−)-α-phenylethanol. [Adapted with permission from L. Verbit, *J. Am. Chem. Soc.*, **87**, 1617 (1965). Copyright 1965 American Chemical Society.]

$$\begin{array}{c} H \\ \equiv \\ C_6H_5 \blacktriangleright C \blacktriangleleft CH_3 \\ \equiv \\ C(CH_3)_3 \end{array}$$

10-8

185 nm. The spectra are shown in Figure 10-16. The Cotton effects, all of which are positive, correspond to the ^1B, ^1L_a, and ^1L_b transitions of the aromatic ring. The ^1L_b band in the 260 nm region, $[\theta]_{max}$ +5,000 is more than an order of magnitude more intense than those displayed by typical open chain phenyl compounds (compare α-phenylethanol) and suggests the possibility of restricted conformational mobility. Similar large ellipticity values are found in rigid aromatic systems such as alkaloids. Evidence that chiral 2-phenyl-3,3-dimethylbutane is constrained to a single conformation because of steric effects imposed by the bulky *tert*-butyl group was obtained by measurement of the 260 nm Cotton effect at −100° C (ref. 10-7). The almost complete absence of CD changes over this temperature range indicates that the compound is already in its lowest energy conformation at room temperature, at which the spectra in Figure 10-16 were determined.

The polycyclic hydrocarbon hexahelicene is the classic example of an inherently chiral chromophore. Its chirality extends in a helical manner throughout the entire molecule, and its enormous D line rotation, $[\alpha]_D$ 3750°, is characteristic of this class of compounds. The UV and CD spectra of (+)-hexahelicene are shown in Figure 10-17. In 1972, Lightner and co-workers (ref. 10-18) reported an x-ray study of optically active

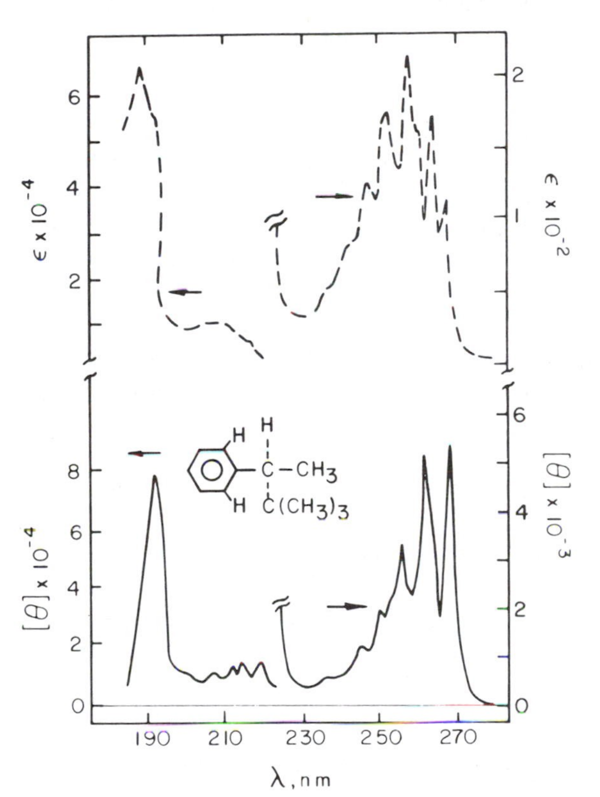

FIGURE 10-16 The UV (- - -) (heptane) and CD (——) (methylcyclohexane-isopentane, 1:3) spectra of (*S*)-(+)-2-phenyl-3,3-dimethylbutane. The arrows indicate the respective scales. [Adapted with permission from P. Salvadori, L. Lardicci, R. Menicagli, and C. Bertucci, *J. Am. Chem. Soc.*, **94**, 8598 (1972). Copyright 1972 American Chemical Society.]

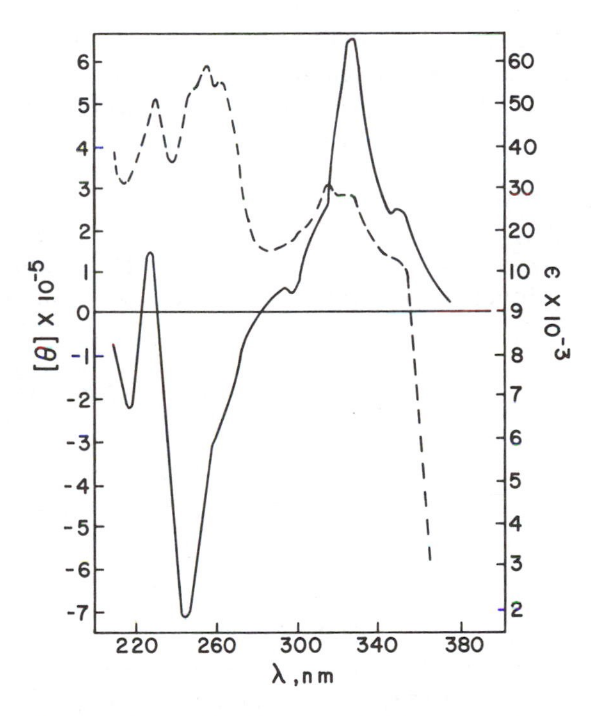

FIGURE 10-17 The UV (- - -) and CD (——) in methanol of (+)-hexahelicene. [Adapted with permission from M. S. Newman, R. S. Darlak, and L. Tsai, *J. Am. Chem. Soc.*, **89**, 6191 (1967). Copyright 1967 American Chemical Society.]

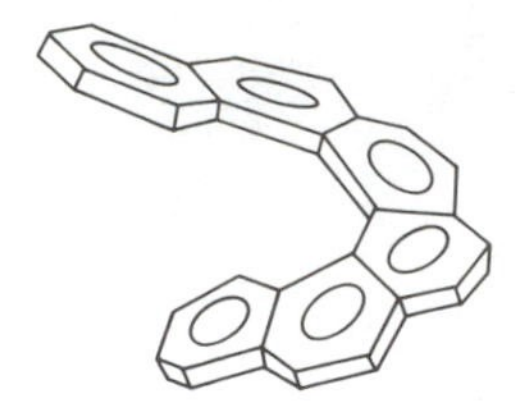

FIGURE 10-18 Absolute configuration of (+)-hexahelicene.

2-bromohexahelicene. Their results show that (+)-hexahelicene has right-handed helicity or the *P* configuration in the *RS* nomenclature system (Figure 10-18).

10-4

The Exciton Chirality Rule

A powerful rule for predicting absolute stereochemistry was advanced recently by Harada and Nakanishi (ref. 10-19): the exciton chirality rule. It is based on coupling of the locally excited states of two or more chromophores that are situated sufficiently near to one other to lead to splitting of the electronic excited states (exciton splitting), as in Figure 10-19. If the electronic transitions of the independent chromophores are strongly allowed, e.g., $\pi \rightarrow \pi^*$, and the orientations of the associated electric transition moment dipoles and the vector connecting them are known, then the absolute stereochemical relationship of the chromophores can be determined with a very high degree of certainty. Thus, in a bichromophoric molecule, electronic excitations to the two split energy levels (Figure 10-19) yield two absorption bands (Figure 10-20, lower, dotted lines), whose λ_{max} are separated by the Davydov splitting energy ($\Delta\lambda$) and which appear in the UV–vis spectrum as a composite, summed curve (solid lines). The composite exciton absorption band is broader than its component parts, and its profile depends on the magnitude of $\Delta\lambda$ and the shapes and intensities of the component bands. If the *bichromophoric* molecule is also optically active, the two component (UV) electronic transitions have counterpart CD transitions with mutually opposite Cotton effect signs (Figure 10-20, middle, dotted lines). Summation of the component CD curves gives net *bisignate* Cotton effects (solid lines) with the λ_{max} of the bisignate components flanking the λ_{max} of the composite UV curve, as is characteristic of exciton splitting. The signed order of the bisignate CD Cotton effects [either long wavelength positive (+) followed by short wavelength negative (−), or vice versa] is determined by the

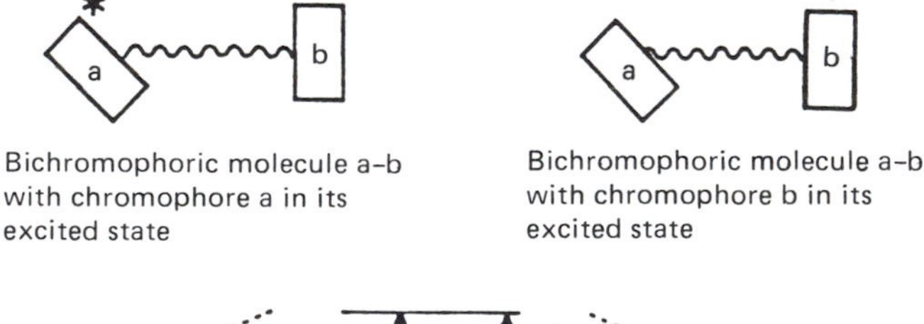

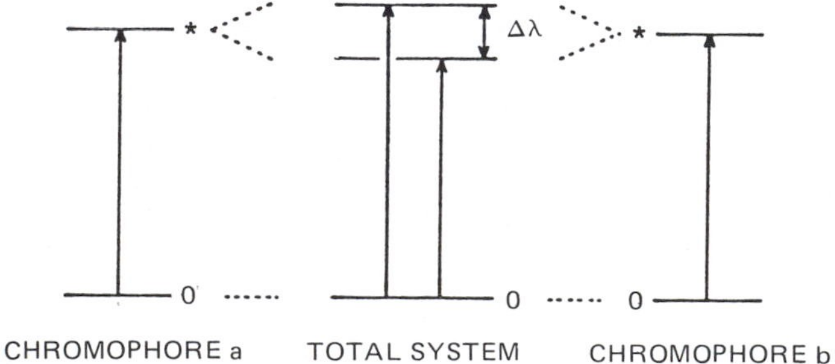

FIGURE 10-19 (Upper) The excited state (exciton) of a bichromophoric molecule a–b is delocalized between the component chromophores a and b. (Lower) Exciton coupling of the excited states of chromophores a and b splits the excited state of the system into two energy levels. The energy gap, Δλ, is called the Davydov splitting. [Adapted with permission from H. Harada and K. Nakanishi, *Circular Dichroism Spectroscopy—Exciton Coupling in Organic Stereochemistry*, University Science Books, Mill Valley, CA, 1983.]

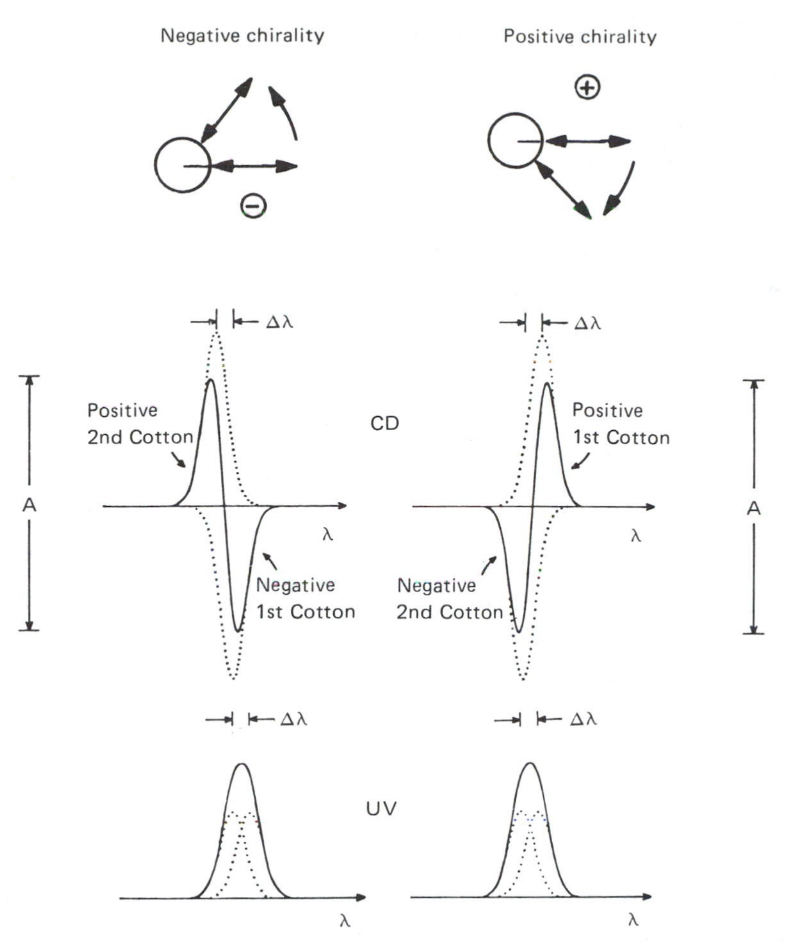

FIGURE 10-20 Upper: Negative and positive chirality of the electric dipole transition moments of chromophores involved in exciton coupling in a bichromophoric molecule. The chirality is determined by rotating the nearer transition moment vector into coincidence with the farther along the shortest path. Anticlockwise rotation corresponds to negative chirality; clockwise rotation corresponds to positive chirality. Middle: Summation CD curves (solid lines) of two Cotton effects (dotted lines) of opposite signs separated by Davydov splitting, $\Delta\lambda$. Positive chirality is associated with positive long wavelength followed by negative short wavelength Cotton effects, whereas negative chirality is associated with negative long wavelength followed by positive short wavelength Cotton effects. The amplitude, A, of split CD Cotton effects is defined as $A = \Delta\epsilon_1 - \Delta\epsilon_2$, where $\Delta\epsilon_1$ and $\Delta\epsilon_2$ are the intensities of the long wavelength and short wavelength Cotton effects, respectively. Lower: In the UV-vis spectrum, summation of the two exciton splitting bands (dotted lines) leads to a single maximum (solid lines) when $\Delta\lambda$ is small. [Adapted with permission from H. Harada and K. Nakanishi, *Circular Dichroism Spectroscopy—Exciton Coupling in Organic Stereochemistry*, University Science Books, Mill Valley, CA, 1983.]

chirality (or screw sense) of the dihedral angle made by the intersection of the component chromophores' relevant electric dipole transition moment vectors with the vector connecting them. As illustrated in the Newman-type projection diagrams of Figure 10-20, the transition moment vectors make a left-handed [negative (−)] dihedral angle, called negative chirality, when the nearer transition moment vector can be made to eclipse the farther transition moment vector by rotating along the shortest path in an anticlockwise direction. A positive (+) chirality is associated with a clockwise rotation. Clearly, the rule becomes difficult to apply when the two transition moment vectors cannot be defined accurately, or when the dihedral angle is close to 0° (nearly eclipsed, synperiplanar) or 180° (antiperiplanar). Under these special circumstances, use of the exciton chirality rule to assign absolute configuration is rendered tenuous. Generally, however, the signed order of the bisignate Cotton effects gives the chirality and hence the absolute configuration, as shown in Figure 10-20.

The requirements for the exciton chirality rule are summarized below, and the rule is expressed diagrammatically in Figure 10-20.

1. Large extinction coefficient values in UV spectra.
2. Isolation of the band in question from other strong absorptions.
3. Established direction of the electric transition moment in the geometry of the chromophore (if the skew angle made by the transition moment vectors with the vector connecting them is approximately 0° or 180°, then the directions must be determined exactly (refs. 10-6 and 10-17).
4. Unambiguous determination of the exciton chirality in space, inclusive of configuration and conformation.
5. Negligible molecular orbital overlap or homoconjugation between the chromophores, if any.

To illustrate exciton splitting and application of the exciton chirality rule to determine absolute configuration, consider a case in which the two chromophoric units (a and b of Figure 10-19) are each anthracene. Anthracene has a very intense ($\epsilon \sim 200{,}000$) UV absorption (^1B_b in the Platt notation, Section 10-3) near 250 nm polarized along the long axis of the molecule, as shown in Figure 10-21. Anthracene also has a much weaker ($\epsilon \sim 7{,}000$) absorption (designated ^1L_a) near 350 nm polarized along the short axis of the

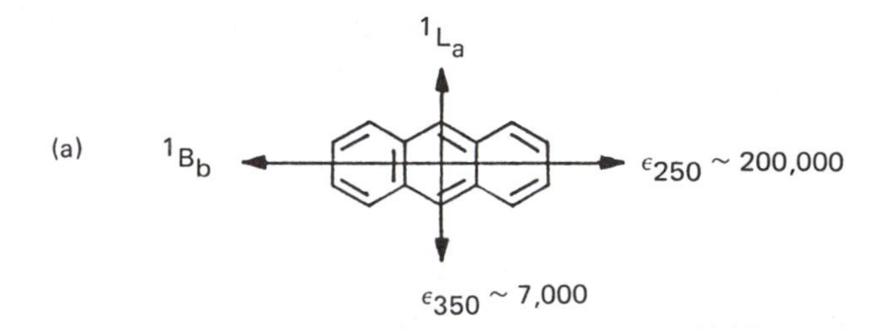

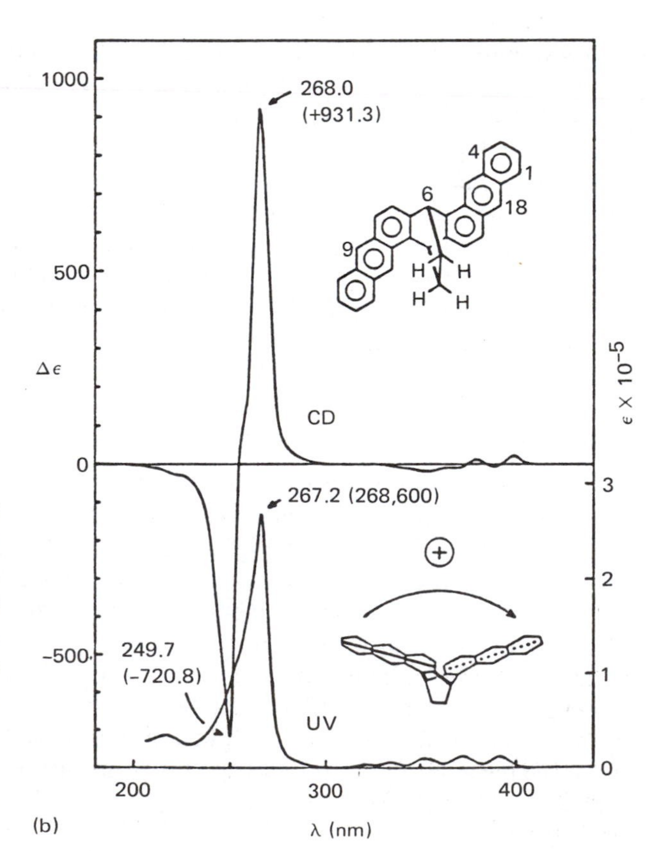

FIGURE 10-21 **(a) Orientation (polarization) of the UV transitions of anthracene. (b) CD and UV spectra of (6*R*,15*R*)-(+)-6,15-dihydro-6,15-ethanonaphtho[2,3-*c*]pentaphene. The positive sign of the first Cotton effect is in agreement with the positive exciton chirality between the long axes of the two anthracene moieties.** [Reproduced with permission from H. Harada and K. Nakanishi, *Circular Dichroism Spectroscopy—Exciton Coupling in Organic Stereochemistry*, University Science Books, Mill Valley, CA, 1983.]

molecule. The ^1B_b transition is ideal for observing exciton coupling in the CD. When the anthracene chromophores are held in a fixed orientation by the bicyclo[2.2.2]octane framework of 6,15-dihydro-6,15-ethanonaphtho[2,3-*c*]pentaphene, two enantiomeric configurations are possible. The 6*R*,15*R* absolute configuration has a positive (+) chirality for the anthracene ^1B_b transition moments; the 6*S*,15*S* configuration has a negative (−) chirality. If exciton splitting takes place in the molecule, a knowledge of the CD spectrum of an optically active sample and application of the exciton chirality rule should be sufficient to determine the absolute configuration of the major enantiomer present in the CD sample. Evidence for exciton coupling in a sample of the (+)-6,15-dihydro-6,15-ethanonaphtho[2,3-*c*]pentaphene can be found in its UV spectrum (Figure 10-21), where the UV band is broader than the component anthracene ^1B_b bands, and its ϵ_{max} is less than the sum of the component ϵ_{max}. More obvious evidence for exciton coupling can be seen in the CD spectrum, where a very strong bisignate Cotton effect, with (+) long wavelength and (−) short wavelength components, is centered on the UV transition. The signed order of the Cotton effects corresponds to a (+) chirality of the ^1B_b transition moment vectors, hence the 6*R*,15*R* absolute configuration for the major enantiomer of the CD sample.

Although there are many instances where the exciton chirality rule can be applied fruitfully to obtain stereochemical information that is otherwise difficult to acquire, it has been applied with particular success to the stereochemistry of 1,2-glycol systems. The glycol is usually derivatized as its dibenzoate or bis(*p*-dimethylaminobenzoate) in order to provide two chromophores with strong ($\pi \rightarrow \pi^*$) UV absorptions whose associated electric transition dipole moments lie roughly along the axis connecting the ester group and the para benzene carbon, as shown for the two mirror image structures in Figure 10-22. The exciton chirality rule predicts the sign of the dihedral angle and hence the absolute configuration of the glycol, made by the intersection of the transition moments with the (vector) carbon–carbon bond connecting the two hydroxyl groups of the glycol.

An example of an application of the exciton chirality rule to determine the absolute stereochemistry of a vicinal glycol may be seen in cholest-5-ene-3β,4β-diol derivatized as its bis(*p*-dimethylaminobenzoate) ester, whose CD spectrum is shown in Figure 10-23. Here, the interacting chromophores are the *p*-dimethylaminobenzoates, and the spatial arrangement of the coupled electric dipole moments leads to a predicted negative (−) chirality [Figures 10-20 (left) and 10-22 (left)]. The exciton splitting may be seen in the strong bisignate Cotton effect: (−) at 320.5 nm ($\Delta\epsilon = -63.1$) followed by a (+) at 295.5 nm ($\Delta\epsilon = +39.7$) (Table 10-7 and Figure 10-23). Therefore, the absolute configuration of the steroid must be 10*R*—as shown—rather than

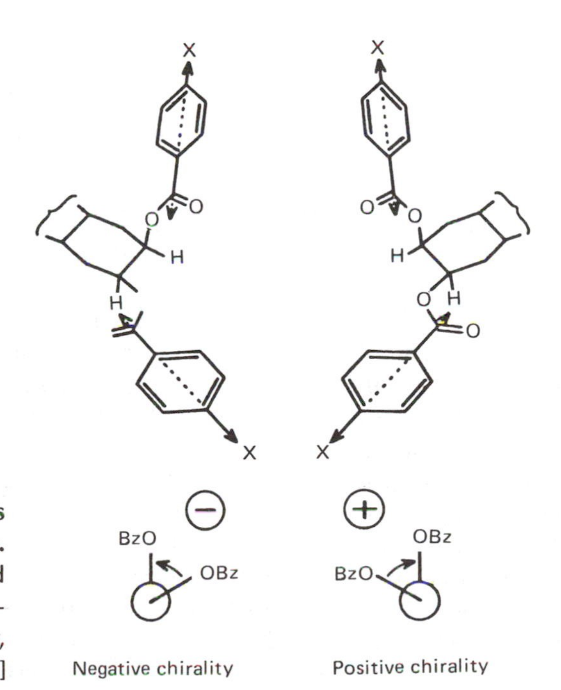

FIGURE 10-22 Chiralities of α-glycol dibenzoates and their $\pi \rightarrow \pi^*$ electric transition dipole moments. [Reprinted with permission from H. Harada and K. Nakanishi, *Circular Dichroism Spectroscopy—Exciton Coupling in Organic Stereochemistry*, University Science Books, Mill Valley, CA, 1983.]

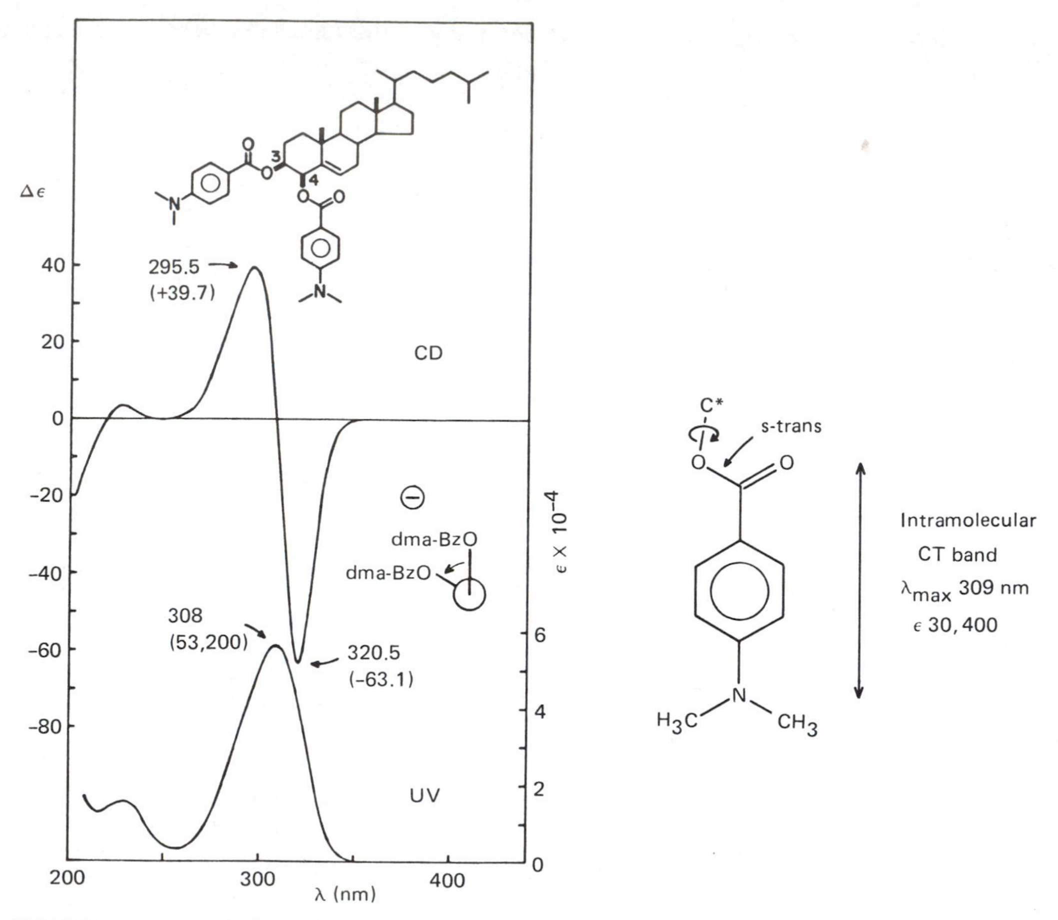

FIGURE 10-23 CD and UV spectra of cholest-5-ene-3β,4β-diol bis(*p*-dimethylaminobenzoate) in ethanol, and the direction of the electric transition dipole moment in the *p*-dimethylaminobenzoate chromophore. [Reprinted with permission from H. Harada and K. Nakanishi, *Circular Dichroism Spectroscopy—Exciton Coupling in Organic Stereochemistry,* University Science Books, Mill Valley, CA, 1983.]

the mirror image absolute configuration, which would have a (+) chirality of the 3β,4β-diol bis(*p*-dimethylaminobenzoate).

Compared with bis(*p*-dimethylaminobenzoate) ester derivatives, dibenzoate esters of 1,2-glycols may also show exciton coupling (Table 10-7). However, the former has a more intense ^1L_a electronic transition, which is also strongly red-shifted to a less crowded region of the UV; hence, bis(*p*-dimethylaminobenzoates) are the derivatives of choice.

The exciton chirality rule is not limited to proximal chromophores or 1,2-glycol derivatives. However, the *intensity* of interaction (hence the splitting, $\Delta\lambda$) falls off with distance, as illustrated for the 1,8-glycol derivative in Figure 10-24. It is also highly geometry dependent (Figure 10-25), with a ~90° orientation of the interacting electric dipole transition moments leading to maximum splitting, but with only weak splitting being observed for an ~0° dihedral angle or an ~180° dihedral angle. For example the hydroxyl groups of (2*S*,3*R*)-camphane-2,3-diol are eclipsed (synperiplanar) and thus the bis(*p*-dimethylaminobenzoate) derivative is predicted to show no exciton coupling. A very weak bisignate CD is observed, nonetheless (Figure 10-26), and the signed order of the component Cotton effects may be used to infer a (+) chirality. The exciton chirality rule should not, however, be used here to predict the absolute configuration because a very small change in relative orientation of the electric transition moments (as might accompany a change in solvent or temperature) can invert the signs and hence the chirality.

The exciton chirality rule is not limited to aromatic chromophores. An important application may be

TABLE 10-7 UV and CD Data for Cholest-5-en-3β,4β-diol Bis(p-dimethylaminobenzoate) (upper) and 5α-Cholestan-2α,3β-diol Dibenzoate (lower) Showing Exciton Splitting

Compound	Exciton Chirality	UV λ_{max} (nm)	UV ϵ	CD λ_{ext} (nm)	CD $\Delta\epsilon$
Cholest-5-en-3β,4β-diol bis(p-dimethylaminobenzoate)	$\ominus$	308	53,200	320.5	−63.1
				308	0.0
				295.5	+39.7
		227	13,400		
5α-Cholestan-2α,3β-diol dibenzoate	$\ominus$	229.3	26,700	234	−13.9
				228	0.0
				219	+14.6

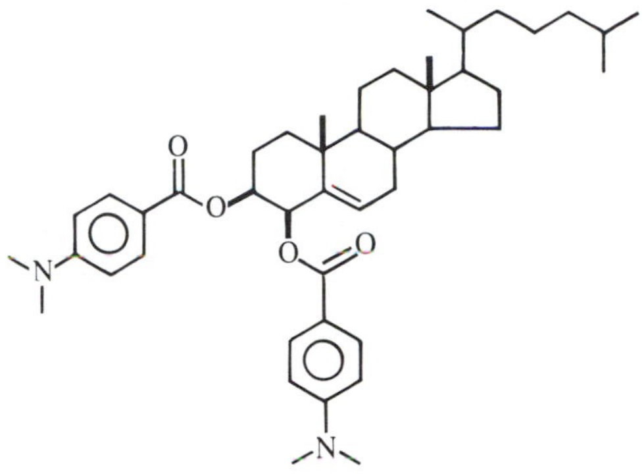

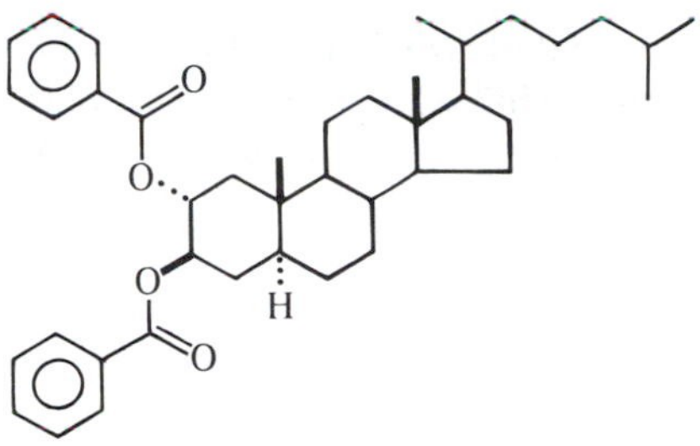

Data from H. Harada and K. Nakanishi, *Circular Dichroism Spectroscopy—Exciton Coupling in Organic Stereochemistry*, University Science Books, Mill Valley, CA, 1983.

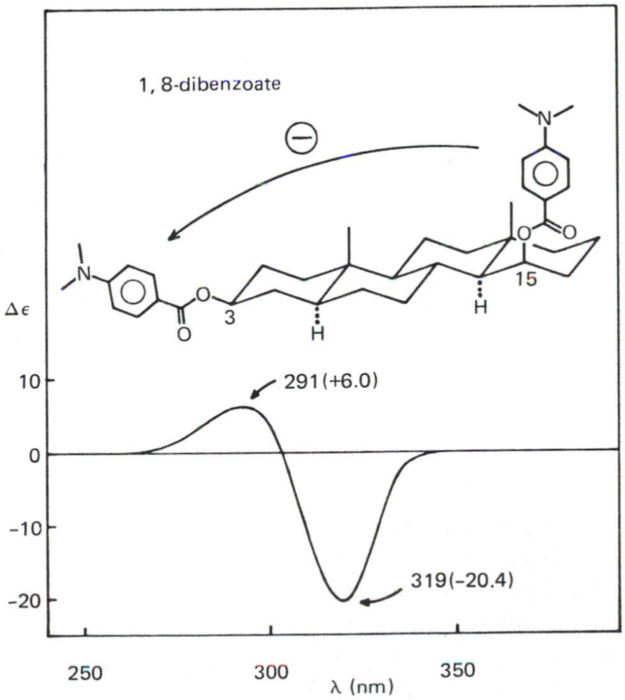

FIGURE 10-24 Coupled Cotton effects of a remote dibenzoate system (1,8-glycol); CD spectrum of D-homo-5α-androstane-3β,15β-diol bis(p-dimethylaminobenzoate) in ethanol. [Reproduced with permission from H. Harada and K. Nakanishi, *Circular Dichroism Spectroscopy—Exciton Coupling in Organic Stereochemistry*, University Science Books, Mill Valley, CA, 1983.]

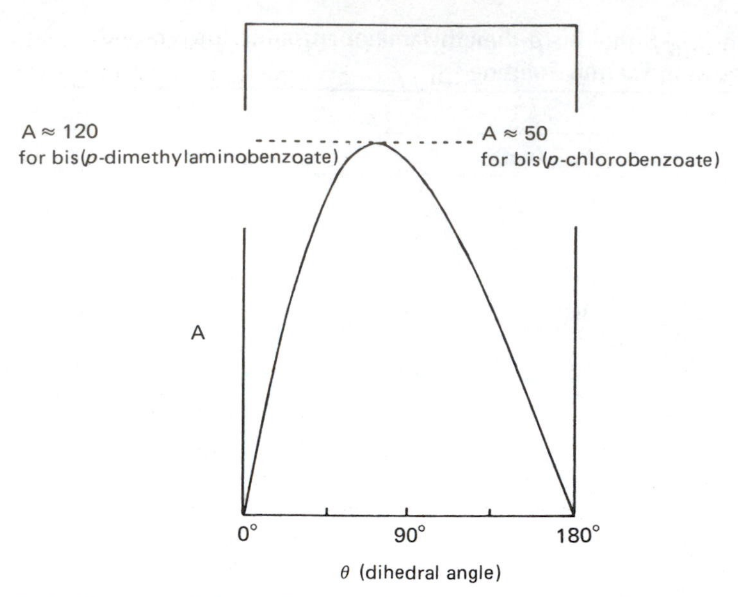

FIGURE 10-25 Relation between the interchromophoric (induced electric dipole moment) dihedral angle θ and the CD amplitude (Figure 10-20) A ($= \Delta\epsilon_1 - \Delta\epsilon_2$) of vicinal bis(*p*-substituted benzoates). [Reproduced with permission from H. Harada and K. Nakanishi, *Circular Dichroism Spectroscopy—Exciton Coupling in Organic Stereochemistry,* University Science Books, Mill Valley, CA, 1983.]

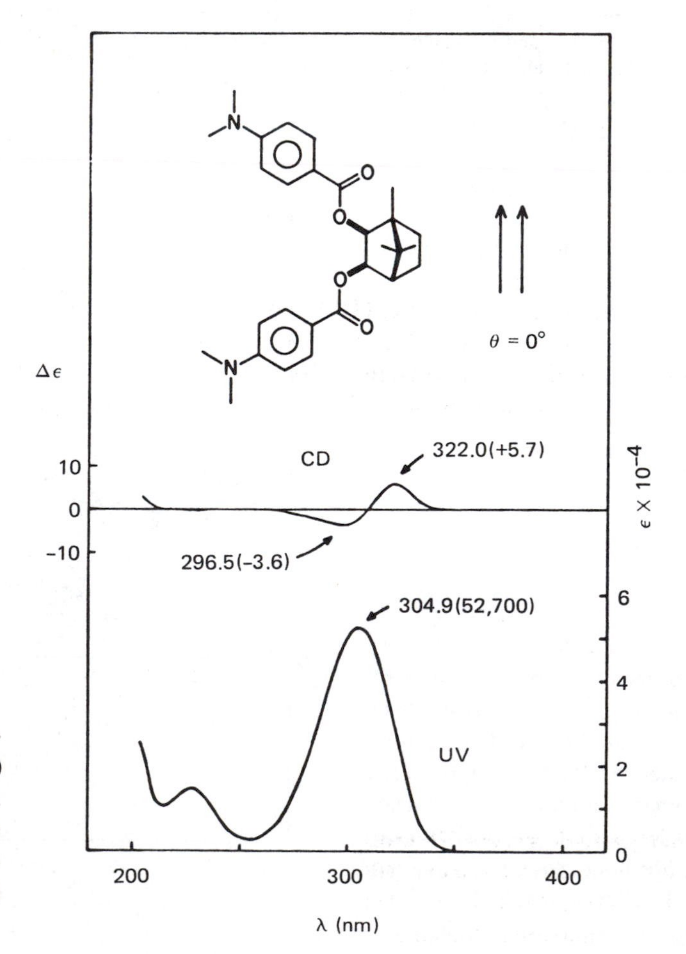

FIGURE 10-26 CD and UV spectra of (2*S*,3*R*)-camphane-2,3-diol bis(*p*-dimethylaminobenzoate) in ethanol. [Reproduced with permission from H. Harada and K. Nakanishi, *Circular Dichroism Spectroscopy—Exciton Coupling in Organic Stereochemistry*, Univerisity Science Books, Mill Valley, CA, 1983.]

found in the determination of the absolute configuration of the plant growth regulator (+)-abscisic acid (below). The ORD and CD spectra of this compound showed typical Davydov-type split Cotton effect

(+)-abscisic acid

bands. This splitting was interpreted as due to the transition dipole–dipole coupling between the enone and diene–carboxylic acid systems. Quantitative application of the exciton chirality method indicated that the *S* configuration should be assigned to (+)-abscisic acid. A chemical correlation of (+)-abscisic acid with (*S*)-malic acid had been carried out simultaneously and independently of this work; happily, the results led to the same absolute configuration.

10-5

Solvent and Temperature Effects on Conformation and Equilibria

Solvent-induced conformational changes have been detected for (+)-*trans*-6-chloro-3-methylcyclohexanone (eq. 10-2). Thus, in isooctane solvent, a negative $n\rightarrow\pi^*$ Cotton effect is observed, corresponding to a

10-9	10-9a	⇌	10-9e	(10-2)
Octant rule predictions:	(-)		(+)	Cotton effect

predominance of the conformer (**10-9a**) with the methyl and chloro groups oriented diaxially (Figure 10-27). In a more polar solvent, EPA (5/5/2 diethyl ether–isopentane–ethanol), the Cotton effect sign begins to become reversed with the bisignate form at 25°C. This observation suggests that more of the conformer (**10-9e**) with the equatorial methyl and chloro groups is present in the equilibrium. At −192°C in EPA, very

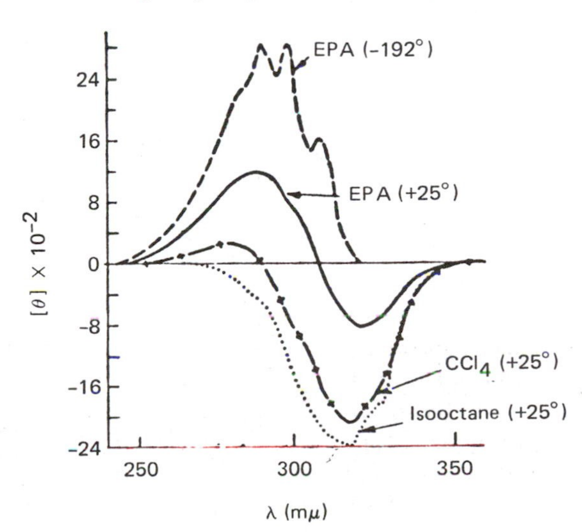

FIGURE 10-27 Temperature-dependent CD curves of (+)-*trans*-6-chloro-3-methylcyclohexanone (10-9). [From C. Djerassi, *Proc. Chem. Soc.*, 314 (1964), reproduced by permission of the editor.]

little of the diaxial form is present. Apparently **10-9** attempts to minimize its net dipole moment in the more hydrocarbon (low dielectric) solvents by orienting the C=O and C—Cl bond dipoles as nearly opposite to one another as can be achieved, i.e., in the conformation with an axial Cl.

Solvent effects on conformational equilibria have also been detected for (−)-menthone (**10-10**), as detected

10-10

by CD spectroscopy (Figure 10-28). The octant rule predicts a (+) n→π^* Cotton effect for the diequatorial conformer (**10-10e**) and a (−) Cotton effect for the diaxial conformer (**10-10a**). However, the twist boat conformer (**10-10tb**), which can be expected to relieve the serious axial–alkyl interactions of **10-10a**, is also predicted to have a (−) Cotton effect (eq. 10-3). From the data of Figure 10-28, one might tend to think in

10-10e ⇌ **10-10a** ⇌ **10-10tb** (10-3)

terms of the increasing presence of **10-10a**–**10-10tb** as the solvents change from methanol to isooctane. The bisignate nature of the CD curves is explained by overlapping a shorter wavelength negative n→π^* Cotton effect (due to **10-10a** or **10-10tb**) with a longer wavelength positive n→π^* Cotton effect (due to **10-10e**).

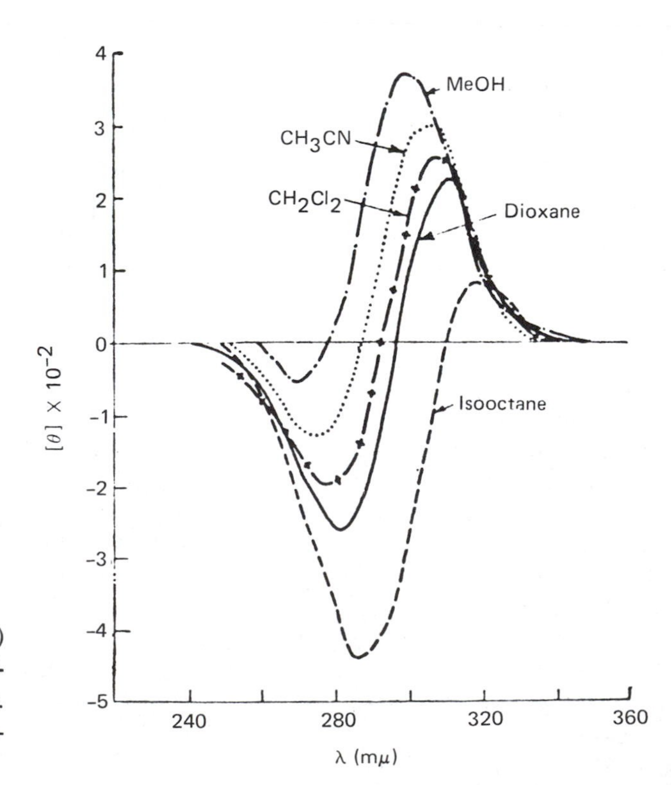

FIGURE 10-28 CD curves of (−)-menthone (10-10) in methanol, acetonitrile, methylene chloride, dioxane, and isooctane. [From C. Djerassi, *Proc. Chem. Soc.*, 314 (1964), reproduced by permission of the editor.]

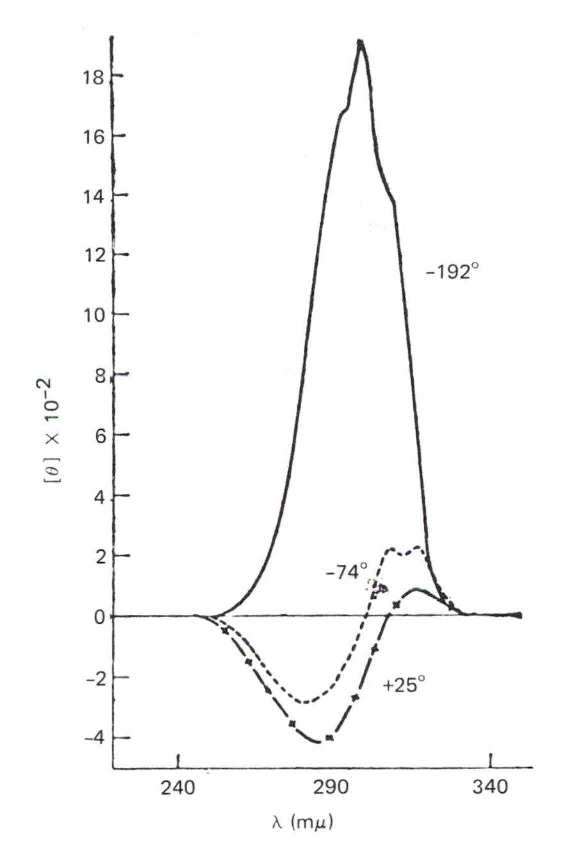

FIGURE 10-29 CD curves of (−)-menthone (10-10) in isopentane-methylcyclohexane at −192°C and in decalin at −74°, +25°, and +162°C. [From C. Djerassi, *Proc. Chem. Soc.*, 314 (1964), reproduced by permission of the editor.]

Interestingly, the CD curves of (−)-menthone (**10-10**) measured in the glass-forming hydrocarbon solvent, isopentane–methylcyclohexane, change sign between +25° and −192° C (Figure 10-29). These data suggest that the conformational equilibrium (eq. 10-3) is sensitive to temperature and that at very low temperatures, high energy conformers with (+) Cotton effects, e.g., **10-10a** and **10-10tb**, are less prevalent and that the equilibrium lies largely in favor of the (expectedly more stable) diequatorial conformer, **10-10e**.

When a ketone is dissolved in methanol and a drop of hydrochloric acid added, an equilibrium is established in which the ketone is converted to a dimethyl ketal, as shown in eq. 10-4. (An analogous reaction

$$RR'C{=}O \xrightleftharpoons{CH_3OH/H^+} R{-}\underset{OCH_3}{\overset{OCH_3}{\underset{|}{\overset{|}{C}}}}{-}R' + H_2O \tag{10-4}$$

can be written for conversion of an aldehyde to an acetal.) Although the ketal can be isolated only after removal of the acid catalyst, its formation in solution is readily monitored by UV or chiroptical spectroscopy, since ketal formation causes the carbonyl group and its associated $n \rightarrow \pi^*$ transition to disappear. The reaction of cholestan-3-one with acidified methanol was investigated by Zalkow and co-workers, who found that the ketone–ketal equilibrium is strongly dependent upon the amount of water present. This result is illustrated in Figure 10-30, which also shows the use of CD as a kinetic tool. The Cotton effect at 289 nm is due to the concentration of free ketone present; the ketal absorbs at much shorter wavelength. Successive additions of small amounts of water shift the ketone–ketal equilibrium toward the free ketone. The same workers also found that ketal formation depends upon the structure of the alcohol, as well as on stereochemical factors, such as the size of groups in the molecule near the carbonyl function. Thus, cholestan-3-one gives 96% of the dimethyl ketal, 84% of the diethyl ketal, and only 25% of the diisopropyl ketal.

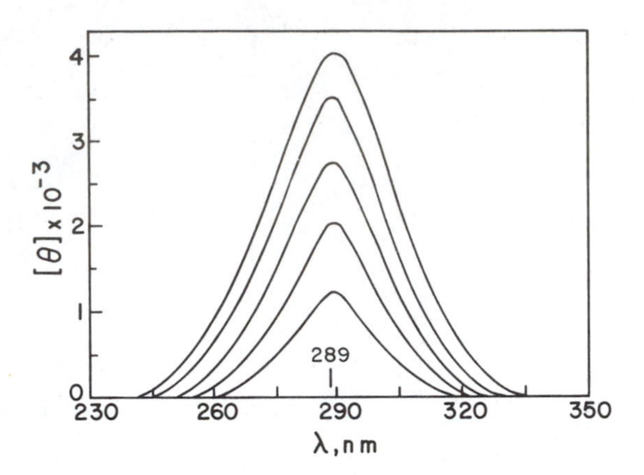

FIGURE 10-30 A CD investigation of the ketone-ketal equilibrium for cholestan-3-one in acidified methanol. The Cotton effect due to free ketone increases with successive additions of water to the ketal. [Plotted from data of L. H. Zalkow, R. Hale, K. French, and P. Crabbé, *Tetrahedron,* **26,** 4947 (1970).]

10-6

Applications of UV and CD Spectroscopy in Organic Structure Determination

10-6a Woodward's Rules

The isomeric diterpene natural products, abietic acid (**10-12**) and levopimaric acid (**10-13**), can be distinguished by their UV spectra. Because **10-13** has a homoannular diene, it is expected to show a longer wavelength $\pi \rightarrow \pi^*$ λ_{max} than **10-12**, which has a heteroannular diene. Application of Woodward's rules (Table 10-3) to **10-12** predicts $\lambda_{max} = [214 \text{ (parent)} + 4 \text{ (alkyl)} \times 5 + 1 \text{ (exocyclic C=C)} \times 5] = 239$ nm; and to **10-13**, $\lambda_{max} = [214 \text{ (parent)} + 4 \text{ (alkyl)} \times 5 + 1 \text{ (exocyclic)} \times 5 + 1 \text{ (homoannular)} \times 39] = 278$ nm. The predicted values are in good agreement with those observed. Note that ϵ_{max} is significantly smaller for **10-13** than for **10-12**. A rough rule of thumb is that the greater the distance between the ends of the diene, the greater is ϵ_{max}.

HOOC HOOC

	10-12	**10-13**
Observed λ_{max} (nm)	238	273
ϵ_{max}	16,000	7000

10-6b Steric Hindrance

Stilbene can be isolated as two different configuration isomers: trans (**10-14**) and cis (**10-15**). Only the trans isomer can easily adopt a conformation with maximum coplanarity of the aromatic ring π systems with the central C=C. Consequently, one expects a greater λ_{max} and ϵ_{max} for the long wavelength electronic transition.

10-14		**10-15**	
λ_{max} **(nm)**	ϵ_{max}	λ_{max} **(nm)**	ϵ_{max}
296	29,000	280	10,500
228	16,500	224	24,000

As explained earlier (Section 10-3), the coupling of two benzene chromophores in biphenyl lowers the energy of the allowed benzene ${}^{1}L_a$ transition from 204 to 248 nm. However, when coplanarity, and hence maximum π overlap of the two rings of biphenyl, is severely sterically inhibited, the biphenyl derivative behaves more like the sum of the two independent aromatic chromophores. Compare the data for benzene and biphenyl with data for *m*-xylene and 2,2′,6,6′-tetramethylbiphenyl.

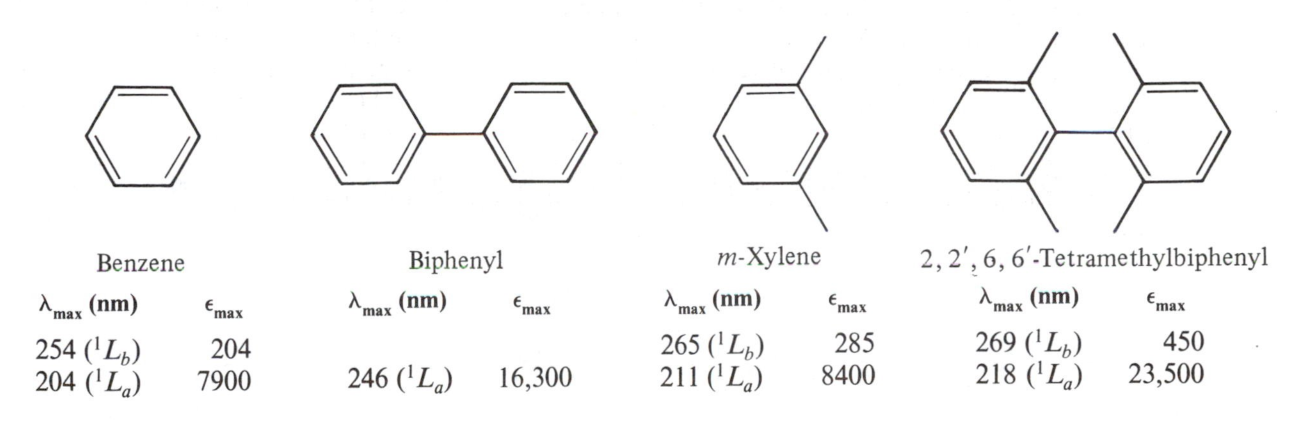

Benzene		Biphenyl		*m*-Xylene		2,2′,6,6′-Tetramethylbiphenyl	
λ_{max} **(nm)**	ϵ_{max}	λ_{max} **(nm)**	ϵ_{max}	λ_{max} **(nm)**	ϵ_{max}	λ_{max} **(nm)**	ϵ_{max}
254 (${}^{1}L_b$)	204			265 (${}^{1}L_b$)	285	269 (${}^{1}L_b$)	450
204 (${}^{1}L_a$)	7900	246 (${}^{1}L_a$)	16,300	211 (${}^{1}L_a$)	8400	218 (${}^{1}L_a$)	23,500

Examples of steric hindrance to conjugation are not limited to biphenyls and find expression in substituted nitroanilines and acetophenones, among others. *p*-Nitroaniline (**10-16**) shows ϵ_{max} 16,000 at λ_{max} 375 nm, but its *o,o*′-dimethyl analogue (**10-17**) shows ϵ_{max} 4800 at λ_{max} 385 nm. Similarly, *p*-methylacetophenone (**10-18**) shows ϵ_{max} 15,000 at λ_{max} 252 nm, but the *o,o*′-dimethyl analogue (**10-19**) shows ϵ_{max} 3200 at λ_{max} 262 nm. The methyl groups of **10-17** inhibit coplanarity (hence maximum p orbital overlap) of the nitro π system with the aromatic ring π system leading to a greatly diminished ϵ_{max} and a red shift of the absorption. As in **10-17**, the ortho methyl groups of **10-19** inhibit coplanarity of the carbonyl π system with the aromatic ring π system.

10-16 **10-17** **10-18** **10-19**

10-6c Environmental Factors and Tautomerism

Cyclohexane-1,3-dione in cyclohexane solvent exhibits weak absorption near 295 nm (ϵ_{max} ~50), but in ethanol the UV–vis spectrum changes: λ_{max} 255 nm (ϵ_{max} 12,500). In basified ethanol the spectrum changes again: λ_{max} 280 nm (ϵ_{max} 20,000). The diketone can exhibit the equilibrium shown in eq. 10-5. In the hydro-

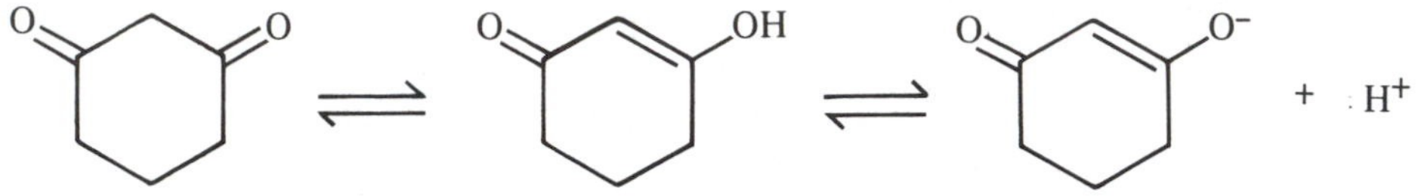

+ H^+ (10-5)

carbon solvent, the equilibrium lies largely in favor of the diketo tautomer, which exhibits the weak λ_{max} = 295 n→π* absorption. In the polar, protic solvent, ethanol, the β-hydroxyenone tautomer is favored, and this form shows the λ_{max} at 255 nm, as predicted by Woodward's rules [215 (parent) + 12 (β-alkyl) + 30 (β-OH) = 257 nm] for the π→π* absorption. Deprotonation of this enol gives the enolate anion, which absorbs more intensely and at longer wavelengths than the enol form.

Tautomerism (eq. 10-6) of 2-hydroxypyridine (**10-20**) leads to a structure (**10-21**) that has a different

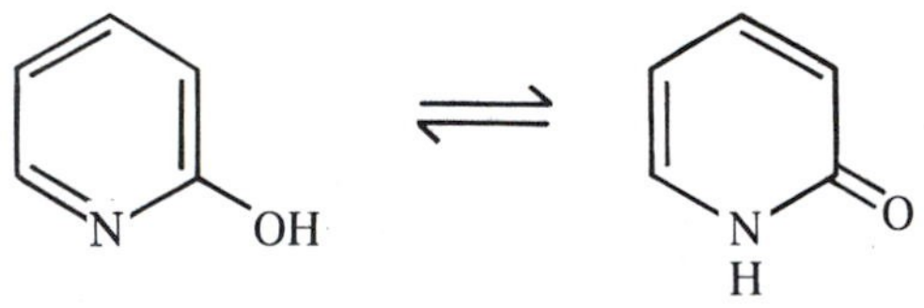

(10-6)

10-21 **10-22**

chromophore. The position of equilibrium can be determined from the UV–vis spectra of **10-22a** (λ_{max} 224 nm, ϵ_{max} 7200) and **10-22b** (λ_{max} 293 nm, ϵ_{max} 5900) from the following models.

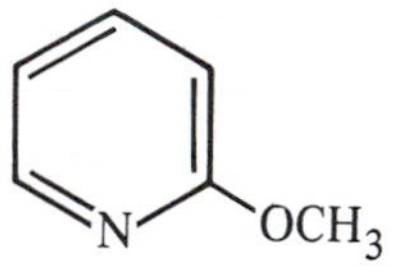

10-22a

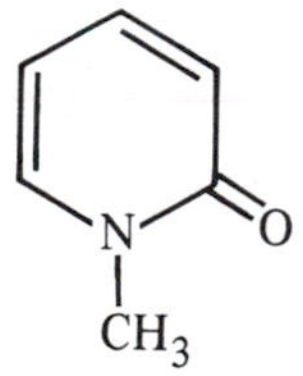

λ_{max} (nm)	ϵ
269	3200
<205	>5300

10-22b

λ_{max} (nm)	ϵ
293	5900
226	6100

10-6d Absolute and Relative Configuration

The recent widespread research efforts directed toward the total synthesis of prostaglandins have made important use of chiroptical techniques. The ORD and CD curves of natural prostaglandin E_1 (PGE) of

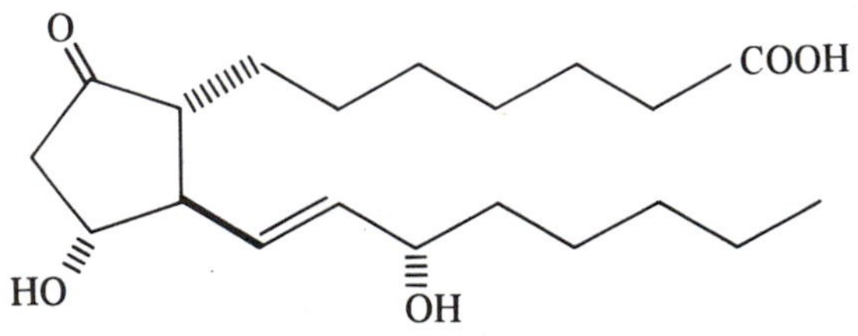

Prostaglandin E_1 (PGE)

established configuration have been reported by Korver (ref. 10-20). A recent investigation by Miyano and Dorn used chiroptical methods to correlate the absolute configuration of PGE with several intermediates involved in prostaglandin synthesis. They resolved 7-(2-*trans*-styryl-3-hydroxy-5-oxocyclopentenyl)heptanoic acid (**10-23**), the key intermediate in their total synthesis of racemic prostaglandins, and converted the resolved acids into (8*S*,12*S*,15*S*)-dihydroPGE (**10-24**) and its diastereomer (8*S*,12*S*,15*R*)-dihydroPGE (**10-25**) by an unambiguous series of reactions. Chiroptical studies in the region of the n→π* transition

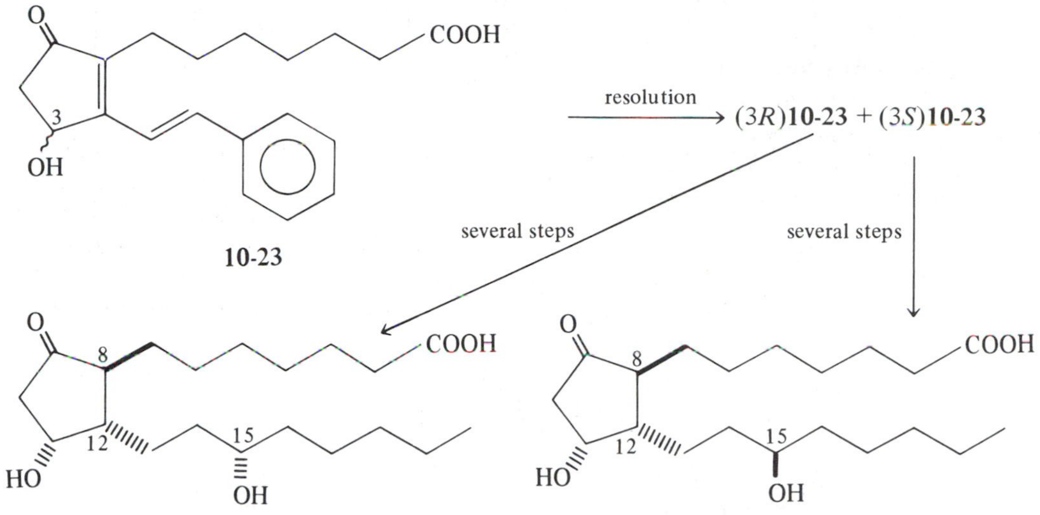

O COOH 3 OH

resolution → (3*R*)**10-23** + (3*S*)**10-23**

several steps several steps

10-23

O 8 COOH HO 12 15 OH

O 8 COOH HO 12 15 OH

10-24 10-25

(Table 10-8) showed that the ORD and CD curves of **10-24** and **10-25** are both mirror images of those of natural PGE. In agreement with octant rule projections, the signs of the Cotton effects indicate that the stereochemistry about the cyclopentanone ring must be the same for both diastereomers and must be enantiomeric to that of natural PGE.

An example of the use of model compounds of known configuration to deduce stereochemical information is provided by cafestol, a diterpene isolated from the coffee bean. Degradation of cafestol (**10-26**) gave the ketone (**10-27**) whose ORD curve was found to be almost the mirror image of that of

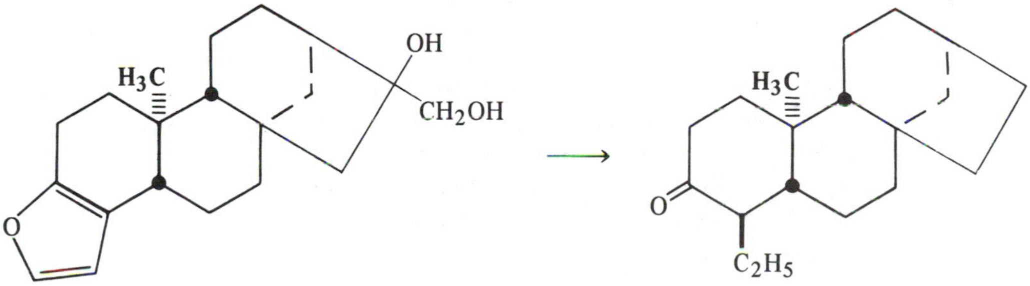

H_3C OH CH_2OH O → H_3C O C_2H_5

10-26 10-27

TABLE 10-8 Chiroptical Data for the n→π* Absorption Region of PGE, 10-24, and 10-25

	ORD				CD	
	Peak		Trough			
	[Φ]	λ (nm)	[Φ]	λ (nm)	[Φ]	λ (nm)
PGE	+7,200	272	−6,200	314	−11,000	296
10-24	+3,800	315	−4,100	273	—	—
10-25	+3,700	315	−5,600	274	+ 7,600	295

Data for methanol solution. See M. Miyano and C. R. Dorn, *J. Am. Chem. Soc.*, **95**, 2664 (1973).

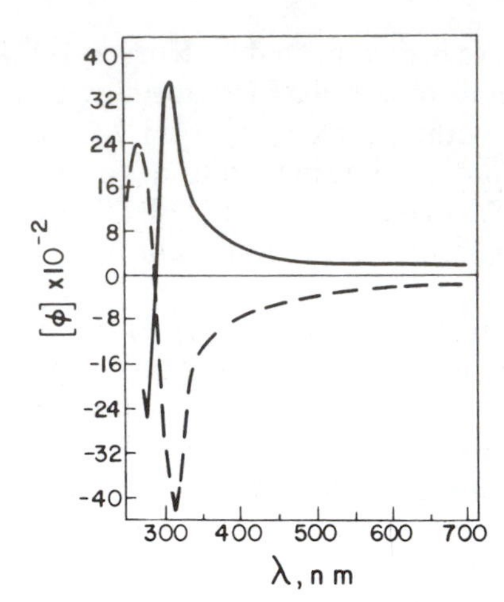

FIGURE 10-31 The ORD curves of the ketone (10-27) from cafestol (- - -) and of 4α-ethylcholestan-3-one (10-28) (———). [Adapted with permission from C. Djerassi, M. Cais, and L. A. Mitscher, *J. Am. Chem. Soc.*, **81**, 2386 (1959). Copyright 1959 American Chemical Society.]

4α-ethylcholestan-3-one (**10-28**). The ORD curves are shown in Figure 10-31. Since the observed Cotton

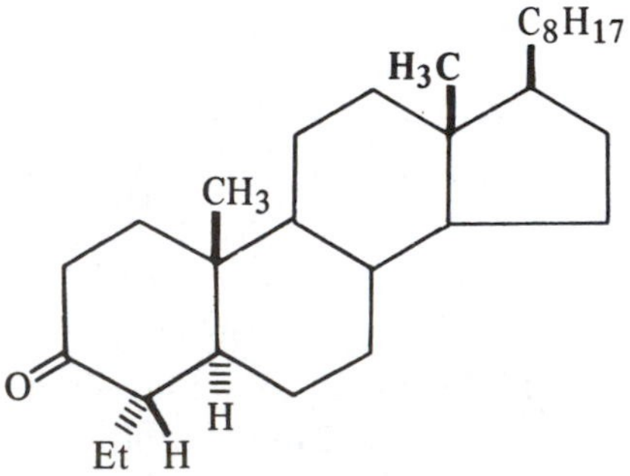

10-28

effect curves are essentially mirror images and the two ketones **10-27** and **10-28** are structurally the same in the vicinity of the carbonyl group, the conclusion was reached that **10-27**, and hence cafestol, from which **10-28** was obtained, possessed enantiomeric stereochemistry at the A/B ring junction, as depicted in Figure 10-32.

We have seen that ORD and CD methods may be used for the assignment of *relative configuration* to a molecule, that is, relative to some model compound whose *absolute configuration* has been determined by other means. A few theoretical treatments exist that are useful in certain cases for nonempirical assignments of absolute configuration. One of these is the coupled oscillator theory of Werner Kuhn, which is applicable to rigid, inherently chiral systems, in particular those that contain two identical but noncoplanar chromophores. With knowledge of the direction of the electric dipole transition moment for the chromophore, one can predict the sign of the Cotton effect arising from that transition for a chosen absolute configuration. There is no need for a comparison of chiroptical data with compounds of known absolute configuration, so the method is nonempirical.

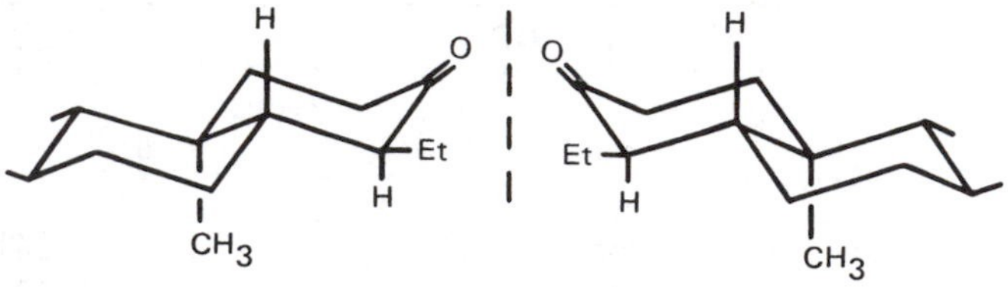

10-27 **10-28** (partial structure)

FIGURE 10-32 Conformation of the A-B ring portion of the cafestol-derived ketone (10-27) and of the model compound 4α-ethylcholestan-3-one (10-28) showing the mirror image relationship. See Figure 10-30 for the ORD curves.

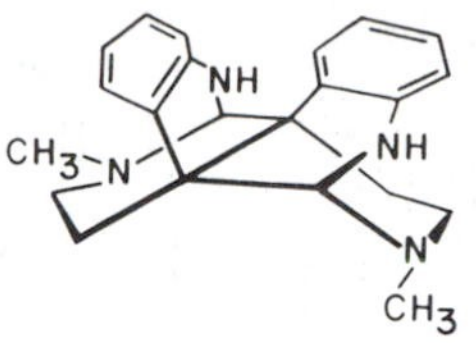

FIGURE 10-33 Absolute configuration of the indole alkaloid (+)-calycanthine. [After S. F. Mason and G. W. Vane, *J. Chem. Soc.* (*B*), **370** (1966).]

A number of natural products are essentially dimeric. An example is the indole alkaloid calycanthine, whose absolute configuration was determined by Mason and Vane (Figure 10-33). From the point of view of coupled oscillator theory, calycanthine consists of two aniline chromophores. Planes through the two aromatic rings make an angle of 61° with each other. The C_2 axis through each aniline residue makes an angle of 28° with the intersection of these planes. The rotational strengths of the two longest wavelength Cotton effects were determined theoretically from consideration of the absorption spectrum of aniline and of the calculated transition dipole interactions between the aniline chromophores in each of the two optical isomers of calycanthine. The theory predicted that for the absolute configuration of (+)-calycathine shown in Figure 10-33, the CD spectrum should show a long wavelength positive Cotton effect followed by a negative one. The experimental CD spectrum was in agreement with this prediction: the longest wavelength transition occurred near 320 nm with a molar ellipticity of about 100,000.

PROBLEMS

10-1[1] Calculate λ_{max} for the following and explain the lower ϵ_{max} for the middle diene.

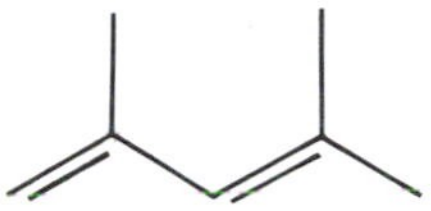

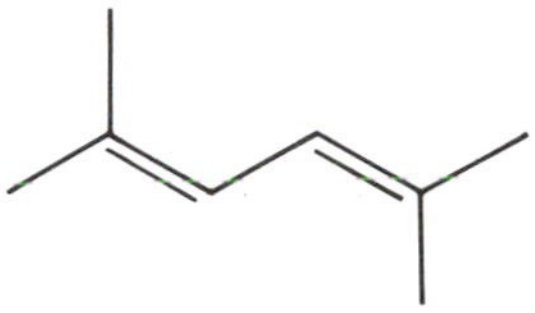

Observed				
	λ_{max} (nm)	217	228	241
	ϵ_{max}	21,000	8500	23,000

10-2[1] Calculate λ_{max} for the following.

a.

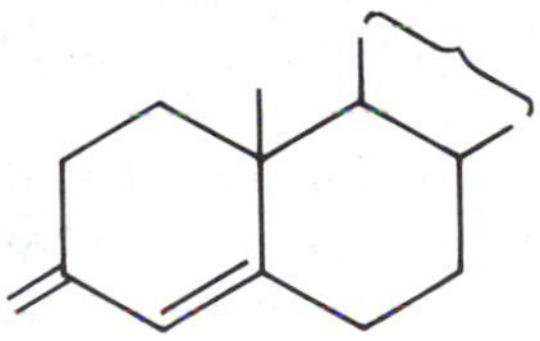

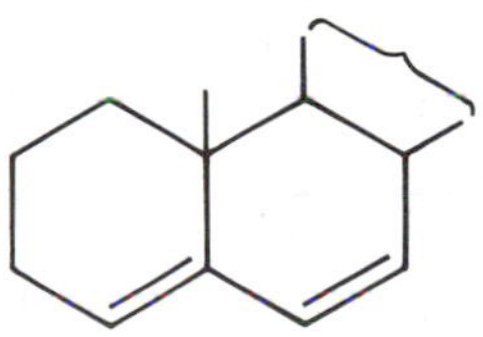

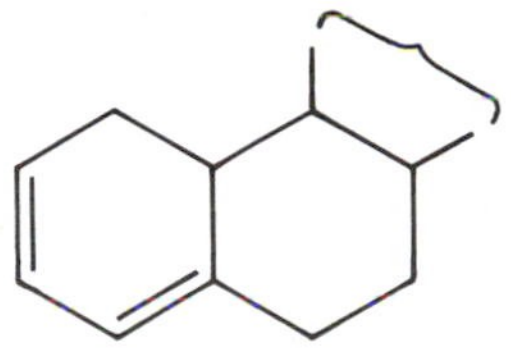

Observed				
	λ_{max} (nm)	239	235	275
	ϵ_{max}	17,300	19,000	10,000

b.

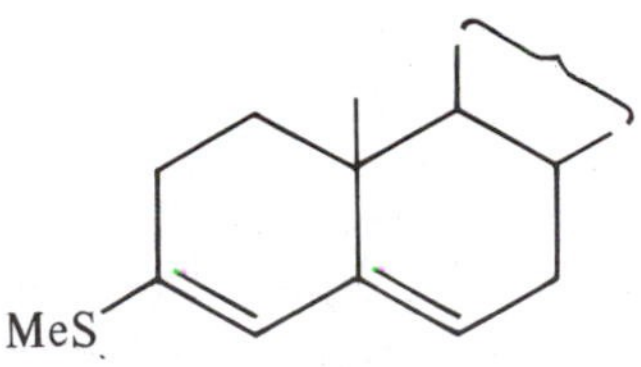

MeS 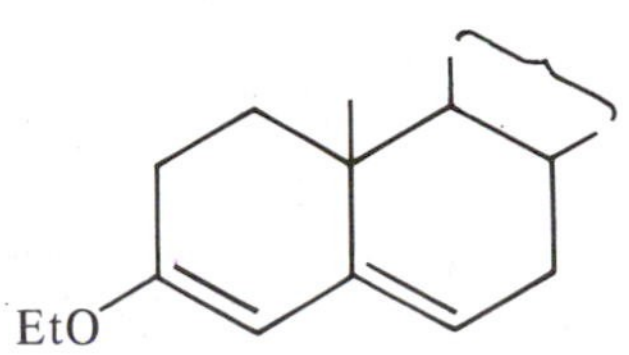 EtO 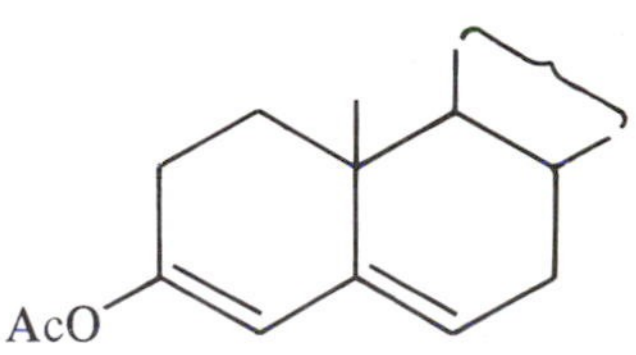 AcO

Observed				
	λ_{max} (nm)	268	241	235
	ϵ_{max}	22,600	22,600	19,000

[1]Observed data from A. I. Scott, *Interpretation of the Ultraviolet Spectra of Natural Products*, New York, 1962.

c.

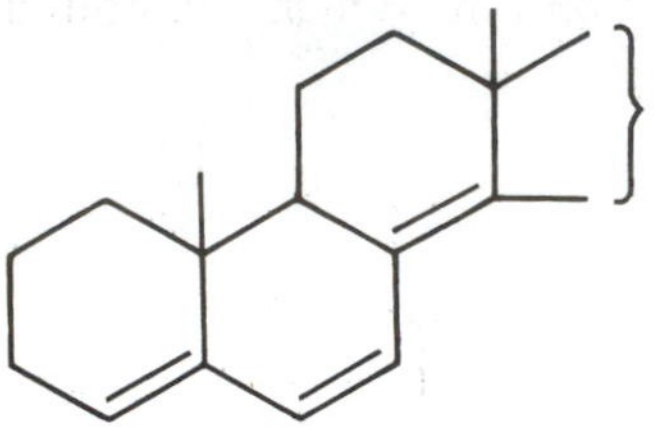

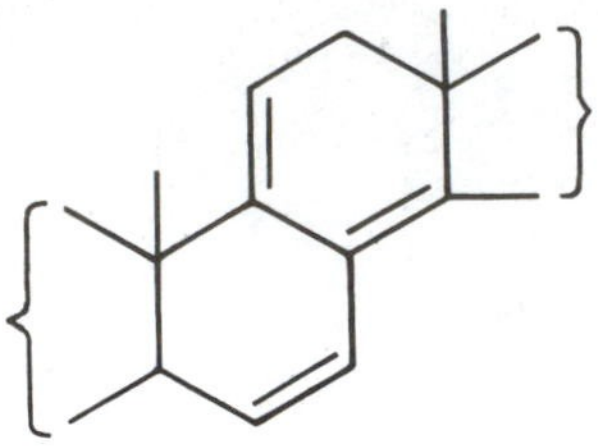

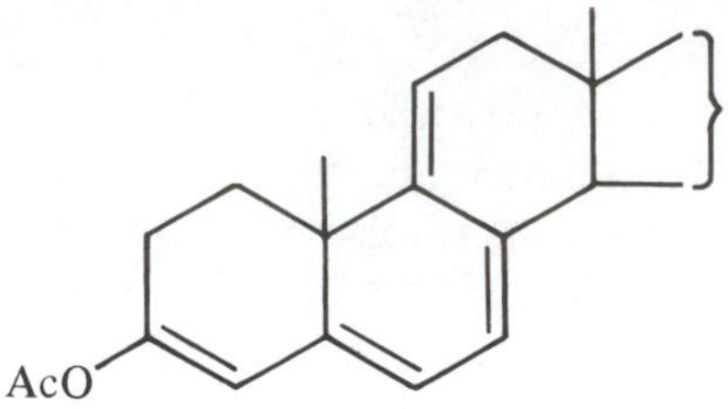

Observed				
	λ_{max} (nm)	283	285	355
	ϵ_{max}	33,000	9100	19,700

d.

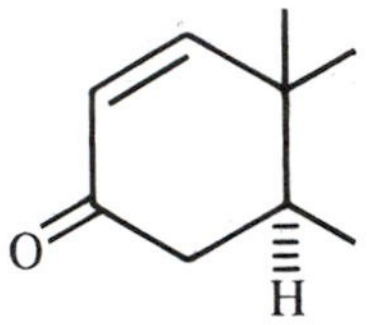

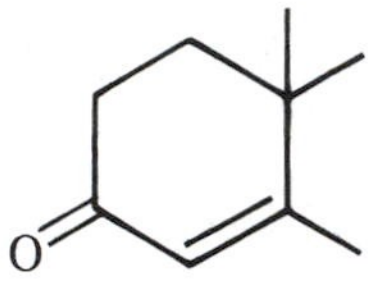

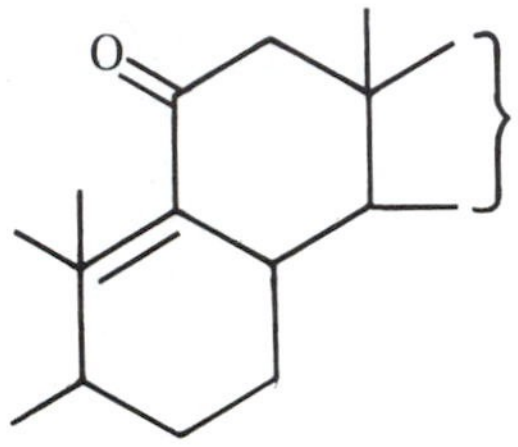

Observed				
	λ_{max} (nm)	230	241	254
	ϵ_{max}	10,000	16,600	9100

e.

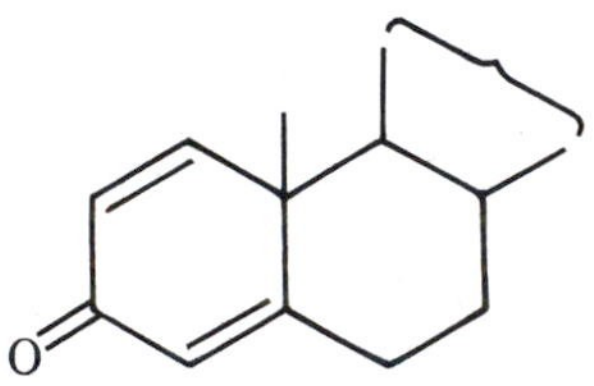

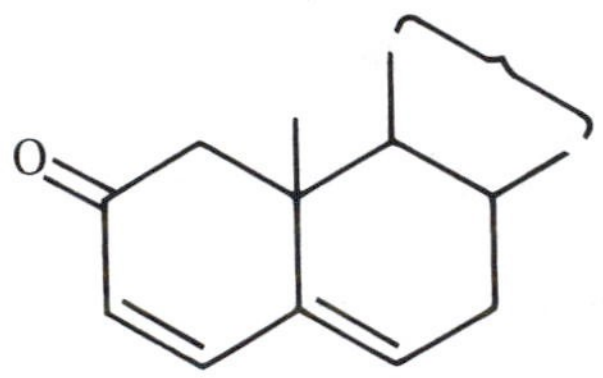

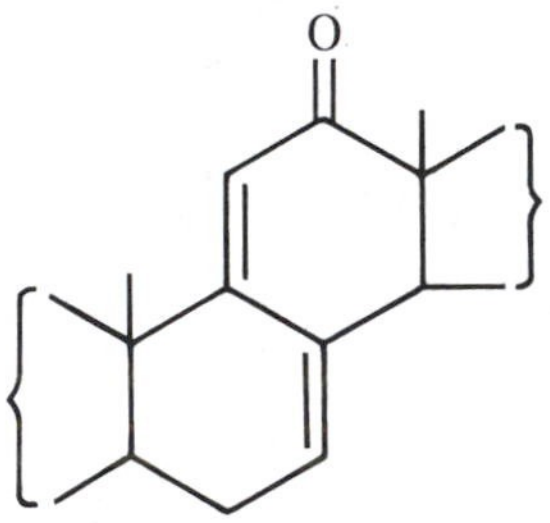

Observed				
	λ_{max} (nm)	244	290	292
	ϵ_{max}	15,000	12,600	13,000

f.

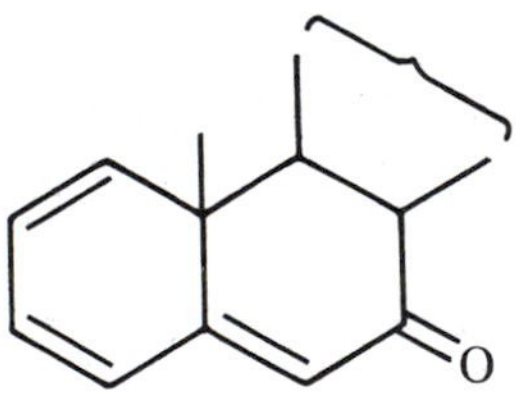

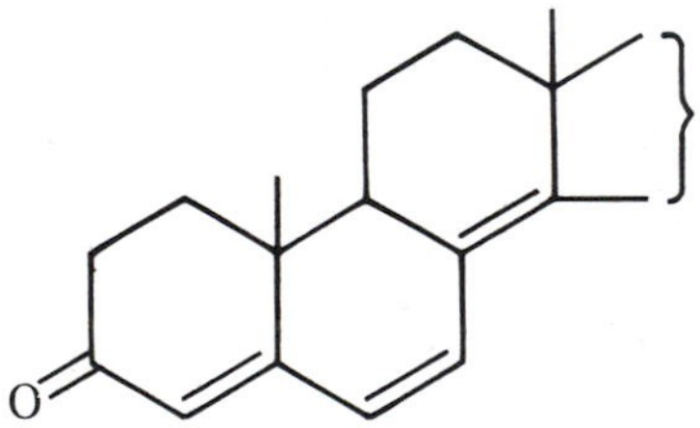

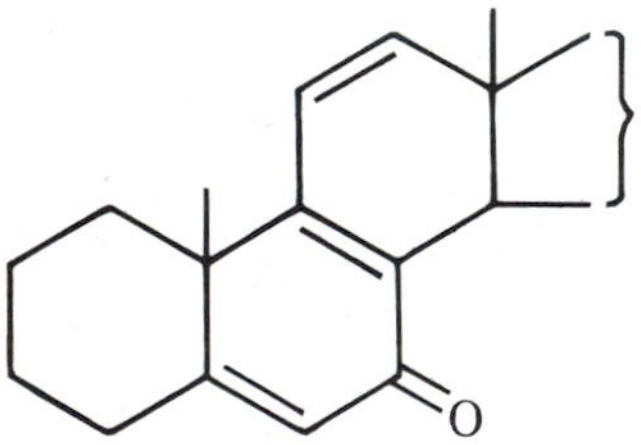

Observed				
	λ_{max} (nm)	348	348	327
	ϵ_{max}	11,000	26,500	

10-3 Calculate the $\pi \rightarrow \pi^*$ transition λ_{max} for cholesta-4,6-diene-3-one (A) and its enol acetate (B).

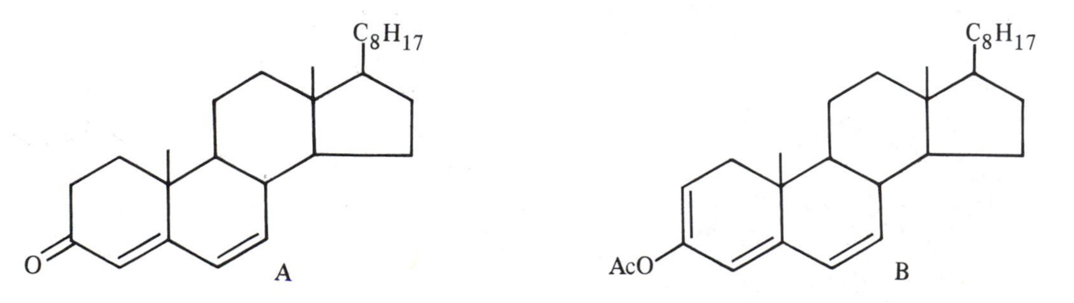

10-4 The compounds below have λ_{max} 303, 274, and 283 nm. Which compound has which absorption?

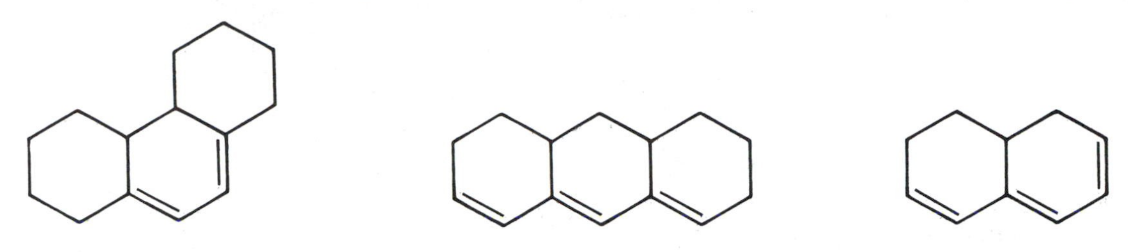

10-5 The compounds below have λ_{max} at 305, 349, and 360 nm. Which compound has which absorption?

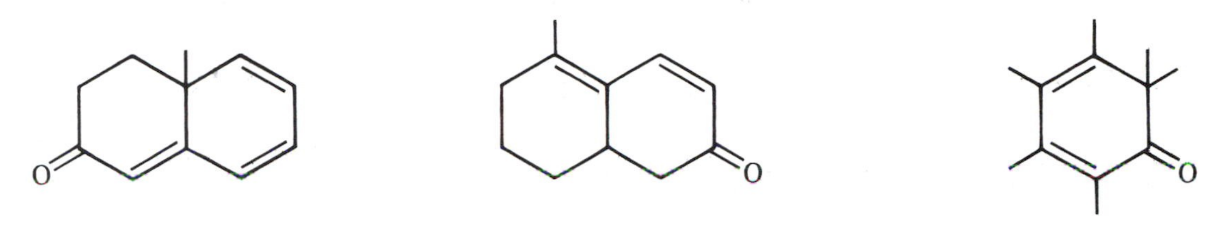

10-6 What UV–vis λ_{max} would you predict for the $\pi \rightarrow \pi^*$ transitions of the following compounds?

10-7 The following polyenes have ϵ_{max} values of 382, 294, 249, 318, 234, and 244 nm. Assign the λ_{max} to each structure.

10-8 Calculate the approximate λ_{max} for the $\pi \rightarrow \pi^*$ transition of each of the following compounds.

10-9 The following unsaturated ketones have λ_{max} values of 254, 239, 280, 249, 244, and 407 nm. Assign the λ_{max} to the structure.

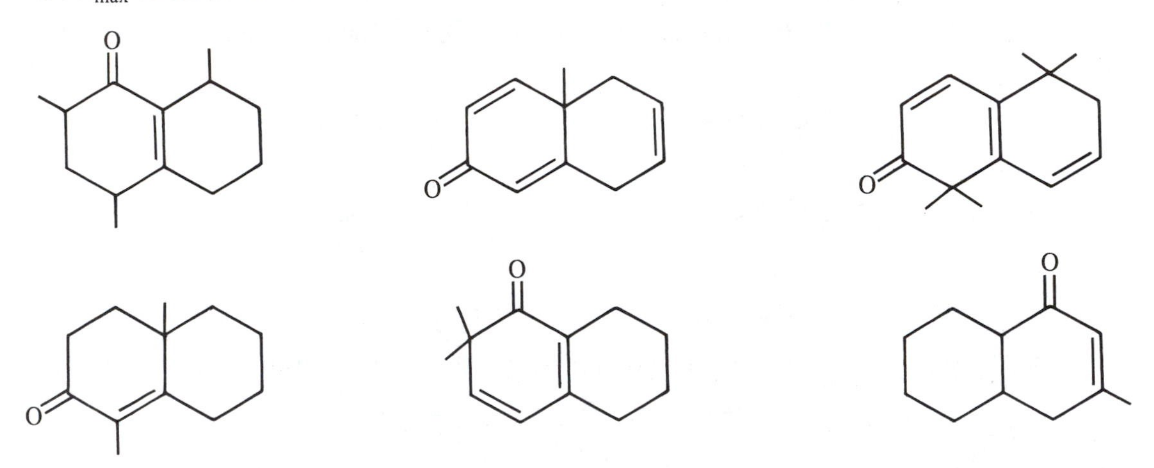

10-10 Calculate the approximate λ_{max} for the $\pi \rightarrow \pi^*$ transitions of each of the following compounds.

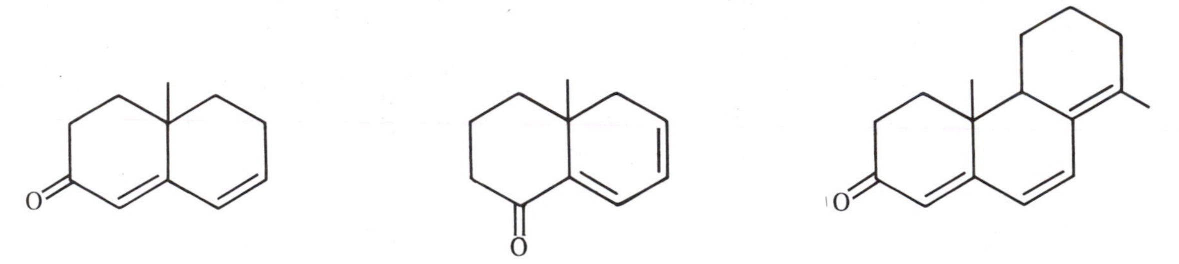

10-11 The following compounds absorb at 283, 227, 234, and 249 nm. Which compound has which absorption?

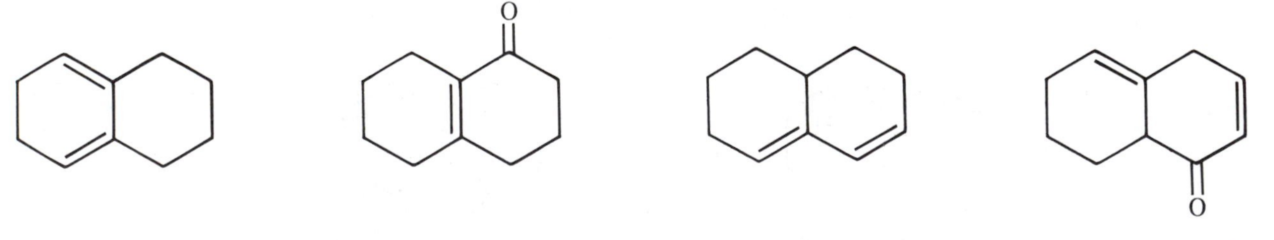

10-12 An enol acetate of cholest-4-ene-3-one (A) is prepared and has $\lambda_{max} = 238$ nm with log $\epsilon_{max} = 4.2$. Is the enol acetate B or C?

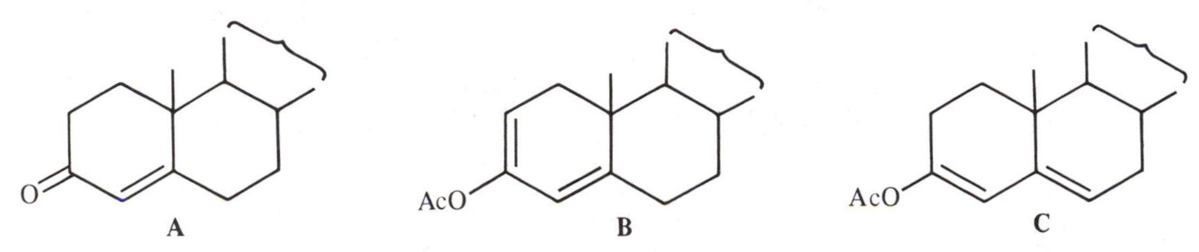

10-13 Spiroenones were prepared of structures A and B. One showed an intense λ_{max} at 247 nm, the other at 241 nm. Assign the structures.

10-14 Account for the following observations.

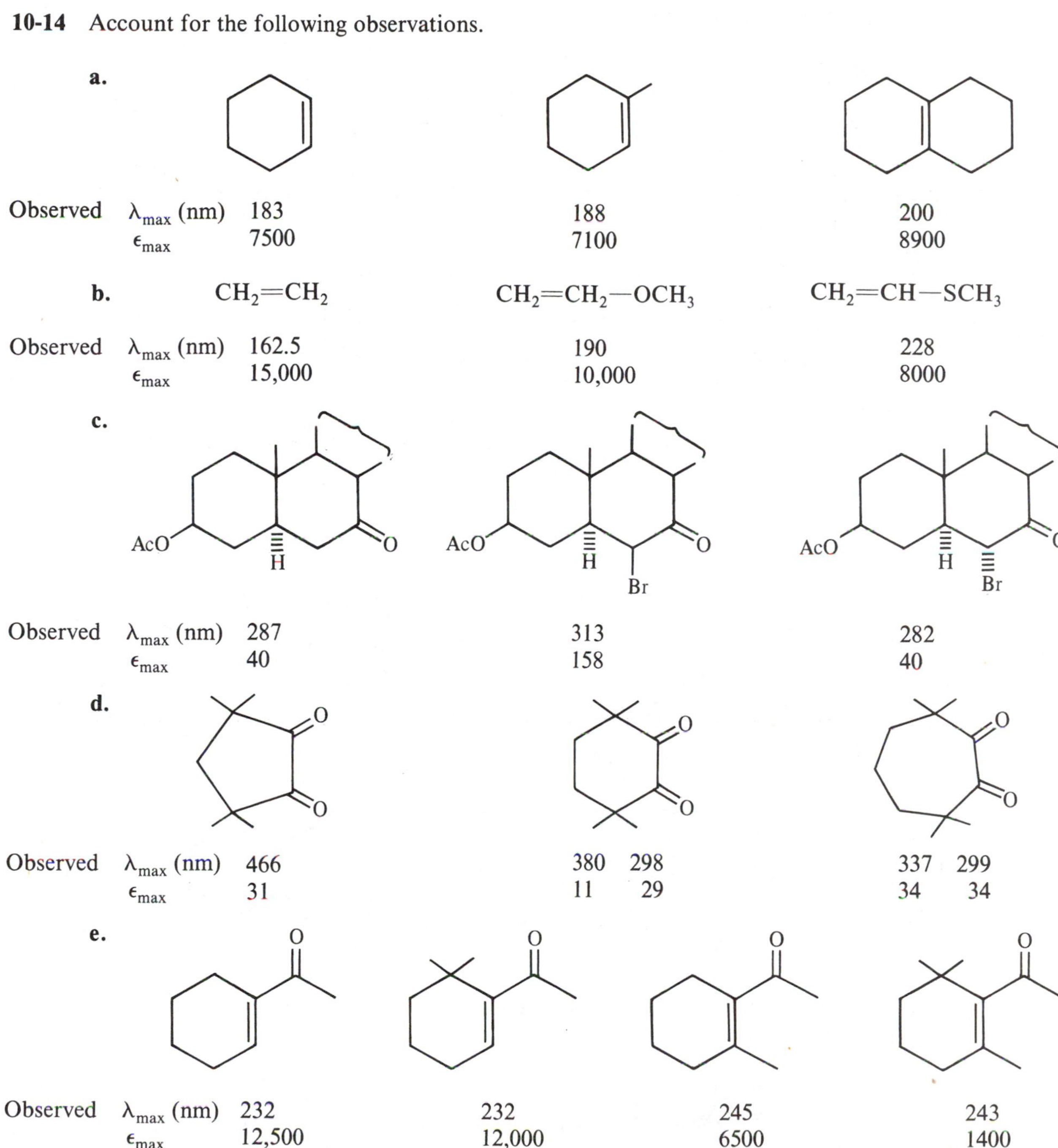

a.

Observed			
λ_{max} (nm)	183	188	200
ϵ_{max}	7500	7100	8900

b.

Observed	$CH_2{=}CH_2$	$CH_2{=}CH_2{-}OCH_3$	$CH_2{=}CH{-}SCH_3$
λ_{max} (nm)	162.5	190	228
ϵ_{max}	15,000	10,000	8000

c.

Observed			
λ_{max} (nm)	287	313	282
ϵ_{max}	40	158	40

d.

Observed			
λ_{max} (nm)	466	380 298	337 299
ϵ_{max}	31	11 29	34 34

e.

Observed				
λ_{max} (nm)	232	232	245	243
ϵ_{max}	12,500	12,000	6500	1400

10-15 Calculate the λ_{max} for each of the following compounds.

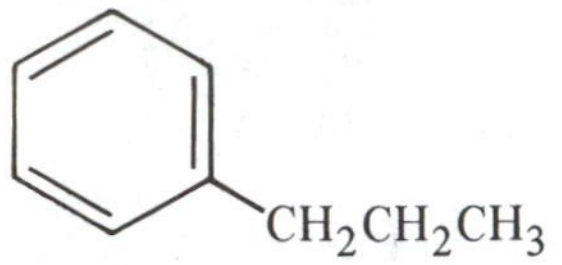

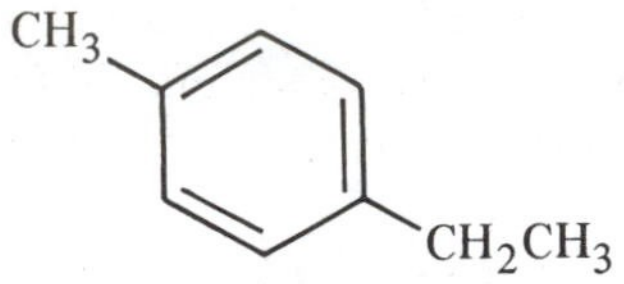

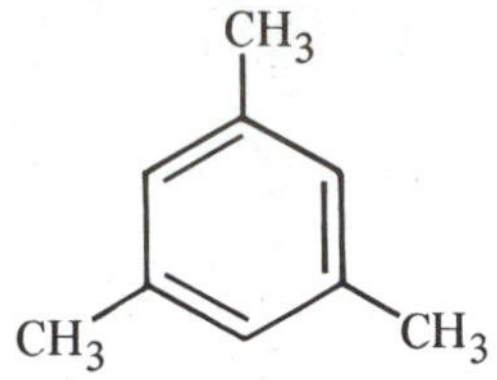

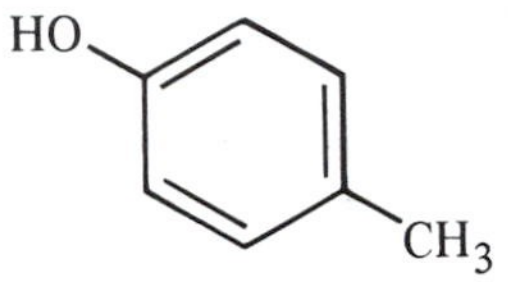

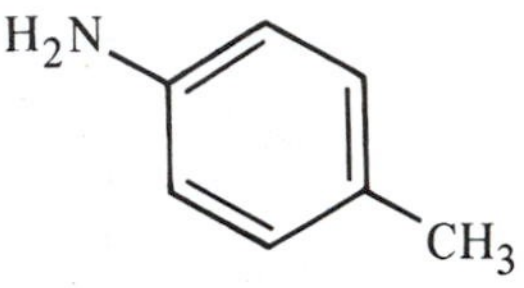

$CH_2CH_2CH_3$ CH_3 CH_2CH_3 CH_3 CH_3 CH_3

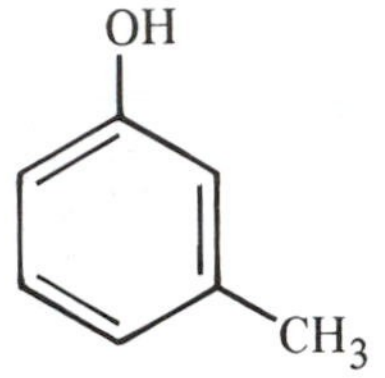

OH CH_3 HO CH_3 H_2N CH_3

10-16 Predict and explain whether UV–vis spectroscopy can be used for distinguishing members of the isomeric pairs.

a. 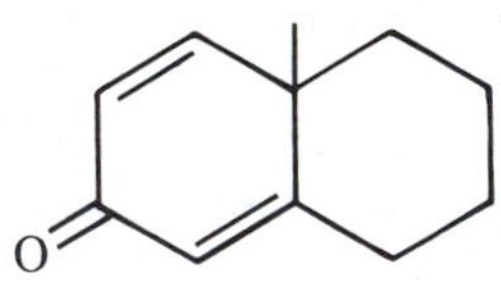 O and O

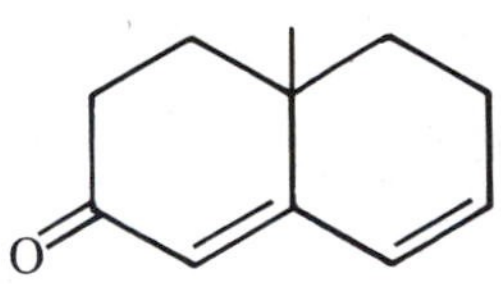

b. 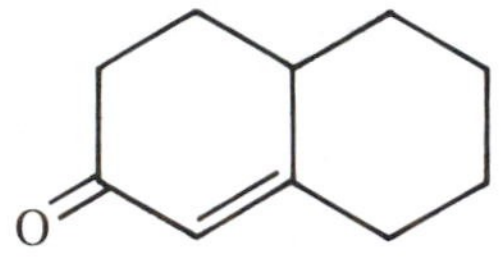 O and O

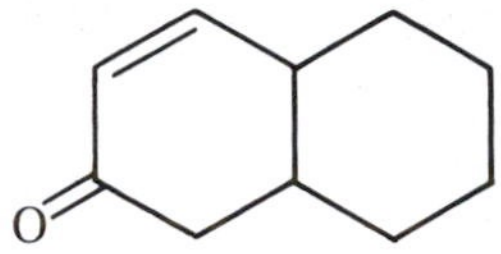

c. 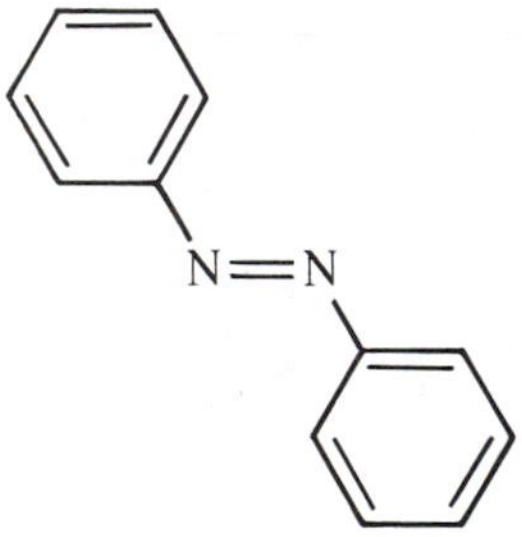 N=N and N=N

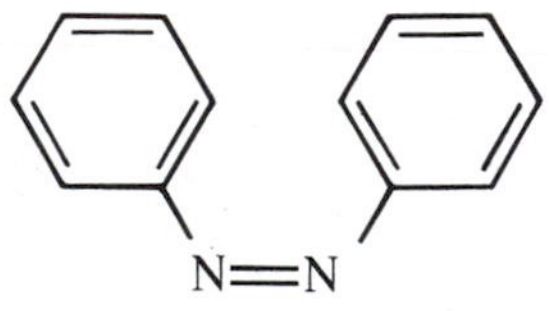

d. 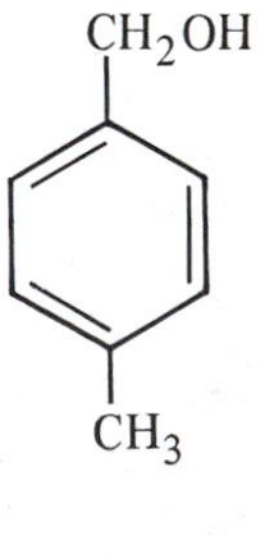 CH_2OH CH_3 and OH CH_2CH_3

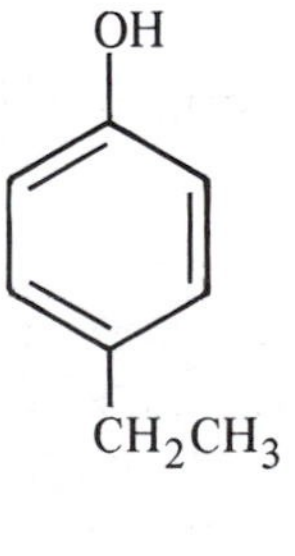

e. 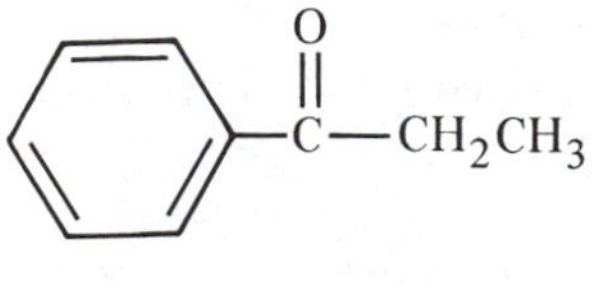 $\overset{O}{\overset{\|}{C}}$—$CH_2CH_3$ and —$CH_2\overset{O}{\overset{\|}{C}}$—$CH_3$

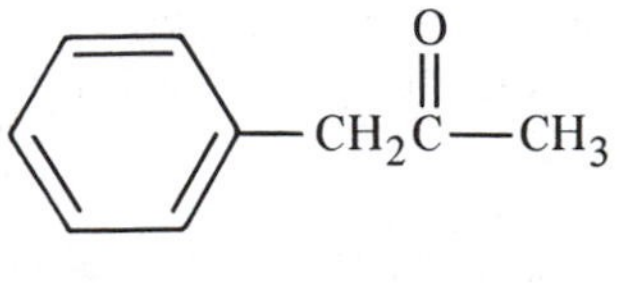

f. $CH_3CH_2COOCH_3$ and $CH_3COOCH_2CH_3$

g. $CH_3{-}CH{=}CH{-}CH_2{-}CH{=}CHCH_3$ and $CH_3CH_2{-}CH{=}CH{-}CH{=}CH{-}CH_3$

10-17 Compounds X (C_6H_8) and Y (C_6H_8) each take up 2 mole equivalents of H_2 in the presence of Pd/C to give cyclohexane. What are the structures of X and Y when X has λ_{max} 267 nm and Y has λ_{max} 190 nm?

10-18 A compound (A), $C_{11}H_{16}$, has λ_{max} 288 nm. On treatment with Pd/C (which dehydrogenates cyclic compounds completely to aromatic compounds without rearrangement), α-methylnaphthalene (below) is produced. What is the structure of A?

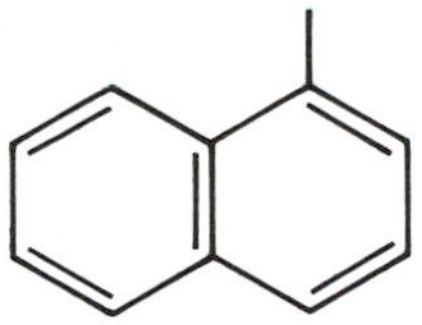

10-19 An unknown monocyclic hydrocarbon A, C_8H_{14}, has a λ_{max} at 234 nm and could be selectively ozonized to yield B, $C_7H_{12}O$, which has a λ_{max} at exactly 239 nm. Reduction of B with $LiAlH_4$ and careful elimination of water (dehydration) gave C, C_7H_{12}, which has a λ_{max} at 267 nm. Give structures for A, B, and C.

10-20 An optically active allylic alcohol, A, $C_8H_{14}O$, was treated with hot H_3PO_4, and three new products, B, C, and D, were distilled. B and C were optically active, and their formation did not involve rearrangement. B, C, and D could be further dehydrogenated (Pd/C + heat) to give *p*-xylene (1,4-dimethylbenzene). The λ_{max} for B was 229, for C 268, and for D 273 nm. Provide the structures for B–D and deduce the structure of A from the information given.

10-21[2] Explain the following observations.

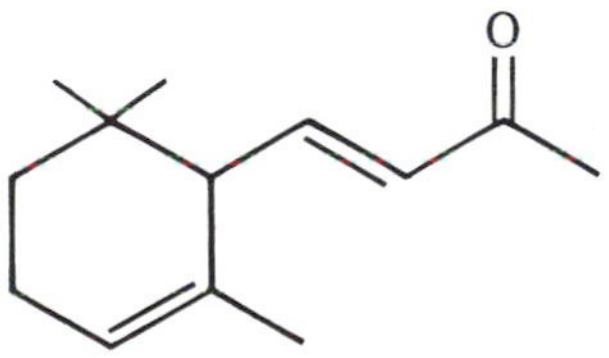

	α-Ionone	β-Ionone	ψ-Ionone
Observed λ_{max} (nm)	228	281	291
ϵ_{max}	14,300	9500	21,800

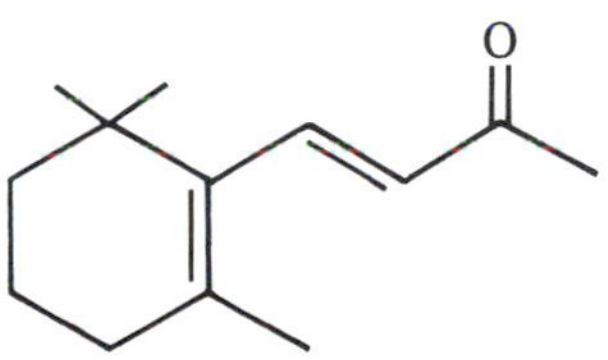

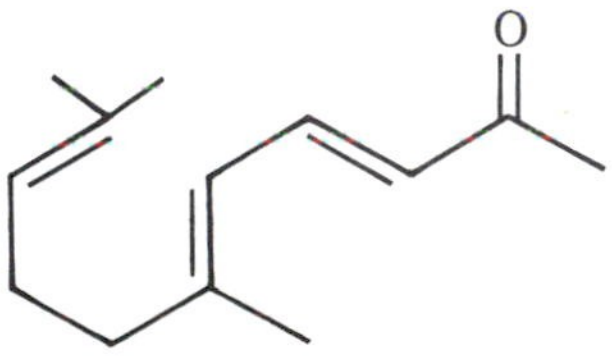

10-22 A diene, $C_{11}H_{16}$, was thought to have the structure

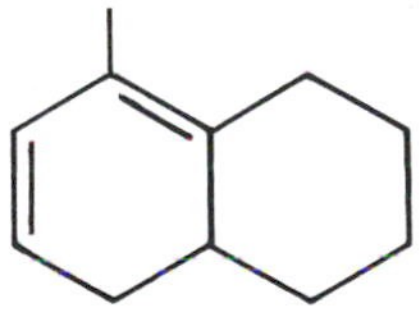

Its ultraviolet spectrum showed λ_{max} 245 nm with $A = 1.47$.

a. Can the suggested structure be correct? (Apply Woodward's rules to it.)

b. Write a structure that satisfies the ultraviolet spectrum and λ_{max} and has the same carbon skeleton.

c. If the recorded spectrum was determined in a 1 cm cell with 3 mg of compound in 250 ml of solvent, determine the molecular extinction coefficient ϵ.

d. From your knowledge of transition allowedness, is the ϵ determined in **c** reasonable?

[2]Observed data from E. S. Stern and C. J. Timmons, *Gillam and Stern's Introduction to Electronic Absorption Spectroscopy in Organic Chemistry*, 3rd ed., St. Martin's Press, New York, 1970.

e. What would the value of ϵ be if the measurement in **c** had been determined in a 1.0 m cell and gave the recorded spectrum? Is this a reasonable value?

10-23 A diene, $C_{11}H_{16}$, was thought to have the structure

Its ultraviolet spectrum showed λ_{max} 261 nm with $A = 1.92$.

a. Can the suggested structure be corrected? (Apply Woodward's rules to it.)
b. Write a structure that satisfies the ultraviolet spectrum and λ_{max} and has the same carbon skeleton.
c. If the recorded spectrum was determined in a 1 cm cell with 3 mg of compound of 250 ml of solvent, determine the molecular extinction coefficient ϵ.
d. From your knowledge of transition allowedness, is the ϵ determined in **c** reasonable?
e. What would the value of ϵ be if the measurement in **c** had been determined in a 1.0 m cell and gave the recorded spectrum? Is this a reasonable value?

10-24 An unsaturated ketone, $C_9H_{12}O$, was thought to have the structure

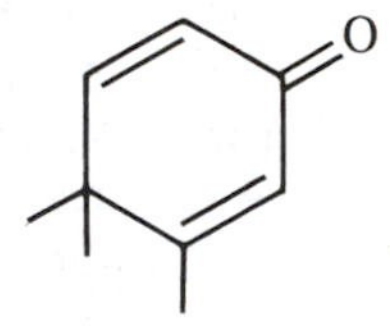

Its ultraviolet spectrum showed λ_{max} 300 nm with $A = 1.34$.

a. Can the suggested structure be corrected? (Apply Woodward's rules to it.)
b. Write a structure that satisfies the ultraviolet spectrum and λ_{max} and has the same carbon skeleton.
c. If the recorded spectrum was determined in a 1 cm cell with 4 mg of compound in 200 ml of solvent, determine the molecular extinction coefficient ϵ.

10-25 Optically pure 3-methylcyclohexanone shows a CD Cotton effect $\Delta\epsilon = +2.0$ at $\lambda = 295$ nm. To what electronic transition does this correspond? Using the octant rule, determine the absolute configuration.

10-26 Predict the sign and approximate $\Delta\epsilon$ for the $n \rightarrow \pi^*$ CD Cotton effects of (*S*)-9-methyl-*trans*-decalone-2 and cholestan-2-one.

10-27 The compounds below have negative (−) Cotton effects. Draw the absolute configuration and conformation of each.

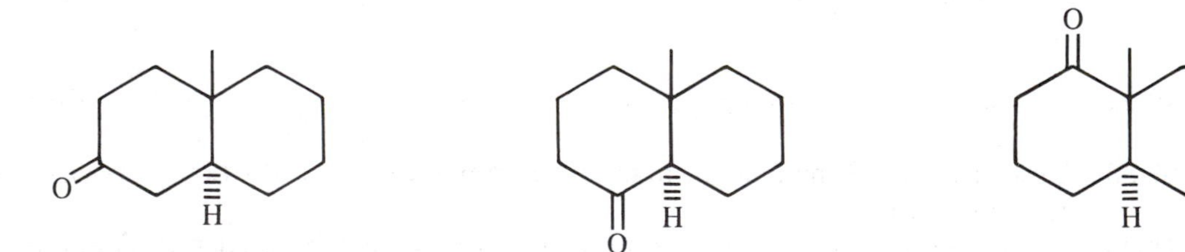

10-28 The compounds below exhibit negative (−) Cotton effects. On treatment with base, a (+) Cotton effect is produced. Explain fully.

CH₃ CH₃ O O H Br

10-29 An isomer of the compound below has a weakly positive (+) Cotton effect. On treatment with base, another isomer is obtained with $\Delta\epsilon = -0.8$. Draw the configurations and conformations of the two isomers and explain the transformation by drawing the intermediate between the two. Indicate (R,S) the absolute configuration.

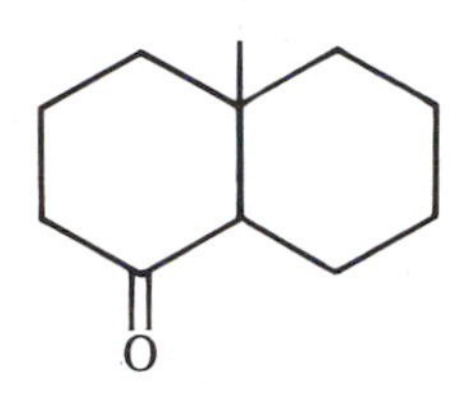

10-30 The partial structure and experimentally determined Cotton effect (CE) signs of cholestanone (**A**), lanostanone (**B**), β-amyrone (**C**), and 2,2-dimethylcholestanone (**D**) are given below. The partial structures given are sufficient for octant rule projection diagrams. Explain the experimental CE signs or rationalize any difference in terms of conformational structures.

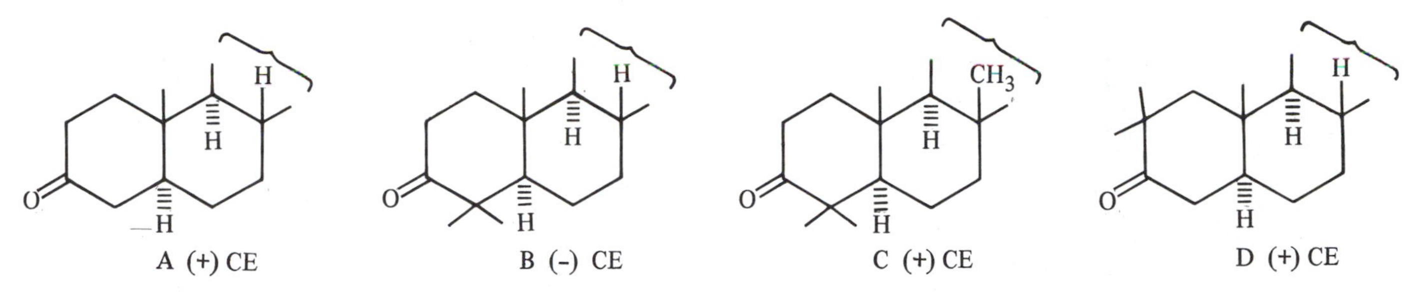

A (+) CE B (-) CE C (+) CE D (+) CE

10-31 Explain the following data in terms of the octant rule (the absolute configurations are as given).

CH$_3$ H O H CH$_3$ O

equatorial CH$_3$ axial CH$_3$

	equatorial CH$_3$	axial CH$_3$
λ_{max} (nm) (dioxane)	300	305
$\Delta\epsilon$	−0.8	+0.09

10-32 Optically pure 2-methylcyclopentanone shows $\Delta\epsilon = -1.8$; draw its conformation and absolute configuration.

10-33 Optically pure 3-methylcyclopentanone shows $\Delta\epsilon = +2.1$; draw its conformation and absolute configuration.

10-34 Treatment of (+)-α-pinene with $Fe(CO)_5$ gives two isomeric ketones, A and B. A has a (−) Cotton effect ($\Delta\epsilon$ −13.1, λ_{max} 292 nm); B has a (+) Cotton effect ($\Delta\epsilon$ +14.1, λ_{max} 292 nm). Assign the absolute configurations of A and B.

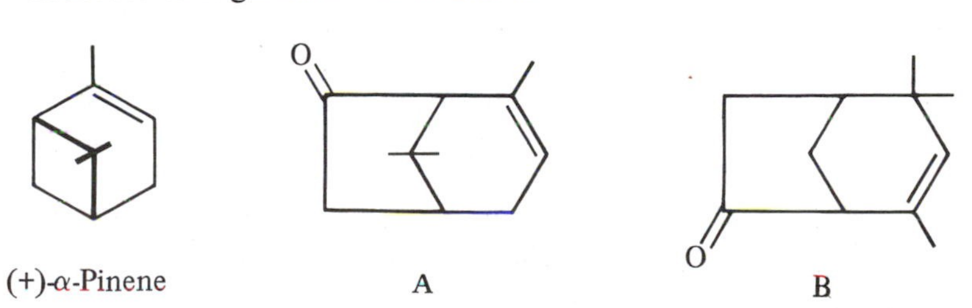

(+)-α-Pinene A B

10-35 The figure shows the CD curves of (+)-*trans*-6-chloro-3-methylcyclohexanone in methanol (*M*, ———) and isooctane (*I*, - - -). Rationalize in terms of conformations for the absolute configuration shown.

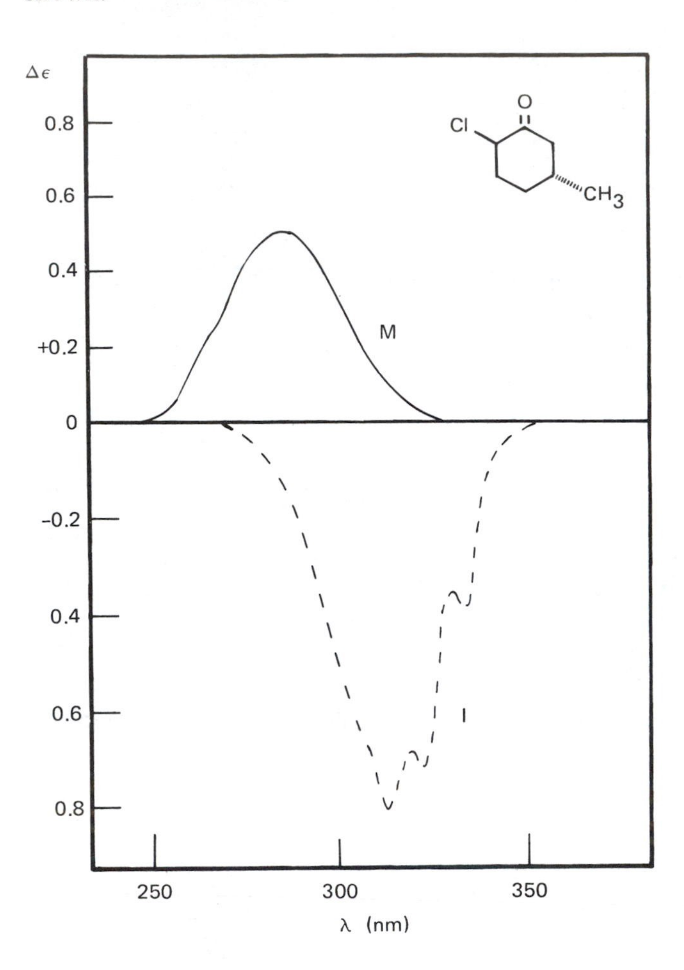

10-36 A 4-hydroxycholesterol (below) di-2-naphthoate shows a bisignate CD Cotton effect with $\Delta\epsilon_{242}$ −214.5, $\Delta\epsilon_{229}$ +191.9 for UV $\epsilon_{max,232}$ 124,500. What is the configuration at C-4?

10-37 A 6-hydroxy-5α-dihydrocholesterol bis(*p*-dimethylaminobenzoate) shows a bisignate CD Cotton effect with $\Delta\epsilon_{320}$ −37.6, $\Delta\epsilon_{295}$ +19.2 for UV $\epsilon_{max,309}$ 53,300. What is the configuration at C-6?

10-38 The absolute configuration of succinic anhydride *trans*-2,3-diol was determined by measuring the CD Cotton effects of its dibenzoate ester: $\Delta\epsilon_{240}$ +27.5, $\Delta\epsilon_{223}$ −5.7 for UV $\epsilon_{max,233}$ 29,500. Draw the structure indicating its absolute configuration.

10-39 (+)-Camphorquinone was reduced to give two different trans diols. The di(*p*-chlorobenzoate) of one gave CD Cotton effects: $\Delta\epsilon_{248}$ −35, $\Delta\epsilon_{230}$ +9. Draw the structure of the diol indicating absolute stereochemistry.

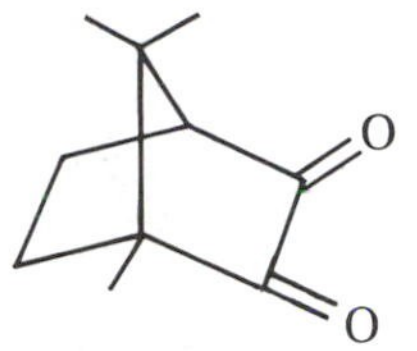

10-40 The absolute configuration of *trans*-7,8-dihydroxy-7,8-dihydrobenzo[*a*]pyrene, a carcinogenic metabolite of benzo[α]pyrene, was determined by the CD spectrum of its bis(*p*-dimethylaminobenzoate): $\Delta\epsilon_{322}$ −78, $\Delta\epsilon_{292}$ +74. Indicate the absolute configuration of the diol.

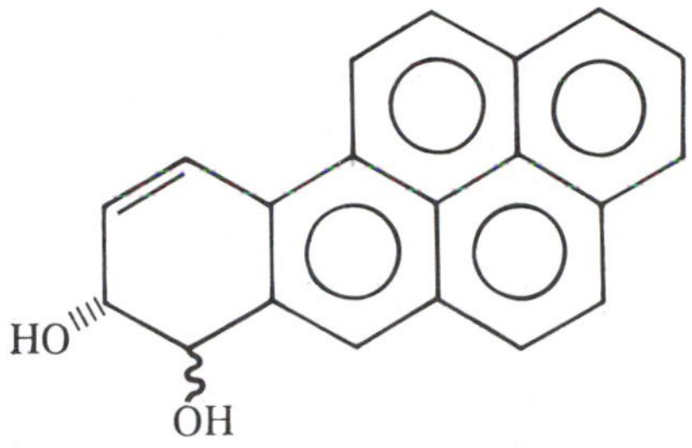

10-41 The absolute configuration of spiro[4.4]nonan-1,6-dione can be determined by CD on the bis(*p*-dimethylaminobenzoate) of the cis,trans diol (reduction product): $\Delta\epsilon_{321}$ −37, $\Delta\epsilon_{295}$ +16. Draw the dione, showing its absolute configuration.

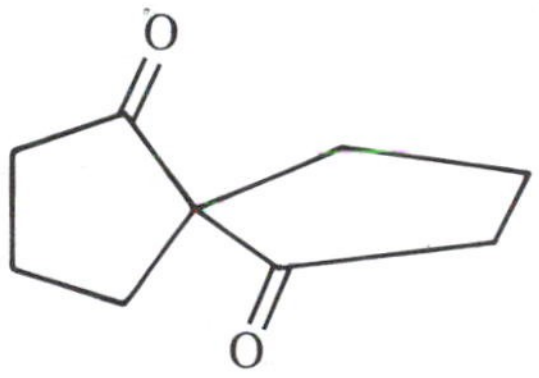

10-42 The absolute configuration of *trans*-2-aminocyclohexanol was determined by CD of its benzoate benzamide derivative: $\Delta\epsilon_{238}$ −17, $\Delta\epsilon_{222}$ +10 for UV $\epsilon_{max,228}$ 4,400. Draw the absolute configuration of the aminoalcohol.

10-43 1,1′-Binaphthyl exhibits a bisignate CD Cotton effect, $\Delta\epsilon_{225}$ +250, $\Delta\epsilon_{214}$ −179 for UV $\epsilon_{max,220}$ 108,000. Draw a structure of binaphthyl that shows its absolute configuration.

10-44 The benzoate ester of 7-hydroxycholest-4-en-3-one shows a bisignate CD Cotton effect: $\Delta\epsilon_{250}$ +13, $\Delta\epsilon_{230}$ −3. What is the configuration at C-7?

10-45 In the presence of Ni(AcAc)$_2$, butane-2,3-diol shows a bisignate CD Cotton effect: (+) at 395 nm, (−) at 370 nm. Draw a structure of the diol showing its absolute configuration.

REFERENCES/BIBLIOGRAPHY

10-1 For discussion and leading references see D. A. Lightner, T. D. Bouman, W. M. D. Wijekoon, and A. E. Hansen, *J. Am. Chem. Soc.,* **108**, 4484 (1986).

10-2 H. H. Jaffé and M. Orchin, *Theory and Applications of Ultraviolet Spectroscopy*, Wiley, New York, 1962.

10-3 C. Djerassi, *Optical Rotatory Dispersion*, McGraw-Hill, New York, 1960.

10-4 W. Moffitt, R. B. Woodward, A. Moscowitz, W. Klyne, and C. Djerassi, *J. Am. Chem. Soc.*, **83**, 4013 (1961).

10-5 T. D. Bouman and D. A. Lightner, *J. Am. Chem. Soc.*, **98**, 3145 (1976).
10-6 A. E. Hansen and T. D. Bouman, *Adv. Chem. Phys.*, **44**, 545 (1980).
10-7 D. A. Lightner and D. E. Jackman, *J. Am. Chem. Soc.*, **96**, 1938 (1974), and references therein.
10-8 C. Djerassi and W. Klyne, *J. Am. Chem. Soc.*, **79**, 1506 (1957).
10-9 A. J. Merer and R. S. Mulliken, *Chem. Rev.*, **63**, 639 (1969).
10-10 A. Yogev, J. Sagiv, and Y. Mazur, *Chem. Commun.*, 411 (1972).
10-11 T. D. Bouman, A. E. Hansen, B. Voigt, and S. Rettrup, *Int. J. Quantum Chem.*, **23**, 595 (1983).
10-12 J. K. Gawroński, *Tetrahedron*, **38**, 3 (1982), and references therein.
10-13 D. A. Lightner, D. E. Jackman, and G. D. Christiansen, *Tetrahedron Lett.*, 4467 (1978).
10-14 A. Moscowitz, K. Mislow, M. A. W. Glass, and C. Djerassi, *J. Am. Chem. Soc.*, **84**, 1945 (1962).
10-15 D. J. Sandman, K. Mislow, W. P. Giddings, J. Dirlam, and G. C. Hanson, *J. Am. Chem. Soc.*, **90**, 4877 (1968).
10-16 J. R. Platt, *J. Chem. Phys.*, **17**, 484 (1949).
10-17 P. Salvadori, L. Lardicci, R. Menicagli, and C. Bertucci, *J. Am. Chem. Soc.*, **94**, 8598 (1972).
10-18 D. A. Lightner, D. T. Hefelfinger, T. W. Powers, G. W. Frank, and K. N. Trueblood, *J. Am. Chem. Soc.*, **94**, 3492 (1972).
10-19 H. Harada and K. Nakanishi, *Circular Dichroism Spectroscopy—Exciton Coupling in Organic Stereochemistry*, University Science Books, Mill Valley, CA, 1983.
10-20 O. Korver, *Rec. Trav. Chim Pays-Bas*, **88**, 1070 (1969).

Part IV

MASS SPECTROMETRY

11

Basics of Mass Spectrometry

11-1

Characteristics of Mass Spectrometry

Mass spectrometry, which is the most sensitive method of molecular analysis, is fundamentally different from other forms of spectroscopy (nuclear magnetic resonance, microwave, Raman, emission, infrared, etc.) because electromagnetic radiation is not involved. Mass spectrometry is a form of spectroscopy in the sense that a distribution of masses is analogous to a plot of radiation intensity versus wavelength.

A mass spectrum is shown in Figure 11-1. As is usually done, the plot is normalized so that the pattern of abundances is independent of the number of items measured in recording the distribution. In this example, mass is given in kilograms and the items measured are individual people. There is, however, a strict analogy between this spectrum and that measured for organic ions in the gas phase, for which atomic mass units (daltons, d) are a more appropriate unit and in which the number of items used to record the distribution is typically from 10^8 to 10^{10} (although even single ions can be detected in appropriate experiments).

A mass spectrum can be used to characterize a *population* of ions on the basis of their mass distribution. If the set of ions all arise from one type of molecule, then this *compound* is characterized by the distribution. The mass spectrum of the compound then serves as a fingerprint. Unlike the other spectra given by a compound, its mass spectrum is a *product distribution*, and chemical reactions necessarily occur in the course of obtaining the spectrum.

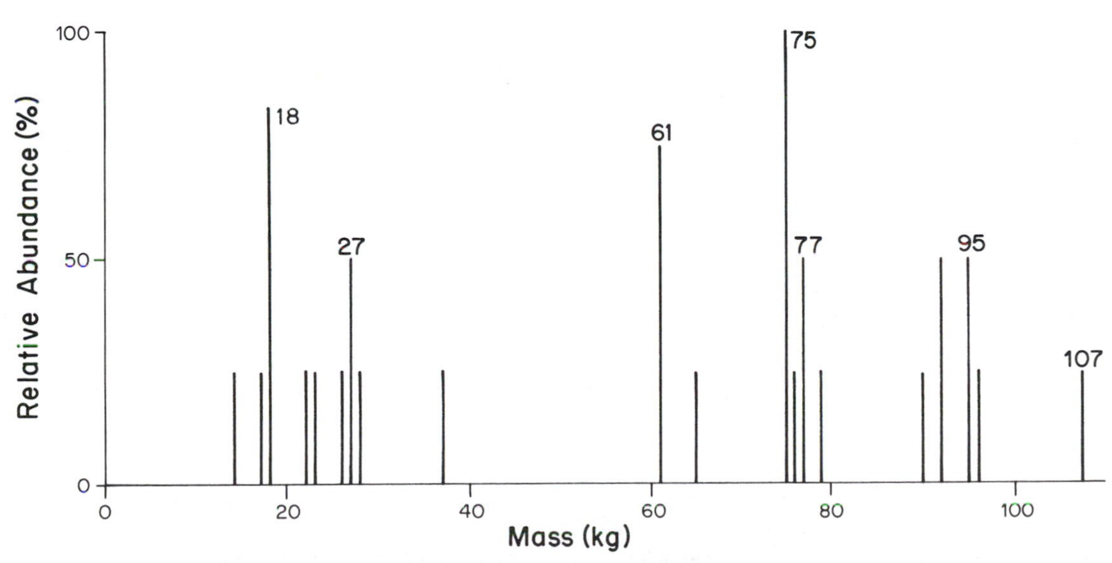

FIGURE 11-1 A mass spectrum is a normalized abundance distribution expressed as a function of mass. The mass data used in this plot are for a population of 31 people—a soccer team being welcomed by a group of preschool children.

It will help to appreciate the various experiments possible with mass spectrometry if one recognizes three facts.

1. Mass spectrometry, as already noted, is a form of spectroscopy.
2. Mass spectrometry shares features with chromatography.
3. Mass spectrometry is a means of carrying out chemical reactions.

The relationship to chromatography is particularly evident in the technique of tandem mass spectrometry (Section 12-5b) where the compound to be characterized by its mass spectrum is first ionized and separated from other species by an initial stage of mass analysis. The third aspect of mass spectrometry is, for many chemists, its most exciting attribute. A mass spectrometer is a device in which chemical reactions are invariably required to produce the ions from which mass spectra are recorded. So close is this relationship that, to many chemists, gas phase ion chemistry and mass spectrometry are synonymous. Not only can one follow reactions by mass spectrometry, but chemical reactions are intrinsic to obtaining the spectrum itself.

TABLE 11-1 Nomenclature of Mass Spectrometry

Ionization methods	see Table 11-4
Types of analyzers	see Table 11-5
Mass-to-charge ratio (m/z)	daltons/electronic charge
Mass spectrum	plot of ion abundance vs. mass-to-charge ratio normalized to most abundant ion
Breakdown curve	plot of ion abundances vs. ion internal energy normalized at each energy
Molecular ion ($M^{\cdot +}$ or $M^{\cdot -}$)	ion derived from the neutral molecule by loss or gain of an electron
Quasimolecular ion	ion that is simply related to a neutral molecule, e.g., $(M+H)^+$, $(M+Cl)^-$, $(M+Ag)^+$
Fragment ion	ion not generated by direct ionization of a neutral molecule
Parent ion (m_1^+)	any ion (including negatively and doubly charged ions) that gives fragments
Daughter ion (m_2^+)	any ion generated by fragmentation as opposed to direct ionization
Metastable ion	ion that fragments slowly and not in ion source; in sector instruments the products are focused at a unique mass-to-charge ratio and serve to identify the parent and daughter ions
Base peak	the most abundant ion in the spectrum
Relative abundance (RA)	normalization relative to the base peak
Isotopic peak (ion)	due to other isotopes of the same composition
Isobaric peak (ion)	of the same nominal mass
Ionization energy (IE)	endothermicity of process $M \rightarrow M^{\cdot +}$
Appearance energy	endothermicity of process $AB \rightarrow A^+ + B$
Electron affinity (EA)	endothermicity of process $M^- \rightarrow M + e^-$
Proton affinity (PA)	endothermicity of process $(M+H)^+ \rightarrow M + H^+$
Ion internal energy (ϵ)	total electronic, vibrational, and rotational energy referenced to ground state of the ion
Distribution of internal energy ($P(\epsilon)$)	analogue of Boltzmann distribution for molecules not in thermal equilibrium
Unimolecular rate constant ($k(\epsilon)$)	in s^{-1}, dependent upon ϵ, which is shown explicitly
Collision-activated dissociation (CAD) } Collision-induced dissociation (CID)	process whereby a mass-selected ion is excited and caused to fragment, especially in MS–MS
Charge exchange	electron transfer process used as a form of chemical ionization (eq. 11-12)
Gas chromatography–mass spectrometry (GC–MS)	combined techniques for mixture analysis
Liquid chromatography–mass spectrometry (LC–MS)	combined technique for mixture analysis
Tandem mass spectrometry (MS–MS)	combined technique for mixture analysis; also used to study chemistry of selected ions
Selected ion monitoring (SIM)	experiment in which mass analyzer is used to detect one or a few ions as a function of time

TABLE 11-2 Conversion Factors in Mass Spectrometry

Charge	1 electronic charge = 1.60×10^{-19} coulomb
Current	1 ion s^{-1} = 1.60×10^{-19} A
Energy	1 eV = 23.06 kcal mol^{-1}; 1 kcal mol^{-1} = 4.18 kJ mol^{-1}
Mass	1 dalton = 1.66×10^{-24} g
Pressure	1 Torr = 133 Pascal = 1 mm Hg = 1.33 mbar
	1 Torr = 3.2×10^{16} molecules cm^{-3}
Rate constant	
unimolecular	s^{-1}
bimolecular	1 liter mol^{-1} s^{-1} = 2×10^{-21} cm^3 $molecule^{-1}$ s^{-1}
Velocity	particle of 1 dalton accelerated to 1 eV = 1.1×10^6 cm s^{-1}

The nomenclature of mass spectrometry requires some attention, and a number of terms, acronyms, and definitions are collected in Table 11-1. Many of these refer to concepts or experiments that will only be encountered later in this chapter. The table is arranged by grouping together closely related material. Table 11-2 lists quantities and conversion factors important in mass spectrometry.

11-2

Use of Mass Spectrometry by the Organic Chemist

The organic chemist uses mass spectrometry for the following purposes.

1. *Molecular weight determination*, either on pure compounds, synthetic or isolated, or on mixtures as in the cases of reaction products or samples isolated from natural sources. Often both separation and identification are achieved by mass spectrometry, for example by using the combined gas chromatography–mass spectrometry (GC–MS) instrument. Molecular weight determination is the most common use of mass spectrometry made by organic chemists. Unit mass resolution, the ability to distinguish ions that differ in mass by one dalton, is readily available over a mass range of several thousand and is adequate for this task. Interpretation of the results requires *recognition* of ions that pinpoint the molecular weight, and this simple act of interpretation is the key step in determining molecular weight by mass spectrometry.

2. *Molecular formula determination.* If masses can be measured with sufficient accuracy, it is possible to determine the molecular formula for any ion. In practice, interest centers on the molecular ion, i.e., the ion with the same formula as the neutral molecule, indicated as $M^{\cdot+}$, or a closely related species such as the protonated molecule $(M+H)^+$. This determination requires high resolution mass spectrometry. For example, a resolution from 10^4 to 10^5 suffices to distinguish between simple organic molecules containing the common elements C, H, N, and O. The required resolution quickly increases and can reach unattainable levels as mass increases or as more elements are considered. Even in such circumstances, however, the measurement is extremely valuable in *excluding* certain formulas and in obtaining data consistent with a particular composition. In each of these respects it is analogous to conventional elemental analysis.

3. *Structural analysis*, including both confirmation of structural assignments and identification of unknown structures by interpretation of the fragmentations occurring in the spectrometer. The former task can be accomplished by comparing data with those for known compounds included in libraries of spectra or by measuring the spectrum of an authentic compound. The identification of unknowns by spectrum–structure correlation is accorded considerable space in Chapter 12.

4. *Isotopic incorporation*, including both the extent and the site of incorporation of stable isotopes in molecules that are isotopically enriched. Special instruments having more than one detector are used when ratios of isotopes are to be measured with ppm precision, especially in solving biochemical and geochemical problems. Incorporation of isotopes in organic molecules can be quantified less precisely using conventional instruments as described in Section 12-3.

11-3

Essentials of Instrumentation

The *mass spectrometer* accomplishes far more than the separation of ions according to mass, and later sections will provide some indications of the range of the mass spectrometry experiment. However, reduced to essential elements, three operations occur in all mass spectrometers: *ionization*, *mass analysis*, and *detection* (Table 11-3).

TABLE 11-3 Components and Operations in Mass Spectrometry

	Component		
	Ion source	**Mass analyzer**	**Detection system**
Operation	ionization and fragmentation	separation	recording spectrum
Common types	chemical ionization	magnetic sector	electron multiplier
	desorption ionization	quadrupole mass filter	image current
	electron ionization	time-of-flight	
		ion cyclotron resonance	

The *ion source* is the region where gas phase ions are generated and also where reactions of the initially formed ions occur. These reactions may be ion–molecule reactions, resulting from collisions between ions and neutral molecules, or they may be unimolecular processes in which energetic ions fragment spontaneously. The latter are our chief concern in the common technique of electron ionization mass spectrometry. On the other hand, reactive collisions form the basis for chemical ionization, a more versatile ionization method that often provides molecular weight information missing in electron ionization experiments. In the new desorption ionization techniques, condensed phase samples are examined directly through the input of energy. These methods, typified by fast atom bombardment (FAB), provide the ability to ionize nonvolatile and thermally labile compounds. The various ionization methods are briefly summarized in Table 11-4. All are discussed at greater length below (Section 11-5), while hardware is presented in Section 11-7c.

TABLE 11-4 Ionization Methods

Abbreviation	Method	Energy supplied by	Typical reaction
CI	chemical ionization	chemical reagent	$BH^+ + M(\text{vapor}) \longrightarrow MH^+ + B$
DI	desorption ionization	particle or photon impact	$M(\text{solution}) \xrightarrow[5\text{ keV}]{Xe^0} MH^+$
EI	electron ionization	electron impact	$M(\text{vapor}) \xrightarrow[70\text{ eV}]{e^-} M^{\cdot +} + 2\,e^-$

The *mass analyzer* achieves separation of charged particles on the basis of mass-to-charge ratio or a property related to it. Detailed treatment is reserved for Section 11-7d, but four of the principal methods employed are illustrated schematically in Figure 11-2, and their properties are summarized in Table 11-5. The

TABLE 11-5 Characteristics of Mass Analyzers

Method	Quantity measured	Equation	Mass range (d)	Resolution at 1000 d	Dynamic range
Sector magnet	momentum	11-28	$>10^4$	10^5	10^7
Quadrupole	filters for m/z		$>10^3$	10^3	10^5
Time of flight	time	11-24	$>10^4$	10^3	10^5
Cyclotron resonance	frequency	11-29	$>10^4$	10^6	10^4

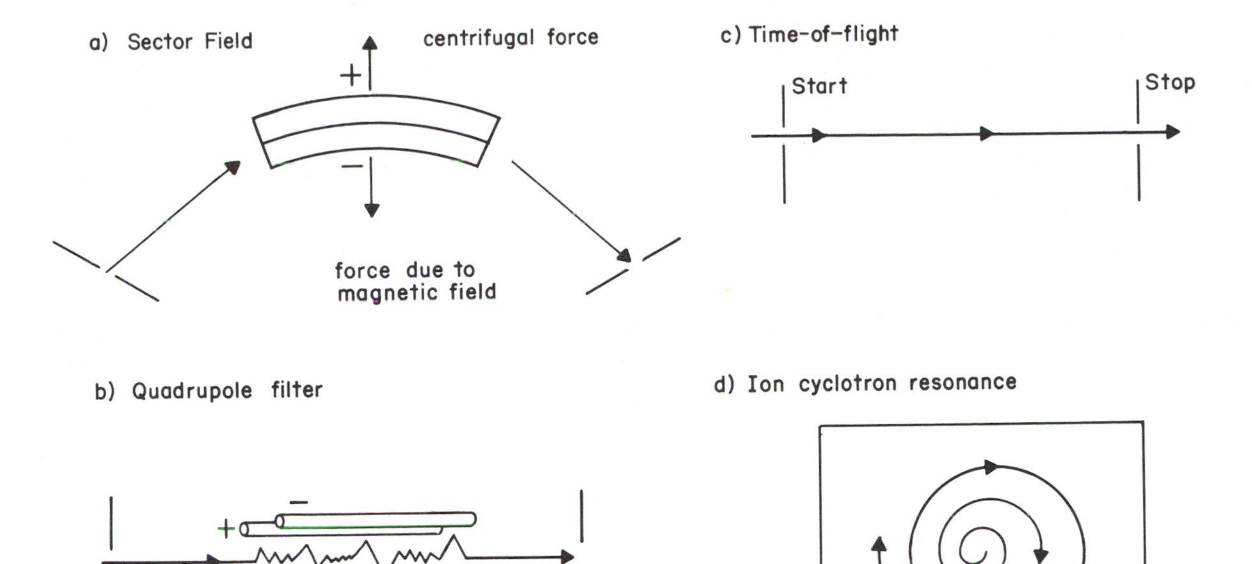

FIGURE 11-2 Methods of mass analysis.

commonly used sector magnet technology measures mass-to-charge ratios of ions via deflection of an ion beam through a fixed angle. More and less massive ions bearing the same charge will strike the outer and inner walls, respectively, while the chosen ions pass cleanly through the device. A completely different method of mass separation employs alternating electric fields that can be arranged to cause unstable trajectories for ions having mass-to-charge ratios that fall outside a given range. Quadrupoles and their three-dimensional counterpart, the quistor or ion trap, utilize this principle. Flight time of an ion to cover a fixed distance depends upon its mass, the number of charges it bears, and the potential through which it is accelerated. A simple mass analyzer based chiefly on timing circuitry utilizes this principle. Finally, the cyclotron frequency of an ion in a magnetic field is the basis of the cyclotron resonance method of mass analysis.

Ion detection is usually achieved with an electron multiplier. This device is the analogue of a photomultiplier, except that the first stage requires an ion → electron rather than a photon → electron conversion (compare Figure 11-22). In cyclotron resonance instruments one measures the image current induced in a metal by the gas phase ion as it moves by. For precise isotope ratio measurements one employs a Faraday cup, rather than an electron multiplier.

11-4

Applications Outside Organic Chemistry

Since this treatment of mass spectrometry emphasizes heavily the deduction of molecular structures and the behavior of gaseous organic ions, it seems desirable to establish some perspective by broadly sketching other uses of the technique. The roots of mass spectrometry lie in the experimental work of W. Wien and J. J. Thomson done around the end of the nineteenth century. Development of the instrument has both depended on and contributed to improvements in vacuum technology, ion optics, and recently data acquisition and reduction.

In the years 1919–1939, after Thomson's discovery of the stable neon isotopes ^{20}Ne and ^{22}Ne, the search for *new isotopes* and the determination of their exact masses and natural abundances were pursued because of their basic importance in physical science. A later more technical application of mass spectrometry arose from the needs of the Manhattan project. Large mass spectrometers known as calutrons neutralized and deposited separated ions in special collectors, allowing macroscopic amounts of isotopes to be prepared. Mass spectrometry is the standard method for the analysis of the results of isotopic labeling experiments and the determination of *stable isotope incorporation* at high precision, which is proving to be an increasingly powerful method of following chemical processes in the biosciences, where it has obvious advantages over radioisotopic tracer work.

The accurate determination of isotopic abundances now constitutes the basis for important branches of geoscience and archaeology. Various dating methods, including the new carbon dating method in which ^{14}C is *directly* measured, at ratios relative to ^{12}C as small as 1 in 10^{14}, depend on mass spectral data. This procedure is an improvement over fission counting methods. In a similar vein, variations in natural abundances of stable isotopes with the nature of the source can be measured accurately. From the measurement of this historical isotope effect, the physical, chemical, or biological conditions that effected the original enrichment can sometimes be inferred. For instance, the average temperature and even the seasonal temperature variation during the lifetimes of Cenozoic marine crustaceans can be found from the $^{16}O/^{18}O$ ratio in the calcium carbonate of their shells, since this ratio depends on the water temperature in which the organisms lived.

Following intensive effort on ion mass and abundance measurements, rapid development occurred in the utilization of *thermochemical data* obtainable from the mass spectrometer. The initial focus of this effort was the ionization efficiency curve, a plot of the abundance of an ion against the ionizing energy (e.g., electron energy). From this plot the fundamental properties of electron affinity (EA), ionization energy (IE), and appearance energy (AE) can be derived (Table 11-1). When these values are combined with known thermochemical data on neutral species, *ionic heats of formation* and *bond dissociation energies* can be determined. A valuable feature of the mass spectrometric approach is the fact that unstable species, such as free radicals, can be studied and bonds that cannot otherwise be cleaved can be investigated. Thus the α-haloacetophenone molecular ion undergoes C—C cleavage in a process that releases about 29 J mol^{-1} (6.9×10^{-3} kcal mol^{-1}) of internal energy as translational energy. This is of the order of magnitude of rotational energies. Another mass spectrometric procedure, field ionization kinetics, can be used to follow ionic reactions occurring on the *picosecond* time scale.

A rich source of thermochemical information is available from studies of *ionic equilibria* in the gas phase. These methods, which employ high pressure chemical ionization types of sources or long reaction times, as in ion cyclotron resonance measurements, provide thermochemical data that are typically more accurate than those available for neutrals with calorimetry. Similar experimental methods allow one to study the chemistry of ions solvated by as few as one or two solvent molecules.

Recent years have seen the application of the mass spectrometer in many new ways as well as the development of new types of instrumentation. Remarkably, much of this growth originally derived from the role mass spectrometers once played in a single area of industrial production, namely, quality control in petroleum refining. This application provided a powerful stimulus for structural analysis and led to the introduction of commercial instruments in the 1940s. Mass spectrometers, fitted with a wide range of ion sources, can now be used to create ions representative of aqueous solutions, refractory solids, and gases, to name some types of samples. The method is now of importance to chemists and biochemists and an essential tool in geoscience, and it is also widely used in the chemical and pharmaceutical industries. Even in nonchemical industries such as semiconductors, it is used for trace element determination, microscopic imaging of surfaces, and doping of devices. Similarly, *surface science* experiments often incorporate mass spectrometers. For example, the outermost monolayers of a surface are sputtered away and the ejected ions are mass analyzed to provide surface compositional data in the technique of secondary ion mass spectrometry.

The huge amount of data obtained when a mass spectrometer is run on the components emerging from a gas chromatograph or run under high resolution conditions has caused a great deal of effort to be devoted to *data acquisition and processing*. The availability of many thousands of reference mass spectra has led to the development of efficient library search routines that match unknown and library spectra in seconds. In addition, the complexity of mass spectra and the highly individual approach used in solving structural problems have contributed to major progress in automated spectral interpretation and most recently in *artificial intelligence*.

Perhaps the most notable development since the systematic application of mass spectrometry to organic chemistry in the early 1960s has been the enormous increase in its use in solving problems in the *biological sciences*. These applications have followed directly upon advances in the mass spectrometer that have made possible the examination of higher molecular weight compounds (above 10,000 and being extended), as well as nonvolatile and unstable samples. Particularly noteworthy has been the *coupling of gas and liquid chromatographs to the mass spectrometer* (Section 11-7a), leading to the routine study of complex mixtures. These techniques have greatly extended the range of applications of mass spectrometry in many fields, some of which, e.g., drug metabolism, have been greatly affected by these instrumental developments. Natural products chemistry, among other subjects, has been influenced by a related procedure, tandem mass spectrometry, which utilizes two mass spectrometers to obtain qualitative and quantitative information on individual constituents in complex mixtures (Section 12-5b). The development of a suite of new methods for ionizing nonvolatile, fragile molecules (such as peptides, phospholipids, and oligonucleotides) has completed the opening of the biological sciences to mass spectrometry. The individual methods of ionization are discussed in Section 4-5. The accessible molecular weight range is well in excess of 10,000 d.

Another area in which mass spectrometry is proving essential is *environmental monitoring*. As requirements for compound-specific detection at ultratrace levels have increased, so mass spectrometry has become the standard method for these measurements. Modern quadrupole mass spectrometer systems (gas chromatograph–quadrupole mass filter–data system) have developed in parallel with the requirements for environmental analysis. Toxic waste analysis (e.g., chlorinated dioxins that have been measured down to the 20 parts per quadrillion level) has relied almost exclusively on gas chromatography–mass spectrometry.

Elemental analysis, particularly of refractory materials, has long been accomplished using special arc and spark source mass spectrometers. The sensitivity to both trace and major elements and applicability across the periodic table are characteristics of these methods. Recently improved vaporization–ionization procedures have made these determinations more convenient. Particularly powerful is the combination of the *inductively coupled plasma* with the *mass spectrometer* (ICPMS). This device allows rapid, multielement trace anaysis of solution samples with few matrix effects.

11-5 Ionization

Three methods of ionization are discussed: electron ionization (EI), chemical ionization (CI), and desorption ionization (DI). There are desiderata (Table 11-6) that apply to all methods and can be used to compare them. CI, DI, and EI have complementary properties that make access to all three highly desirable (see Part V, "Integrated Problems"). Even when only one method is available, it is worthwhile to record spectra under several conditions in order to establish the molecular weight of the sample firmly.

11-5a Electron Ionization

In electron ionization the sample is first vaporized, and the vapor is subjected to bombardment by an energetic electron beam. Some electrons are simply reflected (elastically scattered) by sample molecules,

TABLE 11-6 Desiderata for Ionization Methods

Ion current	should be high for sensitivity, stable for quantitative analysis; values of 10^{-9} A are typical for all three methods
Ionization efficiency	conversion of introduced molecules to detected ions; values of 10^{-8} coulomb/μg, i.e., 10^{-2}%, are typical for all three methods
Polarity	equal performance in generating, analyzing, and detecting positive and negative ions is desirable; achieved for CI and DI, not EI
Internal energy distribution	ideally, should be single valued and settable; in practice, should be variable; achieved for CI and EI, not DI
Matrix effects	undesirable; avoided completely only in EI
Physical state of sample	mass spectrometers should be able to handle samples in any physical state

whereas others undergo collisions that cause the molecules to be electronically excited. A few collisions cause complete removal of an electron from the molecule (eq. 11-1). Note that the products of this interaction are

$$AB + e^- \rightarrow AB^{\cdot +} + 2\,e^- \tag{11-1}$$

the radical ion, a charged species that is also a free radical, and two electrons, one the bombarding electron itself. The radical cation, $AB^{\cdot +}$, is termed the *molecular ion* because of its relationship to the original neutral molecule. It is desirable that this ion retain the structure of the neutral precursor, since structural inferences are based on its behavior. However, the reactivity of free radicals is such that intramolecular rearrangement can occur and can complicate the interpretation of electron impact mass spectra.

The molecular ion is the ion formed by loss of an electron from the molecule. A *parent ion* is any ion that fragments to produce a product or *daughter ion*. The molecular ion is just one example of a parent ion. Fragment ions formed directly from the molecular ion are primary fragments. The whole network of competing and consecutive reactions leading to the set of ions that is ultimately sampled to provide a mass spectrum is termed the *fragmentation pattern*.

Ions may be *singly* or *multiply* and *positively* or *negatively* charged. The properties of a charged species will depend strongly on whether or not it is a free radical, hence the important distinction between *odd*- and *even-electron ions*. (The free radical or odd-electron ion is always indicated, as in $H_2O^{\cdot +}$ and $AB^{\cdot +}$.) Ions containing a sequence of atoms not present in the molecule are *rearrangement ions*.

Equation 11-1 represents the dominant process whereby ionization is effected by electron bombardment. Multiply charged or negatively charged ions may be formed as shown in eqs. 11-2 and 11-3, although usually

$$AB + e^- \rightarrow AB^{2+} + 3\,e^- \tag{11-2}$$

$$AB + e^- \rightarrow AB^{\cdot -} \tag{11-3}$$

in low abundance. Contributions from ion pair production (eq. 11-4) and predissociation (eq. 11-5) are detectable only in particular cases.

$$AB + e^- \rightarrow A^+ + B^- + e^- \tag{11-4}$$

$$AB + e^- \rightarrow A^+ + B^{\cdot} + 2\,e^- \tag{11-5}$$

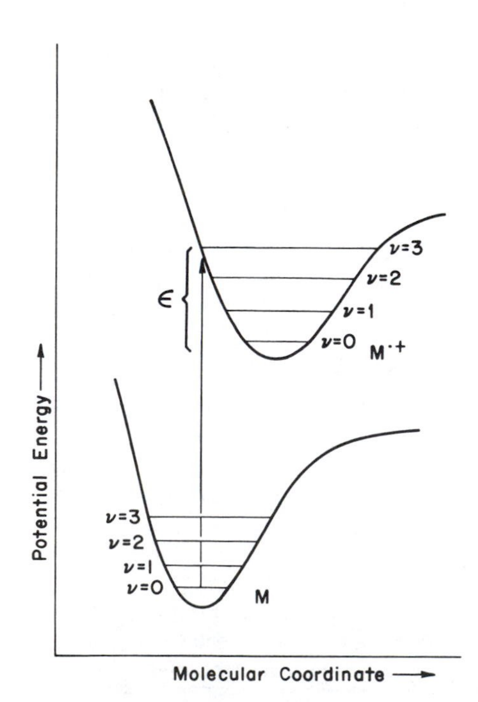

FIGURE 11-3 Vertical transitions associated with electron ionization ($M \rightarrow M^{\cdot +}$) showing deposition of internal energy ϵ in the ion $M^{\cdot +}$.

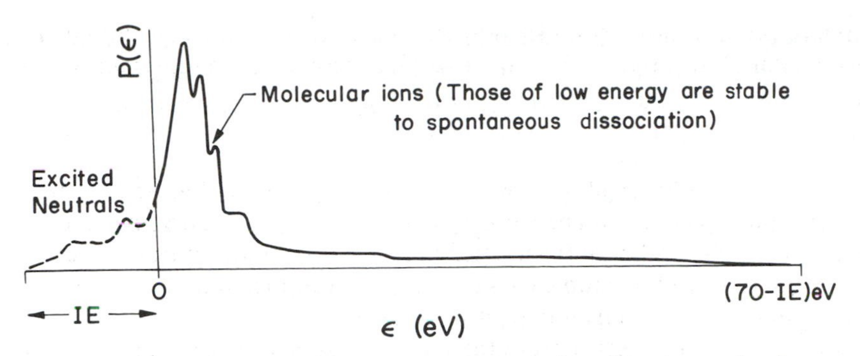

FIGURE 11-4 Internal energy distribution $P(\epsilon)$ resulting from impact by electrons of 70 eV kinetic energy upon an organic molecule. The maximum internal energy that an ion can acquire is 70 eV, minus the ionization energy (IE). Most ions acquire small internal energies.

Ionization by electron impact is the result of an electronic transition in the molecule. Because the velocity of the bombarding electron (5×10^8 cm s^{-1} for 70 eV electrons) greatly exceeds the rate of intramolecular atomic motion (typical bond vibrations require $>10^{-12}$ s, whereas a 70 eV electron transits a 10 Å molecule in 2×10^{-16} s), the molecule remains frozen as the electronic excitation occurs, a condition described as the *Franck–Condon* principle. Figure 11-3 illustrates such an ionizing event.

Ions are generated with a distribution of internal energies upon electron ionization. Transfer of the full kinetic energy of the electron to the molecule is unlikely; the amount of energy transferred depends on the nature of each individual electron–molecule collision. When a population of molecular ions is considered, there will be an associated distribution of ion internal energies, termed $P(\epsilon)$ (Figure 11-4). Since electron ionization is usually achieved under high vacuum, i.e., collision-free conditions, the distribution of internal energies of molecular ions $M^{\cdot+}$ will *not* be altered by collision and it will certainly not be a Boltzmann distribution. The shape of the distribution associated with electron impact depends on the probability of excitation to various excited states of the ion and is difficult to characterize in detail in larger organic ions. It can be controlled to some extent by varying the ionizing electron energy since the maximum energy the ion $M^{\cdot+}$ can have is the difference between the electron energy and the minimum energy required for ionization, namely, the ionization energy.

The internal energy distribution $P(\epsilon)$ is important because it controls the subsequent unimolecular dissociation behavior of the ion population, just as the temperature of a system with a Boltzmann distribution of internal energies controls its chemistry. In Section 11-6a the relationship between fragment ion abundances and the internal energy distribution is explored further.

Table 11-7 provides an indication of the ability of electron ionization to generate abundant molecular ions. The compounds considered all give useful electron impact mass spectra (many compounds are not suitable for analysis by this procedure). Abundant molecular ions are favored both by small energy transfer and resistance of the ion to dissociation. In neither regard does electron impact have the appropriate characteristics for abundant molecular ion formation.

TABLE 11-7 Molecular Ion Abundances in EI Spectra[a]

Relative abundance of $M^{\cdot+}$ (% of base peak)	100–50	49–6	5–1	<1
Fraction of sample (%)	30	32	8	30

[a]A collection of 120 compounds, molecular weight 220. Data from the EPA–NIH library of mass spectra.

To summarize some of the features of electron ionization,

1. A radical cation is produced (negative ion formation is much less probable).

2. This cation is generated with a broad distribution $P(\epsilon)$ of internal energies that contributes to the observation of both intact molecular ions and fragment ions formed by unimolecular dissociation.
3. The energy deposited in the ion can be varied, to a limited extent, by varying the ionizing electron energy.
4. The efficiency of ionization is of the order of 1 part per 1000–10,000.

The main advantage of EI is that the spectra obtained are highly reproducible, largely because collisions between ions and molecules are avoided, thus allowing only electron–molecule and unimolecular processes to occur. The reproducibility of EI spectra is such that they form the basis for the only extensive collections of mass spectra. There are three main disadvantages of electron ionization. First, the sample must be presented in the vapor state. This requirement precludes work with many thermally labile compounds and is an especially severe limitation for the many heat-sensitive compounds of biological origin. The second disadvantage concerns the energy deposited upon ionization, which is often large enough to cause extensive fragmentation. The net result is that the molecular ion, from which information on the molecular weight and formula are obtained, is often of low relative abundance. Third, the radical character of the molecular ion means that some fragmentation products may arise as a result of rearrangement reactions that can make the interpretation of the spectrum more difficult.

11-5b Chemical Ionization and Desorption Ionization

The newer methods of ionization, chemical ionization and desorption ionization, address the three main disadvantages of EI. Chemical ionization still requires that the sample be vaporized, but it allows a wide choice with regard to the type of ion formed, e.g., $(M+H)^+$ and $(M-H)^-$. Even more significantly, the energy deposition accompanying ionization can be controlled through choice of a particular reagent. By making an appropriate choice, one can achieve efficient ionization with small degrees of internal excitation (so-called "soft" ionization). One can therefore select conditions that give abundant ions from which molecular weights can be deduced. Such control is not available to the same extent in electron ionization or in desorption ionization. Desorption ionization is a relatively soft method compared to 70 eV electron impact, but its more significant advantage is that either solids or solutions can be examined. Thus, DI does not share the limitation of EI mass spectrometry to thermally stable, volatile samples.

Interestingly, both CI and DI have ionization efficiencies similar to that of EI, and all three methods are suitable for use in high resolution mass spectrometers. CI and DI spectra do not have the reproducibility of EI spectra, but they do have the additional advantage of generating negative ions in comparable quantities to positive ions. A second advantage provided by CI and DI is that the ion formed directly from the molecule (sometimes termed the quasimolecular ion) is not normally a radical ion and this results in simpler fragmentation than observed in EI. Before examining the chemical basis for these methods, it is useful to generalize, in a simple qualitative form, the relative merits of the three most widely used ionization techniques (Table 11-8).

TABLE 11-8 Performance Comparison of Ionization Methods

Property	Method		
	CI	DI	EI
Ion current	$\geq 10^{-9}$ A	$\geq 10^{-9}$ A	$\geq 10^{-8}$ A
Polarity	±	±	+
Simplicity, reliability	good	good	excellent
Quantitative accuracy	good	good	excellent
Reproducibility of spectra	good	fair	excellent
Information on molecular weight	good–excellent	excellent	good
Labile samples	good	excellent	poor
Applicability	selective	selective	universal
Information on molecular structure	good	good	good
Insensitivity to interfering ions	good	poor	excellent

Chemical Ionization. Chemical ionization utilizes electron ionization to generate a reagent ion from a reagent gas. For example, if ammonia is chosen as reagent gas, electron ionization occurs to give the molecular ion of ammonia (eq. 11-6). In a chemical ionization source ion–molecule collisions are induced by

$$NH_3 + e^- \rightarrow NH_3^{\cdot+} + 2\,e^- \qquad (11\text{-}6)$$

raising the pressure to ca. 0.3 Torr (as opposed to 10^{-5} Torr in EI). When ammonia is the reagent, the dominant ion–molecule reaction is hydrogen atom abstraction (eq. 11-7), which yields a stable population of

$$NH_3^{\cdot+} + NH_3 \rightarrow NH_4^+ + NH_2^{\cdot} \qquad (11\text{-}7)$$

NH_4^+ reagent ions. These ammonium ions do not react with ammonia itself except by the switching process (eq. 11-8), which serves to remove any excess internal energy they may have.

$$NH_4^+ + NH_3 \rightarrow NH_3 + NH_4^+ \qquad (11\text{-}8)$$

However, if a *small* amount of sample is introduced into the CI source, it can react with the reagent ion even though its partial pressure makes direct electron ionization negligible. For example, pyridine at 10^{-3} Torr can accept a proton from NH_4^+ in a proton transfer reaction that is slightly exothermic and that occurs on virtually every collision (eq. 11-9). The product ion, protonated pyridine, has a small internal

$$NH_4^+ + Py \rightarrow NH_3 + PyH^+ \qquad (11\text{-}9)$$

energy and hence does not fragment. Molecular weight determination is therefore straightforward.

The thermochemistry that underlies chemical ionization reveals much about the process. To understand this approach, consider the particular case of proton transfer. The proton affinity (PA) of a molecule M is defined as the exothermicity of protonation (eq. 11-10). Table 11-9 reproduces some proton affinity values.

$$PA(M) \equiv \Delta H_f(M) + \Delta H_f(H^+) - \Delta H_f(MH^+) \qquad (11\text{-}10)$$

If a proton is transferred between two species, A and B, then the enthalpy of reaction is simply the difference in proton affinities ΔPA. This relationship is readily shown by applying the definition in eq. 11-10 to a reaction such as 11-9. The excess energy must appear as internal energy of the products. If this energy (ΔPA) is large, extensive fragmentation will be seen, a condition that often occurs with electron ionization. If it is small, the protonated product molecule will be abundant. Although ΔPA describes the sum of the internal energies of both the ionic and the neutral products of eq. 11-9, the larger ionic product typically takes almost all the excess energy. The result is that the ion internal energy distribution $P(\epsilon)$ for chemical ionization is rather narrow and can be set to high or low values by choice of appropriate reagent gas.

TABLE 11-9 Some Values of Proton Affinities

Base	Proton Affinity kJ mol^{-1}	Proton Affinity kcal mol^{-1}	Base	Proton Affinity kJ mol^{-1}	Proton Affinity kcal mol^{-1}
H_2	422	101	$(CH_3)_2CO$	823	197
O_2	423	101	NH_3	857	205
HF	468	112	$C_6H_5NH_2$	884	211
CH_4	536	128	CH_3NH_2	894	214
H_2O	723	173	$(CH_3)_2NH$	920	220
H_2S	738	177	Pyridine	921	220
HCN	748	179	$(CH_3)_3N$	938	224
i-C_4H_8	823	197	$(CH_3)_2N(CH_2)_2N(CH_3)_2$	996	238

From D. H. Aue and M. T. Bowers in M. T. Bowers (Ed.), *Gas Phase Ion Chemistry*, Academic Press, New York, 1979, Vol. 2, Ch. 9.

Chemical ionization relies on ion–molecule reactions to create quasimolecular ion(s) from the analyte, but fragment ions arise from unimolecular dissociation, just as in EI, which explains the importance of $P(\epsilon)$ in both experiments. Consideration of the ion residence time in a CI source (1–10 μs) confirms that unimolecular reactions will predominate. At 0.3 Torr, ions undergo approximately 100 collisions in 3 μs. Hence, the first collision will occur approximately 3×10^{-8} s after ionization. This time is long compared to the rates of many unimolecular dissociations (Section 11-6a).

Proton transfer processes are just one type of ion–molecule reaction useful in CI. It is a simple matter to change the chemical ionization reagent and so generate a different type of molecular ion. For example, methyl chloride as reagent gas, at an appropriate pressure, yields as reagent ion the dimethylchloronium cation. This species readily transfers a methyl group in an exothermic reaction leading, in the case of pyridine, to methylated pyridine (eq. 11-11). This form of quasimolecular ion is just as useful as the pro-

$$\mathrm{Py} + \mathrm{CH_3ClCH_3^+} \rightarrow \mathrm{PyCH_3^+} + \mathrm{CH_3Cl} \qquad (11\text{-}11)$$

tonated molecule in determining the molecular weight of the sample.

Chemical ionization has an additional useful characteristic that is illustrated by considering the consequences of introducing a pyridine–benzene mixture into the ionized ammonia gas. The reagent ion NH_4^+ does not transfer a proton to benzene (this reaction is endothermic) and hence the spectrum of the pyridine–benzene mixture is identical to that of pyridine itself. In other words, chemical ionization using ammonia as reagent gas is transparent to some constituents of mixtures, i.e., CI is a *selective method* of ionization. On the other hand, because all vapor phase compounds can be ionized by an electron beam, EI is a *universal ionization* method. Both characteristics can be advantageous in particular circumstances. Because it depends upon chemical reactions to create ions of the analyte, CI is capable of almost infinite variation. Reagent ions can be Brønsted or Lewis acids or bases, redox reagents, or other species. Reagent ions can be selected to deposit large internal energies in the sample ions resulting in extensive fragmentation or small internal energies that minimize fragmentation and facilitate molecular weight determination.

Table 11-10 lists some chemical ionization reagent gases, including the major reagent ions formed from each. For example, hydrogen, which yields an ionizing agent H_3^+ and has a proton affinity of only 422 kJ mol^{-1} (101 kcal mol^{-1}), will transfer a proton to almost any compound and in doing so will transfer relatively large amounts of energy and cause extensive fragmentation. The ammonium ion, by contrast, is a selective protonating agent that gives abundant quasimolecular ions. [The proton affinity of ammonia is 857 kJ mol^{-1} (205 kcal mol^{-1}); hence the difference in energy deposited by these reagent ions is 435 kJ mol^{-1} (104 kcal mol^{-1}) or 4.5 eV.]

The ability to control the extent of dissociation is evident in a form of chemical ionization that depends upon the process of charge exchange (eq. 11-12). Figure 11-5 shows the molecular ion (m/z 78) decreases and

TABLE 11-10 Some Chemical Ionization Reagents

Reagent Gas	Reagent Ion	Analyte Ion	Comment
H_2	H_3^+	$(M+H)^+$	very energetic, considerable fragmentation
CH_4	CH_5^+	$(M+H)^+$	energetic protonating agent
CH_4	$C_2H_5^+$, $C_3H_5^+$	$(M+C_2H_5)^+$, $(M+C_3H_5)^+$	form adduct ions
i-C_4H_{10}	$C_4H_9^+$	$(M+H)^+$	mild protonating agent, ionizes all nitrogen bases
NH_3	NH_4^+	$(M+NH_4)^+$	selective; little fragmentation
NH_3–CH_4	NH_4^+	$(M+H)^+$	selective protonating agent
Biacetyl	CH_3CO^+	$(M+CH_3CO)^+$	acetylating agent
Ar	$Ar^{\cdot+}$	$M^{\cdot+}$	energetic charge exchange agent
CS_2	$CS_2^{\cdot+}$	$M^{\cdot+}$	mild charge exchange agent
CH_3ONO–CH_4	CH_3O^-	$(M-H)^-$	mild proton abstraction reagent
NF_3	F^-	$(M-H)^-$	proton abstraction reagent
$CHCl_3$–CH_4	Cl^-	$(M+Cl)^-$	chloride addition reagent

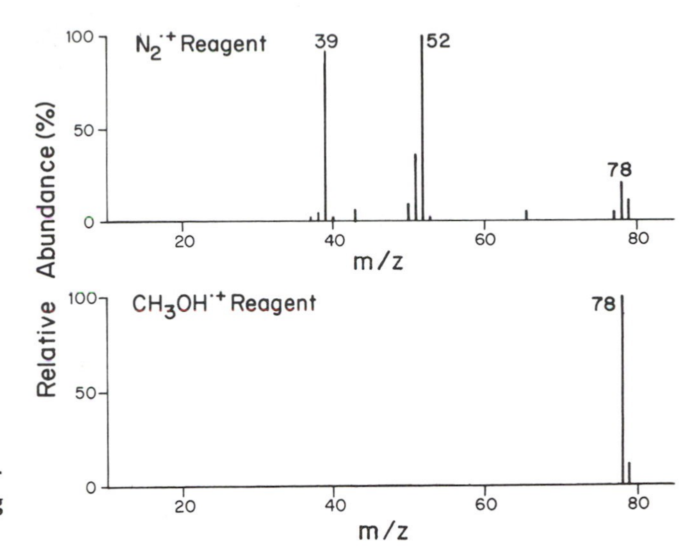

FIGURE 11-5 Control of molecular ion abundance in charge exchange of benzene using $CH_3OH^{\cdot+}$ and $N_2^{\cdot+}$ as reagent gases.

$$M + A^{\cdot+} \rightarrow M^{\cdot+} + A \tag{11-12}$$

the extent of fragmentation increases as more energetic charge exchange reagents are used to ionize benzene. The ionization energies of benzene, methanol, and N_2 are 9.25, 10.85, and 15.6 eV, respectively. From these observations it is evident that the internal energy available when benzene is ionized by methanol is at most $10.85 - 9.25 = 1.6$ eV, while $N_2^{\cdot+}$ can deposit $15.6 - 9.25 = 6.35$ eV. This larger energy accounts for the difference in molecular ion to fragment ion abundances in the two spectra. Note that this particular CI method *does* generate radical ions as the primary product of ionization. (Note also that the methanol molecular ion used as reagent in the above experiment was mass selected and introduced into the ion source; it is not the product of CI with methanol.)

Desorption Ionization. The desorption ionization methods share many characteristics with chemical ionization since ion–molecule reactions are often a part of these methods, too. The distinguishing feature of DI is that condensed phase samples are examined directly. The sample is energized by a particle or photon beam, and the resulting gas phase ions are extracted from the ion source, mass analyzed, and detected. A variety of experiments and conditions fall into the DI category and Table 11–11 summarizes some of these procedures.

TABLE 11-11 Procedures Used in Desorption Ionization (DI)

Agent	Method	Flux	Matrix	Mass Analyzer	Comments
keV ions (Xe^+, Cs^+, Ar^+)	secondary ion mass spectrometry (SIMS)	10^{-10} A cm^{-2} (10^9 particles cm^{-2} s^{-1})	none, solid	any	surface sensitive, low signal, low background
		10^{-6} A cm^{-2} (10^{13} particles cm^{-2} s^{-1})	liquid	any	higher signal, lasts minutes
keV atoms (Xe^0)	fast atom bombardment (FAB)	10^{-5}–10^{-7} A cm^{-2} (10^{12}–10^{14} particles cm^{-2} s^{-1})	liquid	any	high signal, background
MeV ions (e.g., ^{252}Cf fission fragments)	plasma desorption (PD)	10^3 particles cm^{-2} s^{-1}	none	time of flight	best ionization efficiency, low signals
Photons (any wavelength)	laser desorption (LD)	$\geq 10^6$ watt cm^{-2}	none, solids or liquids	any	CW lasers can give thermal degradation

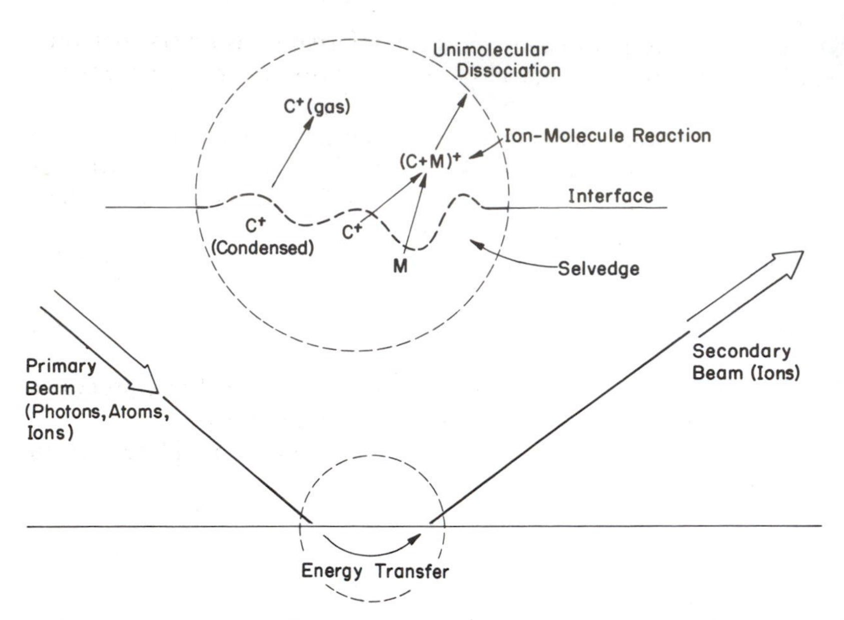

FIGURE 11-6 Processes that occur in DI (C^+ represents a preformed ion). Unimolecular dissociation occurs after the ion has left the interfacial high pressure region known as the selvedge.

In each of these methods, energy deposited into the surface of the sample can cause ionization reactions as well as provide the ions with sufficient kinetic energy to leave the surface. The mass spectra are, to a remarkable extent, independent of the means by which energy is originally deposited. A temporary, charged, local region of high pressure forms at the condensed phase–vacuum interface. Ion–molecule reactions occur in this region, which is known as the selvedge. The energetic ions that result can fragment either in the selvedge or after they have left the interfacial region. Figure 11-6 is a schematic representation of the processes that occur in desorption ionization. Note the occurrence of both ion–molecule reactions and of unimolecular fragmentation of energetic ions. This accounts for two characteristic features of DI spectra: (1) the appearance of several different ions derived directly from the molecule, e.g., $(M+K)^+$ and $(M+Na)^+$, and (2) the occurrence of rather abundant fragment ions. As might be expected, the types of compounds present, the nature of the matrix, and the rate of energy input all affect ion internal energy and hence the degree of fragmentation.

In the simplest of cases desorption ionization merely transfers material from one phase to another. Consider a solution of 1-methylpyridinium chloride in a glycerol–water solvent. Pyridinium ions must be transferred from solution to the gas phase to allow a mass spectrum to be recorded. This phase transfer can be effected by rapid heating of the solution via irradiation with a laser or high energy particle beam (eq. 11-13). The gas phase ions have internal energy distributions $P(\epsilon)$ that depend upon, among other things, the

$$\text{[1-methylpyridinium]}^+ \, Cl^- \text{ (solvated)} \longrightarrow \text{[1-methylpyridinium]}^+ \text{ (gas)} + Cl^- \text{ (gas)} \qquad (11\text{-}13)$$

method of energization and the matrix (solvent) from which the ion is liberated. Soft ionization conditions are accessible. The large ion currents measured in the phase transfer experiments just described often lead to attempts to derivatize samples chemically for DI to generate *pre-formed* ionic compounds. This objective can be contrasted with that when derivatizing samples for gas chromatography (and GC–MS) where the aim is to convert polar compounds to more volatile nonpolar derivatives (such as trimethylsilyl derivatives). For this reason, derivatization procedures used for desorption ionization have been referred to as reverse derivatizations. The conversion of pyridine to the methylpyridinium chloride derivative, prior to mass spectrometry, exemplifies this process.

Instead of examining a solution of methylpyridinium chloride, pyridine can simply be irradiated in an acidic solution in order to generate high yields of pyridinium ions (eq. 11-14). Alternatively, in the presence of salts such as NaCl, the natriated form of the gaseous molecule is generated (11-15). In experiments such as

$$\text{Py (solvated)} \longrightarrow \text{PyH}^+ \text{(vapor)} \qquad (11\text{-}14)$$

$$\text{Py (solvated)} \longrightarrow \text{PyNa}^+ \text{(vapor)} \qquad (11\text{-}15)$$

these, the quasimolecular ion may be generated by ion–molecule reactions in the selvedge. These processes are less efficient than simple desorption of preformed ions and yield lower ion abundances.

The mechanisms by which ions arise in DI are summarized in Table 11-12, which should be read in conjunction with Figure 11-6.

TABLE 11-12 Ionization Mechanisms in Desorption Ionization

Direct emission:	C^+ (condensed) → C^+ (gas)
Ion–molecule reactions:	$C^+ + M$ (selvedge) → $(C{+}M)^+$ (gas)
	$M + C^{\cdot +} \rightarrow M^{\cdot +} + C$
Electron ionization:	$M + e^- \rightarrow M^{\cdot +} + 2e^-$

Negative Ions. This short section summarizes and extends observations on negative ion formation made elsewhere. There are two principal modes of negative ion formation, those that depend upon ion–molecule interactions and those that utilize electron–molecule reactions (eqs. 11-16 and 11-17 give examples of

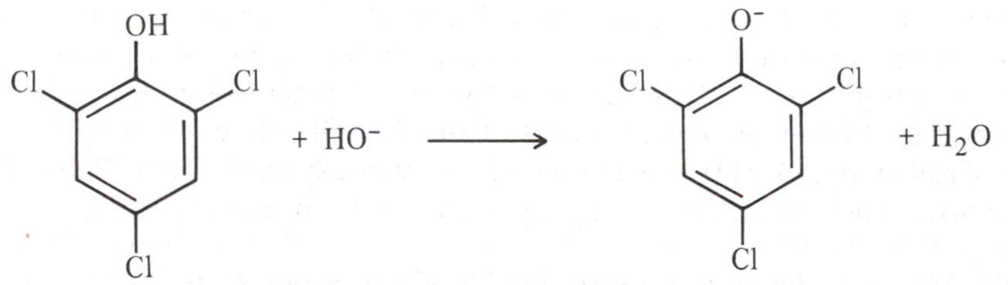

[2,4,6-trichlorophenol: OH, Cl, Cl, Cl] + HO^- ⟶ [2,4,6-trichlorophenoxide: O^-, Cl, Cl, Cl] + H_2O (11-16)

[2,4,6-trichlorophenol: OH, Cl, Cl, Cl] + e^- ⟶ [2,4,6-trichlorophenol: OH, Cl, Cl, Cl]$^{\cdot -}$ (11-17)

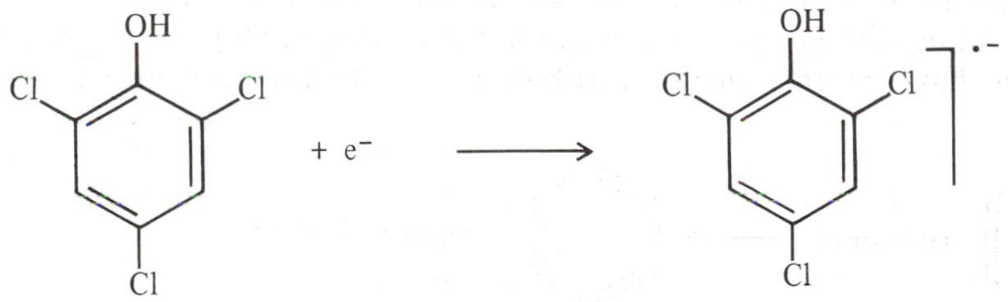

each in the specific case of 2,4,6-trichlorophenol). The former mode shares many of the features of positive ion chemical ionization (see Section 11-5b). The characteristics of several reagent ions are given in Table 11-10. Negative ion formation by electron–molecule interactions is a straightforward process that requires collisions between the molecule and slow electrons. The experiment is normally only successful in CI sources where the relatively high pressure moderates the electron energy *and* removes excess energy from the radical anion formed upon electron attachment (stable $M^{\cdot -}$ ions have positive electron affinities, i.e., they are of lower enthalpy than the reagents, $M{+}e^-$). Electron attachment (eq. 11-17) is a sensitive selective ionization method. Dissociative attachment (eq. 11-18) and ion pair formation (eq. 11-4) often compete with simple electron attachment.

$$\text{2,4,6-trichlorophenol (OH; Cl, Cl, Cl)} + e^- \longrightarrow \text{2,4,6-trichlorophenoxide } (O^-;\ Cl, Cl, Cl) + H\cdot \qquad (11\text{-}18)$$

Certain groups, notably fluorinated groups such as pentafluorobenzyl, have high electron capture cross sections and hence are introduced into molecules to achieve efficient electron capture. Chemical derivatization in this fashion is often worthwhile if trace analysis is required. The high velocities of electrons make the rates of their reactions with neutral molecules much greater than the rates of ion–molecule reactions. Large electron capture cross sections lead to the 100-fold increase in sensitivity sometimes observed for electron capture, and dissociative electron capture, over other ionization methods that yield positive or negative ions.

11-6

Ionic Reactions

11-6a Unimolecular Fragmentation

Thermochemical Control. The use of mass spectrometry as a structural tool (as opposed to its use simply for molecular weight measurements) depends on observation of the unimolecular fragmentations that ions undergo. In this section the principles that underlie ionic reactions in the *isolated* phase environment are discussed. The controlling factors are seen to be the distribution of internal energies the ions possess and the thermochemistry of the dissociation processes available to them. Kinetic considerations can be relegated to a secondary role. The following discussion should be read carefully, given the familiar association (from solution chemistry) of activation energies with kinetic control and their applicability here to thermodynamic control.

Fragmentation occurs unimolecularly for ions whose internal energy ϵ equals or exceeds a minimum value, ϵ_0. Ions with less than this internal energy are stable indefinitely, while those with greater energies fragment at a rate that rapidly increases with energy in excess of the activation energy. We have already seen how vertical electronic transitions can convert neutral molecules into molecular ions with a range of internal energy states (Section 11-5a). In some collisions the minimum energy required to cause ionization, the ionization energy IE, is transferred to the molecule. This situation leads to formation of the ion $AB^{\cdot+}$ in its ground state, with zero internal energy. In other collision events the energy transferred might exceed the minimum energy required to cause ionic dissociation, namely, the appearance energy AE. Figure 11-7 illustrates these quantities in addition to their difference, the *activation energy*, ϵ_0 (= AE − IE). Illustrated are both the potential energy surfaces themselves and the internal energy distribution of the ionic population.

It is possible to divide the internal energy distribution into two regions, based on whether the initially formed ions have or do not have sufficient energy to fragment unimolecularly. This division provides an indication of the molecular ion abundance relative to fragment ion abundances recorded in the mass spectrum. Note that those ions with sufficient energy to fragment may nevertheless not do so on the time scale of the experiment. This result does not introduce serious errors in the approximate treatment suggested here. (The reaction rate typically increases very rapidly with excess internal energy; see next section.) Consider the behavior of acetophenone upon electron ionization. Energetic electrons interact with the vapor phase molecules producing the molecular ion as a result of electronic transitions (Figure 11-3). In a few cases, the minimum energy required for ionization (the ionization energy) may be delivered to the molecules giving molecular ions in their ground electronic and vibrational states. In many more cases each molecular ion will

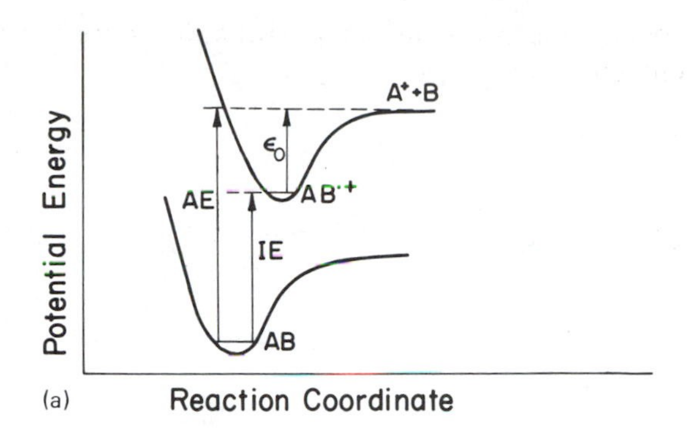

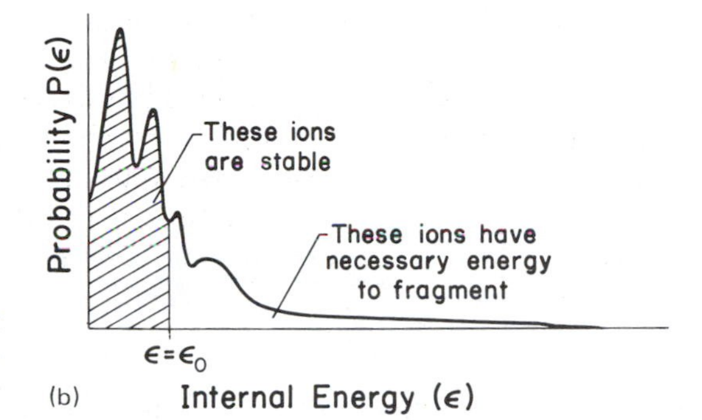

FIGURE 11-7 **(a) Thermochemical quantities that control unimolecular dissociation and (b) their mapping in the internal energy distribution.**

have an excess internal energy ϵ and the population will have an energy distribution $P(\epsilon)$ (Figure 11-4). The more highly excited ions will rapidly fragment to give ionic fragments that are themselves excited and can fragment further. Ions of lower energy may fragment but by different routes, the favored fragmentation mode depending upon the degree of internal excitation of each ion. A network of competing and consecutive unimolecular reactions exists and leads to a set of products that evolves with time. The mass spectrum is simply a sampling of this product distribution at a particular time after ionization, often a time of about 1 μs.

The main electron impact chemistry of acetophenone is shown in eq. 11-19 together with the energy requirements for each reaction referenced to the neutral molecule. A linear sequence of reactions is involved

EI chemistry:	M $\xrightarrow{e^-}$	$M^{\cdot+}$ m/z 120 $\longrightarrow$	$(M-CH_3)^+$ m/z 105 $\longrightarrow$	$C_6H_5^+$ m/z 77 $\longrightarrow$	$C_4H_3^+$ m/z 51
Chemical species:	$C_6H_5\overset{O}{\overset{\|}{C}}CH_3$ $\longrightarrow$	$[C_6H_5\overset{O}{\overset{\|}{C}}CH_3]^{\cdot+}$ $\longrightarrow$	$C_6H_5C{\equiv}O^+$ $\longrightarrow$	$C_6H_5^+$ $\longrightarrow$	$C_4H_3^+$ (11-19)
Energy requirements:	0 eV	9 eV	10 eV $\epsilon_0 = 1$ eV	13 eV $\epsilon_0 = 4$ eV	16 eV $\epsilon_0 = 7$ eV

in this particular molecule. If one assumes that a fragment ion will be generated from all ions that have sufficient energy to make this reaction accessible, then the mass spectrum can be inferred from a knowledge of the molecular ion internal energy distribution $P(\epsilon)$ and the thermochemistry of fragmentation. To a first approximation the mass spectrum *is* determined by such thermochemical factors. Figure 11-8 illustrates this principle for acetophenone. For example, all molecular ions with internal energies between 1 eV, the onset of 105^+ formation, and 4 eV, the onset of 77^+ formation, can be assigned as fragmenting to yield stable 105^+ ions. The ion abundance in the mass spectrum is taken simply as this area in the $P(\epsilon)$ distribution. Figure 11-8

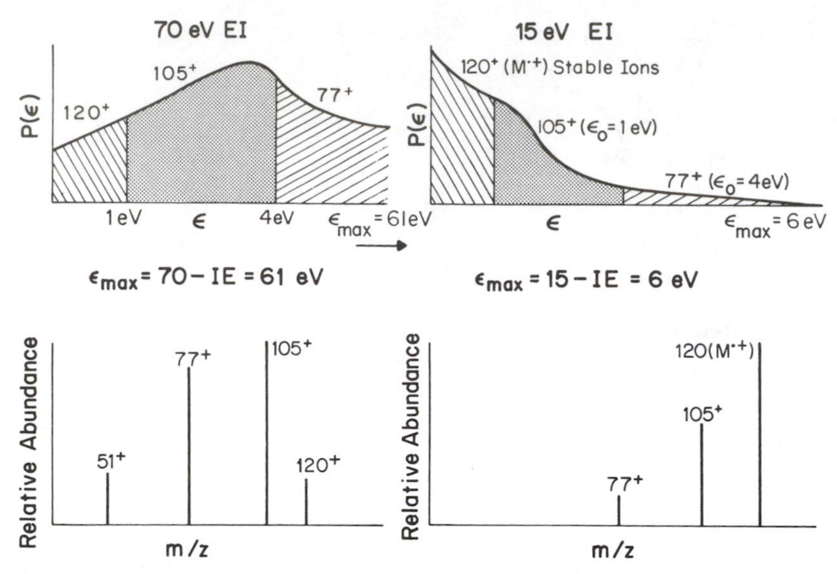

FIGURE 11-8 Thermochemical factors control mass spectrum of acetophenone.

also shows the increase in molecular ion abundance as the ionizing electron energy is reduced. (The practical advantage of this relative increase in molecular ion abundance is offset by a decrease in ionization efficiency.)

Those molecular ions in the distribution $P(\epsilon)$ that are generated with insufficient internal energy to surmount the barrier for even the lowest energy fragmentation process *must* appear in the mass spectrum as molecular ions. Those with greater internal energies will fragment.

Acetophenone behaves straightforwardly because competitive processes are not important in the breakdown pattern. All molecular ions of energy 1–4 eV behave the same way and lose CH_3 to give benzoyl fragment ions of mass 105. When competitive reactions are possible, it is still often true that thermochemical factors dominate and only the lowest energy processes available to a given ion are observed.

The control of thermochemistry over the type and degree of fragmentation can be illustrated by considering a set of monosubstituted benzenes, each of which undergoes a simple cleavage reaction as its main fragmentation pathway. Table 11-13 collects the thermochemical data and illustrates how the abundance of the fragment ion increases with decreasing activation energy (ϵ_0). If the molecular ions of all these compounds have roughly similar internal energy distributions, this inverse dependence of ion abundance on ϵ_0 is exactly the behavior expected from a consideration of Figure 11-7. Summarizing this approach, if one knows (1) the energy requirements for fragmentation and (2) the internal energy distribution $P(\epsilon)$, one can

TABLE 11-13 Thermochemical Control of Fragmentation of Substituted Benzenes

Substituent	Neutral Fragment	IE (eV)	AE (eV)	ϵ_0 (eV)	Fragment Ions/$M^{\cdot+}$ [a]
$COCH_3$	CH_3	9.3	10.0	0.7	2.2
$C(CH_3)_3$	CH_3	8.7	10.3	1.6	2.1
$(CO)OCH_3$	OCH_3	9.3	10.8	1.5	1.5
C_2H_5	CH_3	8.8	11.3	2.5	1.0
NO_2	NO_2	9.9	12.2	2.3	0.7
CH_3	H	8.8	11.8	3.0	0.5
I	I	8.7	11.5	2.8	0.4
Br	Br	9.0	12.0	3.0	0.4
Cl	Cl	9.1	13.2	4.1	0.2

Adapted from I. Howe and D. H. Williams, *Principles of Organic Mass Spectrometry*, McGraw-Hill, New York, 1972, p. 55.

[a] Ratio of fragment to molecular ion abundance, 70 eV EI.

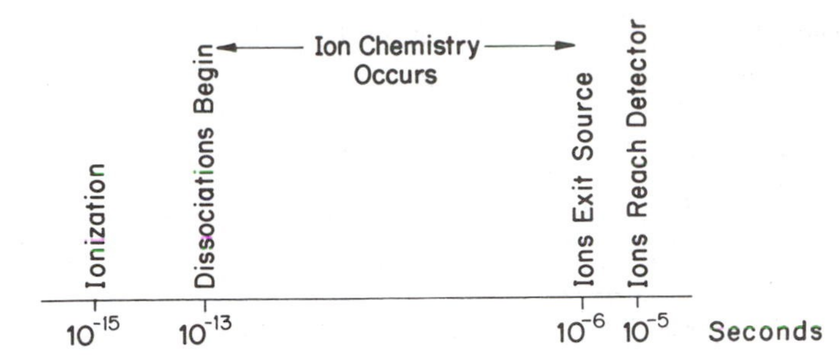

FIGURE 11-9 Time scale for electron impact mass spectrometry.

approximate the mass spectrum without considering the kinetics of individual reactions. *This approximate thermochemical approach to mass spectra works best when complex rearrangements are not important.* These processes, which often have slow rates compared with higher activation energy simple cleavages, must now be considered.

Kinetics. The mass spectrometer time scale is illustrated in Figure 11-9. The mass spectrum is a time-integrated picture of all reactions occurring after ionization and before ions leave the source. Although this time period is short ($\sim 10^{-6}$ s), it is long compared to bond vibrational periods from 10^{-10} to 10^{-12} s. In almost all excited molecules, fragmentation does not occur within a vibrational period; rather internal energy is shuttled around the excited, isolated ion until it collects in the appropriate modes for fragmentation to occur. The rate at which fragmentation occurs increases rapidly with excess internal energy above the activation energy.

Unimolecular fragmentations are normally endothermic reactions, and, even when exothermic, they often have substantial activation energies. Potential energy surfaces are shown schematically in Figure 11-10 for the two main types of unimolecular reactions. These are *simple bond cleavages*, which have small or *no reverse activation energies*, and fragmentations that involve *rearrangement of bonds* and in which reverse activation energies can be substantial. In the former case relatively little energy in excess of the activation energy for fragmentation is required to observe products on the time scale of the mass spectrometer. In the latter case, the additional energy required for the fragmentation to proceed at an observable rate can be substantial. The different forms of the kinetic curves associated with these reaction types are illustrated also

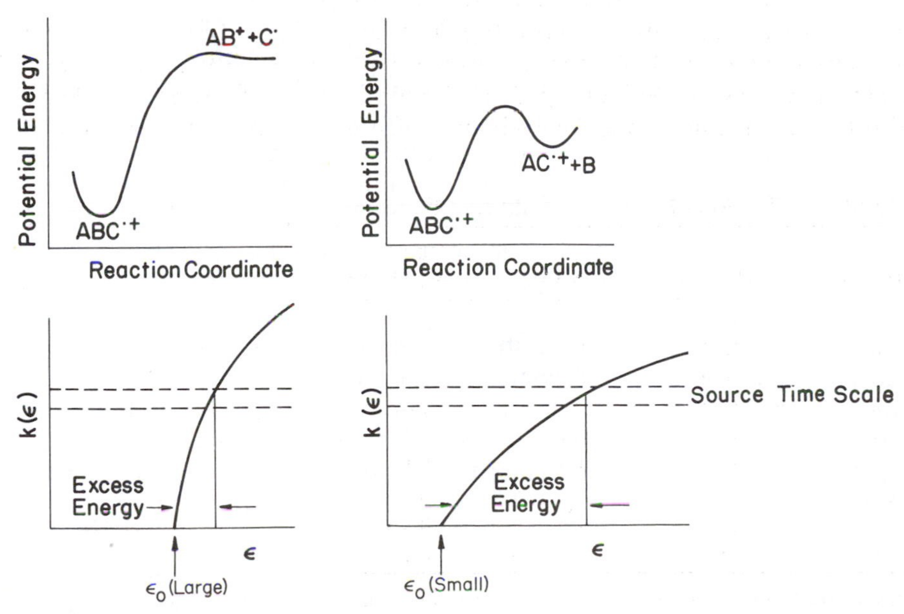

FIGURE 11-10 Potential energy diagrams and energy dependence of unimolecular rate constants for simple cleavage (left) and rearrangement reactions (right).

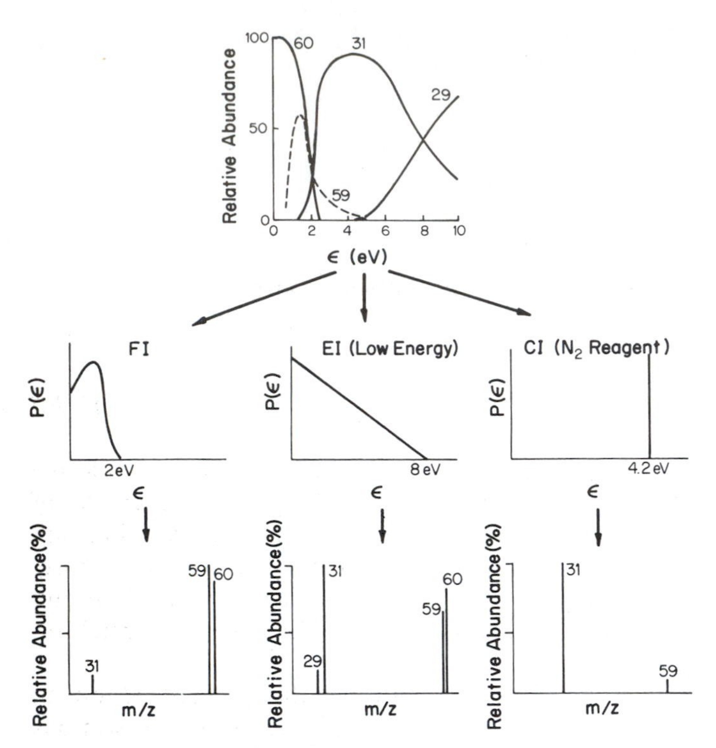

FIGURE 11-11 Convolution of breakdown curve (an intrinsic property) with internal energies $P(\epsilon)$ due to different ionization techniques to produce mass spectra of 1-propanol.

in Figure 11-10. Note the dependence of rate constants upon internal energy for (a) simple cleavage and (b) a rearrangement process. The important, practical conclusion is that only when rearrangements have substantially lower activation energies can they compete with simple cleavages, because of the greater excess energy required to cause observable fragmentation on the time scale of the mass spectrometer. The lower part of Figure 11-10 illustrates competition between rearrangement and simple cleavage reactions and shows that the former only compete if they have low activation energies *and* if the ions examined are of low internal energy.

Fragmentation behavior is controlled by the thermochemical factors (activation energies) already discussed as well as the kinetic factors just referred to. These can be considered together to produce a *breakdown curve*. Such a curve shows the dependence of the fragmentation behavior upon the internal energy of an ion. Figure 11-11 (top) illustrates the breakdown curve of 1-propanol. At low internal energies fragmentation is precluded and only the molecular ion is observed. As the internal energy is increased, the lowest energy fragmentation, $H^{\cdot}$ loss to give m/z 59, occurs, only to be replaced in turn by consecutive reactions yielding m/z 31 and 29. One can read from the breakdown curve how molecular ions with particular internal energies will behave. For example, ions that have exactly 4.4 eV of internal energy yield 96% of 31^+ and 4% of 59^+. Such ions can be generated by charge exchange with $N_2^{\cdot +}$ reagent (IE = 15.6 eV, IE propanol = 10.2 eV).

A mass spectrum is simply a breakdown curve with appropriate weighting of the internal energy axis. In other words, depending upon the method and conditions of ionization, ions will have a particular internal energy distribution $P(\epsilon)$. By convoluting this distribution with the breakdown curve, one obtains the mass spectrum (Figure 11-11). This figure, which should be studied carefully, illustrates how different methods of generating a set of molecular ions result in different unimolecular product distributions, that is, different mass spectra. FI indicates field ionization, a soft ionization method used for vapor phase samples.

11-6b Ion-Molecule Reactions

When organic ions collide with neutral molecules, reorganization of chemical bonds can occur (eq. 11-20).

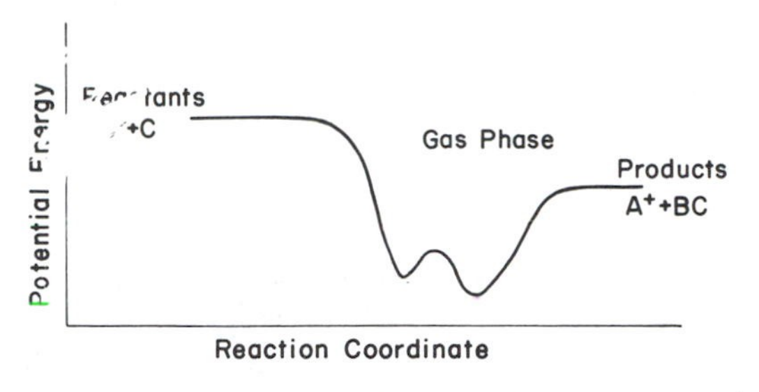

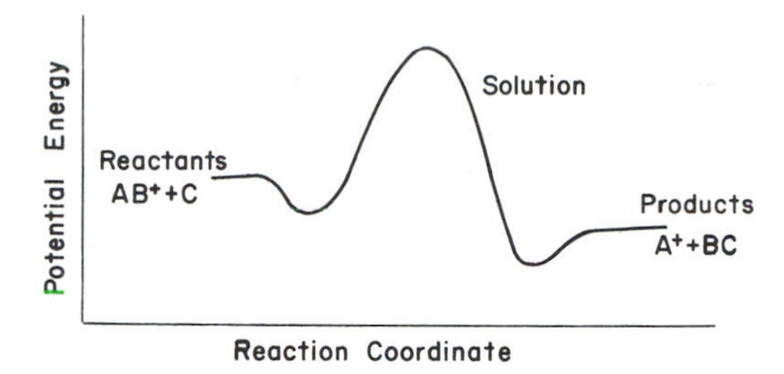

FIGURE 11-12 Potential energy diagrams of exothermic ion–molecule reactions in the gas phase and in solution.

$$AB^+ + C \rightarrow A + BC^+ \qquad (11\text{-}20)$$

Although electron impact mass spectrometry is carried out under collision-free conditions, ion–molecule reactions are important in the CI and DI procedures. Section 11-5b should be referred to for examples of ion–molecule reactions associated with these ionization methods. Note that ion–molecule reactions are similar to conventional organic reactions in a number of ways, including the types of reactions (acid–base, nucleophilic substitution, elimination, etc.), the fact that they often occur for systems at thermal equilibrium, and the fact that they proceed through reaction intermediates. The differences, which can be substantial, are ascribed to solvent effects that therefore become accessible for study.

Most exothermic ion–molecule reactions occur on every collision. In other words they proceed without an activation barrier (Figure 11-12). Given the fact that ion–dipole attractive forces raise collision cross sections compared with neutral molecule collisions, the rates of these reactions are very high (10^{-9} cm^3 molecule^{-1} s^{-1} $= 5 \times 10^9$ liter mol^{-1} s^{-1}). Endothermic ion–molecule reactions can also occur and lead to structural rearrangement and fragmentation. These processes are not often encountered under the low energy conditions that prevail in most ion sources.

A key to gas phase ion chemistry is that gaseous ions have high enthalpies and are stabilized by interaction with *any* molecule. Polarization stabilizes the charge, and ion–molecule association complexes are normally more stable than the free reagents (Figure 11-12). However, if the collision complex does not have an available reaction channel, it will revert to reactants if excess energy is not removed, usually by collision with a third body. The very different forms of the potential energy surfaces for the same ion–molecule reaction in solution and in the gas phase are illustrated in Figure 11-12. The differences are due to solvent effects. In solution the ionic reagent (AB^+) is already stabilized by solvation, but its gas phase counterpart is not. Hence the interaction with C is strongly stabilizing in the latter but not the former case. In both systems the structural changes necessary to achieve the activated complex configuration are endothermic with respect to the surrounding regions of the potential energy surface. Large energy changes occur across the gas phase reaction surface because energy differences between unsolvated and monosolvated ions are large.

The above discussion on ion–molecule collisions refers to processes that occur at low translational energies, typically but not always under conditions of thermal equilibrium. There is a second broad class of ion–molecule interactions that is important in mass spectrometry. These interactions occur at much higher translational energy (10–10^4 eV) than used to study reactive collisions. (These experiments involve ions that have been accelerated from the ion source and encounter neutral gas targets in the analyzer of the mass spectrometer.) Because of the high relative velocities of the collision partners, collision complexes cannot be formed. Interactions are therefore limited to the exchange of electrons, or to energy transfer. Among the many processes that occur in this regime, collisional activation (eq. 11-21) is most important. In this process, some of the translational energy of the collision partners is converted to internal energy and leads to subsequent unimolecular dissociation of the ion (eq. 11-22). This process is the basis for the method of tandem mass spectrometry (Section 12-5a).

$$m_1{}^+ + N \rightarrow m_1{}^{+*} + N \qquad (11\text{-}21)$$

$$m_1{}^{+*} \rightarrow m_2{}^+ + m_3 \qquad (11\text{-}22)$$

11-7

Techniques and Components

Although mass spectrometers are capable of subnanogram sensitivity, samples are normally required to be in the microgram range. Ease of sample transfer from the sample vial to the mass spectrometer will require dissolution in some cases, so an appropriate solvent must be specified. Mass spectrometers are operated under vacuum, often 10^{-6} Torr or less, which is maintained by a combination of mechanical forepumps with diffusion pumps, turbomolecular pumps, or ion pumps. Many systems can be vented to atmosphere and working vacuum re-achieved in a matter of minutes. Samples are introduced into the ion source without breaking vacuum by one of the methods described in Table 11-14. These are the vapor reservoir system, direct insertion probe, gas or liquid chromatograph, and the membrane interface.

TABLE 11-14 Sample Introduction Systems

System	Sample Type	Minimum Sample	Characteristics	Ionization	Figure
Batch (reservoir)	gas, liquid, low-melting solid	<1 mg	steady sample delivery for long periods	CI, EI	—
Direct probe	less volatile samples	<1 μg	sample delivery varies with probe temperature	CI, DI, EI	11-15
Gas chromatograph	mixtures	<1 μg	more volatile samples	CI, EI	11-13
Liquid chromatograph	mixtures in solution	<1 μg	less volatile samples	CI, DI	11-13
Membrane	mixtures in solution	$<10^{-6}$ M	widely varying selectivity, volatiles only	CI, EI	11-14

11-7a Sample Introduction Systems

In the *reservoir inlet* system, sample vapor is bled into the source from a vessel at a higher pressure. This type of inlet system requires more sample than the other methods and also demands greater sample volatility and thermal stability. It was once the most common method of introducing organic samples, and it has the advantage that a steady sample pressure is maintained for long periods, allowing slow scanning or extensive signal averaging. *Direct insertion probes* transfer the sample into the ion source. They are used with all the ionization methods and minimize required sample sizes. Probe temperature is used to control the rate of sample evaporation, and rapid scan techniques are often employed to monitor fractionation during evaporation. *In-beam ionization methods* use modified probes to introduce the sample directly into the electron beam (EI) or the CI plasma. This procedure allows samples of low volatility to be examined by EI and CI. The performance of in-beam ionization methods, for all but the least volatile compounds, is comparable to that of the DI techniques.

Mixtures, or pure substances in solution, are often introduced via a *chromatograph*. The key to the success of gas and liquid chromatography–mass spectrometry lies in the separators that enrich the eluant in analyte. Figure 11-13 illustrates the methods of jet separator, thermospray, and moving belt. The *jet separator* removes helium carrier gas with high efficiency and is widely used for GC–MS. The *thermospray* and moving belt systems represent contrasting approaches to coupling LC and MS. The former method creates ions directly in the course of solvent evaporation. No external ionization agent is required, although for nonionic samples it may be advantageous to use electron ionization. The eluant is first vaporized to a fine aerosol, and further evaporation completely removes solvent and yields gas phase ions directly. Electrolytes are normally added to the eluant, and the types of ions seen include cation-attachment species such as $(M+Na)^+$, which is also observed in DI (Section 11-5b). The thermospray method is particularly successful in the analysis of polar compounds present in aqueous solutions. The *moving belt* procedure is more successful with normal phase liquid chromatography; solvent is removed from the eluant and the dried material is vaporized and ionized by EI, CI, or, sometimes less successfully, in situ by DI. Other LC–MS interfaces are based on condensation of the mobile phase on cold surfaces in the ion source. Direct coupling of the chromatograph

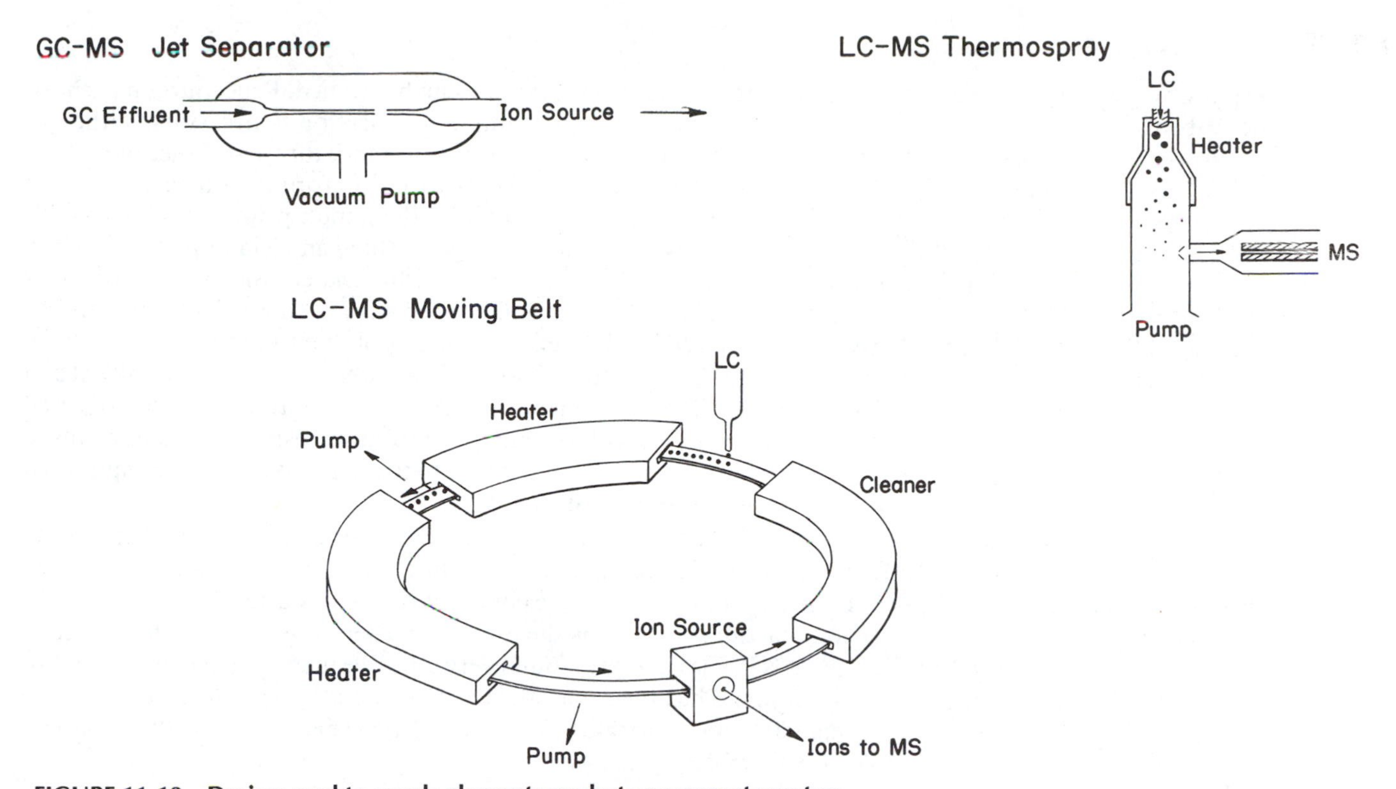

FIGURE 11-13 Devices used to couple chromatographs to mass spectrometers.

and mass spectrometer (i.e., without using an interface) is also possible with LC and GC capillary columns. Data can be collected continuously or the sample can be rerun and data collected only in the regions of interest. Section 12-5 should be consulted for further discussion of chromatography–mass spectrometry.

A *membrane* can be used to sample a flowing fluid stream with a mass spectrometer. One interface is shown in Figure 11-14, which illustrates the coupling of a chemical reactor (a round bottom flask) directly to a mass spectrometer for continuous monitoring of the product distribution. The use of silicon rubber membranes allows low molecular weight nonpolar compounds to be examined in aqueous solution. Although the method is extremely convenient and has very high sensitivity for small nonpolar compounds, differences in membrane permeation rates are very large. Development of membranes that transmit larger and more polar compounds would greatly increase the usefulness of this type of introduction method.

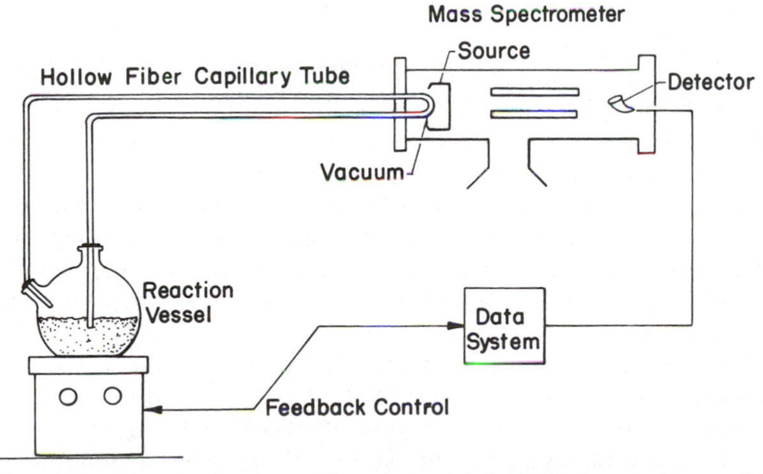

FIGURE 11-14 Membrane capillary for sample introduction used to monitor a chemical reaction continuously.

11-7b Sample Preparation

It is a strength of mass spectrometry that a wide range of sample types can be studied. Procedures have been described in the preceding section. In most cases little or no sample preparation is necessary, although *desorption ionization* methods do provide opportunities to improve performance through choice of matrix, as well as manipulation of the chemical form of the analyte. The fact that preformed ions are desorbed efficiently has already been noted (Section 11-5b) as a reason to derivatize the sample prior to analysis by DI. Alternatively, improved performance is achieved by admixture of reagents that can yield ionic forms of the analyte in situ. For example, acidification can increase $(M+H)^+$ production, whereas the presence of metal salts often yields intense adduct ions with the analyte. Silver and copper have been particularly useful in generating adducts that are then readily recognized by their characteristic isotopic signatures.

In the desorption ionization techniques the sample is often dissolved in a lower vapor pressure solvent and the solution is bombarded by an ion beam (SIMS) or an atom beam (FAB). The solution is typically carried on (or continuously fed to) a direct insertion probe. Loss of solvent by evaporation restricts measurements to a period of some minutes. Table 11-15 lists some of the solvents that are used. The nature of the sputtered species depends upon the compounds present at the solvent surface.

The solvent contributes ions to the mass spectrum that can obscure the low mass region of a DI spectrum. In other DI methods a solid matrix is employed (e.g., in californium fission fragment bombardment, in laser desorption, and in some SIMS experiments). In these cases the choice of the matrix affects the types of ions observed (through analyte–matrix ion reactions) as well as the energy transferred and hence the degree of fragmentation. Simple salt matrices (e.g., NH_4Cl) have been found very effective in increasing ion yields. For many larger biologial compounds, glutathione has the same effect. The substrate also affects ion yields in these experiments, presumably through analyte–substrate interactions. Nitrocellulose has proved particularly useful in DI of large biolmolecules from the solid state.

TABLE 11-15 Some Solvents and Reagents for Desorption Ionization

Solvents	Cations	Derivatizing Reagents
glycerol	alkalis	Girard's reagent (steroids)
thioglycerol	Th	pyrylium salts (primary amines)
sulfolane	Ag	$SbCl_3$ (aromatics)
tetraglyme	Cu	acid–base reactions
diethanolamine	K	charge transfer reagents
fomblin	Li	
threotol	Na	
triethanolamine		

11-7c Ion Sources

The heart of the mass spectrometer is the ion source. Not only is ionization effected in this region but most of the ionic reactions occur here, too. We describe first the most widely used source, the electron impact source.

Electron Ionization In order to achieve maximum sensitivity with a limited amount of sample, the *ionization chamber* is a small, relatively gas-tight unit to which sample is supplied and from which ions are extracted. Pressures of $\sim 2 \times 10^{-5}$ Torr are standard in electron ionization; much lower pressures will often give insufficient signal while ion–molecule reactions (which change the character of the mass spectrum) can become important at higher pressures. While the efficiency of this type of ion source is relatively high, only about one molecule in 10^3 that enter the ion chamber is ionized; most are simply pumped away.

The total ion current produced by an electron impact source is of the order of 10^{-8} A. After acceleration, passage through the source slit, and mass analysis, beams of 10^{-9} A (6×10^9 ions s^{-1}) represent a common upper limit. Fortunately, very weak signals ($<10^{-17}$ A) can be monitored accurately with an electron multiplier (see below). Next to ionization efficiency, the second major parameter for judging an ion source is the energy spread of the accelerated ion beam. A large energy spread is particularly undesirable in high

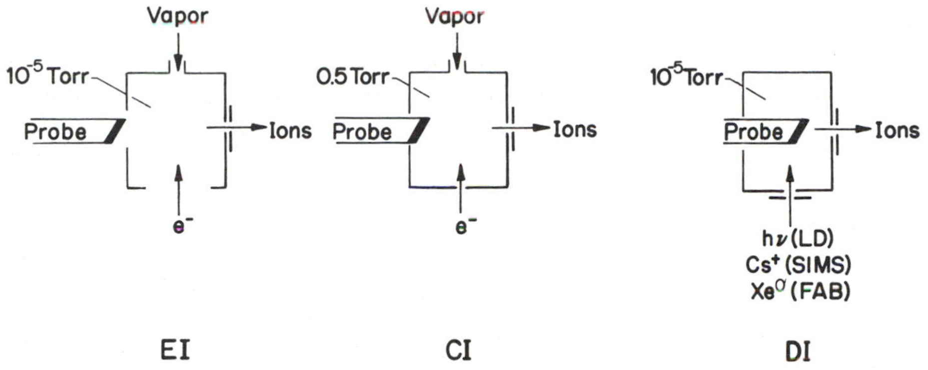

FIGURE 11-15 Ion sources (schematic). Sample either is carried on the probe or introduced as a vapor (EI and CI).

resolution work with sector instruments and in time-of-flight analysis. A spread of about a volt in several thousand is commonly obtained from an electron impact source.

A schematic diagram of the ion source is given in Figure 11-15. The source slit is narrow but long in mass spectrometers that analyze in one plane (e.g., sector magnets), whereas an aperture is used when analysis is spatially symmetrical (e.g., time-of-flight). Electrons are generated by *thermionic emission* from an incandescent rhenium or tungsten filament. The temperature of the metal ($\sim$1800° C) is responsible for the distribution in electron energy, the average energy being determined by the potential difference between the filament and the ion chamber. (In the usual arrangement the filament is external to the chamber, and the electrons enter through a slit.) The *ionizing electron beam*, which usually takes the form of a wide but shallow band, traverses the chamber under the influence of a small external magnetic field. This field constrains the electrons to a shallow equipotential region in the chamber (thereby reducing the energy spread in the accelerated ion beam) and causes them to move in helical paths (thereby increasing the possibilities of ionizing collisions). The electron beam passes through the ion chamber and is collected at a trap or anode that is positive with respect to the chamber. Quantitative work requires precise ion abundance measurements, for which the ionizing electron current must be kept constant. A typical value is 100 μA. The current reaching the trap is controlled by a feedback loop to the filament circuit.

Chemical and Desorption Ionization. Chemical ionization and desorption ionization sources are adaptations of the electron impact type. The chemical ionization source is much more gas tight than the EI source, consistent with operation at a higher pressure (Figure 11-15). Sample is often, but not necessarily, introduced via a direct insertion probe. In the "in-beam" method of desorption chemical ionization (DCI), a version of CI, the probe carries the sample directly into the CI plasma. The result is performance for nonvolatile compounds that is intermediate between that of CI and the DI methods. In the DI methods, the sample, again carried on a probe, is bombarded by an externally generated ion, particle, or photon beam. By reversing the polarity of the potentials applied to the source, negatively charged ions are extracted. In CI and DI these will often result in ion currents comparable to those for positive ions, namely, $\sim 10^{-9}$ A.

The similarities in ion sources should not mask the sharp differences in the chemical and physical phenomena associated with the several ionization techniques. It is worth emphasizing that the major ionization methods (EI, CI, and DI) typically yield complementary information based as they are on different physical principles and different ion chemistry. Positive vs. negative ion spectra also provide complementary information. For these reasons it is extremely useful to obtain more than one type of spectrum when characterizing an unknown compound.

11-7d Mass Analyzers

The four devices in common use are based on sector fields, quadrupoles, time-of-flight, and cyclotron resonance. All depend on the interactions of charge particles with electric or magnetic fields. Table 11-5 summarizes the methods and their principal characteristics.

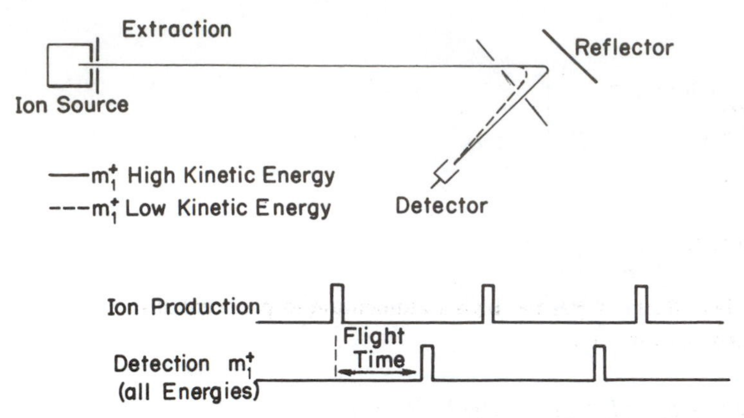

FIGURE 11-16 Time-of-flight mass spectrometer (schematic) showing focusing in time of ions of different kinetic energy.

Time-of-Flight The time-of-flight (TOF) method is exceedingly simple in concept. A beam of ions is accelerated through a known potential V, and the time t taken to reach a detector at a distance d, in a linear flight tube, is measured. All ions fall through the same potential V, so their velocities v must be inversely proportional to the square roots of their masses m (eq. 11-23). Hence, flight time is related to the mass-to-

$$\tfrac{1}{2}mv^2 = zV \tag{11-23a}$$

or

$$v = \left(\frac{2zV}{m}\right)^{1/2} \tag{11-23b}$$

charge ratio m/z by eq. 1-24. The source must be pulsed in order to avoid simultaneous arrival of ions of

$$t = \left(\frac{m}{2zV}\right)^{1/2} \times d \tag{11-24}$$

different mass-to-charge ratios. Typical pulsing frequencies vary from hundreds to thousands of Hertz.

The simplicity of the experiment and the wide availability of devices that measure events on the microsecond time scale have made TOF popular. An important advantage is that *all* ions are detected, rather than only those that satisfy some transmission condition. This factor greatly improves sensitivity (Felgett advantage), but it makes demands for rapid data acquisition that cannot yet be fully satisfied. Mass measurements in excess of 10,000 d were first made on biological molecules with TOF instruments. The resolution available with time-of-flight methods is modest, although it can be improved by methods that focus ions so as to compensate for small differences in kinetic energies. Such focusing often takes the form of electrostatic repulsion arranged so that more energetic ions of the same mass spend more time in the reflecting lens than do their slower, less energetic counterparts (Figure 11-16).

Sector Fields. The widely used sector magnet analyzer also requires ions of fixed kinetic energy on which to perform mass-to-charge analysis. When the force exerted by the magnetic field B balances the centrifugal force (eqs. 11-25 and 11-26), ions are transmitted by the magnet (Figure 11-17). The device functions there-

$$Bzv = \frac{mv^2}{r} \tag{11-25}$$

or

$$\frac{mv}{z} = Br \tag{11-26}$$

fore as a momentum-to-charge analyzer. If all ions have the same kinetic energy, it acts as a mass analyzer.

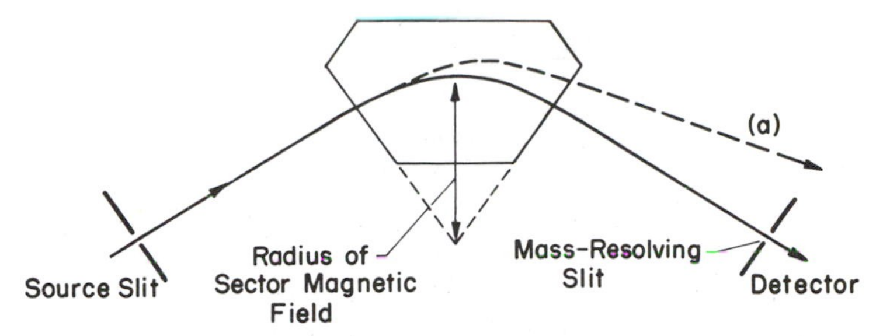

FIGURE 11-17 Momentum analysis using a sector magnetic field. Beam (a) has too high a momentum to pass centrally through the sector and be transmitted to the detector by the mass-resolving slit.

Eliminating v from eqs. 11-23 and 11-26 yields eqs. 11-27 and 11-28.

$$Bz = \frac{m}{r}\left(\frac{2zV}{m}\right)^{1/2} \quad (11\text{-}27)$$

$$\frac{m}{z} = \frac{B^2r^2}{2V} \quad (11\text{-}28)$$

The mass range available depends on the maximum field strength and the radius of curvature and inversely on the accelerating voltage employed. Magnets of 2 tesla and large dimension give mass ranges of some 10,000 d for ion energies of 10,000 eV. Much higher mass ranges are available by decreasing the accelerating voltage, although this objective is only achieved with some loss of resolution and sensitivity.

Improved resolution in mass analysis is possible if the inevitable distribution in ion kinetic energy is compensated for by using a second analyzer in series with the magnetic sector. An electrostatic analyzer that selects for ion kinetic energy fulfills this role. While it is possible to use this device simply as a prefilter for kinetic energy before ions enter the magnetic sector, the loss in signal would be severe. Instead, the concept of *velocity focusing* is used. The geometry of the instrument is chosen so that ions that have the appropriate mass but an incorrect kinetic energy suffer off-setting deflections due to the two sector fields. In this way high resolution is achieved without excessive attenuation in signal strength. Because beams that diverge in direction are also brought to a point of focus, such instruments are termed *double focusing* mass spectrometers. One arrangement for achieving double focusing is illustrated in Figure 11-18. This particular arrangement focuses ions moving in the z direction as well as in the usual xy plane, thereby increasing system sensitivity.

Quadrupoles and Ion Traps. Methods of mass analysis that are proving increasingly valuable are based on inducing periodic motion in an ion. Two important applications of this principle are the *quadrupole mass filter* and the *ion trap*. The quadrupole has good ion transmission (10–50% at unit mass resolution). The ion

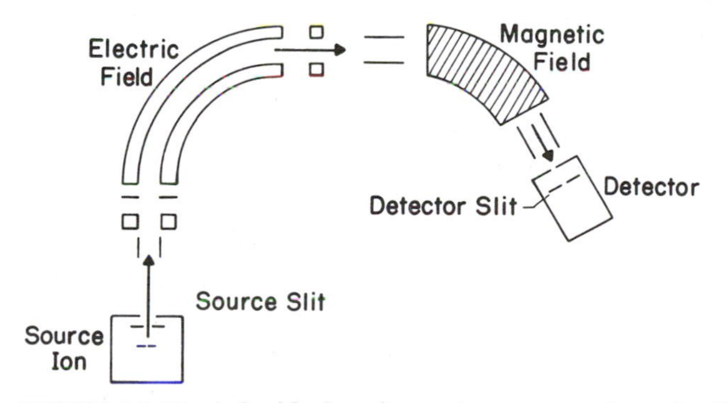

FIGURE 11-18 A double-focusing sector mass spectrometer. The magnetic field is normal to, and the electric field is in the plane of, the figure.

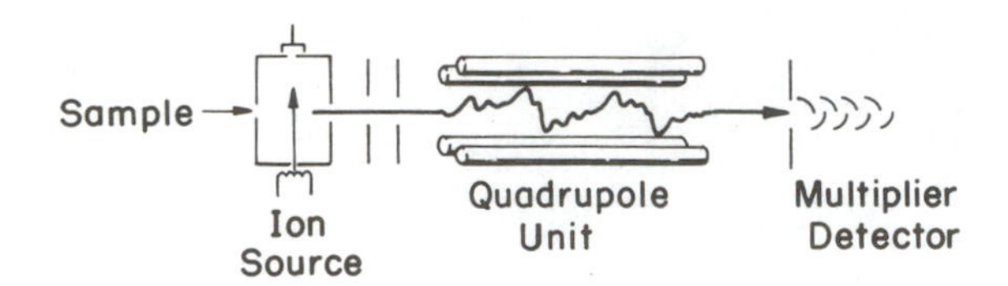

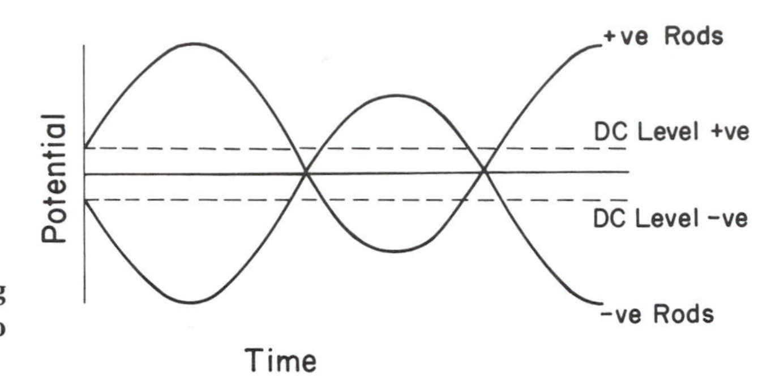

FIGURE 11-19 Quadrupole mass filter, showing an ion trajectory and the potentials applied to opposite pairs of rods.

drifts into an assembly of four parallel rods (Figure 11-19), which are charged by dc currents of opposite sign and are subjected to an oscillating rf field, also applied to opposite rods, that successively reinforces and then overwhelms the dc field. An ion moving through the array experiences a force only in the plane normal to the rod length. Provided its oscillations in this plane are stable, the ion will drift down the rod assembly and reach the collector. Stable oscillations are only achieved by ions of given mass-to-charge ratio for a given rod assembly, oscillation frequency, rf voltage, and dc voltage. Mass scanning is usually achieved by sweeping the dc and rf voltages, keeping their ratio and the oscillator frequency constant. Typical operating parameters include rf voltages of several thousand volts in the 10^6 Hz range. High mechanical precision and strict control of operating voltages are needed.

The ease with which electric fields (as opposed to magnetic fields) can be controlled is an important advantage of this instrument. The mass scale is linear, and after calibration any mass can be electrically selected. Computer control and data collection are much simplified, and the instruments are very reliable. Although a resolution of 20,000 has been reported with a quadrupole mass spectrometer, high resolution is not normally available and the mass range is somewhat limited.

Newer devices that employ similar principles are the three-dimensional *ion traps*. Electric fields are used to select ions from an "electric bottle" in which ion–molecule reactions can conveniently also be carried out prior to mass analysis. These inexpensive devices are built to low mechanical tolerances, operate at relatively high pressures (1×10^{-3} Torr of He) and can trap mass-selected ions for many seconds. An rf voltage is applied to the ring electrode, the end cap electrodes being grounded (Figure 11-20). Mass selection is achieved by raising the rf voltage so as to cause ions to become unstable and be ejected from the trap. Detection is with an external electron multiplier. Ion traps have many features in common with ion cyclotron resonance instruments, including the ability to study sequences of reactions by MS–MS methods. The former are much simpler devices, which are (as yet) incapable of the extremely high mass resolutions attainable in the latter.

Cyclotron Resonance. The advantage of ready computer control, without the disadvantage of precise mechanical tolerances, also characterizes the final type of mass analyzer, that based upon *ion cyclotron resonance*. The principles of ion cyclotron resonance can be derived from a consideration of the forces experienced by ions in magnetic and electric fields. An ion of mass m and velocity v, in the plane normal to a magnetic field of strength B, experiences a force Bzv normal to both the field direction and to its direction of motion (eq. 11-25). Hence, the ion describes a circle of radius r in the plane normal to the field, so that the cyclotron frequency is given by eq. 11-29. Thus every ion has a characteristic cyclotron frequency that is

$$\omega_c = \frac{v}{r} = \frac{Bz}{m} \tag{11-29}$$

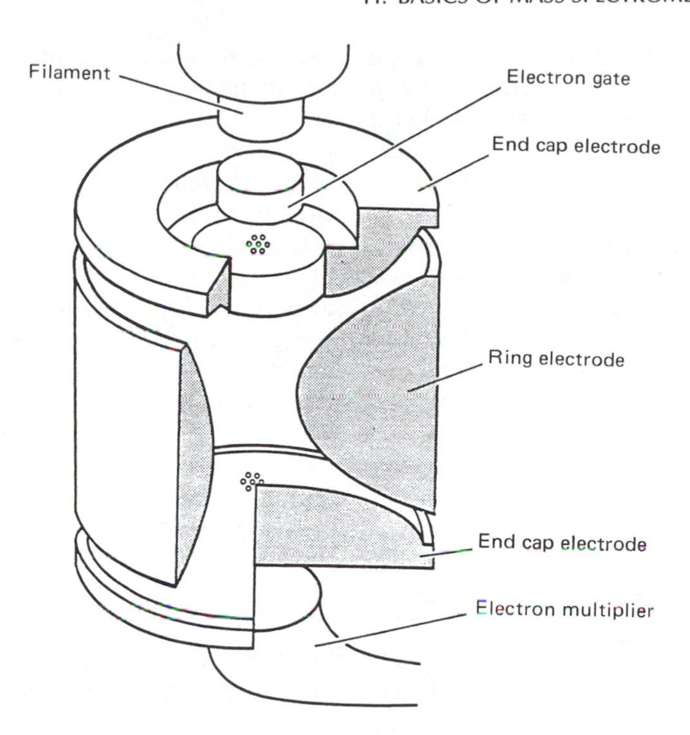

FIGURE 11-20 Ion trap mass spectrometer utilizes an rf field to confine mass-selected ions. [Courtesy Finnigan Corp.]

inversely proportional to its mass-to-charge ratio and directly proportional to the applied magnetic field. While ω_c is independent of the velocity of the ion, the radius of the orbit is directly proportional to ion velocity. Because frequencies can be selected and measured with great accuracy, the mass resolution obtainable using eq. 11-29 is very large indeed, especially at low mass, for which high frequencies are employed. This capability is illustrated in Figure 11-21, which shows nominal mass 28 split into the CO^+ and N_2^+ contributions.

In early work an oscillating electric field, frequency ω, was applied normal to B; then when $\omega = \omega_c$, energy is absorbed by the ions, thereby increasing their velocities and orbital radii. The absorption of energy is

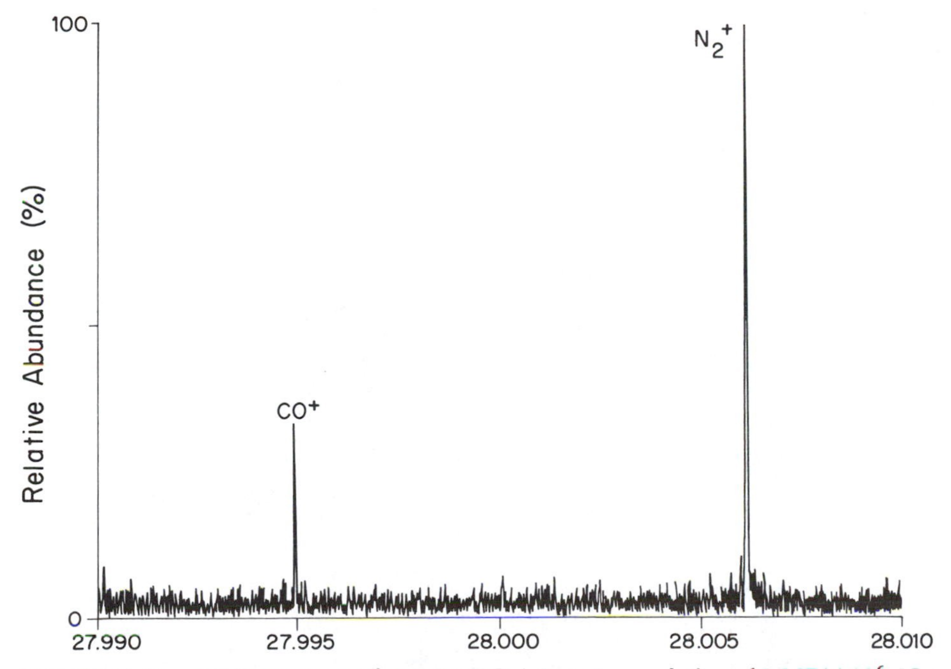

FIGURE 11-21 FTMS spectrum of m/z 28 ($CO + N_2$) at a resolution of 1.147×10^6. [Courtesy of Alan G. Marshall.]

signaled by a change in the power required by the oscillator. In order to achieve mass analysis, the *oscillator frequency* is conveniently *kept constant* and the magnet strength varied. Since m/z is proportional to B, the mass spectrum so obtained has a linear mass scale. More recently, ion detection has employed *variable frequency oscillators* to excite the trapped ion population. The resulting coherent motion of ions of a given frequency (mass) is detected as an image current of appropriate frequency in the walls of the vacuum cell. The mass spectrum is recorded by *Fourier transformation* of the signal from the time domain into the frequency domain, and the individual masses are recognized by their characteristic frequencies. This experiment shares many of the properties of Fourier transform NMR, including the ability to monitor a broad range of signals simultaneously. FTMS yields exceptionally high resolution mass spectra, and it is electronically complex and mechanically simple. Moreover, it allows sequential events to be followed including series of ionic reactions in which particular products are selected, collisionally excited, and reacted further. Resonance is lost by collisions so the cell must be maintained at high vacuum, $\sim 10^{-9}$ Torr. This requirement limits the dynamic range and places restrictions on chromatographic sample introduction and some desorption ionization methods.

At present, rapid development is occurring in each of the main types of mass analyzers. These improvements are extending the mass range, decreasing scan times, improving the resolution, and lifting limitations with regard to ionization methods. Capabilities for tandem mass spectrometry (Section 12-5b) have been developed for each type of analyzer and for hybrid instruments that combine more than one type.

Ion Detection. The availability of devices that allow the detection of ions arriving at the rate of one per second is an underlying reason for the excellent sensitivity of mass spectrometers. Electron multipliers convert ion beams to greatly amplified electron beams through a cascade effect, which results in gains of 10^5 to 10^7. This procedure allows single ion counting, although chemists usually employ analog detection. In this mode of operation ion currents caused by the arrival of relatively large numbers of ions give a signal (current, voltage) that is further amplified and processed. Note that negative ion detection can employ an electron multiplier operated at the same potentials as for positive ion detection, provided a conversion dynode is employed. Positive ions are generated from collisions occurring at this electrode prior to ion-to-electron conversion and electron multiplication (Figure 11-22).

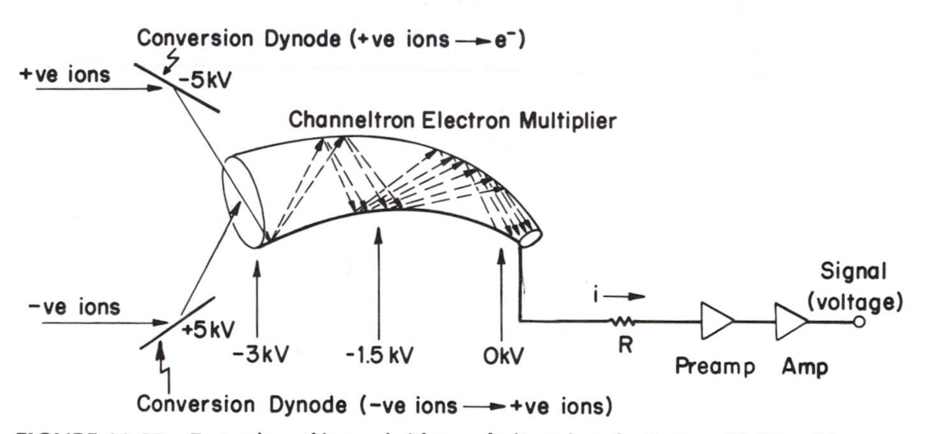

FIGURE 11-22 Detection of ions of either polarity using electron multiplier with conversion dynodes.

Bibliographic references for this chapter are given at the end of Chapter 12.

12

Problem Solving with the Mass Spectrometer

12-1

Molecular Weight Determination

12-1a Electron Impact

The single most valuable item of information that the mass spectrometer can provide is the molecular weight of a compound. Mass spectrometry is the preferred procedure for this determination because of its speed and accuracy (nearest mass number for low resolution spectrometers). A convenient starting point in making this measurement is to record a conventional 70 eV EI spectrum. Figure 12-1 illustrates the results of this approach for a number of cases. When the sample is pure and the compound is sufficiently volatile, a molecular ion is often observed and the molecular weight is thus known. This is the case for the derivatized pentasaccharide shown in Figure 12-1a. This assignment can be checked for its consistency with the fragment ions that appear in the spectrum. For example, in spectrum (b), an ion 4 mass units higher than the probable molecular ion of an organic compound indicates an impurity or an incorrect molecular ion assignment. (The ion I^+ from an iodo compound gives rise to this peak at m/z 127.) In other cases, illustrated by (c), the thermal

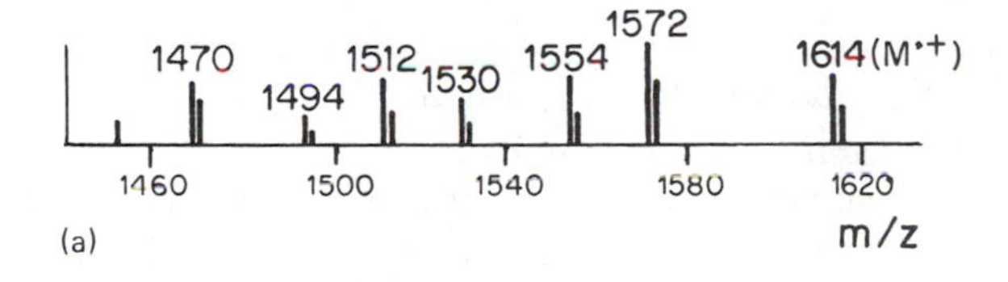

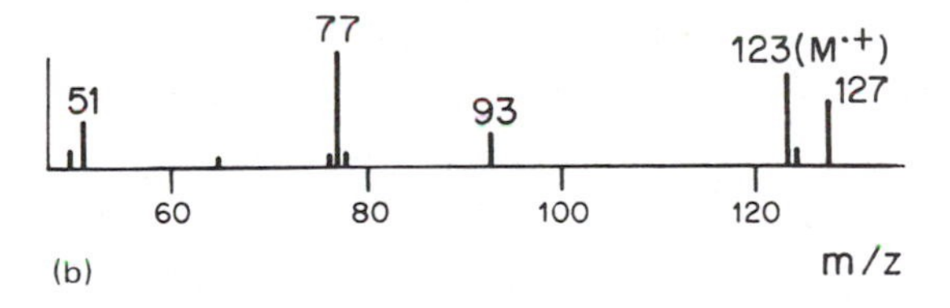

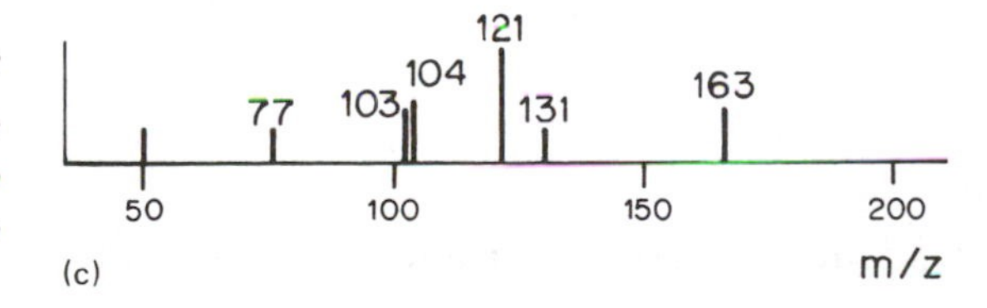

FIGURE 12-1 Molecular weight determination by EI. (a) Maltopentose 1-phenylflavazole peracetate, molecular weight 1614 daltons, (b) nitrobenzene, molecular weight 123, (c) $C_6H_5CHBrCH_2CO_2CH_3$, molecular weight 242–244.

lability of the sample or ease of fragmentation results in no ions in the molecular ion region. This case is difficult to interpret without additional data.

12-1b Confirmation by Other Methods

Unless mass spectrometry is being used only to confirm a suspected structural assignment, additional experiments may be necessary to secure the molecular weight. Figure 12-2 starts from the standard electron impact experiment and suggests additional avenues that can be taken. For each ionization method there exist choices regarding the chemical state of the analyte, as well as the ionization conditions. Among the many possibilities, a selection can be made based on the nature of the sample. For example, acidic substances often give good negative ion spectra. High molecular weight compounds should probably be examined by desorption ionization. Easily fragmented molecules can be examined by reducing the ionizing energy (electron energy in EI, matrix selection in DI, ionizing reagent acidity or basicity in CI) to increase the yield of molecular ions relative to fragment ions.

The variety of experiments can be used to determine molecular weight is illustrated by the case of the amino acid arginine. Figure 12-3 shows arginine mass spectra of several types, any one of which could be used to characterize the molecular weight as 174 d. Three of the experiments employ DI (Section 11-5b), achieved by laser desorption, fast atom bombardment, and field desorption, the last being a procedure in which the solid sample is heated in a high electric field. It is worth noting that even for this simple compound, conventional electron impact and methane or isobutane chemical ionization do not provide molecular ions.

It is because *chemical reactions* underlie mass spectrometry that the arginine mass spectra display such variety. The consequences of different choices of ionization method for the nature and intensity of the peaks that characterize molecular weight deserve further notice. The case of chemical ionization allows chemical effects to be illustrated clearly. The form in which the ionized molecule occurs in CI includes $M^{\cdot +}$, $M^{\cdot -}$, $(M+H)^+$, $(M-H)^+$, and $(M+C)^+$. Electron and proton transfer, as well as adducts with metal and other ions (C^+), generate these species. Consider the use of ammonia as a chemical ionization reagent gas. The major reagent ion is NH_4^+. This ion can react either by proton transfer (eq. 12-1), or by simple adduct generation

$$M + NH_4^+ \rightarrow MH^+ + NH_3 \tag{12-1}$$

(eq. 12-2). While the conditions that dictate the balance between these products cannot be dealt with here, the

$$M + NH_4^+ \rightarrow MNH_4^+ \tag{12-2}$$

effectiveness of molecular weight determination often depends on the choice of CI reagent. This is illustrated in the case of D-glucopyranose in Figure 12-4; the sample only yields a quasimolecular ion when the *mild* protonating reagent NH_4^+ is employed.

The abundance of the ion that provides molecular weight information in chemical ionization is controlled by the internal energy deposited during the ionization event. We have already discussed in Section 11-3 how the internal energy depends on the difference in proton affinities (Table 11-9) between the analyte and the

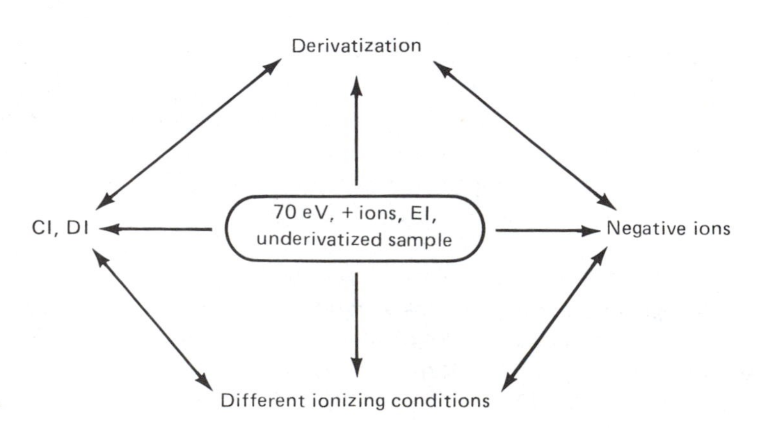

FIGURE 12-2 Options for confirming molecular weight.

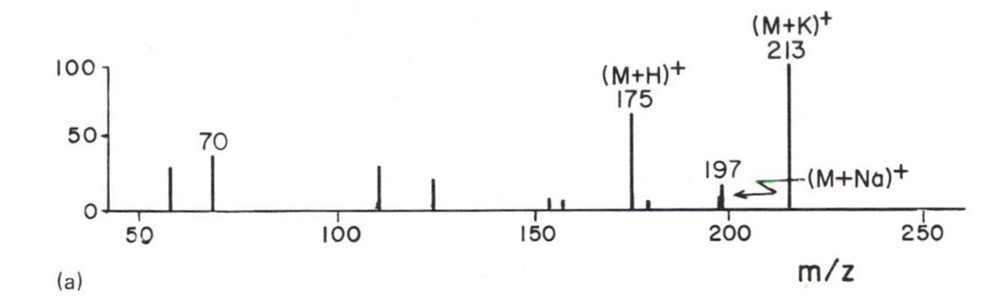

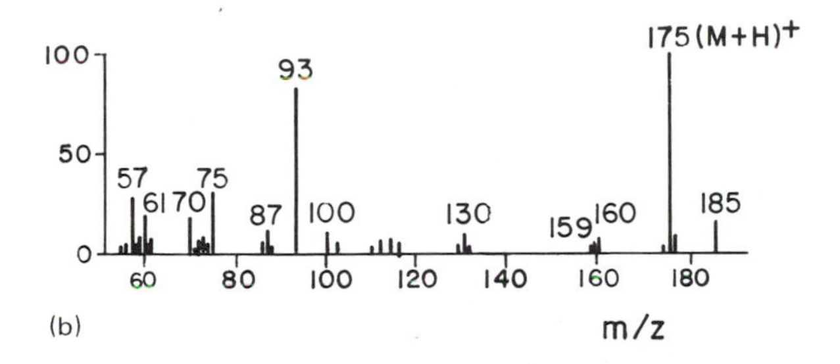

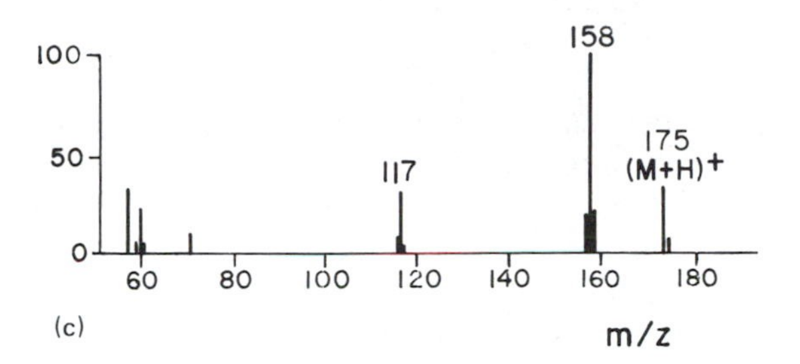

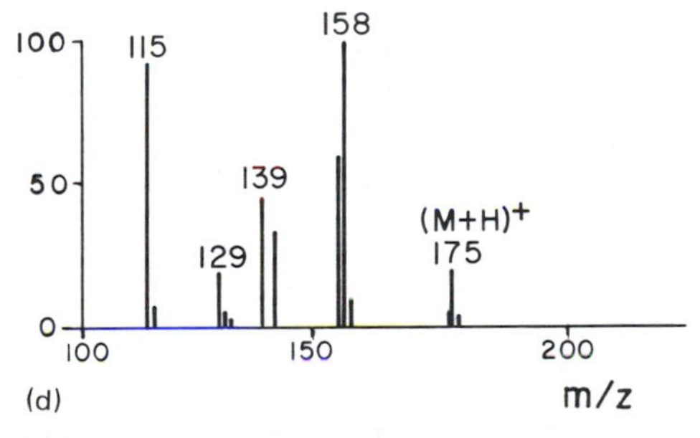

FIGURE 12-3 Mass spectra of arginine. (a) Laser desorption mass spectrum (CO_2 laser, 0.1 J pulse^{-1}, 0.15 μs pulse width). (b) Fast atom bombardment (argon, 5 keV, sample in glycerol). (c) Field desorption (emitter temp. 220° C). (d) Chemical ionization (NH_4^+ reagent, mass-selected). [(a) reprinted with permission from M. A. Posthumus et al., *Anal. Chem.*, **50**, 985 (1978), copyright 1978 American Chemical Society; (b) from J. J. Zwinselman et al., *Org. Mass Spectrom.*, **18**, 525 (1983), copyright 1983 John Wiley & Sons Ltd., reprinted by permission of John Wiley & Sons Ltd.; (c) from H. U. Winkler and H. D. Beckey, *Org. Mass Spectrom.*, **6**, 655 (1972), copyright 1972 John Wiley & Sons Ltd., reprinted by permission of John Wiley & Sons Ltd.; (d) reprinted with permission from R. J. Beuhler et al., *J. Am. Chem. Soc.*, **96**, 3990 (1974), copyright 1974 American Chemical Society.]

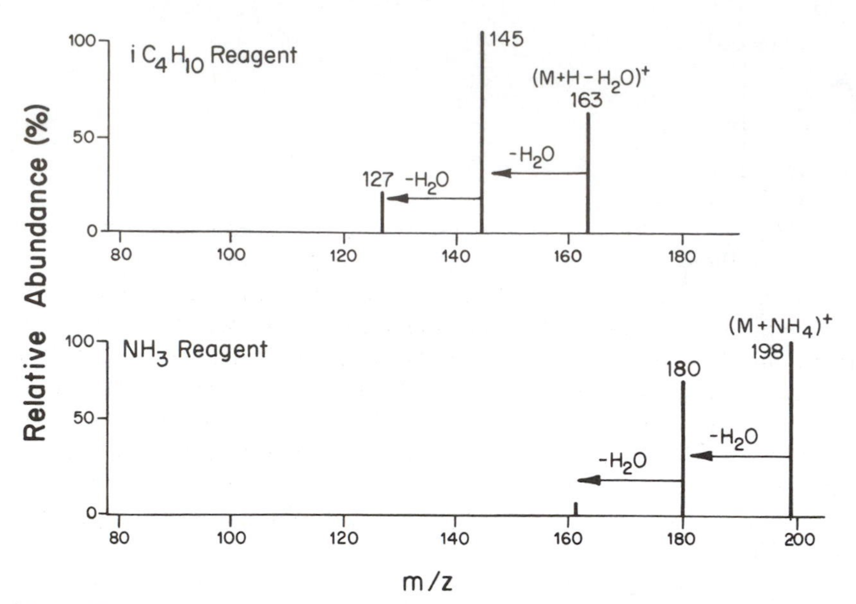

FIGURE 12-4 Glucopyranose examined with different CI reagents. [Reprinted with permission from A. G. Harrison, *Chemical Ionization Mass Spectrometry*, 1983, p. 119. Copyright CRC Press, Inc., Boca Raton, FL.]

conjugate base of the reagent ion (in the case of proton transfer chemical ionization). The consequences for determination of molecular weight are evident in the isobutane and ammonia chemical ionization spectra of glucopyranose. The reagent ions generated from isobutane and ammonia under chemical ionization conditions are $C_4H_9^+$ and NH_4^+, respectively. The proton affinity of C_4H_8 is 823 kJ mol^{-1} (197 kcal mol^{-1}) that of NH_3 is 857 kJ mol^{-1} (205 kcal mol^{-1}); hence the energy available when isobutane is used as reagent gas is 34 kJ mol^{-1} (8 kcal mol^{-1}) greater than that for when ammonia is used. Ammonia, which yields the softer protonating agent, gives a spectrum that shows the more abundant protonated molecule. Note also that the spectra shown in Figure 12-4 show fragment ions that arise only by simple elimination reactions. Simple spectra are not always the case in CI, but it stands in contrast to electron ionization, where the free radical character of the molecular ion typically results in a more complex fragmentation chemistry.

Molecular weight determination by DI is illustrated in Figure 12-5, which shows the $(M+H)^+$ ion of human proinsulin. The measurement is accurate to a few tenths of a dalton. Note the use of readily desorbed inorganic cluster ions $(Cs_{n+1}I_n)^+$, which are easily recognized and serve as standards to establish the mass scale.

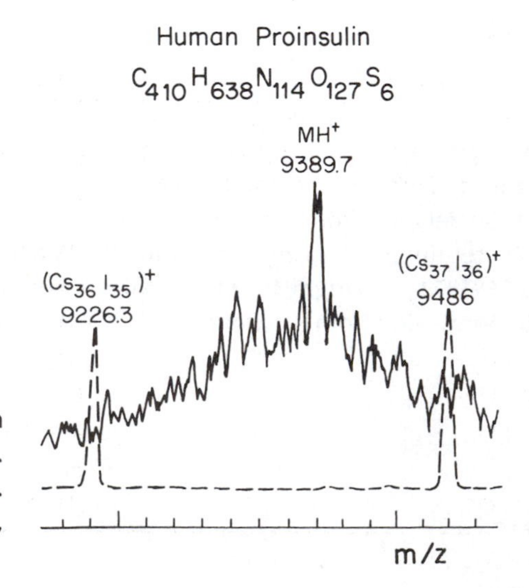

FIGURE 12-5 The molecular ion region of human proinsulin ionized by FAB. [Data from M. Barber, R. S. Bordoli, G. J. Elliott, N. J. Horoch, and B. N. Green, *Biochem. Biophys. Res. Commun.*, **110**, 753 (1983).]

12-2

Molecular Formula

12-2a Exact Mass Measurement

A molecular formula is normally arrived at from an exact mass measurement. Exact mass measurements frequently can be made to an accuracy of better than 10^{-3} d with an internal standard of known exact mass. Even with this accuracy, however, a unique fit is seldom obtained when all possible elemental compositions at any nominal mass are considered. The number of possibilities increases with the number of elements that may be present, with the number of atoms of each element possible, and with the molecular weight. For organic compounds that have only the common "heteroatoms" (O, N, F, Cl, Br, I, S, and Si) and molecular weights less than 500 d, only a few possibilities need be considered (see Appendix IV.A for masses and abundances of the isotopes). Thus, while exact mass measurement seldom gives a single formula, it will often give a single *reasonable* formula, especially when other information on the sample is considered. Some of this ancillary information may come from the mass spectrum itself or from isotopic abundance distributions, which are considered below (Section 12-2b).

Exact mass measurements are performed on mass spectrometers capable of high mass resolution, namely double focusing sector or Fourier transform instruments. High resolution measurements can be made with all the common types of ionization methods. The resolution required to separate the common H_2/D doublet (mass difference 1.7×10^{-3} d) is approximately 10^6 at 1700 d. This resolution somewhat exceeds available capabilities and illustrates the need for supplementary data to assign molecular formulas confidently. Frequently, only the composition of the molecular ion and several of the more abundant fragment ions will be measured in solving structural problems.

To illustrate the application of high resolution measurements to the mass spectrum of a natural product available in small amounts, a portion of the high resolution spectrum of the ergot alkaloid dihydroelmyoclavine ($C_{16}H_{20}N_2O$) is reproduced in Table 12-1. The mass scale is calibrated using ions from perfluorokerosine, an internal standard. The assignments shown in the table are not the only ones possible within the uncertainties of the experiment. They are, however, the only reasonable ones (for example, the mass of the molecular ion, 256.1579 d, also fits the less reasonable composition $C_{14}H_{18}N_5$).

12-2b Isotope Cluster Abundances

Even without high resolution data, it is often possible to arrive at a probable molecular formula before undertaking a detailed interpretation of the low resolution mass spectrum. This information comes from analysis of isotopic clusters. Natural carbon, in addition to its ^{12}C component, contains 1.1% of the stable

TABLE 12-1 Part of the High Resolution Mass Spectrum of an Alkaloid[a]

Measured	Calculated	Error ($\times 10^3$ d)	Composition
235.9866	235.98722	−0.61	C_7F_8
237.1388	237.13917	−0.35	$C_{16}H_{17}N_2$
238.1464	238.14700	−0.55	$C_{16}H_{18}N_2$
241.1347	241.13409	0.59	$C_{15}H_{17}N_2O$
242.1430	242.14191	1.13	$C_{15}H_{18}N_2O$
242.9856	242.98562	0.02	C_6F_9
243.1445	243.14527	−0.80	$C_{14}(^{13}C)H_{18}N_2O$
254.9856	254.98562	0.02	C_7F_9
256.1579	256.15756	0.37	$C_{16}H_{20}N_2O$

[a]The fluorine-containing ions are due to an internal standard.

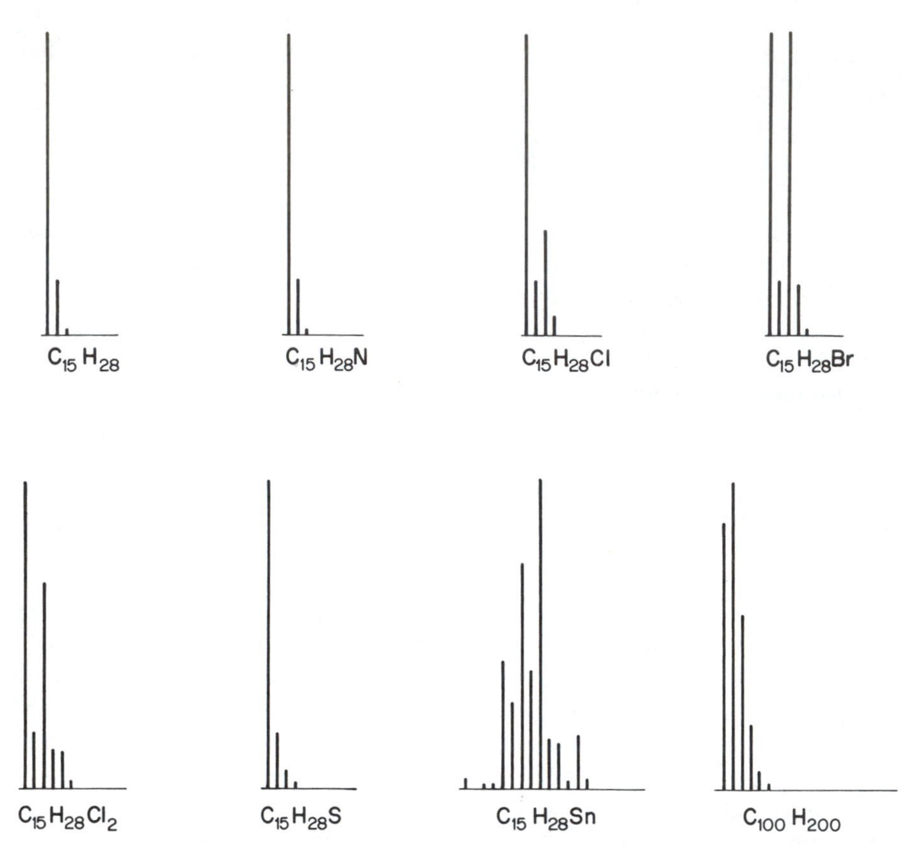

FIGURE 12-6 Typical isotopic patterns.

isotope ^{13}C. This means that a molecule containing 20 carbon atoms has a 100:22 distribution of $^{12}C_{20}$ and $^{13}C^{12}C_{19}$ constituents. The presence and abundance of isotopic peaks can be exceedingly valuable in interpreting mass spectra. The cases of most interest, besides carbon, are those in which the isotopic patterns are sufficiently distinctive to allow the ready recognition of their presence and *number* in organic compounds. These cases include chlorine, bromine, boron, sulfur, silicon, and many of the metals. Figure 12-6 illustrates some typical isotopic patterns. Appendix IV.A contains the masses and abundances of the isotopes commonly encountered in organic chemistry. Remember that any mass spectrum will show characteristic isotopic patterns for fragment ions as well as for the molecular ion, although the former are more likely to be obscured by interfering ions with different compositions.

The data shown in Figure 12-6 are obtained by calculations based on binomial expansions. Programs are commonly available that calculate isotopic patterns for given molecular formulas. Appendix IV.B reproduces one of these. Note that in addition to the common distinctive patterns, monoisotopic elements can often be recognized by the fact that the $(M+1)^+$ ion has a much *lower* abundance, relative to the molecular ion, than can be explained on the basis of the presence of C, H, N, and O alone. Such a situation can be illustrated by the case of an unknown compound, molecular weight 322, which shows an $[(M+1)^+]/[M^{\cdot+}]$ ion abundance ratio of only 17%. Since the natural abundance of ^{13}C is 1.1% of ^{12}C, the number of carbon atoms must be less than 15 and, as a maximum, the CH content can only account for 212 ($C_{15}H_{32}$) of the 322 d mass. A 6% $(M+2)^+$ ion suggests the presence of one sulfur atom, but >78 mass units are still left to be accounted for by heteroatoms. Such a situation suggests the possible presence of fluorine, iodine, or phosphorus atoms, and indeed in this case the actual molecular formula was $C_{14}H_8F_6S$ (measured 322.032; expected 322.033).

Just as in NMR, where many of the naturally abundant isotopes do not have appropriate spin states, so the species most difficult to determine from isotope abundances are the elements C, H, O, and N. Since the presence of a single nitrogen atom can usually be recognized from the odd mass molecular weight, and that of two or more nitrogen atoms will usually be indicated by other evidence, the problem reduces most often to deciding between C and O. It is preferable to make this decision by estimating the upper limit of the number of carbon atoms from eq. 12-3 and then to consider also the possibilities in which one or two carbons are replaced by oxygen.

$$\frac{[(M+1)^+]}{[M^{\cdot+}]} = 0.11 \text{ (no. of C atoms)} + 0.004 \text{ (no. of N atoms)} \qquad (12\text{-}3)$$

For example, a compound of molecular weight 162 showed an $M^{\cdot+}/(M+1)^+/(M+2)^+$ ratio of 100/13.0/1.2. The maximum possible number of carbon atoms is 11 since this requires $(M+1)^+ \geq 12.1\%$; formulas with 10 carbon atoms, $(M+1)^+ \geq 11.0\%$, should also be considered. Hence, the probable molecular formulas are $C_{11}H_{14}O$, $C_{10}H_{14}N_2$, and $C_{10}H_{10}O_2$. Examination of the fragmentation behavior, or exact mass measurement, is needed to decide between these possibilities. It is extremely valuable to consider the units of unsaturation (rings and double bonds) when deciding between possible formulas. This is simply done by considering the corresponding hydrocarbons, which in this case are $C_{11}H_{14}$, $C_{10}H_{12}$, and $C_{10}H_{10}$, respectively. These formulas have 5, 5, and 6 units of unsaturation, respectively.

For organometallics the *isotope cluster method* can even become competitive with *exact mass measurements* for determining elemental compositions. As an illustration of the measure of agreement that can be achieved, the molecular ion of triferrocenylborane ($C_{30}H_{27}BFe_3$) gives the calculated and observed isotope pattern shown in Table 12-2. The procedure is only applicable, however, when there are no interfering ions, especially ions such as $(M-H)^+$.

TABLE 12-2 Calculated and Measured Isotope Abundance Pattern for Ferrocene

	Relative Abundance	
Mass, *m/z*	Calculated	Observed
562	1.1	1.1
563	4.7	4.4
564	18.8	19.0
565	29.6	28.5
566	100	100
567	39.2	39.2
568	8.7	7.6
569	1.3	1.1

12-3 Isotopic Incorporation

The preceding discussion leads us naturally to the problem of determining the extent of incorporation of stable isotopes into molecules. Although recent improvements in NMR sensitivities have made the NMR method competitive in the case of some nuclei, mass spectrometry is still the standard technique for these measurements. In the following discussion, emphasis is placed on determination of the *extent* of incorporation, but it should be noted that the *site* of incorporation also can be determined from a knowledge of the fragmentation pattern of the compound.

The most common stable isotopic labels are ^2H, ^{15}N, ^{13}C, and ^{18}O. The mass spectrometric method of determining isotopic incorporation requires reasonable abundance and absence of interference in the ion chosen for measurement. This will frequently be the molecular ion in EI or a fragment known to be formed without loss of isotopic label. Ions that can interfere with $M^{\cdot+}$ include $(M+H)^+$ and $(M-H)^+$, since isotopic forms of these species have the same nominal mass as the several isotopic forms of $M^{\cdot+}$.

The analytical method involves obtaining spectra of the labeled compound and the unlabeled analogue under as nearly identical conditions as possible. Since ion abundances are required, multiple slow scans are taken at low resolution. The spectrum of the unlabeled compound includes contributions from natural

TABLE 12-3 Calculation of the Mass Spectrum of a Labeled Compound

	m/z					
	138	139	140	141	142	143
Experimental data: d_0	100	9.6	0.2			
Hence for d_3				100	9.6	0.2
for d_2			100	9.6	0.2	
for d_1		100	9.6	0.2		
80% d_3				80	7.7	0.2
15% d_2			15	1.4	0	
5% d_1		5	0.5			
Sum, renormalized (80% d_3, 15% d_2, 5% d_1)		6.1	19.0	100	9.5	0.2

isotopic species, ion–molecule reaction products, and $(M-H)^+$ and $(M-H_2)^{\cdot+}$ ions. If these last two can be completely removed, the determination becomes trivial and should be accurate to 1%, e.g., 25 ± 1% incorporation. The determination of deuterium incorporation may have appreciable uncertainty if processes leading to $(M-H)^+$ cannot be removed, since (1) accurate correction for primary isotope effects cannot be made and (2) the site of origin of the protium atom lost from the unlabeled ion is not always known, so its distribution between protium and deuterium loss in the labeled compound is uncertain.

Consider the molecular ion region of a pure compound, *m*-dimethoxybenzene, molecular weight 138 d, ionized by electron impact. Ions occur at *m/z* 138, 139, and 140 in relative abundances of 100, 9.6, and 0.2%. Even in this simple case, each nominal mass does not consist of a single isobaric ion (e.g., *m/z* 139 includes the ^{13}C and ^{2}H forms of the molecular ion as well as a contribution from the small amount of the protonated molecule generated in EI). The simplest possible assumption is that the pure monodeuterated compound will contain exactly the same relative abundance ratios, each shifted by one dalton to higher mass. On this basis, the mass spectrum of a sample that is 80% d_3, 15% d_2, and 5% d_1 is calculated as shown in Table 12-3.

To determine isotopic compositions from a mass spectrum one must reverse the procedure just described. For example, consider an experiment in which catechol is exchanged in NaOD–D_2O at 80°C and subsequently methylated to give 1,3-dimethoxybenzene-2,4,6-d_3. The problem is to determine the isotopic purity of this product. The molecular ion is the base peak in the mass spectrum and the unlabeled compound shows no $(M-1)^+$ ions at low ionizing energy. The molecular ion regions of the unlabeled (d_0) and labeled compounds are measured as

m/z:	138	139	140	141	142	143
unlabeled:	100	9.6	0.2			
labeled:		0.5	2.6	100	9.6	0.2

Compositions are assigned to the ions in the d_3 spectrum, heavy isotope and protonated species are subtracted in the ratio found for the d_0 compound, and the isotopic composition is calculated as shown in Table 12-4.

Now consider a case in which the contributions of an $(M-H)^+$ ion cannot be removed. In such cases there will be additional uncertainty in the measurement because the assignment of H relative to D loss in the labeled compound will not usually be known. Butyl ethyl ether is generated from 1-butanol, and the isotopic incorporation in the ether synthesized from butanol-1,1-d_2 is required. At 70 eV the mass spectrum of the unlabeled ether shows an $(M-H)^+$ ion that has 13% the abundance of the molecular ion. Even at 16 eV the $(M-H)^+$ ion is of substantial abundance, and the following experimental data are obtained for the labeled (d_2) and unlabeled (d_0) compounds.

m/z:	100	101	102	103	104	105	106
unlabeled:	0.2	4.6	100	6.5	0.3		
labeled:			1.9	3.1	100	6.3	0.3

TABLE 12-4 Calculation of Isotopic Incorporation from Mass Spectra

	m/z					
	138	139	140	141	142	143
Experimental, d_0	100	9.6	0.2			
Experimental, labeled		0.5	2.6	100	9.6	0.2
Assigning 139 to d_1		0.5	0.05			
Subtracting d_1 (0.5 unit)		0	2.55	100	9.6	0.2
Assigning 140 to d_2			2.55	0.24		
Subtracting d_2 (2.55 units)			0	99.76	9.6	0.2
Assigning 141 to d_3				99.76	9.58	0.2
Subtracting d_3 (99.76 units)				0	0.02	0

Conclusion: Labeled compound is 99.76 units d_3, 2.55 units d_2, 0.5 unit d_1; i.e., 97.0% d_3, 2.5% d_2, and 0.5% d_1.

From the d_0 spectrum, m/z 100 (0.2%) can only be due to loss of H_2 from the molecular ion. Its ^{13}C isotope will contribute 6.6% (<0.01 unit) to m/z 101 and can be ignored. The signal at m/z 101 (4.6 units) must be entirely due to H$^\cdot$ loss from the molecular ion, m/z 102, since H_2 loss from m/z 103 will be negligible, namely, 0.2% of 6.5 units or 0.01 unit. The ^{13}C isotope of $C_6H_{13}O^+$ (m/z 101) will contribute 6.6% of 4.6 units, or 0.3 unit, to m/z 102. Hence m/z 102 represents 99.7 units of $C_6H_{14}O$ and 0.3 unit of $^{13}CC_5H_{13}O$. The ions at masses m/z 103 and m/z 104 due to higher isotopes of $C_6H_{14}O$ have abundances of 6.6 units and 0.4 unit, respectively (considering ^{13}C and ^{18}O). Hence within experimental error (0.1 unit), m/z 103 and 104 are due entirely to isotopic contributions of $C_6H_{14}O$. Thus 99.7 units of $C_6H_{14}O.^+$ gives 4.6 units of $C_6H_{13}O^+$ and 0.2 unit of $C_6H_{12}O^{\cdot+}$, or, normalizing, 100 units of $C_6H_{14}O^{\cdot+}$ gives 4.6 units of $C_6H_{13}O^+$ and 0.2 unit of $C_6H_{12}O^{\cdot+}$.

In analyzing the d_2 ether, assumptions regarding the relative losses of H$^\cdot$ and D$^\cdot$ must be made. To a good approximation all H$^\cdot$ loss from alkyl ethers occurs from the α position (Section 12-6b). Hence, if we assume a k_H/k_D isotope effect of unity for the fast ion source reactions under study, the labeled compound should suffer equal H$^\cdot$ and D$^\cdot$ loss. Now if the labeled compound were 100% butyl-1,1-d$_2$ ethyl ether, the molecular ion region would show

m/z	102	103	104	105	106
	0.2+2.3	2.3	100	6.3	0.3

In this computation all molecular hydrogen loss is assigned to H_2, since loss of D_2 or HD is statistically much less likely, and $(M+1)^+$ and $(M+2)^+$ ion abundances are found experimentally for the d_0 compound rather than calculated (in this case the difference between the two is negligible). Agreement with experiment is optimized if the calculation includes some d_1 ether. The calculation for 99.0% d_2–1.0% d_1 gives

m/z	102	103	104	105	106
	2.4	3.1	100	6.1	0.3

which is in good agreement with the experimental data.

12-4

Quantitative Analysis

The basis for quantitative work in mass spectrometry is the accurate measurement of ion abundances. Difficulties in exactly reproducing ion source conditions mean that quantitation is best achieved by using an internal standard. External standard and standard addition methods both require more than one sample introduction. While the internal standard may simply be a closely related compound, best accuracy is obtained if one uses an isotopically labeled form of the analyte. Once this has been added to the sample, separations and chemical manipulations can be performed without requiring total sample recovery. It is not even necessary to know the sample recovery since losses of the analyte and internal standard can be considered to be equally likely. It is particularly convenient to choose the labeled compound so that there is a separation of a few daltons between peaks due to the labeled and unlabeled compounds. In this way contributions from naturally occurring isotopic forms are avoided. For example, to quantitate 1,3-dimethoxybenzene (molecular weight 138), the d_3 analogue (molecular weight 141), could be prepared using methyl-d_3 iodide, and it would be an excellent choice as internal standard. The ions at mass 141 and 138 represent labeled and unlabeled compounds, and their ratios allow one to quantitate the analyte using eq. 12-4, in which Q is the amount and S the signal due to analyte (A) and to the internal

$$\frac{Q_A}{Q_I} = \frac{S_A}{S_I} \tag{12-4}$$

standard (I). The quantity of internal standard used in eq. 12-4 *must* refer to the amount of the particular form measured, in this case the trideuterated compound. Hence, if the labeled compound was the 97.0% d_3 sample discussed above (Table 12-4), Q_I in eq. 12-4 will be 97% of the total amount of labeled standard added to the sample.

There are two advantages to this procedure that go beyond the obvious merits of taking ratios of isotopic ions in order to minimize analytical errors. First, the total quantity of analyte is increased by addition of the labeled version, which assists in minimizing losses in the handling and chromatographic steps (the labeled analyte can be added in considerable excess without prejudice to the determination). Second, once the labeled compound has been mixed with the naturally occurring analyte, only the ratio is significant. Any losses by whatever process, chemical or physical, as long as they are not subject to significant isotope effects, can be ignored.

Consider the quantitation, in a mixture including other DNA hydrolysates, of the methylated nucleoside 3-methylcytidine (m^3C). This measurement is made with liquid chromatography (for separation) and secondary ion mass spectrometry (one of the desorption ionization procedures) with the synthetic d_3-methyl analogue added as internal standard. The nucleoside can be determined by the abundance ratio of the ions corresponding to protonated methylcytosine (the free base) at m/z 126 and its d_3-analogue at m/z 129 (Figure 12-7). Note that the experiment uses a fragment ion in quantitation; the molecular ion is of lower abundance and the fragment ion is free of interference and arises by a well-defined process. Note also that linear calibration curves need not apply (or even be examined) for this method to be used. The experiment illustrated in Figure 12-7 employed a 1 μg aliquot of a mixture of 10.5 mg of enzymatically degraded oligonucleotide and 2.08 mg of labeled methylcytidine, namely, $(1/12.58) \times 10.5\ \mu$g of nucleotide mixture and $(1/12.58) \times 2.8\ \mu$g of the labeled compound. This standard consisted of 99% methyl-d_3-cytidine. The ratio of m/z 126 to m/z 129 is 1.22 ± 0.17. Hence, from eq. 12-4, $S_A/S_I = Q_A/Q_I = 1.22$:

$$Q_A = 1.22 \cdot 0.99 \cdot 2.08 \cdot 10^{-6}\ \text{g} = 2.02 \times 10^{-7}\ \text{g}$$

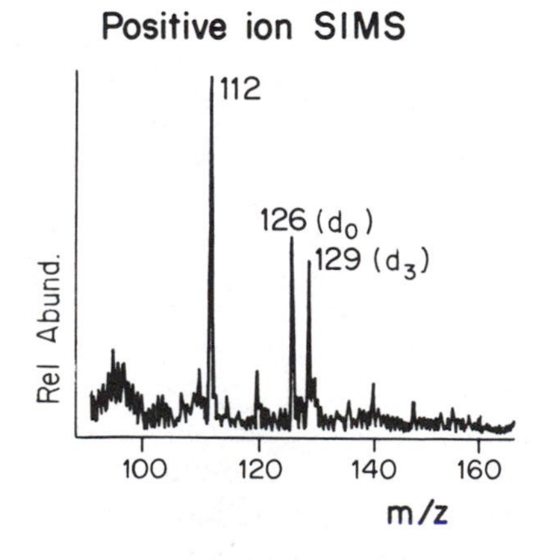

FIGURE 12-7 Quantitation of a nucleoside, 3-methylcytidine. Positive ion SIMS spectrum taken in the presence of the d_3-labeled analyte as internal standard. Sample size 10^{-6} g. [Reprinted with permission from D. J. Ashworth, C-J. Chang, S. E. Unger, and R. G. Cooks, *J. Org. Chem.*, **46**, 4770 (1981). Copyright 1981 American Chemical Society.]

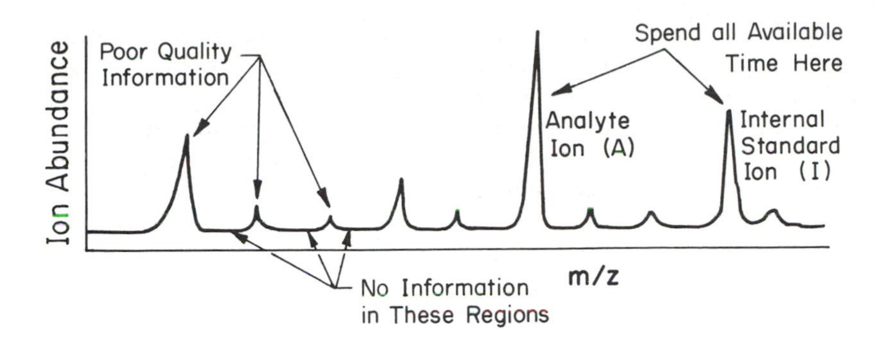

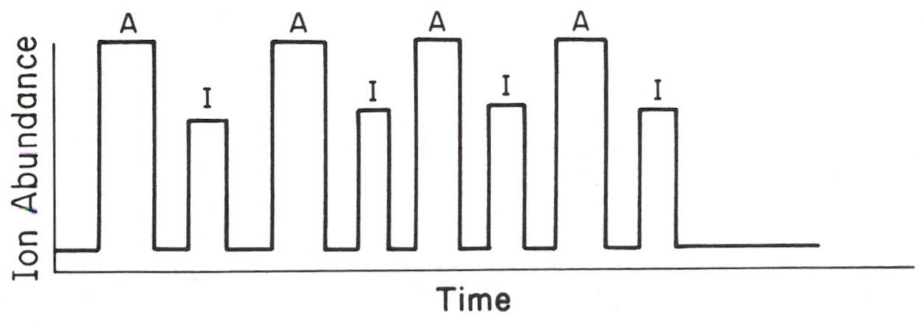

FIGURE 12-8 Selected ion monitoring: (upper) the mass spectrum, (lower) output of SIM experiment in which just two ions are monitored and analyte flux is not changing with time.

$$\% \text{ analyte in the sample} = \frac{2.02 \times 10^{-7}\text{ g}}{10^{-6}\text{ g}} \cdot \frac{12.58}{10.5} \cdot 100 = 24\%$$

In experiments of the type just described, it is not necessary to record the entire mass spectrum since only selected ions are used in the calculation. Hence *selected ion monitoring* is used; this procedure increases signal-to-noise ratios (often by 10^3 times) by having the mass spectrometer monitor only the ion or ions in which the information is concentrated. The procedure is illustrated in Figure 12-8. Most routine quantitative work by mass spectrometry employs selected ion monitoring.

12-5

Mixture Analysis

12-5a GC-MS and LC-MS

Qualitative and quantitative analysis of mixtures are typically performed on samples introduced into the mass spectrometer via a gas or liquid chromatograph. The instrumentation has been presented in Section 11-5a. The mass spectrometer can be thought of as a detector for the chromatograph, and it is operated in either of two modes. In one mode full spectra of each component are obtained by rapidly scanning the spectrometer (typically 10 spectra per chromatographic peak, which can elude in less than a second). In the second mode, the spectrometer monitors selected ions (Figure 12-8) that are diagnostic for a particular compound(s). In this procedure specificity is traded away for enhanced sensitivity.

The output of a GC–MS or LC–MS run can be presented as a chromatograph in which the total ion current is displayed against time, or the signal due to selected ions can be displayed to allow recognition of particular compounds or types of compounds present in the mixture. Figure 12-9 illustrates each of these possibilities in the case of a mixture of barbiturates examined by GC–MS. The upper trace is a *total ion chromatogram*, which responds to all constituents of the mixture. The lower *mass chromatogram* shows only the constituents that yield m/z 195, a characteristic fragment of allyl-substituted barbiturates.

Because enormous amounts of data are collected in GC–MS systems, it is normal to save (and hardcopy) only a small fraction of it. For example, background subtraction might be done after each run, data compacted,

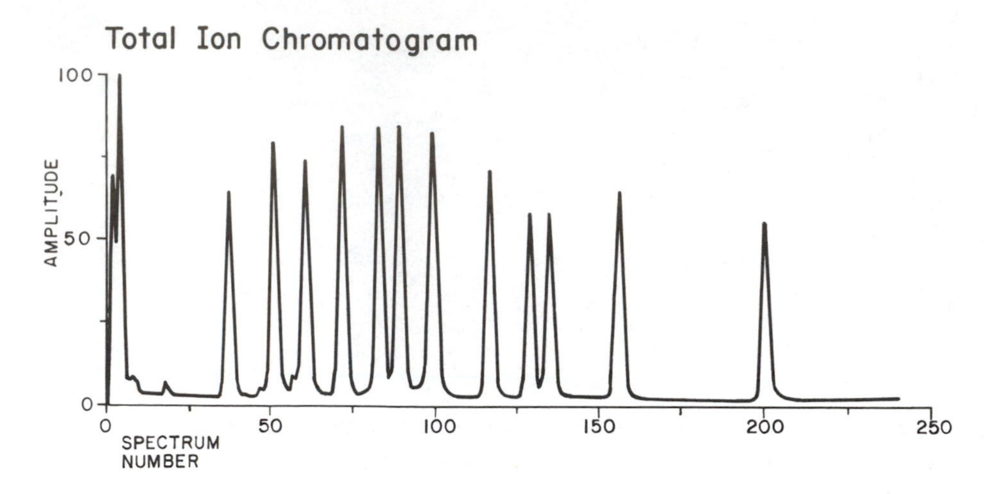

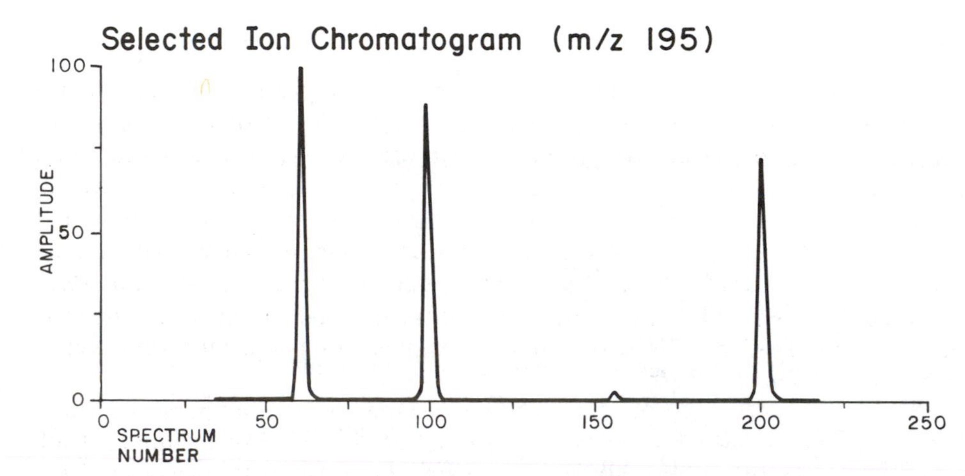

FIGURE 12-9 Universal and selective detection in GC–MS of barbiturate mixture, 2×10^{-7} g per component: (upper) total ion chromatogram, (lower) mass chromatogram for *m/z* 195 recorded by selected ion monitoring. (Data from Finnigan MAT.)

and the excess raw data erased. Consider the number of items of information that must be stored and manipulated in a GC–MS experiment. Commonly, GC runs will last 10 min or more, a complete mass spectrum will be taken each second during the chromatograph (required by the width of capillary column peaks), and relative abundance information is accurate to at least 1%. These conditions result in 4×10^7 items of information. Note that mass spectrometer scan speeds can be increased by at least a factor of 10, that abundances are often stored to six figures, and increases in the other items are also possible. Nevertheless, these data are collected and processed to yield chromatograms of individual masses and other items of information virtually on demand. (Real-time data processing is employed to keep the volume of data within reasonable bounds.)

Among the chromatography–mass spectrometry procedures, the thermospray LC–MS method (Figure 11-13) is rather unusual in acting as both an interface and an ion source. Its performance is illustrated in Figure 12-10, which shows full spectra obtained from relatively small amounts of sample. Note that nonvolatile samples can be examined and that negative ion data can be obtained either by negative ion–molecule reactions or (as illustrated for the benzophenone) by electron attachment (eqs. 11-3 and 11-17). Enhanced sensitivity is again available by selected ion monitoring.

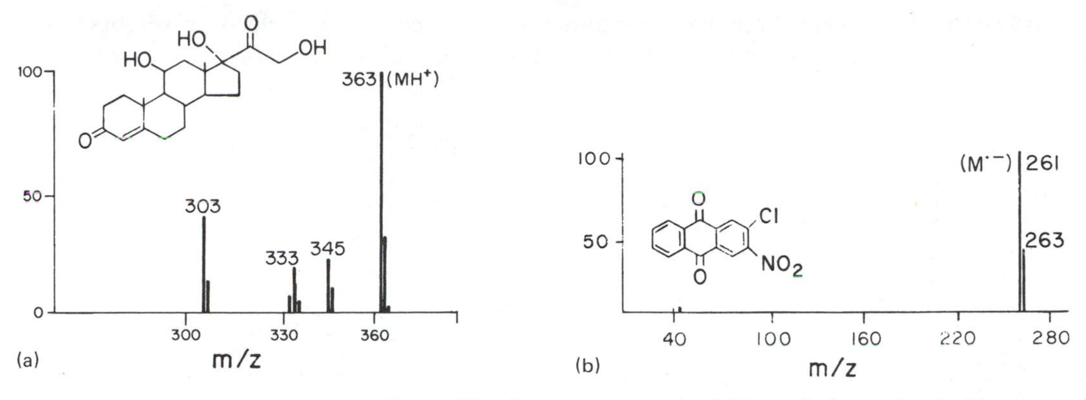

FIGURE 12-10 LC–MS spectrum obtained by thermospray method: (a) cortisol examined without supplementary ionization; (b) benzophenone with an auxiliary filament to facilitate electron capture. [Data from Vestec Corp.]

12-5b Tandem Mass Spectrometry

Tandem mass spectrometry (MS–MS) is a method of mixture analysis that complements the chromatography–mass spectrometry methods. This procedure employs two stages of mass spectrometry, the first to isolate the ion of interest and the second to characterize it via the spectrum of fragments generated in a collision process. Figure 12-11 illustrates the concept. An ion, which retains the structure of the neutral molecule being analyzed, is selected from other components of the mixture by mass analysis. It is then excited by collision and dissociates to yield characteristic fragments (eq. 11-18). A second mass analyzer is employed to record the mass spectrum of these fragments. When operated in this way, the device records a *daughter spectrum* that serves to characterize targeted, individual components of the mixture. In this scan mode the first analyzer is not scanned but is used to select ions of a particular mass-to-charge ratio (and hence compounds of a particular molecular weight).

An alternative scan mode can be used to characterize all those mixture constituents that possess certain functional groups. This scan, the *parent scan*, also employs collisional activation. For example, if all alkylnaphthols in a mixture are to be identified, the second mass analyzer can be set to a characteristic

Mixture —Ion Source→ Mixture of Ions —1st MS→ Single Ion —Collision→ Daughter Ions —2nd MS→ MS/MS Spectrum

Surrogate for a Particular Molecule

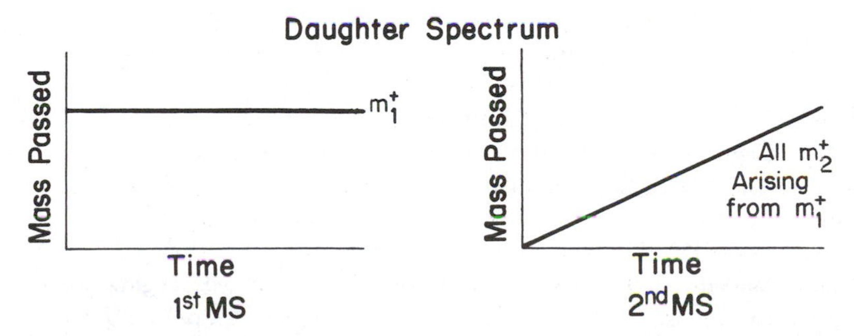

FIGURE 12-11 Concept of tandem mass spectrometry, showing scan (daughter spectrum) that characterizes particular compounds in mixtures.

fragment ion (m/z 157), and the first analyzer can be scanned so as to record the molecular weights of all compounds that fragment by the process shown in eq. 12-5 and therefore fulfill this test for alkylnaphthols.

CH_2R $]^{\bullet+}$ → CH_2^+ (12-5)

OH OH

m/z 157

Individual constituents can then be identified by recording their daughter spectra as already described. Interpretations of each of these daughter spectra, or comparison with library spectra, is the final step in identifying the set of compounds with the designated functional group. Alternative tests for the same group of compounds can also be done, for example, dehydration of $(M+H)^+$ ions generated by CI could serve to recognize the alkylnaphthols. An MS–MS scan that responds to all ions that fragment by loss of a particular neutral fragment (here mass 18) is available to implement this experiment.

OH_2^+ → $^+$ + H_2O (12-6)

R R

A significant advantage of MS–MS is the improvement in signal-to-noise ratios often observed compared with conventional mass spectrometry (Figure 12-12). The ion of interest (m/z 166), the free base derived from protonated nucleoside, is obscured by other ions in the mass spectrum (left). When mass-selected and dissociated, it gives a daughter spectrum with good signal-to-noise characteristics (right).

In a further parallel between GC–MS and MS–MS, quantitation by MS–MS uses a procedure, selected reaction monitoring, that is the analogue of selected ion monitoring in GC–MS. Reactions characteristic of the analyte and internal standard are monitored in turn by setting the first and second mass analyzer to monitor the appropriate ions, first for one reaction, then for the other.

Tandem mass spectrometry is useful in the characterization of pure compounds as well as of mixtures, because background ions are removed and fragmentation can be controlled. Removal of contributions from the liquid matrix is particularly useful in FAB experiments. LC–MS techniques often give insufficient

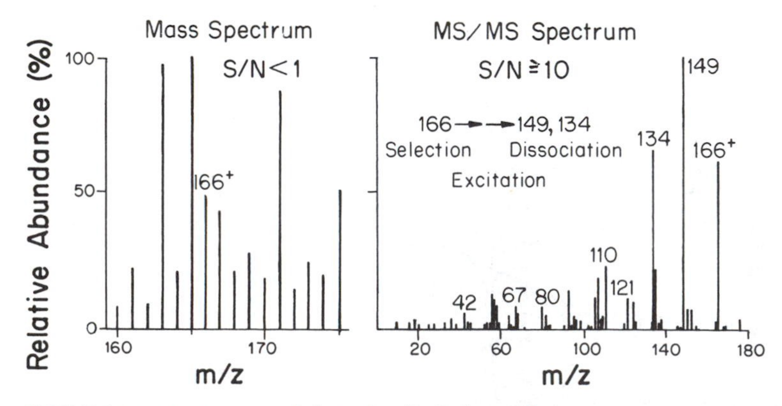

FIGURE 12-12 Improved detection limits in tandem mass spectrometry. Mass spectrum and MS-MS daughter spectrum (of *m/z* 166) for O^6-methyldeoxyguanosine. [From D. J. Ashworth, W. M. Baird, C. J. Chang, J. D. Ciupek, and R. G. Cooks, *Biomed. Mass Spectrum.*, **12**, 309 (1985). Copyright 1985 John Wiley & Sons Ltd. Reprinted by permission of John Wiley & Sons Ltd.]

fragmentation to allow structures to be deduced, so tandem mass spectrometry can be useful in promoting fragmentation. Figure 12-13 illustrates the structural information obtainable from a cyclic peptide ionized by FAB and dissociated in an MS–MS experiment.

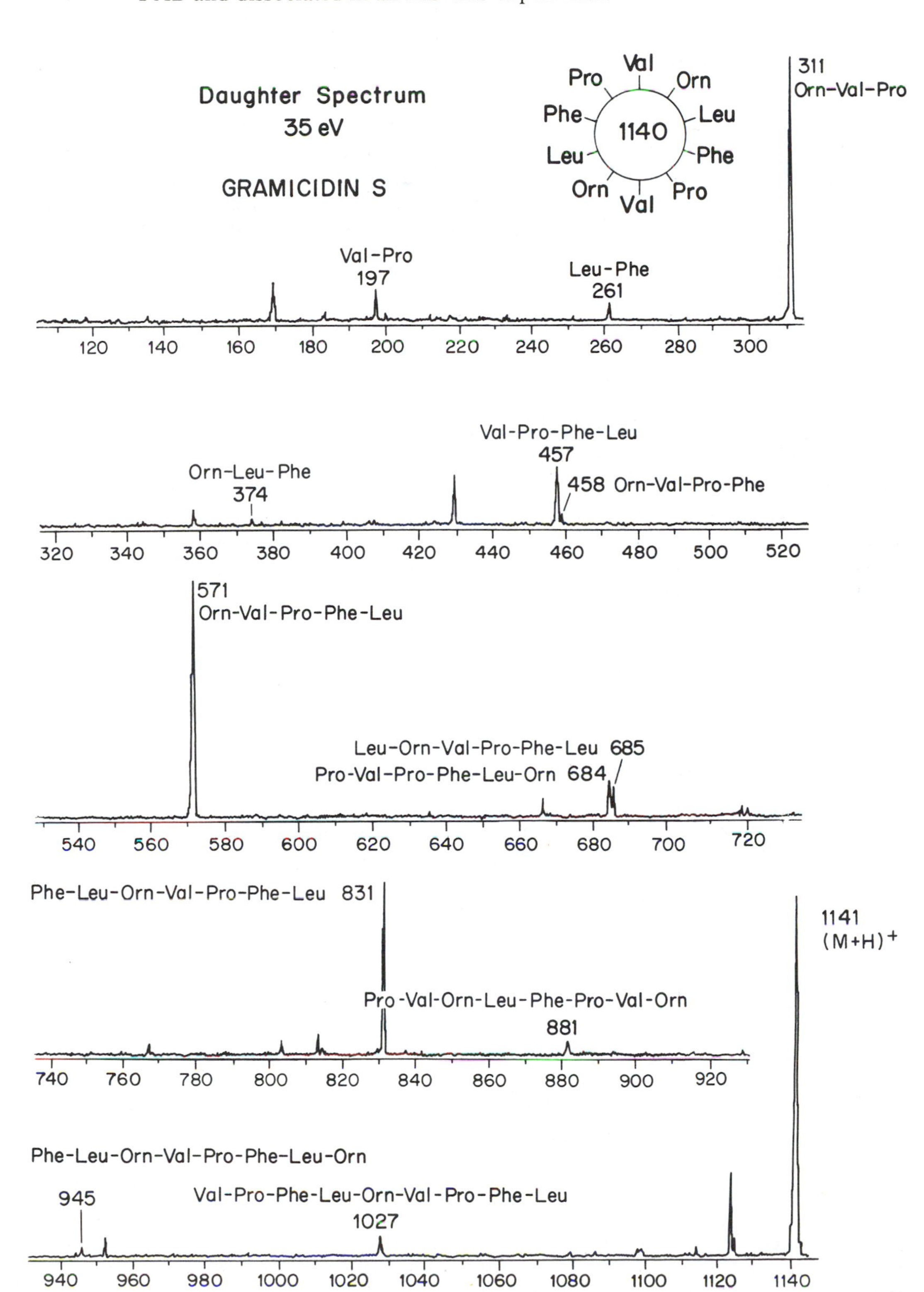

FIGURE 12-13 Use of MS-MS to sequence a cyclic peptide. [Data from A. E. Schoen and R. Pesch, Finnigan MAT.]

12-6

Structure Elucidation

12-6a Generalizations Regarding Fragmentation

Even in experiments where a single molecular ion with a particular internal energy is initially generated, a complex mixture of ions normally will be formed within a microsecond of ionization, a typical ion source residence time (Figure 11-9). These ions will differ in mass and they will have been formed from a network of competing and consecutive individual reactions. The mass spectrum approximates an instantaneous picture of this ion collection as described in Section 11-6a. It is fortunate that only a few reactions are usually dominant and that these processes can often be rationalized from a knowledge of ionic and free-radical solution chemistry.

Energetics of Dissociation. The energy available in a fragmenting ion is normally quite small. Consequently, weak bonds in the ion (usually the same bonds that are weak in the corresponding neutral molecule) are more likely to cleave than are strong bonds. Reactions that lead to electron unpairing are relatively unlikely, but reactions that result in conjugation of charge or radical or otherwise lead to stabilized products are favored. Since the ionic rather than the neutral product is the high enthalpy species, one can generalize that mass spectral fragmentation is *governed by product ion stability*.

Assessment of the relative stabilities of different ions is straightforward, based as it is on chemical principles that are also applicable in solution. Maintenance (whenever possible) of an octet of electrons, localization of charge on the most favorable site available, resonance delocalization, and absence of electron unpairing are fundamental. These principles are worth illustrating by several examples that, taken together, form a basis from which much of ion functional group chemistry can be extrapolated.

1. Dialkyl ethers tend not to give RO^+ or $R{-}O{-}CH_2{-}CH_2^+$ ions. They do give $R{-}O{-}CH_2^+ \leftrightarrow R{-}\overset{+}{O}{=}CH_2$.
2. Dialkyl thioethers, however, do give RS^+ ions.
3. Aryl ethers give ArO^+.
4. Alkyl arenes give $ArCH_2^+$ in strong preference to Ar^+ or $Ar(CH_2)_n^+$.
5. The bifunctional compound $R_1{-}O{-}CH_2{-}NR_2R_3$ gives $CH_2{=}\overset{+}{N}R_2R_3$ rather than $R_1{-}\overset{+}{O}{=}CH_2$.

Naturally, tertiary carbenium ion formation is favored over that of secondary or primary ions. Which, however, will be lost more readily—a large or a small radical? The answer for 70 eV EI is illustrated in eq. 12-7

$$\left[\text{2-}C_4H_9\text{-2-}C_5H_{11}\text{-1,3-dioxolane}\right]^{\cdot+} \xrightarrow{-C_5H_{11}^{\cdot}} \text{2-}C_4H_9\text{-1,3-dioxolan-2-ylium} \quad \text{100 parts}$$

$$\left[\text{2-}C_4H_9\text{-2-}C_5H_{11}\text{-1,3-dioxolane}\right]^{\cdot+} \xrightarrow{-C_4H_9^{\cdot}} \text{2-}C_5H_{11}\text{-1,3-dioxolan-2-ylium} \quad \text{84 parts} \qquad (12\text{-}7)$$

for a case in which the difference in radical size is minimal. Note that the favored product ion is *less* stabilized by polarization forces of the alkyl group; hence it is intrinsically slightly less stable. However, its fewer available fragmentation modes account for its greater abundance.

Some of the above points, including reasons for the preferred stability order, can be summarized as follows.

$R_3C^+ > R_2CH^+ > RCH_2^+$ polarizability

$R_4N^+ > R_3O^+ > R_2F^+$ electronegativity

$R_4N^+ > R_3C^+$ octet rule

$CH_2{=}CH{-}CH_2^+ > CH_2{=}C^+{-}CH_3$ resonance delocalization

Odd- and Even-Electron Ions. A useful consideration, when accounting for the effects of energetics on fragmentation behavior, is the closed shell or radical character of the dissociating ion. Molecular and other odd-electron ions may eliminate either a radical or an even-electron neutral species (eqs. 12-8 and 12-9), but

$$\text{odd}^{\cdot+} \rightarrow \text{even}^{+} + R^{\cdot} \tag{12-8}$$

$$\text{odd}^{\cdot+} \rightarrow \text{odd}^{\cdot+} + N \tag{12-9}$$

an even-electron ion will not usually eliminate a radical to form an odd-electron species (eq. 12-11); its normal fragmentation yields another even-electron ion (eq. 12-10). Exceptions to this last rule occur when the

$$\text{even}^{+} \rightarrow \text{even}^{+} + N \tag{12-10}$$

$$\text{even}^{+} \rightarrow \text{odd}^{\cdot+} + R^{\cdot} \tag{12-11}$$

bond cleaved in the even-electron ion is particularly weak (polybrominated compounds, for example, can undergo multiple successive $Br^{\cdot}$ losses) or when a particularly stable odd-electron ion results (eq. 12-12). As a

$$[C_6H_5CH{=}CHCOCl]^{\cdot+} \xrightarrow{-Cl^{\bullet}} C_6H_5CH{=}CHC^+{=}O \xrightarrow{-H^{\bullet}} [\text{indenone}]^{\cdot+} \tag{12-12}$$

consequence of the odd–even electron generalization, the important even-electron fragment ions $R{-}C{\equiv}\overset{+}{O}$, $ArCH_2^+$, and $R{-}\overset{+}{O}{=}CH_2$ undergo the further fragmentations shown in eqs. 12-13, 12-14, and 12-15.

$$RC{\equiv}\overset{+}{O} \longrightarrow R^+ + CO \tag{12-13}$$

$$ArCH_2^+ \longrightarrow C_5H_5^+ + C_2H_2 \quad (Ar = C_6H_5) \tag{12-14}$$

$$R{-}\overset{+}{O}{=}CH_2 \longrightarrow R^+ + H_2CO \tag{12-15}$$

Note in each case that the neutrals lost not only are closed shell molecules but also are particularly stable.

Stevenson's Rule. The control of product ion enthalpy over the extent of dissociation by two competitive pathways is summarized in Stevenson's rule. This rule states that for a simple bond cleavage, the fragment with the lowest ionization energy will preferentially take the charge. Since competition for the charge is involved, no consideration need be given to entropic factors. For the competitive reactions 12-16 it is readily

$$A{-}B^{\cdot+} \nearrow A^{\cdot} + B^{+} \quad \searrow A^{+} + B^{\cdot} \qquad (12\text{-}16)$$

shown, from the definition of ionization energy (Table 11-1), that the difference in activation energies equals the difference $IE(A^{+}) - IE(B^{+})$. Hence, the rule operates successfully, provided fragment ion abundances are governed by the energetics of fragmentation as repeatedly suggested in Section 11-6a. Note that elimination reactions also follow the rule. For example, ionized cyclohexene undergoes a retro Diels–Alder reaction (eq. 12-17) and yields $C_4H_6^{\cdot+} + C_2H_4$ in preference to $C_4H_6 + C_2H_4^{\cdot+}$. (The ionization energy of 1,3-butadiene is 9 eV, that of ethylene is 10 eV.)

(12-17)

Rearrangement vs. Simple Cleavage. When ions of very low internal energy are examined, as in electron impact experiments done at low ionizing energy, the importance of reactions of low activation energies that generate stable products becomes even more marked than it is for higher internal energy ions. Under low energy conditions some entropically favored processes such as simple bond cleaveages will no longer be allowed energetically, and lower energy but intrinsically slower processes, often involving molecular rearrangement, will increase in importance. The choice of low energy conditions will have marked effects on the mass spectra (compare Figure 11-10).

The conditions used for the measurement will tend to favor either simple bond cleavages or rearrangements. Most analytical applications, particularly of EI, which is emphasized here, employ ions of relatively high internal energy, for which only the simpler rearrangements need be considered. These will often involve the elimination of a stable neutral molecule (H_2O, N_2, CO_2, CO, an alkene, or an alcohol) via a simple four-, five-, or six-membered cyclic transition state. For example, diaryl carbonate molecular ions eliminate CO_2 to give ions that show all the properties of the diaryl ether molecular ions (eq. 12-18). (See Example 12-13 for another example of aryl migration.) Similarly, 1-hexanol eliminates water via a six-centered intermediate (eq. 12-19).

$$(Ar{-}O)(Ar{-}\dot{O}^{+})C{=}O \longrightarrow Ar{-}\overset{\cdot+}{O}{-}Ar + CO_2 \qquad (12\text{-}18)$$

$$H{-}O^{\cdot+}{-}CH_2{-}CH_2{-}CH_2{-}CH(H)(C_2H_5) \longrightarrow C_6H_{12}^{\cdot+} + H_2O \qquad (12\text{-}19)$$

Under low energy conditions more than one new bond can be formed, as in the double hydrogen rearrangement shown in eq. 12-20, which generates a protonated enol.

$$CH_3\overset{\overset{O^{\cdot+}}{\|}}{C}-C_5H_{11} \longrightarrow CH_3-\overset{\overset{H_2O^+}{|}}{C}=CH_2 + C_4H_7^{\cdot} \qquad (12\text{-}20)$$

Ions of High Internal Energy. It should be apparent from the internal energy distributions considered in some detail in Section 11-6a (e.g., Figure 11-4) that a small fraction of the ions generated by electron impact will have very high internal energies indeed. These ions are not well behaved and do not dissociate by familiar chemical processes; for example, they undergo fragmentations leading to $C_n^{\cdot+}$, $C_nH^{\cdot+}$, and other high enthalpy products. The fact that signals are observed at almost every mass in electron impact mass spectra is partly because of the presence of these ions that undergo indiscriminate, high energy bond cleavages.

Proximate vs. Remote Fragmentation. Most fragmentations of gas phase ions are triggered by the charge or radical site. Bonds in the vicinity of the atoms carrying low electron density (or high unpaired spin)are weakened and dissociation is promoted. In some ions, however, the charge is a spectator to dissociation, which therefore occurs by a process that resembles thermolysis. An example of such a high energy, remote fragmentation is C—C cleavage in the *n*-alkyl carboxylate anion (eq. 12-21). The reaction occurs with approximately equal facility at every bond along the chain. Similar reactions occur in quarternary ammonium cations.

$$RCH_2CH_2(CH_2)_nCO_2^- \longrightarrow CH_2=CH(CH_2)_nCO_2^- + RH \qquad (12\text{-}21)$$

Charge Localization. Much of what has been said above can be summarized by this statement: dissociation of an ion can be rationalized as proceeding from its most stable, charge-localized structure. The thermodynamically most stable structure will also have the maximum internal energy (total energy is fixed upon ionization) and hence it should also be most reactive. According to this view, the charge is associated with the group best able to stabilize it. If this group is retained in the product ion, this is also a particularly stable product, making the charge localization concept similar to that of control of fragmentation by product ion stability. A further advantage of the charge localization idea is that it forces one to do accurate electron bookkeeping in rationalizing dissociation behavior.

Consider the α,ω-amino alcohol shown in eq. 12-22. The two charge-localized forms of the molecular ion differ in internal energy by approximately 1.2 eV (the difference between typical alkylamine and alkanol ionization energies). Hence the structure with charge localized on the more electropositive atom (N) should be more reactive, no matter what the total energy of the system.

$$\overset{\cdot+}{N}H_2-(CH_2)_n-OH \nearrow \overset{+}{N}H_2=CH_2 + {}^{\cdot}(CH_2)_{n-1}OH$$

$$NH_2-(CH_2)_n-\overset{\cdot+}{O}H \searrow NH_2(CH_2)_{n-1}^{\cdot} + CH_2=\overset{+}{O}H \qquad (12\text{-}22)$$

As illustrations of the hypothesis in rationalizing fragmentation, consider reactions 12-23, 12-24, and 12-25, each of which is a major process in the normal EI mass spectrum. Note the use of single- and double-headed arrows to show the movements of single electrons and electron pairs, respectively.

$$C_2H_5-\underset{\underset{CH_3}{|}}{CH}-\overset{\cdot+}{O}-CH_3 \longrightarrow CH_3CH=\overset{+}{O}-CH_3 + C_2H_5^{\cdot} \qquad (12\text{-}23)$$

$$C_6H_5-\overset{\cdot+}{S}-CH_3 \longrightarrow C_6H_5S^+ + CH_3^{\cdot} \qquad (12\text{-}24)$$

$$\text{C}_6\text{H}_5^{\cdot+}\text{–CH(Cl)–C(=O)Br} \xrightarrow{-\text{Cl}^\bullet} \text{C}_6\text{H}_5\text{–}\overset{+}{\text{C}}\text{H–C(=O)Br} \longrightarrow \text{C}_6\text{H}_5\text{–}\overset{+}{\text{C}}\text{H–Br} + \text{CO} \qquad (12\text{-}25)$$

Metastable Ions. Interpretation of fragmentation pathways is facilitated by the observation of metastable peaks, which serve to establish parent–daughter ion relationships. Ions of low internal energy will fragment slowly, a process that can occur after full acceleration but prior to magnetic sector mass analysis. These ions will have the velocity v_1 of the parent ion (m_1^+) but the mass m_2 of the daughter ion. They therefore have a unique momentum-to-charge ratio, and in magnetic sector instruments they give rise to a signal at an apparent mass $m^* = (m_2)^2/m_1$. The positions of these metastable peaks therefore provide information on the masses of both the parent ion, m_1^+, and the daughter ion, m_2^+. For example, benzene (see Figure 11-6 for a mass spectrum) shows a metastable peak at m/z 34.7 due to the favored reaction 12-26.

$$\underset{m/z\ 78}{C_6H_6^{\cdot+}} \longrightarrow \underset{m/z\ 52}{C_4H_4^{\cdot+}} + C_2H_2 \qquad (m^* = 34.7) \qquad (12\text{-}26)$$

The process of collision-induced dissociation (eq. 11-18) has similar utility to metastable ion dissociation, but signals are greater and the fragmentation of any chosen ion can be examined (Section 12-5a).

12-6b Typical Functional Groups

The rather abstract principles just described can be reduced to practice by considering, first, the behavior of just two functional groups. The two functional groups chosen, ethers and ketones, are in many ways archtypical and their study should prepare the reader for more difficult problems. In the course of the discussion, characteristics of other functional groups will be mentioned in a comparative fashion. The examples in Section 12-8 are to be considered an integral part of this presentation.

Ethers: *Primary fragmentations.* Alkyl ethers typify functional groups containing singly bonded heteroatoms. Consider first their EI mass spectra. *Primary fragmentations* are (1) α-cleavage, (2) C—O cleavage, and (3) simple H rearrangement.

Figure 12-14 shows the spectrum of ethyl 1-phenylethyl ether, which displays ions due to each of these primary reactions of the molecular ion.

1. α-Cleavage, i.e., cleavage of the first C—C bond counting from the functionality, gives rise to m/z 135, the base peak in the spectrum.
2. C—O cleavage with charge retention by the hydrocarbon and formation of an oxy radical produces m/z 105.
3. The third primary reaction is represented by the rearrangement process in which ethane is lost and the low abundance ion, m/z 120, is produced.

These processes are typical of ethers, although the charge-stabilizing character of the aryl ring has evident effect on ion abundances.

If α-cleavage can occur at several alternative positions, the process leading to the most stable carbenium ion will occur preferentially. If different radicals can be lost by α-cleavage, then the more stable radical (tertiary > secondary > primary) will preferentially be lost, provided the rule concerning product ion stabilities is not contradicted. Thus ethyl isobutyl ether loses the $C_3H_7^{\cdot}$ radical to give an ion of 8% relative abundance and the $CH_3^{\cdot}$ radical to give an ion of less than 1% relative abundance. If formation of a stable ion and formation of a stable radical are competitive processes, then ionic stabilization is more important. Thus, isobutyl isopropyl ether loses a secondary propyl radical to give ion **12-1** to a smaller extent than it loses a

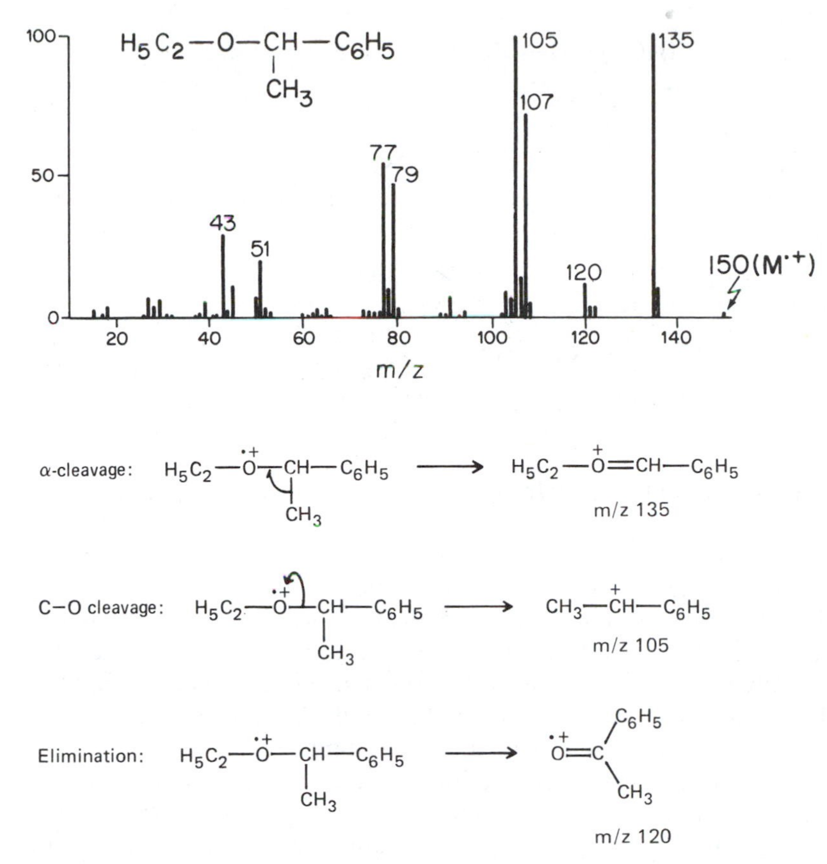

FIGURE 12-14 EI (70 eV) mass spectrum and primary fragmentations of ethyl 1-phenylethyl ether. [Spectrum from E. Stenhagen, S. Abrahamsson, and F. W. McLafferty (Eds.), *Atlas of Mass Spectral Data*, John Wiley & Sons, New York, 1969.]

primary methyl radical to give ion **12-2**. Each of these generalizations covers other functional groups as well as ethers.

$$\overset{+}{C}H_2—O—i\text{-Pr} \leftrightarrow CH_2{=}\overset{+}{O}—i\text{-Pr} \qquad i\text{-Bu}—O—\overset{+}{C}H—CH_3 \leftrightarrow i\text{-Bu}—\overset{+}{O}{=}CH—CH_3$$

12-1 **12-2**

In cases where alternative fragmentation reactions give radical and ionic products that have identical degrees of substitution, ion abundance is controlled by the size of the alkyl radical lost: the loss of the larger alkyl radical gives the more abundant fragment ion in 70 eV spectra (compare eq. 12-7). The operation of this rule is seen in the fact that ionized butyl propyl ether fragments by loss of the larger radical ($C_3H_7^{\cdot}$) to give an ion of 24% relative abundance, while loss of the smaller ethyl radical yields an ion of 13% relative abundance.

Secondary fragmentations. Secondary fragmentations of ethers occur because some primary fragment ions retain sufficient internal energy to dissociate further. For the α-cleavage products, formation of an even-electron product ion with elimination of an even-electron neutral will be favored on energetic grounds (see eq. 12-10). The tendency to form the most stable product possible will usually mean retention of the charge on the heteroatom. As so often happens when these two considerations obtain, a neutral hydrocarbon molecule is lost. Thus, the α-cleavage product, m/z 135, in ethyl 1-phenylethyl ether (Figure 12-14) fragments further by loss of ethylene to yield the abundant ion of mass m/z 107. Similarly, ethyl isobutyl ether forms an ion with m/z 31 of 76% relative abundance, by the sequence of reactions shown in eq. 12-27. As an alternative mode of secondary fragmentation, neutral formaldehyde may be eliminated in the process shown in eq. 12-15. Naturally, this pathway will be most important when R^+ is a stable ion, such as benzyl cation. (Direct formation of R^+ from the molecular ion also occurs.)

$$
\begin{array}{c}
CH_3CH_2\overset{\bullet +}{O}CH_2CH(CH_3)_2 \ (m/z\ 102) \xrightarrow{-C_3H_7\bullet} CH_3CH_2-\overset{+}{O}=CH_2 \ (m/z\ 59) \xrightarrow{-C_2H_4} H\overset{+}{O}=CH_2 \ (m/z\ 31) \\
CH_3CH_2\overset{\bullet +}{O}CH_2CH(CH_3)_2 \ (m/z\ 102) \xrightarrow{-CH_3\bullet} CH_2=\overset{+}{O}-CH_2-CH(CH_3)_2 \ (m/z\ 87) \xrightarrow{-C_4H_8} H\overset{+}{O}=CH_2 \ (m/z\ 31)
\end{array}
\quad (12\text{-}27)
$$

Those secondary fragment ions that retain enough energy for further fragmentation will again expel a neutral molecule. Thus $CH_2{=}\overset{+}{O}{-}H$ (m/z 31) loses H_2 to give $H{-}C{\equiv}O^+$ (m/z 29). Usually, however, it is unnecessary to consider these higher order fragmentations because of the time and energy limitations imposed upon ionic fragmentation and because the low mass region of the spectrum provides little specific information on molecular structure. Nevertheless, low mass ions, especially homologous series of ions, can indicate the type(s) of *functional groups* present. Some of this information is summarized in Table 12-5. Note, however, that mass spectra are usually interpreted by considering the *neutral fragments* that are lost from the molecular ion. In other words, the information-rich region of a mass spectrum is the high and not the low mass region. Table 12-6 lists some neutral losses and their structural implications.

Returning to our discussion of ethers, we note that the relative importance of ions with and without heteroatoms in ethers is intermediate between that for compounds containing more electropositive heteroatoms such as nitrogen, where hydrocarbon ions are of low total abundance (the fragmentation patterns of amines therefore resemble those of ethers, but are simpler), and less electropositive atoms such as the halogens (but also sulfur), where there is a strong tendency to form hydrocarbon ions since fragmentations are less strongly directed by the heteroatom. (See Examples 12-2 and 12-29 for further discussion of α-cleavage.) Naturally, C—O cleavage is most significant when a tertiary, benzylic, allylic, or other stabilized carbocation can be formed. Further fragmentation of the resulting cation follows a pattern typical of the spectra of alkanes and other species in which alkyl cations are formed: H_2 loss is ubiquitous, and CH_4 and larger alkane losses also can be observed. Highly unsaturated alkyl cations like $C_7H_7^+$ tend to lose carbon-containing fragments (eq. 12-28) rather than H_2. The spectrum

TABLE 12-5 Some Characteristic Ions

Mass	Ion	Possible Functionality
15	CH_3^+	methyl, alkane
29	$C_2H_5^+$, HCO^+	alkane, aldehyde
30	$CH_2{=}NH_2^+$	amine
31	$CH_2{=}OH^+$	ether or alcohol
39	$C_3H_3^+$	aryl
43	$C_3H_7^+$, CH_3CO^+	alkane, ketone
45	CO_2H^+, CHS^+	carboxylic acid, thiophene
47	CH_3S^+	thioether
50	$C_4H_2^{\cdot +}$	aryl
51	$C_4H_3^+$	aryl
77	$C_6H_5^+$	phenyl
83	$C_6H_{11}^+$	cyclohexyl
91	$C_7H_7^+$	benzyl
105	$C_6H_5C_2H_4^+$	substituted benzene
	$CH_3{-}C_6H_4CH_2^+$	disubstituted benzene
	$C_6H_5CO^+$	benzoyl

TABLE 12-6 Some Characteristic Neutral Losses

Mass	Composition	Possible Functionality
14	impurity, homologue	
15	CH_3	methyl
16	CH_4	methyl
	O (rarely)	nitrogen oxide
	NH_2	amide
17	NH_3	amine (CI)
	OH	acid, tert. alcohol
18	H_2O	alcohol, aldehyde, acid (CI)
19	F	fluoride
20	HF	fluoride
26	C_2H_2	aromatic
27	HCN	nitrile, heteroaromatic
28	CO	phenol
	C_2H_4	ether
	N_2	azo
29	C_2H_5	alkyl
30	CH_2O	methoxy
	NO	aromatic nitro
	C_2H_6	alkyl (CI)
31	CH_3O	methoxy
32	CH_3OH	methyl ester
33	$H_2O + CH_3$	alcohol
	HS	mercaptan
35	Cl	chloro compound
36	HCl	chloro compound
42	CH_2CO	acetate
43	C_3H_7	propyl
44	CO_2	anhydride
46	NO_2	aromatic nitro
50	CF_2	fluoride

$$\underset{m/z\ 91}{C_7H_7^+} \rightarrow \underset{m/z\ 65}{C_5H_5^+} + C_2H_2 \qquad \searrow \underset{m/z\ 39}{C_3H_3^+} + C_2H_2 \tag{12-28}$$

of ethyl 1-phenylethyl ether (Figure 12-14) shows many features from hydrocarbon ions. For example, the 1-methylbenzyl cation (m/z 105) decomposes via loss of C_2H_4, and then C_2H_2, to give m/z 77 and 51, respectively.

Rearrangements. Some bond-forming reactions, particularly simple eliminations occurring in the molecular ion, are of value in characterizing the analyte. These processes often give rather low abundance ions in the 70 eV mass spectra, but become more important at lower electron energies. In ethers there are two such reactions: loss of an alkene and loss of an alcohol. The reactions are complementary; only the location of the charge is changed (eq. 12-29). Retention of the charge by the alcohol, the species with the lower ioniza-

$$\text{H—C—C—}\overset{\cdot +}{\text{O}}\text{—R} \rightarrow \begin{cases} \text{C=C} + \text{ROH}^{\cdot +} \\ \text{C=C}^{\cdot +} + \text{ROH} \end{cases} \tag{12-29}$$

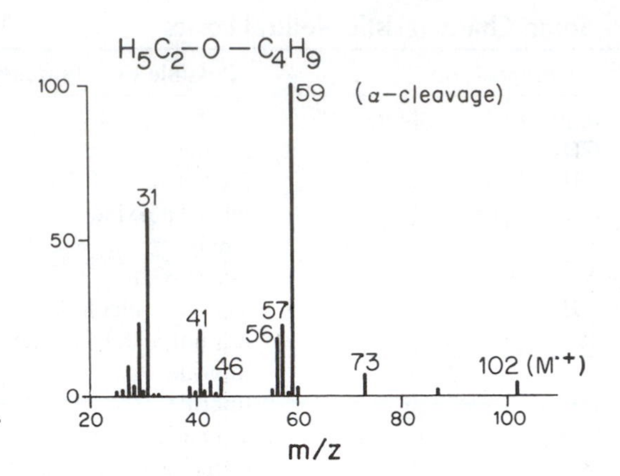

FIGURE 12-15 EI (70 eV) mass spectrum of *n*-butyl ethyl ether.

tion energy, is favored. This result can be thought of as a consequence of the charge localization hypothesis and is known as *Stevenson's rule* (see above). Both reactions are evident in the butyl ethyl ether spectrum (Figure 12-15), although the ion at *m/z* 56, composition $C_4H_8^{\cdot+}$, as expected, is more abundant than $C_2H_5OH^{\cdot+}$ at *m/z* 46. (Ionization energies are 9.2 and 10.5 eV, respectively.)

The species ROH_2^+ is another fragment ion that should be mentioned, not because it is abundant in 70 eV spectra of ethers (it is usually barely detectable), but because it is the dominant fragment ion at low electron energies. Obviously, several bonds must be cleaved to form this ion, and the absolute value of the entropy of activation is large. On the other hand, the ion is very stable. Similar stable ions formed by multiple hydrogen transfers occur for other functional groups, including $RCO_2H_2^+$ for alkyl esters (compare also eq. 12-20). The spectrum of dihexyl ether taken at 70 and 12 eV is given in Figure 12-16. The 70 eV spectrum is dominated by formation and further fragmentation of the alkyl cation (*m/z* 85) and of the ionized alkene (*m/z* 84). Oxygen-containing ions are almost absent. At low energy further fragmentation of *m/z* 84 and 85 is virtually eliminated, but the relative abundances of the molecular ion and the ROH_2^+ rearrangement ion (*m/z* 103) increase dramatically.

Aryl ethers. Alkyl aryl ethers differ from dialkyl ethers in that they typically show molecular ions of much greater abundance. The stabilizing role of the aryl group is a general phenomenon. Butyl phenyl ether can be compared to dibutyl ether for purposes of illustration. The molecular ion abundance of the aromatic compound is 20%, that of the dialkyl ether <1%. Aryl methyl ethers differ from dialkyl ethers in two of their most important primary fragmentations. One of these reactions is loss of the methyl radical to give the aroxy

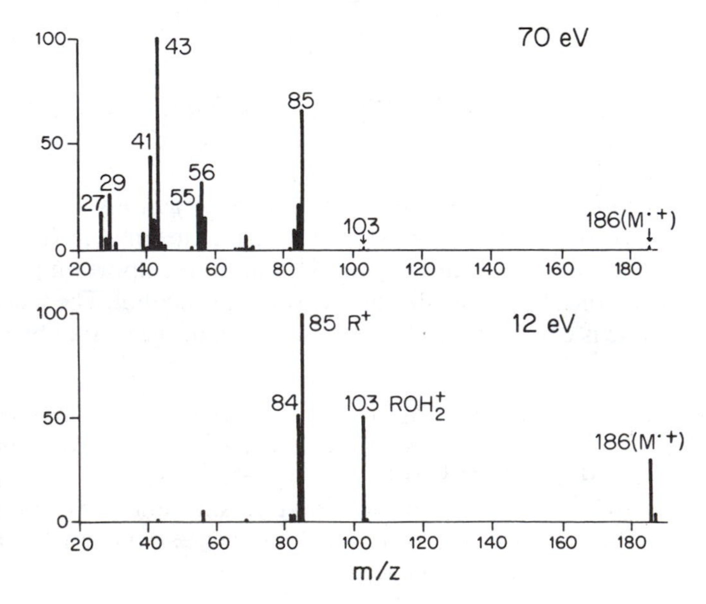

FIGURE 12-16 High energy (70 eV) and low energy (12 eV) EI mass spectra of di-*n*-hexyl ether. [From G. Spiteller and M. Spiteller-Freidmann, in R. Bonnett and J. G. Davis (Eds.), *Some Newer Physical Methods in Structural Chemistry*, 1967.]

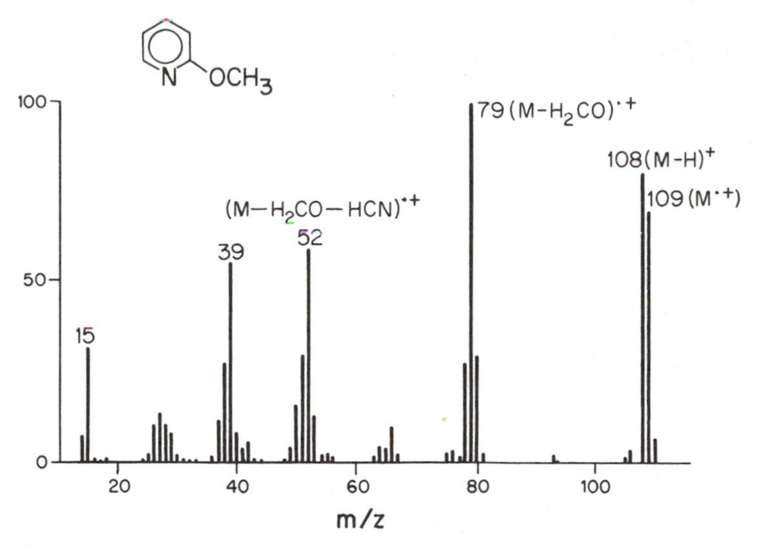

FIGURE 12-17 EI (70 eV) mass spectrum of 2-methoxypyridine. [From E. Stenhagen, S. Abrahamsson, and F. W. McLafferty (Eds.), *Atlas of Mass Spectral Data*, John Wiley & Sons, New York, 1969.]

ion (ArO^+), a far more stable species than are alkoxy cations. Further fragmentation of this even-electron cation occurs by elimination of CO. The second primary fragmentation—loss of the heteroatom as formaldehyde to give the molecular ion of the parent aromatic compound (eq. 12-30)—is an illustration of the fact that *simple* elimination reactions can lead to relatively abundant ions in 70 eV spectra.

$$\left[C_6H_5OCH_3 \right]^{\cdot +} \longrightarrow \left[C_6H_6 \right]^{\cdot +} + H_2CO \qquad (12\text{-}30)$$

The arene fragment ion undergoes the same further fragmentation reactions as do $ArH^{\cdot +}$ molecular ions formed by direct ionization. The subsequent reactions (ignoring those due to substituents that may be present) are $H^{\cdot}$ loss, C_2H_2 loss, and $C_3H_3^{\cdot}$ loss. The spectrum of 2-methoxypyridine (Figure 12-17) shows the fragmentation sequence $M^{\cdot +} \rightarrow (M-H_2CO)^{\cdot +} \rightarrow (M-H_2CO-HCN)^{\cdot +}$, in which the $(M-H_2CO)^{\cdot +}$ ion apparently has the same structure as the pyridine molecular ion. This spectrum is also important because it provides our first example of an *ortho effect*. The molecular ion loses $H^{\cdot}$ to a far greater extent than do most methyl ethers; here $(M-1)^+ > M^{\cdot +}$, whereas in anisole $(M-1)^+$ is 3% of $M^{\cdot +}$. Direct interaction of the amino and methoxy groups occurs, forming a unique stable ion, perhaps **12-3**. In benzene itself (as in anisole), frag-

$\overset{+}{N}$ O
CH_2

12-3

ment ions occur at m/z 77, 52, and 39. The benzene molecular ion also shows losses of two and three hydrogen atoms to give m/z 75 and 76, further fragmentation of m/z 77 to give m/z 51, and of m/z 76 to give m/z 50, as well as other minor reactions.

Another primary reaction of substituted anisoles, this one also observed in diaryl ethers, is loss of the alkoxyl radical. This process is important because the ion Ar^+ (m/z 77 in phenyl ethers) provides a means of recognizing the aromatic group.

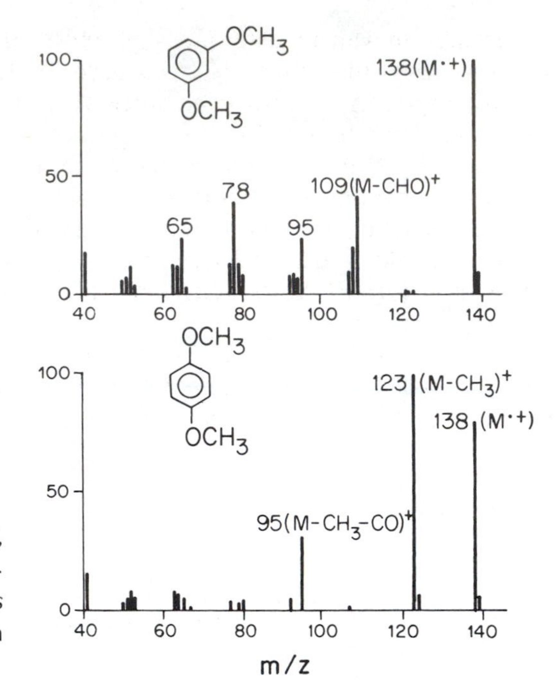

FIGURE 12-18 Contrasting mass spectra (70 eV, EI) of isomeric dimethoxybenzenes. [From H. Budzikiewicz, C. Djerassi, and D. H. Williams, *Mass Spectrometry of Organic Compounds*, Holden Day, San Francisco, 1967.]

In certain substituted anisoles the loss of the formyl radical occurs to yield the species ArH_2^+. Why should the ion undergo a double hydrogen transfer in order to yield this product? As always, product stability is the answer. The process is restricted to anisoles that bear oxygen- or nitrogen-containing substituents to which hydrogen can be transferred. It is also restricted to anisoles in which the competitive methyl elimination process is not dominant. These points are well illustrated by comparing the mass spectra of the *m*- and *p*-dimethoxybenzenes (Figure 12-18). The $(M-CH_3)^+$ ion in the para isomer is resonance stabilized, so that it gives the base peak in the spectrum (37% of the total ion current); formyl radical loss is not observed. In the meta isomer, methyl radical loss barely occurs, giving rise to an ion carrying <1% of the total ion current; formyl radical loss is the dominant fragmentation with 11% of the total ion current. The formyl loss product probably has either of the stable structures **12-4** or **12-5** (eq. 12-31). Either pathway has a requirement of a

OCH3 ·+ OCH3 $\xrightarrow{-CHO^\bullet}$ CH3 O+ H or CH3 O+ H H (12-31)

12-4 **12-5**

second heteroatom, which accounts for the absence of this fragmentation in anisole itself. This process is comparable to ROH_2^+ formation from long chain ethers (compare also eq. 12-20)

The third isomer, *o*-dimethoxybenzene, exhibits a mass spectrum that is different from both the others. In particular, interaction of the ortho substituents occurs. Again loss of $CHO^\cdot$ from the molecular ion does not occur. Therefore, in this particular case mass spectrometry can be used to distinguish isomeric disubstituted benzenes. Some other isomeric substituted anisoles give identical mass spectra. Once again we have encountered a general situation that is far more widely applicable than just to this class of compounds. Simple EI mass spectrometry serves to identify isomeric substituted benzenes in perhaps half of all cases.

The fragmentation behavior of a disubstituted aromatic compound will frequently be dominated by one group, and substituents can be ranked in order of their reactivity (Table 11-13). This order also applies to monosubstituted benzenes, in which it determines the extent of fragmentation relative to the molecular ion abundance. Table 11-13 gives this crude reactivity order and also the major reaction of each functional group. The activation energies (AE–IE) for each process are seen to correlate inversely with the reactivity of the substituent, as expected (Section 11-6a).

Turning from anisoles to other alkyl aryl ethers, one finds that alkene elimination to give the ionized phenol is the dominant reaction. Diaryl ethers are notable for their stable molecular ions and for the occurrence of a complex skeletal rearrangement resulting in CO elimination (see Example 12-12). While this particular rearrangement could hardly have been predicted, guidelines have been recognized (see above) that rationalize most skeletal rearrangements. Let us not forget, however, that these are minor processes in analytical 70 eV spectra.

Chemical ionization. The behavior of ethers under chemical ionization conditions can now be considered. The main expectation is that simple elimination reactions will be significant. For example, the loss of a neutral alcohol molecule gives the carbenium ion (eq. 12-32) and alkene elimination gives the protonated

$$R_1{-}\overset{+}{\underset{H}{O}}{-}R_2 \rightarrow R_1OH + R_2^+ \qquad (12\text{-}32)$$

alcohol (eq. 12-33). Both reactions follow the even-electron → even-electron rule (eq. 12-10), as does alkane

$$R_1{-}\overset{+}{\underset{H}{O}}{-}R_2 \rightarrow R_1OH_2^+ + (R_2{-}H) \qquad (12\text{-}33)$$

loss, a formal 1,2-elimination process that is important in some ethers. In addition, alkyl radical loss, an electron unpairing reaction, is observed, especially under high energy conditions (Table 12-7).

TABLE 12-7 Reactions of Protonated *i*-Butyl Ethyl Ether

Reaction	Reactant	Products
Alkyl loss	$H_9C_4^i{-}\overset{+}{O}(H){-}C_2H_5$	$\longrightarrow C_4H_9OH^{\cdot +} + C_2H_5^{\cdot}$
Alkane loss	$H_7C_3^s{-}CH(H){-}\overset{+}{O}(H){-}C_2H_5$	$\longrightarrow C_3H_7CH{=}\overset{+}{O}H + C_2H_6$
Alkene loss	$CH_3{-}C(CH_3)(H){-}CH_2{-}\overset{+}{O}(H){-}C_2H_5$	$\longrightarrow C_2H_5{-}\overset{+}{O}H_2 + C_4H_8$
Alcohol loss	$H_9C_4^i{-}\overset{+}{O}(H){-}C_2H_5$	$\longrightarrow C_2H_5^+ + C_4H_9OH$
	$H_9C_4^i{-}\overset{+}{O}(H){-}C_2H_5$	$\longrightarrow C_4H_9^+ + C_2H_5OH$

From M. L. Sigsby, R. J. Day, and R. G. Cooks, *Org. Mass Spectrom.*, **14**, 278 (1979).

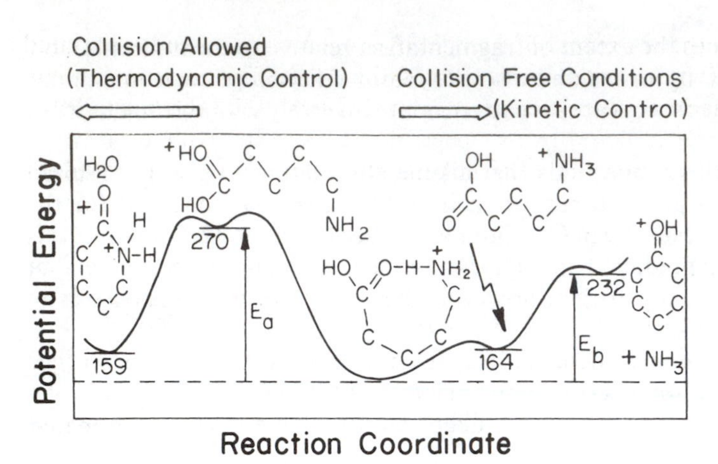

FIGURE 12-19 Potential energy surface (kJ mol^{-1}) for H_2O and NH_3 loss from protonated 5-aminopentanoic acid. [From V. H. Wysocki, D. J. Burinsky, and R. G. Cooks, *J. Org. Chem.*, **50**, 1287 (1985). Copyright 1985 American Chemical Society.]

In interpreting CI spectra it is important to recognize that the molecules are not isolated. Rather, the neutral product of unimolecular dissociation may have a higher proton affinity than the ionic product, and the charge might be transferred in a subsequent exothermic reaction. This principle is nicely illustrated by the CI mass spectrum of 5-aminopentanoic acid. There are only two primary reactions of significance, loss of H_2O and loss of NH_3 (eq. 12-34). The latter is favored on kinetic grounds (Figure 12-19) and is the dominant product when *isolated* protonated molecules fragment (namely, in MS–MS, Section 12-5b). However, the normal CI mass spectrum shows almost exclusive formation of the more stable protonated lactam.

$$O{=}C(-HO\cdots H^+\cdots NH_2)-CH_2-CH_2-CH_2-CH_2 \quad \text{(12-34)}$$

Product	Mass spectrum	MS-MS spectrum
$-NH_3$ → HO^+ ring (O)	1	5
$-H_2O$ → O ring ($\overset{+}{N}H_2$)	5	1

Ketones. Ketones are representative of the second major functional type, those possessing a multiple bond to a heteroatom. Once again EI mass spectra are emphasized. A major primary process is acylium ion formation, and the rule regarding large vs. small radical loss is applicable here too. In other words, R_LCOR_S will give R_SCO^+ with the greater abundance at 70 eV, even though R_LCO^+ has a slightly lower appearance potential. If R_S is secondary and R_L primary, then of course this factor becomes dominant. Further fragmentation of the acylium ions is facile, and the ions $R_L{}^+$, $R_S{}^+$, R_LCO^+, R_SCO^+ are easily recognized in almost all ketone spectra.

The second major primary fragmentation is the six-membered cyclic hydrogen transfer and associated γ-cleavage leading to loss of an alkene molecule. The reaction is termed the *McLafferty rearrangement.* Although some hydrogen exchange occurs in ketone molecular ions even at 70 eV, the rearrangement is highly specific. It is by far the most studied rearrangement in mass spectrometry, and the evidence favors a stepwise reaction initiated by the radical site on oxygen (eq. 12-35). The product ions are unusually abundant

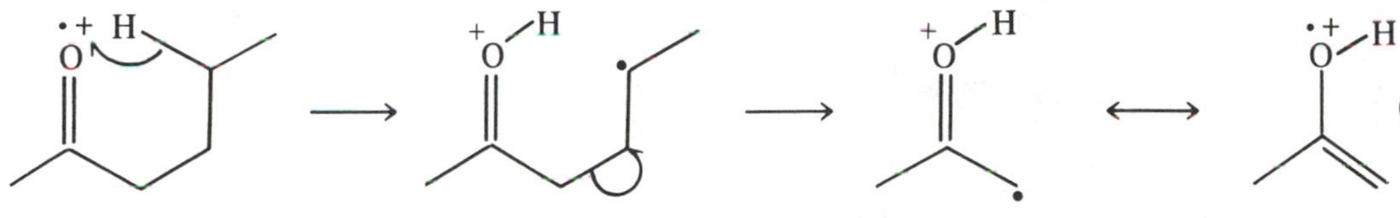

(12-35)

for a rearrangement fragmentation. The McLafferty rearrangement product is the enolic species shown. (Evidence for the structure comes from isotopic labeling and metastable ion studies.) Further fragmentation of the ion product occurs by α-cleavage, after ketonization, or by alkene elimination.

There are several other notable features of the McLafferty rearrangement.

1. A methyl group does not rearrange when substituted for hydrogen.
2. A formally analogous process can occur in even-electron ions.
3. Related reactions occur in other C=X systems, for example, oximes.
4. Secondary hydrogen atoms in the γ position are more readily abstracted than are primary.
5. If both alkyl groups can undergo the McLafferty rearrangement, the more abundant product is due to reaction from the larger group. In addition, the product of two successive alkene eliminations (double McLafferty rearrangement) will usually appear as an abundant ion.

Many of the major features of ketone mass spectra are illustrated in the spectra of butyrophenone (Figure 12-20) and 4-nonanone (Figure 12-21).

The butyrophenone spectrum is very simple, the base peak m/z 105 ($C_6H_5CO^+$) being formed by methyl radical loss from the McLafferty rearrangement product, m/z 120, as well as by α-cleavage in the molecular ion. It is interesting to note that methyl loss involves the CH_2 group of the enol ion and an ortho ring hydrogen, *not* the hydroxyl hydrogen. In other words, a hydrogen atom is abstracted via a five- rather than a

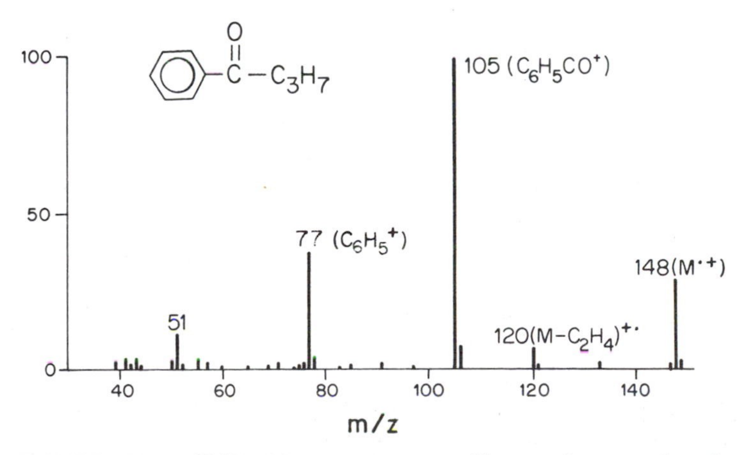

FIGURE 12-20 EI (70 eV) mass spectrum of butyrophenone (phenyl *n*-propyl ketone).

four-membered cyclic transition state (eq. 12-36). The acetophenone spectrum (Figure 11-8) is even simpler, being dominated by the benzoyl ion (m/z 105) and its fragmentation products m/z 77 and 51.

$$m/z\ 148,\ M^{\cdot+} \xrightarrow{-C_2H_4} m/z\ 120 \longrightarrow \xrightarrow{-CH_3^{\cdot}} C_7H_5O^+\ (m/z\ 105) \qquad (12\text{-}36)$$

4-Nonanone fragments by α-cleavage to give the acylium ions at m/z 71 and 99. These ions fragment further by CO loss to give the alkyl cations at m/z 43 and 71 (the latter is composed of both $C_5H_{11}^+$ and $C_3H_7CO^+$). Formation of the alkyl cations directly from the molecular ion is unimportant. Of the three possible odd-electron ions due to McLafferty rearrangement, that associated with elimination of the smaller alkene is not observed (loss of C_2H_4 would give m/z 114), but loss of C_4H_8 gives the ion at m/z 86 and the double McLafferty rearrangement gives m/z 58. Both of these ions fragment further by α-cleavage and then CO elimination, and so contribute to the acylium and alkyl ions already mentioned. The only other ion that appears in the spectrum besides hydrocarbon fragments at m/z 55, 41, etc., is a low abundance ion, m/z 113, from loss of $C_2H_5^{\cdot}$ from the molecular ion. This ion comes from a rearrangement reaction and probably has the stable structure **12-6**, although a cyclized structure cannot be excluded. Although not evident in the 70 eV

$$C_3H_7{-}C({=}\overset{+}{O}H){-}CH{=}CH{-}CH_3$$

12-6

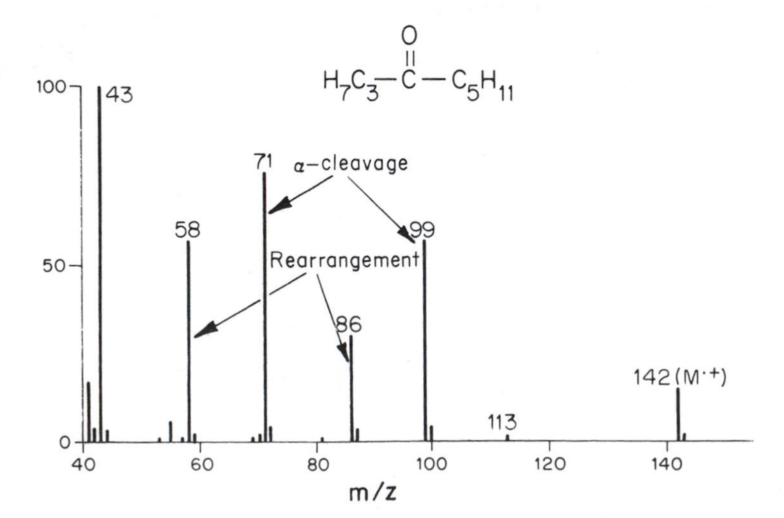

FIGURE 12-21 EI (70 eV) mass spectrum of 4-nonanone. [From G. Eadon, C. Djerassi, J. H. Beynon, and R. M. Caprioli, *Org. Mass Spectrom.*, **5**, 917 (1971).]

mass spectrum of 4-nonanone, several other low energy rearrangement processes do appear in ketone spectra. One of these is analogous to the McLafferty rearrangement, except that two hydrogen atoms are transferred to the carbonyl group, probably to give the stable ion **12-7**. It will be recognized that this is analogous to ROH_2^+ formation from ethers (see eq. 12-20).

$$C_3H_7{-}\overset{\overset{+}{O}H_2}{\overset{|}{C}}{=}CH_2$$

12-7

Thus these rather complex rearrangements do not interfere with the analytical applications of mass spectrometry, nor are they haphazard in their occurrence. Rather, they are controlled by product ion stability, and they increase the structural information available in a mass spectrum, particularly when low energy metastable reactions are examined.

Cyclic ketones show behavior that is typical of other substituted cycloalkanes. They undergo α-cleavage but, because of the cyclic nature of the ion, this process leads to an isomeric form of the molecular ion rather than directly to a fragment ion. Hydrogen rearrangement is then often followed by loss of a methyl radical. Related reactions are illustrated in Examples 12-17 and 12-19 (Section 12-8).

The chemical ionization mass spectra of ketones (proton transfer reagents) show alkane and alkene eliminations analogous to those observed for ethers. For example, diisopropyl ketone fragments via loss of propane or of a propyl radical. Note that fragment ion abundances in CI depend on the nature and pressure of the reagent gas (compare Figure 12-4). The figure in Example 12-18 (Section 12-8) illustrates the CI fragmentation of a substituted benzophenone.

12-6c Other Functional Groups

This section summarizes the fragmentation behavior of several functional groups. The ketone and ether discussions, above, should be read first to understand the reasons for particular reactions. Strict rules are no more possible for mass spectral fragmentation than they are for describing solution phase reactions. What follows are therefore only guidelines. More complex compounds are dealt with in Section 12-8. All data are for EI unless otherwise indicated.

Aliphatic Hydrocarbons. These compounds show numerous intense, low mass ions, but the molecular ions have low abundance (even in CI where $(M-H)^+$ is common; see Example 12-19). Characteristic ions are separated by 14 d in $C_nH_{2n+1}^+$ and $C_nH_{2n-1}^+$ series. Isomers, including positional isomers of alkenes, are often difficult to distinguish (extensive carbon and hydrogen rearrangement is common). See Figure 12-16 for typical hydrocarbon features, including the homologous series 27, 41, 55, 69, 83, ... and 29, 43, 57, 71, 85, ... Cyclic compounds often show retro Diels–Alder fragmentation, as seen in eq. 12-17.

Aromatic Hydrocarbons. Abundant molecular ions and a relatively few abundant fragment ions dominate the spectra, especially for CI and DI. Fragmentation is β to the ring, e.g., to give $C_7H_7^+$ from alkyl-substituted benzenes. Loss of $H^{\cdot}$ and H_2 from $M^{\cdot+}$ is common, as is loss of C_2 fragments (especially C_2H_2) and C_3 fragments (see the benzene spectrum in Figure 11-5). Characteristic ring cleavage ions occur at m/z 50 and 51 for substituted benzenes. Table 11-13 illustrates the more important substituent cleavage modes of substituted aromatics. Some rearrangement ions occur, e.g., $C_7H_8^{\cdot+}$ (m/z 92) in alkylbenzenes. Figures 12-17 and 12-18 display some of these characteristics. Problem 12-1 is a typical spectrum of an aromatic compound, and Examples 12-8, 12-12 and 12-16 also cover aromatic compounds.

Halides. Weak molecular ions with loss of the halogen to give carbenium ions and hydrocarbon-like spectra are typical of bromides and iodides. Chlorides lose HCl to enter the alkene manifold. Carbon–carbon bond cleavage is rare (even α-cleavage) except when γ- and δ-cleavage can lead to cyclic halonium ions, which are then favored. Aryl chlorides and bromides show facile loss of the halide, iodides fragment less readily, and fluorides very little by this route. Fluorinated compounds undergo unique reactions including CF_2 loss with F rearrangement (CH_2 loss rarely occurs) and formation of fluorinated carbenium ions. Acid halides undergo ready halogen loss (eq. 12-12). Many halogenated compounds are readily recognized by their characteristic isotopic signature, a feature evident in Problems 12-2, 12-4, and 12-9.

Alcohols. Molecular ions are of only moderate abundance. Characteristic fragmentation is α-cleavage to form oxonium ions (m/z 31, 45, 59, ...). In addition, tertiary alcohols tend to undergo $HO^{\cdot}$ loss, while primary alcohols eliminate H_2O (eq. 12-18). Further fragmentation follows the behavior of aliphatic hydrocarbons (R^+) and alkenes $(R-H)^{\cdot+}$. Phenols eliminate CO and $HCO^{\cdot}$ as their characteristic fragmentations. Negative ion spectra give abundant $(M-H)^-$ ions for alcohols and phenols. The 1-propanol EI spectrum (Figure 11-11) illustrates some of these features.

Aldehydes. Aldehydes display moderate $H^{\cdot}$ loss to yield stable acylium ions. In addition, McLafferty rearrangements occur. (Both reactions parallel ketone behavior.) An additional primary fragmentation is dehydration, which gives abundant fragment ions.

Carboxylic Acids. Carboxylic acids are best examined through their negative ion spectra, where $(M-H)^-$ is dominant (e.g., by CI using HO^- as reagent). Electron impact generates positive molecular ions of moderate abundance. Fragmentation by $HO^{\cdot}$ loss is not facile except in aromatic acids, nor is H_2O loss a general process. The McLafferty rearrangement gives abundant fragments, as does C—C cleavage some distance from the functionality, e.g., δ-cleavage. Aromatic acids yield benzoyl ions by $HO^{\cdot}$ loss, while ortho effects can result in dehydration. See Example 12-1 (Section 12-8) for further discussion of a particular aromatic acid.

Esters. Molecular ions are of moderate abundance. Fragmentation of the ether oxygen substituent is by McLafferty rearrangement (here O—C cleavage) to give the ionized carboxylic acid or, commonly, by a double hydrogen rearrangement to yield the protonated carboxylic acid (see Figure 12-16 for an analogous double hydrogen transfer). Fragmentation via the substituent on the carbonyl group is by McLafferty rearrangement (β-cleavage, with H transfer, to give m/z 74 in methyl esters) and by simple δ and higher cleavages (m/z 87, 101, ... in methyl esters). Acylium ion formation is also observed. In esters of aromatic acids this is a dominant process. The CI mass spectra of two esters are discussed in Examples 12-20 and 12-21 (Section 12-8) while Problem 12-6 presents EI data.

Amines. As already noted in connection with ethers, amines fragment predominantly by α-cleavage. Characteristic ions are observed in a series m/z 30, 44, 58, ... Aromatic amines (aniline and pyridine types) show HCN loss as a significant cleavage mode. Under CI conditions, protonated amines can lose NH_3. Other reactions typical of protonated and quaternary amines are discussed in Section 12-6e. Example 12-8 and Problem 12-1 provide more information on the EI spectra of aromatic nitrogen compounds. Example 12-14 illustrates the dominance of α-cleavage in the spectrum of an amino acid, and Example 12-29 follows the fragmentation behavior of a cyclic amine.

Amides. This group of compounds shows good molecular ion intensity. The C—N bond resists cleavage except in *N*-aryl amides, where it is accompanied by hydrogen rearrangement. Fragmentation is by α-cleavage to nitrogen rather than to carbonyl. McLafferty rearrangement in the carbonyl substituent is facile.

12-6d Negative Ion Fragmentations

For appropriate compounds, *negative ions* are generated in large quantities, especially in CI and DI experiments (see Section 11-5b). They can be generated by electron attachment (eq. 11-3), a gas phase resonance process that yields a radical anion, $M^{\cdot-}$. Anion radicals show similarities in their fragmentation behavior to that of radical cations. For example, ketone anion radicals undergo α-cleavage to yield acyl anions RCO^-, esters fragment to yield RCO_2^-, and ethers yield alkoxy or aryloxy anions. Of special interest are the highly electronegative functional groups present in aromatic nitro, phosphorus, and sulfur compounds. Electron capture in these cases leads to the formation of very reactive radicals that undergo hydrogen atom abstraction and initiate quite complex fragmentation reactions. For example, *o*-nitroacetanilide undergoes HO˙ loss as a prominent process (eq. 12-37). The ortho effect, evident here, is of course well established in positive ions too.

NO_2, $NHCOCH_3$ $\xrightarrow{+e^-}$ O^-, N, O˙, H, N, $COCH_3$ $\longrightarrow$

O^-, N, OH, $NCOCH_3$ $\xrightarrow{-\dot{O}H}$ O^-, N, N—C—CH_3, O (12-37)

Conjugate anions of acidic compounds, $(M-H)^-$, are commonly generated by CI and DI. These even-electron anions tend to fragment by simple elimination processes, as illustrated by the negative ion DI (SIMS) spectrum of nicotinamide (Figure 12-22), which shows loss of HCN and HNCO from $(M-H)^-$ to yield m/z 94 and 78, respectively. Other examples of negative ion fragmentation already encountered include remote fragmentations of carboxylates (eq. 12-21). Example 12-11 illustrates negative ion fragmentations in anisole.

12-6e Fragmentation in Desorption Ionization

Fragmentation in desorption ionization is largely due to gas phase processes resembling those that occur in other ionization methods. This phenomenon is illustrated by the nicotinamide spectrum (Figure 12-22) and by the behavior of the quaternary alkaloid candicine, which has been examined in the solid state and from

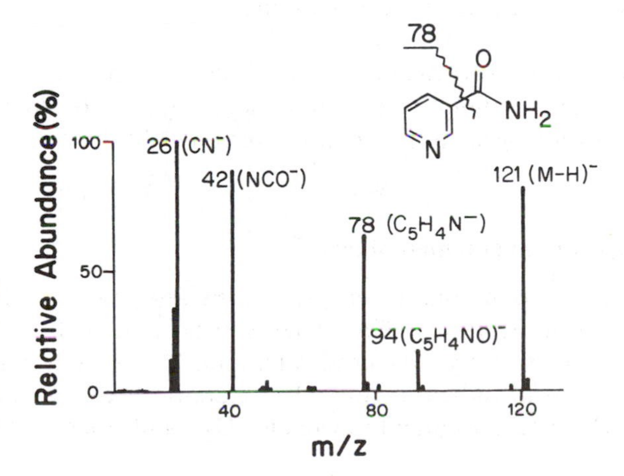

FIGURE 12-22 Negative ion (DI) spectrum of nicotinamide (Ar^+ bombardment, 5 keV, solid sample). [From S. E. Unger, R. J. Day, and R. G. Cooks, *Int. J. Mass Spectrom. Ion Phys.*, **39**, 231 (1981), as redrawn in J. H. Bowie, *Mass Spectrom. Rev.*, **3**, 161 (1984), Fig. 16.]

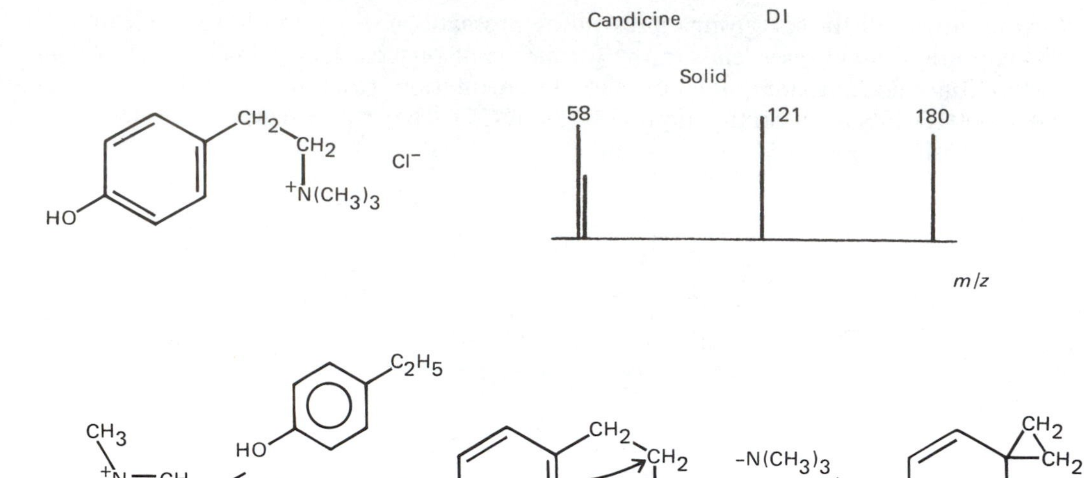

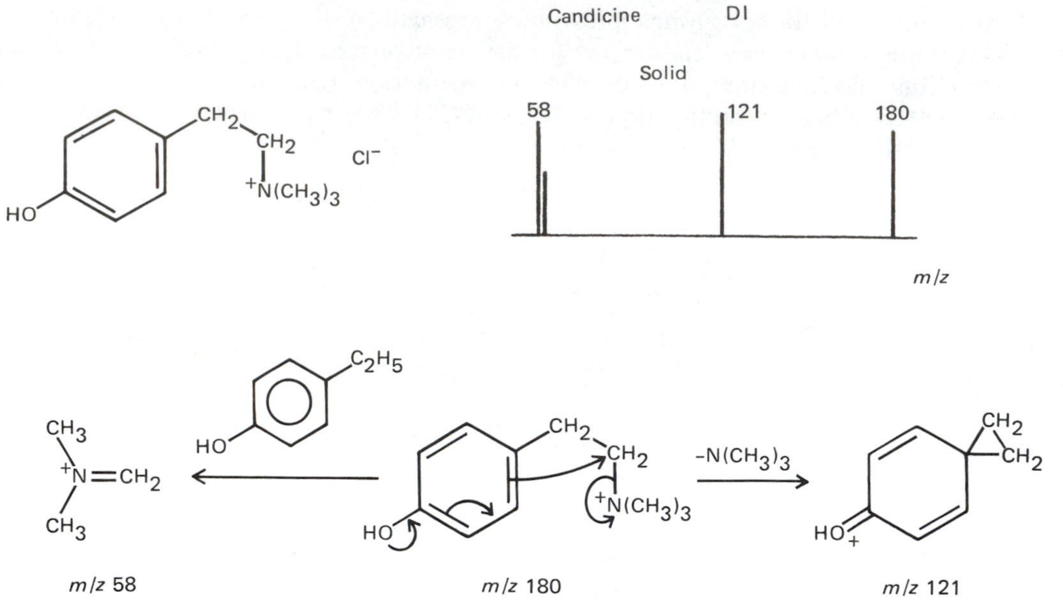

FIGURE 12-23 Fragmentation in DI. Elimination of neutral molecules from the intact cation (m/z 180) desorbed by ion bombardment of candicine. [From K. L. Busch and R. G. Cooks, *Science*, **218**, 247 (1982). Copyright 1982 by the American Association for the Advancement of Science.]

glycerol solution using Ar^+ bombardment. The data (Figure 12-23) show the intact cation (m/z 180) as the most intense peak as well as fragmentations from this even-electron ion to give even-electron daughter ions with elimination of stable neutral molecules. This behavior is seen also in the spectra of related compounds.

The DI spectra of simple quaternary ammonium compounds illustrate many of the rules for fragmentation encountered previously. For example, *N*-ethyl-*N*-methylpiperidinium iodide is readily ionized by any of the DI techniques. If the intact cation is selected, excited by collisions, and allowed to fragment (an MS–MS experiment, see Section 12-5b), it yields the products shown in eq. 12-38 (ref. 12-27). Note that radical loss occurs under the conditions of this particular experiment, in contravention of the even → even electron rule (eq. 12-10), which holds less rigidly for CI and DI than for EI.

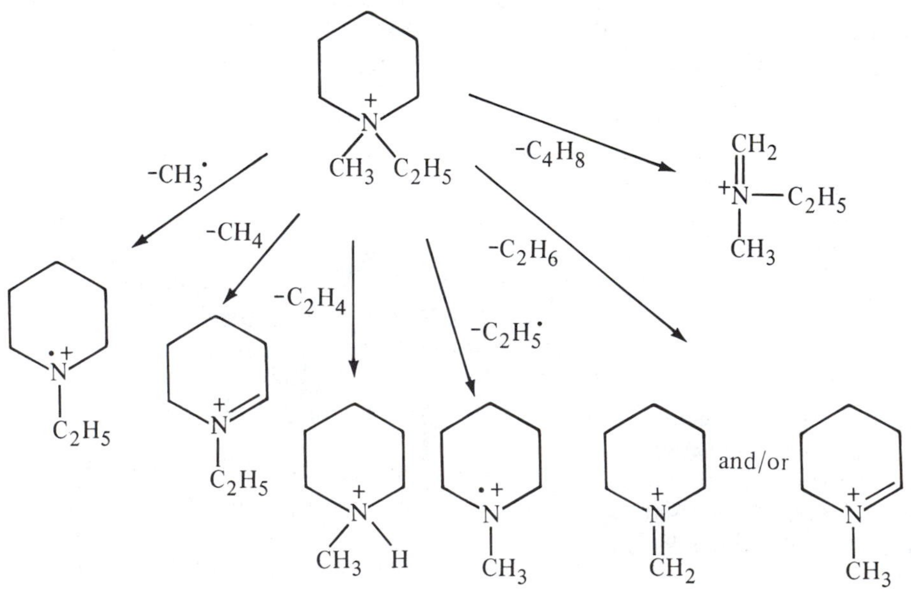

(12-38)

It must be remembered that in DI the molecule can be ionized in a variety of ways (see, e.g., arginine in Figure 12-3) and each quasimolecular ion will display its own set of fragment ions. A more serious problem in interpreting DI spectra is that intermolecular reactions can sometimes occur. These range from simple transmethylations, giving peaks 14 mass units greater than expected, to more unexpected reactions like the intermolecular N atom transfer (eq. 12-39) that occurs in the spectrum of benzotriazole when examined from

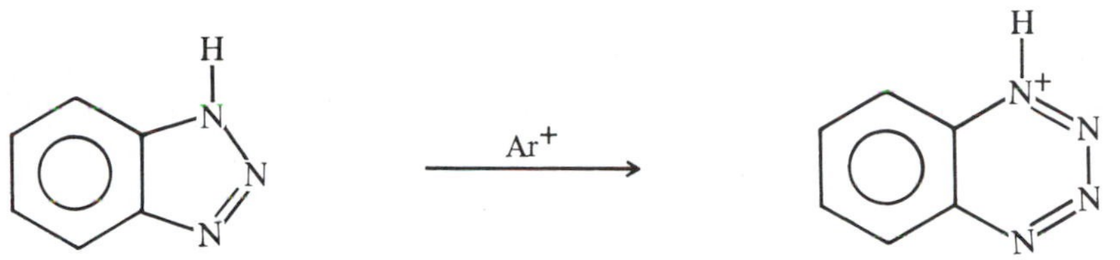

$$\text{benzotriazole (solid), M} \xrightarrow{Ar^+} (M+N)^+ \qquad (12\text{-}39)$$

the solid state, though not from solution. Further examples of fragmentation in DI can be found in Figure 12-23 and Section 12-8.

12-7

Some Ion Chemistry

Gas phase ion chemistry is extremely rich, both in structure and reactivity, and relatively easily accessible experimentally. In a treatment such as the present one, we have *rationalized* mass spectra rather than enquiring deeply into fragmentation mechanisms. This short section attempts to indicate some highlights of ion chemistry without reference to the methods needed to reach the conclusions that follow.

1. The molecular ion of methyl acetate fragments, as expected, to yield the stable acylium ion, m/z 43 (eq. 12-40). However, the neutral fragment is not $CH_3O^{\cdot}$; it is the isomer $^{\cdot}CH_2OH$.

$$CH_3C(=\ddot{O})OCH_3^{\cdot +} \longrightarrow CH_3C{\equiv}O^+ + COH_3 \qquad (12\text{-}40)$$

2. Acylium ions dissociate, as expected, by loss of the stable neutral molecule CO to yield carbenium ions (eq. 12-13). Isomerization of R^+ occurs in the course of this apparently simple bond cleavage, e.g., n-RCO^+ yields *sec*-R^+. (This type of result has been interpreted as suggesting that unimolecular dissociations actually occur via ion–molecule complexes of the type encountered in low energy ion–molecule reactions—Section 11-6b.)
3. Doubly charged ions can dissociate to give two singly charged fragments. From the coulombic energy of repulsion the distance between the charges in the dissociating doubly charged ion can be measured. Cyclic vs. ring-opened structures are readily distinguished.
4. The methanol molecular ion $CH_3OH^{\cdot +}$ is much less stable than its isomer, $^{\cdot}CH_2{-}^{+}OH_2$, which is formally an ionized ylid. Both species exist in deep potential wells, as predicted by calculation. The neutral ylid $^{-}CH_2{}^{+}OH_2$ is unstable.
5. The ketonic ion $CH_3COCH_3^{\cdot +}$ is much less stable than the ionized enol $CH_2C(OH)CH_3^{\cdot +}$.
6. Internal energy is incompletely randomized in some ions, which fragment unimolecularly with rate constants greater than $10^{10}\ s^{-1}$.
7. Mass spectral fragmentations show close analogies with solution behavior. For example, the Fischer

indole synthesis (eq. 12-41) and many other well-known acid- and base-catalyzed reactions occur in the

$$C_6H_5NHNH_2 + \begin{matrix} R \\ \;\;\diagdown \\ C{=}O \\ \diagup \\ H_3C \end{matrix} \xrightarrow{CH_5^+} \text{(indole ring: CH, –R, N–H)} \qquad (12\text{-}41)$$

gas phase. The indole synthesis proceeds by deamination of the protonated hydrazone, exactly as in solution. Such similarities should not obscure the fact that the unique conditions of the mass spectrometer provide access to uniquely interesting molecules. Hypervalent species, including CH_5^+, were known for many years in mass spectrometry before being encountered in the condensed phase.

12-8

Worked Examples

EXAMPLE 12-1

Figure 12-24 shows a 70 eV electron impact spectrum of an organic compound. (a) Is the compound pure? (b) What is the compound's molecular weight? (c) Are heteroatoms present? (d) Is the compound aromatic or aliphatic? (e) What is the probable molecular formula? (f) Why is there a peak at m/z 77.5?

Answer: (a) There is no evidence of impurities, i.e., ions that cannot be connected to each other by reasonable fragmentations.

(b) Probably 154. This ion could be a fragment, in which case the compound does not display a molecular ion at all. See Figure 12-2 for further steps to confirm the molecular weight.

(c) The ion m/z 156 is 1.3% relative abundance, i.e., 5% of the abundance of m/z 154. This value suggests the presence of one sulfur atom.

(d) The high intensities of the molecular ions and of the high mass fragments and the presence of m/z 77 suggest an aromatic structure.

(e) The even mass of the molecular ion suggests the compound does not contain nitrogen. Hence likely formulas are $C_9H_{14}S$ and $C_7H_6O_2S$. The degrees of unsaturation in these cases (see Section 12-2b) are four and five, respectively. The second formula is indicated because oxygen is clearly present (loss of 17 and 18 to give m/z 137 and 136) and because it allows an aromatic structure (at least four units of unsaturation).

(f) The peak is due to the ^{13}C isotope of M^{2+}. The abundance of the doubly charged ion supports the suggestion that the molecule is aromatic.

The compound is 2-mercaptobenzoic acid.

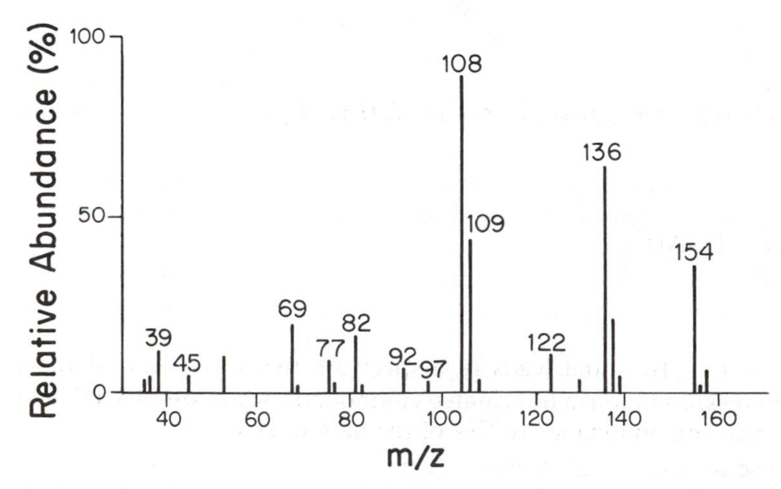

FIGURE 12-24 Electron impact (70 eV) mass spectrum of an unknown compound.

EXAMPLE 12-2

(a) Using the ionization energies of the simple alkyl derivatives given below, write charge localized

	Ionization energy (eV)
CH_3OH	10.85
CH_3NH_2	8.97
CH_3Cl	11.27

structures (Section 12-6a) for the molecular ions of the 1,3-propane derivatives **12-8** through **12-11**. (b) Write structures for the primary α-cleavage products. (c) What types of nitrogen-containing compounds might yield fragment ions of mass 44?

$NH_2(CH_2)_3OH$	$NH_2(CH_2)_3CH_3$	$NH_2(CH_2)_3Cl$	$HO(CH_2)_3Cl$
12-8	**12-9**	**12-10**	**12-11**

Answer:

(a) $^{\cdot+}NH_2{-}(CH_2)_3OH$
$^{\cdot+}NH_2{-}(CH_2)_3CH_3$
$^{\cdot+}NH_2{-}(CH_2)_3Cl$
$^{\cdot+}HO{-}(CH_2)_3Cl$

(b) $^{+}NH_2{=}CH_2$
$^{+}NH_2{=}CH_2$
$^{+}NH_2{=}CH_2$
$HO^{+}{=}CH_2$

(c) $CH_3\overset{\cdot+}{N}H{-}CH_2{-}R \rightarrow CH_3\overset{+}{N}H{=}CH_2$ $\quad$ (m/z 44)

$\overset{\cdot+}{N}H_2{-}\underset{R}{\underset{|}{C}}HCH_3 \rightarrow H_2N^{+}{=}CHCH_3$ $\quad$ (m/z 44)

EXAMPLE 12-3

1,2,3,4-Tetrafluorobenzene is ionized by charge exchange with argon. If the ionization energy of the organic molecule is 9.6 eV and that of argon is 15.8 eV, what is the internal energy of the molecular ion? (See Section 11-5b for an analogous problem in chemical ionization.)

Answer: Charge exchange proceeds as follows:

$$Ar^{\cdot+} + C_6F_4H_2 \longrightarrow Ar + C_6F_4H_2^{\cdot+}$$

$$\Delta H\,(\text{reaction}) = \Delta H_f(Ar) - \Delta H_f(Ar^{\cdot+}) + \Delta H_f(C_6F_4H_2^{\cdot+}) - \Delta H_f(C_6F_4H_2)$$

but $IE(M) = \Delta H_f(M^{\cdot+}) - \Delta H_f(M)$

$$\therefore \Delta H\,(\text{reaction}) = IE(C_6F_4H_2) - IE(Ar) = 9.6 - 15.8\text{ eV} = -6.2\text{ eV}$$

(This 6.2 eV appears as internal energy of $C_6F_4H_2^{\cdot+}$ and leads to a spectrum that shows considerable fragmentation. The base peak is due to CF_2 loss, common in fluorinated compounds, while loss of CHF and CHF_2 also gives intense peaks. The molecular ion abundance is 22% of the base peak.)

EXAMPLE 12-4

Use Stevenson's rule (Section 12-6a) to assign charge to one or the other of the products of each of the fragmentations shown.

$$\left[\text{3-methylpent-4-en-2-ol}\right]^{\bullet+} \longrightarrow \text{2-butene} + CH_3CHO$$

IE = 9.0 eV IE = 10.2 eV

$$HO{-}CH_2{-}CH_2{-}NH_2\,]^{\bullet+} \longrightarrow HO{-}CH_2^{\bullet} + {}^{\bullet}CH_2{-}NH_2$$

IE = 7.6 eV IE = 6.2 eV

Answer: 2-Butene will tend to take the charge in the rearrangement reaction, and the nitrogen-containing fragment ($CH_2{=}\overset{+}{N}H_2$) will be charged in the simple cleavage. (Charge preferentially goes with the low ionization energy fragment.)

EXAMPLE 12-5

Referring to Figure 11-10, explain why metastable ions are more likely to occur for rearrangement reactions than for simple bond cleavages. At what apparent mass should aniline show a metastable peak (Section 12-6a) in its EI spectrum?

Answer: Metastable peaks arise from the slow fragmentation of ions of low internal energy. Rearrangements are characterized by large entropies of activation. Therefore, they must have low activation energies to compete with simple bond cleavages. Ions of low internal energy consequently tend to undergo low activation energy reactions, including rearrangements. At higher internal energies entropic factors dominate; hence rearrangements become less competitive. Aniline molecular ion (m/z 93) fragments by loss of HCN, a slow rearrangement process, to give m/z 66. The resulting metastable peak will have apparent mass $m^* = m_2^2/m_1 = 66^2/93 = 46.8$.

EXAMPLE 12-6

Show that perpendicular magnetic (B) and electric (E) fields act as a mass analyzer, termed a Wien filter. *Hint:* Use eq. 11-23.

Answer: Perpendicular B and E fields cause parallel deflections. Hence, an ion beam moving through such a field will suffer no net deflection provided that the forces due to the fields cancel. This occurs when the force $Bzv = Ez$, i.e., $E/B = v$. If all the ions examined fall through a potential V, they have kinetic energy zV, and their velocities are given by eq. 11-23.

$$\tfrac{1}{2}mv^2 = zV \qquad v = \left(\frac{2zV}{m}\right)^{1/2} \tag{11-23}$$

Eliminating v gives

$$\frac{2zV}{m} = \frac{E^2}{B^2}$$

and

$$m/z = \frac{2B^2V}{E^2}$$

EXAMPLE 12-7

A sector mass spectrometer of radius 30 cm and maximum field strength 0.8 Tesla is normally operated at an accelerating voltage of 8000 V. What is its mass range for singly charged ions? What accelerating potential should be used in order to reach a mass range of 6000 d?

Answer: Eq. 11-28 can be expressed in units of daltons/electronic charge, tesla, cm, and electron

$$m/z = \frac{B^2 r^2}{2V} \tag{11-28}$$

volts as follows.

$$m/z = \frac{4.83 \times 10^3\, B^2 r^2}{V}$$

Hence, for $B = B_{max} = 0.8$ T and $z = 1$ (singly charged ions) $m_{max} = 348$ d. To achieve $(m/z)_{max} = 6000$ d, V must be reduced to 348/6000 of its initial value, i.e, V must be 464 V. Mass spectrometers can be operated at these low accelerating voltages but only with considerable loss in sensitivity and resolution. Higher fields are the preferred way to achieve these high masses.

EXAMPLE 12-8

Predict the extent to which compounds **12-12** and **12-13** will lose (a) $H^{\cdot}$ and (b) $CH_3^{\cdot}$ upon 70 eV electron impact. Give reasons.

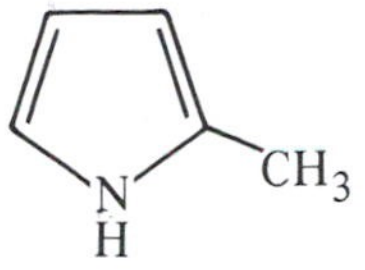

12-12 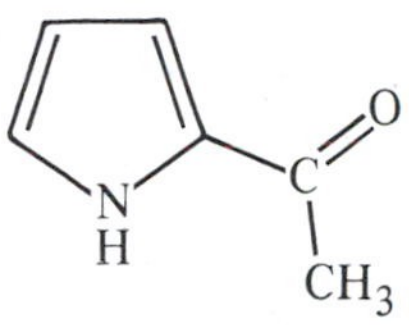 12-13

Answer: (a) **12-12** should lose $H^{\cdot}$ by analogy with the behavior of toluene and other aromatic compounds that bear a hydrogen on the α-carbon. **12-13** should lose little $H^{\cdot}$ since loss of the NH hydrogen gives an unstable nitrenium ion and loss of the methyl hydrogen gives an α-keto carbenium ion.

(b) **12-12** should not lose $CH_3^{\cdot}$ since this process would give the simple aryl cation that competes ineffectively with the aryl-carbenyl ion just discussed. Conversely, methyl loss from **12-13** generates a stable acylium-type ion and hence is strongly favored. (Note that $(M-H)^+$ is the base peak in the spectrum of **12-12** and $(M-CH_3)^+$ is the base peak for **12-13**.)

EXAMPLE 12-9

How many isotopic forms of the molecular ion of $C_{11}Cl_{11}^+$ exist? How many nominal masses are represented? Carbon has natural isotopes ^{12}C and ^{13}C, chlorine has ^{35}Cl and ^{37}Cl (see Appendix IV.A).

Answer: C_{11} exists in 12 forms ranging from all-^{12}C through ^{12}C–^{13}C mixtures to all-^{13}C. Similarly Cl_{11} must exist in 12 isotopic forms. All combinations of the C_{11} and Cl_{11} species are allowed, i.e., $12!/2!10! = (12 \cdot 11)/2 = 66$. This is the total number of isotopic forms. Some of these have the same nominal masses. *All* integral masses between the two extremes $^{12}C_{11}{}^{35}Cl_{11}^+$ and $^{13}C_{11}{}^{37}Cl_{11}^+$ must be represented since there is a 1 dalton interval between the masses of the carbon isotopes and a 2 dalton interval between the chlorine isotope masses. Hence, the integral masses range from m/z 517 to m/z 550 and include 34 nominal masses.

EXAMPLE 12-10

Predict the results of attempted negative chemical ionization of nitrobenzene and biacetyl (electron affinities of 0.4 and 1.1 eV, respectively) (a) if the experiment is done under conditions that promote electron attachment (e^- reagent) or (b) if the reagent ion is OH^- (see Section 11-5b).

Answer: (a) Electron attachment is often successful when the compound has a positive electron affinity. Both compounds should yield molecular anions, $M^{\cdot -}$

(b) Biacetyl is relatively acidic and should yield the deprotonated molecule, $(M-H)^-$, with hydroxyl reagent. Nitrobenzene will not be ionized by this reagent.

EXAMPLE 12-11

Figure 12-25 shows the negative ion spectrum of anisole taken under electron impact conditions. (See discussion in Section 11-5 on negative ion formation.) Suggest ionization mechanisms and structures for the major ions.

Answer: Two ionization mechanisms are possible, dissociative electron attachment and ion pair production (eq. 11-18 and eq. 11-4). In the case of the formation of m/z 93⁻, the two reactions are

$$C_6H_5OCH_3 \xrightarrow{e^-} [C_6H_5OCH_3]^{\cdot -} \longrightarrow C_6H_5O^- + CH_3^{\cdot}$$

$$C_6H_5OCH_3 \xrightarrow{e^-} C_6H_5O^- + CH_3^+ + e^-$$

It is not necessary that the hydrocarbon ions be generated as fragments of 93⁻. They could result from independent ionizing events, e.g.,

$$C_6H_5OCH_3 \xrightarrow{e^-} C_6H_5^- + \dot{O}CH_3$$

$$C_6H_5^- \xrightarrow{-H_2} C_6H_3^- \xrightarrow{-H_2} C_6H^-$$

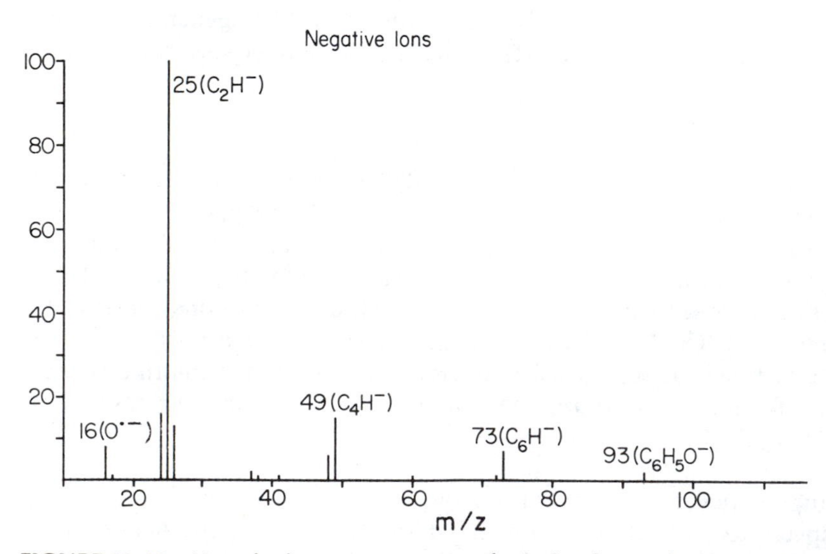

FIGURE 12-25 Negative ion mass spectrum of anisole taken under conventional EI conditions. [Reprinted with permission from R. T. Aplin, H. Budzikiewicz, and C. Djerassi, *J. Am. Chem. Soc.*, **87**, 3180 (1965). Copyright 1965 American Chemical Society.]

EXAMPLE 12-12

Ionized diphenyl ether undergoes two competitive primary reactions that yield product ions of comparable abundance at 70 eV.

$[C_6H_5OC_6H_5]^{\cdot+}$ $\xrightarrow{-CO}$ m/z 142

$[C_6H_5OC_6H_5]^{\cdot+}$ $\xrightarrow{-C_6H_5O^{\cdot}}$ m/z 77

(a) Predict the result of decreasing the ionizing electron energy to 15 eV. (Compare Figures 11-11 and 12-16.)

(b) Draw curves of the internal energy dependence of the rate constant for these processes. (Compare Section 11-6a.)

Answer: (a) The complex rearrangement process will compete with the simple cleavage more favorably in ions of lower internal energy.

(b) See Figure 12-26.

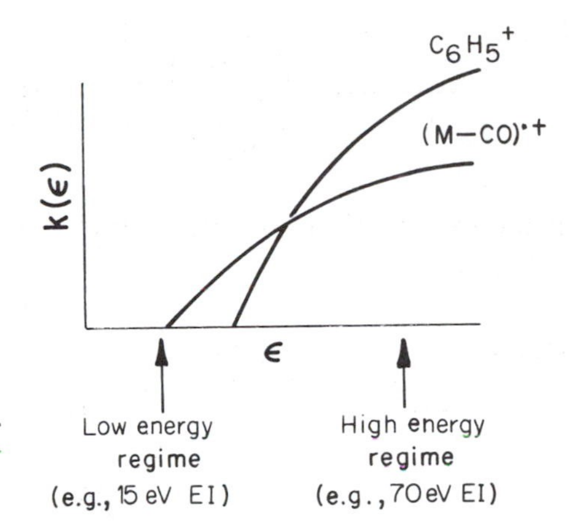

FIGURE 12-26 Energy dependence of the unimolecular rate constants for CO loss and $C_6H_5^+$ formation from diphenyl ether.

EXAMPLE 12-13

Sulfur and phosphorus compounds are particularly prone to molecular rearrangements in EI mass spectrometry. Account, in detail, for the observation of ions at m/z 109 and 111 (100/5 ratio) in the spectrum of **12-14.**

$CH_3-P(=S)(CH_3)-OC_6H_5$

12-14

Answer: The ion abundance ratio is suggestive of sulfur ($^{32}S/^{34}S = 100/4.4$). If so, the formula could be $C_6H_5S^+$. This result would require aryl migration from O to ionized S (charge-localized structure).

$(CH_3)_2P(=S^{\cdot+})-OC_6H_5 \longrightarrow (CH_3)_2P(={}^{+}SC_6H_5)-O^{\cdot} \longleftrightarrow (CH_3)_2P(-{}^{+\cdot}SC_6H_5)=O \longrightarrow C_6H_5S^+$

m/z 109 and 111

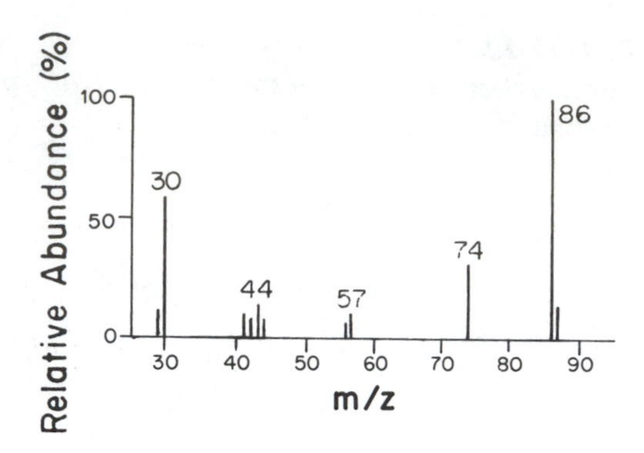

FIGURE 12-27 EI (70 eV) mass spectrum of leucine.

EXAMPLE 12-14

The electron impact mass spectrum of leucine is shown in Figure 12-27.

(a) Account for the major ions.

(b) Suggest three specific methods of obtaining a spectrum that shows an ion characteristic of the molecular weight (Section 12-1).

Answer: (a)

$$C_3H_7-CH_2-CH(\overset{\cdot+}{N}H_2)-CO_2H \longrightarrow C_3H_7-CH_2-CH{=}NH_2^+ \quad (m/z\ 86)$$

$$C_3H_7-CH_2-CH(\overset{\cdot+}{N}H_2)-CO_2H \longrightarrow (H)(CO_2H)C{=}\overset{+}{N}H_2 \quad (m/z\ 74) \longrightarrow (H)(H)C{=}\overset{+}{N}H_2 \quad (m/z\ 30)$$

(b) (i) Chemical ionization, with ammonia or isobutane reagent. (The amino acid is strongly basic, so almost any gas phase acid will protonate it.) (ii) Electron impact after derivatization to form the methyl ester. (iii) Desorption ionization, recording either positive or negative ions, e.g., negative ion FAB.

EXAMPLE 12-15

Methyl *m*- and *o*-methylbenzoates give identical 70 eV EI mass spectra with the exception of an ion at m/z 118. Account for this behavior. Of what general circumstance is it a particular case?

Answer: The ion m/z 118 is due to loss of 32 mass units. In a methyl ester this loss is normally due to CH_3OH (Table 12-6). Hence, one compound loses CH_3OH, the other does not, an example of an ortho effect.

$$o\text{-}CH_3C_6H_4-C(=O)-\overset{\cdot+}{O}-CH_3 \longrightarrow \cdot CH_2C_6H_4-C(=O)-\overset{+}{O}(H)-CH_3 \xrightarrow{-CH_3OH} [CH_2{=}C_6H_4{=}C{=}O]^{\cdot+}$$

EXAMPLE 12-16

The EI mass spectrum of butylbenzene recorded at 14 eV ionizing energy shows three ions of significant relative abundance, the base peak at 134, an ion of 70% relative abundance at m/z 92, and one of 40% relative abundance at m/z 91. Considering the effects of internal energy distributions on mass spectra (Section 11-5a), predict how the spectrum will change as the ionizing energy is raised.

Answer: Raising the ionizing electron energy will create relatively more molecular ions of high internal energy. The molecular ion, m/z 134, will therefore decrease in abundance relative to fragment ions. The simple cleavage fragment m/z 91 will increase relative to the rearrangement ion (see Examples 12-5 and 12-12). Ions due to further fragmentation will appear. These should include m/z 65, formed by acetylene loss from m/z 91.

EXAMPLE 12-17

Hernandulcin, an intensely sweet compound discovered by reviewing conquistador literature [*Science*, **227**, 417 (1985)], gives a high resolution mass spectrum that includes a molecular ion at 236.18005 d and prominent fragment ions at m/z 110 ($C_7H_{10}O^+$), 95 ($C_6H_7O^+$), and 82 ($C_5H_6O^+$). Discuss the claim that these data are consistent with structre **12-15**. ($^{12}C = 12.0000$, $^{1}H = 1.0078$, $^{16}O = 15.9949$).

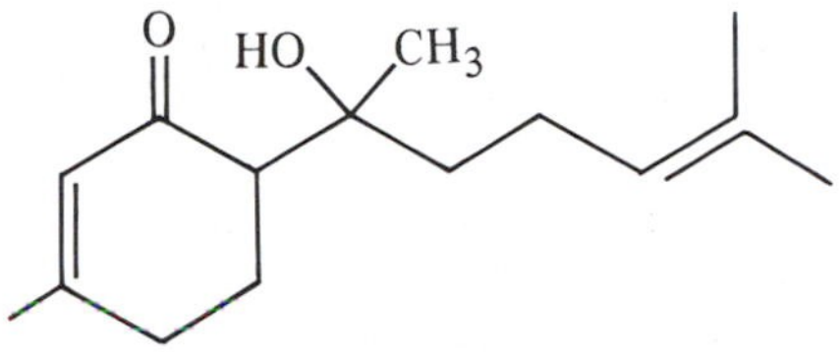

O HO CH3

12-15

Answer: The formula of **12-15** is $C_{15}H_{24}O_2$, which has an exact mass of 236.1795 d. The experimental error of the measured value is 0.5 ppm, which is well inside the error range normally encountered. In fact, several other formulas fit the experimental value, among them $C_{13}H_{22}N_3O$, which requires 236.1781. The mass spectral fragmentation pattern and the measured compositions of the fragment ions, however, demand the $C_{15}H_{24}O_2$ formula. The fragmentation behavior is rationalized as follows.

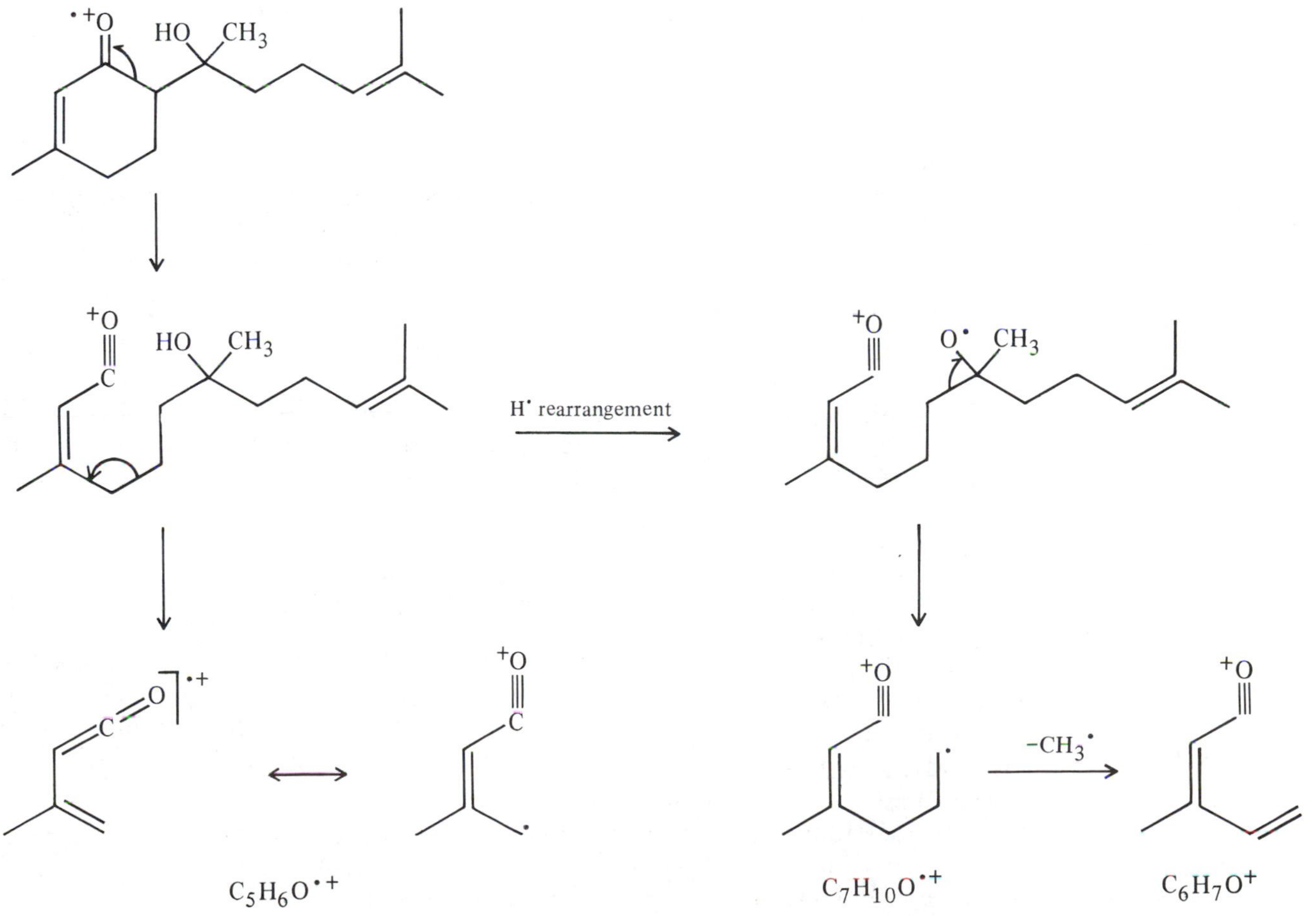

•+O HO CH3

+O HO CH3 C

H• rearrangement

+O O• CH3

C=O]•+

+O C

+O

$-CH_3^{\bullet}$

+O

$C_5H_6O^{\bullet +}$

$C_7H_{10}O^{\bullet +}$

$C_6H_7O^+$

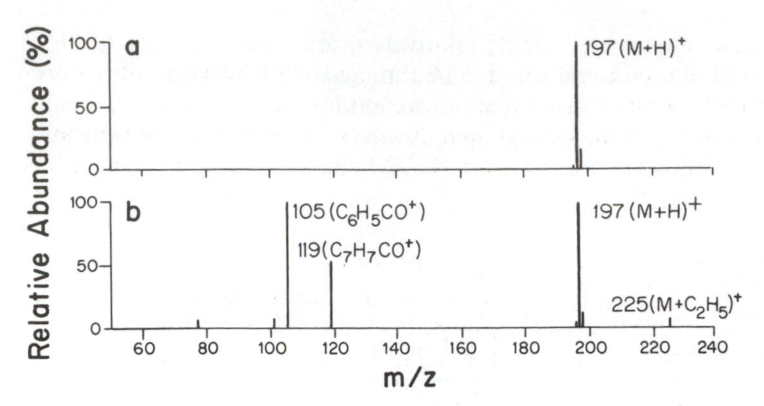

FIGURE 12-28 CI mass spectra of 4-methylbenzophenone using (a) isobutane and (b) methane as reagent gas. [Reproduced with the permission of Heyden and Son from J. Michnowitz and B. Munson, *Org. Mass Spectrom.*, **4**, 481 (1970).]

EXAMPLE 12-18

The CI (isobutane) mass spectrum of 4-methylbenzophenone (**12-16**) shows only the protonated molecule and its ^{13}C isotopic peak. When methane is employed as reagent, ions occur at both lower and high mass (Figure 12-28). Explain their origin and usefulness. The proton affinities of methane, isobutene, and **12-16** are 536, 823, and 857 kJ mol^{-}1 (128, 197, and 205 kcal mol^{-1}). Compare Table 11-9.

Answer: The protonation of **12-16** by CH_5^+ is more exothermic by $823 - 536 = 287$ kJ mol^{-1} (69 kcal mol^{-1}) than is protonation by $C_4H_9^+$. The ions below $(M+H)^+$ are due to unimolecular fragmentation of $(M+H)^+$ because of its large internal energy, $857 - 536 = 321$ kJ mol^{-1} (77 kcal mol^{-1}). The fragments are indicators of structure.

^+OH CH_3 **12-16** $\xrightarrow{-C_6H_6}$ $CH_3C_6H_4C{\equiv}O^+$ *m/z* 119

$\xrightarrow{-C_7H_8}$ $C_6H_5C{\equiv}O^+$ *m/z* 105

The ion of greater mass than $(M + H)^+$ is the result of the ionization reaction $M + C_2H_5^+ \rightarrow (M + C_2H_5)^+$ (methane gives CH_5^+, $C_2H_5^+$, and $C_3H_5^+$ as reagent ions, Table 11-10). The presence of *two* quasimolecular ions, $(M + H)^+$ and $(M + C_2H_5)^+$, separated by 28 mass units, helps in checking the molecular weight assignment.

EXAMPLE 12-19

The CI (methane) mass spectra of alkanes show prominent $(M-H)^+$ ions in the molecular ion region. Explain this observation.

Answer: The major methane reagent ion is CH_5^+ (Table 11-10). The reaction $RH + CH_5^+ \rightarrow RH_2^+ + CH_4$ is only slightly exothermic for alkanes (it is thermoneutral for $R = CH_3$). Hence $(R-H)^+$ does *not* arise by fragmentation of the usual quasimolecular ion, $(M+H)^+$. Rather, other ion–molecule reactions can compete with protonation. In particular, hydride abstraction occurs readily for alkanes.

$$CH_5^+ + RH \rightarrow CH_4 + H_2 + R^+$$

$$C_2H_5^+ + RH \rightarrow C_2H_6 + R^+$$

EXAMPLE 12-20

The CI mass spectrum of methyl cinnamate, recorded with a protonating reagent ion, shows fragments at m/z 121 and 131. Account for these with charge-localized structures.

Answer: While protonation is favored at the most basic site, chemical ionization is not normally done under conditions of chemical equilibrium. Fragmentation from several charge-localized structures is therefore possible. Methyl cinnamate illustrates this principle.

$$C_6H_5\text{—CH=CH—C(=O)—}\overset{+}{O}(CH_3)H \xrightarrow{-CH_3OH} C_6H_5\text{—CH=CH—C}\equiv\overset{+}{O}$$

m/z 131

$$C_6H_5\text{—}\overset{+}{C}H\text{—CH}_2\text{—C(=O)OCH}_3 \xrightarrow{-CH_2CO} C_6H_5\text{—}\overset{+}{C}HOCH_3$$

m/z 121

EXAMPLE 12-21

What is the structure of the compound for which positive ion CI spectra are shown in Figure 12-29? Note that H_2 and CH_4 yield reagent ions (H_3^+ and CH_5^+) that differ by 114 kJ mol^{-1} (27 kcal mol$^-$) in proton affinity (Table 11-9).

Answer: The main difference between the spectra is in the degree of fragmentation, consistent with the formation of very energetic $(M+H)^+$ ions on protonation by H_3^+. The molecular weight is therefore firmly established as 100. The major fragments, m/z 73 and 69, arise by losses of 28 and 32 d, respectively. From Table 12-6 these are likely to be C_2H_4 or CO and CH_3OH, respectively. We conclude that at least one oxygen is present. Possible molecular formulas are $C_6H_{12}O$, which has one unit of unsaturation and $C_5H_8O_2$, which has two. The ion at m/z 41 suggests the presence of the $C_3H_5^+$ group. The loss of C_3H_6 (giving m/z 59) confirms this conclusion. The loss of 32, CH_3OH, is suggestive of a methyl ester (acids lose H_2O on CI), and the structure methyl methacrylate is suggested.

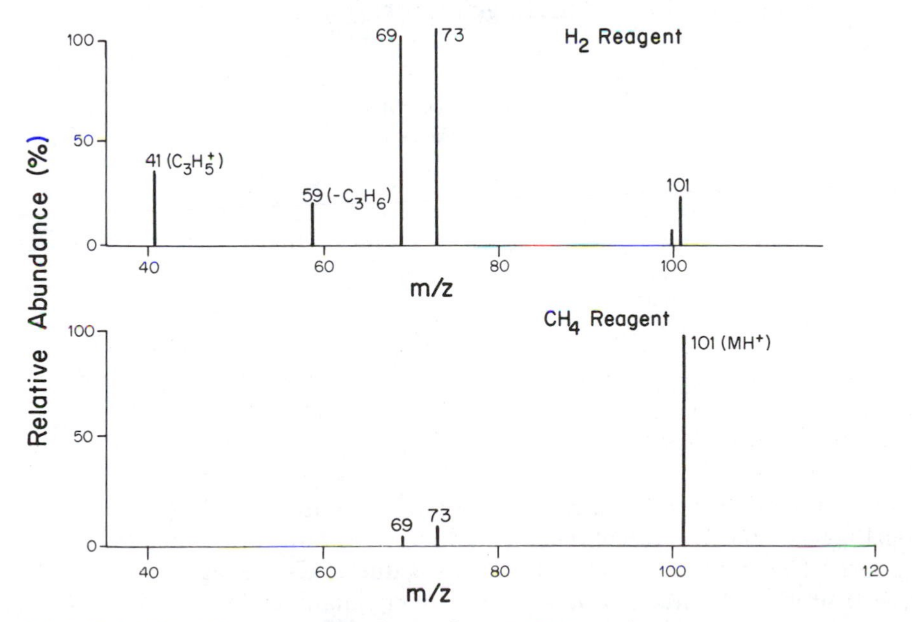

FIGURE 12-29 CI mass spectra of an unknown compound.

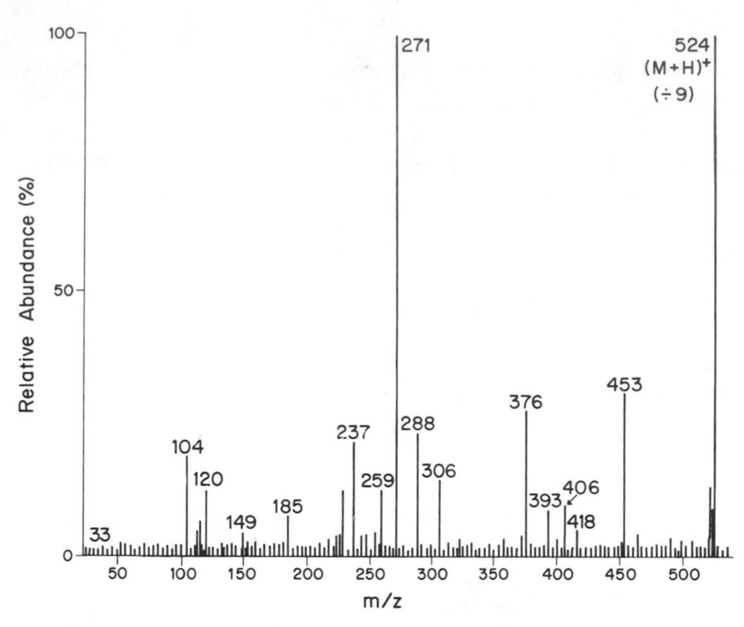

FIGURE 12-30 Daughter (MS-MS) spectrum of protonated Met-Arg-Phe-Ala (m/z 524) ionized by FAB (10 keV Xe^0; 25 eV collisions upon argon). [Data from Finnigan MAT.]

EXAMPLE 12-22

Peptides undergo fragmentations at the C—C and C—N bonds, with and without hydrogen rearrangement. These cleavages yield ions classed by Roepstorff as A, B, C, X, Y, and Z types. Figure 12-30

$$H_2N\text{—}\overset{R}{CH}\text{—}\overset{O}{\overset{\|}{C}}\text{—}NH\text{—}\overset{R}{CH}\cdots\overset{R}{CH}\text{—}\overset{O}{\overset{\|}{C}}\text{—}NH\text{—}\overset{R}{CH}\text{—}CO_2H$$

A B C X Y Z

shows the daughter MS–MS spectrum (Section 12-5b) of protonated Met-Arg-Phe-Ala, a small peptide ionized directly by FAB. Comment on the origin of some of the fragment ions, given that internal peptide residues have masses as follows.

	Internal Residue Mass		Internal Residue Mass
glycine	57	serine	87
alanine	71	threonine	101
phenylalanine	147	methionine	131
valine	99	tryptophane	186
leucine	113	aspartic acid	115
isoleucine	113	glutamic acid	129
proline	97	arginine	156
cystine	222	histidine	137
tyrosine	163	lysine	128
hydroxyproline	113	asparagine	114

Answer: The DI (FAB) mass spectrum gives the molecular weight but relatively little information on structure because of the high background due to glycerol. The MS–MS spectrum shows better signal-to-noise ratios and structurally diagnostic fragment ions. The ion at 453^+ is due to Z-cleavage of terminal alanine. Ion 306^+ belongs to the same series and is due to further loss of phenylalanine ($453 - 147 = 306$). A-type cleavage yields 104^+, 259^+, and 406^+. B-type cleavage is represented by m/z 288, while B-cleavage with charge retention by the carboxylic acid unit is represented by m/z 393 and 237. These fragments define the sequence. See Figure 12-13 for another example of a sequence determination.

EXAMPLE 12-23

Consider the data in Figure 12-31 and describe the key steps that would be needed in a quantitative assay for the pesticide DDD at subnanogram levels (compare Section 12-4).

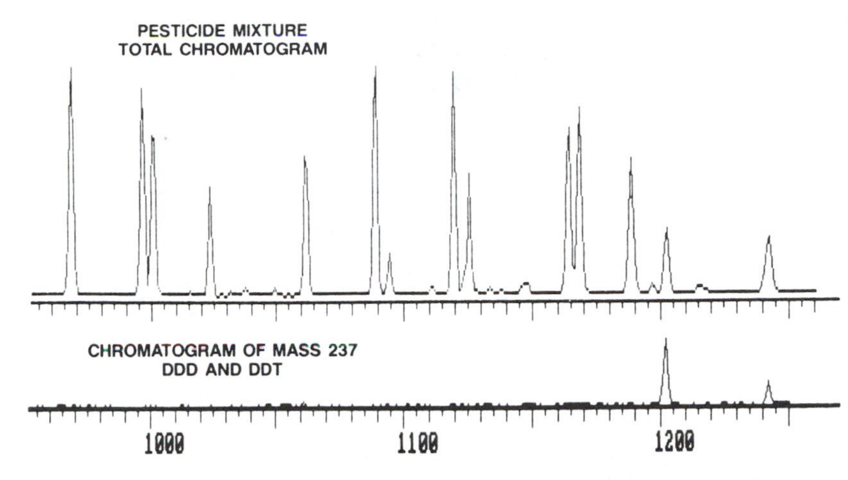

FIGURE 12-31 Reconstructed gas chromatogram of a pesticide mixture (~20 ng per component) and ion chromatogram of *m/z* 237 (characteristic DDD fragment). Data obtained by EI in an ion trap. [Data from Finnigan MAT.]

Answer: The data show a gas chromatogram recorded using mass spectrometric detection (full mass spectra). The data are also replotted so as to show chromatographic profiles for a selected ion (m/z 237), namely, a mass chromatogram. Additional sensitivity is available if the latter experiment is done while spending all the available time measuring a few selected ions instead of the full mass spectrum. Such a selected ion monitoring experiment would be necessary to achieve subnanogram detection limits. (SIM is 100–1000 times more sensitive than scans of full mass spectra.) For quantitative accuracy a labeled analogue of DDD should be synthesized. The ring d_4- or d_8-compound would be suitable. If the former were chosen SIM data would be taken for m/z 237 and 241. From the known amount of the labeled compound, the analyte would be quantitated by using eq. 12-4.

EXAMPLE 12-24

The region of a CI mass spectrum containing the protonated molecule shows relative abundances as follows.

m/z 321	2.03%
m/z 322	100.00%
m/z 323	18.88%
m/z 324	34.06%
m/z 325	3.83%

Deduce as much as possible about the molecular formula (compare Section 12-3).

Answer: The presence of one Cl atom is evident from the 3:1 (100:34) ratio of m/z 322 and 324. The presence of an odd number of N atoms is inferred from the even mass after protonation. The number of carbons can be determined from the ratio of 322:323 after appropriate isotopic corrections have been made. The ion at 323 must be corrected for H$^{\cdot}$ loss from 324 [^{37}Cl form of $(M+H)^+$]. That at 322 must be corrected for ^{13}C contributions from 321$^+$. The ^{13}C contribution of 321 will be approximately the same as that of 322, which has a heavy isotope peak of 18.88%. Hence, taking 19% of 2.03%, i.e., 0.39%, as the approximate value of ^{13}C-M$^{\cdot+}$ at m/z 322, one is left with 99.61% of ^{12}C,^{35}Cl-$(M-H)^{\cdot+}$. The ^{37}Cl-M$^{\cdot+}$ contribution to m/z 323 will be in the same proportion to 324 as the ^{35}Cl-M$^{\cdot+}$ ion at 321 is to 322, i.e., 2.03% (ignoring isotope effects and second order corrections). Hence, 2.03% of 34.06%, i.e., 0.69%, is the correction, leaving 18.88 − 0.69%

or 18.17% as the magnitude of $^{35}Cl,^{13}C\text{-}(M+H)^{\cdot+}$. This ion therefore occurs in its ^{12}C and ^{13}C forms in the ratio 99.61 to 18.17 units, i.e., 18.24%. The $^{13}C/^{12}C$ ratio is 1.1% per atom, i.e., the maximum number of carbon atoms in the molecule is 16 (requires 17.6%, found 18.24%). The partial formula is therefore C = 16, Cl = 1, N = 1. Even with the maximum complement of hydrogen allowed (34) a considerable fraction of the mass is missing, suggesting the presence of oxygen or monovalent elements. The actual formula is $C_{16}H_{17}NO_2PCl$.

EXAMPLE 12-25

Account for the major fragment ions at m/z 304 and 207 in the negative ion FAB spectrum of the antibiotic ampicillin **12-17** (cf. Sections 12-6d and e).

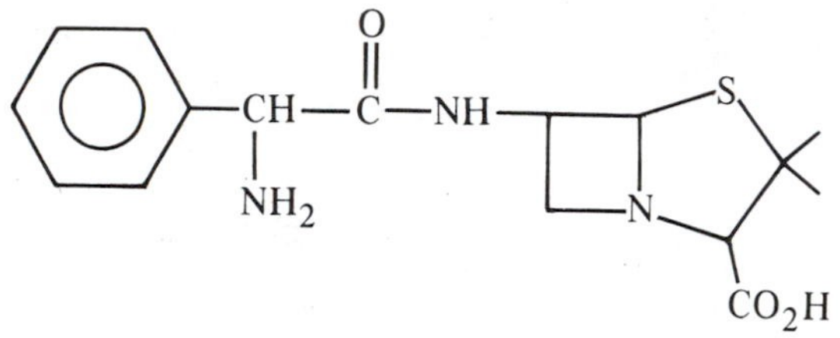

12-17

Answer: This acidic compound yields an abundant $(M-H)^-$ ion. This ion is readily desorbed from solution into the gas phase, where it fragments in a straightforward fashion. (Gas phase fragmentation is a characteristic of DI spectra, Section 11-5b.) Simple decarboxylation yields m/z 304. The ion at m/z 207 is the result of cleavage of the β-lactam ring.

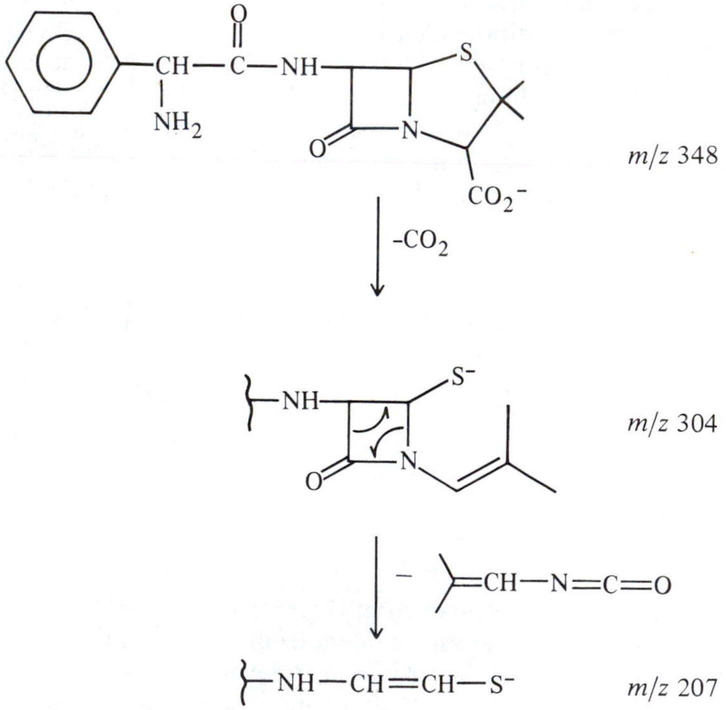

m/z 348

$-CO_2$

m/z 304

$-$ $(CH_3)_2C{=}CH{-}N{=}C{=}O$

$\sim NH{-}CH{=}CH{-}S^-$ m/z 207

EXAMPLE 12-26

Figure 12-32 shows results of a selected ion monitoring experiment to quantitate 2,3,7,8-tetrachlorodibenzodioxin (2,3,7,8-TCDD) in an industrial dust sample. The internal standard is the $^{13}C_8$-labeled analogue, which is 85% pure (isotopic and chemical impurities both considered). If the 2 μl sample injected

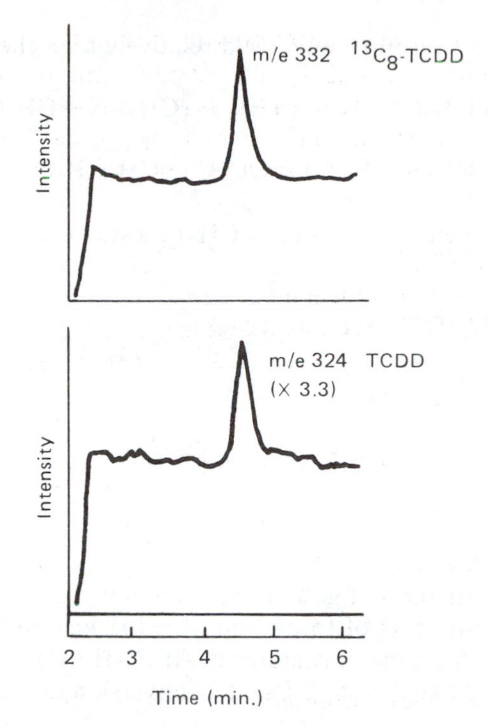

FIGURE 12-32 Selected ion monitoring (EI, 70 eV) for *m/z* 322 ($M^{\cdot+}$ of TCDD) and *m/z* 324 ($M^{\cdot+}$ of $^{13}C_8$-TCDD) in an enviromental sample. [Reproduced with permission from L. C. Lamparski and T. J. Nestrick, *Anal. Chem.*, **52**, 2045 (1980). Copyright 1980 American Chemical Society.]

into the GC–MS contained 500 pg/μl of the labeled compound, what is the concentration of 2,3,7,8-TCDD in this sample? If this 2 μl sample represents a 10% aliquot of the extract of a 5 g flyash sample, what is the concentration in the flyash?

Answer: Responses for m/z 332 and 324 are in the ratio 10:3; hence 2,3,7,8-TCDD is present at 500 pg/μl $\times 0.85 \times 0.3 = 128$ pg/μl in the extract. Hence, total amount of TCDD in 5 g of flyash is 2560 pg. Concentration in the original flyash is therefore 510 pg/g, i.e., *510 ppt*. [Note that this type of experiment can quantitate to the 200 ppq level! See, for example, L. C. Lamparski and T. J. Nestrick, *Anal. Chem.*, **52**, 2045 (1980).] The combination of excellent preseparation, chromatography, and mass spectrometry is needed to achieve this level of performance, which corresponds to locating, identifying, and measuring a few molecules per microgram of sample.

EXAMPLE 12-27

The neurotransmitter acetylcholine is a quaternary ammonium ester (**12-18**) that can be characterized by its DI spectrum. The SIMS spectrum shows the intact cation at m/z 146 and thus provides molecular weight information. Structural information is available from fragment ions, the most abundant of which corresponds to loss of 59 d. There are two groups in the molecule that have this mass. Which is lost?

$(CH_3)_3\overset{+}{N}CH_2CH_2OCOCH_3\ Cl^-$

12-18

Answer:

$N(CH_3)_3 = 59$

not $\quad O\overset{\overset{O}{\|}}{C}CH_3 = 59$

Simple cleavage of either group would yield relatively high energy products.

$$(CH_3)_3\overset{+}{N}{-}CH_2CH_2{-}OCOCH_3 \rightarrow (CH_3)_3\overset{+}{N}{-}CH_2CH_2^{\cdot} + {}^{\cdot}OCOCH_3$$

$$(CH_3)_3\overset{+}{N}{-}CH_2{-}CH_2{-}OCOCH_3 \rightarrow \overset{+}{C}H_2CH_2OCOCH_3 + N(CH_3)_3$$

However, a rearrangement with loss of $N(CH_3)_3$, a stable neutral molecule, can yield a stable product.

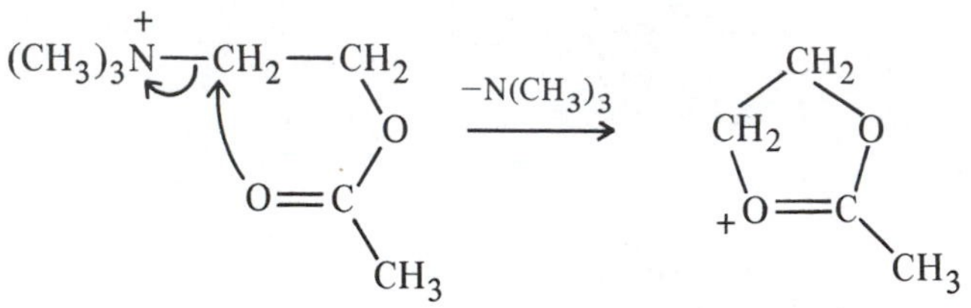

$$(CH_3)_3\overset{+}{N}{-}CH_2{-}CH_2{-}O{-}C(CH_3){=}O \xrightarrow{-N(CH_3)_3} \text{cyclic } CH_2{-}CH_2{-}O{-}C(CH_3){=}\overset{+}{O}$$

EXAMPLE 12-28

Heats of formation of gaseous ions are one means of characterizing them. Secondary alcohols yield $C_2H_5O^+$ ions with heats of formation of ~141 kcal mol^{-1}. Does the $C_2H_5O^+$ ion generated from dimethyl ether have this same structure, if $AE(C_2H_5O^+) = 10.7$ eV, $\Delta H_f(H^{\cdot}) = 52$ kcal mol^{-1}, and $\Delta H_f(CH_3OCH_3) = -44$ kcal mol^{-1}? Use the conversion factors in Table 11-2.

Answer

$$CH_3OCH_3 \xrightarrow{e^-} C_2H_5O^+ + H^{\cdot}$$

$$\Delta H_{\text{reaction}} \equiv AE(C_2H_5O^+) = 10.7 \text{ eV} = 247 \text{ kcal mol}^{-1} \qquad (1 \text{ eV} = 23.06 \text{ kcal mol}^{-1})$$

but

$$\Delta H_{\text{reaction}} = \Delta H_f(C_2H_5O^+) + \Delta H_f(H^{\cdot}) - \Delta H_f(CH_3OCH_3)$$

Therefore,

$$\Delta H_f(C_2H_5O^+) = 247 - 55 - 44 = 151 \text{ kcal mol}^{-1}$$

The 10 kcal mol^{-1} (42 kJ mol^{-1}) difference suggests that the two structures are *not* identical. In fact, the ether gives $CH_3{-}\overset{+}{O}{=}CH_2$, the alcohols $CH_3{-}CH{=}\overset{+}{O}H$.

EXAMPLE 12-29

A compound of molecular weight 113 gives ions in its 70 eV electron impact mass spectrum at m/z 84, 70, 56, 41, 20, 44, and 98 (decreasing abundance). Give a structure that is consistent with the data.

Answer: The odd mass value of the molecular ion implies that the compound contains an odd number of N atoms, probably one. The ions at m/z 30 and 44 are characteristic of alkylamines (30 is $CH_2{=}\overset{+}{N}H_2$, 44 either $CH_3CH{=}\overset{+}{N}H_2$ or $CH_2{=}\overset{+}{N}HCH_3$). The neutrals lost are 15 (giving 98^+), 29 (giving 70^+), and 43 (giving 56^+). The last may be a combination of 15 and 28. These fragmentations suggest that the compound contains a C_2 hydrocarbon chain at least, and, to account for ethyl loss, probably C_3. Reasonable formulas are $C_7H_{15}N$ (one unit of unsaturation) and $C_6H_{11}NO$ (two units of unsaturation). Dealing only with the former, we probably have either a cyclic or an unsaturated amine. There are a number of structures that account for most but not all the data. For example, allylmethylpropylamine does not explain methyl loss. (*N*-Ethyl)cyclopentylamine does (perhaps surprisingly) fit the data. The key fragmentation is ethyl loss. Of course, cleavage of the C—N bond does *not* occur, but the normal α-cleavage reaction yields a form of the molecular ion that rearranges and fragments to give stable products. (Note that analogous processes occur in cyclic alcohols and ketones, e.g., cyclohexanone eliminates $CH_3^{\cdot}$ by an analogous mechanism.)

H, $N^{\bullet +}$, CH_2CH_3 —α-cleavage→ H, N^+, C_2H_5 —1,4-H migration→

CH_3, H, N^+, C_2H_5 —$-C_2H_5^{\bullet}$→ H, N^+, C_2H_5

PROBLEMS

The ten problems presented here are intended for practice in deducing molecular structures from mass spectra. The examples used, while simple, all represent real structural problems, and the mass spectra were determined to provide structural information. No source of information besides mass spectrometry was employed in solving these problems. Of course, this is seldom the case in practice, so that much more complex structures than these examples are actually amenable to analysis.

Molecular ions are assigned to facilitate problem solving. This assignment was trivial in almost all cases. It was, nevertheless, checked by the methods discussed in Section 12-1. Finally, the spectra are restricted to ions with greater than 1% relative abundance.

12-1

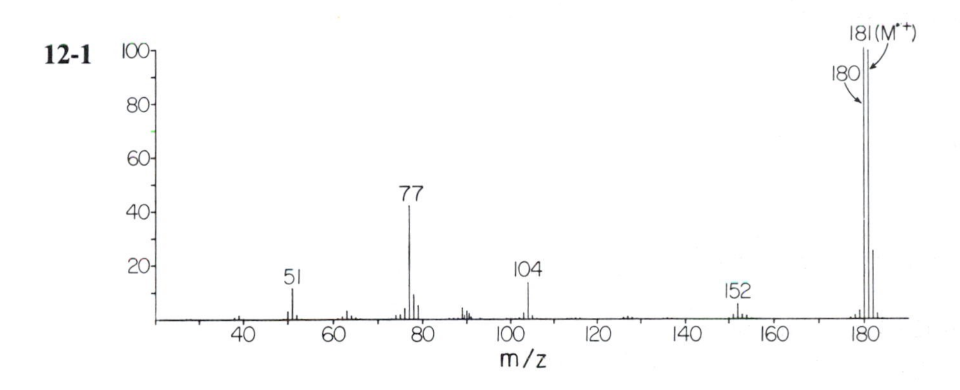

12-2

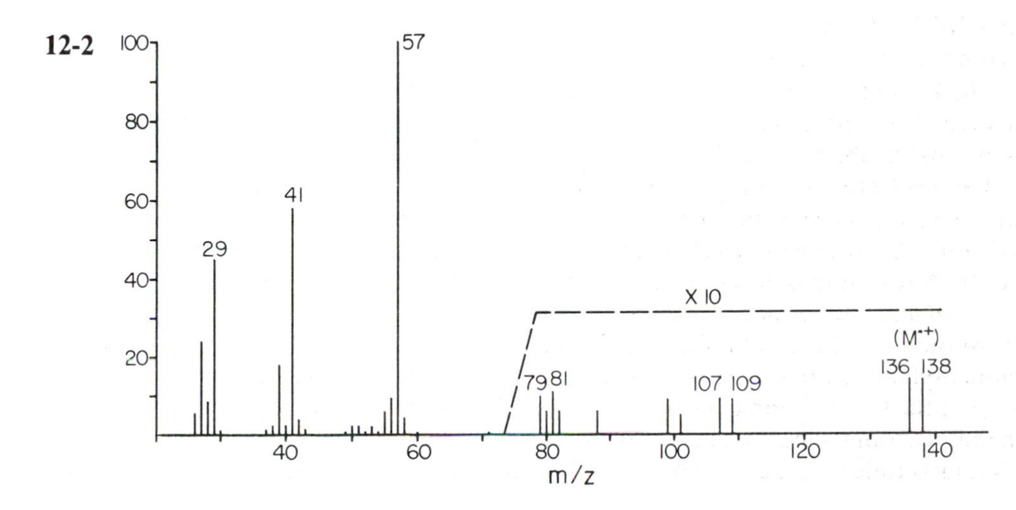

12-3

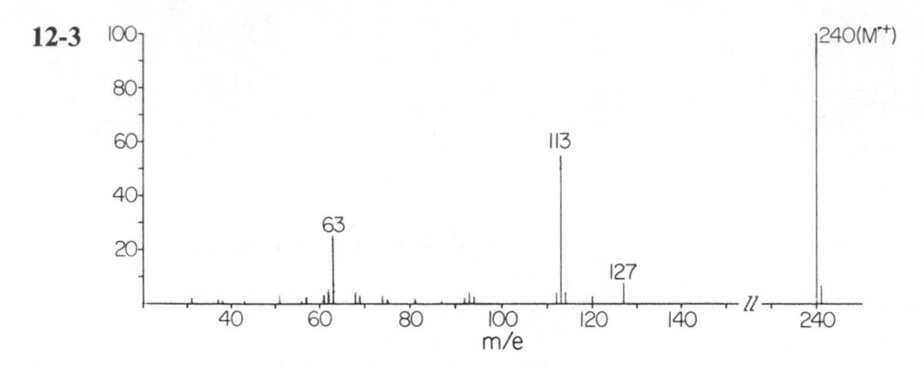

12-4

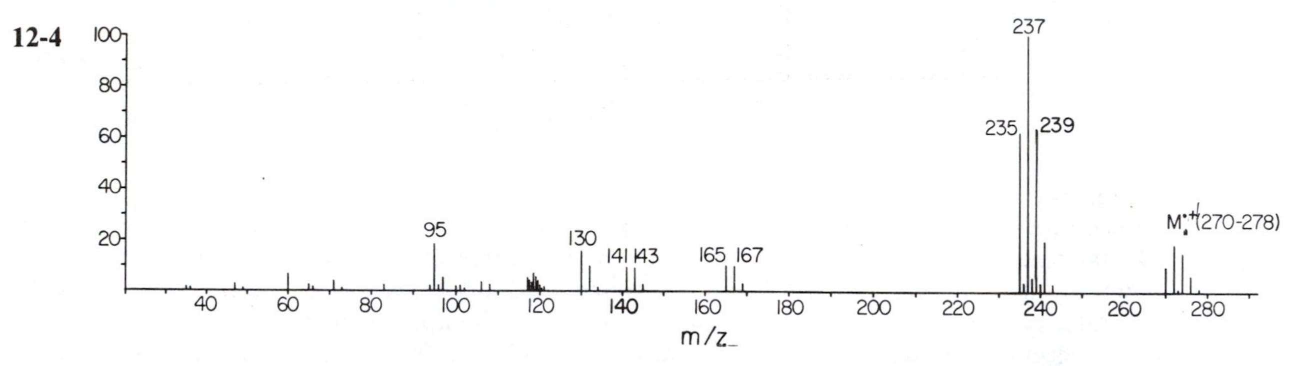

12-5

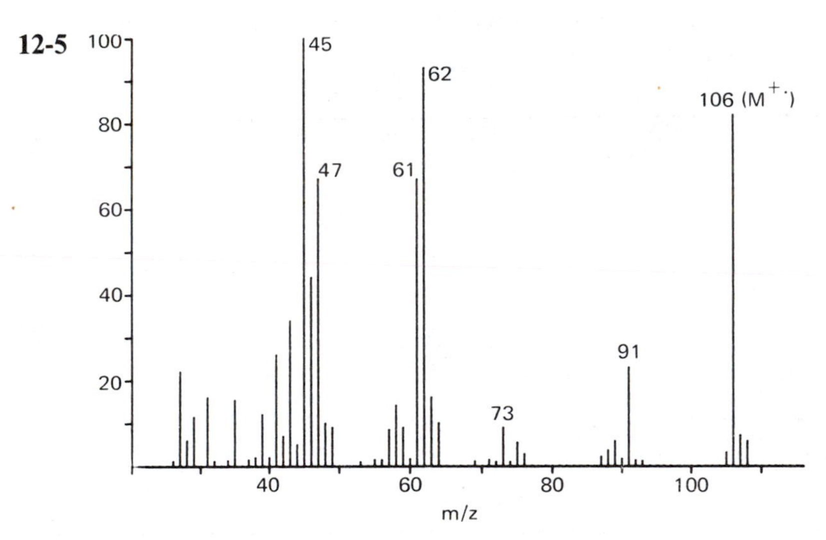

12-6

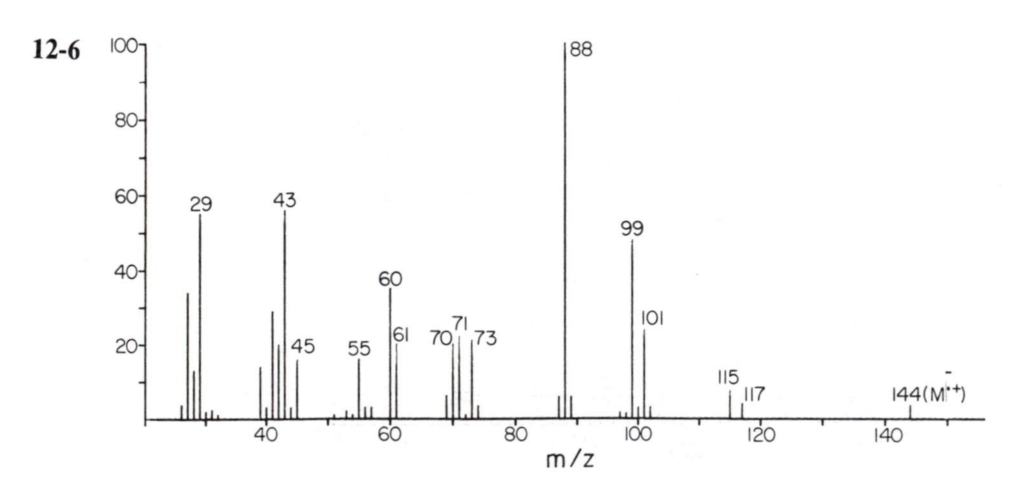

12-7

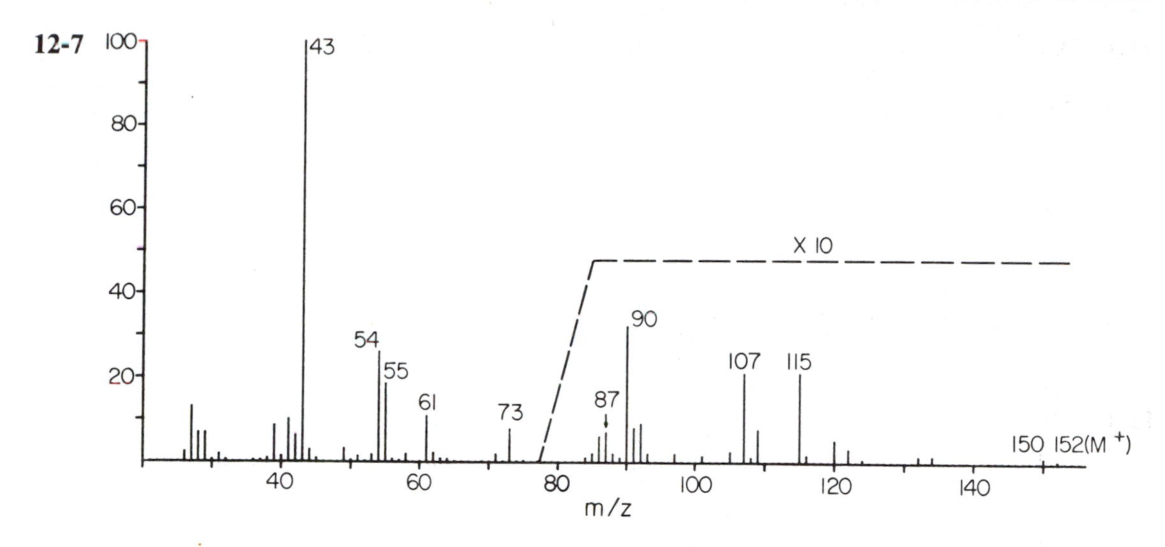

12-8

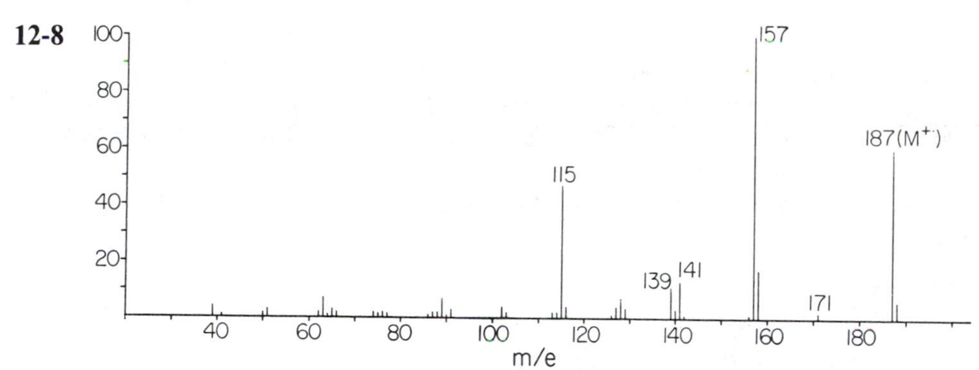

12-9

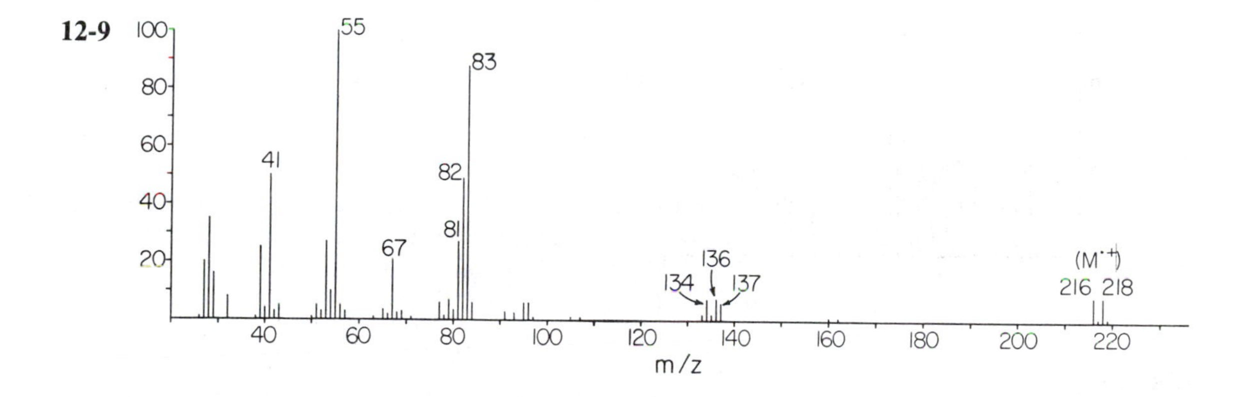

12-10

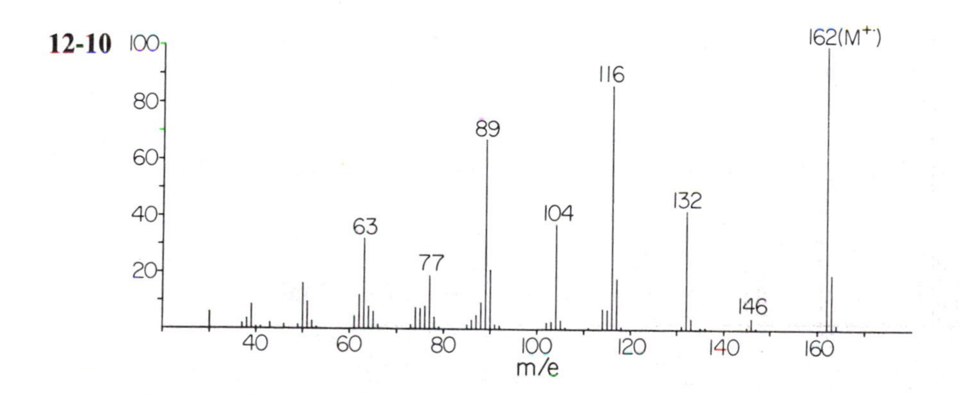

REFERENCES/BIBLIOGRAPHY

Historical

12-1 F. W. Aston, *Mass Spectra and Isotopes*, 2nd ed., Edward Arnold, London, 1942.
12-2 R. W. Kiser, *Introduction to Mass Spectrometry and Its Applications*, Prentice-Hall, Englewood Cliffs, NJ, 1965.
12-3 J. J. Thomson, *Rays of Positive Electricity*, Longmans, London, 1913.

Recent Developments

12-4 R. G. Cooks, K. L. Busch, and G. L. Glish, *Science*, **222**, 273–291 (1983).

Industrial and Engineering Aspects of MS

12-5 F. A. White and G. Wood, *Mass Spectrometry in Science and Technology*, Wiley, New York, 1985.

Peptides by MS

12-6 K. Biemann, *Int. J. Mass Spectrom. Ion Phys.*, **45**, 183 (1982).

Biological Applications

12-7 V. N. Reinhold and S. A. Carr, *Mass Spectrom. Rev.*, **2**, 153–221 (1983).
12-8 A. L. Burlingame, K. M. Straub, and T. A. Baillie, *Mass Spectrom. Rev.*, **2**, 331–387 (1983).
12-9 M. E. Rose and R. A. W. Johnstone, *Mass Spectrometry for Chemists and Biochemists*, Cambridge University Press, Cambridge, 1982.
12-10 G. Waller (Ed.), *Biochemical Applications of Mass Spectrometry, 1st Supplemental Volume*, Wiley, New York, 1980.

Fundamentals and Theory

12-11 K. Levsen, *Fundamental Aspects of Organic Mass Spectrometry*, Verlag Chemie, Weinheim, 1978.
12-12 I. Howe, D. H. Williams, and R. D. Bowen, *Mass Spectrometry, Principles and Applications*, 2nd ed., McGraw-Hill, London, 1981.

Metastable Ions

12-13 R. G. Cooks, J. H. Beynon, R. M. Caprioli, and G. R. Lester, *Metastable Ions*, Elsevier, Amsterdam, 1973.

Ion–Molecule Reactions

12-14 M. T. Bowers (Ed.), *Gas Phase Ion Chemistry*, Vols. 1 and 2, Academic Press, New York, 1979.
12-15 R. G. Cooks (Ed.), *Collision Spectrometry*, Plenum, New York, 1978.
12-16 R. J. Ausloos (Ed.), *Kinetics of Ion–Molecule Reactions*, Plenum, New York, 1979.
12-17 T. H. Morton, *Tetrahedron*, **38**, 3195 (1982).

Spectral Interpretation Also see ref. 12-12.

12-18 F. W. McLafferty, *Interpretation of Mass Spectra*, 3rd ed., University Science Books, Mill Valley, CA, 1980.
12-19 H. Budzikiewicz, D. Djerassi, and D. H. Williams, *Mass Spectrometry of Organic Compounds*, Holden-Day, San Francisco, 1967.

Chemical Ionization

12-20 A. G. Harrison, *Chemical Ionization Mass Spectrometry*, CRC Press, Boca Raton, FL, 1983.

Negative Ions

12-21 J. H. Bowie, *Mass Spectrom. Rev.*, **3**, 161–207 (1984).
12-22 C. E. Melton, *Principles of Mass Spectrometry and Negative Ions*, Marcel Dekker, New York, 1970.

Desorption Ionization

12-23 R. G. Cooks and K. L. Busch, *J. Chem. Educ.*, **59**, 926 (1982).
12-24 P. A. Lyon (Ed.), *Desorption Mass Spectrometry*, ACS Symposium Series, #291, American Chemical Society, Washington, DC, 1985.
12-25 A. Benninghoven (Ed.), *Ion Formation from Organic Solids*, Springer-Verlag, New York, 1983.
12-26 I. W. Wainer, M. Ahashi, R. P. Barron, and W. Benson, *Biomed. Mass Spectrom.*, **11**, 532 (1984).

General and Instrumental

12-27 M. L. Gross (Ed.), *High Performance Mass Spectrometry: Chemical Applications*, ACS Symposium Series, #70, American Chemical Society, Washington, DC, 1978.
12-28 J. Roboz, *Introduction to Mass Spectrometry, Instrumentation and Techniques*, Wiley–Interscience, New York, 1968.
12-29 B. S. Middeditch (Ed.), *Practical Mass Spectrometry*, Plenum, New York, 1979.
12-30 P. Dawson, *Quadrupole Mass Spectrometry and Its Applications*, Elsevier, Amsterdam, 1976.
12-31 C. A. McDowell (Ed.), *Mass Spectrometry*, McGraw-Hill, New York, 1963.
12-32 T. A. Lehman and M. M. Bursey, *Ion Cyclotron Resonance Spectrometry*, Wiley, New York, 1976.

Tandem Mass Spectrometry (MS–MS)

12-33 F. W. McLafferty (Ed.), *Tandem Mass Spectrometry*, Wiley–Interscience, New York, 1983.
12-34 R. G. Cooks and K. L. Busch, *J. Chem. Educ.*, **59**, 926–933 (1982).
12-35 R. A. Yost and D. D. Fetterolf, *Mass Spectrom. Rev.*, **2**, 1–45 (1983).

Quantitative Analysis

12-36 B. I. Millard, *Quantitative Mass Spectrometry*, Heyden, London, 1978.

LC–MS

12-37 C. R. Blakley and M. L. Vestal, *Anal. Chem*, **55**, 750 (1983).

GC–MS

12-38 W. H. McFadden, *Techniques of Combined Gas Chromatography–Mass Spectrometry*, Wiley, New York, 1973.
12-39 W. L. Budde and J. W. Eichelberger, *Organic Analysis Using GC–MS*, Ann Arbor Science, Ann Arbor, MI, 1979.

Interfaces and Data Processing

12-40 J. R. Chapman, *Computers in Mass Spectrometry*, Academic Press, New York, 1978.

Ion Chemistry See any recent issue of the American Society for Mass Spectrometry's Annual Conference on Mass Spectrometry and Allied Topics.

Remote and Proximate Fragmentation

12-41 N. Jensen, K. Tomer, and M. L. Gross, *J. Am. Chem. Soc.*, **107**, 1863 (1985).

Stereochemistry by MS

12-42 M. M. Green, *Top. Stereochem.*, **9**, 35–110 (1976).

Data Collections

12-43 E. Stenhagen, S. Abrahamsson, and F. W. McLafferty, *Registry of Mass Spectral Data*, Wiley, New York, 1974.
12-44 S. R. Heller and G. W. A. Milne, *EPA–NIH Mass Spectral Data Base*, Vols. 1–3, NSRDS-NBS 63, National Bureau of Standards, Washington, DC, 1978.

Appendix

IV.A

Table of Natural Isotopic Masses and Abundances

Element	Isotope	Mass	Natural Abundance
Hydrogen	^{1}H	1.0078	99.985
	^{2}H	2.0140	0.015
Carbon	^{12}C	12.0000	98.89
	^{13}C	13.0033	1.11
Nitrogen	^{14}N	14.0031	99.63
	^{15}N	15.0001	0.37
Oxygen	^{16}O	15.9949	99.759
	^{17}O	16.9991	0.037
	^{18}O	17.9992	0.204
Fluorine	^{19}F	18.9984	100
Silicon	^{28}Si	27.9769	92.21
	^{29}Si	28.9765	4.70
	^{30}Si	29.9738	3.09
Phosphorus	^{31}P	30.9738	100
Sulfur	^{32}S	31.9721	95.0
	^{33}S	32.9715	0.76
	^{34}S	33.9679	4.22
Chlorine	^{35}Cl	34.9689	75.53
	^{37}Cl	36.9659	24.47
Bromine	^{79}Br	78.9183	50.54
Iodine	^{127}I	126.9004	100

From CRC Handbook of Chemistry & Physics, 60th ed., R. C. Weast (Ed.) and M. J. Astle (Assoc. Ed.), CRC Press, Boca Raton, FL 33431, 1979.

Appendix
IV.B
Program for Calculation of Isotopic Distributions from Molecular Formula

This program, written for an IBM PC by Werner Stolz, Rick Korzenowski, and others, allows the calculation of the isotopic distributions given the molecular formula.

```
10 REM                              ***** ISOTOPE *****
20 REM                        WRITTEN BY RICHARD W. KORZENIOSKI
30 REM                           REVISED BY WERNER LEE STOLZ
40 REM                                   JULY 1986
50 REM                              REQUIRES 20K OF MEMORY
60 HOME
70 PRINT "PLEASE WAIT...LOADING DATA"
80 DIM ST(63,13)
90 DIM NU%(200)     : REM ATOMIC NUMBER OF NTH ATOM IN MOLECULE
100 DIM F$(40)      : REM ATOMIC NUMBER OF NTH ATOM IN MOLECULE
110 DIM MW%(92)     : REM BASE ATOMIC MASS OF ELEMENT
120 DIM AT%(40)     : REM ATOMIC NUMBER ELEMENTS
130 DIM N$(95)      : REM SYMBOL FOR EACH ELEMENT
140 DIM IS(92,13)   : REM ABUNDANCE OF ISOTOPES OF ELEMENT
150 DIM NI(92)      : REM SIZE OF RANGE OF ISOTOPES
160 FOR I = 1 TO 92
170 READ MW%(I),N$(I),NI(I)
180 FOR J = 1 TO NI(I)
190 READ IS(I,J)
200 NEXT J
210 NEXT I
220 REM LOCATE 25,25
230 REM PRINT "Mass Spec Isotope Peak Simulation"
240 REM LOCATE 1,1
250 PRINT "This program generates a simulated mass spectrogram which indicates"
260 PRINT "the relative peak heights due to natural isotope abundancies."
270 PRINT
280 PRINT "The program will print a list of the elements with their atomic"
290 PRINT "numbers if you wish.  Next you must specify how many elements are"
300 PRINT "in the molecule you are investigating."
310 PRINT
320 PRINT "The program will then ask for a range of mass units, you must then"
330 PRINT "specify how large a mass range the program should calculate"
340 PRINT "abundancies for.  Too small a range will not show all the peaks due"
350 PRINT "to isotopes, and too large a range will slow down execution time"
360 PRINT "considerably and generate a larger than neccessary printout with "
370 PRINT "zeros for all the abundancies at higher masses."
380 PRINT
390 PRINT "Before the program begins to calculate the peaks, it will show an"
400 PRINT "estimated output time, this is only an approximation."
410 PRINT
420 PRINT "It would be advisable not to get a printou of the results until"
430 PRINT "you have discovered just how large a mass range you need.  This"
440 PRINT "will save paper.  Good luck!"
450 PRINT
460 PRINT "Press RETURN to continue"
470 LINE INPUT Z$
480 DIM MP(63)      : REM RELATIVE STRENGTH OF PEAK FOR A MASS N
490 DIM M1(63)      : REM RELATIVE STRENGTH OF PEAK AS A PERCENT
500 DIM M2(63)      : REM TEMPORARY STORAGE FOR PEAK STRENGTHS
```

```
510 HOME
520 PRINT "Would you like a list of the elements and their atomic ";
530 INPUT "numbers? (Y or N) ";A$
540 IF A$ = "Y" OR A$ = "y" THEN 660
550 IF A$ <> "L" OR A$ = "l" THEN 720
560 PRINT : PRINT
570 FOR I = 1 TO 92
580 PRINT I;":";N$(I)
590 SM = 0
600 FOR J = 1 TO NI(I)
610 SM = SM + IS(I,J)
620 PRINT IS(I,J);",";
630 NEXT J
640 PRINT TAB(65);"*,>";SM;"*("
650 NEXT I
660 FOR I = 1 TO 19
670 PRINT N$(I);I;TAB(12);N$(I+19);I+19;TAB(25);N$(I+38);
680 PRINT I+38;TAB(38);N$(I+57);I+57;TAB(52);N$(I+76);
690 IF N$(I+76) = "" THEN PRINT ELSE PRINT I+76
700 NEXT I
710 PRINT
720 PRINT "Enter the number of elements in the molecule : (1-40) ";
730 LINE INPUT A$
740 PRINT
750 NE = VAL(A$)
760 IF NE <= 0 OR NE > 39 THEN 2430
770 PRINT "Enter the range of mass units to cover : (1-50) ";
780 LINE INPUT A$
790 PRINT
800 UC = VAL(A$)
810 IF UC <= 0 OR UC > 50 THEN 2430
820 PRINT "Enter the atomic symbols of the elements that occur in the molecule"
830 PRINT "and the number of times that each is present. (one pair per line, ";
840 PRINT "separated by a comma)"
850 NI(0) = 0
860 FOR I = 1 TO NE
870 INPUT S$,SU%(I)
880 J = 1
890 WHILE S$ <> N$(J) AND J < 93
900 J = J + 1
910 WEND
920 IF J > 92 THEN 2450
930 AT%(I) = J
940 IF NI(J) > NI(0) THEN NI(0) = NI(J)
950 NEXT I
960 PRINT
970 N=0
980 FOR KE=1 TO NE
990 IF SU%(KE)=0 THEN 1050
1000 FOR K=1 TO SU%(KE)
1010 N=N+1
1020 IF N>200 THEN 2400
1030 NU%(N)=AT%(KE)
1040 NEXT K
1050 NEXT KE
1060 MP(1)=1
1070 FOR KA = 2 TO UC
1080 MP(KA) = 0
1090 NEXT KA
1100 TP = (UC * ((N * (2 + NI(0)) * 1.312) * .878) + (NE * 4/3)) * .03156
1110 TH = INT(TP/3600)
1120 TM=INT((TP - TH * 3600)/60)
1130 TS=INT(TP - TH * 3600 - TM * 60)
1140 HOME
1150 PRINT "Estimated output time : ";
```

```
1160 PRINT TH;" hours, ";TM;" minutes, ";TS;" seconds"
1170 PRINT
1180 LINE INPUT "Continue? ";A$
1190 IF MID$(A$,1,1)="N" OR MID$(A$,1,1) = "n" THEN 2340
1200 PRINT
1210 P = 0
1220 LINE INPUT "Output to printer? (Y or N)";P$
1230 P$=MID$(P$,1,1)
1240 IF P$ = "Y" OR P$ = "y" THEN P = 1
1250 FOR L = 1 TO N
1260 F = NU%(L)
1270 IF F <= 0 THEN 1520
1280 FOR I = 1 TO UC
1290 M2(I) = 0
1300 NEXT I
1310 FOR I = 1 TO UC
1320 FOR J = 1 TO NI(0)
1330 ST(I,J) = IS(F,J) * MP(I)
1340 IF ST(I,J)<.0000001 THEN ST(I,J) = 0
1350 M2(I + J - 1) = M2(I + J - 1) + ST(I,J)
1360 ST(I,J) = 0
1370 NEXT J
1380 NEXT I
1390 FOR I=1 TO UC
1400 MP(I) = M2(I)
1410 NEXT I
1420 NEXT L
1430 PRINT
1440 TL = 0
1450 FOR I = 1 TO UC
1460 TL = TL + MP(I)
1470 NEXT I
1480 NS = 0
1490 FOR I = 1 TO NE
1500 G = AT%(I)
1510 IF G <= 0 THEN 1540
1520 F$(I) = N$(G)
1530 NS = NS + 1
1540 NEXT I
1550 PRINT
1560 PRINT "PARENT/PEAK FRAGMENTATION ANALYSIS OF THE COMPOUND :"
1570 IF P = 1 THEN LPRINT "ISOTOPE"
1580 IF P THEN LPRINT "PARENT/PEAK FRAGMENTATION ANALYSIS OF THE COMPOUND :"
1590 PRINT "ATOMIC #";
1600 IF P THEN LPRINT "ATOMIC #";
1610 FOR I = 1 TO NS
1620 PRINT TAB(I * 5 + 5);AT%(I);
1630 IF P THEN LPRINT TAB(I * 5 + 5);AT%(I);
1640 NEXT I
1650 PRINT
1660 IF P THEN LPRINT
1670 PRINT "SYMBOL";
1680 IF P THEN LPRINT "SYMBOL";
1690 FOR I=1 TO NS
1700 PRINT TAB(I * 5 + 6);F$(I);
1710 IF P THEN LPRINT TAB(I * 5 + 6);F$(I);
1720 NEXT I
1730 PRINT
1740 IF P THEN LPRINT
1750 PRINT "# ATOMS";
1760 IF P THEN LPRINT "# ATOMS";
1770 FOR I=1 TO NS
1780 PRINT TAB(I * 5 + 5);SU%(I);
1790 IF P THEN LPRINT TAB(I * 5 + 5);SU%(I);
1800 NEXT I
```

```
1810 PRINT
1820 IF P THEN LPRINT
1830 TE = 1 - TL
1840 PRINT "TOTAL ERROR IN CALCULATION :";TE
1850 IF P THEN LPRINT "TOTAL ERROR IN CALCULATION :";TE
1860 PRINT
1870 PRINT "       AMU         % OF 1.0000                           % W.R.T. ";
1880 IF P THEN LPRINT "       AMU         % OF 1.0000                           % ";
1890 PRINT "LARGEST PEAK"
1900 IF P THEN LPRINT "W.R.T. LARGEST PEAK"
1910 MO = 0
1920 FOR KJ = 1 TO NE
1930 I=AT%(KJ)
1940 IF I = 0 THEN I = 93
1950 MO = MO + SU%(KJ) * MW%(I)
1960 NEXT KJ
1970 AM = 0
1980 FOR KK = 1 TO UC
1990 IF MP(KK) > AM THEN AM = MP(KK)
2000 NEXT KK
2010 FOR KL = 1 TO UC
2020 M1(KL) = MP(KL) / AM * 100
2030 NEXT KL
2040 MO = MO - 1
2050 FOR KM = 1 TO UC
2060 MO = MO + 1
2070 PRINT TAB(6);MO;TAB(17);MP(KM);TAB(45);M1(KM)
2080 IF P THEN LPRINT TAB(6);MO;TAB(17);MP(KM);TAB(45);M1(KM)
2090 NEXT KM
2100 PRINT
2110 PRINT
2120 PRINT
2130 IF P THEN LPRINT : LPRINT : LPRINT
2140 IF P AND UC > 21 THEN LPRINT CHR$(12)
2150 PRINT " AMU       SIMULATION OF PARENT/FRAGMENT PEAK REGION"
2160 PRINT "-----     -------------------------------------------"
2170 IF P THEN LPRINT " AMU       SIMULATION OF PARENT/FRAGMENT PEAK REGION"
2180 IF P THEN LPRINT "-----     -------------------------------------------"
2190 MO = MO - UC
2200 FOR KM = 1 TO UC
2210 MO = MO + 1
2220 MM = M1(KM) * .7
2230 MM = INT(MM + .5)
2240 PRINT MO;TAB(7);"I";
2250 IF P THEN LPRINT MO;TAB(7);"I";
2260 IF MM < 1 THEN 2310
2270 FOR KX = 1 TO MM
2280 PRINT "*";
2290 IF P THEN LPRINT "*";
2300 NEXT KX
2310 PRINT
2320 IF P THEN LPRINT
2330 NEXT KM
2340 PRINT
2350 IF P THEN LPRINT CHR$(12)
2360 LINE INPUT "ANOTHER PROBLEM? ";A$
2370 A$ = MID$(A$,1,1)
2380 IF A$ = "Y" OR A$ = "y" THEN ERASE MP,M1,M2 : GOTO 480
2390 END
2400 PRINT
2410 PRINT "TOO MANY ATOMS (LIMIT = 200).  CHOP OFF HYDROGENS"
2420 GOTO 2340
2430 PRINT "**** ERROR ****"
2440 GOTO 2340
2450 PRINT "YOU HAVE ENTERED AN INCORRECT ATOMIC SYMBOL, PLEASE TRY AGAIN"
```

```
2460 GOTO 870
2470 DATA 1,H,2,.99985,1.5E-4
2480 DATA 3,He,2,1.3E-6,.999987
2490 DATA 6,Li,2,.075,.925
2500 DATA 9,Be,1,1
2510 DATA 10,B,2,.2,.8
2520 DATA 12,C,2,.9889,.0111
2530 DATA 14,N,2,.9964,.0036
2540 DATA 16,O,3,.9976,4.0E-4,.002
2550 DATA 19,F,1,1
2560 DATA 20,Ne,3,.9051,.0027,.0922
2570 DATA 23,Na,1,1
2580 DATA 24,Mg,3,.7899,.1,.1101
2590 DATA 27,Al,1,1
2600 DATA 28,Si,3,.9233,.0467,.031
2610 DATA 31,P,1,1
2620 DATA 32,S,5,.95,.0076,.0422,0,2.0E-4
2630 DATA 35,Cl,3,.7577,0,.2423
2640 DATA 36,Ar,5,.0034,0,7.0E-4,0,.9959
2650 DATA 39,K,3,.9326,1.0E-4,.0673
2660 DATA 40,Ca,9,.96941,0,.00647,.00135,.02086,0,4.0E-5,0,.00187
2670 DATA 45,Sc,1,1
2680 DATA 46,Ti,5,.08,.075,.737,.055,.053
2690 DATA 50,V,2,.0025,.9975
2700 DATA 50,Cr,5,.0435,0,.8379,.095,.0236
2710 DATA 55,Mn,1,1
2720 DATA 54,Fe,5,.058,0,918,.021,.003
2730 DATA 59,Co,1,1
2740 DATA 58,Ni,7,.6827,0,.261,.0113,.0359,0,.0091
2750 DATA 63,Cu,3,.692,0,.308
2760 DATA 64,Zn,7,.486,0,.279,.041,.188,0,.006
2770 DATA 69,Ga,3,.6,0,.4
2780 DATA 70,Ge,8,.205,0,.274,0,.078,.365,0,.078
2790 DATA 75,As,1,1
2800 DATA 74,Se,9,.009,0,.09,.076,.235,0,.498,0,.092
2810 DATA 79,Br,3,.5069,0,.4931
2820 DATA 78,Kr,10,.0035,0,.0225,0,.116,.115,.57,0,0,.173
2830 DATA 85,Rb,3,.7217,0,.2783
2840 DATA 84,Sr,5,.005,0,.099,.07,.826
2850 DATA 89,Y,1,1
2860 DATA 90,Zr,7,.514,.112,.171,0,.175,0,.028
2870 DATA 93,Nb,1,1
2880 DATA 92,Mo,9,.148,0,.093,.159,.167,.096,.241,0,.096
2890 DATA 98,Tc,1,1
2900 DATA 96,Ru,9,.055,0,.019,.127,.126,.171,.316,0,.186
2910 DATA 103,Rh,1,1
2920 DATA 102,Pd,9,.01,0,.11,.222,.273,0,.267,0,.118
2930 DATA 107,Ag,3,.5183,.0,.4817
2940 DATA 106,Cd,11,.012,0,.009,0,.124,.128,.24,.123,.288,0,.076
2950 DATA 113,In,3,.043,0,.957
2960 DATA 112,Sn,13,.01,0,.007,.004,.146,.077,.243,.086,.324,0,.046,0,.056
2970 DATA 121,Sb,3,.573,0,.427
2980 DATA 120,Te,11,.001,0,.025,.009,.046,.07,.187,0,.317,0,.345
2990 DATA 127,I,1,1
3000 DATA 124,Xe,13,.001,0,.001,0,.019,.264,.041,.212,.269,0,.104,0,.089
3010 DATA 133,Cs,1,1
3020 DATA 130,Ba,7,.001,0,.024,.066,.079,.112,.717
3030 DATA 138,La,2,9.0E-4,.9991
3040 DATA 136,Ce,7,.002,0,.003,0,.884,0,.111
3050 DATA 141,Pr,1,1
3060 DATA 142,Nd,9,.272,.122,.238,.083,.172,0,.057,0,.056
3070 DATA 145,Pm,1,1
3080 DATA 144,Sm,11,.031,0,0,.151,.113,.139,.074,0,.266,0,.226
3090 DATA 151,Eu,3,.478,0,.522
3100 DATA 152,Gd,9,.002,0,.022,.148,.205,.157,.248,0,.218
```

```
3110 DATA 159,Tb,1,1
3120 DATA 156,Dy,9,6.0E-4,0,.001,0,.0234,.189,.255,.249,.282
3130 DATA 165,Ho,1,1
3140 DATA 162,Er,9,.001,0,.016,0,.334,.229,.27,0,.15
3150 DATA 169,Tm,1,1
3160 DATA 168,Yb,9,.001,0,.031,.143,.219,.162,.317,0,.127
3170 DATA 175,Lu,2,.974,.026
3180 DATA 174,Hf,7,.002,0,.052,.185,.271,.138,.352
3190 DATA 180,Ta,2,1.2E-4,.99988
3200 DATA 180,W,7,.001,0,.263,.143,.307,0,.286
3210 DATA 185,Re,3,.374,0,.626
3220 DATA 184,Os,9,2.0E-4,0,.0158,.016,.133,.161,.264,0,.41
3230 DATA 191,Ir,3,.373,0,.627
3240 DATA 190,Pt,9,1.0E-4,0,.0079,0,.329,.338,.253,0,.072
3250 DATA 197,Au,1,1
3260 DATA 196,Hg,9,.002,0,.101,.169,.231,.132,.297,0,.068
3270 DATA 203,Tl,3,.295,0,.705
3280 DATA 204,Pb,5,.014,0,.241,.221,.524
3290 DATA 209,Bi,1,1
3300 DATA 209,Po,1,1
3310 DATA 210,At,1,1
3320 DATA 222,Rn,1,1
3330 DATA 223,Fr,1,1
3340 DATA 226,Ra,1,1
3350 DATA 227,Ac,1,1
3360 DATA 232,Th,1,1
3370 DATA 231,Pa,1,1
3380 DATA 234,U,5,5.0E-5,.0072,0,0,.99275
```

Part
V

INTEGRATED PROBLEMS

In the following problems all four spectroscopic methods are to be used for the derivation of structures. The problems are arranged in approximate order of increasing difficulty, and spectra are presented just as they might be obtained from the respective instruments. Thus elemental formulas are not provided from the mass spectra, integrals are not calculated from the ^{1}H NMR data, and molar absorptivities are not given for the electron spectra.

One possible procedure for approaching these problems is as follows. Obtain the molecular weight from the chemical ionization mass spectrum. Compare this value with the relative numbers of protons and carbons from the NMR spectra in order to get an initial idea of the numbers of carbon and hydrogen atoms in the molecular formula. Consider the possibility of nitrogen, sulfur, or silicon by examination of the mass spectrum. Use the infrared spectrum to determine the nature of the functional groups in the molecule. At this stage suggest a complete molecular formula and determine whether it is consistent with all the data, including the UV–Vis spectrum. Calculate the unsaturation number from the formula and proceed to develop structural units from all the spectra. Finally, bring the units together in possible structures and reexamine all the data to determine if they are consistent with the structures. Circular dichroism spectra are given when the unknown is optically active, and they should be used to determine absolute configuration.

If you reach an impasse after due effort, we provide three tables at the end of the problems. In the first are given the molecular formulas for all the compounds. If you are still not able to deduce a structure, proceed to the second table, which lists the functional groups in the molecules. Finally, in the third table the full structures are given so that you may check your conclusions.

Use the spectra as further exercises on the principles developed in this text. Identify every peak in the NMR spectra with the appropriate structural component, and justify chemical shifts and coupling constants. Find infrared absorptions for all the functional groups and hydrocarbon components in the molecules. Rationalize the location of the wavelength maxima and the intensities of the ultraviolet absorbances. Calculate ϵ, assuming that the pathlength is always 1.0 cm. Explain the differences among the various mass spectra (EI, CI, and DI) and identify important fragmentation pathways.

PROBLEM 1

Mass spectrum

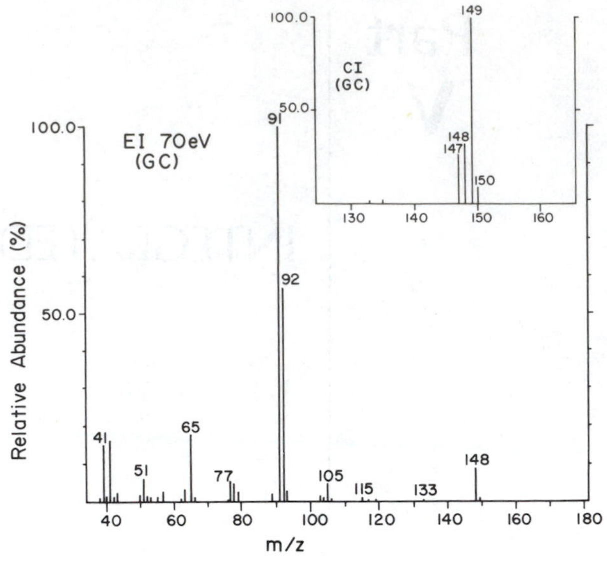

Proton NMR spectrum
$(CDCl_3)$

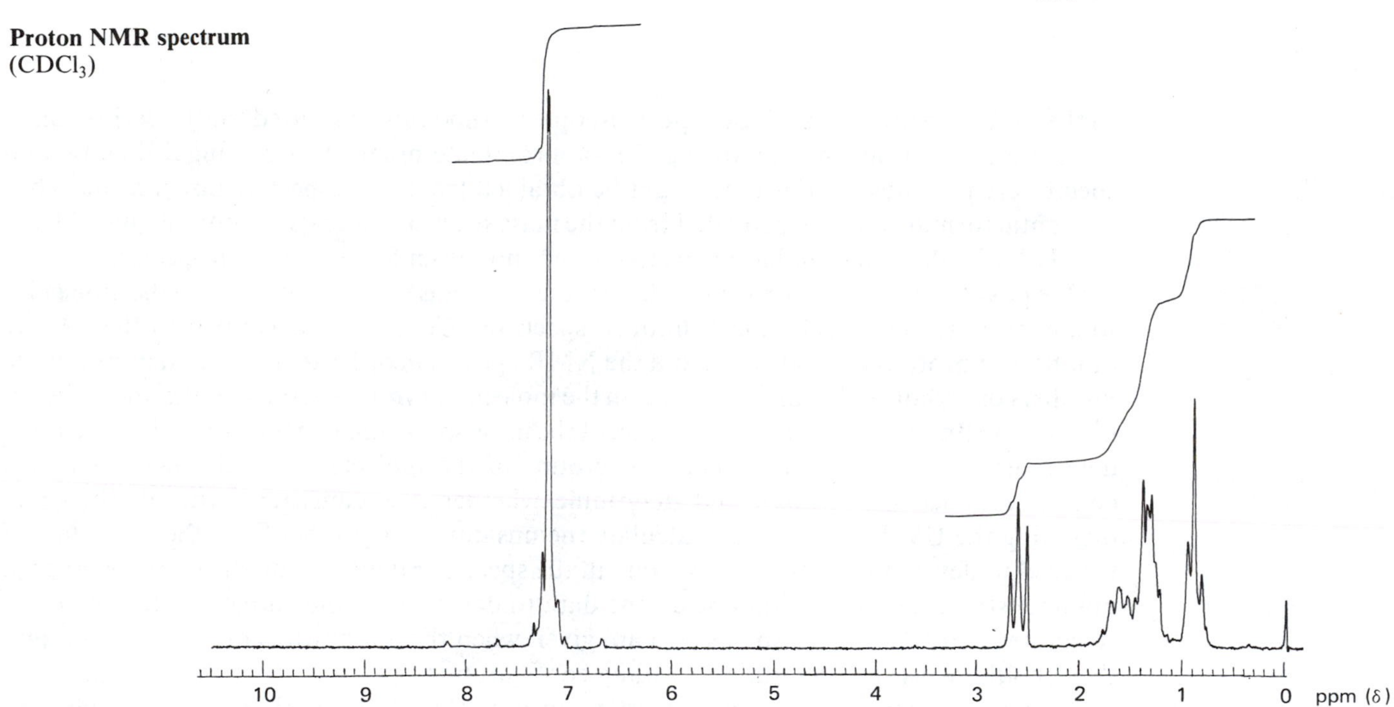

Carbon-13 NMR spectrum
$(CDCl_3)$

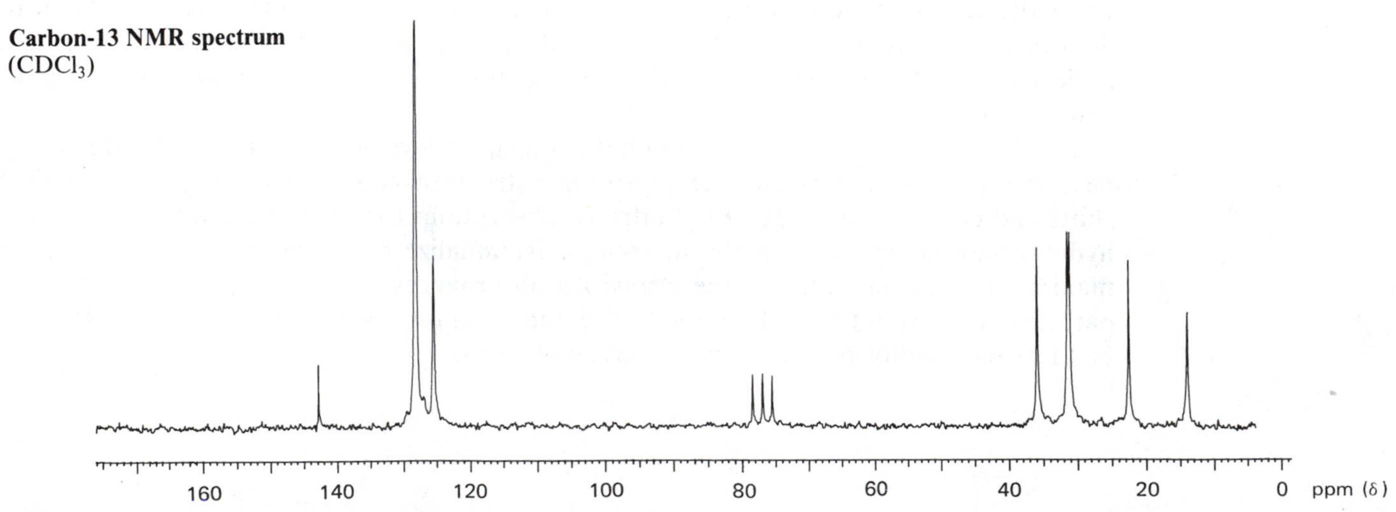

Infrared spectrum (neat)

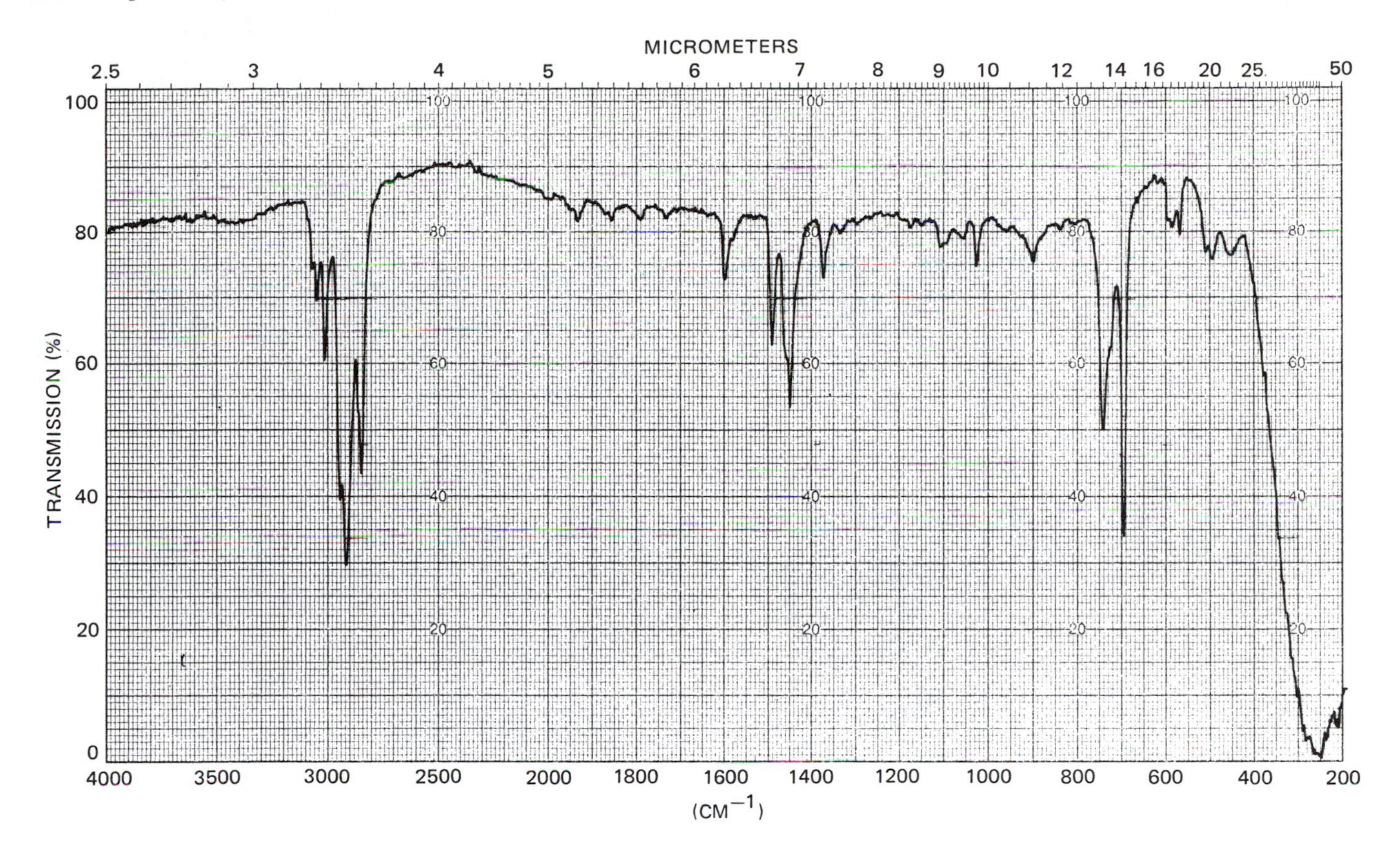

Ultraviolet-visible spectrum (CH_3OH, 2.33×10^{-3} M)

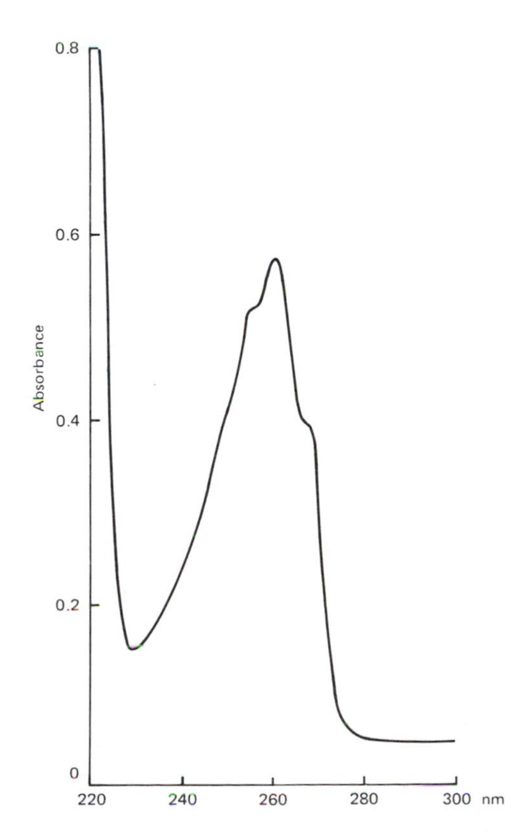

PROBLEM 2

Mass spectrum

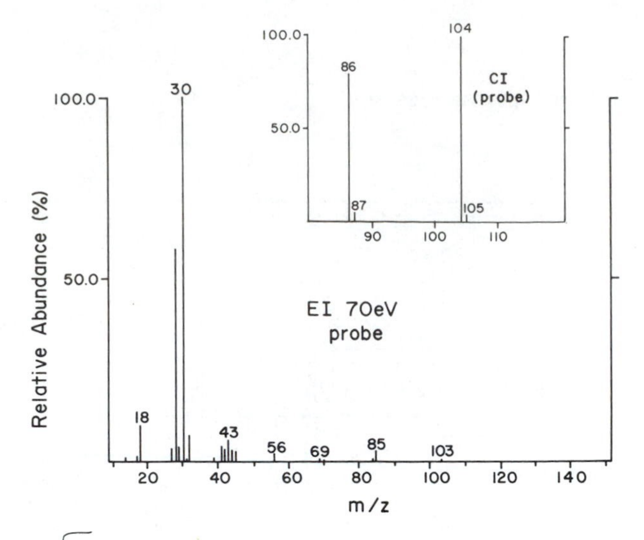

Proton NMR spectrum (D_2O)

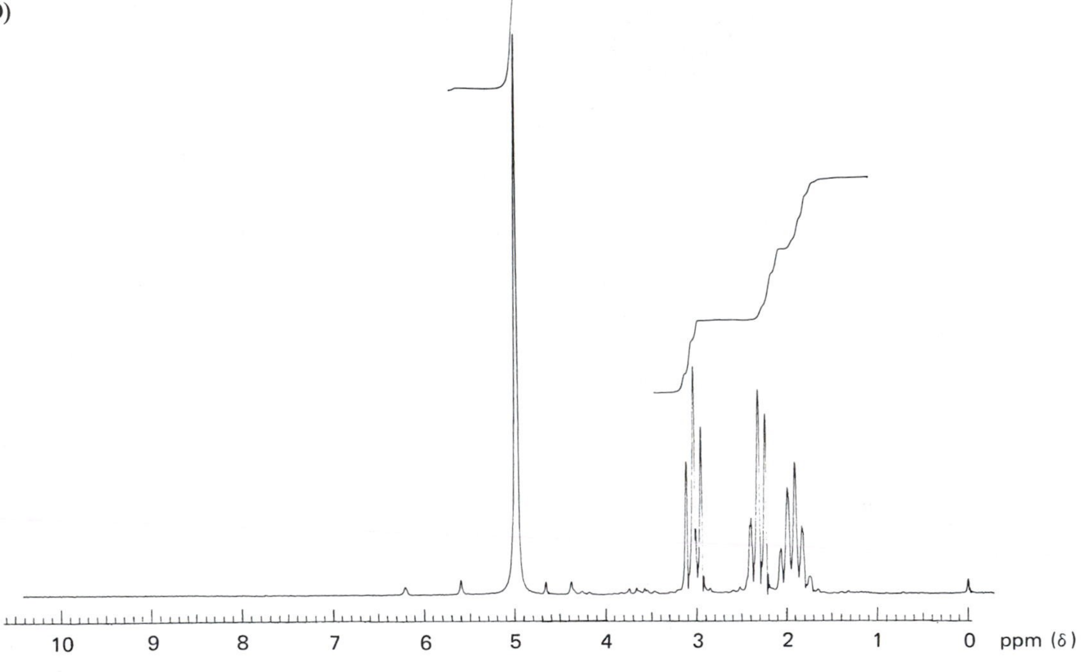

Carbon-13 NMR spectrum (D_2O)

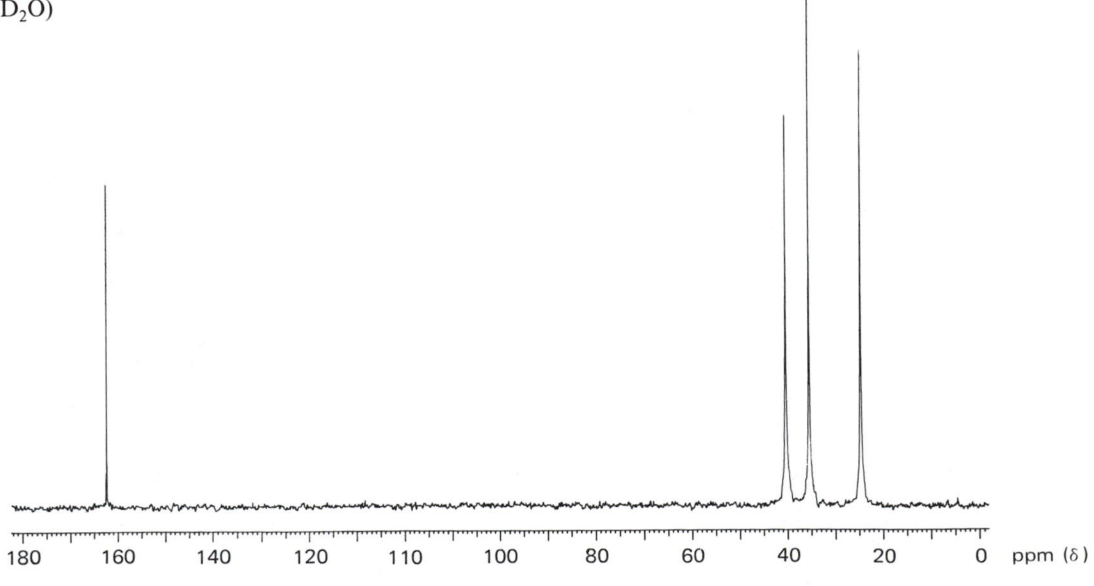

Infrared spectrum (KBr disc)

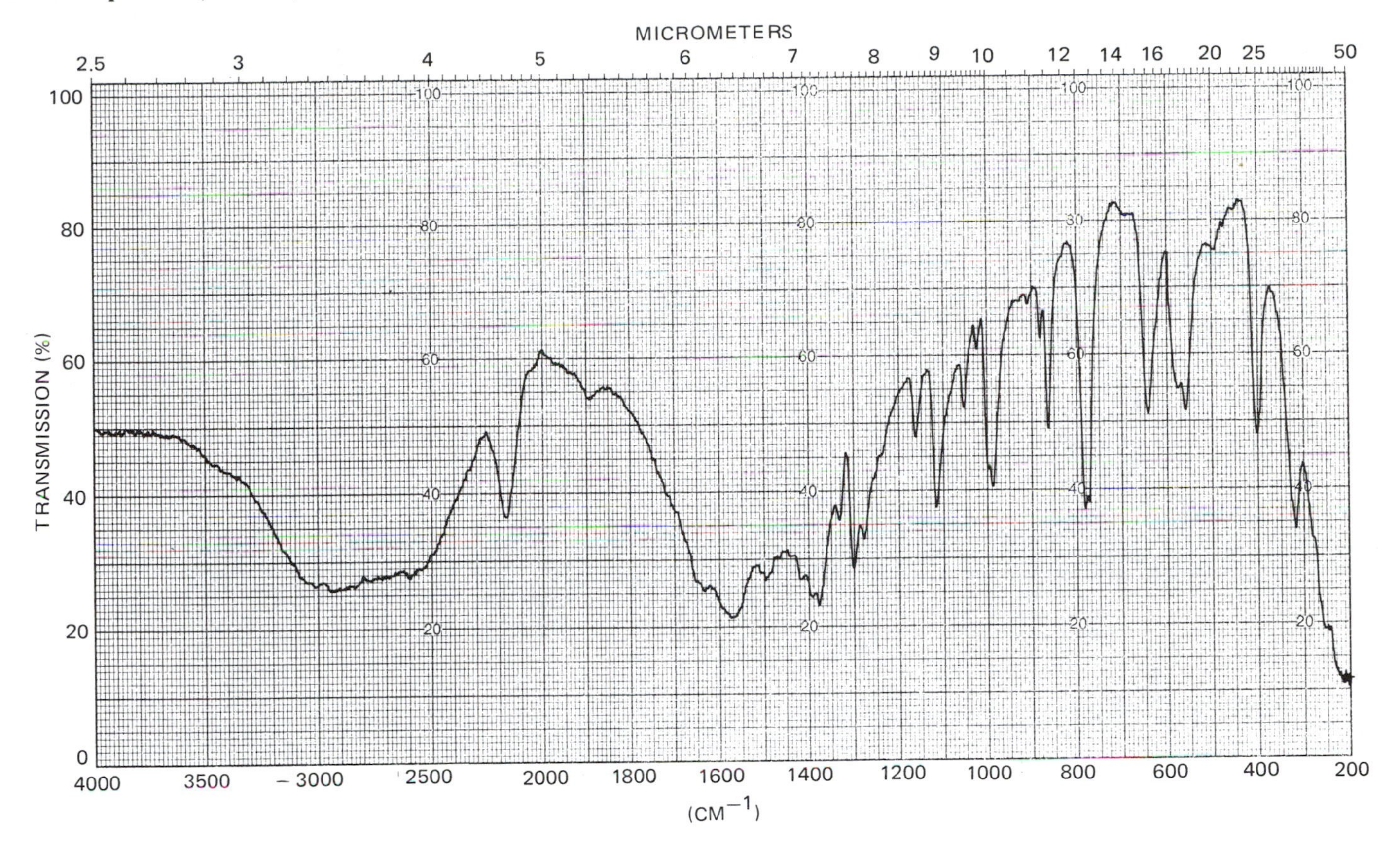

Ultraviolet-visible spectrum (CH_3OH, 0.027 M)

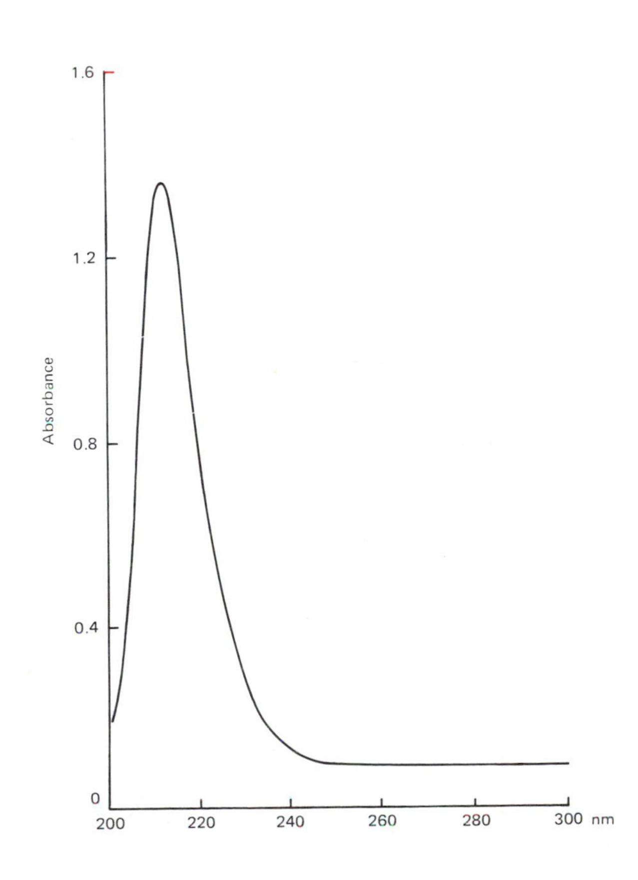

PROBLEM 3

Mass spectrum

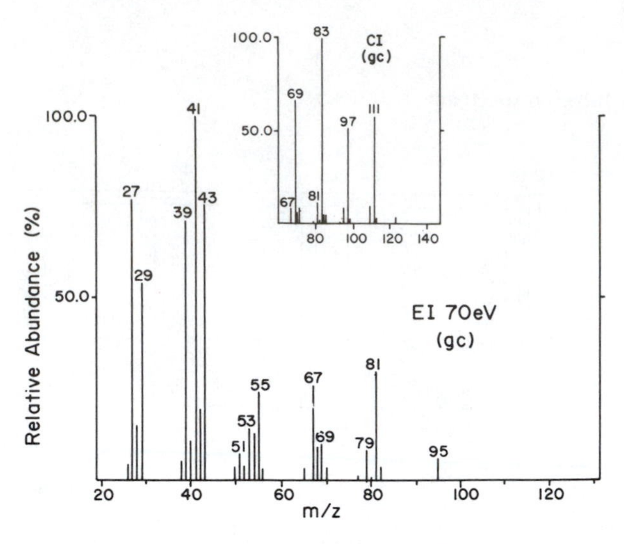

Proton NMR spectrum ($CDCl_3$)

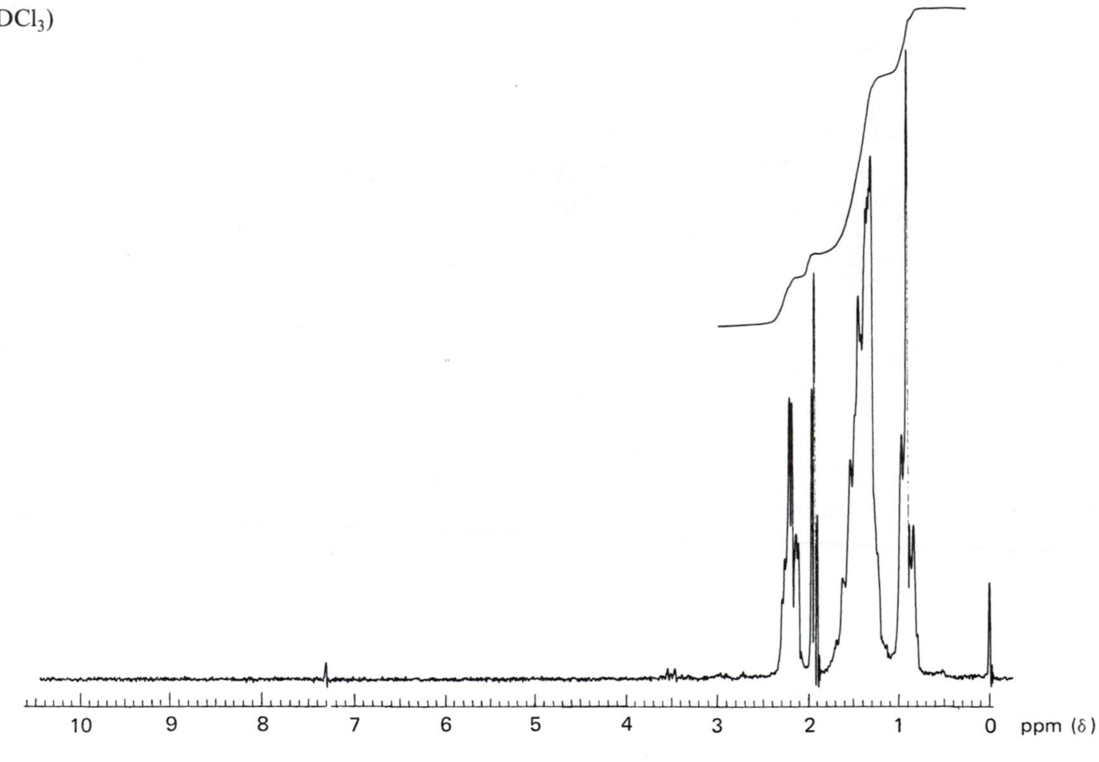

Carbon-13 NMR spectrum ($CDCl_3$)

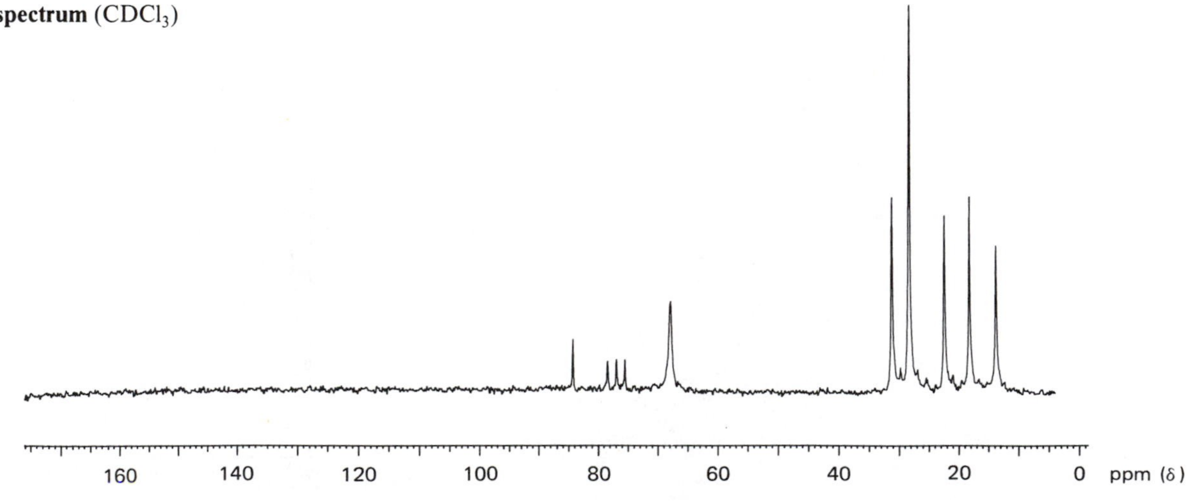

Infrared spectrum (neat)

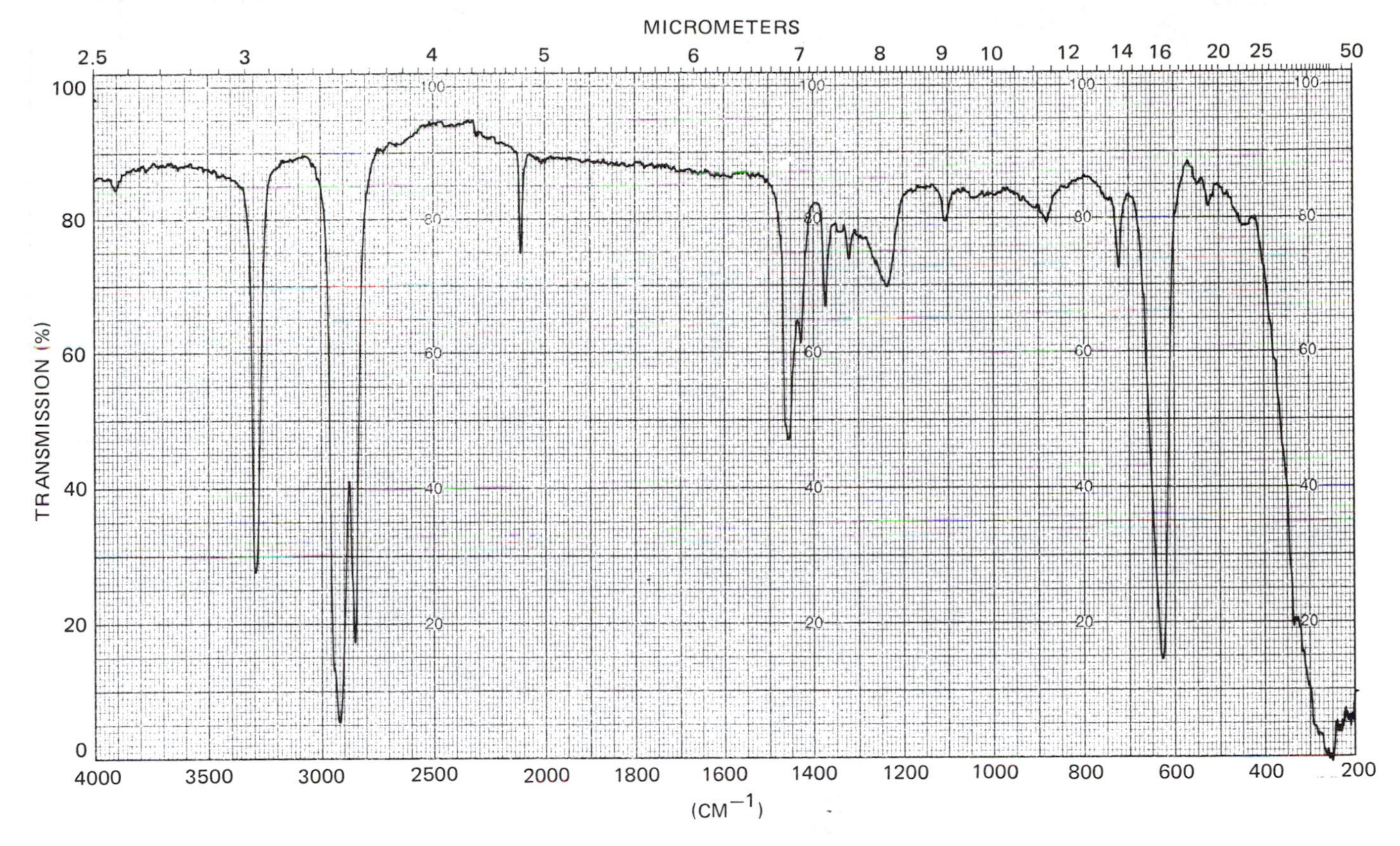

Ultraviolet–visible spectrum (heptane, 0.02 M)

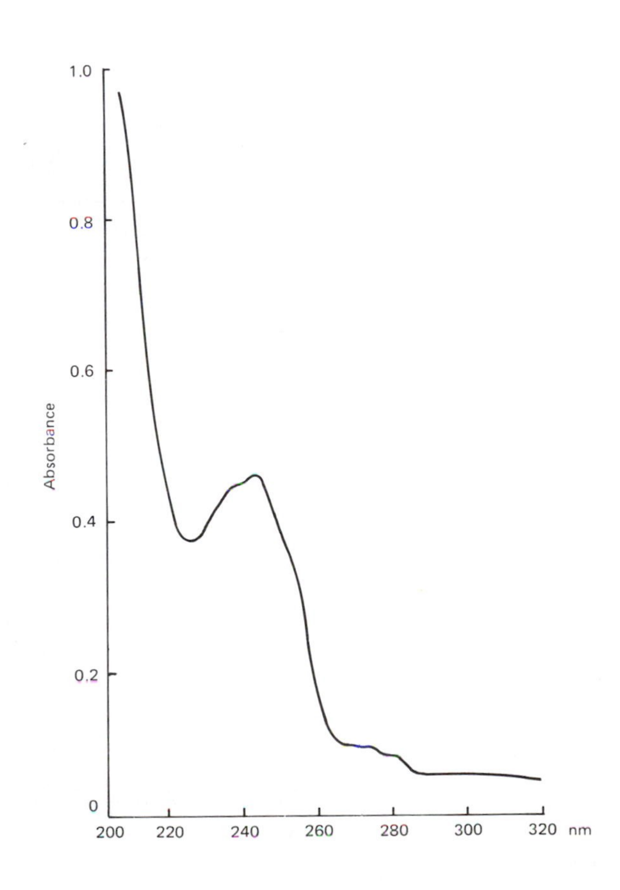

PROBLEM 4

Mass spectrum

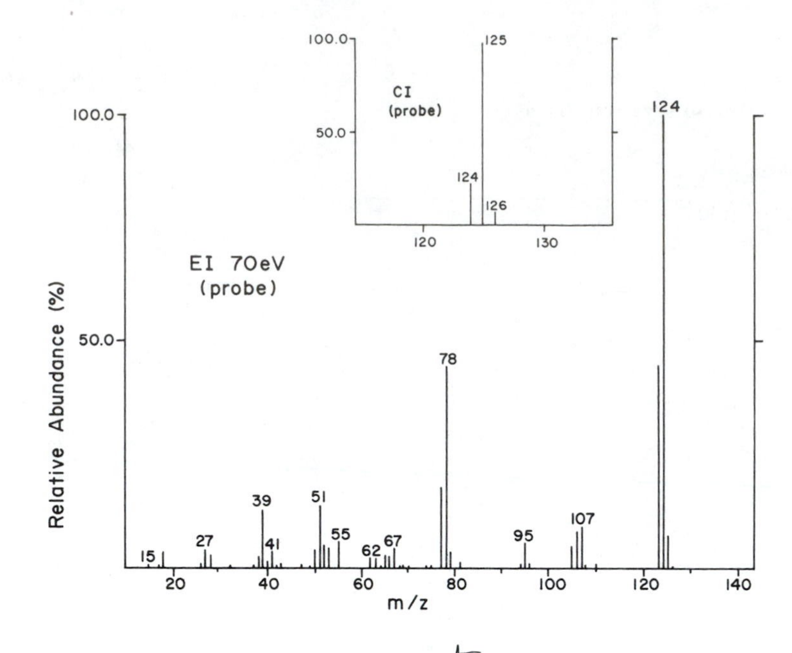

Proton NMR spectrum ($CDCl_3$)

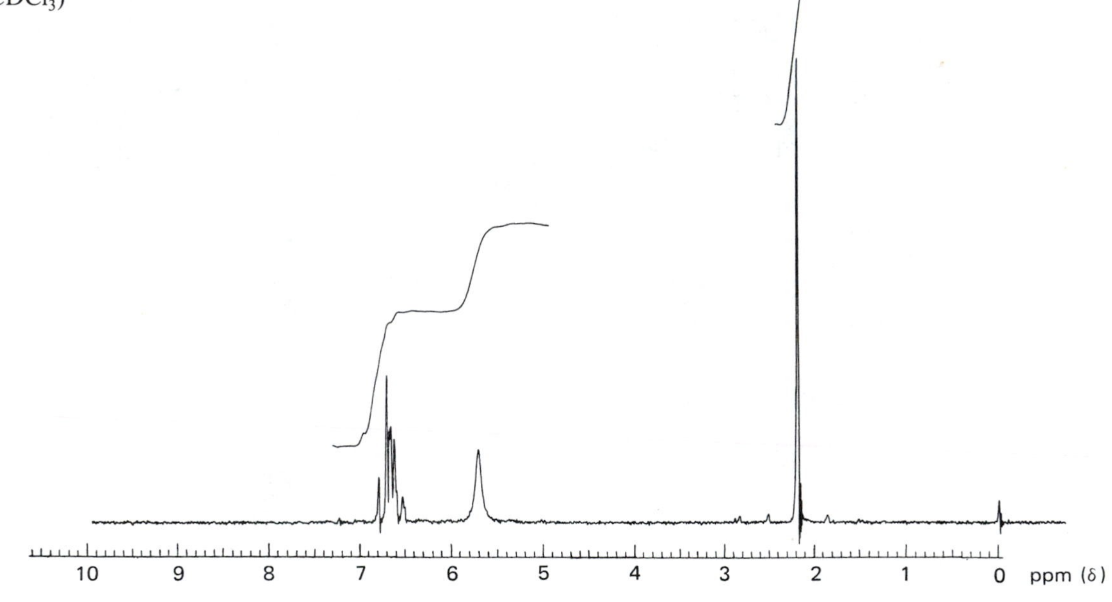

Carbon-13 NMR spectrum ($CDCl_3$)

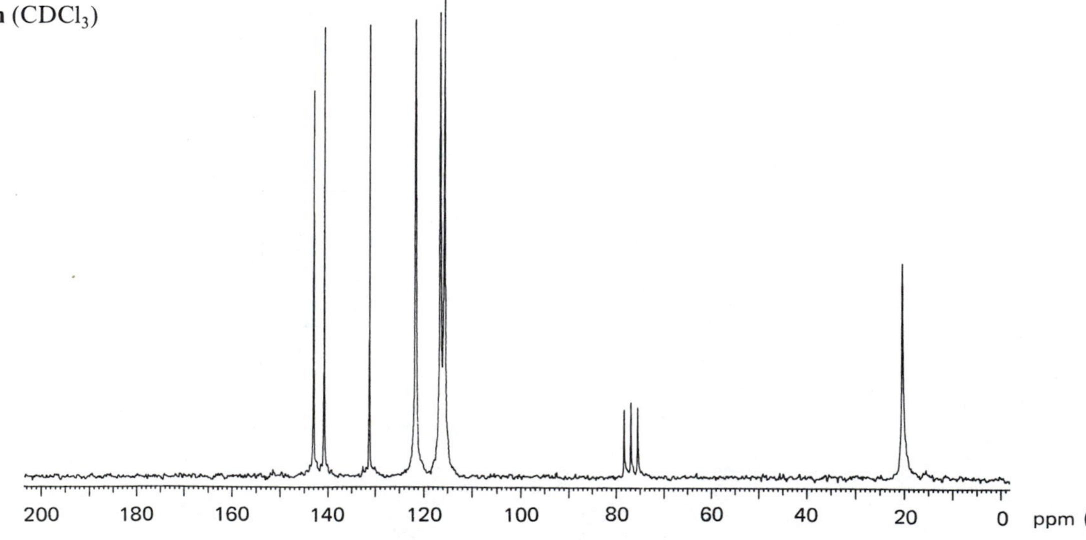

Infrared spectrum (KBr disc)

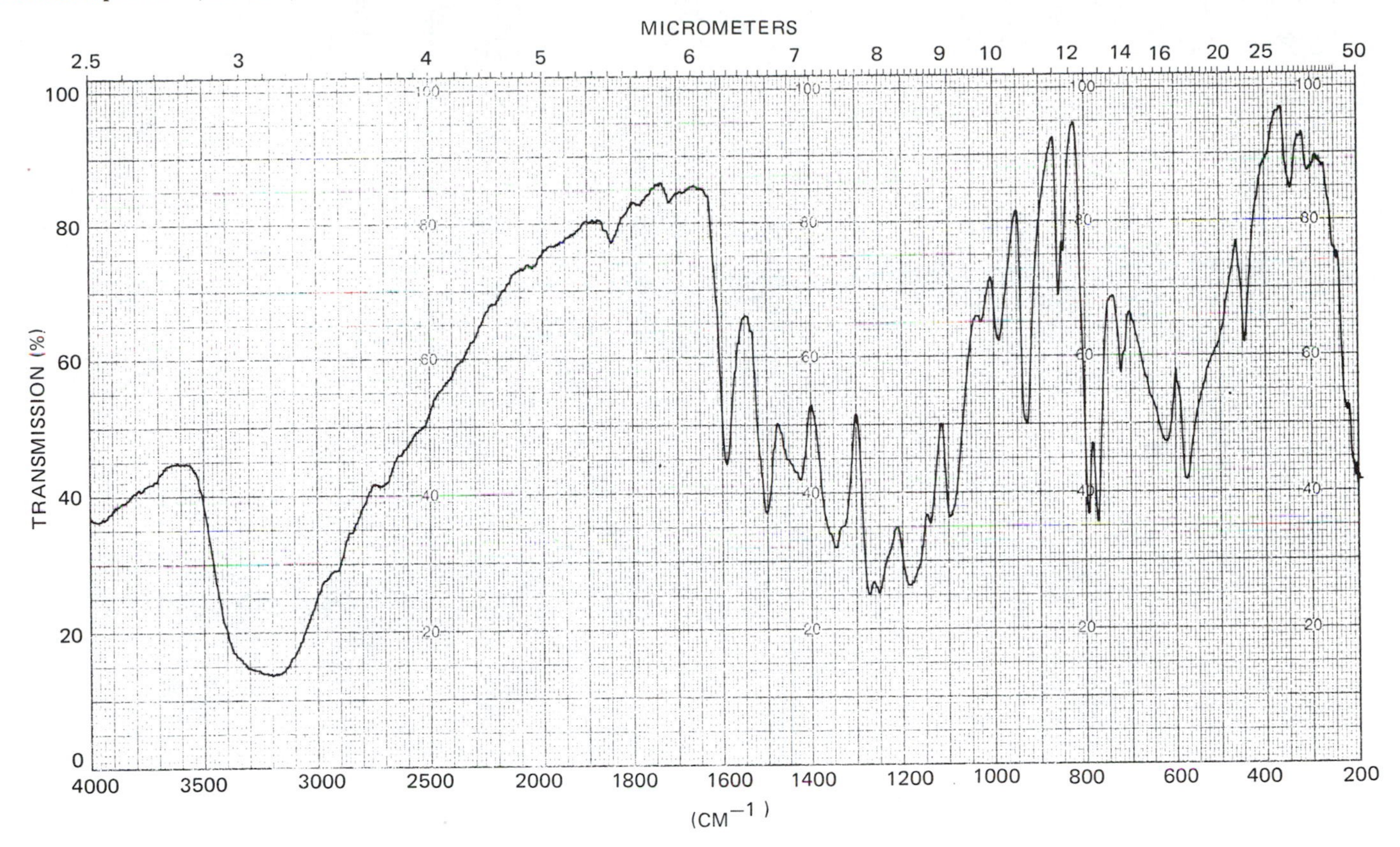

Ultraviolet-visible spectrum (CH_3OH, 9.4×10^{-5} M)

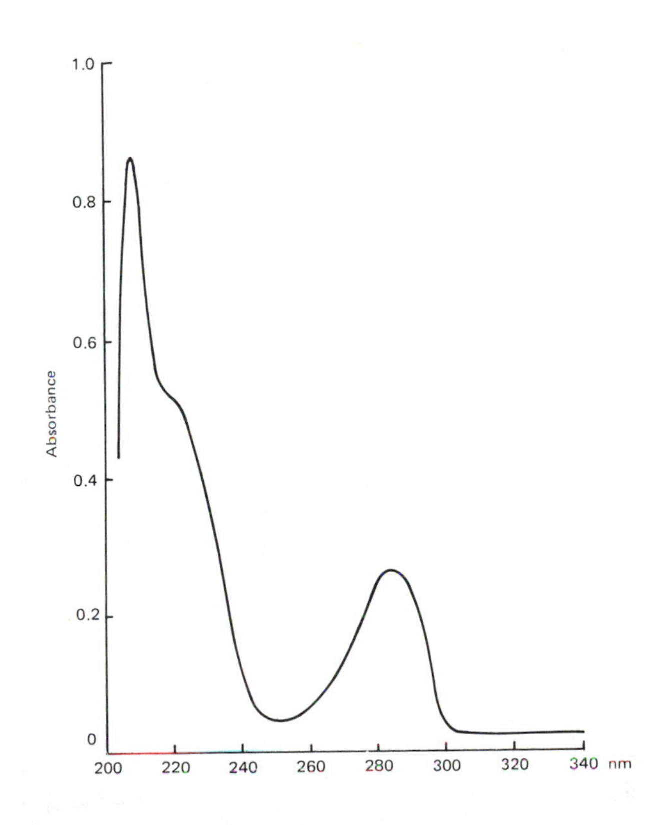

PROBLEM 5

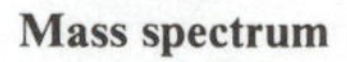

Mass spectrum

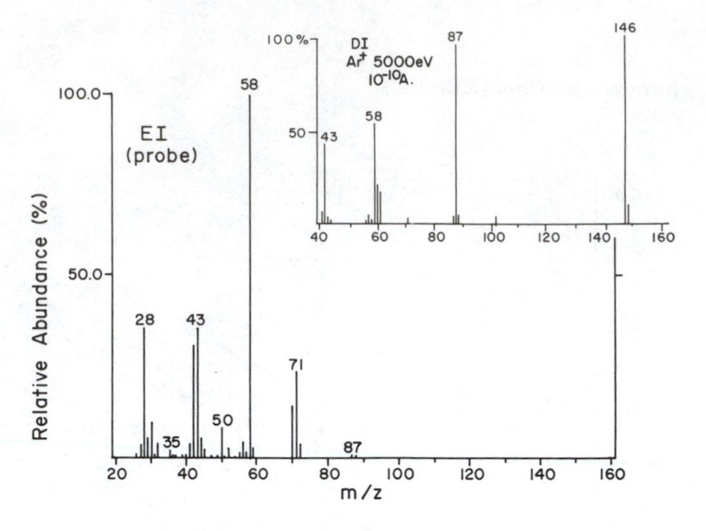

Proton NMR spectrum (D_2O)

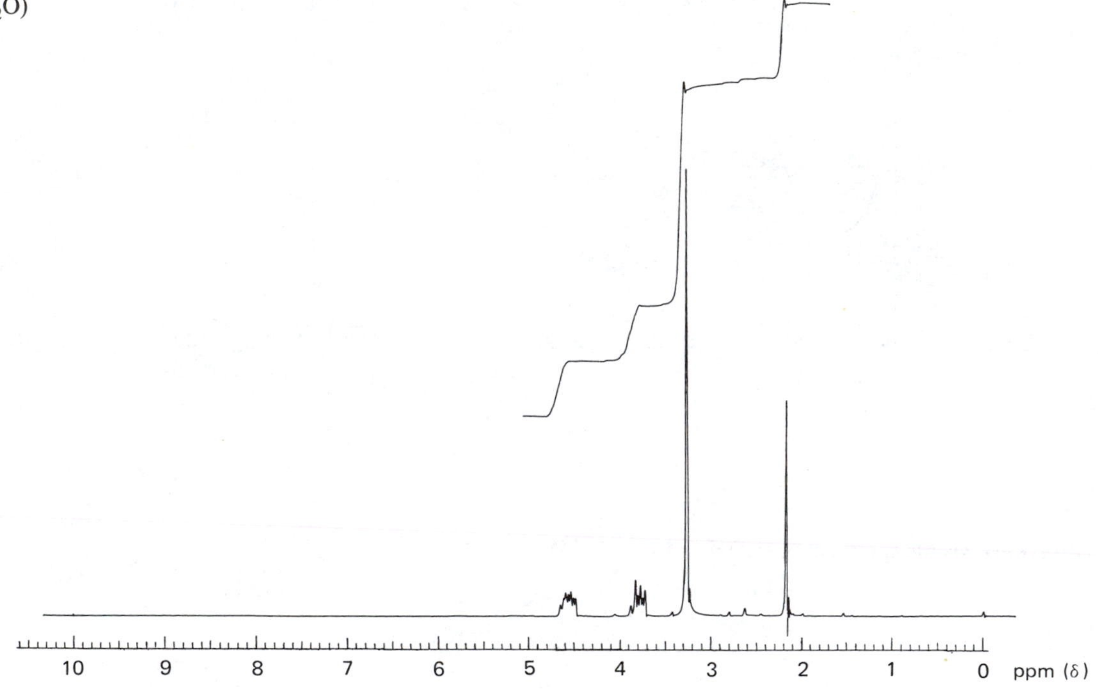

Carbon-13 NMR spectrum (D_2O)

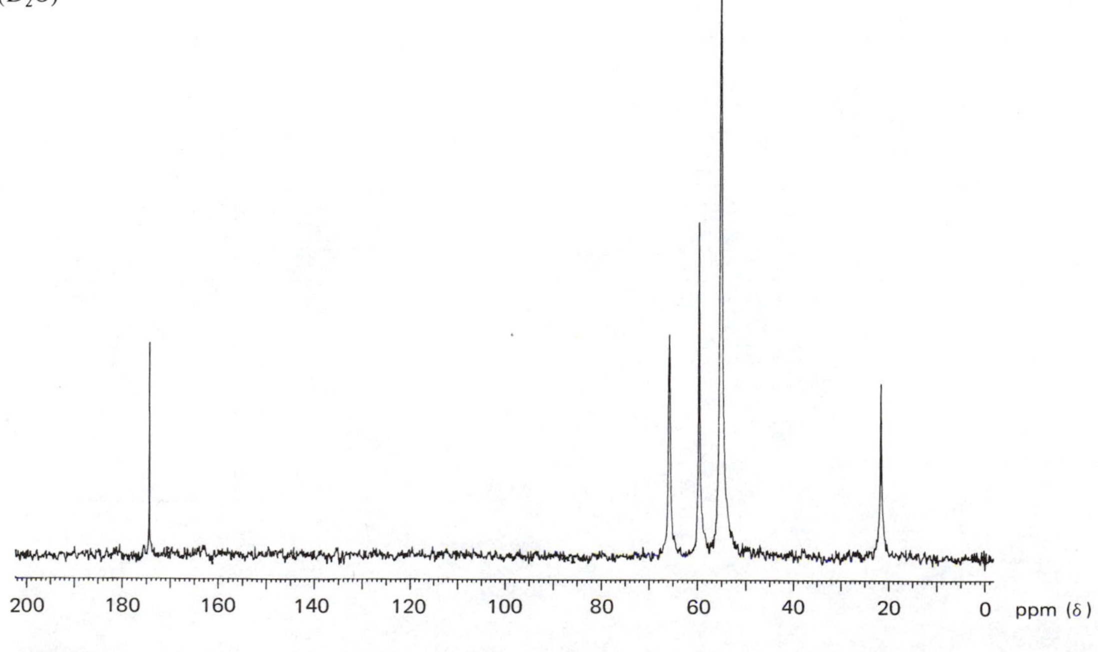

Infrared spectrum (KBr disc)

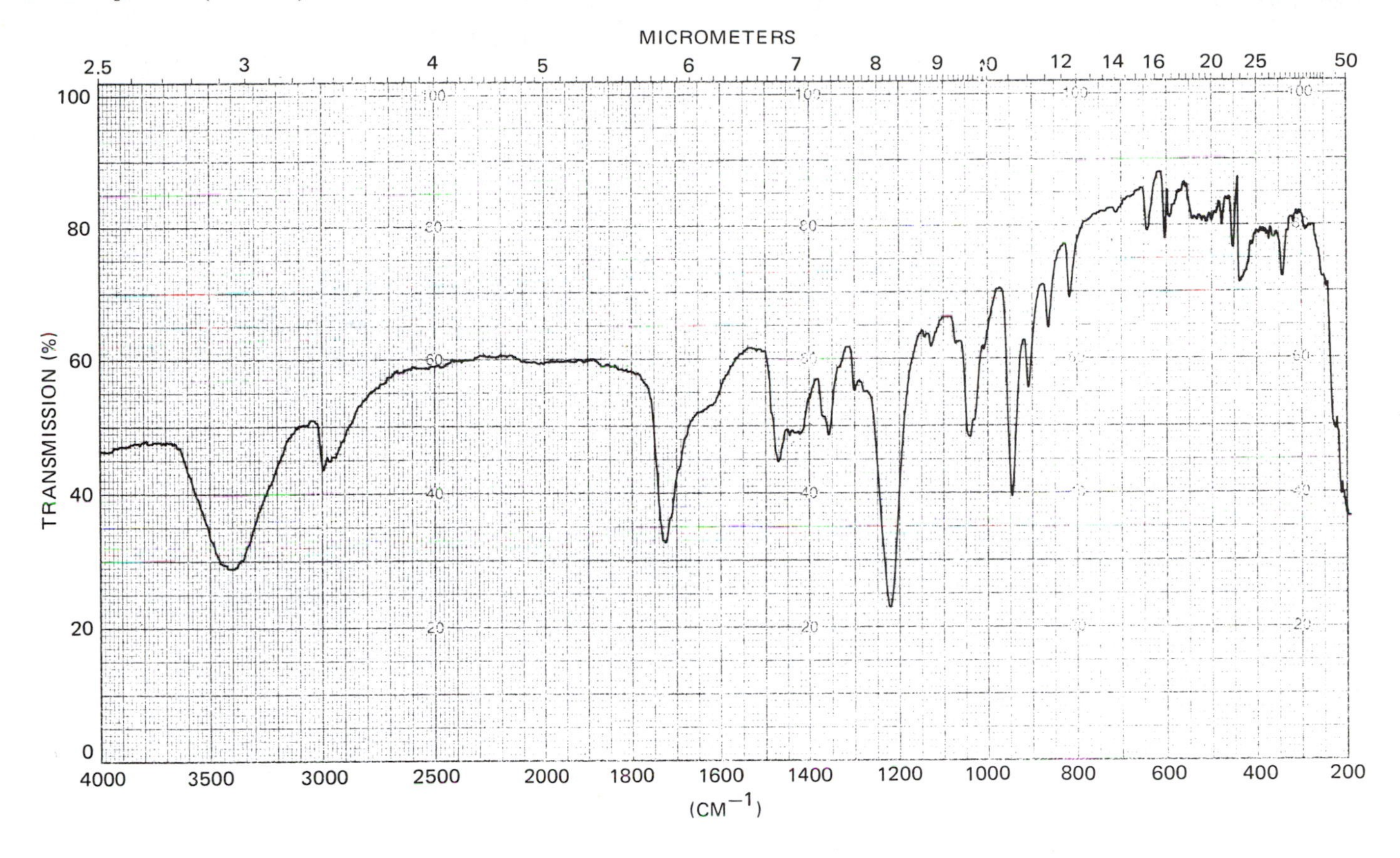

Ultraviolet-visible spectrum (CH_3OH, 0.016 M)

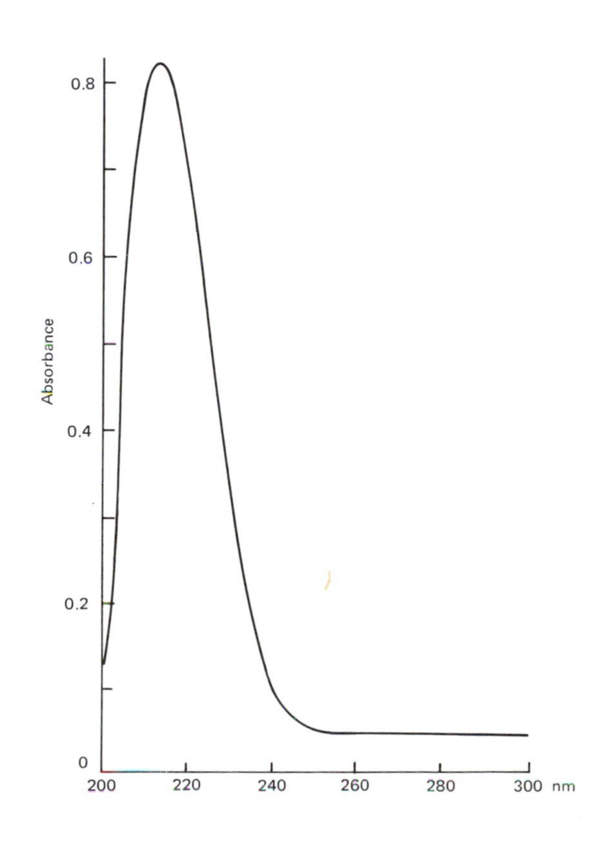

PROBLEM 6

Mass spectrum

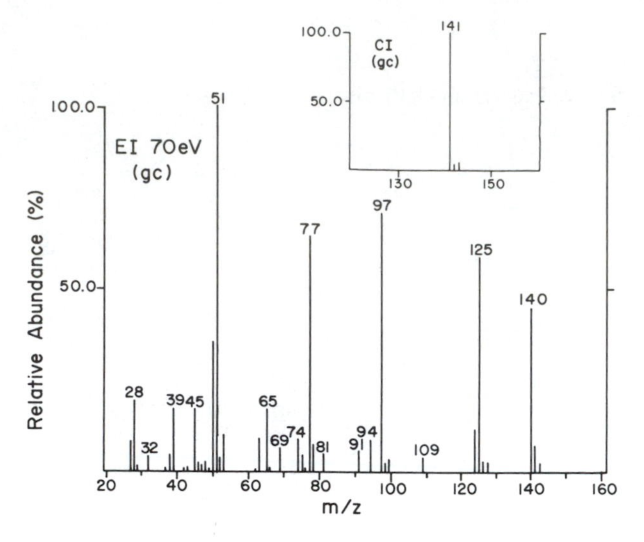

Proton NMR spectrum ($CDCl_3$)

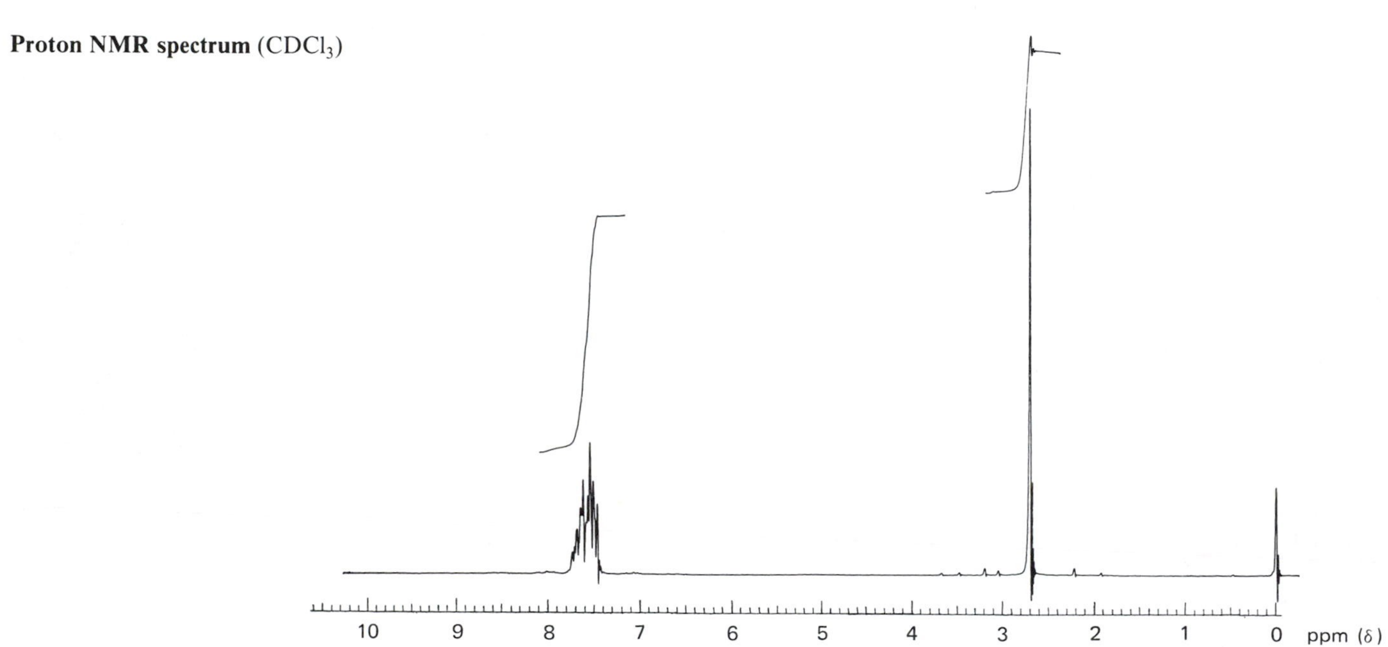

Carbon-13 NMR specturm ($CDCl_3$)

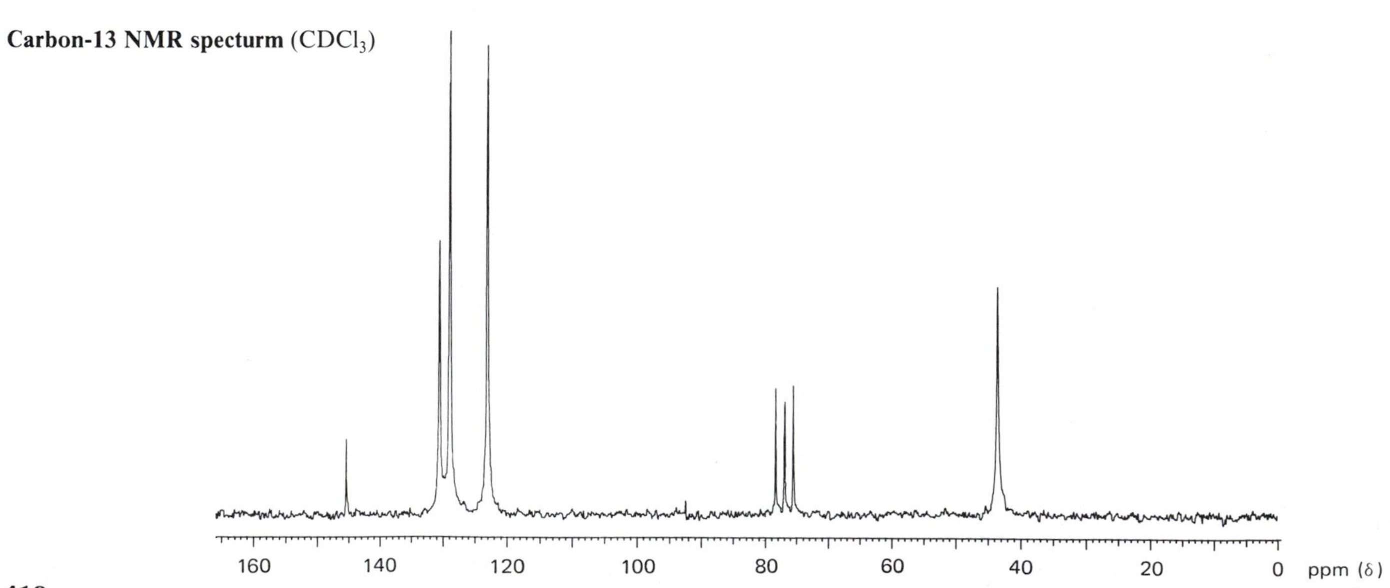

Infrared spectrum (neat)

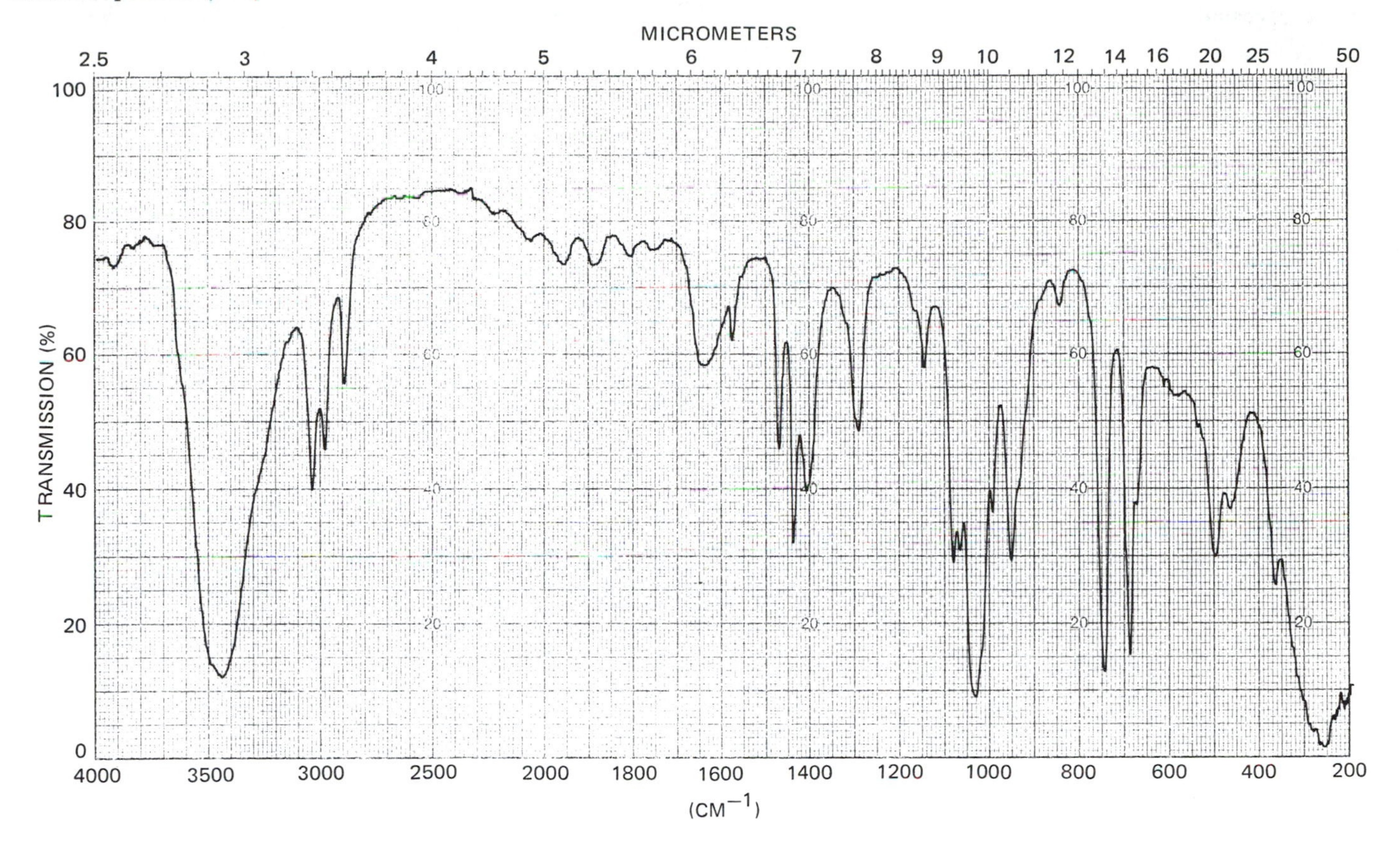

Ultraviolet–visible specturm (1.027×10^{-5} M)

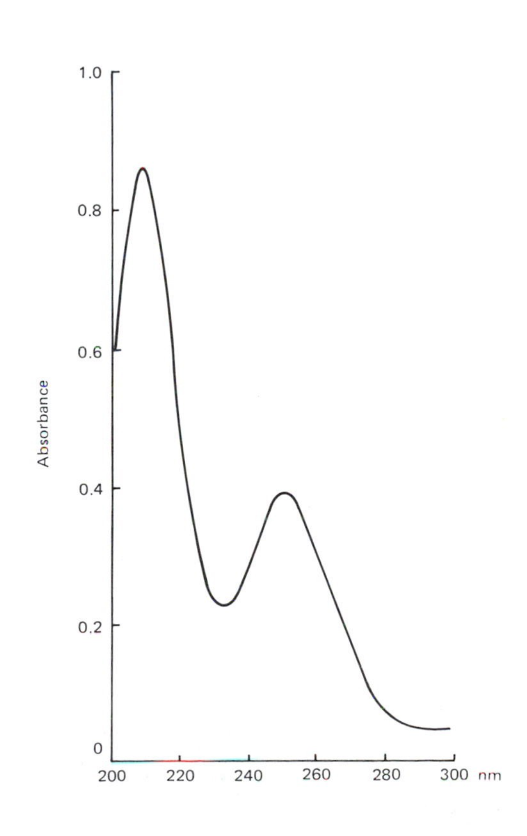

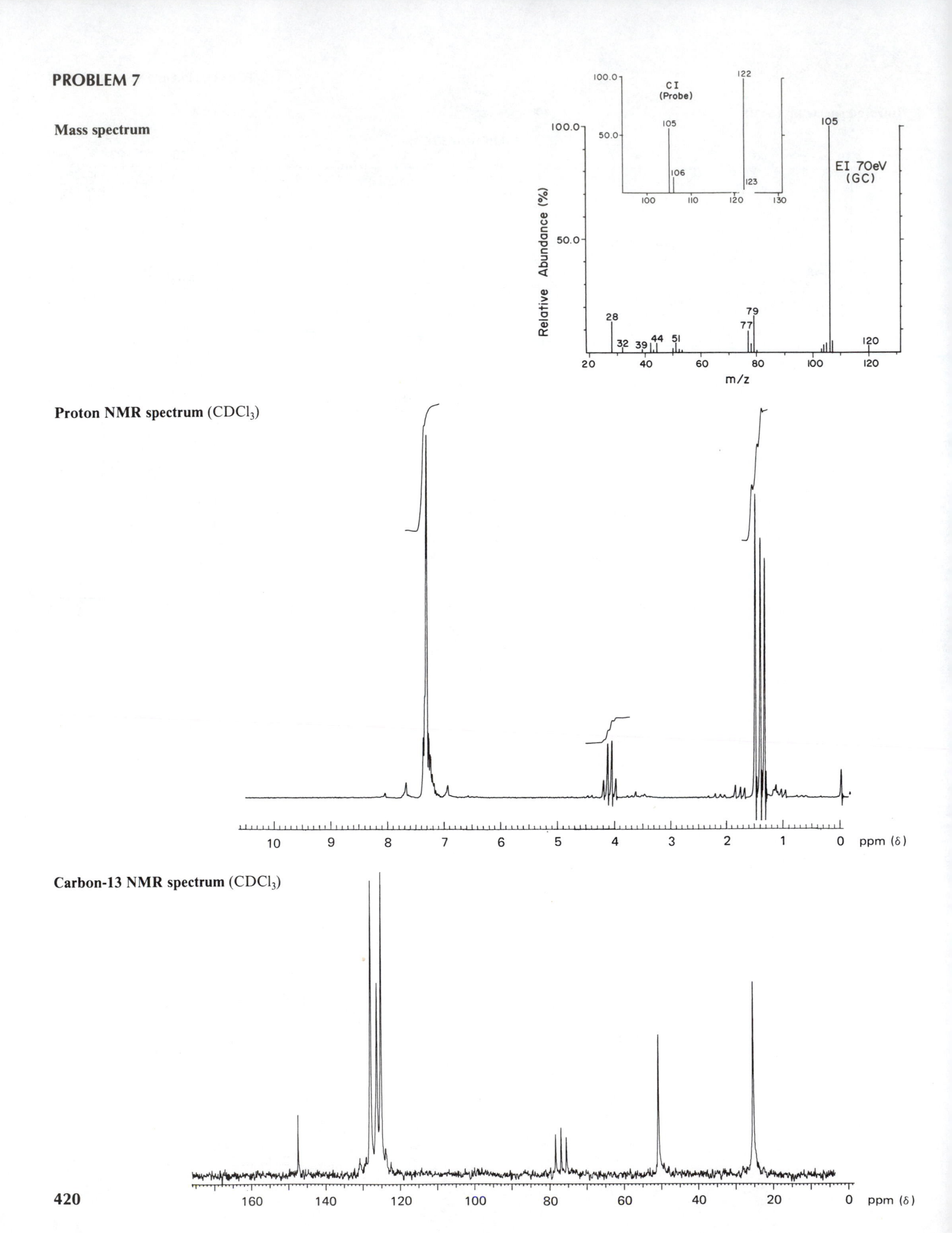
PROBLEM 7
Mass spectrum
100.0
CI
(Probe)
122
105
50.0
106
123
100
110
120
130
100.0
105
EI 70eV
(GC)
Relative Abundance (%)
50.0
28
32
39
44
51
77
79
120
20
40
60
80
100
120
m/z
Proton NMR spectrum (CDCl3)
10
9
8
7
6
5
4
3
2
1
0
ppm (δ)
Carbon-13 NMR spectrum (CDCl3)
160
140
120
100
80
60
40
20
0
ppm (δ)

Infrared spectrum (neat)

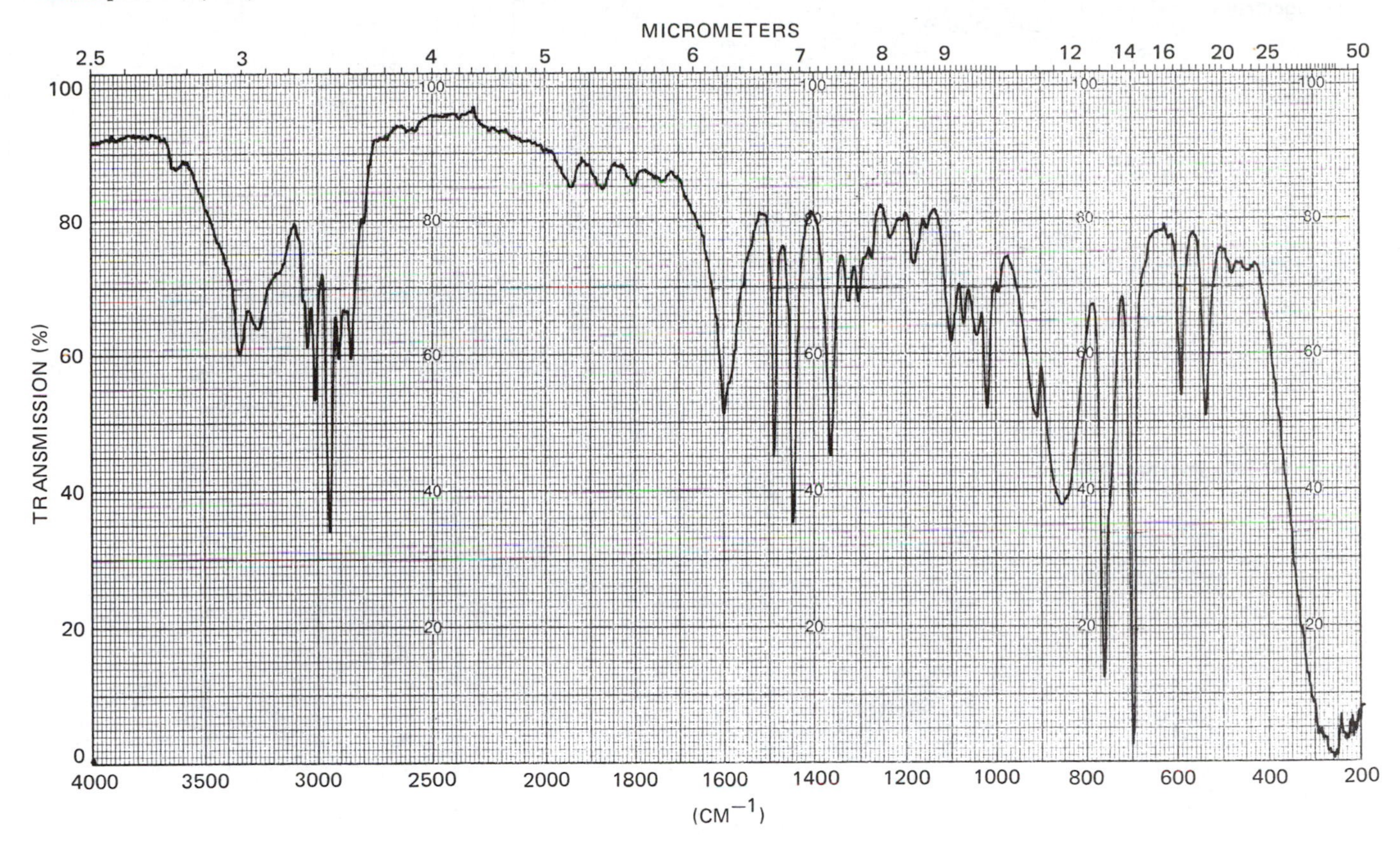

Ultraviolet-visible spectrum (CH_3OH, 2.33×10^{-3} M)

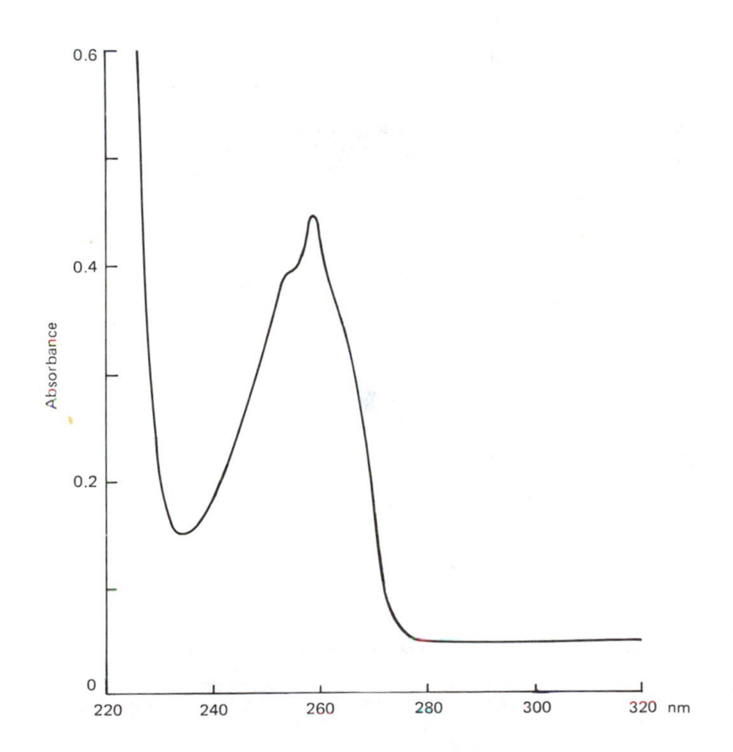

Circular dichroism spectrum (CH_3OH, 2.33×10^{-3} M)

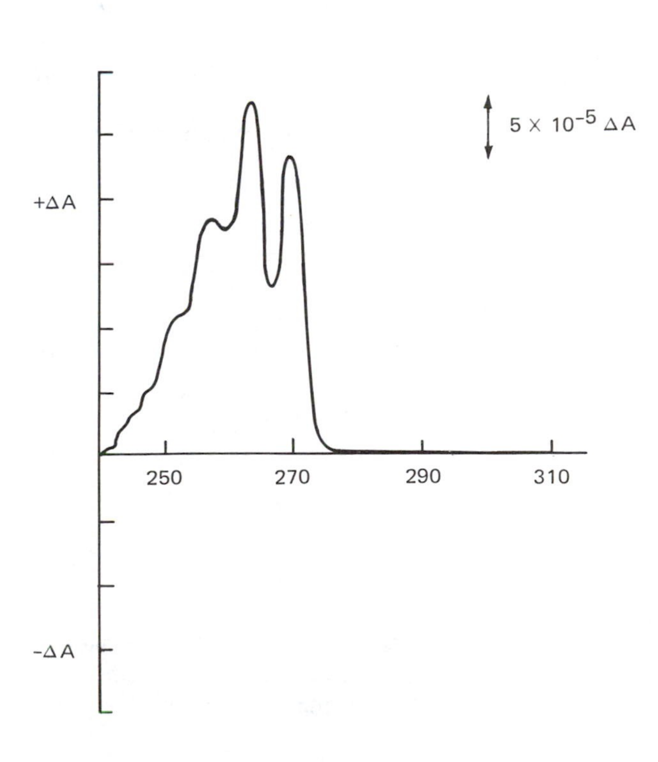

PROBLEM 8

Mass spectrum

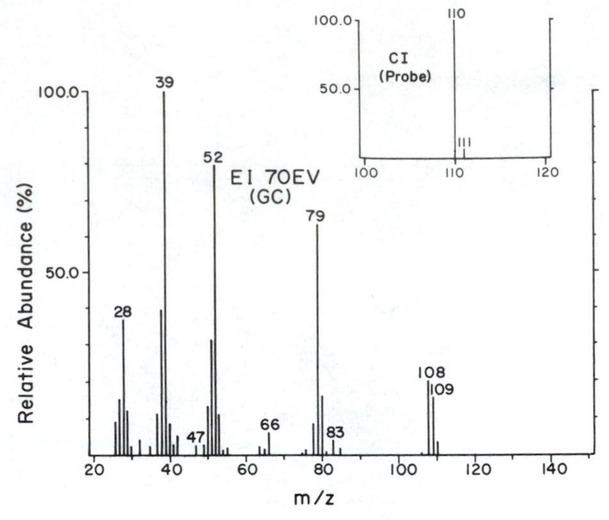

Proton NMR spectrum ($CDCl_3$)

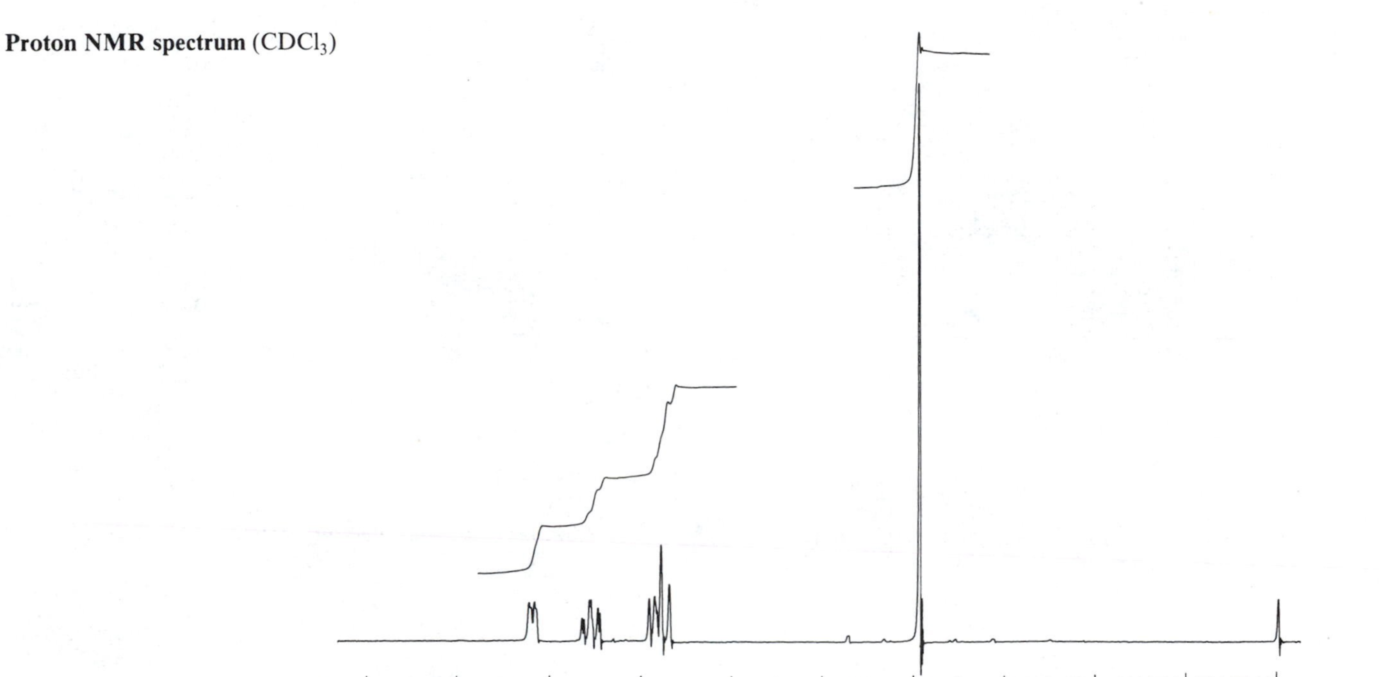

Carbon-13 NMR spectrum ($CDCl_3$)

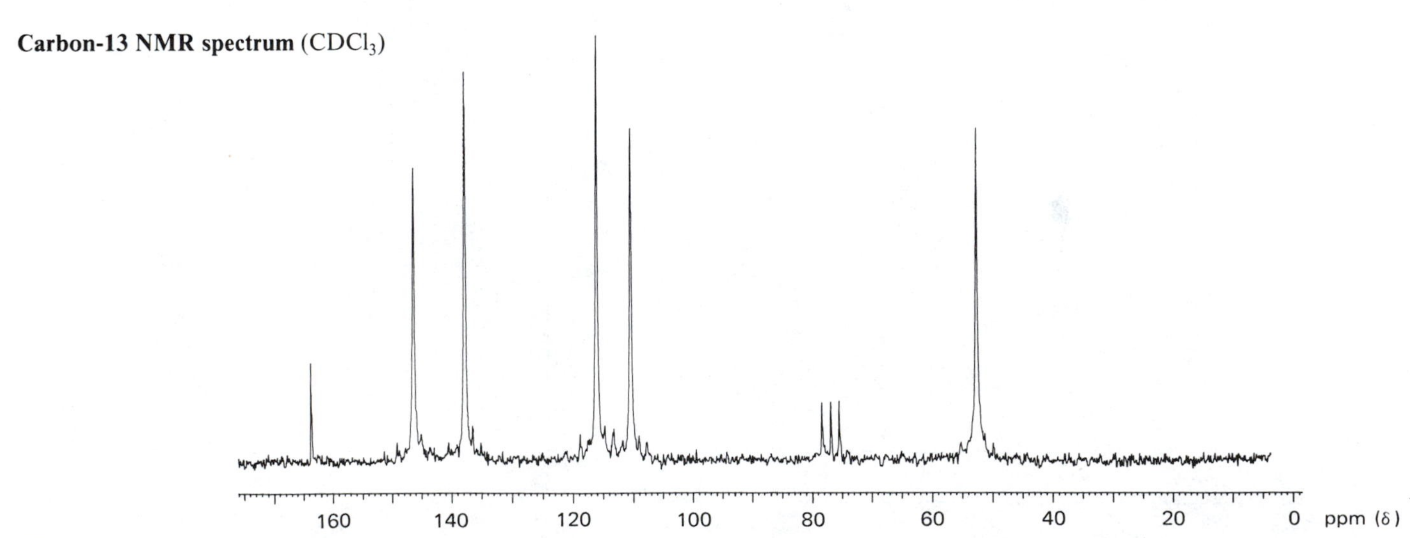

Infrared spectrum (neat)

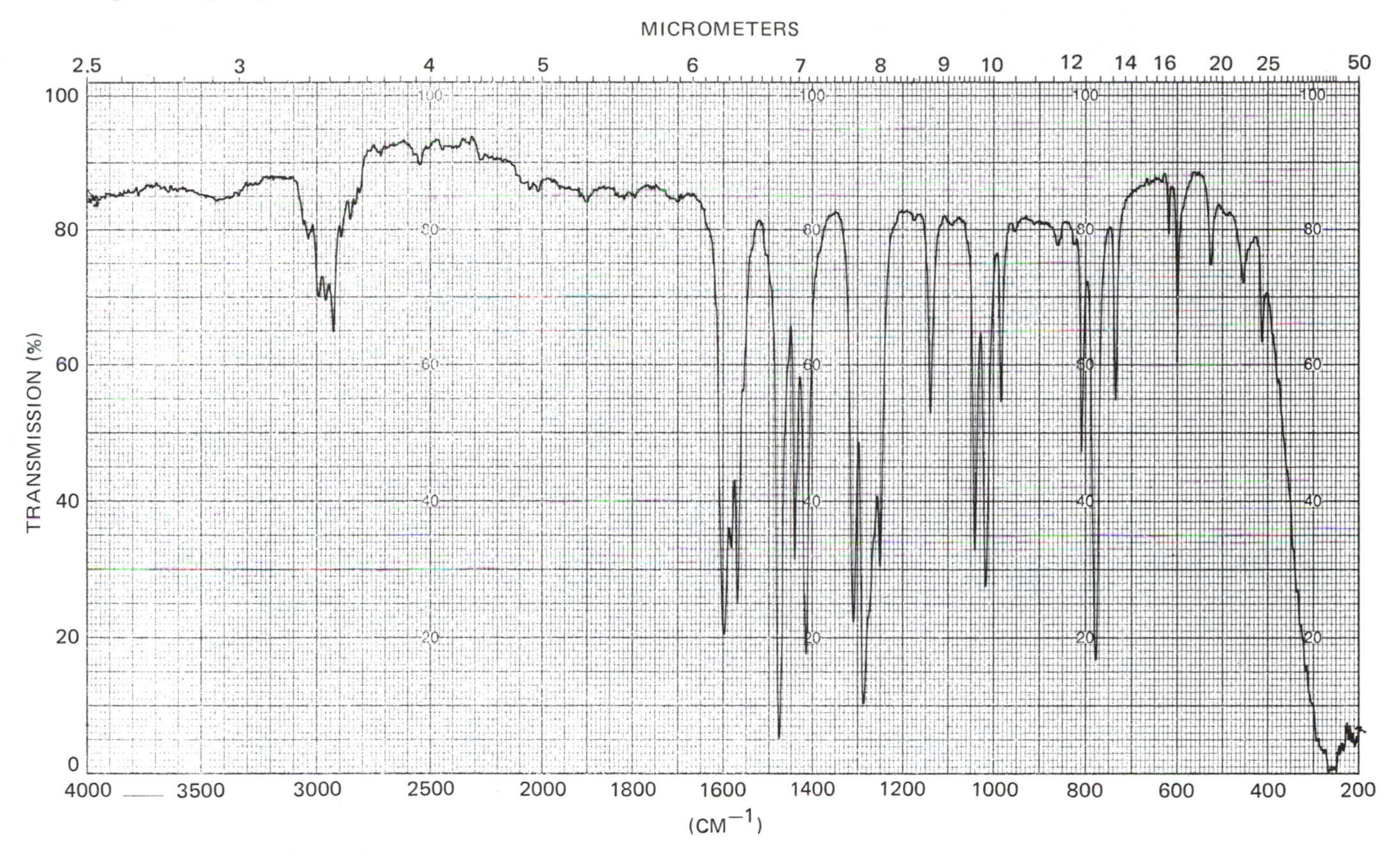

Ultraviolet-visible spectrum (CH_3OH, 7.98×10^{-5} M)

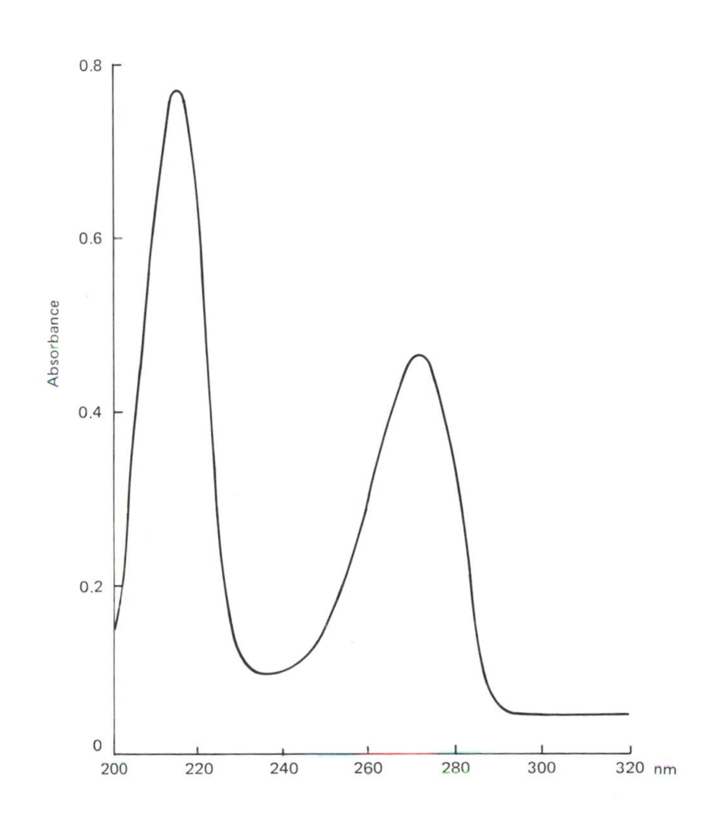

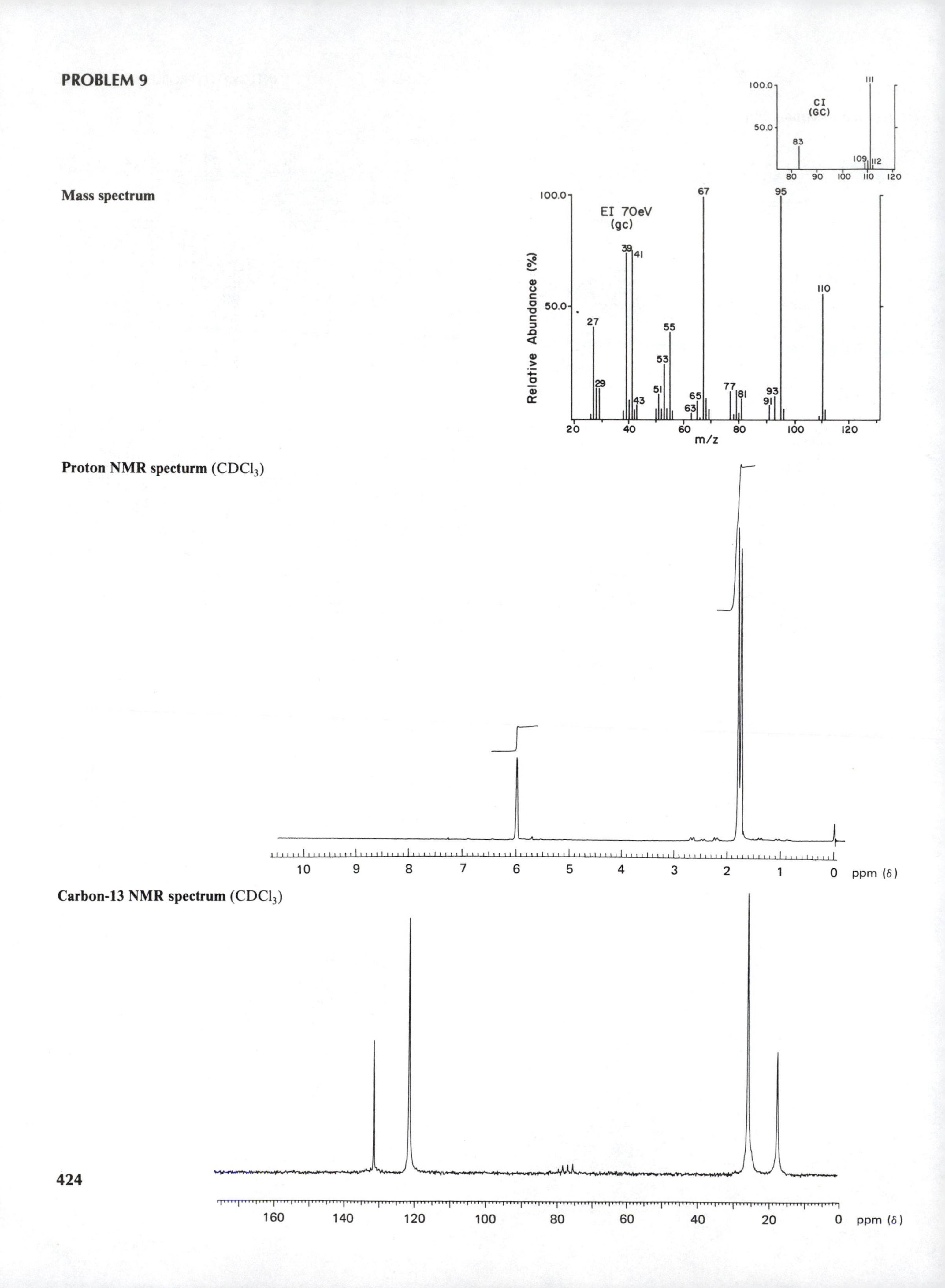
PROBLEM 9
Mass spectrum
CI
(GC)
100.0
50.0
83
109
111
112
80
90
100
110
120
EI 70eV
(gc)
Relative Abundance (%)
100.0
50.0
27
29
39
41
43
51
53
55
63
65
67
77
81
91
93
95
110
20
40
60
80
100
120
m/z
Proton NMR specturm (CDCl3)
10
9
8
7
6
5
4
3
2
1
0
ppm (δ)
Carbon-13 NMR spectrum (CDCl3)
160
140
120
100
80
60
40
20
0
ppm (δ)

Infrared spectrum (neat)

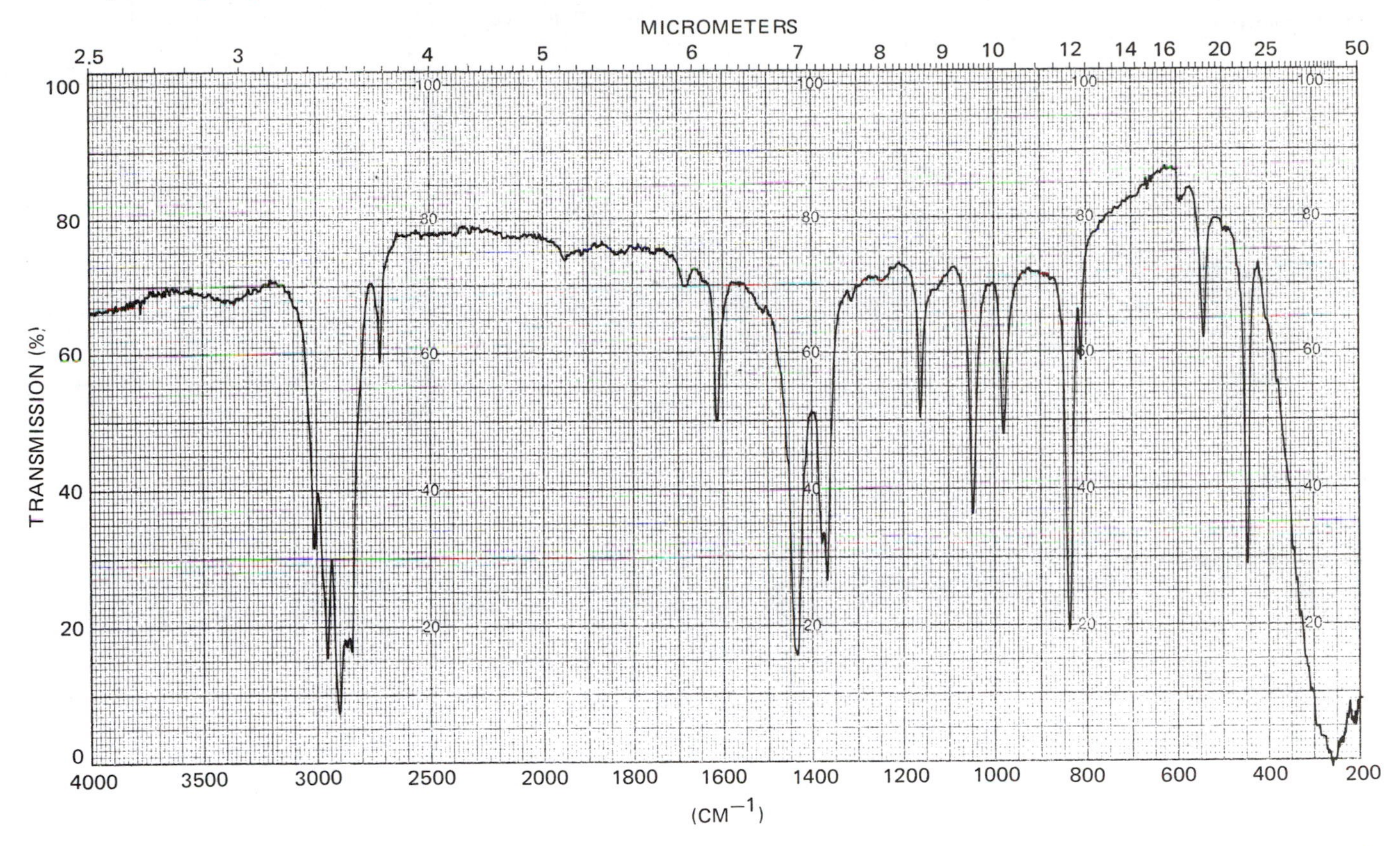

Ultraviolet-visible spectrum (heptane, 1.26×10^{-5} M)

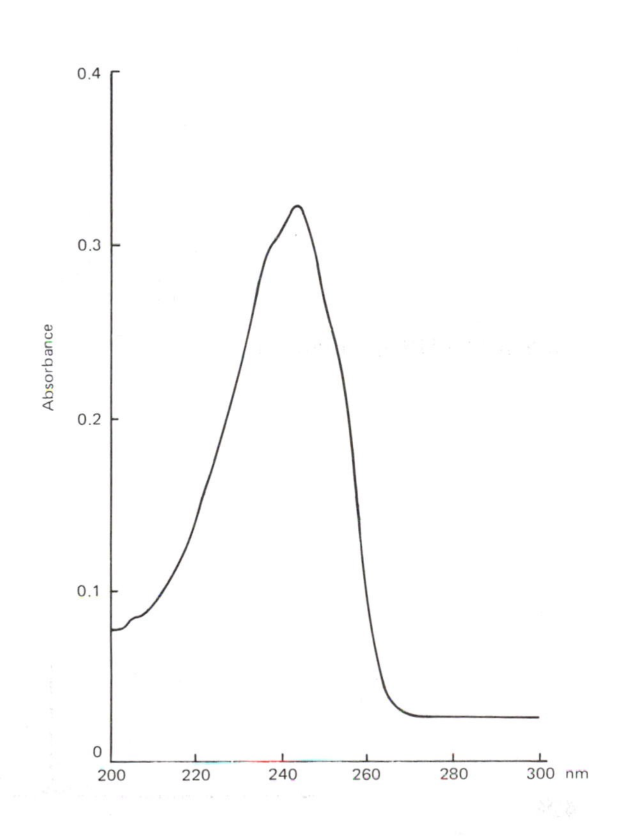

PROBLEM 10

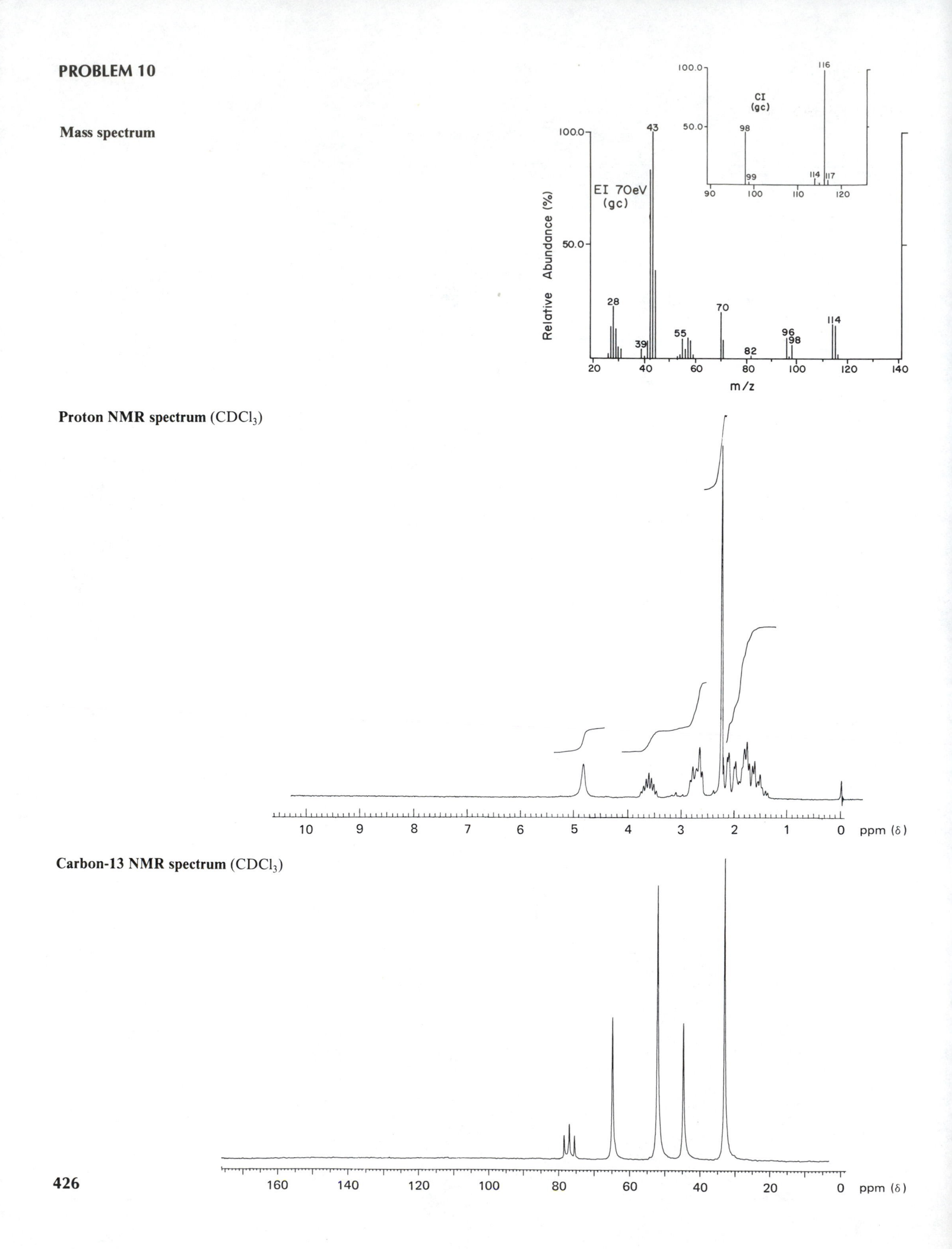

Mass spectrum
100.0
50.0
Relative Abundance (%)
EI 70eV
(gc)
43
28
39
55
70
82
96
98
114
20
40
60
80
100
120
140
m/z
CI
(gc)
116
98
99
114
117
90
100
110
120
Proton NMR spectrum (CDCl3)
10
9
8
7
6
5
4
3
2
1
0
ppm (δ)
Carbon-13 NMR spectrum (CDCl3)
160
140
120
100
80
60
40
20
0
ppm (δ)

Infrared specturm (neat)

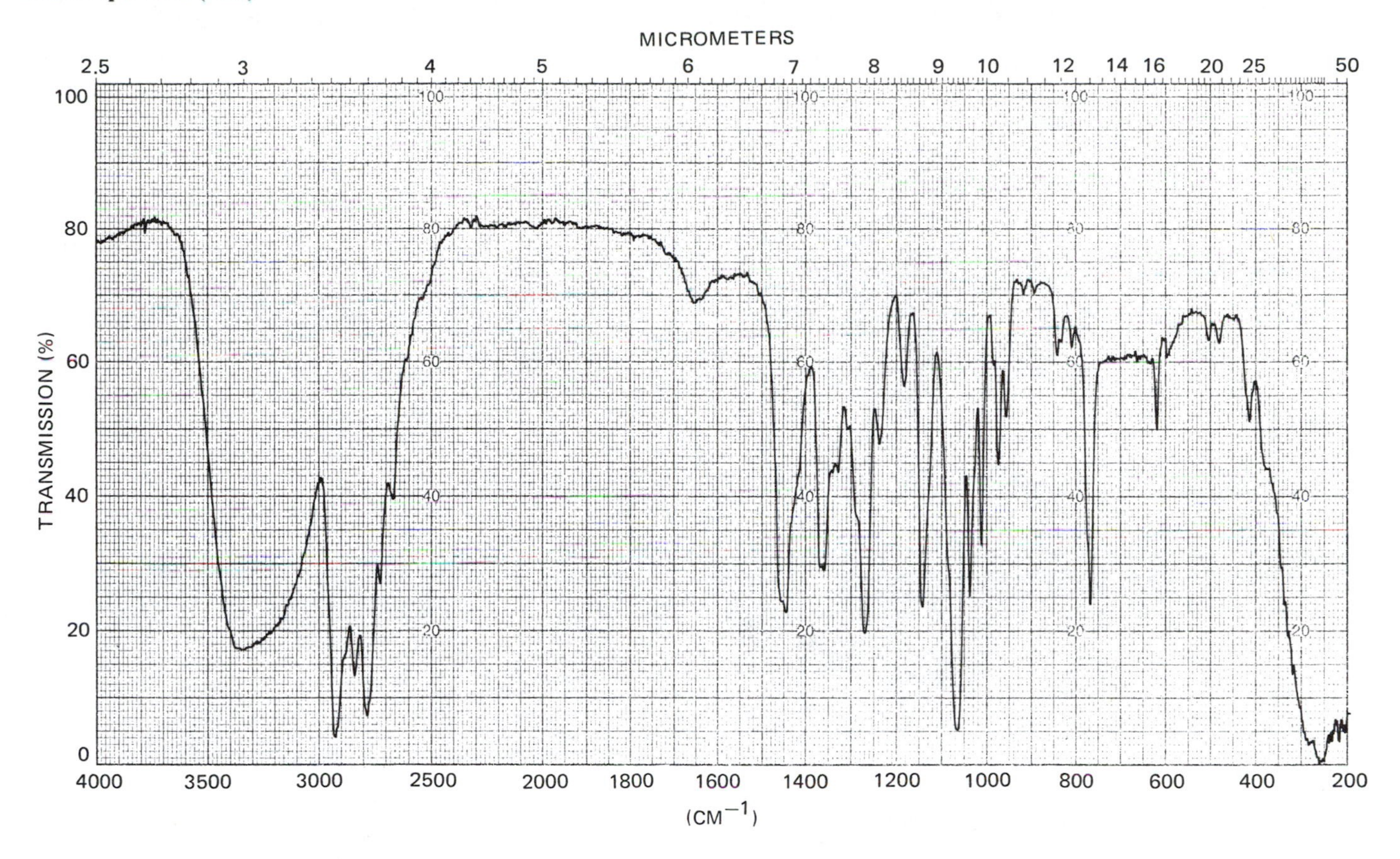

Ultraviolet-visible spectrum (end absorption only)

PROBLEM 11

Mass spectrum

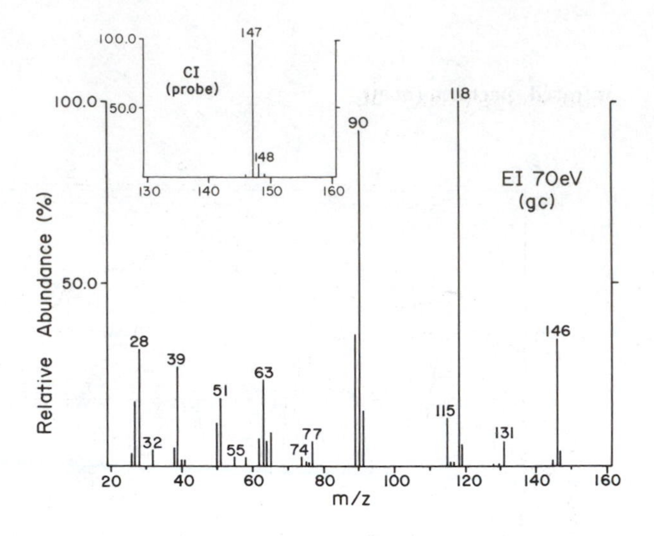

Proton NMR spectrum ($CDCl_3$)

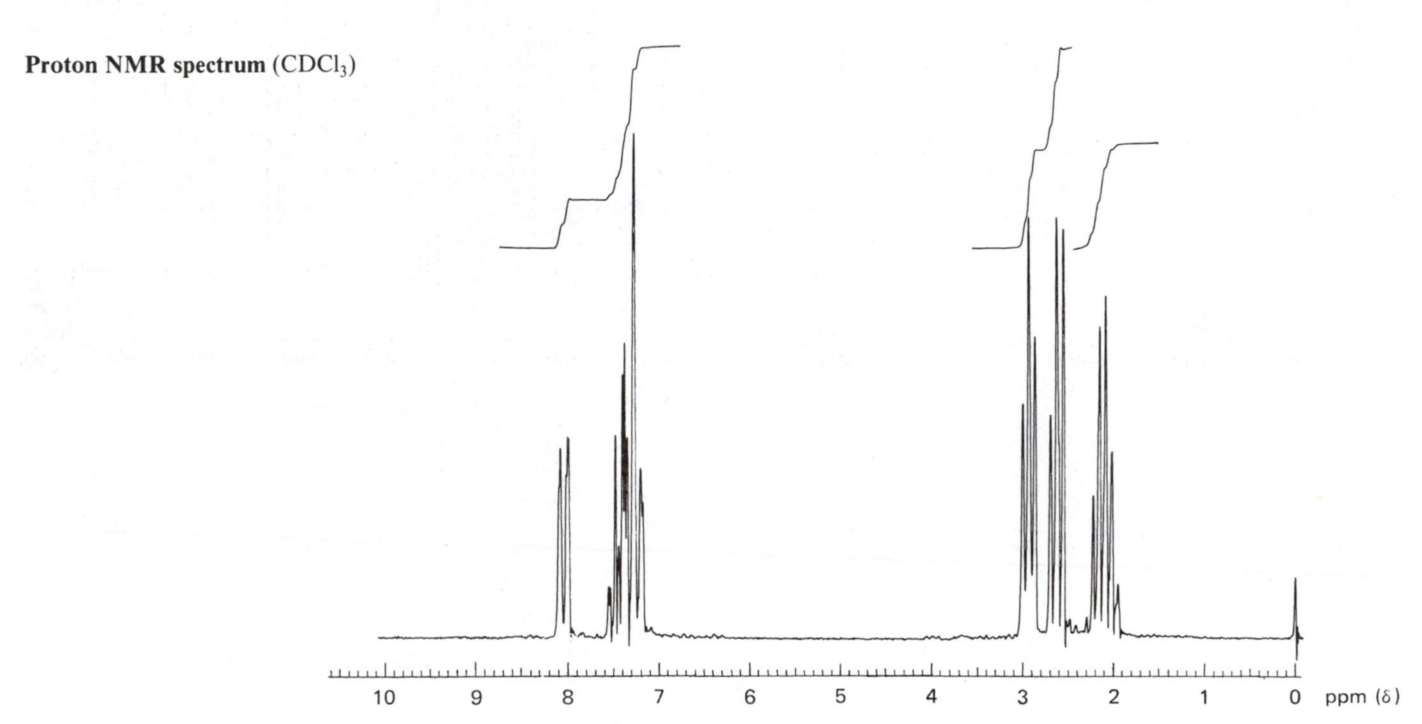

Carbon-13 NMR spectrum ($CDCl_3$)

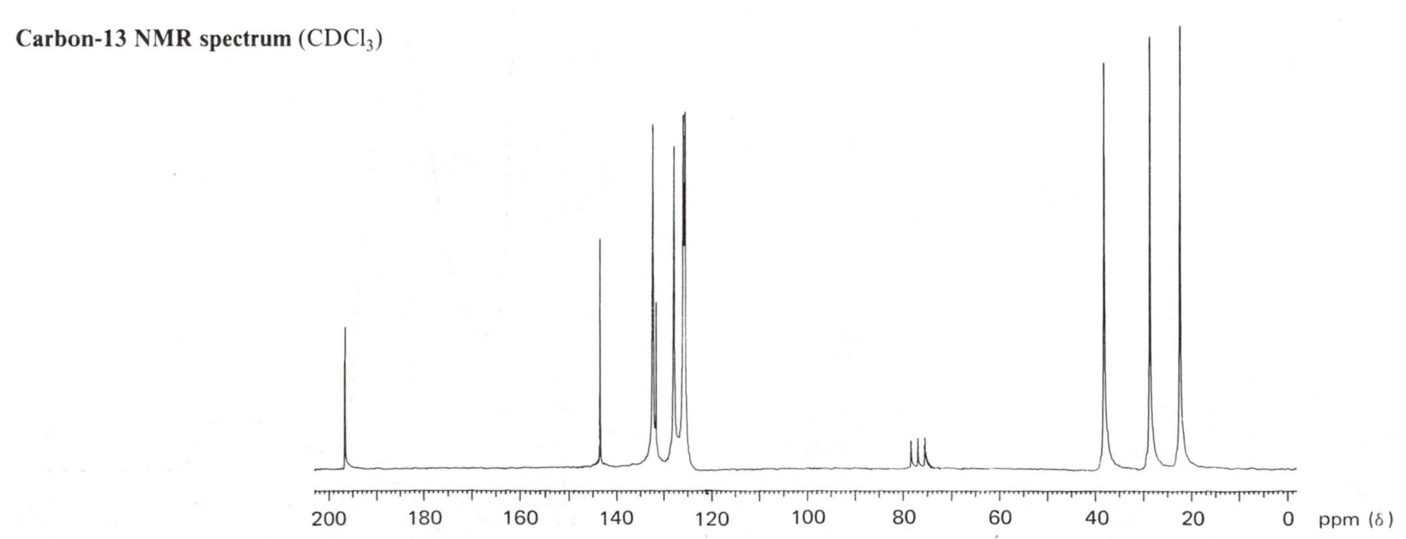

Infrared spectrum (neat)

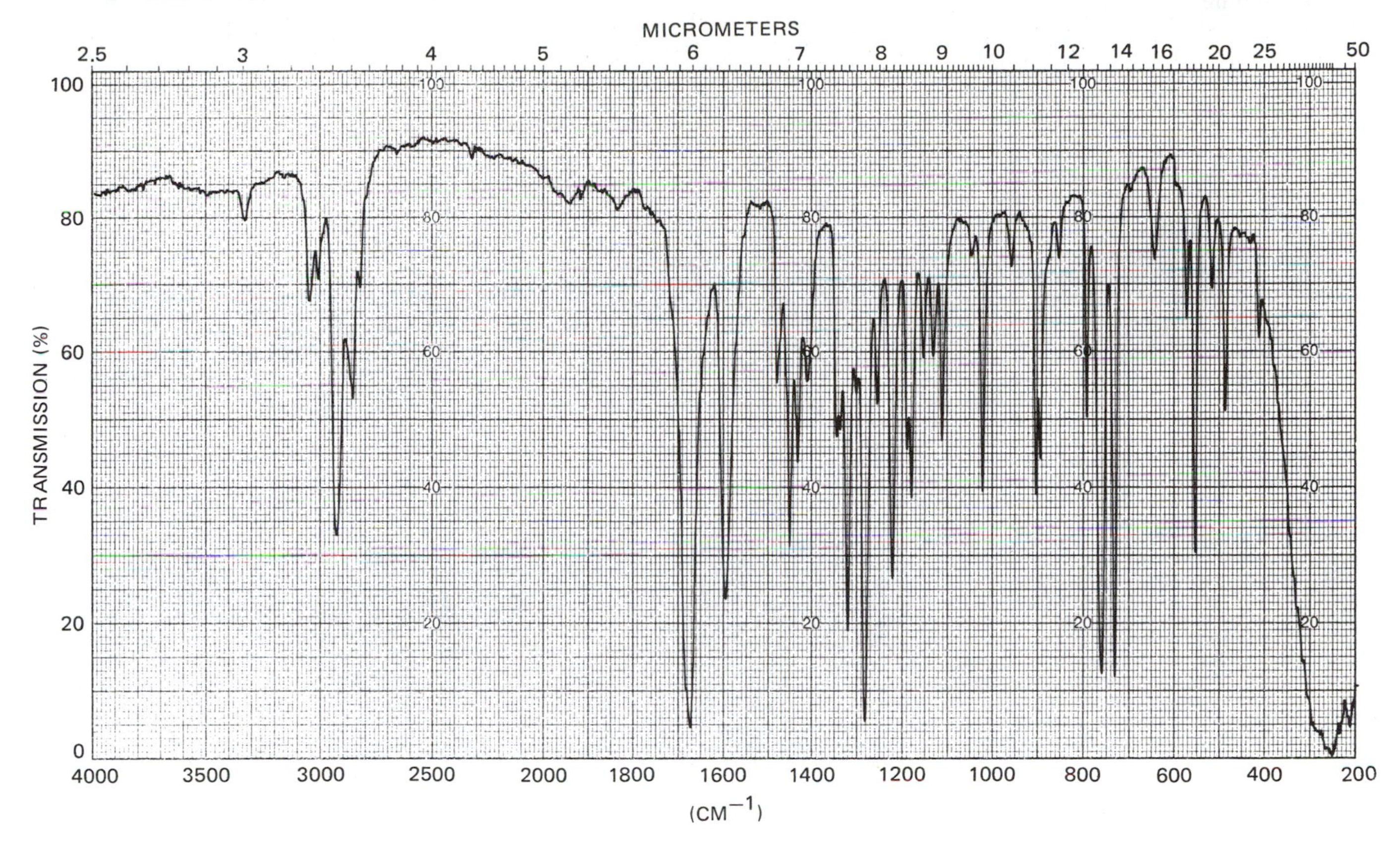

Ultraviolet-visible spectrum (heptane, 4.51×10^{-5} M)

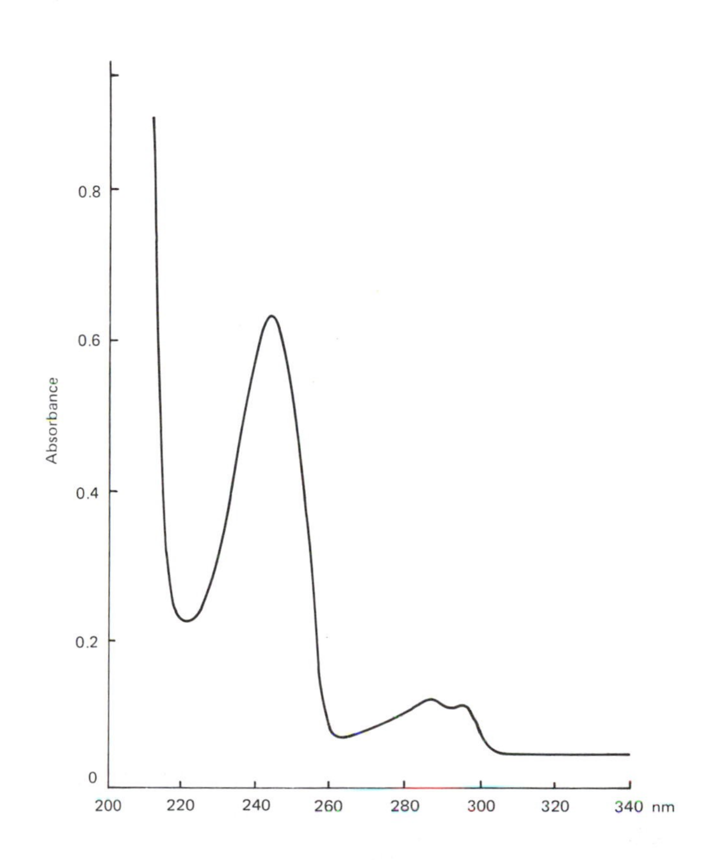

PROBLEM 12

Mass spectrum

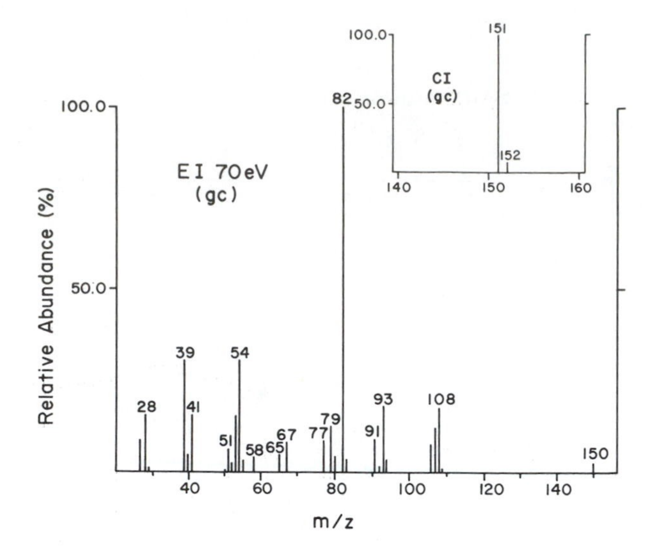

Proton NMR spectrum ($CDCl_3$)

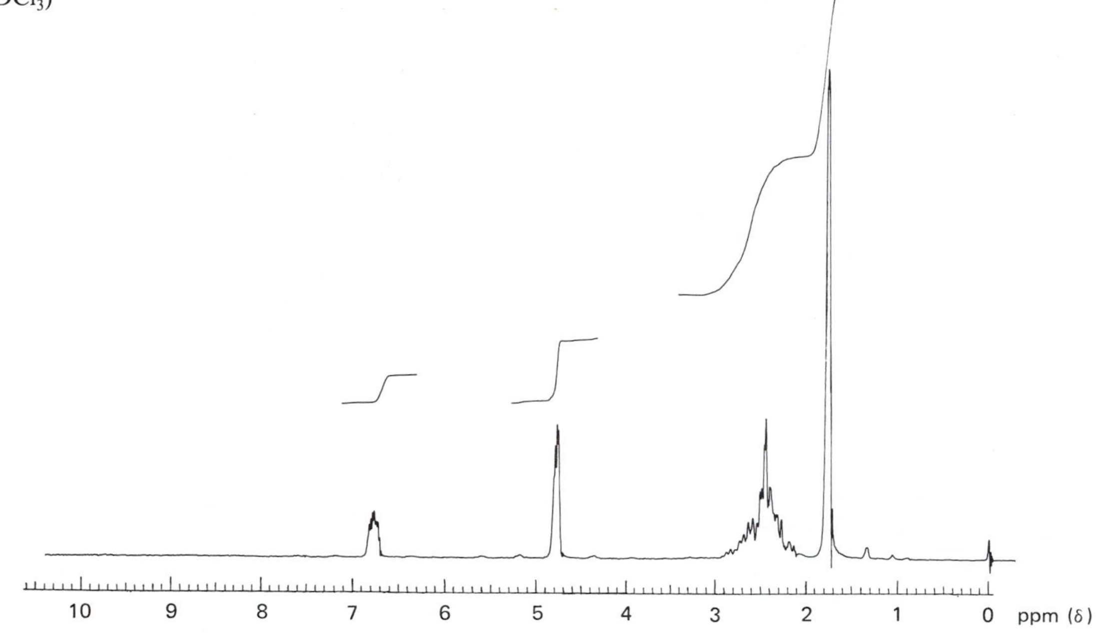

Carbon-13 NMR spectrum ($CDCl_3$)

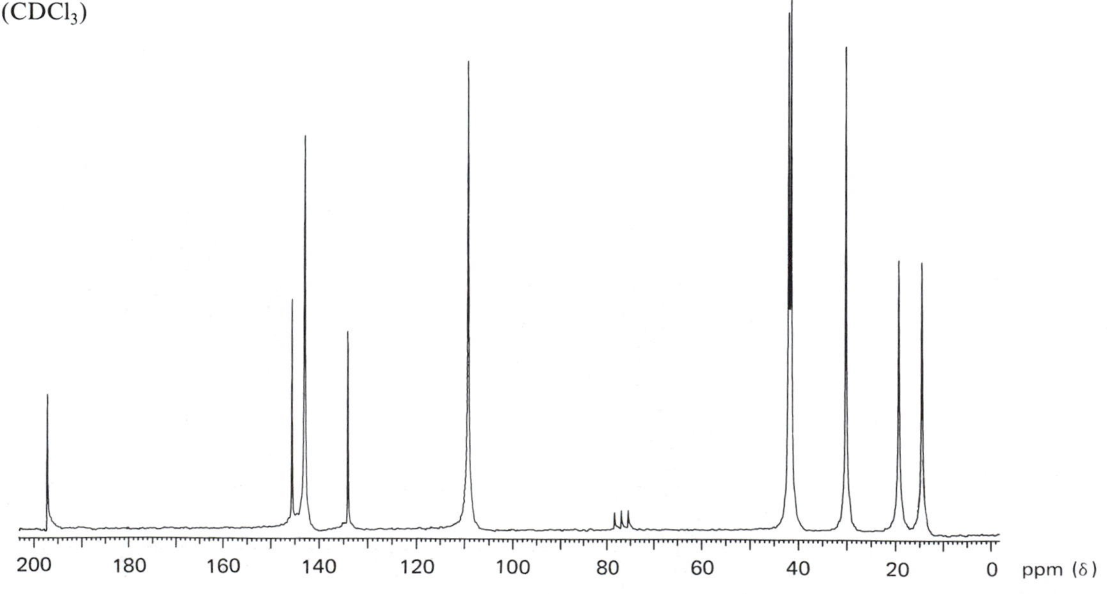

Infrared spectrum (neat)

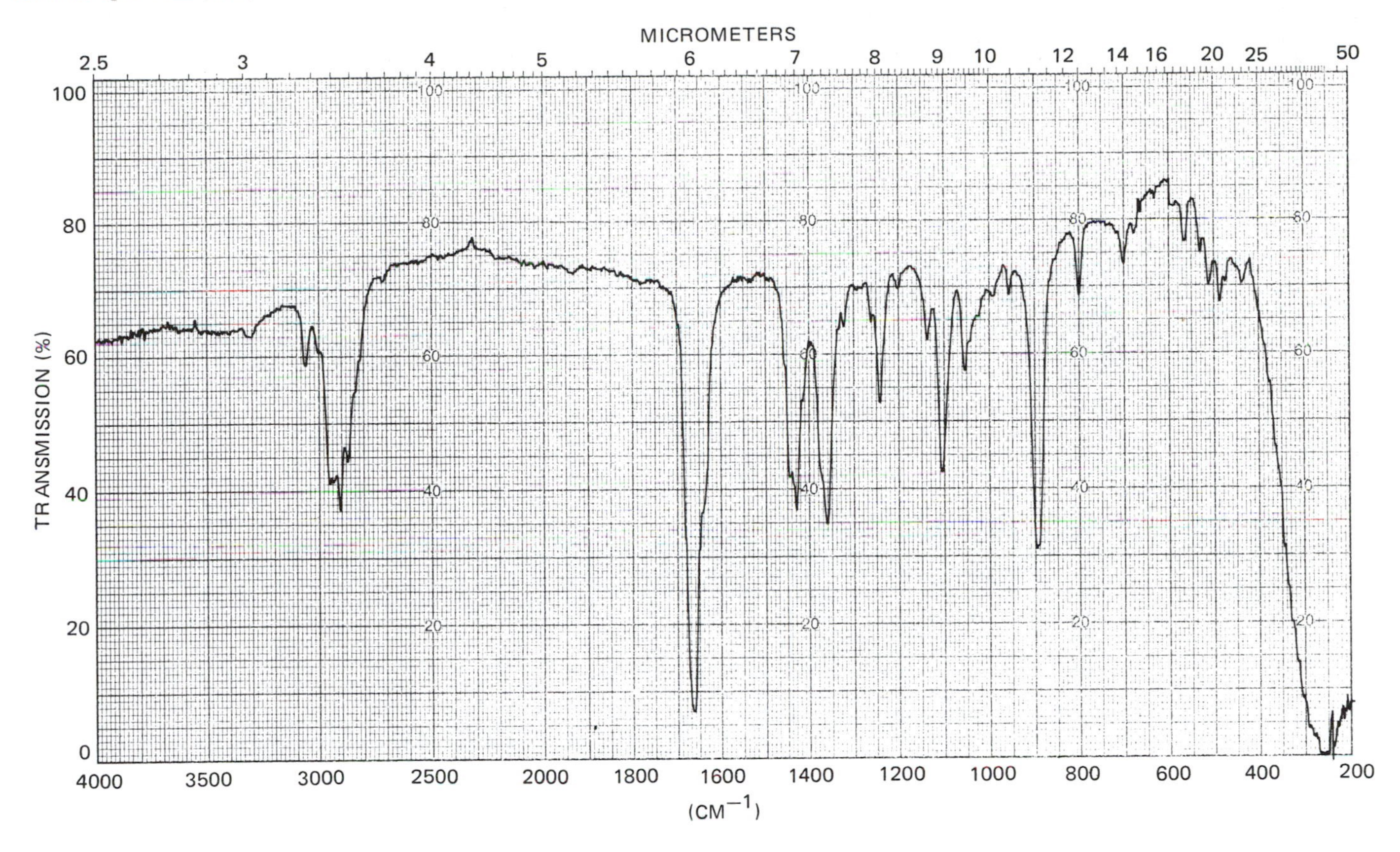

Ultraviolet-visible spectra (CH_3OH, 9.0×10^{-5} M, left, 7.66×10^{-3} M, right)

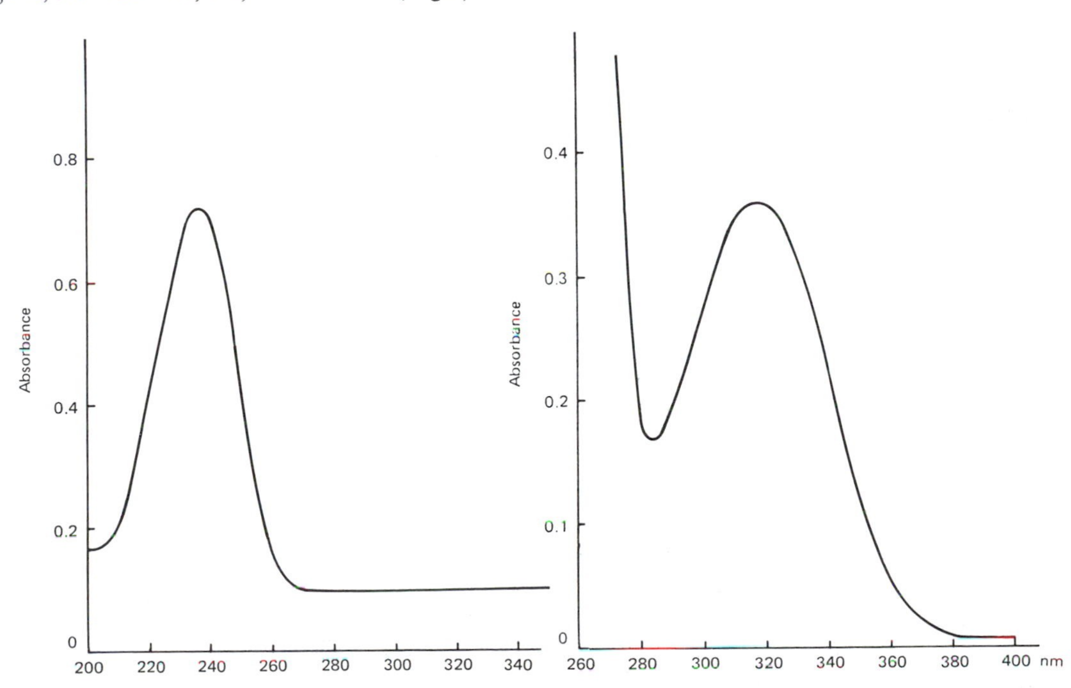

Circular dichroism spectra (CH_3, 9.0×10^{-5} M, left, 7.66×10^{-3} M right)

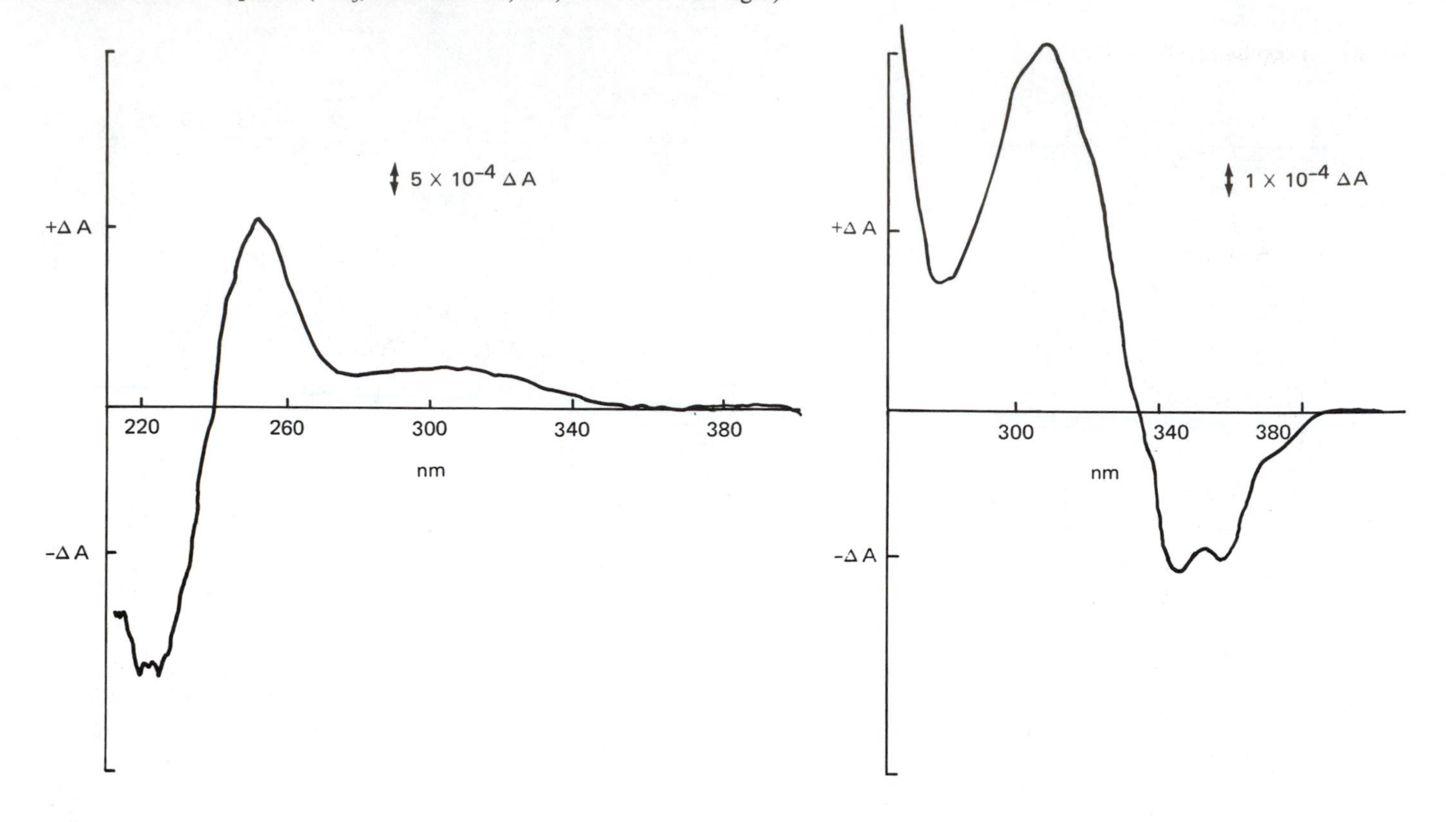

Molecular Formulas

1. $C_{11}H_{16}$
2. $C_4H_9O_2N$
3. C_8H_{14}
4. $C_7H_8O_2$
5. $C_7H_{16}O_2NCl$
6. C_7H_8OS
7. $C_8H_{11}N$
8. C_6H_7ON
9. C_8H_{14}
10. $C_6H_{13}ON$
11. $C_{10}H_{10}O$
12. $C_{10}H_{14}O$

Functional Groups

1. Monosubstituted aryl
2. Carboxyl, primary amino
3. Alkyne
4. Trisubstituted aryl, hydroxyl
5. Ester, quaternary ammonium chloride
6. Monosubstituted aryl, sulfoxide
7. Monosubstituted aryl, primary amino
8. Monosubstituted pyridinyl, methoxy
9. Conjugated diene
10. Piperidine, hydroxyl
11. Fused aryl, ketone
12. Unconjugated diene, α,β-unsaturated ketone

Molecular Structures

1. Amylbenzene, $C_6H_5CH_2CH_2CH_2CH_2CH_3$
2. 4-Aminobutyric acid, $NH_2CH_2CH_2CH_2CO_2H$
3. 1-Octyne, $HC{\equiv}CCH_2CH_2CH_2CH_2CH_2CH_3$
4. 4-Methylcatechol,

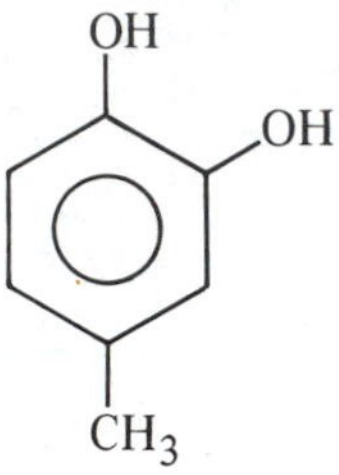

OH
OH
CH_3

5. Acetylcholine chloride, $CH_3\overset{O}{\overset{\|}{C}}OCH_2CH_2\overset{+}{N}(CH_3)_3\ Cl^-$

6. Methyl phenyl sulfoxide, $CH_3\overset{O}{\overset{\|}{S}}Ph$

7. (*S*)-α-Methylbenzylamine, NH_2 C H, CH_3, Ph

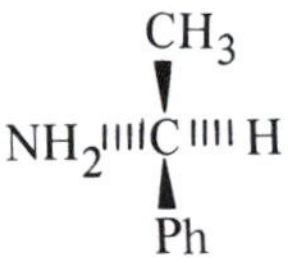

8. 2-Methoxypyridine,

N
OCH_3

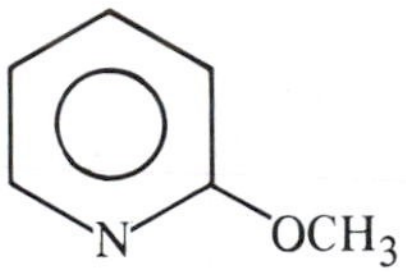

9. 2,5-Dimethyl-2,4-hexadiene,

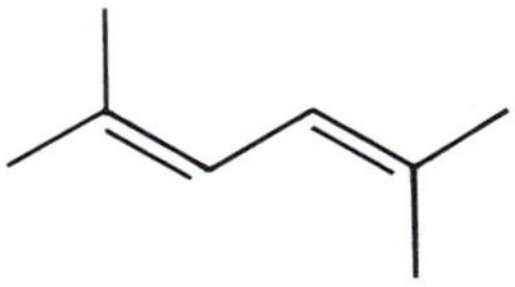

10. 4-Hydroxy-1-methylpiperidine,

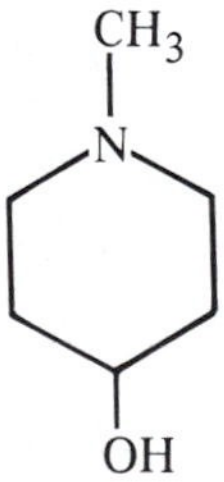

CH_3
N
OH

11. α-Tetralone,

O

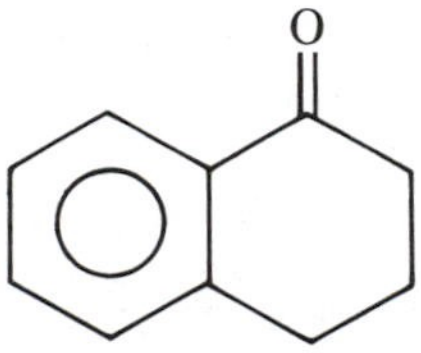

12. (*R*)-Carvone,

O

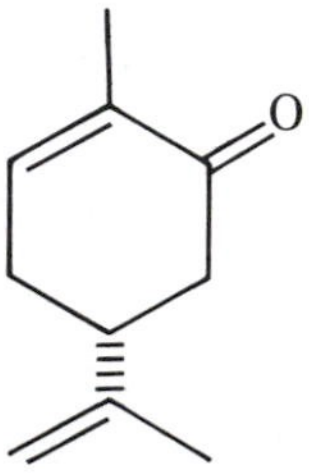

Answers to Problems

Chapter 1

1-1 **a, b, d** structural isomers **c, e** stereoisomers

1-2 **a.** C_7H_{12}, 87.42% C, 12.58% H; $C_5H_{11}N$, 70.53% C, 13.02% H, 16.45% N; C_8H_{16}, 85.63% C, 14.37% H; $C_7H_{10}O_2$, 71.98% C, 6.71% H, 21.31% O; $C_7H_{10}O$, 76.32% C, 9.15% H, 14.53% O

b. 2, 1, 1, 5, 3

c. double bond; amine, cyclopropane; none; phenyl, ester; double bond, hydroxyl

1-3 *1-1a, left:* 7 distinct carbons; 4 pairs of methylene hydrogens, 1 methyl trio of hydrogens, 1 alkenic hydrogen

1-1b, right: 5 different kinds of carbon in the ratio 2/2/1/1/1. The 3 unique carbons (a, b, e) are on the axis of symmetry; the two pairs (c, d) are off it. 4 kinds of hydrogen in the

c a
d CH_2
b
e c
d

ratio 4/4/2/2. The 2 pairs (a, e) are on the axis, the 2 sets of four (c, d) are off it. This analysis assumes rapid ring flipping.

1-1d, both: 7 kinds of carbon in the ratio 2/2/1/1/1/1/1 (ortho and meta carbons on the phenyl ring are doubled); 5 kinds of hydrogen in the ratio 3/2/2/2/1.

1-1e, both: 4 kinds of carbon in the ratio 2/2/2/1 (note the mirror plane in the molecule); 6 kinds of hydrogen in the ratio 2/2/2/2/1/1 (the hydrogens of the CH_2CH_2 portion are made up of 2 exo and 2 endo types).

1-4 **a.** $C_5H_8O_2$ **b.** $C_7H_{13}ON$

Chapter 2

2-1 **a.** quintet (1/4/6/4/1)

++++	+++−	++−−	+−−−	−−−−
	++−+	+−+−	−+−−	
	+−++	−++−	−−+−	
	−+++	+−−+	−−−+	
		−+−+		
		−−++		

b. septet (1/6/15/20/15/6/1)

c. singlet

d. doublet of doublets if couplings to cis and trans protons are different; in cyclopropanes, in fact, J_{cis} is usually larger than J_{trans} (Chapter 4)

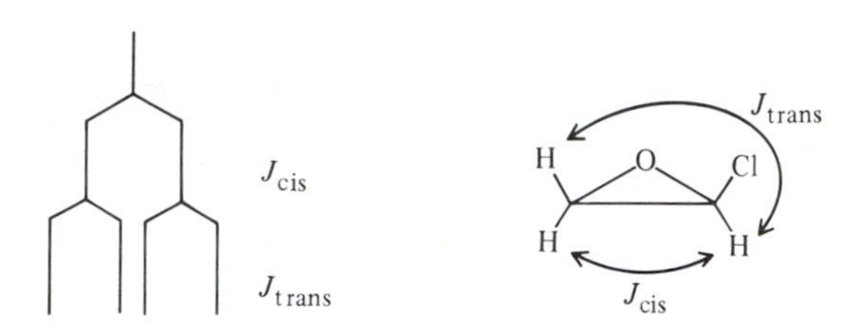

2-2 [s = singlet, d = doublet, t = triplet, q = quartet, m = complex multiplet]

a. $CH_3—CH_2—CH_2—O—C(=O)—CH_3$

	CH_3	CH_2	CH_2	C=O	CH_3
^1H	t	sextet	t	—	s
^{13}C	q	t	t	s	q

b. $C_6H_5—CH_2—CH_2Br$

^1H * t t

^{13}C d d d s t t

*The ^1H aromatic resonances closely coincide so little splitting is observed.

c. (tetrahydropyran, O)

^1H * * t

^{13}C t t t

*quintets or complex multiplets

d. $(CH_3—CH_2)_3N$

^1H t q

^{13}C q t

e. $CH_3—CH=CH—CO_2CH_3$

^1H d m d s

^{13}C q d d s q

2-3 **a.** $CH_3OCH{=}CHC(=O)CH_3$

b. $NH_2-C_6H_4-CO_2CH_2CH_3$ (para) **c.** $CH_3CH_2OCH_2CH_2CN$

2-4 CH_3, 1.22 ppm, 109.8 Hz; CH_2, 3.61 ppm, 324.9 Hz; CH, 5.15 ppm, 463.5 Hz (all Hz at 90 MHz)

CHAPTER 3

3-1 Both H_α and H_β are deshielded the same amount by the ring to which they are attached. They are differentially shielded, however, by the remote ring. Both are deshielded with respect to benzene (δ 7.3), but H_α (8.4) more so than H_β (7.7), since H_α is closer to the remote ring.

H_α H_β

3-2 The hydroxyl of phenol is intermolecularly hydrogen bonded. In dilute $CDCl_3$ many of the hydrogen bonds are broken, so the resonance is shifted upfield. In *o*-nitrophenol the hydroxyl is intramolecularly hydrogen bonded, so even in dilute $CDCl_3$ the anisotropy of the lone pair in the hydrogen bond maintains the resonance at a low-field position.

3-3 The α carbon of pyridine is shifted downfield from that of benzene (δ 129) by the standard effect of an electron-withdrawing group on the radial term. Removal of electrons permits the remaining electron shell to move closer to the nucleus, increasing σ^p and causing a downfield shift. In pyrrole this effect is offset by electron donation through resonance. The formal negative charge on the α carbon

N H ⟷ - N+ H

implies a higher electron density, moving the electron shell further from the nucleus, thereby decreasing σ^p and causing a small upfield shift.

3-4 **a.** δ 3.10 **b.** δ 3.14 **c.** δ 4.42 **d.** δ 4.00

3-5 The value calculated for the cis isomer is δ 6.87, for the trans isomer δ 6.72. The order of the calculated positions suggests that the cis isomer resonates at α 7.03 and the trans isomer at δ 6.65.

3-6 There are six possible trisubstituted isomers with two groups the same. Two are eliminated because they would give only two resonances by symmetry (2-bromo-1,3-dimethoxybenzene and 5-bromo-1,3-dimethoxybenzene). The calculated shifts for the remaining four isomers (A–D) are given below. 1-Bromo-2,4-dimethoxybenzene (D) is in good accord with the observed chemical shifts, even without analysis of splitting patterns. The δ 7.32 shift, caused by the downfield effect of ortho bromine coupled with the small effect of meta methoxyl, is particularly diagnostic.

A: OCH_3, OCH_3, Br; 6.73, 6.68, 7.03

B: OCH_3, OCH_3, Br; 6.63, 7.04, 6.97

C: OCH_3, Br, OCH_3; 6.63, 6.45, 6.98

D: Br, OCH_3, OCH_3; 7.32, 6.34, 6.28

3-7 **a.** $CH_3{-}CH_2{-}CH(CH_3){-}CH(CH_3)_2$
11.0 28.4 51.9 16.8 40.0 9.9
b. $CH_3{-}CH_2{-}CH(NO_2){-}CH_3$
11.5 28.1 86.1 16.5
c. $ICH_2{-}CH_2{-}CH_2Br$
10.0 37.6 35.0
d. $(CH_3)_3C{-}CN$
30.0 28.0 142.0

A base of δ 117.2 from CH_3CN is used in eq. 3-5 in place of -2.5 to calculate the nitrile carbon position ($117.2+3\times9.4-3.4$).

3-8 **a.** *trans*-$CH_3CH{=}CHCO_2H$; the trans stereochemistry is derived from coupling constant analysis described in Chapter 4.

b. $CH_3CHClCH_2CH_2Cl$ **c.** $ClCH_2CH_2C(=O)CH_2CH_3$

CHAPTER 4

4-1 **a.** homotopic, magnetically nonequivalent
b. enantiotopic, magnetically equivalent
c. diastereotopic, magnetically nonequivalent*
d. magnetically nonequivalent*
e. diastereotopic, magnetically nonequivalent*
f. CH_3 on ring: diastereotopic, magnetically nonequivalent*; CH_3 on side chain: diastereotopic, magnetically nonequivalent
g. enantiotopic, magnetically equivalent
h. diastereotopic, magnetically nonequivalent*
i. enantiotopic (different plane of symmetry from **g**), magnetically equivalent

[*Nuclei with different chemical shifts are magnetically nonequivalent by definition.]

4-2 **a.** $A_3MXX'YY'$ (CH_3, A; P, M; CH_2CH_2, X, Y)
b. $A_3A_3'XX'$ (CH_3, A; P, X)
c. ABX_2 (AB, CH_2; X, CHCl)
d. $(AA'BB')_2$(A, axial H; B, equatorial H)

4-3 The Hund's rule argument of Section 4-2 is applied twice: $(H_A{}^+)(e^-)(e^+)C(e^+)(e^-)C(e^-)(e^+)(H_B{}^-)$. The signs represent nuclear and electron spin ($+\frac{1}{2}$ or $-\frac{1}{2}$), not charge. The polarization of H_B to be of opposite spin to H_A is the definition of a positive coupling constant.

4-4 The 142 Hz one bond coupling constant is not consistent with a cyclopropane structure (^1J ~160 Hz). The cyclopropane spontaneously tautomerizes to an annulene. The one bond coupling constant for the bridge structure of the annulene should be close to 140 Hz, as observed.

4-5 The coupling in the axial oxide is −13.7 Hz, in the equatorial oxide −11.7 Hz. The more positive (less negative) coupling constant in the latter case is caused by π donation of the axial lone pair to the α-axial proton in the α CH_2 group, in accord with the Pople–Bothner-By approach. This π donation is maximal for the axial–axial relationship between the donor lone pair and the acceptor C—H bond.

4.6 There are three planar, zigzig four bond couplings in the trans form, but only one in the cis form.

4-7 The large vicinal coupling in B indicates a smaller H—C—C—H dihedral angle, and hence a flatter ring, by standard Karplus considerations. The pseudoaxial 7-trifluoromethyl group (R) = CF_3) must force the ring into a more nearly planar conformation.

4-8 H4, δ 7.47; H2, 6.63; H5, 6.17; H6, 5.37; H1, 5.15.
Coupling constant analysis: (i) Only δ 6.17 has two ^3J (large), so it must be H5. (ii) With H5 identified, J_{56} should be greater than J_{45} because of higher bond order. Therefore $J_{45} = 7.0$ Hz and $J_{56} = 10.0$ Hz. It follows that δ 7.47 is H4 to match J_{45} and δ 5.37 is H6 to match J_{56}. (iii) The remaining vicinal coupling must be $J_{12} = 10.8$ Hz (δ 5.15 and 6.63). (iv) H6 has a large four bond zigzig coupling.
Chemical shift analysis: (i) H1 and H6 should be at highest field, since they are adjacent to a saturated carbon. Hence δ 5.15 is H1 and δ 6.63 is H6. (ii) H2 amd H4 are shifted downfield by the polar effect of CO_2CH_3. (iii) H4 and H6 are shifted downfield by the resonance effect of CO_2CH_3.

4-9 The spectrum is AMX or ABC, almost first order. The high-field peaks have two ^3J(ortho); the midfield and low-field peaks have only one ^3J(ortho) each. Either the 1,5 or 1,8 isomer is consistent with these observations.

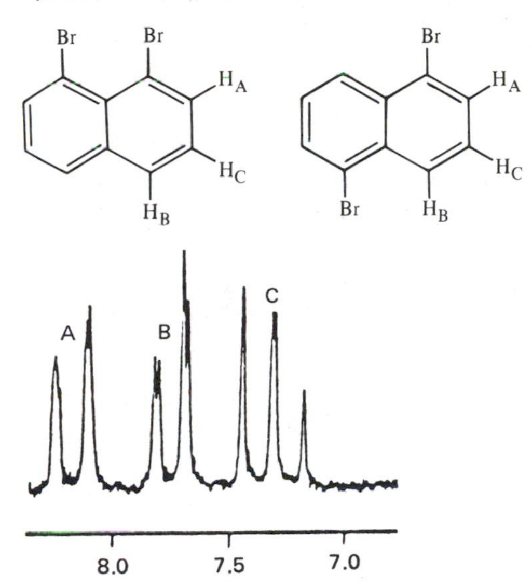

4-10 **a.** The spectrum is AB_2 in the aromatic region, and there is only one type of CH_3. Only the 2,6 and 3,5 isomers would give these patterns. The 3,5 isomer, however, should give typical low-field (δ ~8.5) resonances for protons next to nitrogen. In their absence, the compound must be 2,6-lutidine. The B part has two ^3J(ortho) and the A part only one, consistent with 2,6-lutidine (see structure at end).
b. There is only one proton ortho to nitrogen (δ 8.4). This proton exhibits no large ^3J(ortho). Only 2,5-lutidine has these structural characteristics. Ortho couplings are clearly seen in the high-field multiplet. The methyls are nonequivalent.
c. The spectrum is an AB quartet with a singlet superimposed on the high-field doublet. There is an α proton, with an ortho neighbor. The nonequivalent methyls must be 2,4 in order to isolate the proton that gives the singlet.
d. The methyls are equivalent. There are two isolated α protons (no ortho coupling). Thus the compound is 3,5-lutidine.

2,6 2,5 2,4 3,5

CHAPTER 5

5-1 **a.** A_3 (fast and slow) **b.** A_3X (fast), A_2BX (slow) **c.** A_3X_2 (fast), AA′BXX′ (slow) **d.** A_2X (fast), A_2X and ABX (slow)

5-2 The seven-membered ring is boat shaped, and the major conformer is that with the methoxy group axial. When CH_3O is equatorial, H7 has a NOE from H12. When CH_3O is axial, H7 has none because it is too far from H12. [From the work of P. T. Lansbury.]

5-3 There are two structural isomers. Compound A gives the nitrogen singlet at −72°C, compound B gives the nitrogen AX. The methyl protons produce two unequal singlets, respectively, from the two isomers at −72°C. [See J. E. Bercaw, E. Rosenberg, and J. D. Roberts, *J. Am. Chem. Soc.*, **96**, 612 (1974).]

A: cp′₂Ti(η²-N≡N) ⇌ B: cp′₂Ti—N≡N

a. Look down the C—C axis (of the S-bridged C—C).

The presence of a chiral center makes H_A and H_B inherently nonequivalent and hence diastereotopic. No symmetry element relates H_A and H_B even at fast rotation.

b. Rotation will not interconvert H_A and H_B. A chemical process is required.

$PhCH_2SCHPh(Cl) \rightleftharpoons PhCH_2S—\overset{+}{C}HPh + Cl^-$ ↕ $PhCH_2\overset{+}{S}=CHPh + Cl^-$

Chloroform is an acidic solvent, so exchange of H_A and H_B through a carbanion is not responsible. [See M. Ōki, A. Shimizu, H. Kihara, and N. Nakamura, *Chem. Lett.*, 607 (1980).]

5-5 A glance at the shift table shows that C3, C4, C5, and C17 are off the chart. The two lowest field peaks are off-resonance doublets and must come from C9 and C14. The next highest field peaks remain singlets and must come from the quaternary carbons C10 and C13. The methyl carbons are at highest field (C18 and C19), as they give off-resonance quartets. The remaining peaks all give off-resonance triplets (some are very broad) and must be the methylene carbons at C1, C2, C6, C7, C11, C12, C15, and C16. The C11 and C15 peaks are the highest field of these because they are shifted upfield by the γ-gauche effect of the angular methyl groups. The C8 methinyl resonance simiarly is shifted upfield into the methylene region by γ-gauche effects from both methyl groups. The C2 resonance occurs at midfield because of the offsetting downfield effect by carbonyl and the upfield gauche effect by methyl.

5-6 Acetonitrile has an sp-hybridized nitrogen that is surrounded by an unsymmetrical electron cloud, so that quadrupolar relaxation averages the coupling to zero. The isonitrile ($CH_3—^+N\equiv C:^-$) has a quaternary nitrogen with higher symmetry and slower quadrupolar relaxation, so the coupling is not averaged to zero.

5-7 In general, shorter values of $T_1(Q)$, the quadrupolar spin–lattice relaxation, lead to broader lines. Large electric field gradients (EFG) cause broad lines. The EFG in $^+NMe_4$ is zero, so the line is sharp. In nitromethane the nitrogen is quaternary ($Me—\overset{+}{N}(=O)O^-$), so the EFG is small. The p-orbital lone pair in pyrrole has much lower symmetry and a higher EFG. The lone pair in aniline also is nearly p, but unlike in pyrrole it is not used for aromaticity. Thus the EFG is even larger. Since T_1 is dominated by $T_1(Q)$, not $T_1(DD)$, the closeness of the neighboring protons is not relevant.

5-8 From chemical shift analysis, C1 and C3 are at lowest field (quaternary). They also have the longest T_1 because there are no attached protons. Methyl donates electrons (upfield effect) to the ortho and para positions, so that C4 and C6 are at highest field. C4 and C6 have the fastest relaxation time because they are relaxed by the attached protons, by the ortho protons on C5, and by the methyl protons. C2 and C5 have identical T_1. C2 is probably shifted to lower field by a van der Waals interaction.

5-9 Hydroxyl acts as an even stronger anchor than bromine. Aggregation by hydrogen bonding brings about a large effective size. The nT_1 of CH_3 (9.3 s) indicates that methyl rotates more than twice as fast as ethyl (nT_1 of the first CH_2 is 4.4 s).

5-10 Axial protons should have the shorter T_1, since the nearby 1,3 axial protons can cause dipolar relaxation. T_1 is shorter for H1 in A because it is axial. T_1 of H4 is larger than normal in A and B because there is only one syn-axial neighboring proton. T_1 of H3 is shorter for A because of the additional syn-axial proton (H1). T_1 of H2 is shorter for B because of the added gauche proton (H1). [From C. W. M. Grant, L. D. Hall, and C. M. Preston, *J. Am. Chem. Soc.*, **95**, 7742 (1973).]

5-11 **a.** α is 1, β is 2. In α, H1′ is cis (close) to H2, which provides dominant dipolar relaxation. H3′, H5′, and H5″ are too far away to experience an effect from H1′. Therefore their T_1s are independent of the C1′ configuration. H5′ and H5″ have very short T_1s because they relax each other.

[See S. Tran-Dinh, J.-M. Neumann, J.-M. Thiery, T. Huynh-Dinh, J. Ilgolen, and W. Guschlbauer, *J. Am. Chem. Soc.* **99**, 3267 (1977).]

b. The NOE would be best. Irradiation of H2′ would give enhancement of H1′ only in the α form.

CHAPTER 8

8-1 **a.** The absorption both above and below 3000 cm^{-1} indicates the presence of both saturated and unsaturated groups. The very strong doublet at 990 and 905 cm^{-1} is characteristic of a vinyl group, $-CH{=}CH_2$ (see Section 8-9d). The strong, sharp band at 1630 cm^{-1} supports the presence of an alkenic double bond, and the weak band at 1810 cm^{-1} confirms the vinyl group. The sharp doublet centered near 1370 cm^{-1} indicates the presence of an isopropyl or *tert*-butyl group (see Section 8-7b). Since the molecular weight of the compound is 84, the only possible structures containing a vinyl group and more than one methyl group attached to a single carbon atom are $(CH_3)_2CHCH_2CH{=}CH_2$ or $(CH_3)_3CCH{=}CH_2$. Examination of spectra (15F and 18E) in the Aldrich Library of Infrared Spectra reveals that the compound is 4-methyl-1-pentene.

b. The formula suggests either an alkene or a cyclic saturated hydrocarbon. The absence of absorption above 3000 cm^{-1} and between 2000 and 1600 cm^{-1} rules out alkenes or small rings. The single band at 1370 cm^{-1} indicates a methyl group, or perhaps more than one methyl group, but attached to different carbon atoms. Possible compounds of formula C_7H_{14} are methylcyclohexane, dimethylcyclopentane, and ethylcyclopentane. The compound is, in fact, ethylcyclopentane (Aldrich spectrum 26B).

c. Absorption above 3000 cm^{-1} indicates unsaturation, and the sharp bands at 1600 and 1500 cm^{-1} suggest a benzene ring. The very strong doublet at 740 and 695 cm^{-1} together with the pattern of weak bands between 2000 and 1650 cm^{-1} indicates monosubstitution (Table 8-7). If this is the case, then C_6H_5 accounts for 77 of the 120 molecular weight, leaving 43, which can only be C_3H_7 (*n*-propyl or isopropyl). The presence of a single band at 1380 cm^{-1} favors the *n*-propyl side chain. The compound is propylbenzene (Aldrich spectrum 561E).

d. The hydrocarbon is totally unsaturated (no absorption below 3000 cm^{-1}). It contains an alkenic double bond (sharp band at 1630 cm^{-1}) and a benzene ring (sharp bands at 1600 and 1500 cm^{-1}). It is a monosubstituted benzene derivative (two strong bands at 775 and 695 cm^{-1}). This structure is supported by the pattern of weak bands between 2000 and 1650 cm^{-1} (see Section 8-6b). The unsaturated aliphatic side chain contains a vinyl group (two very strong bands at 985 and 905 cm^{-1}). The compound is styrene, $C_6H_5CH{=}CH_2$ (Aldrich spectrum 569C).

8-2 **a.** The spectrum shows no bands attributable to OH or CO stretching, so an ether is suspected. Absorption both above and below 3000 cm^{-1} indicates both unsaturated and saturated groups. The sharp bands at 1595 and 1505 cm^{-1} suggest the presence of an aromatic ring, and the single strong peak at 745 cm^{-1} indicates that the compound is an ortho-substituted benzene derivative (see Table 8-7). The sharp band at 2850 cm^{-1} is characteristic of a methoxy group. The compound is *o*-dimethoxybenzene (Aldrich spectrum 625B).

b. There is no absorption above 3000 cm^{-1}, and the C=O stretching band at 1725 cm^{-1} is in the region expected for an aliphatic aldehyde or ketone. However, the sharp band of medium intensity at 2725 cm^{-1} is characteristic of an aldehyde (see Section 8-4f). The weak band at 730 cm^{-1} could be due to a chain of four or more $-CH_2-$ groups (see Section 8-8d). There are no other characteristic features in the spectrum, so a reasonable structure for an aliphatic aldehyde of molecular weight 100 is $CH_3(CH_2)_4CHO$ (hexanal, Aldrich spectrum 277E).

c. The strong band at 3350 cm^{-1} suggests an alcohol, and the strong band at 1040 cm^{-1} indicates a primary alcohol. There is no CH absorption above 3000 cm^{-1}, so the compound is aliphatic. The sharp doublet centered at 1380 cm^{-1} indicates the presence of two methyl groups attached to the same carbon atom. The compound is isobutyl alcohol, $(CH_3)_2CHCH_2OH$ (Aldrich spectrum 68B).

d. Again, the strong band at 3380 cm^{-1} indicates an OH group, but the CH absorption above and below 3000 cm^{-1} means both unsaturated and saturated groups are present. The sharp bands at 1610 and 1510 cm^{-1} suggest the presence of a benzene ring, and the single strong band at 830 cm^{-1} together with the two bands between 1900 and 1700 cm^{-1} indicates para disubstitution (Table 8-7). The strong broad band at 1230 cm^{-1} is in the center of the range of the C—O stretching mode of phenols (Table 8-10). Subtracting the molecular weight of C_6H_4OH (93) from the molecular weight of the compound (122) leaves 29, which is exactly that of an ethyl group. The compound is 4-ethylphenol (Aldrich spectrum 647F).

e. The band at 1710 cm^{-1} is clearly due to a carbonyl group and is in the expected range of an aliphatic ketone (Table 8-5). The unsaturation number is 2 (see page 3). One unsaturation is due to the carbonyl group. As there is no absorption above 3000 cm^{-1}, the other unsaturation must be due to a ring. Thus, the compound is an aliphatic cyclic ketone. There is no methyl group present (no band near 1375 cm^{-1}). Hence the compound is probably cyclohexanone (Aldrich spectrum 256B).

f. The very strong broad band extending from 3500 down to 2400 cm^{-1} is characteristic of a hydrogen-bonded carboxylic acid (Table 8-2). The weaker broad bands centered near 930 cm^{-1} together with the strong carbonyl stretching band at 1700 cm^{-1} (Table 8-5) also indicate the presence of a COOH group (Table 8-2). The sharp bands at 1600 and 1495 cm^{-1} suggest a benzene ring, and the doublet at 745 and 695 cm^{-1} indicates that the ring is monosubstituted (Table 8-7). Peaks are seen on top of the broad OH stretching band at 3035 and 2940 cm^{-1}, indicating the presence of both aromatic CH and aliphatic

CH_2 or CH_3 groups. The combined molecular weight of the C_6H_5— and —COOH groups is 122. The remaining mass of 42 could be due to C_3H_6, which would be present in any of the three structural isomers 2-, 3-, and 4-phenylbutyric acid. The absence of a band near 1375 cm^{-1} indicates that there is no methyl group in the molecule. Thus it is concluded that the compound is 4-phenylbutyric acid (Aldrich spectrum 932A).

g. The weak peak at 3055 cm^{-1} and the sharp peak at 1600 cm^{-1} are due to a benzene ring, but it is not possible to establish the substitution pattern from the single strong band at 710 cm^{-1} (Table 8-7). The very strong absorption at 1715 cm^{-1} could be due to an ester or an aldehyde (Table 8-5). The latter is ruled out by the absence of a band near 2740 cm^{-1} (see Section 8-4f). The former is supported by the very strong band at 1275 cm^{-1} (Table 8-10) and the weak band at 3410 cm^{-1} (Table 8-3). Thus it is concluded that the compound is an aryl ester. The compound is, in fact, ethyl benzoate (Aldrich spectrum 1027D).

8-3 **a.** The compound is aromatic (no absorption below 3000 cm^{-1} and sharp peaks at 1595 and 1490 cm^{-1}). It is a nitrile (strong sharp band at 2220 cm^{-1}), and it contains a monosubstituted benzene ring (two very strong bands at 755 and 685 cm^{-1} and the pattern of weak bands between 2000 and 1650 cm^{-1}). The structure is benzonitrile, $C_5H_5C{\equiv}N$ (Aldrich spectrum 1128C). The very strong sharp band at 550 cm^{-1} is due to an out-of-plane ring deformation mode.

b. The most striking feature of the spectrum is the strong broad band centered at 800 cm^{-1}. Examination of Table 8-2 shows that the compound is probably a primary aliphatic amine. The doublet at 3380 and 3310 cm^{-1} confirms this assignment, and the absence of any CH stretching bands above 3000 cm^{-1} indicates that the molecule is saturated. Another clue is the weak band at 725 cm^{-1}, which indicates a chain of at least four $-CH_2-$ groups. The band at 1385 cm^{-1} is due to a methyl group, so that a possible structure of molecular weight 101 is $CH_3(CH_2)_5NH_2$. The compound is hexylamine (Aldrich spectrum 165H).

c. Since the compound does not contain oxygen, the single band at 3395 cm^{-1} suggests a secondary amine (Table 8-11). Absorption in the CH stretching region indicates that both saturated and unsaturated groups are present. The unsaturated group is probably a monosubstituted benzene ring (sharp bands at 1595 and 1495 cm^{-1}, the doublet at 745 and 690 cm^{-1}, and the weak bands between 2000 and 1650 cm^{-1}). Secondary aliphatic amines have a broad band centered near 730 cm^{-1} (Table 8-2 and Figure 8-4e). The absence of this band suggests that the compound is an aniline, C_6H_5NH—, derivative. The remaining hydrocarbon residue is C_4H_9, and the compound is *N*-butylaniline.

d. The compound is known to contain only C, H, and N. However, there are no bands attributable to nitrile or primary or secondary amines. The remaining possibilities are a tertiary amine or a compound in which the nitrogen is doubly bonded to a carbon atom in a $C{-}N{=}C$ group. Further examination of the spectrum reveals that the molecule is aromatic. There is no absorption below 3000 cm^{-1} in the CH stretching region, and there are sharp bands at 1585 and 1485 cm^{-1}. The very strong doublet at 745 and 705 cm^{-1} suggests five adjacent hydrogen atoms on an aromatic ring, as is found in a monosubstituted benzene derivative. Subtracting one nitrogen atom from the molecular weight of 79 leaves 65, which can be attributed to C_5H_5. Thus the molecule is pyridine, C_5H_5N (Aldrich spectrum 1306A).

8-4 **a.** The compound is clearly an amino acid (very strong broad bands centered near 3000 and 1600 cm^{-1} and the weak broad band at 2020 cm^{-1}; Tables 8-2 and 8-3). There is also an OH group (the strong broad band centered at 3450 cm^{-1}). The doublet at 1005 and 915 cm^{-1} resembles that of a vinyl group, but the intensities of the two bands are reversed, and in any case a molecule with only three carbon atoms could not contain both an amino acid and a vinyl group. A possible structure is CH_2OH—$CHNH_2$—COOH, serine (Aldrich spectra 346B, C, and D).

b. The strongest bands in the spectrum are the broad feature centered at 3390 cm^{-1} and the very strong peaks at 1530 and 1355 cm^{-1}. The first of these can be assigned to an OH group, while the other two suggest the presence of a nitro group in the molecule (Table 8-11). The weak peaks at 3105, 2940, and 2950 cm^{-1} suggest that both unsaturated and saturated CH groups are present. The sharp peaks at 1485 cm^{-1} could be due to a benzene ring, and in that case the strong doublet at 800 and 730 cm^{-1} indicates a 1,3-disubstituted benzene derivative. A possible structure with the appropriate molecule weight is *m*-nitrobenzyl alcohol (Aldrich spectrum 808A).

c. The strong doublet at 3355 and 3205 cm^{-1} is due to an NH_2 group. The very strong band at 1630 cm^{-1} could be due to the C=O stretching coupled with the NH_2 deformation of an amide (Tables 8-5 and 8-11). The band near 1140 cm^{-1} could be attributed to a C—N stretch of a primary aliphatic amine or amide (Table 8-11). The very broad absorption centered between 600 and 650 cm^{-1} is characteristic of an amide (Table 8-2). The Nujol bands obscure the CH stretching and bending regions, so nothing can be said about the hydrocarbon part of the molecule. However, if the compound is an amide, the remaining molecular weight is 43, which is that of a propyl group. Hence the structure is $C_3H_7CONH_2$, butyramide (Aldrich spectrum 437D).

8-5 The spectrum has four very strong bands centered at 1205, 1010, 890, and 715 cm^{-1}. Consultation of Table 8-11 indicates that the 1205 cm^{-1} band is probably due to a single S=O group (SO_2 groups would have a second very strong band at higher frequencies), while the 1010 and 890 cm^{-1} bands could be the two S—O—C stretching modes of a dialkyl sulfite. The molecule is saturated (no bands above 3000 cm^{-1}) and contains a methyl group (a band at 1385 cm^{-1}). The broad band centered at 715 cm^{-1} may be due to O—S stretching. The compound is diethyl sulfite (Aldrich spectrum 532E).

8-6 The strongest band in the infrared spectrum is centered at 1110 cm^{-1}. This is the center of the region of the C—O—C stretching mode of an aliphatic ether. The absence of strong absorption above 1500 cm^{-1} rules out carbonyl, nitro, and nitroso groups. The sharp band at 2840 cm^{-1} is characteristic of a methoxy group (Section 8-4e). This peak supports the assignment of the 1110 cm^{-1} band to an aliphatic ether. A tertiary amine is not entirely ruled out, since the band at 1195 cm^{-1} could be due to the C—N stretch and there are no other characteristic bands for this functional group. The problem of the location of the nitrogen atom is solved by the Raman spectrum, in which a strong band is observed near 2250 cm^{-1}. This band can only be due to a nitrile group. The molecule is then identified as methoxyacetonitrile, $CH_3OCH_2C{\equiv}N$ (Aldrich spectrum 508D).

CHAPTER 9

9-1 n–π^*, 99 kcal mol^{-1}; π–π^*, 157 kcal mol^{-1}
9-3 OCH CHO
9-5 $\epsilon = 10{,}500$
9-7 $\Delta\epsilon = +0.6$
9-9 $A = 0.00553$, 1.54 mol l^{-1} cm^{-1}
9-11 **a.** Phenol absorbs at 190 nm, phenoxide at 430 nm. **b.** $pK_a = 8.2$ **c.** MW = 108 **d.** a cresol
9-13 0.0412 mol l^{-1} cm^{-1}
9-15 $K_a = 49 \times 10^{-8}$
9-17 $K = 1.00 \times 10^{-3}$

CHAPTER 10

10-1 Left to right: 213, 229, 234 nm
10-3 A, 280 nm; B, 303 nm
10-5 Left to right: 349, 305, 360 nm
10-7 1st row, left to right; 249, 244, 407 nm; 2nd row, left to right: 254, 285, 239 nm
10-9 1st row, left to right: 249, 244, 382 nm; 2nd row, left to right: 234, 249, 239 nm
10-11 Left to right: 283, 249, 234, 227 nm
10-13 A, 241 nm; B, 247 nm
10-15 1st row, left to right: 261, 264, 271 nm; 2nd row, left to right: 271, 277, 291 nm

10-17 A: B:

10-19 A: CH_2 B: O C:

10-21 Steric hindrance inhibits effective conjugation.

10-23 **a.** no; Woodward's rules predict 239 nm
b.
c. $\epsilon = 23{,}700$ **d.** yes **e.** 237; not reasonable
10-25 n–π^*; 3*R*
10-27 Left to right: 9*R*, 9*S*, 9*S*

10-29 (+):

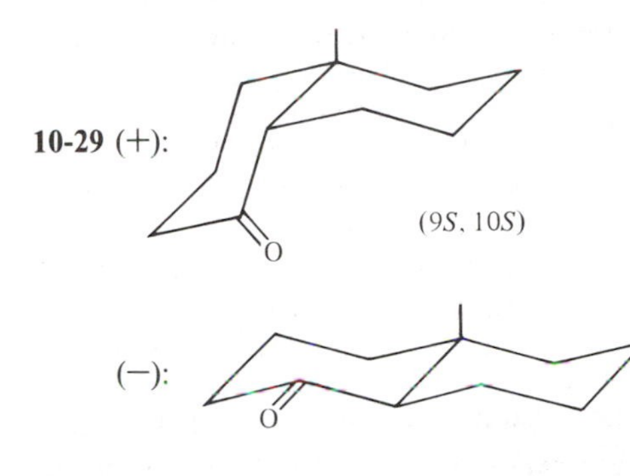

10-31 The axial CH_3 lies in a front octant.

10-33 H CH_3 O CH_3 O

10-35 methanol: Cl O CH_3
isooctane: Cl H H CH_3 O

10-37 β-axial

10-39 OH OH

10-41 O O (*S*)

10-43

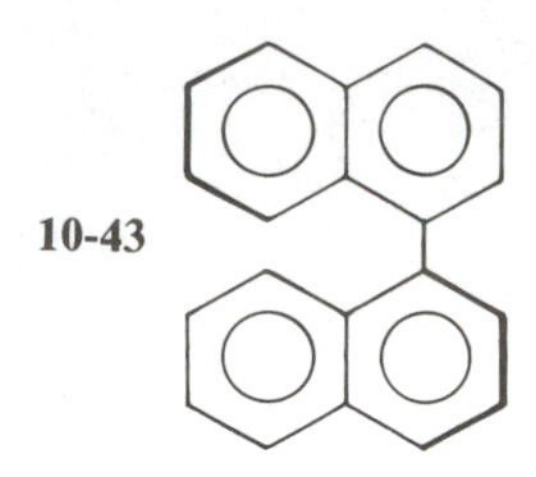

10-45

H

HO CH3

HO CH3

H

CHAPTER 12

12-1

1. This is a very simple spectrum, with the characteristics of an aromatic compound, namely, an abundant molecular ion and only a few abundant ions.
2. In particular, the abundances of the ions at m/z 77 and 51 indicate a monosubstituted benzene derivative. The loss of 77 mass units to give m/z 104 tallies with this assignment.
3. The odd molecular weight indicates the presence of an odd number of nitrogen atoms.
4. Identification of a phenyl group and a nitrogen atom still leaves a massive portion of the molecule unaccounted for. Most significantly, there seems to be no indication of the nature of this fragment in the spectrum. Such a circumstance often suggests that a unit is repeated, in this case the phenyl ring.
5. A clue that this diagnosis is correct is provided by the low abundance ions at m/z 152 and 154, which correspond to ionized biophenyl (m/z 154) and biphenylene (m/z 152). The former species could be generated by the common rearrangement in which two aryl groups are bonded together with the elimination of a stable neutral molecule. In the present case the neutral has mass 27 and includes a nitrogen atom. Hence, the unknown is the imine $C_6H_5CH{=}NC_6H_5$.
6. It is noteworthy that the $(M+1)^+$ ion is far more abundant than the ^{13}C isotopic contribution requires. This is a frequent observation for nitrogen-containing compounds and is due to protonation in the EI source.

12-2

1. The molecular ion abundance is very low, but the isotope distribution clearly indicates the presence of a single bromine atom in the molecule. The m/z 79 and 81 doublets confirm this conclusion.
2. Loss of bromine from the molecular ion occurs with great ease to give the base peak m/z 57. This ion and the presence of other hydrocarbon ions at m/z 27, 29, 39, and 41 indicate a molecular formula C_4H_9Br.
3. The molecular ion loses an ethyl radical to give the bromine-containing ions at m/z 107 and 109. Hence the compound is 2-bromobutane. Methyl radical loss from the molecular ion does occur, but gives a product of very much lower abundance (not recorded) than that due to loss of the larger radical (compare Section 12-6).
4. The presence of ions at m/z 88, 99, and 101 due to impurities may be noted.

12-3

1. This spectrum shows behavior typical of aromatic compounds in that the molecular ion and two or three fragment ions carry the bulk of the ion current.
2. It is striking that in spite of the high molecular weight (240) the ^{13}C isotope of the molecular ion has an abundance of only 7% of $M^{\cdot+}$, limiting the number of carbon atoms present to a maximum of six. Clearly, a considerable proportion of the mass of the molecule is contributed by monoisotopic elements. Iodine and fluorine are the prime candidates.
3. The atomic weight of iodine is 127; hence the ion at m/z 127 corresponds to I^+. In addition, m/z 113 is due to $(M-127)^+$.
4. The only other fragment ion having even moderate abundance is m/z 63, formed, at least in part, from m/z 113, as shown by the presence of a metastable peak at m/z 35.2. The loss of 50 mass units as a neutral species is common only in fluorinated compounds.
5. Assuming that m/z 63 is $C_5H_3^+$, the formula $C_6H_3F_2I$ is obtained. This formula requires four rings or double bonds. A trisubstituted benzene fits all the above data. In fact, the compound is 1,3-difluoro-5-iodobenzene, but the positions of the substituents cannot be determined from the mass spectrum.

12-4

1. In spite of complex isotope patterns, this spectrum is rather simple, just a few ions carrying the bulk of the ion current.
2. At first glance an organometallic might be suspected, but the isotope pattern of the molecular ion differs from that of the major fragment ions. Indeed, the pattern becomes more and more simple as one moves to low mass.
3. This behavior suggests the presence of several atoms of the same multiisotopic element, and the characteristic chlorine 3/1 pattern with a separation of two mass units is seen in several low mass ions.
4. One concludes that this is a polychlorinated compound. The isotope patterns indicate the number of chlorine atoms in each fragment: for example, m/z 71/73 has one chlorine, 130/132/134 has two chlorines, 141/143/145/147 has three chlorines, 235/237/239/241/243 has five chlorines, and the molecular ion, 270/272/274/276/278/280, has six chlorines (the peak at m/z 280 is not recorded in the spectrum since its relative abundance is <1%).
5. Given a molecular weight of 270 (^{35}Cl) and the presence of six chlorine atoms, only 60 mass units remain, possibly C_5. The complete absence of any hydrocarbon ions (m/z 27, 29, 39, 41, 43, etc.) makes the perchlorinated formula C_5Cl_6 seem reasonable.
6. This formula requires three rings or double bonds, implying a cyclopentadiene structure, a vinylcyclopropene formula, or a pentenyne structure. A decision among these alternatives is not readily made from the mass spectrum. The compound, in fact, is perchlorocyclopentadiene.
7. The extent to which $Cl^{\cdot}$ loss predominates over ring fragmentation is shown by the presence of abundant ions in the series

$C_5Cl_5^+$ (m/z 235), $C_5Cl_3^+$ (165), $C_5Cl_2^{\cdot +}$ (130), and C_5Cl^+ (95) (the mass numbers refer to the ^{35}Cl isotope). The only major ions that involve carbon–carbon bond cleavage are $C_3Cl_3^+$ (141) and $C_3Cl_2^{\cdot +}$ (106).
8. The abundance of the doubly charged (M− Cl)$^{++}$ ion (in the cluster from m/z 117–121) is notable.

12-5

1. The presence of one sulfur atom is suggested by the abundance of the $(M+2)^{\cdot +}$ ion.
2. The ion m/z 47 is unusual since this region of the spectra for compounds containing C, H, N, and O is invariably blank. The ion CH_3S^+ is probably responsible for this peak.
3. The proposed methylthio group may also account for the observed loss of a methyl radical from the molecular ion.
4. The ion m/z 61 is apparently the higher homologue of m/z 47. This result suggests the partial structure CH_3SCH_2—.
5. The remainder of the molecule has mass 45, and indeed an ion m/z 45 forms the base peak in the spectrum. This ion must have the composition CO_2H^+ or $C_2H_5O^+$.
6. The presence of ions due to loss of 17, 18, and 19 mass units from the molecular ion suggest an alcohol. In conjunction with the abundance of m/z 45, this observation requires the structure $CH_3SCH_2CH(OH)CH_3$.
7. The propensity of sulfur to engage in bond-forming reactions in mass spectrometry is exemplified by two ions in this spectrum. First, the loss of $SH^{\cdot}$ occurs to give m/z 73, even though a thiol group is not present in the molecule. As is usual for such complex rearrangements, this ion is of low relative abundance. Second, the formation of m/z 62 by elimination of acetaldehyde is due to the ease with which the ionized sulfur atom can abstract a hydrogen atom via a five-membered cyclic transition state.

12-6

1. Exact mass measurement establishes the formula of this unknown as $C_8H_{16}O_2$. Hence there is just one ring or double bond.
2. The base peak, m/z 88, corresponds to loss of 56 mass units from the molecular ion. This observation indicates that a molecule of butene is eliminated from the molecular ion; the presence of a metastable peak at m/z 53.7 confirms the transition (but not the composition of the neutral).
3. Since the molecule has but one ring or double bond and two oxygen atoms and since the base peak arises by alkene elimination from the molecular ion, it is reasonable to assume that McLafferty rearrangement to a carbonyl group is responsible for butene loss.
4. The second oxygen atom might be present as several functionalities, but the abundant ion at m/z 99 $(M-45)^+$ and the presence of an ion m/z 45 suggest the ethoxy group, and in all probability the unknown is an ethyl ester. An ethyl ester would be expected to give an ion $EtOC{\equiv}O^+$, m/z 73, and this ion is observed.
5. The loss of butene from an ester $C_5H_{11}CO_2C_2H_5$ in a McLafferty rearrangement requires that the α carbon be unsubstituted.
6. The only remaining question concerns whether or not the C_4H_9 unit in $C_4H_9CH_2CO_2C_2H_5$ is branched. The presence of the $(M-43)^+$ ion, which may at first sight seem puzzling, helps to answer this question. Carbonyl compounds undergo a reaction that appears to be simple γ-cleavage. This reaction would give the $(M-43)^+$ ion, m/z 101, if both the α and the β carbons were unsubstituted. The ion m/z 115 is the higher homologue of m/z 101, and its formation requires that the entire amyl chain be linear, so the unknown is ethyl hexanoate.
7. The origin of several other ions may be briefly noted: m/z 71 is the $C_5H_{11}^+$ ion, formed in part by CO loss from the acylium ion; m/z 60 is ionized acetic acid, generated in part by C_2H_4 loss from the McLafferty rearrangement product. The ion at m/z 117, due to $C_2H_3^{\cdot}$ loss, is unexpectedly abundant for an ethyl ester, although the product ion, $RCO_2H_2^+$, is particularly stable (compare Figure 12-16).

12-7

1. Although the molecular ion is of very low abundance, the two isotopic forms suggest that one chlorine atom is present; this conclusion is confirmed by the isotope pattern seen in several other ions, notably m/z 107/109.
2. The fact that the chlorine atom is readily lost (to give m/z 115) suggests that it is present as an alkyl or acyl chloride. The presence of the ion CH_2Cl^+ (m/z 49/51) and the loss of $CH_2Cl^{\cdot}$ from the molecular ion to give m/z 101 clarify the nature of the chloro group.
3. The abundance of m/z 43 suggests the presence of an acetyl group; the loss of acetic acid from the molecular ion to give m/z 90/92 confirms that the compound is an acetate.
4. The units $ClCH_2$— and $CH_3C(O)O$— leave 42 mass units to be accounted for. Only C_3H_6 and C_2H_2O are reasonable possibilities. The structure $Cl(CH_2)_4O{-}COCH_3$ is suggested by the presence of homologous ions belonging to the series 115, 101, 87, and 73.
5. The ion at m/z 54 corresponds to ionized butadiene and is apparently formed by loss of protonated acetic acid from the $(M-Cl)^+$ species. There are several other ions of low abundance that arise by rearrangement reactions.

12-8

1. The presence of just a few dominant ions suggests an aromatic compound.
2. The odd molecular weight implies that there is an odd number of nitrogen atoms (probably one rather than three or more).
3. There is no indication from isotope patterns of the presence of any elements beside C, H, N, and O.
4. The major fragment ion in the high mass region (and coincidentally the base peak) is m/z 157, which corresponds to loss of 30 mass units from the molecular ion (a reaction that is substantiated by the observation of a metastable peak at m/z 131.8). This observation is good evidence for the presence of a nitro group. The low abundance $(M-16)^+$ ion is diagnostic of an N—O group. The loss of 46 mass units from the molecular ion (to give m/z 141) provides further evidence for the presence of the nitro group.
5. The unknown is therefore an aromatic nitro compound, X—NO_2, where X has mass 141 and could have zero, one, or more oxygen atoms. Hence the formulas that must be considered for this unit are $C_{11}H_9$, $C_{10}H_5O$, and C_9HO_2. Only $C_{11}H_9$ seems at all likely, and it is strongly indicted by the fact that m/z 115 (loss of 26, i.e., C_2H_2, from 141) is abundant. In fact, both m/z 141 and 115 are common ions in larger aromatic compounds.

6. The molecular formula is therefore $C_{11}H_9NO_2$. The structure must exclude ring-substituted methyl groups since an $(M-H)^+$ ion is not observed. The actual compound is *p*-(2H-cyclopentadienyl)nitrobenzene.

12-9

1. The presence of a single bromine atom is immediately evident from the isotope distribution in the molecular ion.
2. The spectrum shows the typical features of aliphatic compounds, including a molecular ion of relatively low abundance and abundant fragment ions of low mass.
3. The loss of bromine does not give rise to an abundant fragment ion; hence bromine is not bonded to a saturated carbon, but rather to an alkenic carbon.
4. The low abundance of *m/z* 43, 57, and 71 indicates that the compound does not contain a saturated alkyl chain of more than two atoms.
5. The abundant ion at *m/z* 83 (and its fragment ion *m/z* 55) provides the key to this structural determination. Recognition that the cyclohexyl group has mass 83 allows one to rationalize all the abundant ions in the spectrum. Thus, *m/z* 82 is ionized cyclohexene, the major fragmentation of which is methyl radical loss to give *m/z* 67. Further loss of C_2H_2 should give rise to *m/z* 41.
6. The cyclohexyl group and the bromine atom leave 54 mass units to be accounted for. Possible formulas are C_4H_6 and C_3H_2O. Reasonable structures can be deduced on the basis of both possibilities, but an exact mass measurement proved the hydrocarbon formula to be C_4H_6.
7. The formula $C_6H_{11}-C_4H_6-Br$ requires one ring or double bond in addition to the cyclohexyl ring. The bromine must be attached to the C_4H_6 unit, and it has already been suggested that it is an alkenic bromide. The absence of an $(M-15)^+$ ion precludes an allylic methyl group.
8. A unique structural assignment can be made on the basis of the formation of *m/z* 134/6. This fragment contains bromine and is due to elimination of a neutral molecule of cyclohexene from the molecular ion. This process may occur via a six-membered cyclic rearrangement only for 1-bromo-2-methyl-3-cyclohexyl-1-propene if the above-memtioned structural requirements are also to be satisfied.

12-10

1. The fragment ion of highest mass, *m/z* 146, which corresponds to loss of 16 mass units from the molecular ion, probably results from loss of an oxygen atom. Such loss is diagnostic of an N—O group.
2. This analysis is confirmed by the fact that the molecular ion also loses fragments having masses of 30 and 46. An aryl nitro group is clearly indicated by these processes.
3. Since the molecular weight of the compound is even, it must contain an even number of nitrogen atoms, probably two.
4. Metastable peaks show that, at least in part, the ion at *m/z* 104 arises from 132 and that at *m/z* 89 from 116. The subsequent loss of CO after $NO^\cdot$ loss to give *m/z* 132 is an expected process. The further loss of 27 mass units after loss of $NO_2^\cdot$ almost certainly means that HCN is the neutral.
5. One knows, therefore, that the compound contains an aromatic nitro group and a cyano group, and the presence of *m/z* 77 indicates a benzene ring (although not necessarily monosubstituted). The molecular weight of 162 leaves only 14 mass units unaccounted for, which suggests a methylene group.
6. As there is no loss of $H^\cdot$ from the molecular ion, there can be no ring methyl group. Hence the compound must be a (cyanomethyl)nitrobenzene. The mass spectrum does not allow one to choose among the ortho, meta, and para isomers, but the compound is *p*-(cyanomethyl)nitrobenzene.

Index to Functional Groups

Index